Green Energy and Technology

More information about this series at http://www.springer.com/series/8059

Panagiotis Grammelis
Editor

Energy, Transportation and Global Warming

Springer

Editor
Panagiotis Grammelis
Chemical Process and Energy Resources Institute
Centre for Research and Technology Hellas
Athens, Greece

ISSN 1865-3529 ISSN 1865-3537 (electronic)
Green Energy and Technology
ISBN 978-3-319-30126-6 ISBN 978-3-319-30127-3 (eBook)
DOI 10.1007/978-3-319-30127-3

Library of Congress Control Number: 2016933773

Printed on acid-free paper

This Springer imprint is published by Springer Nature
The registered company is Springer International Publishing AG Switzerland

Preface

This book is associated with the Global Conference on Global Warming: Focus on Energy-Transport-Greenhouse Effects, held in Athens on May 24–27, 2015. Objectives are in accordance with the conference's presentations and mainly concern the exchange of technical information, dissemination of high-quality research results and presentation of new policy and scientific developments, while promoting future priorities for a more sustainable development and energy security.

In particular, energy-related issues in all engineering disciplines for a wide area of applications in the renewables and fossil fuels sectors are described, incorporating cross-cutting effects. It includes main aspects of transportation discipline with emphasis on the elimination of the impact on greenhouse effects.

High-quality technical knowledge and research results from specific test cases around the world are being analysed, providing a holistic view in the main aspects of the Global Warming issue. The latter also concern current policies and emissions from air and maritime transport, in addition to the fossil fuel applications. Novel technologies such as Carbon Capture and Storage are investigated along with process/systems analysis and optimization for mitigating CO_2 emissions. Water resources management and waste water treatment as well as waste management issues are also tackled. Biomass, hydrogen and solar energy technologies are presented along with an insight on green buildings.

The utmost scope of this book is to contribute to the scientific community, since it includes scientific approaches from many organisations around the globe, presenting key issues, challenges and research results in a variety of scientific areas that relate to the Global Warming effects.

Athens, Greece — Panagiotis Grammelis

Contents

Part I
Global Warming and Climate Change: General Issues and Challenges

General Aspects of Global Warming, Current Policies and Challenges Policies and General Aspects of Global Warming with Focus on Specific Examples

Chapter 1
Urban Development Policy and Urban Sprawl in Turkey

Münevver Özge Balta

Introduction

Metropolitan areas are rapidly changing in response to urban development dynamics. Urban sprawl is defined as a form of urbanization which inefficient, low-density, suburban development around the periphery. Sprawl is a leading process at the edges of urban growth and implies deficient and weak planning control on land policy.

Since the beginning of twentieth century, the world population has increased dramatically, and especially metropolitan areas have undergone diverse structural changes. As many other developing countries, Turkey was faced with the fast urbanization which necessitated immediate formulation of new planning policies in metropolitan areas. Metropolitan development areas have spread out into rural areas, so planning policies supposedly regulate this new pattern of development.

In Turkey, local governments frequently use partial plans with a tendency for discontinuity for urban fringe areas. In actuality, the urban space is patchy and the applications can be different from the plan decisions. Partial plans may completely change the urban settlement character. Since 1980, urban development in metropolitan areas in Turkey has been especially piecemeal, rather than holistic. So, in the process, urban development planning has tended to be piecemeal, resulting urban sprawl.

M.Ö. Balta (✉)
Department of Architecture, Faculty of Architecture and Design, Aksaray University, 68100 Aksaray, Turkey
e-mail: ozgearas@gmail.com

P. Grammelis (ed.), *Energy, Transportation and Global Warming*, Green Energy and Technology, DOI 10.1007/978-3-319-30127-3_1

This paper is concerned with the notion of urban sprawl, particularly planning policies in metropolitan areas. It aims to discuss the impacts of planning tools and regulations for urban sprawl. The paper also investigates a research problem that could show the characteristic of metropolitan development and planning procedure in Turkey.

Urban Sprawl

"Urban sprawl" is used to describe low-density, inefficient, suburban development around the periphery. Sprawl is defined as a form of urbanization distinguished by leapfrog development, commercial strips, low density, separated land uses, automobile dominance and a minimum of public open space (Table 1.1).

Squires (2002) defines sprawl as "a pattern of urban and metropolitan growth that reflects low density, automobile-dependent, exclusionary new development on the fringe of settled areas often surrounding a deteriorating city". Urban sprawl is a consequence of many factors, such as the need for industrial establishment for larger and less expensive locations which force them to move out of the city centre where they were previously settled, the increasing mobility of middle and high income groups due to private car ownership and changing socio economic value systems.

Urban sprawl is a consequence of many factors, such as the need for industrial establishment for larger and less expensive locations which force them to move out of the city centre where they were previously settled, the increasing mobility of middle and high income groups due to private car ownership and changing socio economic value systems (Balta and Eke 2011).

European Environment Agency defines the stimulants of urban sprawl as macro economic factors, micro-economic factors, demographic factors, housing preferences, inner city problems, transportation and regulatory frameworks (Table 1.2). Sprawl is the leading edge of urban growth and implies deficient and weak planning control on land subdivision.

Table 1.1 Characteristics of sprawl (Ewing 1997)

Leapfrog or scattered development	It means subdivisions shopping centers and office parks widely spread apart
Commercial strip development	It is characterized by huge roads with shopping centres, gas-stations, fast-food restaurants, banks, parking lots, etc.
Low density	Suburban buildings are often single-story and widely spaced, with intervening parking lots and roadways
Large expanses of single-use development	Housing consist predominantly of single-family homes on individual lots
Poor accessibility (Automobile dominance)	The longer distances between activities
Lack of public open space	In suburban area, public open space can be difficult to find except school yard

Table 1.2 Stimulants of urban sprawl (EEA 2006)

Economic factors	Macro-economic	Globalization Economic development Integration
	Micro-economic	Quality of life Land value Land availability Competition between local governments
Demographic factors	Population growth	
	Increase in household formation	
Physical factors	Housing	Housing preferences More space per person
	Urban problems	Poor air quality Small houses Noise Unsafe urban environment Lack of open and green areas Poor quality of schools
	Transportation	Private car ownership Roads accessibility Fuel cost Lack of public transportation
	Urban planning	Weak land use planning Lack of enforcement of existing plans Lack of coordination and organization

In summary, some scholars (Beauregard 2006; Gillette 2005; Teaford 2006) thus conclude that urban sprawl should be accepted as one of the forms of metropolitan growth while others (Burchell et al. 1998; Cox and Utt 2004; Ewing et al. 2002; Snyder and Bird 1998) condemn urban sprawl because of its detrimental affects such as social segregation, upheave and change or economic prosperity; urban sprawl is one of the evitable spatial attributes of metropolitan development (Balta and Eke 2011).

There are two main reasons of urban sprawl. First reason is lack of laws and regulations supporting planning, second, reason is lack of goodwill and determination of local authorities to keep the urban macroform as planned.

Urban Sprawl and Energy Consumption

In the last years, cities have gone through problems as global climate change. Together with the urban growth, greenhouse gases (GHGs) are receiving increasingly more attention (Glicksman 2007). The rise in population and individual car ownership is mainly attributed to urban sprawl. Suburban areas is often believed to be less energy efficient than urban core. Overall energy consumption of a territory, especially as far as travel energy consumption is concerned with urban sprawl (Ewing 1994).

Urban sprawl is a consequence of many factors, such as the need for industrial establishment for larger and less expensive locations which force them to move out of the city centre where they were previously settled, the increasing mobility of middle and high income groups due to private car ownership and changing socio economic value systems.

Urban sprawl has many effects on urban environment as unplanned expansion of the city may not coordinate with the public transportation system which brings loss of time in journeys towards the city centre, increased energy consumption and increased traffic congestion. Motorization has increased energy demand.

Urban Development Policy in Turkey

Metropolitan growth of Turkey is due to fast urbanization and results in urban sprawl. The formulation and implementation of urban development plans in Turkey is guided by the statutory provisions of the country's planning system. The multiple plans are lack of determined policies and the multiplicity of speculative market forces and their heavy demands in Turkey. Thus, metropolitan growth was transformed into an oil-spill form in most of the Turkish cities.

Metropolitan areas have been attracting a large population from rural areas since the 1960s. Although the rate of migration has slowed down in the last two decades, it has caused a considerable increase in the population of the metropolitan areas of Turkey (Erkip 2005). Turkey was urbanizing fast, changing from an agriculture to an urban-based economy. Two major steps were taken in 1960s. A new ministry of planning was established to deal with planning, housing and infrastructural issues, being responsible to draw urban development strategies, to undertake housing programmes and improvement of squatter housing, and above all to approve all plans prepared and adopted in local municipal councils (Balta and Eke 2011).

The new value systems, procedures and concepts brought to the era by globalization, privatization and neoliberalism affected Turkey as well. Two major reforms were undertaken in the Turkish planning system in 1984. Besides the classical municipal administration for cities, metropolitan municipalities were established responsible for larger cities to plan and control the metropolitan region and the municipalities in that area, the Metropolitan Planning Offices being joined to the metropolitan municipalities. Metropolitan planning offices were established in large cities with the collaboration of both local and central authorities which had the aim of formulating new policies, models and methods to confront the negative consequences of metropolitanism. The second major reform was the decentralization of several administrative powers including planning. The local authorities were now responsible for all the stages of planning practice, several administrative powers including approval (Balta and Eke 2011).

In Turkey, local municipalities is flexible regulation of urban development. Cities can be changed by partial plans or planning decisions (Ozuduru and Varol 2009). This process has led to piecemeal implementation through partial plans and plan amendments in metropolitan areas.

As many other developing countries Turkey was faced with the problem of fast urban growth. However, the globalisation trends coupled with the decentralization demands of the age which also affected Turkey caused the abolishment of the planning units, leaving the ground to local authorities which could not stand against the pressures of market forces so easily and urban sprawl has been occurred.

In Turkey, urban development plans are detailed end-state blueprint plans, which envision that a time would come and the spatial development of any city would be completed in the specific planning period. The allegation of the planning system is to control every detail during urban development. Development plans, of which structure and content are defined in planning legislation, are the main planning control tools in the Turkish system. Urban Development Law (3194), issued in 1985, exists at the very center of this legislation and is the main law directly related to production of the urban built environment (Ünlü 2005).

Urban planning in Turkey are guided by the statutory provisions of the country's planning system. In metropolitan cities, master plans are prepared by metropolitan municipalities, and implementation plans by county municipalities, and are then approved. The process stimulates the development of entities' individual behavior, and individual acts come to the foreground primarily through piecemeal implementations as partial plans and plan amendments. Local governments frequently use partial plans for the purpose of steering the public benefit. In actuality, the urban space is broken into pieces and is privatized through these plans, and public spaces are turned over to the private sector through urban projects/partial plans. To allay reactions that the public may put forth, local governments define the plans in planning regulations and apply the procedures anticipated by the law predominantly in a stylistic manner, thereby rendering the applications different from the plan decisions. Partial plans may completely change the urban settlement character as urban sprawl. Since 1980, urban development in metropolitan areas in Turkey has been primarily incremental rather than holistic. In the process, urban development planning has tended to be piecemeal, resulting in the fragmentation of the urban built environment (Balta et al. 2012).

In 1980s, after the legal arrangements, the central government's role in planning has diminished, and authority has been decentralized to local governments. However, in metropolitan areas, the limited capacity and experience of local governments resulted in urban space to be developed through the private sector. So, urban fringe faced partial developments. The holes in planning legislation and the popularity of partial plans resulted in a urban sprawl in metropolitan areas.

The rapid growth in metropolitan areas resulted in infrastructure, environment, and housing problems. In contrast, density increments and land use changes in urban environment have been realized through plan modifications.

Conclusion

In the twentieth century duc to rapid growth, the structure of urban settlements has changed dramatically. Metropolitan areas were mostly affected by increasing population. Metropolitan growth will either concentrate within the city boundaries, increasing the density, but most likely the city will expand on the periphery of the city through decentralization. The detrimental consequences of urban sprawl can be solved through planned decentralization. The diverse existing urban systems must be coordinated with the new development areas in order to establish the integrity of the macroform.

The new value systems, procedures and concepts brought to the era by globalization affected Turkey as well. Turkey is also similar to other developing countries where metropolitan growth is due to fast population growth and results in urban sprawl. The multiple plans were not effectively implemented due to lack of policies because of speculative market forces and their heavy demands in Turkey.

Consequently, partial plans have resulted in uncontrolled development and metropolitan growth was transformed into an oil-spill form in Turkey. Such development increases infrastructure costs and burdens local authorities. Urban sprawl creates problems, such as incongruity of functions, environmental problems, and higher of public services cost, increased energy consumption. To overcome these problems, the uncontrolled growth should be stopped and development policies that address both public and private sectors should be addressed.

References

Balta, M. Ö., & Eke, F. (2011). Spatial reflection of urban planning in metropolitan areas and urban rent: A case study of Cayyolu, Ankara. *European Planning Studies (SSCI), 19*(10), 1817–1838.

Balta, M. Ö., Tekel, A., & Tekel, İ. (2012). Urban development process of built environments in metropolitan areas in Turkey: A case study of Angora Settlement, Ankara. *Journal of Urban Planning and Development-ASCE (SCI), 138*(1), 70–77.

Beauregard, R. A. (2006). *When America became suburban*. Minneapolis: University of Minnesota Press.

Burchell, R., Shad, N. A., Listokin, D., Phillips, H., Downs, A., Seskin, S., Davis, J. S., Moore, T., Helton, D., & Gall, M. (1998). *The costs of sprawl, revisited. Transportation research board* (pp. 1–40). Washington, DC: National Research Council.

Cox, W., & Utt, J. (2004). *The costs of sprawl reconsidered: What the data really show* (pp. 1–18). Unpublished paper. The Heritage Foundation. Retrieved May 12, 2009, from http://www.heritage.org/research/smartgrowth/.

EEA. 2006. Urban sprawl in Europe—The Ignored Challenge EEA Report No.: 10/2006. Retrieved May 12, 2009, from http://reports.eea.europa.eu/eea_report_2006_10/en/eea_report_10_2006.pdf.

Erkip, F. (2005). The rise of the shopping mall in Turkey: The use and appeal of a mall in Ankara. *Cities, 22*(2), 89–108.

Ewing, R. H. (1994). Characteristics, causes and effects of sprawl: A literature review. *Environmental and Urban Studies, 21*, 1–15.

Ewing, R. (1997). Is Los Angeles—Style sprawl desirable? *Journal of the American Planning Association, 63*(1), 2–4.

Ewing, R., Pendall, R., & Chen, D. (2002). *Measuring sprawl and its impact* (pp. 1–31). Washington, DC: Smart Growth America.

Gillette, H. (2005). *Camden after the fall: Decline and renewal in a post-industrial city*. Philadelphia: University of Pennsylvania Press.

Glicksman, L. R. (2007). Editorial: The energy crisis—The need for more balanced solutions. *HVAC&R Research Journal, 13*(4), 521–523.

Ozuduru, B. H, & Varol, C. (2009). Global restructuring of the marketplace: Local and regional impacts on the spatial distribution of retail activity in Turkey. In *23rd congress of the Association of European Schools of Planning (AESOP)*, Liverpool, İngiltere.

Snyder, K., & Bird, L. (1998). *Paying the Costs of Sprawl: Using Fair-Share Costing to Control Sprawl*. Washington, DC: U.S. Department of Energy.

Squires, G. D. (2002). *Urban sprawl: Causes, consequences & policy responses*. Washington, DC: The Urban Institute Press.

Teaford, J. C. (2006). *The metropolitan revolution: The rise of post-urban America*. New York: Columbia University Press.

Ünlü, T. (2005). *Plan modifications within the contexts of planning control mechanisms Mersin case*. Unpublished PhD thesis, School of Natural and Applied Sciences of Middle East Technical University, Ankara

Chapter 2
Perspectives on the Implementation of Climate Change Public Policies in Brazil

Rommel de Santana Freire, Monica Carvalho, Charles Ulises de Montreuil Carmona, and Alexandre Magno Vieira Gonçalves de Brito

Introduction

As a developing country, Brazil plays an important role in the global scenario for the reduction of greenhouse gas (GHG) emissions. Several public policies have been created towards the reduction of pollution levels. Brazil's participation in international agreements and the creation of a positive agenda for the development of new regulations for environmental presentation direct the country to the construction of a cleaner scenario.

Brazil participates in the Kyoto Protocol (largest international agreement signed for the reduction of GHG emissions), but still does not count with established mandatory reduction levels. However, the country already acts in a pro-active way through the creation of several sectional plans for mitigation and adaptation

R. de Santana Freire (✉)
Department of Accountancy and Finances, Center of Applied Social Sciences, Federal University of Paraíba (UFPB), Cidade Universitária, João Pessoa, 58051-900 Paraíba, Brazil
e-mail: rommelfreire@gmail.com

M. Carvalho
Department of Renewable Energy Engineering, Center of Alternative and Renewable Energy, Federal University of Paraíba (UFPB), Caixa Postal 5115, Cidade Universitária, João Pessoa, 58051-900 Paraíba, Brazil
e-mail: monica@cear.ufpb.br

C.U. de Montreuil Carmona
Department of Management Science, Center of Applied Social Sciences, Federal University of Pernambuco (UFPE), Cidade Universitária, Recife, 50670-901 Pernambuco, Brazil
e-mail: charles.carmona@gmail.com

A.M.V.G. de Brito
Department of Mechanical Engineering, Center of Technology, Federal University of Paraíba (UFPB), Cidade Universitária, João Pessoa, 58051-900 Paraíba, Brazil
e-mail: alexandreh_magno@hotmail.com

P. Grammelis (ed.), *Energy, Transportation and Global Warming*, Green Energy and Technology, DOI 10.1007/978-3-319-30127-3_2

(Brasil 2011). These plans contemplate strategic areas (forests, energy in agriculture, urban mobility) and industry intensive sectors (metallurgy, mining, transformation). The focus is to reduce the levels of pollution and at the same time, improve the performance and competitiveness of companies. The process of implementing these regulations is slow, as several bureaucratic and cultural barriers still remain, which hinders the obtainment of better results.

A positive point that helps the country to maintain acceptable levels of pollution is the constitution of its energy matrix, which is mainly constituted of hydroelectric plants and therefore presents lower levels of pollution. Incentives to energy generation utilizing renewable sources is still very limited. In a different direction, in 2013 in the European Union it was observed that the energy generation sector was responsible for the largest share of CO_2 emissions—70 % of these companies were located in Germany and England (Carbon Market Data 2014).

Despite the existence of policies directed to the strengthening of energy generation from renewable sources, implementation is still expensive and complex, hindering its application for the majority of the population. Developing countries, such as China and India, count with greater incentives for the installation of residential equipment that better utilize renewable energy sources (World Bank 2014). Regarding the logic of reduction of emissions, the European Union created the European Union Emission Trading Scheme (EU ETS), responsible for regulating the operation of the market for reduction of carbon emissions, establishing goals for the reduction of pollution and distributing the allowance quotas for carbon emissions (European Emission Allowances, EUA) among the members of the European community (Freire 2014).

Currently, the European Union is the largest negotiator of EUAs in the world, also gathering the largest number of companies that follow the determinations of the Kyoto Protocol. Globally, this new active (EUA) has been widely negotiated since the mid-2000s, mainly in stock markets that deal with energy-related contracts, taking advantage of the already existent structure and know-how (Knox-Hayes 2009). The EUA market has grown substantially in the recent years, and can reach the position of being one of the largest commodity markets in the world in the next decades (Pirrong 2011).

However, after the drop in value of EUA's during the 2006 and 2007 crisis and re-establishment of carbon markets, similar methodologies to EU ETS have been applied to several countries, such as Japan, Canada, New Zealand, Australia and United Kingdom—the U.S.A. (especially the state of California) and China deserve special mention (Perdan and Azapagic 2011).

The creation of these new markets shows the directions of public policies that aim at finding a way to reduce levels of emission of pollutants as well as contributing to energy and operational efficiencies of companies, creating therefore a new entrepreneurial culture that is directly related to the wellbeing of the society. A companion paper on the promotion of sustainability by quantifying and reducing the carbon footprint lays the foundation for environmental awareness in consumers (Carvalho et al. 2015).

On the basis of an agreement with the United Nations, Brazil established voluntary actions for the mitigation of GHG, with the objective of reducing between 36.1 % and 38.9 % until 2020, fostering national policies on climate change. Among these actions, the development of a Brazilian Market for Reduction of Emissions is stimulated. The possibility of creating a nation-wide carbon market, with its own regulation and interconnection with other markets in the world, can bring positive contributions from economic and environmental viewpoints. This manuscript aims to deepen the discussion on the subject, highlighting the benefits and opportunities than can be generated by the implementation of public policies, as well as the creation of new supplementary regulations.

Public Policies

The oil crisis, in the beginning of the 1970s, led the Brazilian government to create its first study group on alternative sources of energy, being the first step towards the creation of public policies regarding energy efficiency. After crystallization of the first research efforts on conservation of energy, the first national program on conservation of energy was launched in 1985. In the early 1990s, the first programs that aimed at rational production and use of energy were created, along with programs for better use of oil and natural gas derivatives. These programs represented a milestone for environmental and energy efficiency policies.

In the following decades, Brazil continued to create and improve its national policies on climate change, involving different production sectors and governmental organs. The objective is to create synergy between the involved areas, and the main expected result is the sustainable development of the country, balancing economic growth with environmental preservation. The current economic moment presents low growth, hindering therefore the implementation of environmental regulations that entail excessive costs. Laws that stimulate investments in new equipment, processes or green products, through tributary incentives or for the generation of green revenues, are seen as possible solutions for the mitigation of climate change.

Politic guidelines lead the path to be followed by companies and society in general, in the form of incentives such as tax reductions for the companies that reduce their emissions or apply resources to efficiency energy researches. The creation of new mechanisms to this kind of development and GHG control, especially in the most polluting sectors is fundamental to implement a new culture of environmental preservation. In this new environment, the organizations will be required to optimize its processes/products in order to mitigate the effects of pollution generated.

The Brazilian National Policy of Climate Change (PNMC) was established in 2009, through the adoption of Law 12.187 (Ludeña and Netto 2011). PNMC pursues voluntary actions for the mitigation of GHG emissions, including reduction in deforestation (Amazon and Cerrado), reduction in energy consumption (energy

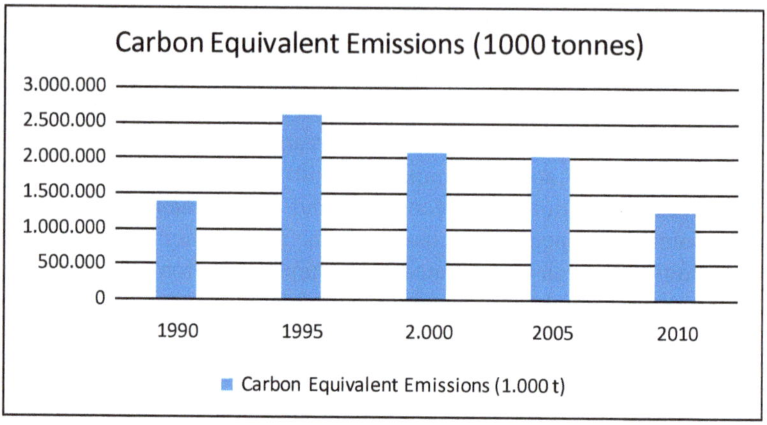

Fig. 2.1 Overall carbon emissions for Brazil (Brasil 2015)

efficiency, alternative energy sources, etc.), stabilization of the share of renewable energy sources in the energy matrix, and increased use of biofuels, among others. Brazil's GHG emissions are calculated every 5 years, and Fig. 2.1 shows the reduction of emissions through the implementation of PNMC, a non-mandatory policy. These data represent the first results after its implementation.

The Brazilian legislation aims equalize sustainable development with economic growth, eradicating poverty and reducing social inequalities. This new interconnected vision drives the country to build a wide platform of emission reduction alternatives. The Brazilian law also encourages the development of a national market of reduce emissions, which in practice is not yet operational, however the main focus is economic and social development compatibility with the protection of the climate system (Brasil 2009, 2010).

The changes in the carbon market could modify the pollution levels defined by the legislation (Montgomery 1972), resulting in greater control of the most polluting activities. The restrictions imposed on companies that exceed the pollution limits could be seen as barriers to competitiveness, which would affect those that pollute more (Perdan and Azapagic 2011).

Several countries are adapting theirs energy and climate policies to impose a new behavior to the companies, one that focuses on cost optimization and reductions of GHG emissions, representing a new era for sustainable development. This represents a new view that economy growth is inextricable from environmental preservation.

The conclusions of the European Council in 2011 (CO EUR 2 CONCL 1 2011) highlight the direction taken by the European Union regarding much necessary environmental policies, with emphasis on sustainable growth, energy efficiency and the necessity of creating policies that are capable of providing more investments in research, development and innovation (Freire 2014). These three pillars are united to create a more modern, safe, sustainable and low carbon energy system (European Council 2011). In 2014, the European Council ratified the relevance to improve the economic growth through the use of energy efficiency and the improvement of competitiveness, maintaining the targets of greenhouse gas emission reduction (CO EUR 2 CONCL 1 2014). Using similar strategy, China has created its own ETS in seven regions of the country. The system will be in adaptation stage until the end of 2015, when the emission reduction plan will be in force for the entire country (Zhang et al. 2014).

Under a regulated market, managers will be enforced to create a new vision of company, using threats as opportunities in the carbon market, creating competitive advantage aims to maximize the shareholders' wealth on behalf of other stakeholders (society and government). Cleary the news ETS needs changes to adapt local/regional issues, but implementation is the first step for developing countries and shows the government provides good conditions for the carbon market start up.

Brazil could follow the same path of other countries, but first needs to implement the existing legislation and adapt/create complementary rules to improve the creation of its own ETS, to create a positive chain between companies-government-society.

Perspectives and Opportunities for Mitigation

Until recently, environment and sustainable development were considered problems and risk factors: however nowadays, these "problems" are seen as opportunities, as possibilities for growth and efficiency improvement (Carvalho et al. 2015). Building upon the work of Carvalho and Freire (2014), mitigation of climate change involves serious reduction in emissions, which can be achieved through the better utilization of natural resources (energy efficiency concept). Obviously, if the same amount of product can be obtained with reduced consumption of resources, lower production costs are also achieved, which is a win-win situation for managers and entrepreneurs.

The rise of new technologies and its learning curve costs contributes to a reduction of transaction costs of organizations (Williamson 1994, 1995), as well to reductions of pollutants gases (Blyth et al. 2009). R&D investments in production models more "clean" and efficient, linked with new environmental policies rise as option to improve energy efficiency consumption in companies, specially to pollution reduction (Gans 2012).

Important opportunity for mitigation of climate change in Brazil include deforestation (Ludeña and Netto 2011) and energy-related carbon emissions (Melo et al. 2013). These mitigation efforts are translated into challenges at federal, state and municipal levels, for both public and private sectors in Brazil.

The reduction of GHG emissions from deforestation and forest degradation is now an important strategy for mitigating climate change, particularly in developing countries with large forests, such as Brazil (Arima et al. 2014). National forest policies designed to slow deforestation on public lands in developing countries have had mixed success. However, there have been successful experiences (IPCC

2007): due to severe environmental and public health consequences, China, the Philippines and Thailand have considerably reduced deforestation rates; in India, there has been effective partnering with communities to reduce forest degradation. The implementation of effective public policies needs backup from government institutions and the general public. Considering the Brazilian scenario of limited budgets, divergent interests and legal obligations, decision makers in Brazil face multiple difficulties in the process of searching for adequate and reliable solutions for the mitigation of GHG emissions (Melo et al. 2013).

In Brazil, buildings were responsible for 48 % of the total electricity consumed in the country in 2010 (Brasil 2011); this important share of electricity consumption could obviously benefit from energy efficiency and conservation strategies aimed at the reduction of energy-related GHG emissions. It is expected that the demand for electricity will increase in Brazil in the next years–there is little wiggle room for hydroelectric power plants to produce more, and therefore thermoelectric power plants based on coal and natural gas will have to increase their contributions to the energy mix. This will lead to higher environmental loads associated with the consumption of electricity, and more overall environmental loads being disposed of into the atmosphere—this figure will definitively be higher than what was predicted in the Brazilian Energy Plan 2030 (Melo et al. 2013). Therefore any decrease in environmental loads will be the result of dissemination efforts of energy efficiency technologies, most probably through public policy mechanisms.

According to Lucon et al. (2014), energy efficiency technologies have been developed at the same time as existing energy efficiency opportunities have been taken up, and therefore the potential for cost-effective energy efficiency improvement is increasing. However, there are barriers to the market uptake of these opportunities, and large potentials will remain untapped without appropriate policies (Lucon et al. 2014): in addition to technologies and architecture, behavior, lifestyle, and culture have a major effect on energy use.

A very promising energy efficiency scenario is linked to the establishment of public policies that highlight well-defined priorities, goals and action plans. A sensitive point involves the education for the development of a collective conscience towards the preservation of the environment. The responsible agencies should widen the education, research and development actions for new technologies and establish partnerships with the private and public sectors. Changes in energy consumers' behavior and practices can lead to considerable reductions in energy demands, which present positives effects on climate-change issues (e.g., reduction of carbon emissions). Conscious energy behavior can be achieved through educational interventions, and by combining interventions with incentives, substantial reductions in energy use can be produced (Kirby et al. 2015).

There is growing acknowledgement on the importance of changing the energy-related behavior of people. Multiple research efforts have been directed to the subject of pro-environmental behavior change (of which energy-related behavior is a component) (Energy Saving Trust 2007). If more consumers agree to a shift towards a more energy-savvy behavior, important reductions in environmental loads will be achieved. Energy-education could help change current and future common routines to save energy, and also encourage consumers to promote similar behavior. One of the main incentives that propel energy users to adopt energy-savvy behaviors and practices is the reduction in consumption, accompanied by the reduction in costs.

Final Remarks

The main conclusion of La Rovere et al. (2014) is that Brazil is in a good position to meet its mitigation goals until 2020. However, after 2020, Brazil will be challenged to combine economic development with low carbon energy-related emissions.

The creation of a system for trading GHG emissions in Brazil would encourage Brazilian companies to reduce their emissions and trade these in a regulated market, achieving double benefits: reduction of emissions (less consumption of resources, lower costs) and benefits realized through the trade of emissions. This model is similar to the model established by the Kyoto Protocol and reproduced internally in many countries.

New environmental policies combined with financial incentives lead company managers to promote the reduction of GHG emissions through innovation/adaptation of processes and products. This action would foster economic development that focuses on environmental preservation, promoting the development of a new culture, where people would vigorously demand the implementation of green policies by companies. Brazil has the potential to maintain sustainable economic growth through the implementation of existing policies and the creation of new mechanisms, permitting the creation of an internal carbon market, which can also be connected to other markets already consolidated in the world.

Acknowledgments The authors wish to acknowledge the support of the Institute for the Development of Paraíba (IDEP-UFPB) and of the National Council for Scientific and Technological Development (CNPq), through project n° 475879/2013-9.

References

Arima, E. Y., Barreto, P., Araújo, E., & Soares-Filho, B. (2014). Public policies can reduce tropical deforestation: Lessons and challenges from Brazil. *Land Use Policy, 41*, 465–473.

Blyth, W., Bunn, D., Kettunen, J., & Wilson, T. (2009). Policy interactions, risk and price formation in carbon markets. *Energy Policy, 37*, 5192–5207.

Brasil. Law n° 12.187, of December 29 2009. Institutes the National Policy on Climate Change—PNMC. Official Diary of the Federative Republic of Brazil (in Portuguese).

Brasil. Decree n° 7.390, of December 9 2010. Provides regulations on paragraphs 6, 11 and 12 of Law 12.187, of December 29 2010, which instituted the National Policy on Climate Change—PNMC. Official Diary of the Federative Republic of Brazil (in Portuguese).

Brasil, Ministry of Environment. *Gas emissions*. Retrieved April 11, 2015, from http://mma.gov.br/mma-em-numeros/emiss%C3%B5es-de-gee.

Carbon Market Data. Retrieved April 11, 2015, from www.carbonmarketdata.com.
Carvalho, M., & Freire, R. S. (2014) Quantificar, reduzir, evidenciar: uma nova prática para empresas sustentáveis. In *ENCONTRO INTERNACIONAL SOBRE GESTÃO EMPRESARIAL E MEIO AMBIENTE*, 16, 2014, São Paulo. Anais... São Paulo: FEA-USP (in Portuguese).
Carvalho, M., Freire, R. S., & Brito, A. M. V. G. (2015) Promotion of sustainability by quantifying and reducing the carbon footprint: New practices for organizations. In *Global conference on global warming*, Athens, 2015.
Energy Saving Trust. (2007). Retrieved April 7, 2015, from http://www.teachshare.org.uk/ExtFiles/EnergyTrust-TheRoleOfEducationAndSchools.pdf.
European Council. CO EUR 2 CONCL 1, February 01, 2011. Brussels, March 8 2011.
European Council. CO EUR 2 CONCL 1, July 01, 2014. Brussels, March 21, 2014.
Freire, R. S. (2014). *Essays on the permissions for carbon emissions (EUA)*. PhD thesis, Graduate Programe in Management (PROPAD), Federal University of Pernambuco (UFPE), Brazil.
Gans, J. S. (2012). Innovation and climate change policy. *American Economic Journal: Economic Policy, 4*(4), 125–145.
IPCC. (2007). Climate change 2007: Working group III: Mitigation of climate change. 9.6.1 Policies aimed at reducing deforestation. Retrieved April 7, 2015, from http://www.ipcc.ch/publications_and_data/ar4/wg3/en/ch9s9-6-1.html.
Kirby, S. D., Guin, A., & Langham, L. (2015). Energy education incentives: Evaluating the impact of consumer energy kits. *Journal of Extension, 53*(1), 10.
Knox-Hayes, J. (2009). The developing carbon financial service industry: Expertise, adaptation and complementarity in London and New York. *Journal of Economic Geography, 9*, 749–777.
La Rovere, E. L., Júnior, A. O. P., Dubeux, C. B. S., & Wills, W. (2014). Climate change mitigation in Brazil. *Climate and Development, 6*, 25–33.
Lucon, O., Ürge-Vorsatz, D., Zain Ahmed, A., Akbari, H., Bertoldi, P., Cabeza, L. F., Eyre, N., Gadgil, A., Harvey, L. D. D., Jiang, Y., Liphoto, E., Mirasgedis, S., Murakami, S., Parikh, J., Pyke, C., & Vilariño, M. V. (2014). Buildings. In O. Edenhofer, R. Pichs-Madruga, Y. Sokona, E. Farahani, S. Kadner, K. Seyboth, A. Adler, I. Baum, S. Brunner, P. Eickemeier, B. Kriemann, J. Savolainen, S. Schlömer, C. von Stechow, T. Zwickel, & J. C. Minx (Eds.), *Climate change 2014: Mitigation of climate change. Contribution of Working Group III to the fifth assessment report of the intergovernmental panel on climate change*. Cambridge/New York, NY: Cambridge University Press.
Ludeña, C., & Netto, M. (2011). *Brazil: Mitigation and adaptation to climate change*. Inter-American Development Bank. Technical note IDB TN 622.
Melo, C. A., Januzzi, G. M., & Tripodia, A. F. (2013). Evaluating public policy mechanisms for climate change mitigation in Brazilian buildings sector. *Energy Policy, 61*, 1200–1211.
Montgomery, W. D. (1972). Markets in licenses and efficient pollution control programs. *Journal of Economic Theory, 5*, 395–418.
Perdan, S., & Azapagic, A. (2011). Carbon trading: Current schemes and future developments. *Energy Policy, 39*, 6040–6054.
Pirrong, C. (2011). *Commodity price dynamics: A structural approach*. Cambridge: Cambridge University Press.
Williamson, O. E. (1994). Strategizing, economizing, and economic organization. In R. Rumelt, D. Schendel, & D. Teece (Eds.), *Fundamental issues in strategy: A research agenda*. Boston: Harvard Business School Press.
Williamson, O. E. (1995). Transaction cost economics and organization theory. In O. E. Williamson (Ed.), *Organization theory: From Chester Barnard to the present and beyond*. New York: Oxford University Press.
World Bank. (2014). *State and trends of carbon pricing 2014*. Washington, DC: World Bank.
Zhang, D., Karplus, V. J., Cassisa, C., & Zhang, X. (2014). Emissions trading in China: Progress and prospects. *Energy Policy, 75*, 9–16.

Chapter 3
30 Years Air Temperature Data Analysis in Athens and Thessaloniki, Greece

T. Slini and K.T. Papakostas

Introduction

The topic of climate change has attracted widespread attention in recent years and is an issue that numerous scientists study on various time and space scales. The reasons for climate change are complex and generate extensive discussion and dispute in the scientific community. The main cause is the greenhouse gas emissions produced by anthropogenic activities, e.g. industry, agriculture, transport. In all these human activities, energy comes mainly from non-renewable sources (liquid, gas and solid fuels) producing emissions that trigger the global warming. It is widely accepted that increasing concentrations of greenhouse gases in the atmosphere is the dominant driver of global warming and climate change. Climate change and its impact are met with various ways: increasing temperature, changing the distribution of rainfall floods, desertification and reduction of water resources, forest fires, shrinking glaciers, rising sea levels.

Over the past 100 years the increase in global average temperature was about 0.74 °C and 11 out of the 12 years in the period 1995–2006 are listed among the top warmest years with reference to the historic beginning of recorded measurements by scientific instruments, i.e. 1880 (IPCC 2007). Moreover, according to recent data (NOAA 2014), on a global scale, the decade 2001–2010 was the warmest ever recorded. Consequently, global warming is projected to persist, if greenhouse gases

T. Slini (✉)
Department of Mechanical Engineering, Laboratory of Heat Transfer and Environmental Engineering, Aristotle University of Thessaloniki, Box 483, 54124 Thessaloniki, Greece
e-mail: lora@aix.meng.auth.gr

K.T. Papakostas
Department of Mechanical Engineering, Laboratory of Process Equipment Design, Aristotle University of Thessaloniki, 54124 Thessaloniki, Greece
e-mail: dinpap@eng.auth.gr

P. Grammelis (ed.), *Energy, Transportation and Global Warming*, Green Energy and Technology, DOI 10.1007/978-3-319-30127-3_3

will continue to be emitted at the same rate. It is estimated that surface air warming in the twenty-first century for a "low emission scenario" will be 1.8 °C, with a likely range of 1.1–2.9 °C, while in a "high emissions scenario", it will be an increase of 4.0 °C with a likely range from 2.4 to 6.4 °C. A temperature increase of 0.1 °C per decade would be expected for the next two decades, even if concentrations of greenhouse gases and aerosols were kept at the levels of the year 2000 (IPCC 2013). In particular, the average global temperature increased after 1978 with an average rate of 0.14 °C per decade, while the average global temperature over land and sea surface reached the record highest temperature (14.47 °C) in April 2010, which is an increase of 0.77 °C as compared to the average of the twentieth century (13.7 °C). This event was characterised as one of the greatest climate anomalies at global scale based on NOAA data (URL 1). According to the same data, the global average temperature in April 2014 over land was 9.45 °C that is 1.35 °C above the average of the twentieth century (8.1 °C), while the average global temperature of April 2014 in the sea was 16.55 °C that is 0.55 °C above the average of the twentieth century (16 °C).

Regarding Greece, according to a report conducted by the Climate Change Impacts Study Committee (Climate Change Impacts Study Committee 2011) and the National Observatory of Athens (WWF 2009), urban centres (e.g., Thessaloniki, Patras, Larissa and Lamia) are expected to face annually a rise at the number of days with maximum temperature above 35 °C, and at the number of "hot" nights, during which the temperature is maintained above 20 °C, with the period 1961–1990 as a reference.

In addition, in recent decades the frequency of occurrence of lower temperature decreases while the frequency of occurrence of higher temperature increases (Papakostas et al. 2013, 2014) and at the same time, the urban heat island effect (Urban Heat Island) is observed in several Greek cities (Santamouris et al. 2001; Kolokotsa et al. 2009; Giannaros and Melas 2012; Vardoulakis et al. 2013; Giannopoulou et al. 2011; Livada et al. 2002). The effects of climate change are of particular interest, as even a slight temperature change is reflected directly in socio-economic sectors such as tourism, agriculture, mining industry, transport, built environment, quality of life and health (Giannakopoulos et al. 2011) as well as in energy consumption in buildings (Papakostas et al. 2010, 2013, 2014; Santamouris et al. 2001).

According to the World Meteorological Organization (WMO) the current official baseline for meteorological models is the 30-year time period from 1961 to 1990. At the same time this reference time period is under review for the "adjusted" evaluation of the latest meteorological data (URL 2). The new proposed baseline is the period 1981–2010, which is already used by several meteorological organizations so as to provide more updated and reliable services for both citizens and policy makers regarding issues such as energy and peak load, civil protection planning and strategic recommendations.

Based on the above, the purpose of this work is to analyze the ambient air temperature in Athens and Thessaloniki for 30 years, namely the period 1983–2012, in order to identify the statistically significant changes and evaluate the current trends in urban sites in Greece.

Methodology

The hourly time series of outdoor air temperature of three decades (1983–2012) for the cities of Athens and Thessaloniki are thoroughly analysed in the frame of the current study. The measurement data are obtained from the meteorological station of the National Observatory of Athens (NOA) (NOA Climatological Bulletin) located on a small hill near the Acropolis (with coordinates 37°58′N, 23°43′E and altitude 107 m), and respectively from the station maintained by the Institute of Meteorology and Climatology of the Aristotle University of Thessaloniki (IMC/AUTh) (Department of Meteorology and Climatology) within the university campus (with coordinates 40°37′N, 22°57′E and altitude 31 m). The measurements were checked for homogeneity (Founda et al. 2009) and the missing values were negligible and did not affect the reliability and quality of results. Specifically, for the Athens area recorded about 1077 out of 263,424 observations (percentage of 0.4 %) and Thessaloniki about 567 in total of 263,520 (percentage of 0.2 %).

For the need of the study descriptive statistics indicators were estimated such as the mean, the standard deviation, the frequency, and quartiles in hourly, daily, monthly, annual and decadal scale. Graphic representation of the data was performed in order to detect and capture any evident trend, seasonality or outlier.

The average for the period 1961–1990 was employed as a benchmark for the identification of outliers and deviations from the "normal" values. Furthermore, according to the Intergovernmental Panel on Climate Change (IPCC), the change in indicators is mainly expressed as deviations from the average or normal mode. The current study adopted the indicators used in the reports of IPCC (IPCC 2007) and the Joint Research Center (2008). In this view, the number of days with daily average of more than 25 °C and daily maximum temperature above 35 °C were estimated.

Finally statistical tests were carried out in order to identify any statistically significant changes in temperature at a significance level of 0.05.

Results

The main results of the analysis are presented in the following graphs exhibiting the average minimum, mean and average maximum monthly values of temperature (°C), the number of days with daily average temperature above 25 °C, the number of days with a maximum value greater than or equal to 35 °C, and the cumulative frequency curves of temperature per decade for the two areas under study. It is noteworthy to point out the intense temperature variations between the decades 1983–1992 and 2003–2012 in all graphs especially during the summer months.

More specifically, there is a gradual increase in average minimum, mean and maximum average monthly values of temperature from decade to decade (Fig. 3.1). Note that in Athens, the differentiation between the temperature per decade is

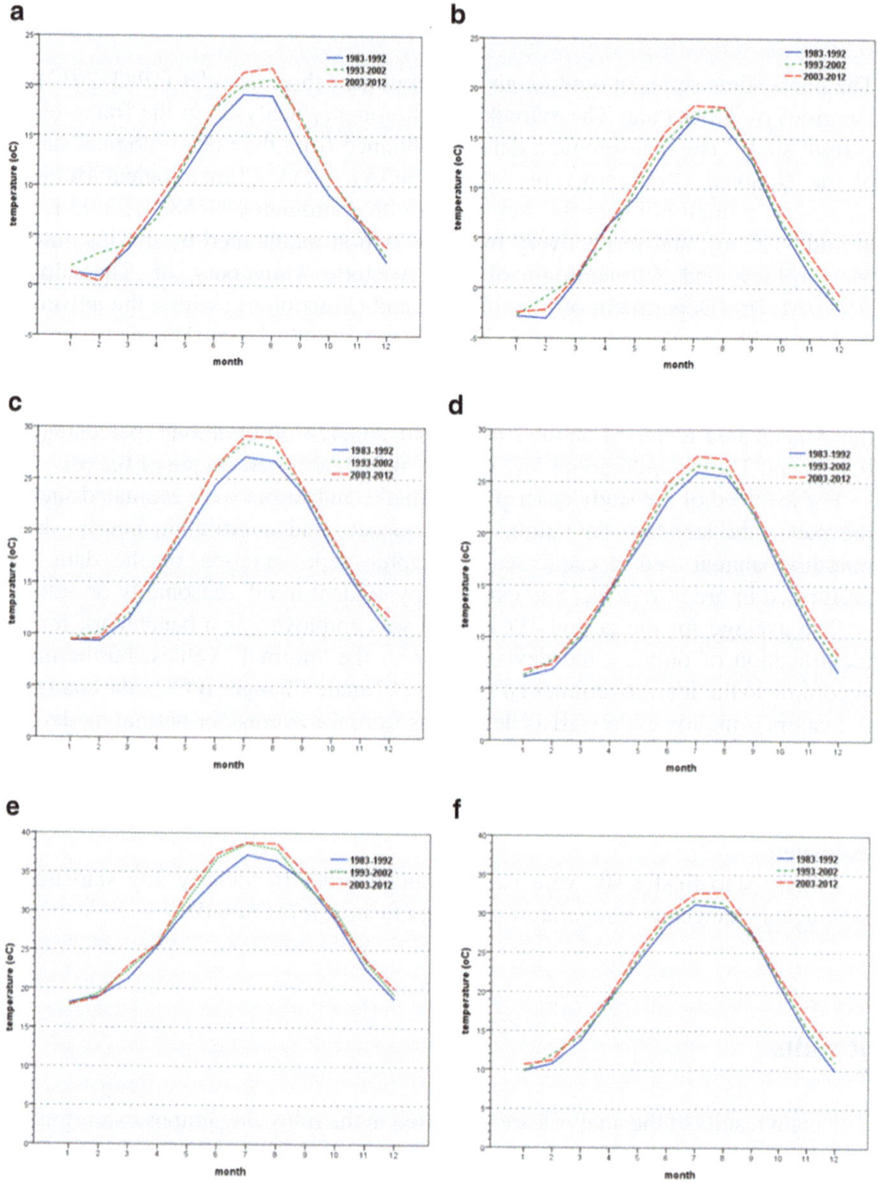

Fig. 3.1 Average minimum, mean, and average maximum monthly temperatures per decade, in Athens (**a**, **c**, **e**) and Thessaloniki (**b**, **d**, **f**): period 1983–2012

statistically significant both in annual scale and individually throughout the summer months, based on the non-parametric Kruskal-Wallis ($p < 0.05$) test, indicating that this variation is not due to a random effect. Respectively the results for Thessaloniki prove that there is a statistically significant difference to both the minimum and maximum values. In particular regarding the variation between the first decade, 1983–1992, and the third decade, 2003–2012, the average minimum monthly temperatures in Athens ranged from 0.1 °C (month of January) to 2.94 °C (month of September), the average monthly temperatures range from 0.04 °C (month of January) to 2.25 °C (month of August), while the average maximum temperatures range from 0.15 °C (month of February) to 2.26 °C (month of August). In Thessaloniki, the change in mean minimum monthly temperatures range from 0.24 °C (month of April) to 1.90 °C (month of August), the average monthly temperatures range from 0.13 °C (month of April) to 1.66 °C (month of August), while the average maximum temperatures ranging from 0.16 °C (month of January) to 2.02 °C (month of August). Moreover, it was observed that the major increase occurred in Athens from the first (1983–1992) to the second decade (1993–2002), while the differentiation of temperatures was much lower from the second to the third decade. Instead, in Thessaloniki greater changes observed between the second and third decade, with the exception of mean minimum temperatures in August, September and October.

Regarding the number of days with daily average temperature above 25 °C, as depicted in the diagrams of Fig. 3.2, a rising trend is exhibited in the number of days for both areas. In detail, there were 717 days in the first decade, 906 in the second and 934 in the third decade in Athens (at least in the measuring range). That is the increase of days with average daily temperature above 25 °C from the first to the third decade was about 30 % and from the first to the second about 26 %. Similarly, in Thessaloniki area (at least in the measurement range), there were 494 days during the first decade, 596 during the second and 748 days in the third decade wherein the increasing tendency was about 51 % from the first to third and 25 % from the first to second decade respectively (Table 3.1).

Higher temperatures increase the amount of moisture that evaporates from land and water, leading to drought in many areas. Generally, lands and cities are affected by drought since they become more vulnerable to flooding once rain falls.

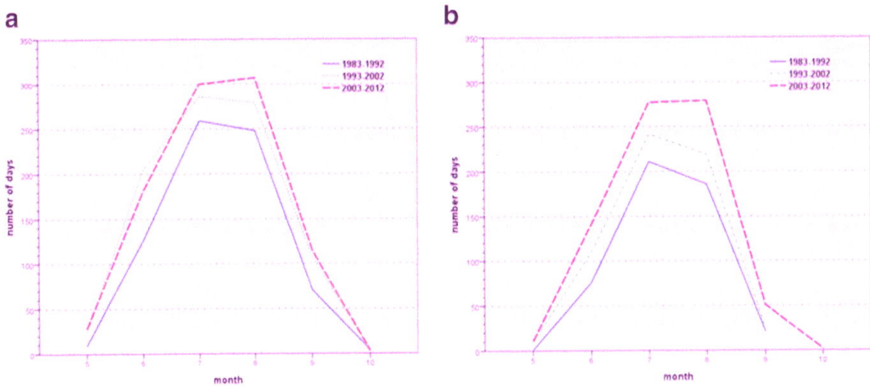

Fig. 3.2 Number of days with daily average temperature above 25 °C per decade (**a**) in Athens and (**b**) in Thessaloniki: period 1983–2012

Table 3.1 Number of days with average daily temperature above 25 °C in Athens and Thessaloniki: period 1983–2012

Athens	May	June	July	August	September	October	Total
1983–1992	9	126	259	248	71	4	717
1993–2002	24	205	286	279	107	5	906
2003–2012	28	181	300	307	115	3	934
Total	61	512	845	834	293	12	2557
Thessaloniki	May	June	July	August	September	October	Total
1983–1992	1	75	211	186	21	0	494
1993–2002	10	107	238	217	24	0	596
2003–2012	11	138	275	273	49	2	748
Total	22	320	724	676	94	2	1838

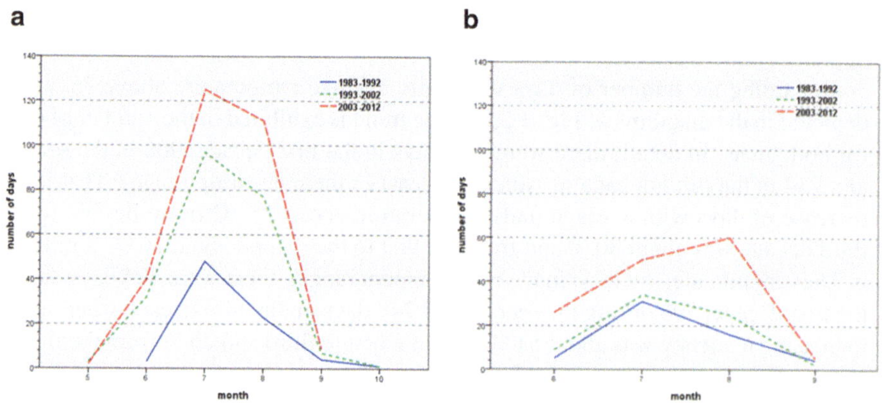

Fig. 3.3 Number of days with maximum temperature equal or greater than 35 °C per decade (**a**) in Athens and (**b**) in Thessaloniki: period 1983–2012

Temperature changes may as well increase the spread of infectious diseases, mainly because warmer temperatures allow disease-carrying insects. On the other hand temperature rise favours the touristic attractiveness during spring and autumn, especially in areas where tourism is an important economic sector.

Regarding the number of the days with a maximum temperature value equal or exceeding the limit of 35 °C, the analysis of data is presented in the graphs of Fig. 3.3 and the conclusions are similar. In the Athens area, the number of days with maximum temperature above or equal to 35 °C was estimated at 79, 217 and 287 respectively for each decade (up 175 % in the second decade and 263 % in the third, relative to the first), while in Thessaloniki on 56, 70 and 141 (an increase of 25 % in the second decade and 152 % in the third, as compared to the first). The most days with this characteristic were monitored during the third decade (2003–2012) as presented in Table 3.2, while their distribution is varying significantly within the three decades only in the Athens area (non-parametric Kruskal-Wallis test).

Table 3.2 Number of days with maximum hourly temperature equal or above 35 °C in Athens and Thessaloniki: period 1983–2012

Athens	May	June	July	August	September	October	Total
1983–1992	0	3	48	23	4	1	79
1993–2002	3	32	97	77	7	1	217
2003–2012	2	40	125	111	9	0	287
Total	5	75	270	211	20	2	583
Thessaloniki	May	June	July	August	September	October	Total
1983–1992	0	5	31	16	4	0	56
1993–2002	0	9	34	25	2	0	70
2003–2012	0	26	50	60	5	0	141
Total	0	40	115	101	11	0	267

Extreme high temperatures perhaps are the most direct effect of climate change on humans. High air temperatures in combination with high humidity on a given day increase the death danger to people with heart and respiratory problems. Higher air temperature also increases the concentration of ozone at ground level. In the lower atmosphere, ozone is a harmful pollutant. It damages lung tissues and causes problems for people with asthma and other lung diseases. Climate change would increase cardio-respiratory morbidity and mortality associated with ground-level ozone. Length, frequency and intensity of high temperatures affects also social life, limits the exposure to open spaces, causes problems to transportation and creates less favorable conditions for tourism in summer. Hot temperatures and dry conditions also increase the likelihood of forest fires, especially in the conifer forests.

Furthermore based on the cumulative frequency temperature curves in Athens (Fig. 3.4a) and Thessaloniki (Fig. 3.4b), it is also evident that the frequency of occurrence of low temperatures decreases, while the frequency of occurrence of high temperatures increases from decade to decade. This has a positive effect on the coefficient of performance of heat pumps (with air as the heating source) in winter, while it has a negative effect on the performance of heat pumps and air-cooled chillers during the summer. Particular analysis is required in order to investigate these deviations in efficiency rates (instantaneous and annual) both in winter and in summer periods. The decrease of low temperatures frequency helps additionally to reduce the demand of buildings for heating while the rise of high temperature frequency increase energy demands for cooling, which may further exacerbate peaks in electricity supply in the summer period. The net local impact of these two effects depends on the current climate in a particular area. From the temperature frequency curves in Fig. 3.4 it is concluded again that the strongest temperature alterations in Athens are monitored from the first to the second decade while in Thessaloniki a smoother change from decade to decade was observed.

Table 3.3 shows the number of weeks per decade, in which days with a maximum temperature value equal or exceeding the limit of 35 °C are observed. The graphs in Fig. 3.5a and b display these weeks during the three decades as well as the number of days per week, in which the temperature is equal to or greater than

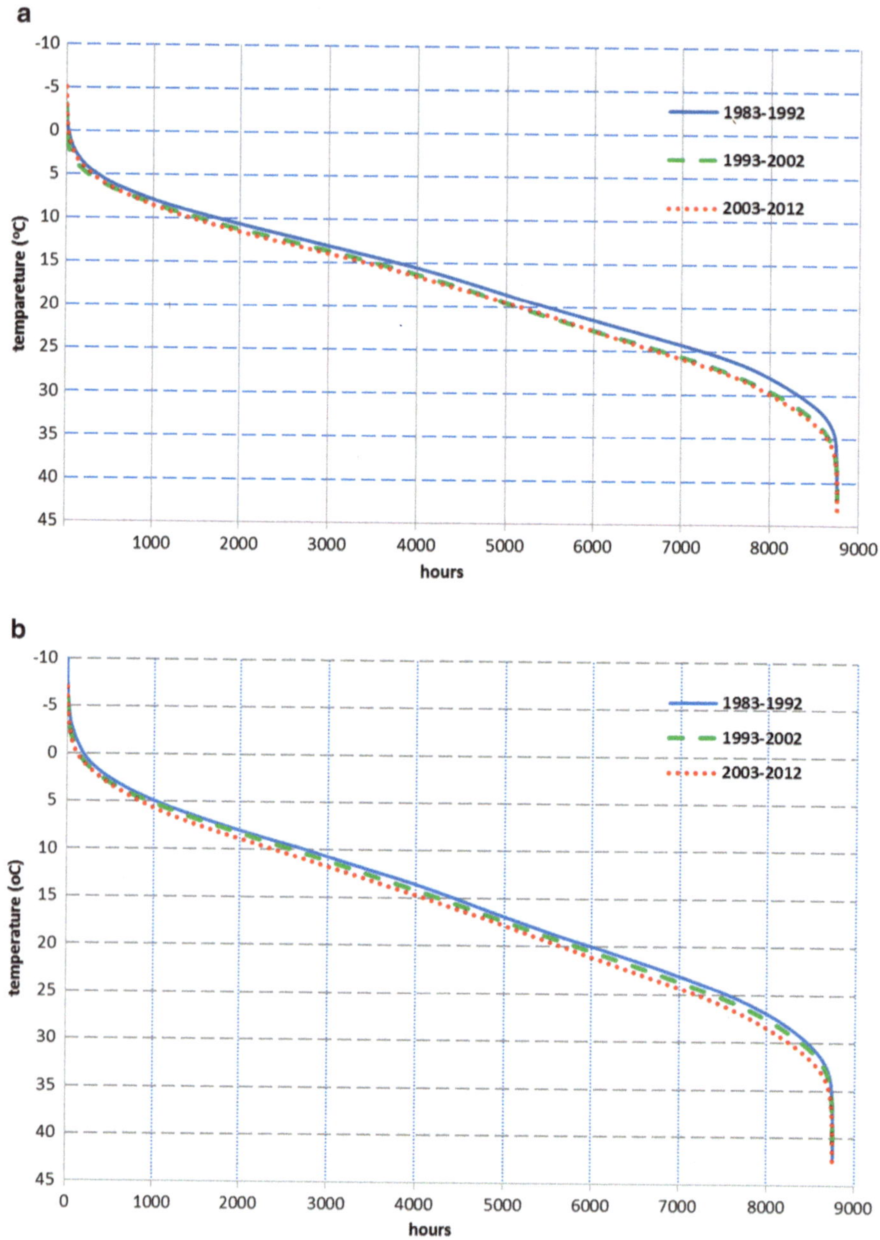

Fig. 3.4 Cumulative frequency curves of the temperature (**a**) in Athens and (**b**) in Thessaloniki for the decades 1983–1992, 1993–2002 and 2003–2012

Table 3.3 Number of weeks with daily maximum temperature equal or above 35 °C in Athens and Thessaloniki for the period 1983–2012

	Athens	Thessaloniki
1983–1992	79	56
1993–2002	217	70
2003–2012	287	141
Total	583	267

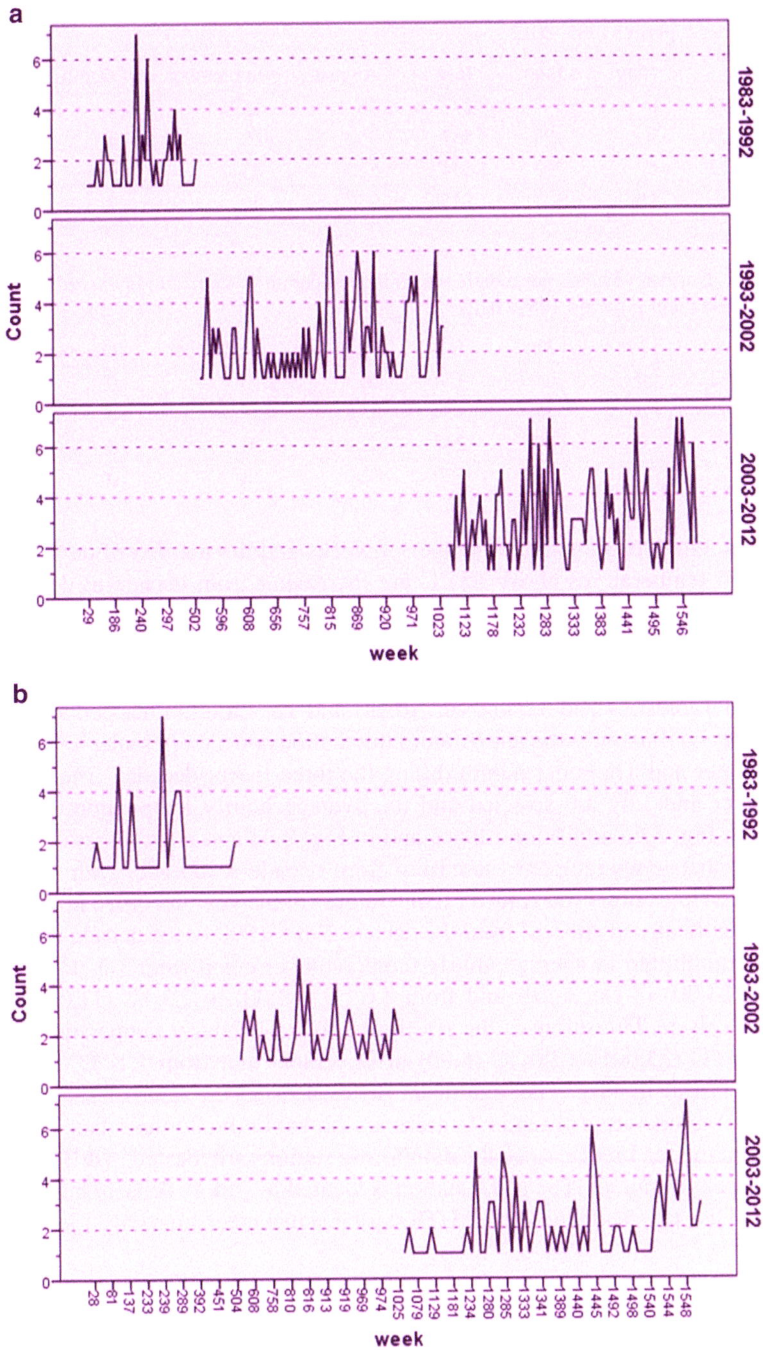

Fig. 3.5 Number of days per week in which the maximum temperature is equal to or exceeds the limit of 35 °C (**a**) in Athens and (**b**) in Thessaloniki for the period 1983–2012

Table 3.4 Number of hours per month, which the temperature is equal to or exceeds the limit of 35 °C. Athens: period 1983–2012

Athens	May	June	July	August	September	October	Total
1983–1992	0	9	217	66	15	1	308
1993–2002	5	122	507	362	16	1	1013
2003–2012	5	203	634	529	25	0	1396
Total	10	334	1358	957	56	2	2717

Table 3.5 Number of hours per month, which the temperature is equal to or exceeds the limit of 35 °C. Thessaloniki: period 1983–2012

Thessaloniki	May	June	July	August	September	October	Total
1983–1992	0	6	137	61	16	0	220
1993–2002	0	37	139	108	2	0	286
2003–2012	0	105	244	231	10	0	590
Total	0	148	520	400	28	0	1096

this limit. Obviously, both the number of weeks and the number of days per week with high temperatures above 35 °C are increasing from decade to decade. The frequency of these events increased in Athens more from the first to the second decade, while in Thessaloniki mainly from the second to the third decade. The number of hours with temperature values equal or exceeding the limit of 35 °C are shown in Tables 3.4 and 3.5 and for Athens and Thessaloniki respectively.

Figure 3.6 illustrates the temperature fluctuations on a daily scale, in both cities, for a winter and a summer month during the three recent decades. The months of December and July are selected and the average hourly temperature curves are shown in Fig. 3.6a and b for Athens and in Fig. 3.6c and d for Thessaloniki. It is apparent that temperature is increasing from decade to decade, with the greater differences to be observed at lower temperatures of the year, recorded in December. More specifically in Athens, from the decade 1983–1992 to the decade 2003–2012 the rise monitored in average hourly temperatures ranged from 1.3 °C (16:00) to 1.9 °C (24.00) in December and from 1.6 °C (18:00) to 2.3 °C (11.00) in July. Respectively in Thessaloniki, the rise in the average hourly temperatures ranged from 1.4 °C (23:00) to 2.0 °C (8:00) in December and from 1.1 °C (at 9:00) to 1.8 °C (17:00) in July. The significant differences in the hour of appearance of maximum temperatures in the two areas are probably due to the urban heat island effect and the sea breeze, since the monitoring stations are located in different built-up urban environment. The NOA station is located in a more densely built-up area, 5 m from the sea, while the IMC/AUTh station is in a green area only 1 km from the sea.

Of particular interest is the fact that the results, the correlations and the variations of the temperature distribution in the three decades reflected in the graphs are confirmed by the non-parametric Kruskal-Wallis test, with the only exception of the number of days with maximum temperature above 35 °C in Thessaloniki area (Table 3.6).

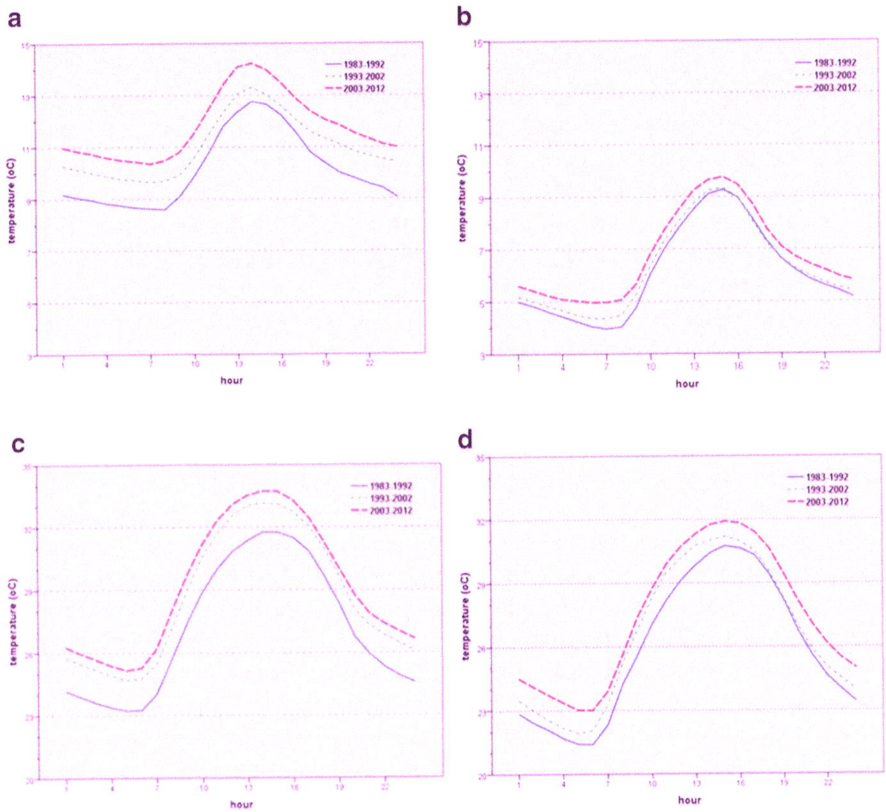

Fig. 3.6 Average hourly temperature for December and July for the period 1983–2012 per decade (**a**, **b**) in Athens and (**c**, **d**) in Thessaloniki respectively

Table 3.6 Results of the Kruskal–Wallis test for the distribution of the studied parameters over the three decades

	Athens	Thessaloniki
Minimum temperatures	**	**
Maximum temperatures	**	**
Number of days with daily temperature above 25°C	**	**
Number of days with maximum temperature equal or above 35°C	**	–
Mean daily temperature of December	**	**
Mean daily temperature of July	**	**

Statistically significant differences marked with **, at significance level of 0.05

Conclusions

Climate change is happening and influences life in cities. Current observations as well as projected changes suggest an increase of the annual mean temperature, a change of precipitation patterns, a rise of sea level, an increase of the number, intensity and duration of heatwaves, as well as an increase of extreme precipitation events and drought. Cities and towns will be affected by the impacts of climate change. The impacts are direct, such as health problems, damages to infrastructure, lower crop productivity, fire risks, decrease in water availability, and indirect such as lower productivity in work, reduction in the use of public spaces, social life restriction, affection to tourism industry, effects on energy consumption and risks in energy supplies. All these impacts are extremely important, their analysis is complex and requires different data. In this work, an analysis of the dry-bulb temperature of the ambient air in Athens and Thessaloniki for the 30 years period 1983–2012, based on hourly measurements of meteorological stations of the National Observatory of Athens and the Aristotle University of Thessaloniki, was performed. The results show a strong and statistically significant upward trend of the temperature values in both cities. More specifically, there is a gradual increase in average minimum, mean and maximum monthly values of air temperature from decade to decade. The differentiation between the temperatures through the decades was demonstrated as statistically significant, based on the non-parametric Kruskal-Wallis test ($p < 0.05$), suggesting the non-randomness of this variation. The cumulative frequency curves of temperature show that the occurrence of low temperatures decreases, whereas the frequency of the high temperatures is increased through the examined period. Specifically an upward trend is shown in the number of days with daily mean temperature above 25 °C and the number of days with maximum value equal to or exceeding 35 °C, for both areas, and suggest a slight shift of the warmest days from July to August. In addition, both the number of weeks and the number of days per week with high temperatures above 35 °C are increasing from decade to decade. Finally, the daily fluctuation in temperature reveals a distinct temperature rise from decade to decade, throughout the day, while at the same time the increase was recorded particularly in low temperatures in winter and in high temperatures in summer. The data that this work presents, may be helpful to support policy development and decision making in urban and built environment design.

References

Climate Change Impacts Study Committee. (2011). *The environmental, economic and social impacts of climate change in Greece*. Athens: Bank of Greece. ISBN 978-960-7032-58-4.

Department of Meteorology and Climatology, Aristotle University of Thessaloniki, Greece. Meteorological observations of Thessaloniki station (1983–2012). Annual edition of Institute of Meteorology and Climatology of the Aristotle University of Thessaloniki (in Greek).

Founda, D., Kambezidis, H. D., Petrakis, M., Zanis, P., & Zerefos, C. (2009). A correction of the recent air-temperature record at the historical meteorological station of the National Observatory of Athens (NOA) due to instrument change. *Theoretical and Applied Climatology, 97*, 385–389.

Giannakopoulos, C., Kostopoulou, E., Varotsos, K., Tziotziou, K., & Plitharas, A. (2011). An integrated assessment of climate change impacts for Greece in the near future. *Regional Environmental Change, 11*(4), 829–843.

Giannaros, T. M., & Melas, D. (2012). Study of the urban heat island in a coastal Mediterranean City: The case study of Thessaloniki, Greece. *Atmospheric Research, 118*, 103–120.

Giannopoulou, K., Livada, I., Santamouris, M., Saliari, M., Assimakopoulos, M., & Caouris, Y. G. (2011). On the characteristics of the summer urban heat island in Athens, Greece. *Sustainable Cities and Society, 1*, 16–18.

IPCC—Intergovernmental Panel on Climate Change. (2007). *Climate change 2007: The fourth assessment report*. Cambridge: Cambridge University Press.

IPCC—Intergovernmental Panel on Climate Change. (2013). *Climate change 2013—The fifth assessment report*. Cambridge: Cambridge University Press.

Joint Research Center. (2008). *Extreme temperatures and precipitation in Europe: Analysis of a high-resolution climate change scenario*. Institute for Environment and Sustainability, European Commission. JRC 44124, EUR 23291 EN, SSN 1018-5593.

Kolokotsa, D., Psomas, A., & Karapidakis, E. (2009). Urban heat island in southern Europe: The case study of Hania, Crete. *Solar Energy, 83*, 1871–1883.

Livada, I., Santamouris, M., Niachou, K., Papanikolaou, N., & Mihalakakou, G. (2002). Determination of places in the great Athens area where the heat island effect is observed. *Theoretical and Applied Climatology, 71*, 219–230.

NOA—National Observatory of Athens. *Climatological bulletin (1983–2012)*. Institute of Meteorology and Physics of the Atmospheric Environment, Athens (in Greek).

NOAA—National Oceanic and Atmospheric Administration. (2014). *U.S. climate extremes index*. Retrieved March 24, 2015, from http://www.ncdc.noaa.gov/extremes/cei/.

Papakostas, K., Mavromatis, T., & Kyriakis, N. (2010). Impact of the ambient temperature rise on the energy consumption for heating and cooling in residential buildings of Greece. *Renewable Energy, 35*, 1376–1379.

Papakostas, K. T., Michopoulos, A., Mavromatis, T., & Kyriakis, N. (2013). Changes of temperature data for energy studies over time and their impact on energy consumption and CO_2 emissions. The case of Athens and Thessaloniki—Greece. *International Journal of Energy and Environment, 4*(1), 59–72.

Papakostas, K. T., Zagana-Papavasileiou, P., & Mavromatis, T. (2014). Analysis of three decades temperature data for Athens and Thessaloniki, Greece—Impact of temperature changes on energy consumption for heating and cooling of buildings. In *International conference ADAPTtoCLIMATE*, Nicosia, Cyprus, 27–28 March.

Santamouris, M., Papanikolaou, N., Livada, I., Koronakis, I., Georgakis, C., Argiriou, A., & Assimakopoulos, D. N. (2001). On the impact of urban climate to the energy consumption of buildings. *Solar Energy, 70*(3), 201–216.

URL 1. Retrieved March 24, 2015, from http://www.ncdc.noaa.gov/cag/.

URL 2. Retrieved March 24, 2015, from http://www.wmo.int/pages/mediacentre/press_releases/pr_997_en.html.

Vardoulakis, E., Karamanis, D., Fotiadi, A., & Mihalakakou, G. (2013). The urban heat island effect in a small Mediterranean city of high summer temperatures and cooling energy demands. *Solar Energy, 70*, 128–144.

WWF Greece. (2009). *Climate change impacts in Greece in the near future*. Athens.

Chapter 4
Mitigation and Adaptation Policies Related to Climate Change in Greece

I. Sebos, A. Progiou, L. Kallinikos, P. Eleni, I. Katsavou, K. Mangouta, and I. Ziomas

Introduction

In response to the emerging evidence that climate change could have a major global impact, the United Nations Framework Convention on Climate Change (henceforth the Convention) was adopted on 9 May 1992 and was opened for signature in Rio de Janeiro in June 1992. Greece signed the Convention in Rio and ratified it in 1994 (Law 2205/94) (Ministry of Environment Energy and Climate Change (MEECC) 2014a).

In that framework, the third meeting of the Conference of the Parties (COP) to the Convention, held in Kyoto (1–11 December 1997), finalised the negotiations related to the establishment of a legal instrument; the Kyoto Protocol on Climate Change. The Kyoto Protocol runs in two commitment periods; the first one started in 2008 and ended in 2012 (United Nations Framework Convention on Climate Change (UNFCCC) 1998), whereas the second started in 2013 and will end in 2020 (European Environment Agency (EEA) 2014). The Protocol provides a foundation upon which future action can be intensified and introduced, for the first time, legally binding commitments for developed countries to reduce emissions of greenhouse gases. The Protocol entered into force on 16 February 2005, after its ratification from 141 Parties (with the exception of USA and Australia—Australia finally ratified the 1997 Kyoto protocol in 2007) including developed countries with a contribution of more than 55 % to global CO_2 emissions in 1990. The detailed rules for the implementation of the Protocol were adopted at COP 7 in Marrakesh, Morocco, in 2001, and are referred to as the "Marrakesh Accords." In Doha,

I. Sebos (✉) • A. Progiou • L. Kallinikos • P. Eleni • I. Katsavou • K. Mangouta • I. Ziomas
Department of Chemical Engineering, National Technical University of Athens,
Zografou Campus, 9 Heroon Polytechniou Street, 15780 Zografou, Greece
e-mail: isebos@mail.ntua.gr; ap@axonenviro.gr; leokalls@central.ntua.gr; peleni@central.ntua.gr; katsavou@central.ntua.gr; kon_maggouta@yahoo.gr; ziomas@chemeng.ntua.gr

P. Grammelis (ed.), *Energy, Transportation and Global Warming*, Green Energy and Technology, DOI 10.1007/978-3-319-30127-3_4

Qatar, on 8 December 2012, the "Doha Amendment to the Kyoto Protocol" was adopted (United Nations Framework Convention on Climate Change (UNFCCC) 1998; United Nations Framework Convention on Climate Change (UNFCCC) 2015), and sets new commitments for Annex I Parties to the Kyoto Protocol who agreed to take on commitments in a second commitment period from 1 January 2013 to 31 December 2020.

With respect to the EU target under the first commitment period of the Kyoto Protocol (i.e. reduction of emissions at 8 % for the period 2008–2012 comparing to 1990 levels), EU has stated that this will be achieved jointly by EU Member-States under the provisions of Article 4 of the Protocol. The Burden-Sharing agreement between all Member States was finalized during the Environment Council in June 1998 and entered into force with Decision 2002/358/EC concerning the approval, on behalf of the European Community, of the Kyoto Protocol. According to this agreement, Greece is committed to limit its GHG emissions increase for the period 2008–2012 to +25 % compared to base year emissions (1990 for CO_2, CH_4 and N_2O emissions—1995 for F-gases). Since the base year emissions of Greece were 106,987,169 t CO_2 eq. (United Nations Framework Convention on Climate Change (UNFCCC) 2007), the assigned amount was calculated to be 668,669,806 t CO_2 eq. (5 * 1.25 * base year emissions). Greece ratified the Kyoto Protocol in 2002 (Law 3017/2002) and adopted a National Programme for achieving its commitment by a decision of the Council of Ministers (DCM5/2003).

Concerning the second period of the Kyoto Protocol, legally binding target trajectories for the period 2013–2020 are enshrined in both the Emissions Trading System (EU-ETS) Directive (Directive 2003/87/EC and respective amendments) and the Effort Sharing Decision (Decision No 406/2009/EC). At the same time the European Union (EU) has set its climate change mitigation objective for 2020, committing itself to reducing its emissions by at least 20 % compared to 1990 levels (30 % subject to the conclusion of a comprehensive international climate change agreement) (Official Journal of the European Union 2013) in the context of the **EU Climate and Energy Package**, which adopted in 2009. These legally binding trajectories not only result in a 20 % GHG reduction in 2020 compared to 1990 but also define the EU's annual target pathway to reduce EU GHG emissions from 2013 to 2020. The EU has also committed to increase to 20 % the share of renewable energies in the EU final energy consumption (with a minimum 10 % share in the transport sector) and to save 20 % of the EU's energy consumption through increased energy efficiency ("20-20-20" objective) (European Environment Agency (EEA) 2014; European Commission/DG Climate Action/ European Environment Agency 2014).

The **EU ETS** is the key tool for cutting industrial greenhouse gas emissions most cost-effectively. The climate and energy package includes a comprehensive revision and strengthening of the legislation which underpins the EU ETS, the Emissions Trading Directive. The revision applies from 2013, the start of the third trading period of the EU ETS. Major changes include the introduction of a single

EU-wide cap on emission allowances in place of the existing system of national caps. Thus, the emission reduction to be achieved from the sectors covered by the EU ETS will be 21 % below 2005 emission levels. In specific, the allowed GHG emissions will start by the mean value of the period 2008–2010 for industries that are included in the ETS and will be decreased by 1.74 % annually until 2020.

The effort to reduce emissions not covered by the EU-ETS, shared between the EU-28 Member States through differentiated annual national GHG targets under the **Effort Sharing Decision** (ESD). The **Effort Sharing Decision** sets annual national emission targets for all Member States for the period 2013–2020 for those sectors not covered by the EU-ETS, expressed as percentage changes from 2005 levels. In March 2013, the Commission formally adopted the national annual limits throughout the period for each Member State. By 2020, the national targets will collectively deliver a reduction of around 10 % in total EU emissions from the sectors covered compared with 2005 levels. The national target of Greece for emissions not included in the ETS (non-ETS) is a 4 % reduction of emissions by 2020 compared to 2005. According to European Commission Decision 2013/162/EU, 4 % reduction of emissions by 2020 for Greece is account for 63,028,265 t CO_2 eq., based on the global warming potential values from the fourth IPCC assessment report (European Environment Agency (EEA) 2014).

National Greenhouse Gas Emissions

Figure 4.1 presents the emissions of greenhouse gases (carbon dioxide, methane, nitrous oxide, HFCs, PFCs), for the years 1990–2012, expressed in carbon dioxide equivalent (Mt CO_2 eq.), while the Land Use, Land Use Change and Forestry (LULUCF) sector is not included (Ministry of Environment Energy and Climate Change (MEECC) 2013).

It should be noted that because the sector Land Use, Land Use Change and Forestry (LULUCF), was a sink of greenhouse gases in 1990 (as in the whole period from 1990 to 2012), emissions from this sector are not taken into account for the calculation of emissions base.

Total emissions of greenhouse gases (without LULUCF) in 2012 amounted to 110.98 Mt CO_2 eq., showing an increase of 5.77 % compared to base year over emissions. Since the objective of the country for the first commitment period of the Kyoto Protocol is to limit the increase of GHG emissions at +25 %, it is obvious that the country not only achieves its goal, but also has surplus of emission allowances (Fig. 4.2).

The contribution of each sector in total emissions is shown in Fig. 4.3. As it is evident from this figure, emissions from the energy sector (emissions from

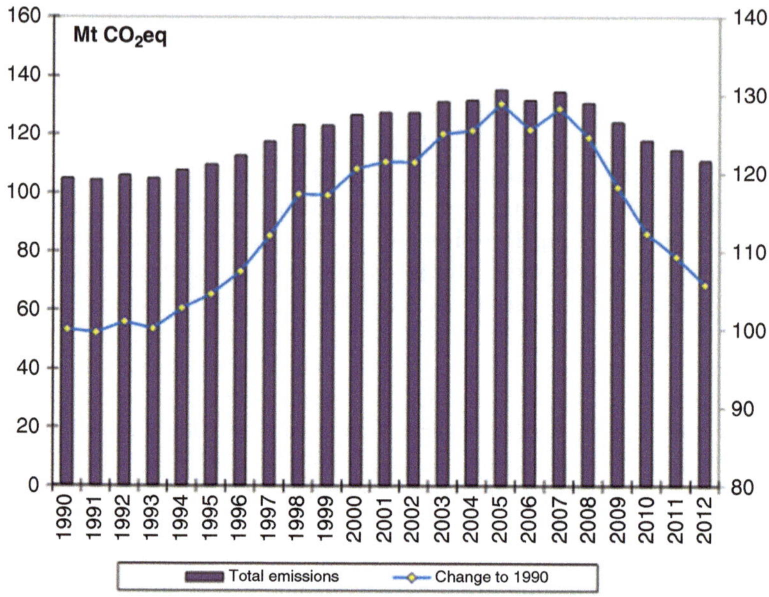

Fig. 4.1 Total GHG emissions time series in Greece in Mt CO_2 eq

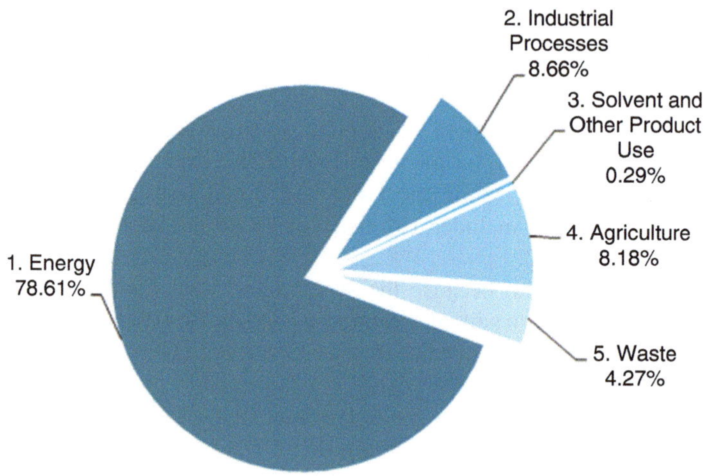

Fig. 4.2 Contribution of each sector to total GHG emissions for 2012

combustion of fossil fuels and fugitive emissions) accounted for 78.61 % of the total emissions for the year 2012.

The key findings are summarized below:

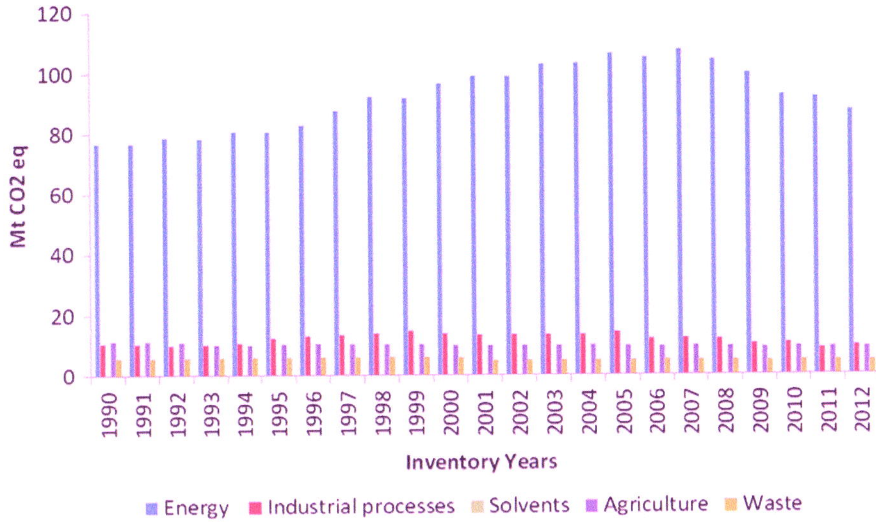

Fig. 4.3 GHG emissions between 1990 and 2012 by sector

Emissions from Energy in 2012 accounted for 78.61 % of total GHG emissions (without LULUCF) and increased by approximately 13.73 % compared to 1990 levels.

The living standards improvement, due to the economic growth, the important growth of the services sector and the introduction of natural gas in the Greek energy system represent the basic factors affecting emissions trends from Energy for the period 1990–2007. The decreasing trend of emissions in all sectors of energy of the years 2008–2012 is attributed among others (i.e. RES, energy efficiency measures, road infrastructure and public transportation improvements, etc.) to the economic recession that the country is facing.

At the same time it should be mentioned that the availability of hydropower has a significant effect to emissions trends. For instance, the significant increase of electricity demand in 1999 was not followed by a similar increase of emissions because of the penetration of natural gas and the high availability of hydropower.

The majority of GHG emissions (64.12 %) in 2012 derived from energy industries, while the contribution of transport, manufacturing industries and construction and other sectors is estimated at 18.63 %, 6.47 % and 10.78 %, respectively. The rest 0.01 % of total GHG emissions from Energy derived from fugitive emissions from fuels. Within the fuel combustion activities, the sector with the greatest increase of emissions for the years 1990–2008 is transport, showing an average rate of increase of 2.48 %. However, for years 2009–2012 a decrease was observed with an average rate of decrease equal to −6.81 %. In addition, energy industries and other sectors (i.e. residential, tertiary and agriculture sectors) presented 1.16 % and 0.94 % average annual rate of increase, respectively. Finally, missions from manufacturing industries and construction emissions had a mean annual rate of decrease of 1.87 %.

Emissions from Industrial Processes in 2012 accounted for 8.66 % of the total emissions (excluding LULUCF) and decreased by 9.42 % compared to 1990 levels. In 2012 mineral products production has continued the decreasing trend of the previous years, following the decrease of all the subcategories of the sector, mainly due to the effects of the economic recession. There is an increase in emissions since 2011 (20.07 %), which followed the intense decrease of 2011 (−36.74 % between 2011 and 2010). Emissions from chemical industry have decreased by 23.67 % since 2011. Emissions from metal production slightly increased with regards to 2011 by 1.05 %, due to the increased production of aluminum and nickel. As regards to fluorinated greenhouse gases (f-gases) emissions, have been increased compared to 2011 (14.48 %). The general increasing trend during the last years of the time-series depicts the continuous substitution of CFCs in the context of the protocol of Montreal.

The contribution of the Solvents and other products use sector to total GHG emissions is minor (0.29 % of the total emissions) and has increased by 3.29 % compared to 1990 level of emissions.

Emissions from Agriculture that accounted for 8.18 % of total emissions in 2012 (without LULUCF), decreased by approximately 20.44 % compared to 1990 levels. Emissions reduction is mainly due to the reduction of N_2O emissions from agricultural soils, because of the reduction in the use of synthetic nitrogen fertilizers. The decrease in the use of synthetic nitrogen fertilizers is attributed to the increase of organic farming, the high price of fertilizers and the impact of initiatives to promote good practice in fertilizer use. The changes of the rest determining parameters of GHG emissions from the sector (e.g. animal population, crops production etc.) have a minor effect on GHG emissions trend.

Emissions from the Waste Sector (4.27 % of the total emissions, without LULUCF), decreased by approximately 19.58 % from 1990. Living standards improvement resulted in an increase of the generated waste and thus of emissions. However, the increase of recycling along with the exploitation of the biogas produced limits the increase of methane emissions. At the same time, emissions from wastewater handling have considerably decreased, due to the continuous increase of the population served by aerobic wastewater handling facilities.

GHG Time Series Trend Assessment

In order to evaluate the trends of GHG emissions time series the indicators presented in Fig. 4.4 were used. Figures present the variation of the indices compared to the base year (base year index value = 100). The base year is 1990 or 2000, respectively, depending on the availability of data (Project: ODYSSEE-MURE 2012).

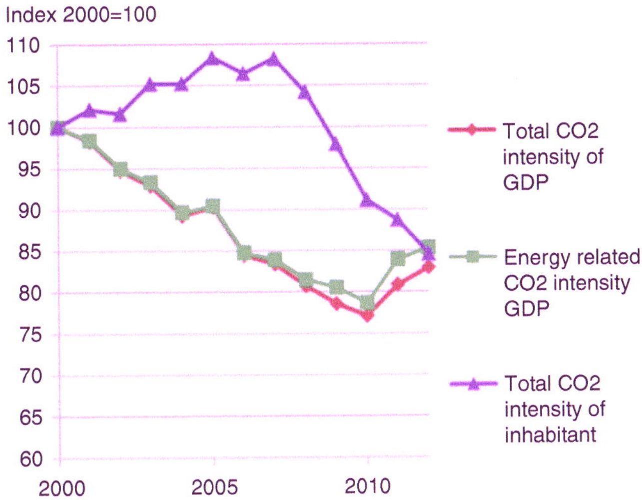

Fig. 4.4 Carbon intensity of GDP and population

Figure 4.4 presents two indices of GHG emissions in relation to the country's economic development and population, respectively. As shown in the Figure, both the total, and the energy related index have decreased significantly compared to 2000 (approximately 17.1 % and 14.7 %, respectively). Thus, the increase in the GDP for the period 2000–2008 was not accompanied by a respective increase of GHG emissions. The entry of natural gas and renewable energy in the energy mix, and other policies and measures, such as energy saving are significant contributions to the above reduction. The increase of the indicators in recent years compared to previous years is probably due to the abrupt decrease in GDP (denominator) due to the economic crisis of recent years.

Figure 4.5 presents two power generation related indicators. Both indicators have decreased significantly compared with 1990 (about 27.6 % in terms of CO_2 specific emissions of public and auto producer power plants and 28.6 % for the electricity carbon intensity of total power generation). The reduction of GHG emissions compared to the generated electricity is the result of policies and measures such as the admission of natural gas in the energy mix, the use of renewable energy sources (RES), the modernization of power plants in combination with energy saving measures etc. The index of total generation presents a lower value because it includes the production of electricity from RES.

Figure 4.6 shows the industry assessment index relative to its growth (gross value added—GVA). The index referring to industry presents a decrement by 27.6 % compared to 2000. Thus, the development of industry (the period 2000–2007, as after 2008 GVA reduced) was not accompanied by a corresponding increase of GHG emissions.

Finally, in Fig. 4.7 the indicators for the residential and tertiary sectors are presented. For households an increase in GHG emissions per household for the

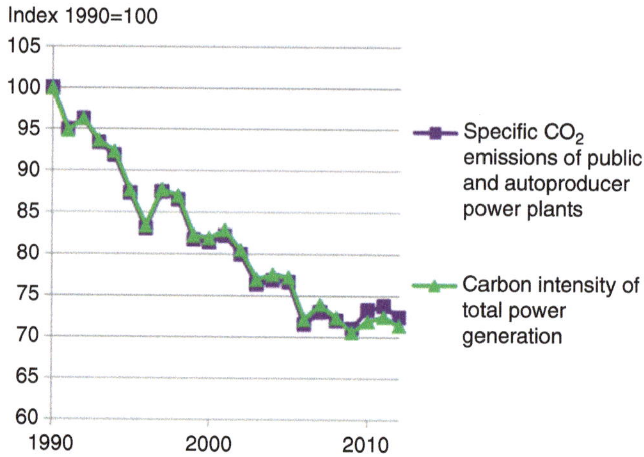

Fig. 4.5 Carbon intensity of power generation

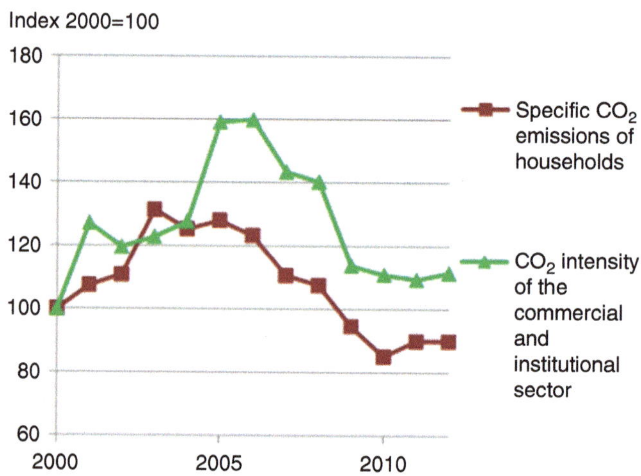

Fig. 4.6 Evaluation indicator of industry (based on gross value added)

period 2000–2003 occurs, which is explained by the improvement of living standards due to the economic growth of the period. Then, the index decreases despite the continued economic growth, mainly due to the gradual replacement of heating diesel by natural gas and the implementation of energy saving measures. Respective conclusions result from the tertiary sector. It should be noted that the above indicators roughly reflect the GHG emissions from burning fossil fuels for heating needs for each home or building in the tertiary sector, so they are strongly influenced by the climatic conditions during winter period of each year (i.e. how hot or cold the winter was every year).

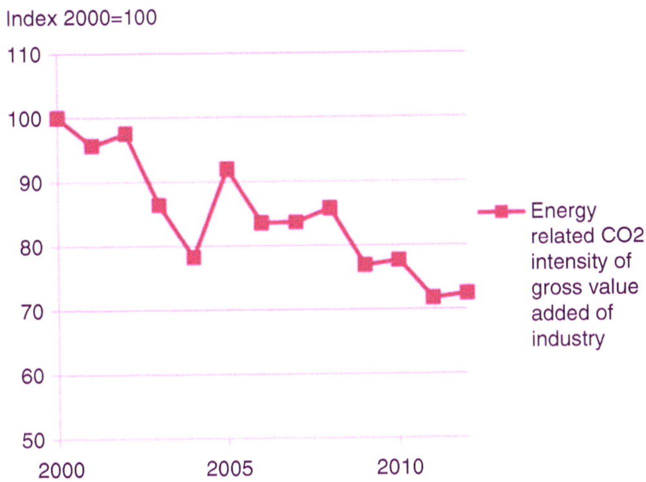

Fig. 4.7 Evaluation indicators of the residential and commercial sectors

Level of Achievement of the National Commitment Under the KP

The GHG emission projections presented in this paper are based on the latest official energy projection scenarios approved by the Ministry of Environment, Energy and Climate Change. In Fig. 4.8 the evolution of GHG emissions and their projections till year 2020, along with the assigned amount of Greece for the first commitment period of Kyoto Protocol are presented.

The fulfillment of the Kyoto Protocol target (first commitment period) is evaluated through a comparison between the total GHG emissions of the years 2008–2012, and the Assigned Amount of Greece for the same period (668,669,806 t CO_2 eq). As it is obvious (Fig. 4.8), Greece achieves Kyoto Protocol target for the first commitment period, on the basis of the domestic policies and measures implemented; and there was a surplus of allowances of GHG emissions of approximately 44 million AAUs.

The presented progress of the country in achieving the emission reduction targets for the period 2013–2020, concerns only the case of non-ETS sectors, since the emissions of sectors covered by the ETS will be regulated by the Emissions Trading System (ETS). Thus, projected emissions from non-ETS sectors and annual emissions allocation for the years 2013–2020, are presented in Fig. 4.9.

Comparing the annual emissions allocation for the years 2013–2020 (Annex 1 of 2013/634/EU), which reflect the non-ETS target of Greece pursuant to European legislation (2013/162/EU and its amendment 2013/634/EU), with the projected emissions from non-ETS sectors, it is concluded that Greece will meet this target, on the basis of the domestic policies and measures. It should be mentioned that this conclusion is based on the comparison of projections and annual emissions

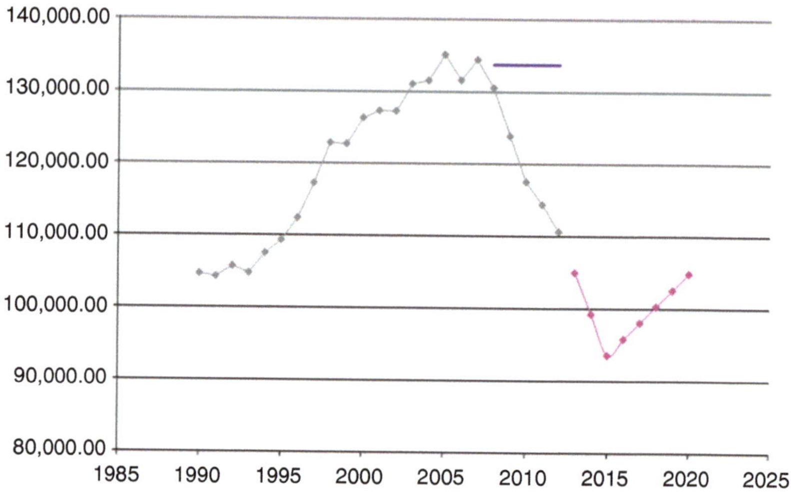

Fig. 4.8 GHG emissions projections

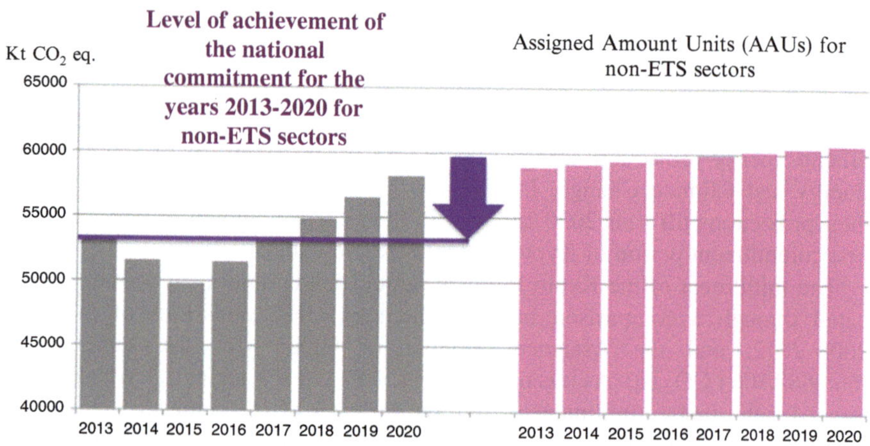

Fig. 4.9 Projected emissions from non-ETS sectors and annual emissions allocation for the years 2013–2020—level of achievement of the national commitment for the period 2013–2020

allocation calculated by applying global warming potential values from the second IPCC assessment report. However, the same conclusion would result if global warming potential values from the fourth IPCC assessment report were considered (Ministry of Environment Energy and Climate Change (MEECC) 2014a; European Commission 2014).

Effect of GHG Mitigation Policies

The most important policies related with the implementation of measures for the mitigation of GHG emissions in Greece are (Ministry of Environment Energy and Climate Change (MEECC) 2014a):

- The National Renewable Energy Action Plan, which sets out Greece' national targets for the share of energy from renewable sources consumed in transport, electricity and heating and cooling in 2020, taking into account the effects of other policy measures relating to energy efficiency on final consumption of energy, and adequate measures to be taken to achieve those national overall targets (raising the share of EU energy consumption produced from renewable resources to 20 %).
- The National Energy Efficiency Action Plans, which covers significant energy efficiency improvement measures and expected and/ or achieved energy savings, including those in the supply, transmission and distribution of energy as well as energy end-use. (Measures for the improvement of energy efficiency.)
- The continuous development of the transmission and distribution network of natural gas, and the promotion of its use.
- The **European emissions trading scheme** (Directive 2003/87/EC) the operation of which started in 2005. In Greece, the trading system for the period 2008–2012 comprises 140 industrial installations (power plants, refineries, cement plants, etc.). An allowance reserve is also created which is intended to cover possible unknown new entrants in the period. In 2013, the EU ETS is now in its third phase, running from 2013 to 2020. A major revision in order to strengthen the system means the third phase is significantly different from phases one and two and is based on rules which are far more harmonized across EU than before.
- The **fiscal measures** that support policies and measures that reduce GHG emissions, such as the tax regime of energy products, the registration tax of vehicles, the Motor vehicle circulation fee (road tax), the income taxation—relief and exemptions.

The total realistic GHG emissions reduction potential from the implemented and adopted policies and measures was estimated to be 33.3 Mt CO_2 eq. for 2015 and 41.0 Mt CO_2 eq. for 2020. The possible interferences between these implemented/adopted measures, which may restrict the estimated GHG emissions reduction potential, were taken into account. Thus, it is obvious that the application of the already implemented and adopted measures for the mitigation of GHG emissions contributes considerably in the restriction of the augmentative trend of emissions (besides the economic recession), leading to the achievement of the 2020 target, exclusively with domestic measures and actions.

Adaptation Measures for Climate Change

Adaptation may include both national and regional strategies as well as measures at community level or by individuals. Although an over-arching adaptation strategy is not yet available in Greece, adaptation measures are currently under implementation as part of a broader network of measures that apply to the specific areas of identified vulnerabilities. It is worth noting that the Overall National Strategy for Adaptation is ongoing and is a key priority of the country (Ministry of Environment Energy and Climate Change (MEECC) 2014a; Bank of Greece 2013).

Greece already has a National Biodiversity Strategy and Action Plan for its implementation, based on a project that started in 1999 and was completed in 2014 (Ministry of Environment Energy and Climate Change (MEECC) 2014b). The aim of the strategy is to stop the loss of biodiversity and degradation of the ecosystems in Greece until 2026, as well as the promotion of biodiversity as a national issue.

Aiming to the adaptation of the country concerning the agricultural sector Greece is participated in the Project **ADAGIO**—ADAptation of AGriculture in European RegIOns at Environmental Risk under Climate Change. The project is focusing on improving awareness and user-orientation of adaptation strategies, improving local representation and considering the main vulnerable regions to Climate Change in Europe. In addition, the **National Action Plan for Combating Desertification** (2001) sets as an objective to combat efficiently the desertification trends in the 35 % of the whole Greek territory that is under direct threat and to prevent the desertification process elsewhere. The main issues in relation to agriculture are **erosion of soils and drought problems.**

The national strategy concerning forest ecosystems is implemented via the **Program of Rural Development (RDP) 2007–2013 "Aleksandros Mpaltatzis"** (MRDF), under which 69 projects have been implemented by the Forestry Services, aiming at the enhancement of forest fires prevention and restoring of forest burnt area, along with anti-erosion, anti-flood works for prevention of flooding in the lowlands and for restoration of forest burnt areas. In addition, a **Life + project** "Adaptation of forest management to climate change in Greece (AdaptFor)" was implemented, aiming at enhancing the capacity of forest services to adapt forest management to climate change and to disseminate the need for adaptation of forest management to other stakeholders and to the general public. **Natura 2000** network has also a key role in protecting and enhancing our natural capital. In addition to safeguarding nature's intrinsic value, investing in Natura 2000 provides multiple benefits to society and the economy at the local, regional, national and EU level. Finally, The **National Action Plan for Combating Desertification** was approved in 2001 and includes measures to reduce the frequency and decrease the spread of forest fires.

The Greek Operational Programme "FISHERIES 2007–2013" was approved by the European Commission, with a strategic objective focusing on "Measures for the adaptation of the fishing fleet". Concerning aquacultures, possible adaptation measures to climate change include institutional measures, policy measures and action planning which are summarized as follows: (a) Insurance aquaculture, (b) Technology transfer and research, (c) Diversification of crop species, (d) Adoption of selection systems for the installation and monitoring of aquacultures.

In the **water resources sector**, according to the **National Action Plan for Combating Desertification**, the suggested measures include: reduction of water loss through the improvement of irrigation efficiency, reduction of water losses and demand in urban and industrial use and increase of water supply through funding of programs for water recycling and reuse, etc. In addition, the **MEDROPLAN** Project: "Mediterranean Drought Preparedness and Mitigation Planning" focuses on developing Guidelines for drought preparedness plans and to setting up a Network for drought preparedness in Mediterranean countries. The Guidelines provide an integrated approach to face droughts from a risk management perspective and therefore minimizing the impacts of drought in the population and resources.

Concerning **coastal zones the General National Framework for Spatial Planning and Sustainable Development** includes priorities that avoid the expansion of existing settlements, especially along the coast; encourage expansion in the areas where population density permits it; protect beaches and natural coastal areas, assure public access etc.

The following measures promoted by the **tourism sector** and especially the Greek National Tourism Organization include adaptation of an appropriate marketing strategy aiming at diversification through the development of new tourism product and distinctive destination brands, promotion of sustainable tourism (eco-tourism, nature based tourism, rural tourism), promotion of Green Destinations, encouragement of sustainable business practices, providing tools and guidance to the tourism industry and investors and new technologies to improve energy efficiency.

The National Action Plan for the **'Response of Environmental Hazards Threatening Health'** for 2008–2012 included a special action dedicated to the 'Exploring of Climate Change Impacts on Health', primarily referring to the identification, research and documentation of the impacts. At the national level, the governments of Europe have been developing actions to address the impacts of climate change: (1) The national health ministries have launched actions to ensure equal access to health services and social justice for all victims of climate change. (2) The national health ministries will also need to design special action plans to address the public health problems. (3) Hospitals will also need proper infrastructure and equipment. (4) Healthcare personnel will need to receive training in environmental epidemiology and the health implications of climate change, etc.

In addition the **General Framework of Spatial Planning and Sustainable Development** has the following specific objectives:

- Constant care for energy-saving;
- Promotion of alternative, and in particular renewable, energy sources;

- Protection and enhancement of natural processes;
- Adaptation to new climate change conditions and mitigation of their consequences (fires, floods, erosion, drought, water salinization, desertification etc.), by putting in place mitigation mechanisms, appropriate infrastructures and plans for action.

The proposed policies and specific policy measures for coping with the impacts on **the transport system include**:

- Development of monitoring systems for crucial infrastructure and use of 'smart' decision-making, risk management and disaster management systems, etc.
- Use of new materials, more resilient to extreme weather conditions.
- Policy measures aimed at reducing transport demand, e.g. carpooling, mobility management, school transport, etc.
- Promotion and support of eco-driving.
- Use of 'smart' technologies and systems with a view to improving freight transport and maximizing capacity use of all means of transport (target: zero empty routes), etc.

References

Bank of Greece. (2013). *Environmental, economic and social impacts of climate change in Greece.*

European Commission. (2014). *European Union transaction log.* http://ec.europa.eu/environment/ets/.

European Commission/DG Climate Action/ European Environment Agency. (2014). *The 2020 climate and energy package.* http://ec.europa.eu/clima/policies/package/index_en.htm.

European Environment Agency (EEA). (2014). *Total greenhouse gas (GHG) emission trends and projections (CSI 010/CLIM 050).* Assessment published November 2014.

European Environment Information and Observation Network (EIONET). (2014). *Central data repository.* http://cdr.eionet.europa.eu/gr.

Ministry of Environment Energy and Climate Change (MEECC). (2013). *Annual inventory submission of Greece under the convention and the Kyoto protocol for greenhouse and other gases for the years 1990–2012.*

Ministry of Environment Energy and Climate Change (MEECC). (2014a). *Sixth national communication and 1st biennial report under the United Nations framework convention on climate change.*

Ministry of Environment Energy and Climate Change (MEECC). (2014b). *National strategy & action plan for biodiversity.*

Official Journal of the European Union. (2013). *COMMISSION IMPLEMENTING DECISION of 31 October 2013 on the adjustments to Member States' annual emission allocations for the period from 2013 to 2020 pursuant to Decision No. 406/2009/EC of the European Parliament and of the Council (2013/634/EU).*

Project: ODYSSEE-MURE. (2012). *Key indicators, final energy intensity.* http://www.indicators.odyssee-mure.eu/online-indicators.html.

United Nations Framework Convention on Climate Change (UNFCCC). (1998). *Kyoto protocol to the United Nations framework convention on climate change.*

United Nations Framework Convention on Climate Change (UNFCCC). (2007). *Report of the review of the initial report of Greece*.
United Nations Framework Convention on Climate Change (UNFCCC). (2015). *Doha amendment to the Kyoto protocol*.

Chapter 5
Assessing Air Quality in the Urban Environment: the Gender Gap

Theodora Slini and Fotini-Niovi Pavlidou

Introduction

Gender Priorities in the European Union

Climate change affects both women's and men's living conditions, welfare and wellbeing, however due to gender roles, women do not affect the environment in the same way as men, and in many countries women's access to resources, and hence their opportunities to manage conditions and adapt are quite limited (IPCC 2014).

At the same time, environmental policies are characterized by lack of sensitivity to women's different economic and social status and needs, having as a result women to be directly and disproportionally affected from environmental degradation. While consumption and lifestyle patterns still differ between two genders, with women to consume less and being more environmentally conscious, "women are clearly under-represented in environmental negotiations, budget deliberations and decisions on achieving a green, sustainable economy" (EC 2012). According to 'Gender aspects of the economic downturn and financial crisis European Parliament resolution of 17 June 2010 on gender aspects of the economic downturn and financial crisis (2009/2204(INI))', the European Parliament "urges the need to

T. Slini (✉)
Laboratory of Heat Transfer and Environmental Engineering, Centre of Space, Technology and Gender, School of Engineering, Aristotle University of Thessaloniki, Box 483, 54124 Thessaloniki, Greece
e-mail: lora@aix.meng.auth.gr

F.-N. Pavlidou
Department of Electrical and Computer Engineering, Centre of Space, Technology and Gender, School of Engineering, Aristotle University of Thessaloniki, Box 489, 54124 Thessaloniki, Greece
e-mail: niovi@auth.gr

P. Grammelis (ed.), *Energy, Transportation and Global Warming*, Green Energy and Technology, DOI 10.1007/978-3-319-30127-3_5

encourage women in local entrepreneurial initiatives in green economy". The European Parliament (2012) resolution of 11 September 2012 'on the role of women in the green economy (2012/2035(INI))' "calls on the Commission and the Member State to introduce gender equality into all environmental policy areas" and introduce gender equality into all environmental policy areas, and at all levels of economic decision-making. On 14 November 2012 the European Commission published its proposal for a Gender Diversity Directive for improving gender balance on company boards, obliging companies with less than 40 % of women non-executive directors to make significant efforts to make appointments in the next 7 years to reach this target.

Furthermore, the Opinion of the European Economic and Social Committee on 'The gender dimension in the Europe 2020 Strategy' 2013/C 76/02 (EU Legislation 2013), the Commission is "highlighting the essential role played by women in sustainable development", while "women can have a key influence on decision-making concerning the environment, particularly with regard to climate change policy. This is a new opportunity for women, who can play a key role and improve their personal and financial situation by getting involved with the new and emerging green economy, which is a crucial sector for development and job creation". After all "women's activism was critical in getting the conversation started," according to historian David Stradling, author of "Smokestacks and Progressives: Environmentalists, Engineers and Air Quality in America, 1881–1951."

In response to these requirements a gender perspective is a critical assumption towards a sustainable environment and a green economy in terms of the ecosystem, consumption, food, growth, transport, energy and citizen welfare.

Gender Priorities in Greece

According to the National Strategic Reference Framework (NSRF) for 2014–2020, the main tools for the implementation of gender policy will be integration of equal opportunities between men and women in all institutions, policies and actions (gender mainstreaming) and the assumption of certain specific activities in the following areas:

- equal participation women in the labor market,
- the participation of women outdoor activities aimed at local development,
- promotion of inclusion of women, prevention and control of female poverty and all forms of gender-based violence,
- mainstreaming of gender issues in social protection and health,
- support family,
- equal participation of women in public life and processes of political, social and economic decisions,
- combat discrimination based on sex and gender stereotypes and
- the integration of gender equality in public policies, monitoring and evaluation.

However, the Greek Legislation does not include the gender perspective as a critical parameter through its development and implementation and in many cases it is characterized as neutral or negative, according to a report by the General Secretariat for Gender Equality (2013). For example,

1. The Law 4067/2012 (OG A79/9-3-2012) the "New Construction Regulation" that includes a number of provisions on accessibility of buildings and built environment. Regarding the motivation for environmental enhancement and the improvement of quality of life in densely populated urban areas, which gender neutral with no comment on specific needs.
2. The master plan for urban planning and environmental protection of the greater Thessaloniki area (L. 1561/1985) refers to the daily needs of citizens ignoring the gender matters. Special requirements due to diverse living habits and routines as well as the different ways in which environment is perceived and used between men and women should be considered for the optimum urban planning and city upgrading.

The most recent L. 3653/2008 on Research and Technology establishes the minimum proportion of male/female participation in national committees of research and development, that is at least 1/3 of each gender, as long the candidates are equally qualified.

Results and Discussion

Gender, Air Quality and Global Warming

Despite it is widely accepted that balanced participation of men and women in decision-making is a prerequisite for improving the operation of democracy and society, the ongoing under-representation of women in political and economic decision-making reflects a basic democratic deficit in Europe and the broader international context. At the same time it is important to highlight that the female population is more vulnerable in effects of environmental degradetation and poor air quality (UN 2014; Fouillet et al. 2007). Furthermore, as suggested by Druckman (2012), the investigation of the concept of carbon footprint as a potential marker for social justice, a higher proportion of an average man's carbon footprint is due to leisure than an average woman's.

There are several studies that prove that poor air pollution conditions can generate negative effects in the female population in both rural and urban sites (Xu et al. 2013; Huisman et al. 2005; Oiamo and Luginaah 2013; Künzli et al. 2010). The proportion of women and young children is mainly affected as these groups of population spend most of their time indoors where solid fuels, fire places and cooking stoves are in extensive use. According to WHO (2014), around three billion people cook and heat their homes using open fireplaces and stoves using various burning materials. These materials are not always the appropriate ones, causing the death of over four million people prematurely from illness due to the poor indoor air quality. The mortality of premature deaths among children under 5 attributed to household air pollution is also extremely high.

Additionally according to The European environment—state and outlook 2010 (EEA 2010), vulnerable population groups (e.g. pregnant women) are more threatened by the bio-accumulation of organic persistent pollutants such as mercury. Chen et al. (2005) and Miller et al. (2007), the associations of $PM_{2.5}$ and cardiovascular morbidity and mortality were greater in women above 60 and at the same time Xu et al. (2013) states that air pollution (referring to air pollutants such as NO_2, SO_2, $PM_{2.5}$, O_3 and CO) increases risk for hypertension in pregnant women during early pregnancy and the full gestational period.

Ambient air pollution has stronger effects on females partly due to autoimmune disorders. Females were more likely to report cardinal symptoms after controlling for income, age and chronic diseases (Oiamo and Luginaah 2013). As Bakke et al. (2007) suggests female gender has reported health symptoms more often than did men and complained more about physical but not psychosocial factors gender, psychosocial, and physical environment factors were related to symptoms and perceived indoor climate. Female gender seems to be highly associated with the sick building syndrome and the symptom of atopia (Magnavita 2015).

Several studies have linked air pollution exposure to autism, including the Volk et al. (2011) study in Environmental Health Perspectives that looked at children in Los Angeles who lived near freeways and the Hallmayer et al. (2011) study that analysed traffic-related air pollution exposure data among pregnant women. The main conclusion of the studies was that a large proportion of the variance in liability can be explained by shared environmental factors in addition to moderate genetic heritability.

In general, studies suggest that health responses to air pollution may differ between women and men, however it is still unclear, whether observed modification is a result of sex-linked biological differences or gender differences in activity patterns or exposure measurement accuracy (Clougherty 2010). Cautious consideration of gender effects and examination in environmental epidemiology may provide critical information on other social factors that can affect population response to air pollution.

In addition, in countries that face the severity of current financial crisis, structural changes occur in European members states as well, including migration and changing of household structures due to energy poverty. For example, in Greece, an exponential increase of wood and pellet employment is monitored primarily for heating purposes (Slini et al. 2014; Santamouris et al. 2013), posing major concern especially for women health (Fig. 5.1), based on the results of a survey conducted in Northern Greece by the Department of Mechanical Engineering, Aristotle University of Thessaloniki (AUTh).

In Greece, due to the economic situation the consumption of oil products, which are the main fuel that is used by the household sector, decreased already by 25.7 % between 2007 and 2010. What is impressive is the exponential increase in the use of wood and pellets, a trend that has begun to have its first implications in matters of air quality, both indoor and outdoor, but also in an enhanced deforestation, with

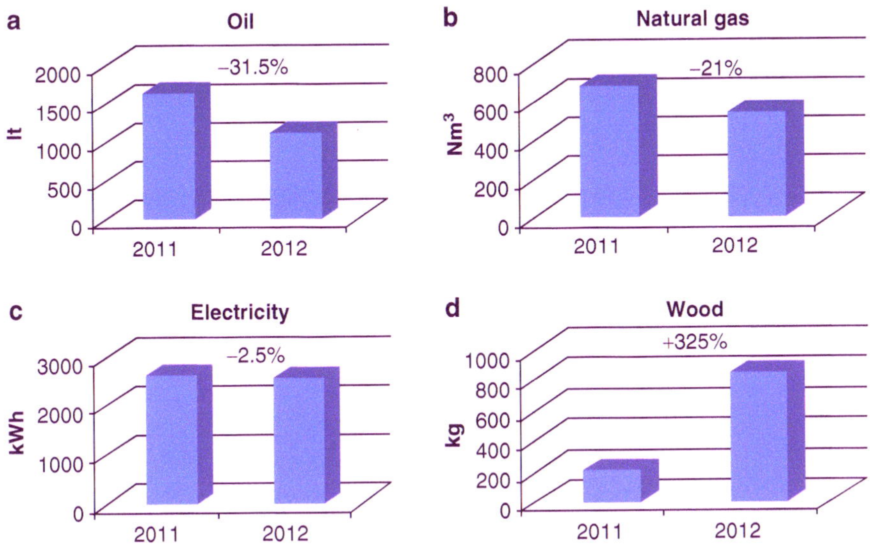

Fig. 5.1 The average annual fuel and electricity consumption for the years 2011–2012 by type: (**a**) oil in litres, (**b**) natural gas in N m^3, (**c**) electricity in kilowatt-hours and (**d**) wood in kilograms

major field studies being currently carried out on those matters (Slini et al. 2014). Towards this direction many countries in Europe are adopting policies and effective precautionary measures in order to reduce the exposure of vulnerable population groups in toxic environments (EEA—European Environment Agency 2015).

At the same time, research on gender-related disparities are mainly focused on energy consumption profiles and transport use. As pointed out at the Report on Gender Equality and Climate Change (EIGE 2012a), the most significant gendered differences in energy consumption were monitored in Greece and Sweden, with men energy consumption reaching a proportion of 39 % and 22 % more than women, respectively. Moreover, it is suggested that women use the public transport or other sustainable means of travel more often than men.

Gender Balance in Environmental Decision Making

As global warming is a multi-fold issue and the abatement policy (e.g. GHG emissions, legislation, and limit values) requires an interdisciplinary approach conducted within environmental authorities. Thus, the gender distribution in authorities responsible for climate change adaptation/mitigation measures is substantial (genderSTE 2014).

According to EU data, concerning environmental governing positions in European States, such as senior and junior ministers in environment, energy and transport at national governments, the women participation was slightly increased

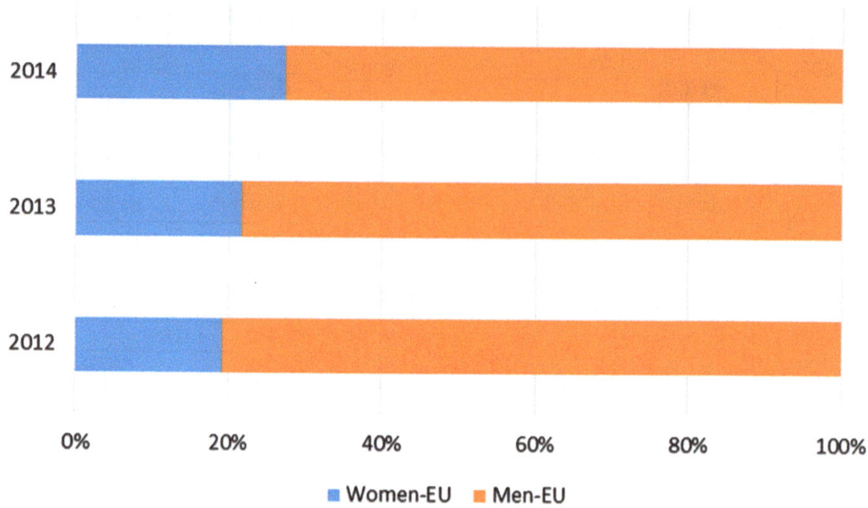

Fig. 5.2 The proportion of men and women in environmental ministers (senior and junior) in European Member States for the period 2012–2014

from 20 % to 27 % during the last 3 years (2012–2014), a proportion that is still significantly lower than male proportion (Fig. 5.2). In Greece during the same period there was no female representation.

However, the national administration positions show a slightly improved distribution with the female proportion increasing from 33 % to 36 % between 2012 and 2014. In Greece the data are considerably more balanced compared to the European figures, though decreased from 48 % to 45 % from 2012 to 2014 (Fig. 5.3).

It is noteworthy to mention that in a broader approach the representation of women in boards of large listed companies is also discouraging, with major gaps and little progress on women representation and career evolution, as presented in Fig. 5.4, in both the European Member States and Greece. More recently, and specifically in October 2014, Greece stands for the proportion of 8.9 % of women in highly ranked position with a European average of 20.2 % (EC 2015).

Having the aforementioned facts under consideration, the European Commission has established a set of indicators in order to determine and evaluate the women's participation in decision making on climate change (EIGE 2012b), that is the following:

- Proportion of women in climate change decision-making bodies at the national level in EU Member State.
- Proportion of women in climate change decision-making bodies at the EU level.
- Proportion of women in climate change decision-making bodies at the international level.
- Proportion of women tertiary graduates of total graduates in natural sciences and technologies at the EU and Member State level.

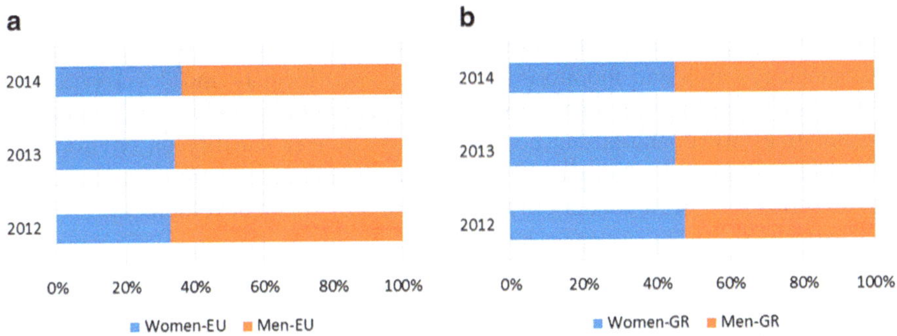

Fig. 5.3 The gender distribution of environmental national administrators (**a**) in the European Member States and (**b**) in Greece for the period 2012–2014

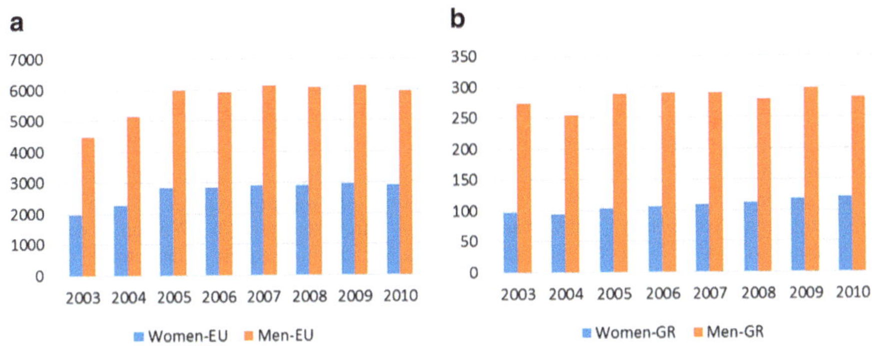

Fig. 5.4 Representation of women and men (absolute numbers) on boards of high listed companies (**a**) in the European Member States and (**b**) in Greece for the period 2003–2010

The development and collection of such data will support the study of gender-climate change issues and may provide the benchmark for the development of new indicators about the correlation between gender, environment and social welfare and justice.

Conclusions

The current study unveil the relations and links between gender and global warming at national and European level. It is proven that there are dissimilarities between males and females regarding global warming attitudes, contributions and capabilities of adaptation in the mitigation measures and the socio-economic impacts of climate change policies. As little progress is achieved, it comes without saying that gender perspective is a critical assumption towards a sustainable environment and a green economy in terms of the ecosystem, consumption, food, growth, transport, energy and citizen welfare. As the engendering of decision making on climate change will reinforce and will improve the efficiency of the suggested policies, the systematic research and monitoring of gender equality data and development of awareness initiatives are on the top of the agenda of policy recommendations. Quantitative targets and strong commitment on gender equality on environment and global warming by national governments, international institutes and authorities are critical areas of concern towards sustainable social development.

References

Bakke, J. V., Moen, B. E., Wieslander, G., & Norbäck, D. (2007). Gender and the physical and psychosocial work environments are related to indoor air symptoms. *Journal of Occupational and Environmental Medicine, 49*(6), 641–650.

Chen, L. H., Knutsen, S. F., Shavlik, D., Beeson, W. L., Petersen, F., Ghamsary, M., et al. (2005). The association between fatal coronary heart disease and ambient particulate air pollution: Are females at greater risk? *Environmental Health Perspectives, 113*(12), 1723–1729.

Clougherty, J. E. (2010). A growing role for gender analysis in air pollution epidemiology. *Environmental Health Perspectives, 118*, 167–176.

Druckman, A., Buck, I., Hayward, B., & Jackson, T. (2012). Time, gender and carbon: A study of the carbon implications of British adults' use of time. *Ecological Economics, 84*, 153–163.

EC—European Commission. (2012, March). Special Eurobarometer 376 "Women in decision-making positions". Conducted by TNS Opinion & Social at the request of Directorate-General Justice.

EC—European Commission. (2014). Report on equality between women and men 2014.

EC—European Commission. (2015, January). *Factsheet March 2015: Gender balance on corporate boards—Europe is cracking the glass ceiling.*

EEA—European Environment Agency. (2010). *The European environment—state and outlook 2010 (SOER 2010).*

EEA—European Environment Agency. (2015). *The European environment—state and outlook 2015 (SOER 2015).*

EIGE—European Institute for Gender Equality. (2012a). *Gender equality and climate change.* ISBN 978-92-9218-126-0.

EIGE—European Institute for Gender Equality. (2012b). *Review of the implementation in the EU of area K of the Beijing platform for action: Women and the environment gender equality and climate change.* ISBN 978-92-9218-026-3.

EU Legislation. (2013). Opinion *of the European Economic and Social Committee on 'The gender dimension in the Europe 2020 Strategy' (own-initiative opinion) 2013/C 76/02.* Retrieved March 26, 2015, from http://eur-lex.europa.eu/legal-content/EN/TXT/?uri=CELEX:52012IE1734&qid=1410262659773.

European Parliament. (2012). Resolution of 11 September 2012 on the role of women in the green economy (2012/2035(INI)). Retrieved March 26, 2015.

Fouillet, A., Rey, G., Jougla, E., Frayssinet, P., Bessemoulin, P., & Hemon D. (2007). A predictive model relating daily fluctuation in summer temperatures and mortality rates. BMC Public Health 7, 114.

genderSTE: A policy-driven targeted network funded by COST (European Cooperation in Science and Technology). Retrieved March 26, 2015, from http://www.genderste.eu/index.php.

General Secretariat. (2013). *Gender impact assessment of 100 laws and regulations.* Retrieved March 26, 2015, from http://www.genderimpact.gr/.

Hallmayer, J., Cleveland, S., Torres, A., Phillips, J., Cohen, B., Torigoe. T., et al. (2011). Genetic heritability and shared environmental factors among twin pairs with autism. *Archives of General Psychiatry, 68*(11), 1095–1102.

Huisman, M., Kunst, A. E., Bopp, M., Borgan, J. K., Borrell, C., Costa, G., et al. (2005). Educational inequalities in cause-specific mortality in middle-aged and older men and women in eight western European populations. *Lancet, 365*(9458), 493–500.

IPCC - Intergovernal Panel on Climate Change. (2014). Climate Change 2014: Impacts, Adaptation, and Vulnerability. Part A: Global and Sectoral Aspects. Contribution of Working Group II to the Fifth Assessment Report of the Intergovernmental Panel on Climate Change. Cambridge University Press, Cambridge, United Kingdom and New York, NY, USA, pp.709–754.

Künzli, N., Jerrett, M., Garcia-Esteban, R., Basagana, X., Beckermann, B., et al. (2010). Ambient air pollution and the progression of atherosclerosis in adults. *PLoS ONE, 5*, e9096. doi:10.1371/journal.pone.0009096.

Magnavita, N. (2015). Work-related symptoms in indoor environments: A puzzling problem for the occupational physician. *International Archives of Occupational and Environmental Health, 88*(2), 185–196. doi:10.1007/s00420-014-0952-7.

Miller, K. A., Siscovick, D. S., Sheppard, L., Shepherd, K., Sullivan, J. H., Anderson, G. L., et al. (2007). Long-term exposure to air pollution and incidence of cardiovascular events in women. *New England Journal of Medicine, 356*, 447–458.

Oiamo, T. H., & Luginaah, I. N. (2013). Extricating sex and gender in air pollution research: A community-based study on cardinal symptoms of exposure. *International Journal of Environmental Research and Public Health, 10*, 3801–3817.

Santamouris, M., Paravantis, J. A., Founda, D., Kolokotsa, D., Michalakakou, P., Papadopoulos, A. M., et al. (2013). Financial crisis and energy consumption: A household survey in Greece. *Energy and Buildings, 65*, 477–487.

Slini, T., Giama, E., & Papadopoulos, A. M. (2014). The impact of economic recession on domestic energy consumption. *International Journal of Sustainable Energy, 34*, 259–270. doi:10.1080/14786451.2014.882335.

UN—United Nations. (2014). Prototype global sustainable development report. Online unedited edition. New York: United Nations Department of Economic and Social Affairs, Division for Sustainable Development, 1 July 2014.

Volk, H. E., Hertz-Picciotto, I., Delwiche, L., Lurmann, F., & McConnell, R. (2011). Residential proximity to freeways and autism in the CHARGE study. *Environmental Health Perspectives, 119*(6), 873–877.

WHO—World Health Organisation. (2014, March). Household air pollution and health. Fact sheet N°292.

Xu, X., Hu, H., Ha, S., & Roth, J. (2013). Ambient air pollution and hypertensive disorder of pregnancy. *Journal of Epidemiology & Community Health, 68*(1), 13.

Chapter 6
Promotion of Sustainability by Quantifying and Reducing the Carbon Footprint: New Practices for Organizations

Monica Carvalho, Rommel de Santana Freire, and Alexandre Magno Vieira Gonçalves de Brito

Introduction

Until recently, companies and organizations had the opinion that the environment and sustainable development were problems and risk factors. However, the progressive development of a generalized, global environmental conscience created a demand for products (processes, services) that are more environmentally-friendly (with less associated impacts), for which there are great advances and developments in the application of eco-efficiency principles by industries and designers.

Nowadays, these "problems" are seen as opportunities, as possibilities for growth and efficiency improvement. According to the World Business Council for Sustainable Development (WBCSD 2000), eco-efficiency is basically the art of doing more with less: better production, with the consumption of less resources. But eco-efficiency is not only defined as the search for energy efficiency, it also stimulates creativity and innovation in the search for the best way to do things (WBCSD 2000). Achieving eco-efficiency does not mean the complete elimination of environmental impacts caused by man, rather lowering these impacts to levels

M. Carvalho (✉)
Department of Renewable Energy Engineering, Center of Alternative and Renewable Energy, Federal University of Paraíba, Caixa Postal 5115, Cidade Universitária, João Pessoa 58051-900, Paraíba, Brazil
e-mail: monica@cear.ufpb.br

R. de Santana Freire
Department of Accountancy and Finances, Center of Applied Social Sciences, Federal University of Paraíba, Cidade Universitária, João Pessoa 58051-900, Paraíba, Brazil
e-mail: rommelfreire@gmail.com

A.M.V.G. de Brito
Department of Mechanical Engineering, Center of Technology, Federal University of Paraíba, Cidade Universitária, João Pessoa 58051-900, Paraíba, Brazil
e-mail: alexandreh_magno@hotmail.com

P. Grammelis (ed.), *Energy, Transportation and Global Warming*, Green Energy and Technology, DOI 10.1007/978-3-319-30127-3_6

that are more compatible to the current degree of environmental conscience. According to Silva (2010), the environmental impacts caused by human activities are also studied in parallel with the development of solutions to increase eco-efficiency, to allow for the perception of solutions.

According to Ribeiro (2011), the oil crisis in the 1970s sparked the concern of society with the depletion of fossil fuel reserves, and as a consequence, the first environmental analyses were carried out. The interest in these analyses was not constant throughout the years, and was rekindled in the 1990s with the creation of the ISO 14040 norms and increase in environmental discussions. In 2006, Lima et al. (2006) mentioned that there was a progressive interest of industry, as well as of environmental specialists, authorities, consumer associations, environmental associations and general public, in the environmental quality of production processes and products.

In this current scenario where emphasis is on the minimization of dependency in fossil fuels along with the integration of alternative energies (progressively increasing participation of renewable energies), quantification of environmental impacts has been established as a powerful tool that creates awareness. Some studies go a step further by searching not only for the optimal solution from an environmental viewpoint (Carvalho et al. 2011a, b), but for a balance or equilibrium between different points of view (environmental and economic perspectives: Carvalho et al. 2011a, b; Buoro et al. 2013). Quantification of environmental impacts have been utilized by developed countries for many years, and the European Union (EU) has even declared Life Cycle Assessment to be a state-of-the-art methodology and included related aspects in the 6th Environmental Action Programme, to assure that environmental legislation is adopted to attack most of the environmental challenges in the EU (EIONET 2003).

Modern companies and institutions have aimed at a balance situation between the expansion of businesses and preservation of the environment, where sustainable growth is the fundamental point for the efficient management of resources. Accurate knowledge on productive processes is essential for the organizations to measure costs and risks; when not measured or correctly controlled, revenue can be significantly decreased to the point of non-viable business in some cases. Correct quantification of the environmental loads of a specific process, or sector, helps the company optimize costs and reduce risks, avoiding future liabilities in function of environmental damages caused. In this way, from the a priori knowledge on the amount of environmental loads generated, companies can establish targets for reduction/mitigation of pollution levels and promote the results to society in a clear manner. The company will publicize the reduction in the pollution generated by its activities as a contribution to the sustainable development of its surroundings. The reduction in environmental loads can be achieved through improvements in processes and products, installation of new equipment, capture and treatment of generated residues for posterior discharge or even sale. Companies can generate non-operational revenues, which can add up to relevant volumes, from the negotiation of processed/treated residues and also from the sale of carbon credits, generated from the elaboration of Clean Development Mechanism (CDM) projects. In these projects, the volume of reduced emissions can be converted into a commodity, and negotiated in the financial market, which has been increasing substantially in recent years (Pirrong 2011). Therefore, the reduction in the levels of emissions brings, besides the environmental benefits, also financial benefits, allowing for the sustainable growth of organizations.

Brazil should, as always, accompany the global trends of environmental conscience. The objective of this manuscript is to explore the relationship between the environmental loads generated by companies, with society and governments. This relationship is based on the tripod QUANTIFY-REDUCE-PROMOTE and can be extended to assess what society itself emits. The idea supporting this project is that the social perception of companies developing studies for the quantification of environmental impacts will be improved. It is important to highlight that the quantification study is the first step: then, points are identified where improvements can be introduced for posterior reduction of emissions. The next step is to promote, disseminate these results, which returns to the initial idea: improvement of social perception and popularity of the company. If the consumers or clients perceive that the company is concerned with the environment, they are more inclined to be loyal to the brand.

Diagnosis and Analysis of the Situation

Due to the progressive development of environmental conscience by the society, along with ever-increasing pressure to reduce environmental loads, emerges the necessity of considering the environment as an additional criterion at the time of designing or building products and processes (Carvalho et al. 2012). Therefore, the evaluation of the performance of products, processes or services depends less on economic criteria and starts to include, progressively, environmental components.

Developed countries committed to reducing their greenhouse gas (GHG) emissions through the Kyoto Protocol, at the end of the 1990s. This commitment is mandatory only for developed countries, reflecting the international consensus that industrialized countries must lead the process of mitigation of climate change (Baer et al. 2000). Developed countries should not only reduce their emissions, but also produce technology and means to help developing countries (Baer et al. 2000; Ashton and Wang 2003). Developing countries insist that the solution for climate change cannot be achieved at the expense of their development (Yale 2014). According to Yale (2014), it is obvious that a reduction in emissions should also be the goal of developing countries, but economic growth and decrease of poverty are connected to the lack of access to energy services. Development seems to be linked, therefore, to an unavoidable increase in the use of fossil fuels and consequently, of GHG emissions. Apart from the multiple schemes for allocation of emissions, the United Nations Framework Convention on Climate Change has already recognized that the global nature of climate change requires the broadest collaboration possible from all countries, and their united participation in an adequate international response, following a common but differentiated responsibility according to the capacity of each country (UNFCCC 2014).

There are also technical-operational arrangements, regulated by the Kyoto Protocol, that can be utilized by companies or countries. These arrangements offer different routes for the countries to achieve emission reduction limits and goals. There are three "flexibilization" mechanisms (United Nations 2014): emission trading (countries that have reduced emissions even beyond the established limit can transfer their reduction excess to another country that has not achieved its limit); CDM, and joint implementation (GHG emission reduction projects in countries that have set goals in the ambit of the Protocol). Only CDM is currently applicable to Brazil. The regulation concerning CDM and carbon credits remains underexplored in Brazil, but even without adequate regulation, in 2007 the first auction sale of carbon credits took place in São Paulo, corresponding to a volume of more than 800 thousand tonners of CO_2-equivalent (CDM from a landfill project) (Perera 2010).

Although GHG emissions are part of the development process of countries, the increasing current environmental conscience demands that environmental impacts are at least diminished (since they cannot be completely eliminated) in the search for eco-efficiency (Carvalho and Freire 2014). During the 2009 United Nations Climate Change Conference, Brazil voluntarily committed to reduce its GHG emissions between 36.1 % and 38.9 % until 2020 (Marques et al. 2010). More recently, in December 2014, during the Conference of the United Nations in Peru, the text "The Lima call for Climate Action" (UNFCCC Newsroom 2014) was announced. Countries can establish their own goals for the reduction of emissions (Intended Nationally Determined Contributions), which, according to Brasil (2014) represent the intended contribution of each country to the global effort of mitigating climate change.

However these reduction efforts should not be limited to mitigation only, including also efforts for adaptation, funding, technology transfer and training (Brasil 2014). Although the word "intended" appears to indicate a non-binding connotation, Brazil defends that the reduction targets should be juridically binding (Brasil 2014). The Lima Call maintains the line that there is differentiated content for developed and developing countries, according to the aforementioned principle of common, but differentiated responsibilities.

Following the idea of sustainable development, consumers can already choose which product to purchase based on the information shown on ecological tags or labels. Ecological labels provide a measure of the environmental performance of the product, informing the consumer about the environmental impacts. The consumer is therefore enpowered, and this action reverberates in the market, as other similar products will suffer pressure to adopt ecological labels; those that already count with this label will also suffer pressure to minimize even more their impacts. In a study carried out in France between 2004 and 2006, it was concluded that the percentage of more selective consumers increased from 15 % to 32 % (Perera 2010).

Still according to Perera (2010), more than 85 % of French citizens believe they play a crucial role in environmental matters, almost 90 % prefer products and brands that are environmentally correct, and 95 % intended to research more on the environmental impacts of consumer products. Ecological labels build on the added value of environmental quality to the consumer (Csutora and Zsóka 2012). The importance of providing environmental information to support more sustainable consumption has been repeatedly highlighted in policy reports and declarations such as UN Agenda 21 United Nations, the EU Sustainable Consumption and Production Action Plan and the UK Sustainable Development Report (Upham et al. 2011).

The Blue Angel (Blaue Engel) was the pioneer in the labeling of products and services, and was instituted in Germany in 1977 (Blaue Engel 2014). The European Ecological Label was created in 1992, and has been raining awareness in companies and consumers ever since (Spain 2014). In a parallel line of thinking, since 1993 Brazil counts with a seal from the National Program for the Conservation of Electrical Energy, which is a simple way of informing the consumer on the efficiency level and consumption of electricity of domestic appliances and equipment (CBIEE 2006). This program was created by the Brazilian Government and although limited, it is evident that Brazil attempts to follow the efforts of other countries.

Entrepreneurship, which can be translated into the search for business opportunities, might appear to be in the opposite direction of sustainability perspectives at the long term—however, maybe because entrepreneurs present the unique ability to combine, in an innovative manner, the individual viewpoint and the circumstances, researchers already start to explore entrepreneurship as a potential mechanism for sustainable development (Kosgaard and Anderson 2010). Despite the increasing interest in this connection between sustainability and entrepreneurship, this relationship is still young but, according to York and Venkataraman (2010), entrepreneurs can contribute to the solution of environmental issues through the creation of more sustainable products or services. From the moment a company or institution manages to concentrate efforts to decrease consumption of energy or increase the efficiency of a process, it provides some "relief" to the environment. The perception of the consumer is positive, returning to the aforementioned environmental conscience process. By giving preference to this product or service, the consumer encourages the competition to decrease their own environmental impacts, closing the cycle: less consumption of resources; less cost with resources; less environmental impact. Gaining commitment and buy-in from the company is crucial. It is also important to highlight the reductions achieved to promote the company as an environmentally-friendly organization. This vision embraces the entrepreneur's potential to supplement: (1) legislation; (2) social corporative responsibility; and (3) activism in the solution of environmental issues (York and Venkataraman 2010).

The Quantify-Reduce-Promote (QRP) Proposal

The overarching intention of the proposal is to implement the concept of eco-efficiency, which involves the extraction of the maximum potential of materials and energy (efficient use), with the consequent minimization of environmental loads and monetary costs (better use of resources means to use less and emit less, with lower costs). International concern about climate change is beginning to positively influence Brazilians, leading to evergrowing interest in the greenhouse gas emissions involved in products, processes and services. The calculation of these emissions is starting to become common in academic environments, however actions directed to reduction and mitigation actions are still in an initial stage.

The reduction of environmental loads can only be confirmed after adequate quantification, and in recent decades, different techniques for global environmental assessment have been developed (e.g., ecological footprint, water footprint, carbon footprint, Life Cycle Assessment—LCA, LEED-NH, to name a few). According to Curran (1999), LCA is one of the most adequate tools to reach objectives directed towards sustainability and is already an internationally accepted and recognized tool, that evaluates the consequences of building and utilizing products and services, through the identification and accounting of material used and disposal of residues into the environment. However, sometimes due to limited resources and data, companies carry out analysis based on simplified LCA approaches or apply the general principals of Life Cycle Thinking (LCT) to specific aspects of the productive system (De Benedetto and Klemeš 2009). LCT takes into consideration all environmental and toxicological impacts associated with a product, process or service, throughout its lifetime.

The first step in the proposal presented herein is to solidify LCT as a methodology applied not only to the quantification of environmental impacts of a product, process or service, but also for the comparison of different alternatives, helping decide on which option is better from an environmental viewpoint. In the light of the above, the necessity of carrying out this project is justified, to pursue eco-efficiency with support from the LCT/LCA methodology, which will help decision makers make informed decisions from an environmental perspective, or even find a commitment between economic and environmental perspectives. With the information generated, it will be possible to establish internal environmental policies that enable maximization of economic-financial results in equilibrium with the preservation of the environment.

Along with the improvement of operational and economic results, there is a concomitant improvement in the image of the company to society, created in function of the reduction in the pollution/emissions generated. This type of publicity (promotion) is already utilized by large companies, especially those that have titles being negotiated in the stock market. In Brazil, the Efficient Carbon Index (ICO2) has sparked the interest of investors worried with environmental issues as well as with revenue expectancy (BM&FBOVESPA 2011). The companies that participate in this index are obligated to publish, in annual reports, the emission of

GHG. However, small and medium sized enterprises do not present any interaction with society, and therefore the existence of a great niche is confirmed for the implementation of this project, especially in energy-intensive industries. In a proactive and positive manner, good practices in environmental education will be stimulated at the same time that environmental conscience is built in a responsible way. Investments in green, positive marketing are important to increase market share.

The traditional saying 'what gets measured gets managed' launches Phase 1, which starts with the QUANTIFICATION of environmental impacts through a comprehensive, detailed environmental analysis. This analysis will allows for the identification of the products, processes and services that are least aggressive to the environment and will help quantify (and identify) critical aspects and even those deserving special attention. This assessment can be extended to comprehend different products, processes or services. Another target for this Phase is the identification of the configurations/options that are most adequate, through the introduction of the environmental component as explicit decision element in synthesis, design or operation stages.

Phase 2 of this project is the REDUCTION of environmental impacts, which occurs after exhaustive analysis of the results of the QUANTIFICATION Phase. This could lead to the formulation of an environmental strategy that could signal a step change in environmental management for the studied company. Opportunities for the REDUCTION of environmental loads can be identified, a priori, in energy efficiency, introduction of renewable energy sources, and minimization of waste (maximization of reuse and recycle). REDUCTION can be extended (but not limited) to supplies and contracts (procurement), transport, construction and temporary activities, and use of water. According to the Ministry for the Environment of New Zealand (2015), these aspects are initially selected because they likely to have the highest environmental impacts, are the most commonly addressed by other major international organisms, and addressing them is consistent with international government environmental policy priorities.

Phase 3 of this project is the PROMOTE, the dissemination of the results. The dissemination, the advertising of the results is essential and therefore a specific Phase of the project is dedicated to this activity, which includes analysis of results, synthesis, conclusions, publicity and dissemination of results. The integration of the previous results will provide a panoramic vision, much more complex and rich than the simple addition of results: environmental evaluation, identification of opportunities and innovative aspects to which R&D efforts must be directed, and recommendations. Due to the high volume of information and data presented in Phases 1 and 2, a great effort of synthesis and creation will be necessary. Phase 3 will influence consumers even when the cost of products is higher (in comparison with competition or similar products). For example, Csutora and Zsóka (2012) remark that hybrid cars are expensive (when the energy saved is accounted for, no cost payback can be expected) but these vehicles still represent prestige or environmental value for some environmentally-conscious consumers and have achieved premium price position and small, but considerable market share.

The company involved in the project will be analyzed from different perspectives, integrating, within an environmental focus, the productive sector with the economic-financial management of the organization. A new focus will be therefore created, generating wealth with an environmentally-responsible basis. Phase 1 is not a simple activity, as it requires the mobilization of a significant amount of data as well as a high consumption of time and computational effort. The product (process, or service) must be described in detail, including inputs (material, energy) and outputs (product, service, residues).

Special care is necessary to avoid any comparisons or relationships with greenwashing claims. Greenwashing is a type of deceptive marketing, used to provide the image that a product is "green". When combined with inexistent or ineffective legislation, greenwashing could decrease the potential of the consumer to drive companies in the direction of better solutions (Dahl 2010). The idea defended herein is that good eco-labeling helps prevent greenwashing, and reduces consumer skepticism.

Opportunities and Advantages

Some of the opportunities identified for the company that adopts the QRP proposal are the development of a more strategic environmental approach, which will position the company to better respond to customer needs, legislative requirements, market changes and stakeholder expectations. Businesses and institutions aim at respectability and need to build a good image, a good reputation and there are two route to achieve these aspects: (1) lead the way, being the pioneer in building this positive, environmentally-conscious image; or (2) be close to whom is a reference, and follow their steps into achieving a good consumer perception.

Regarding the technological-social contribution of the project, it satisfies the priorities of research, development and demonstration of new concepts and technologies to improve energy efficiency and reduce the final consumption of primary energy. The minimization of environmental impacts is a benefit for all Brazilian citizens. This project also presents a strong social bias, as the information on the environmental loads embedded in a product, process or service provides decision power to the consumer, who will be able to choose the least aggressive option, forcing other manufacturers or suppliers to search for more "ecological" ways to maintain competitiveness. The generation of economic-financial benefits represents an extra advantage for the adoption of this project, enabling the company to recover the investments made for its implementation. Already in 2006, Tachizawa predicted that Brazilian consumers would start to give preference to products that were environmentally responsible (Tachizawa 2006; Perera 2010). According to Tan et al. (2014), voluntary carbon labeling schemes have been introduced in several countries, with significant participation from companies that do not want to be seen as lagging behind rivals in cornering the green market.

There are also opportunities for carbon offsetting, where emissions can be mitigated by investing in projects that avoid the production of emissions or remove them from the atmosphere. According to the Ministry for the Environment of New Zealand (2015), commitments to reduce emissions should always be accompanied by offsetting measures in order to be worthy of belief or confidence, as it is more effective to avoid creating the emissions than elaborating strategies for offsetting. Both emission reduction and offsetting can be integrated into an environmental responsibility plan, and it is important to obtain the advice of experts. In Brazil, during the 2014 Soccer World Cup, the Fédération Internationale de Football Association (FIFA) announced a set of carbon offsetting projects in Brazil that were selected to reduce the overall emissions of the tournament, which included emissions from staff travel and accommodation, and extended to officials, teams, volunteers and guests—four certified low-carbon development projects helped achieve the goal (Fédération Internationale de Football Association 2014).

In recent years, the carbon market has progressively expanded and has become a lucrative business for businesses. This market can also include stock market actions and voluntary markets, where a business that reduces pollution generates compensation credits and can sell them. In 2012, Brazil occupied the third position among the countries that participated in the global carbon market, with approximately 5 % of the world total and 268 projects—however, initial estimates indicated that Brazil would absorb 20 % of the carbon credits (Brasil 2012). The CDM is a flexible tool that allows the participation of developing countries (or countries with no reduction commitments) in the carbon market, and has incentivized the creation of new technologies for reduction of GHG gases in Brazil. There is a series of criteria for recognizing carbon credit projects, such as alignment with the sustainable development premises of the host country, defined by a Designated National Authority (DNA)—in Brazil, this authority if the Interministerial Commission of Climate Change, and only after approval of the Commission the project can be submitted to the United Nations for evaluation and registry.

According to Perera (2010), paradoxically, the Brazilian clean energy matrix could hinder CDM mechanisms in Brazil, but the country could become the largest exporter of carbon credits—therefore it is important to be the first in this market to reach leadership in the ability of recruiting international resources for sustainable development.

Final Remarks

Encouraging consumers to make more informed choices requires access to environmental performance information of products. The QRP project presented herein has the ultimate goal of informing consumers about how the purchase of different products within a substitutable range would have differing impacts on the environment. Since Brazilian consumers are starting to demand more data on how their consumption patterns affect the environment, the adoption of the QRP proposal helps businesses and institutions keep up with the competition through the provision of guidance to consumers on the impact of purchases on GHG emissions. According to Tan et al. (2014), the implementation of carbon labeling is likely to have positive effects on the environment.

Transfer of know-how from this project will increase the interest of Brazilian industry and help the sector perceive the effect and importance of quantifying and reducing the GHG emissions, with posterior transference (promotion) of results to the general public as well as other sectors. Technology transfer can be achieved through this proposal, and gaining commitment and buy-in from the company is crucial. Highlighting the reductions achieved has a twofold goal: promote the company as an environmentally-friendly organization and improve its reputation.

The QRP project explained herein brings opportunity, innovation and competitive advantage to the companies and institutions that decide to be on-board. The project will improve the image of the company, even if the initial commitment is only to achieve reduction goals or profit.

Regional integration and articulation with other sectors can be achieved through this project, which will provide specialized services and suggest better technologies or processes, which can be available locally. Integration and articulation will be promoted through the Life Cycle Thinking approach, propelling local and regional economy, and positively influencing the region involved in the project.

The implementation of the activities contained in this project guarantees the rational use of natural resources, decreases social and environmental impacts deriving from their utilization and impels regional economy, resulting in the increase of life quality standards and social standards for local communities. Additionally, the QRP project contributes to sustainable development and improvement of environmental quality.

Acknowledgments The authors wish to acknowledge the support of the Institute for the Development of Paraíba (IDEP-PB) and of the National Council for Scientific and Technological Development (CNPq), through project n° 475879/2013-9.

References

Ashton, J., & Wang, X. (2003). Equity and climate: In principle and practice. In *Beyond Kyoto: Advancing the international effort against climate change*. Pew Center on Global Climate Change working papers series.

Baer, P., Harte, J., Haya, B., Herzog, A. V., Holdren, J., Hultman, N. E., et al. (2000). Climate change: equity and greenhouse gas responsibility. *Science, 289*(5488), 2287–2291.

Blaue Engel. *The blue angel*. Retrieved December 10, 2014, from https://www.blauer-engel.de/sites/default/files/pages/downloads/our-label-environment/be-8-seitiger-eng-web.pdf.

BM&FBOVESPA. (2011). *See how sustainability can guide investments* (in Portuguese). Retrieved January 18, 2015, from http://www.bmfbovespa.com.br/pt-br/noticias/2011/Veja-como-a-sustentabilidade-pode-nortear-os-investimentos-2011-09-22.aspx?tipoNoticia=1&idioma=pt-br.

Brasil—Ministry of Environment. (2012). *Understand how the carbon credit market operates* (in Portuguese). Retrieved January 15, 2015, from http://www.brasil.gov.br/meio-ambiente/2012/04/entenda-como-funciona-o-mercado-de-credito-de-carbono.

Brasil—Ministry of International Affairs. *Process for the preparation of the contribution of Brazil to the new agreement under UNFCCC* (in Portuguese). Retrieved December 30, 2014, from http://diplomaciapublica.itamaraty.gov.br/consultaclima.

Buoro, D., Casisi, M., De Nardi, A., Pinamonti, P., & Reini, M. (2013). Multicriteria optimization of a distributed energy supply system for an industrial area. *Energy, 58*, 128–137.

Carvalho, M., & Freire, R. S. (2014). Quantificar, reduzir, evidenciar: uma nova prática para empresas sustentáveis. In*: ENCONTRO INTERNACIONAL SOBRE GESTÃO EMPRESARIAL E MEIO AMBIENTE*, 16, 2014, São Paulo. Anais... São Paulo: FEA-USP (in Portuguese).

Carvalho, M., Lozano, M. A., & Serra, L. M. (2012). Multicriteria synthesis of trigeneration systems considering economic and environmental aspects. *Applied Energy, 91*, 245–254.

Carvalho, M., Serra, L. M., & Lozano, M. A. (2011a). Geographic evaluation of trigeneration systems in the tertiary sector. Effect of climatic and electricity supply conditions. *Energy, 36* (4), 1931–1939.

Carvalho, M., Serra, L. M., & Lozano, M. A. (2011b). Optimal synthesis of trigeneration systems subject to environmental constraints. *Energy, 36*(6), 3779–3790.

CBIEE—Brazilian Center for Information on Energy Efficiency. *The PROCEL seal* (in Portuguese). Retrieved December 17, 2014, from http://www.procelinfo.com.br/main.asp?TeamID={88A19AD9-04C6-43FC-BA2E-99B27EF54632}.

Csutora, M., & Zsóka, A. (2012). *Effective choice of consumer-oriented environmental policy tools for reducing GHG emissions, greenhouse gases—Capturing, utilization and reduction.* Croatia: Intech.

Curran, M. (1999). The status of LCA in the USA. *International Journal of Life Cycle Analysis, 4* (3), 123–124.

Dahl, R. (2010). Green washing: Do you know what you're buying? *Environmental Health Perspectives, 118*(6), A246–A252.

De Benedetto, L., & Klemeš, K. (2009). The environmental performance strategy map: An integrated LCA approach to support the strategic decision-making process. *Journal of Cleaner Production, 17*, 990–996.

EIONET—European Topic Centre on Sustainable Consumption and Production. (2003). *Life-cycle assessment and life-cycle thinking in resource and waste management*. Retrieved January 18, 2015, from http://scp.eionet.europa.eu/themes/lca.

Fédération Internationale de Football Association—FIFA. (2014). *FIFA to offset all its operational carbon emissions through certified projects in Brazil*. Retrieved January 15, 2015, from http://www.fifa.com/worldcup/news/y=2014/m=6/news=fifa-to-offset-all-its-operational-carbon-emissions-through-certified--2379123.html.

Kosgaard, S., & Anderson, A. R. (2010). Sustainability and entrepreneurial action. In *33rd annual ISBE conference*. London: ISBE.

Lima, A. M. F., Kiperstock, A., & Caldeira-Pires, A. (2006). Evaluation of life cycle in the Brazilian academic environment. In *I Simpósio Nordestino de Saneamento Ambiental—ABES*, Seção Paraíba. João Pessoa (in Portuguese).

Marques, F. R., Magellan, G., & Parente, V. (2010). *The Brazilian carbon market and the absence of a regulatory framework* (in Portuguese). Business School São Paulo, July/2010. Retrieved January 16, 2015, from http://www.revistabsp.com.br/edicao-julho-2010/o-mercado-brasileiro-de-carbono-e-a-ausencia-de-um-marco-regulatorio/.

Ministry for the Environment of New Zealand. (2015). *Major event greening guide: A practical guide to reducing the environmental impact of a major event—Section two: Greening focus areas*. Retrieved January 11, 2015, from http://mfe.govt.nz/publications/sus-dev/major-event-greening-guide/page2.html.

Perera, L. C. J. (2010). Investing in clean development mechanism (CDM) projects in Brazil. *Journal of Operations and Supply Chain Management, 3*(1), 1–14.

Pirrong, C. (2011). *Commodity price dynamics: A structural approach*. Cambridge: Cambridge University Press.

Ribeiro, M. A. J. (2011). Application of life cycle assessment to the search for eco-efficiency in the manufacturing of sliced bread. MSc dissertation. Centro Federal de Educação Tecnológica Celso Suckow da Fonseca, CEFET/RJ (in Portuguese).

Silva, J. C. A. (2010). Life cycle analysis. In: *Theory and practice in sustainable construction in Brazil* . Rio de Janeiro: Environment State Secretary of Rio de Janeiro (in Portuguese). Retrieved January 18, 2015, from http://download.rj.gov.br/documentos/10112/1312245/DLFE-56323.pdf/01_SECAOI_1_INTRODUcaO_docfinal_rev.pdf.

Spain—Ministry of Agriculture, Food and Environment. *What is the European ecological label?* (in Spanish). Retrieved December 17, 2014, from http://www.magrama.gob.es/es/calidad-y-evaluacion-ambiental/temas/etiqueta-ecologica-de-la-union-europea/etiqueta-ecologica-europea/que_es.asp.

Tan, M. Q. B., Tan, R. B. H., & Khoo, H. H. (2014). Prospects of carbon labelling: A life cycle point of view. *Journal of Cleaner Production, 72*, 76–88.

Tachizawa, T. (2006). *Gestão ambiental e responsabilidade social corporativa [Environmental management and corporative social responsibility]* (4th ed.). Atlas: São Paulo [In Portuguese].

UNFCCC Newsroom—United Nations Framework Convention on Climate Change Newsroom. *Decision -/CP.20 Lima call for climate action*. https://unfccc.int/files/meetings/lima_dec_2014/application/pdf/auv_cop20_lima_call_for_climate_action.pdf.

UNFCCC—United Nations Framework Convention on Climate Change. *Full text of the Convention*. Retrieved December 5, 2014, from http://unfccc.int/essential_background/convention/background/items/1349.php.

United Nations. (2014). *The mechanisms under the Kyoto protocol: Emissions trading, the clean development mechanism and joint implementation*. Retrieved January 16, 2015, from http://unfccc.int/kyoto_protocol/mechanisms/items/1673.php.

Upham, P., Dendler, L., & Bleda, M. (2011). Carbon labelling of grocery products: Public perceptions and potential emissions reductions. *Journal of Cleaner Production, 19*, 348–355.

WBCSD—World Business Council for Sustainable Development. (2010). *Eco-efficiency learning module* (231p).

Yale—The Forum on Religion and Ecology at Yale. (2014). *Ethical considerations in reducing global greenhouse gas emissions*. Retrieved December 5, 2014, from http://fore.research.yale.edu/climate-change/ethics/ethical-considerations-in-reducing-global-greenhouse-gas-emissions.

York, J. G., & Venkataraman, S. (2010). The entrepreneur–environment nexus: Uncertainty, innovation, and allocation. *Journal of Business Venturing, 25*(5), 449–463.

Part II
Global Warming and Climate Change: General Issues and Challenges

Climate Change Mitigation Measures and Prospects

Chapter 7
Group Comparison, Trends and Cluster Analysis to Understand Historical Precipitation

Raphael Abrahão

Introduction

In recent years, much research has been carried out regarding climate change and its consequences. The current concentrations of greenhouse gases are already a concern and several scientists predict that the average temperature of the planet can increase between 1.8 and 4.0 °C until the end of this century, which may cause dramatic environmental impacts (IPCC 2007; Malhi et al. 2009; Davidson et al. 2012). According to the IPCC (2014), warming of the climate system is unequivocal, and since the 1950s, many of the observed changes are unprecedented over decades to millennia.

This scenario of increasing temperatures is only one facet of climate change. Shifting in precipitation patterns (frequency and intensity) is another important change that needs to be deeper understood. Such changes in climate may impact ecosystems, trigger plagues and epidemics, threaten urban infrastructure, water and energy supply, as well as agriculture, especially in regions where shortage of water is already an issue (Knapp et al. 2002; Disch et al. 2012; Davidson et al. 2012; Durack et al. 2012; IPCC 2014).

Understanding the nature and extent of these impacts is crucial for the determination of adaptation policies towards avoiding or diminishing the negative impacts of climate change, as well as taking advantage of the positive impacts. However, to understand the potential impacts and plan adequately, we first need to understand how climate is changing.

R. Abrahão (✉)
Department of Renewable Energy Engineering (DEER), Center of Alternative and Renewable Energy (CEAR), Federal University of Paraíba (UFPB), Caixa Postal 5115, Cidade Universitária, João Pessoa 58051-970, PB, Brazil
e-mail: raphael@cear.ufpb.br

P. Grammelis (ed.), *Energy, Transportation and Global Warming*, Green Energy and Technology, DOI 10.1007/978-3-319-30127-3_7

Climate models are useful tools to project climate change scenarios. Although nowadays they are very comprehensive and can take into consideration several parameters and relationships, there is still a high level of uncertainty in the studies using these projections, especially for precipitation (Von Storch et al. 1993; Grillakis et al. 2011; Zhang et al. 2011).

In addition, in many parts of the world, relatively long historical climate datasets are available and little (and in many cases, none) use of this information is made towards a better understanding of climate change impacts. The high costs associated with installation and maintenance of these monitoring stations throughout decades could be better spent and deeper information about the dynamics of regional and local climate change could be obtained from the datasets.

A number of studies used historical data, confirming that climate change is not something that will happen in the future but is a current reality (Magnuson et al. 1997; Vincent and Gullett 1999; Alexander et al. 2006; Wang and Ding 2006). These studies are good examples of how deeper information about climate change can be obtained from real historical data. Furthermore, confidence in the projections from climate models is higher for temperature than for other climate elements such as precipitation; confidence is also higher at global and continental scales than for regional and local scales (AMS 2012).

Thus, can historical data provide sufficient information about climate change in order to understand the potential impacts and develop adaptation strategies? And, more importantly, are the methodologies effective to provide this information? The objective of this paper is to assess the use of cluster analysis, together with other more traditional methodologies such as group comparison and trend analysis, to expand knowledge about climate change on local and regional scales from historical data of precipitation.

Methodology

The methodology was based on three different approaches: (1) a general assessment of annual totals with comparison of past and recent periods, (2) trend analysis, and (3) the application of cluster analysis to analyse monthly and seasonal changes. The intention was not compare these approaches, but evidence the importance of using different approaches, instead of a single one, to assess climate changes that will affect the water cycle and others factors regionally.

Monthly rainfall and snowfall data were obtained from the Sudbury airport station (Canada), from January 1956 to December 2010 (55 complete years). Data was obtained from Environment Canada (Canadian National Climate Data and Information Archive, www.climate.weatheroffice.gc.ca) and it was subjected to a process of quality control before becoming available. The city of Sudbury is part of the province of Ontario and is located at 46°29′N and 81°00′W. The climate of the area is classified as humid continental (Dfb according to the Köppen-Geiger classification).

The 20 first years (1956–1975) and 20 last years of the dataset (1991–2010) were selected for evaluation of annual changes. Since the annual data presented normal distribution, the t-test was applied to compare both periods. The t-test performed a comparison of means to determine whether the means of the two periods where significantly different. A probability of less than a 5 % error ($p < 0.05$) was considered.

The Mann-Kendall test was used for trend analysis. A non-parametric test was selected because the monthly and seasonal data did not present normal distribution (Mann 1945; Kendall 1975). The trend estimation was performed applying the Sen's slope calculation (Sen 1968) and error probabilities between 0.1 % and 10 % were used.

Cluster analysis was carried out to analyze changes in the distribution of rainfall and snowfall throughout the year. The purpose was to place the objects (month-year combinations) into groups (or clusters) suggested by the data, not defined a priori. Thus, data was organized monthly, and each month was considered as a different variable, thereby obtaining data from 660 month-year combinations (12 months in 55 years). Years with similar precipitation distribution were placed into the same cluster. The cluster analysis was performed by standardizing the variables and using the Euclidean square distance as similarity measure (Hair et al. 1998). Due to its simplicity and computational efficiency, the Ward method was used to obtain hierarchical clustering (Hair et al. 1998).

In hierarchical cluster techniques, the objects are progressively aggregated until they form a single cluster. Each object begins in a cluster itself and then the closest clusters are merged to form a new cluster that replaces the two previous clusters. Merging of the two closest clusters is repeated until only one cluster is left (Ramos 2001; Munoz-Díaz and Rodrigo 2004). In some studies, similar results were obtained by applying cluster techniques and principal component analysis (Munoz-Díaz and Rodrigo 2004; Yin et al. 2011).

The "median year" of the objects (month-year combinations) was used as an indicator to compare recent and past periods. All statistical procedures were carried out with Statgraphics 15 and Makesens 1.0 softwares.

Results and Discussion

Changes in Annual Rainfall and Snowfall

The dataset suggests that annual rainfall is increasing in the Sudbury site. Even with the expected variability of rainfall between years, an increasing trend during recent years could be observed (Fig. 7.1). Significant statistical differences ($p < 0.05$) were observed when comparing the 1956–1975 period versus the 1991–2010 period, with average values of 608 and 680 mm/year respectively, which represents a 12 % increase in rainfall.

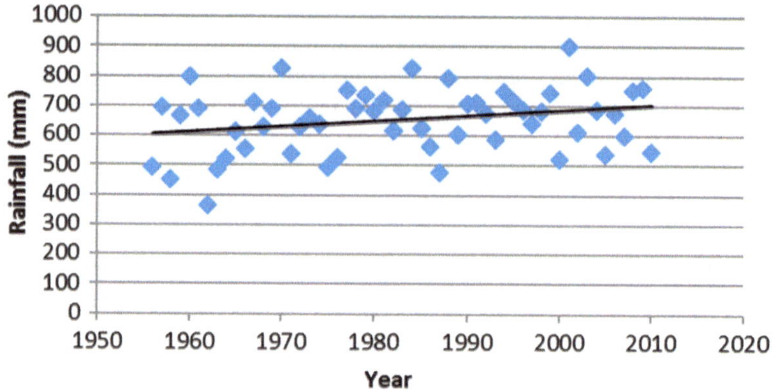

Fig. 7.1 Annual rainfall

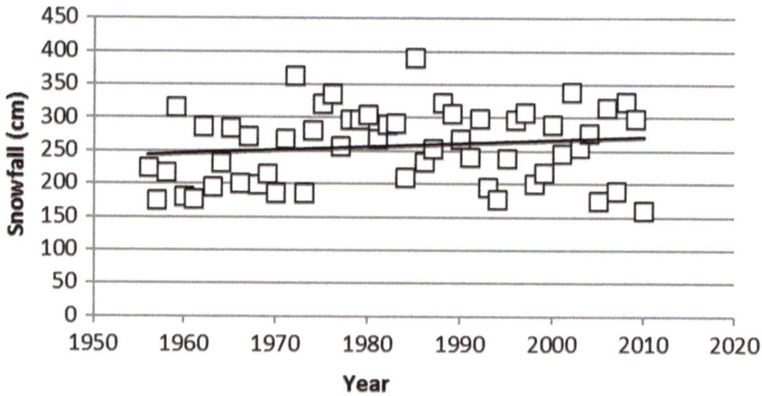

Fig. 7.2 Snowfall in the Sudbury site during the 1956–2010 period

Regarding snowfall, the trend was not that clear. Although a slight increase in snowfall could be observed (Fig. 7.2), there was no statistically significant difference between the means of the two periods ($p < 0.05$).

Changes in Monthly and Seasonal Distribution of Rainfall and Snowfall—Trend Analysis

Trend analysis also detected the annual changes presented in the previous section. A significant increasing trend in rainfall ($p < 0.10$) was quantified by the Sen's slope through a rate of 1.74 mm/year and no significant trend was detected for annual snowfall (Table 7.1).

Table 7.1 Rainfall and snowfall trends detected by the Mann-Kendall test and quantified by the Sen's slope for the Sudbury site during the 1956–2010 period

Period	Rainfall (mm/year)	Snowfall (cm/year)
January	0.05 ns	0.20 ns
February	–	0.17 ns
March	0.28[a]	−0.07 ns
April	0.22 ns	0.07 ns
May	0.58[a]	–
June	−0.13 ns	–
July	0.13 ns	–
August	0.36 ns	–
September	−0.18 ns	–
October	0.49[b]	–
November	0.15 ns	−0.10 ns
December	0.07 ns	0.17 ns
Winter	0.32[a]	0.67 ns
Spring	1.04[b]	0.06 ns
Summer	0.08 ns	–
Fall	0.35 ns	−0.15 ns
Annual	1.74[a]	0.60 ns

ns non significant
[a]$p < 0.10$
[b]$p < 0.05$
[c]$p < 0.01$
[d]$p < 0.001$

The trend analysis could be considered more comprehensive for evaluation of annual changes than the t-test group comparison, because all years of the dataset are included in the analysis and not only the 20 first and 20 last years. However, the expected variability of precipitation could hinder the detection of trends, which was not the case for the Sudbury site during the 1956–2010 period.

Changes in the distribution of precipitation throughout the year were also noted during the studied period. It means that the annual increasing trends observed for rainfall volumes were not uniform over the year. March and May presented significant ($p < 0.10$) positive trends with Sen's slopes of 0.28 and 0.58 mm/year respectively. October presented a more significant ($p < 0.05$) positive trend and a Sen's slope of 0.49 mm/year. The remaining months did not present significant trends.

Seasonally, significant increasing trends were observed in winter ($p < 0.10$) and spring ($p < 0.05$) through Sen's slopes of 0.32 and 1.04 mm/year respectively. The Mann-Kendall test did not detect significant trends in rainfall during summer and fall. Regarding snowfall, no significant trends were detected for any month or season.

Changes in Monthly and Seasonal Distribution of Rainfall and Snowfall—Cluster Analysis

For the monthly rainfall data from the Sudbury site, the cluster analysis applied separated the 55 years into three clusters (Table 7.2). Cluster 1 consisted of years 1956, 1959, 1960, 1961, 1962, 1964, 1970, 1972, 1976, 1978, 1987, 1992, 1997 and 2010. Cluster 2 consisted of years 1957, 1958, 1963, 1965, 1966, 1967, 1968, 1969, 1971, 1973, 1974, 1975, 1977, 1979, 1981, 1982, 1984, 1985, 1986, 1989, 1990, 1991, 1994, 1998, 1999, 2000, 2001, 2002, 2003, 2004, 2005, 2006, 2007 and 2009. Finally, cluster 3 included years 1980, 1983, 1988, 1993, 1995, 1996 and 2008.

Each cluster included similar years regarding precipitation distribution over the year. It is important to note that the median year in Table 7.2 does not necessarily have to be included in the years within the cluster, since it is just an indicator for comparison between clusters representative of past and recent periods.

As the interannual variability of rainfall is often high, there were past and recent years within the three clusters. However, cluster 1 presented a greater concentration of earlier years (median year = 1971), while cluster 2 (median year = 1985) and cluster 3 (median year = 1993) gradually corresponded to more recent years.

The separation into three clusters highlighted the temporal increase in annual rainfall, with cluster 1 adding up to 611 mm/year; cluster 2 to 656 mm/year and cluster 3, representing the most recent years, adding up to 701 mm/year. These results of increased rainfall over time had been indicated by the methods applied previously, however, through cluster analysis, changes in the seasonal distribution of rainfall could also be observed, with little interference from the interannual variability of the data. Thereby, the most interesting result from the cluster analysis was the observation of changes in the distribution of rainfall throughout the year. Increases were observed in most months, but at different scales (Table 7.2). Decreases were observed during the months of June, July and September despite the annual increases. Sincc thcsc months belong to the rainy period of the site (May–October), the shift may have consequences on local water and agricultural management.

Observing these changes through the seasons, it is easier to verify the gradual influence of climate changes on the annual variability of rainfall (Fig. 7.3). Changes were very important during winter and spring, with increases of 309 % and 65 % respectively, when comparing clusters 1 and 3. During the fall, increases were not very important (6 %), while in summer, which is the most important rainy season in Sudbury, the 15 % reduction represented a decrease of 41 mm in rainfall.

For snowfall, cluster analysis separated the 55 years into two clusters. Cluster 1 comprised years 1956, 1957, 1958, 1959, 1960, 1961, 1963, 1964, 1966, 1968, 1969, 1970, 1973, 1977, 1980, 1984, 1986, 1987, 1991, 1993, 1994, 1995, 1998, 2003, 2005, 2007 and 2010, and cluster 2 consisted of years 1962, 1965, 1967, 1971, 1972, 1974, 1975, 1976, 1978, 1979, 1981, 1982, 1983, 1985, 1988, 1989, 1990, 1992, 1996, 1997, 1999, 2000, 2001, 2002, 2004, 2006, 2008 and 2009. The median years of the two clusters were close (1977 and 1987), implying in that the

Table 7.2 Three clusters discriminated by the cluster analysis based on monthly rainfall data in Sudbury site from 1956 to 2010

Cluster	Median Year	January (mm)	February (mm)	March (mm)	April (mm)	May (mm)	June (mm)	July (mm)	August (mm)	September (mm)	October (mm)	November (mm)	December (mm)	Sum (mm)
1	1971	6	0	11	37	67	86	86	95	129	51	35	7	611
2	1985	7	9	30	47	72	86	79	75	94	81	57	19	656
3	1993	36	5	25	69	97	50	70	106	104	82	41	16	701

Median year and annual sum correspondent of each cluster are also provided

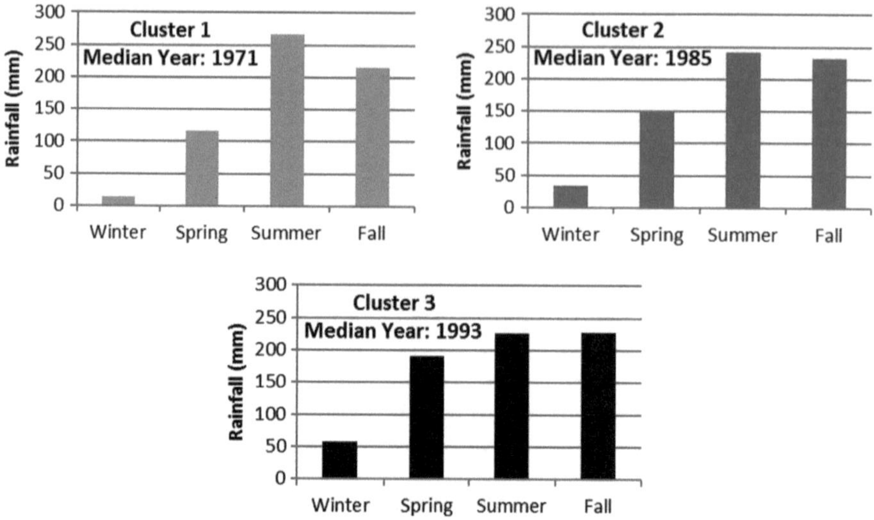

Fig. 7.3 Trends of changes in rainfall distribution throughout the seasons in the Sudbury site from 1956 to 2010, represented by the three clusters discriminated by cluster analysis

trends of change in snowfall distribution over the year were not as clear as for rainfall.

Nevertheless, changes in snowfall distribution could be observed for December, January and February, the 3 months of higher snowfall values in Sudbury. Although the annual sum of snowfall did not change significantly, more recent years presented lower values in December (13 %), and higher values in January (38 %) and especially in February (116 %), when values more than doubled (Fig. 7.4).

The application of cluster and trend analysis evidenced that the increasing annual trends for rainfall volumes in the Sudbury area were not uniform over the year. This annual increase was mainly in winter and spring. Decreases in summer rainfall were detected only through cluster analysis. According to cluster analysis, although years are currently more humid, summers are becoming drier in the location. Previous studies have detected and/or projected changes in rainfall in Sudbury and nearby sites (Magnuson et al. 1997; OCCIAR 2010; IEESC 2012; Charron 2014), however, the changes in seasonality were not considered in depth.

Regarding snow precipitation, while group comparison and trend analysis did not indicate any significant trends, cluster analysis showed clear changes for the months of greatest snowfall (December, January and February). Reductions in December and increases in January and February were observed. This means that the climate of the site is changing towards later winters regarding snowfall, which may also be related to changes in other variables (e.g., temperature).

Although it was not the focus of this study, changes in extreme events should also be investigated. According to Folland et al. (2002), in regions where annual rainfall has increased it is very likely that there have been even more pronounced increases in heavy and extreme rainfall events. And the converse is also true.

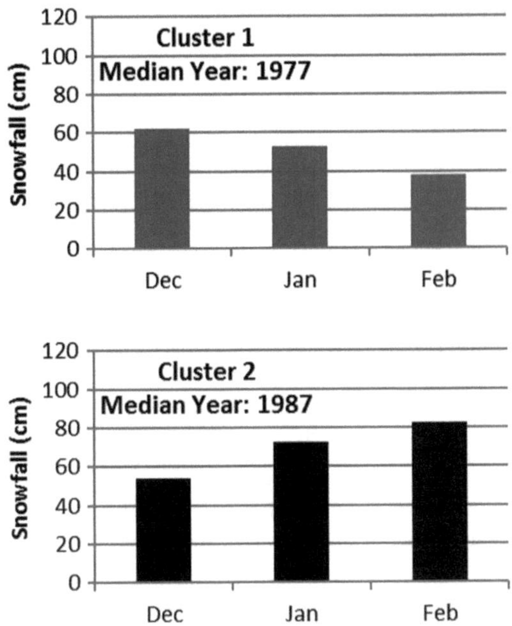

Fig. 7.4 Trends of changes in snowfall distribution throughout December, January and February in the Sudbury site from 1956 to 2010, represented by the two clusters discriminated by cluster analysis

In some regions, heavy and extreme precipitation events have increased, despite the fact that total precipitation has decreased or remained constant.

Conclusions

The methodologies used in this study demonstrated that the detection of annual changes is only the initial step in a more comprehensive understanding of climate change, which also includes complex seasonal and monthly changes. In many areas of the world, the absence of detected changes or significant trends from annual data may give a false idea of the absence of climate change in the location (e.g., snowfall in the site of this study). However, the results presented herein indicate that the inclusion of simple methods, such as cluster analysis, can contribute to a better understanding of seasonal and monthly climate changes.

Acknowledgement This work was supported by the Ontario Centre for Climate Impacts and Adaptation Resources (OCCIAR) and Environment Canada.

References

Alexander, L. V., Zhang, X., Peterson, T. C., Caesar, J., Gleason, B., Klein, A. M. G., et al. (2006). Global observed changes in daily climate extremes of temperature and precipitation. *Journal of Geophysical Research: Atmospheres, 111*, 1–22.

AMS, American Meteorological Society. (2012). *Climate change: An information statement of the American Meteorological Society*. Adopted by AMS Council in 20 August 2012, Boston.

Charron, I. (2014). *A guidebook on climate scenarios: Using climate information to guide adaptation research and decisions*. Montreal: Ouranos.

Davidson, E. A., de Araújo, A. C., Artaxo, P., Balch, J. K., Brown, I. F., Bustamante, M. C., et al. (2012). The Amazon basin in transition. *Nature, 481*(7381), 321–328.

Disch, J., Kay, P., & Mortsch, L. (2012). A resiliency assessment of Ontario's low-water response mechanism: Implications for addressing management of low-water under potential future climate change. *Canadian Water Resources Journal, 37*(2), 105–123.

Durack, P. J., Wijffels, S. E., & Matear, R. J. (2012). Ocean salinities reveal strong global water cycle intensification during 1950 to 2000. *Science, 336*(6080), 455–458.

Folland, C. K., Karl, T. R., & Salinger, M. J. (2002). Observed climate variability and change. *Weather, 57*(8), 269–278.

Grillakis, M. G., Koutroulis, A. G., & Tsanis, I. K. (2011). Climate change impact on the hydrology of Spencer Creek watershed in Southern Ontario, Canada. *Journal of Hydrology, 409*(1–2), 1–19.

Hair, J., Anderson, R., Tatham, R., & Black, W. (1998). *Multivariate data analysis*. Englewood Cliffs, NJ: Prentice-Hall.

IEESC, Institute for Energy, Environment and Sustainable Communities. (2012). *Producing high-resolution (25 km × 25 km) probabilistic climate change projections over Ontario using UK PRECIS*. Regina: University of Regina.

IPCC, Intergovernmental Panel on Climate Change. (2007). *Climate change 2007: The physical science basis. Contribution of working group I to the fourth assessment report of the Intergovernmental Panel on Climate Change*. Cambridge/New York: Cambridge University Press.

IPCC, Intergovernmental Panel on Climate Change. (2014). Summary for policymakers. In *Climate change 2014: Impacts, adaptation, and vulnerability*. Cambridge/New York: Cambridge University Press.

Kendall, M. G. (1975). *Rank correlation methods*. London: Griffin.

Knapp, A. K., Fay, P. A., Blair, J. M., Collins, S. L., Smith, M. D., Carlisle, J. D., et al. (2002). Rainfall variability, carbon cycling, and plant species diversity in a mesic grassland. *Science, 298*(5601), 2202–2205.

Magnuson, J. J., Webster, K. E., Assel, R. A., Bowser, C. J., Dillon, P. J., Eaton, J. G., et al. (1997). Potential effects of climate changes on aquatic systems: Laurentian Great lakes and Precambrian shield region. *Hydrological Processes, 11*(8), 825–871.

Malhi, Y., Aragao, L. E., Galbraith, D., Huntingford, C., Fisher, R., Zelazowski, P., et al. (2009). Exploring the likelihood and mechanism of a climate-change-induced dieback of the Amazon rainforest. *Proceedings of the National Academy of Sciences of the United States of America, 106*(49), 20610–20615.

Mann, H. B. (1945). Nonparametric tests against trend. *Econometrica, 13*, 245–259.

Munoz-Díaz, D., & Rodrigo, F. S. (2004). Spatio-temporal patterns of seasonal rainfall in Spain (1912–2000) using cluster and principal component analysis: Comparison. *Annales Geophysicae, 22*(5), 1435–1448.

OCCIAR, Ontario Centre for Climate Impacts and Adaptation Resources. (2010). *Climate change and conservation authorities in northern Ontario*. Workshop report. OCCIAR, Sudbury.

Ramos, M. C. (2001). Divisive and hierarchical clustering techniques to analyse variability of rainfall distribution patterns in a Mediterranean region. *Atmospheric Research, 57*(2), 123–138.

Sen, P. K. (1968). Estimates of the regression coefficient based on Kendall's tau. *Journal of the American Statistical Association, 63*, 1379–1389.

Vincent, L. A., & Gullett, D. W. (1999). Canadian historical and homogeneous temperature datasets for climate change analyses. *International Journal of Climatology, 19*(1), 1375–1388.

Von Storch, H., Zorita, E., & Cubasch, U. (1993). Downscaling of global climate change estimates to regional scales: An application to Iberian rainfall in wintertime. *Journal of Climate, 6*, 1161–1171.

Wang, B., & Ding, Q. (2006). Changes in global monsoon precipitation over the past 56 years. *Geophysical Research Letters, 33*, 1–4.

Yin, H., Gao, Y., & Fan, C. (2011). Distribution, sources and ecological risk assessment of heavy metals in surface sediments from Lake Taihu, China. *Environmental Research Letters, 6*(4), 1–11.

Zhang, H., Huang, G. H., Wang, D., & Zhang, X. (2011). Uncertainty assessment of climate change impacts on the hydrology of small prairie wetlands. *Journal of Hydrology, 396*(1–2), 94–103.

Chapter 8
Ground Response to Global Warming

Mohamad Kharseh and Mohammed Al-Khawaja

Introduction

Over the period 1880–1985 the mean global temperature increased by 0.5–0.7 °C (Hansen and Lebedeff 1987), bearing in mind that the warming rate at high latitudes of the northern hemisphere during recent past decades was greater (Lachenbruch and Marshall 1986; Peterson et al. 2009). Certainly, the change in surface air temperature (SAT) is more than the change of mean global temperature. A comparison of the averages anomalies of SAT between 1880–1890 and 2000–2010 shows that the SAT has increased 1.2 °C above the preindustrial level (NCDC and NOAA 2010). Figure 8.1 and the meteorological data presented by Hansen and Lebedeff (1987) clearly show that the recorded meteorological data of air temperature varies greatly over time, which might lead to a miscalculation of the warming level (Lachenbruch and Marshall 1986). In other words, the warming that was calculated above might be different if different periods were compared.

The analysis of borehole temperature depth profile (BTDP) has therefore recently become an accepted method for inferring recent climatic changes, rather than measuring the SAT (Deming 1995; Ferguson 2006; Goto 2010; Harris and Chapman 1997). Although in some cases, e.g. in North America, it was shown that warming estimated from the analysis of BTDP is consistent with warming estimated from the analysis of meteorological data over the period 1880–1987 (Hansen and Lebedeff 1987).

Measurements of BTDP evidently show that there are temperature deviations from the linear steady-state ground temperature in the upper sections of boreholes

M. Kharseh (✉) • M. Al-Khawaja
Mechanical & Industrial Engineering Department, Qatar University,
Al Dafna, Jamaa Street, Doha, Qatar
e-mail: kharseh@qu.edu.qa; khawaja@qu.edu.qa

P. Grammelis (ed.), *Energy, Transportation and Global Warming*, Green Energy and Technology, DOI 10.1007/978-3-319-30127-3_8

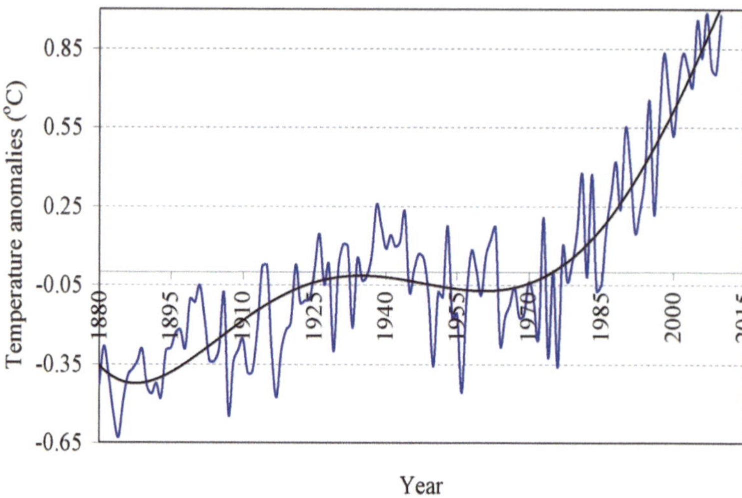

Fig. 8.1 Global mean surface air temperature anomalies (NCDC and NOAA 2010)

(Lachenbruch and Marshall 1986; Goto 2010; Harris and Chapman 1997; Guillou-Frottier et al. 1998).

This deviation in temperature might be theoretically attributed to many different sources (Harris and Chapman 1997):

- There is a disturbance in the energy balance at the Earth's surface on a multi-decade scale. This disturbance might be interpreted by a continental-scale change in the SAT, i.e. recent warming of the mean annual SAT (Deming 1995; Harris and Chapman 1997).
- Change in land use such as more buildings (Lachenbruch and Marshall 1986; Ferguson 2006). Yoshikawa et al. attributed this temperature deviation to increase in albedo and reduction of the leaf area involved in transpiration following deforestation or wildfire (Yoshikawa et al. 2002). It was shown that a recent change in vegetation cover and increased surface temperature over the past century result in a similar shape of BTDP (Nitoiu and Beltrami 2005).
- Changes in thermal conductivity might lead to this temperature departure, that is, "a systematic upward increase in conductivity might cause a systematic upward decrease in gradient and simulate the transient warming" (Lachenbruch and Marshall 1986; Deming 1995).

Thus, using the borehole temperature to analyze the warming is not always easy (Deming 1995). However, in general, the departure of BTDP from the steady state line is considered a direct consequence of a change in SAT (Lachenbruch and Marshall 1986; Deming 1995). Moreover, the effects of other non-climatic sources (e.g. change in land use, variability of thermal conductivity, groundwater movement and so forth) have been shown to be negligible at the noise level of data (Chisholm and Chapman 1992).

It is important to mention that there are temperature measurements of boreholes that show recent cooling trends at the surface which started in the 1970s (Lachenbruch and Marshall 1986; Deming 1995).

However, measurements of borehole temperature generally show smooth enough curves to permit us to approximate it to a mathematical formula (Lachenbruch and Marshall 1986). The current study mainly aims to derive an equation that expresses ground temperature deviation from the steady-state line as a result of warming at ground surface. The derived equation was used to determine the average change of ground temperature over a certain depth and therefore the heat retained by a column of earth during the warming period. This average change of ground temperature is of great importance in the borehole system.

Methodology

Dominating Equation

In order to obtain an equation that expresses the ground temperature change from its linear steady-state profile, the heat conduction equation needs to be solved. For this purpose the transient ground temperature $\theta(z,t)$ at depth z and time t is given as follows:

$$\theta(z,t) = T_s + \gamma \cdot z + T(z,t) \tag{8.1}$$

where T_s is the air temperature at the surface level; γ is the temperature gradient due to geothermal heat flux; z is the depth below the ground surface; t is the time elapsed since the warming started; and $T(z,t)$ is ground temperature deviation from the steady-state profile, i.e. the response of the ground to recent warming of the mean annual surface temperature.

It is worth mentioning that in current study and for the sake of simplicity some assumptions were made as follows. It is well established that urban areas can be 3–6 °C warmer than their surrounding rural areas (Kolokotroni 2009; EPA 2013-01-15) . Certainly, such local temperature changes affect the ground temperature. However, due to the insulating effect of the ground itself (i.e., low thermal conductivity and big heat capacity), it was shown that the effect of change in land use will only extend 50 m beyond the affected area at the surface (Ferguson 2006). In the shallow subsurface, the transfer of heat can have an advection component owing to the transfer of water between the different compartments, but this component might be compensated by phase changes (evaporation). In addition the thermal diffusivity of the ground is assumed to be constant and evaluated at the average over value (Guillou-Frottier et al. 1998; Chisholm and Chapman 1992).

In nutshell, the effect of lateral heat flow and the advection component of heat transfer are neglected and one dimensional heat flow is considered. This simplification of the analysis provides insight into time scales of ground temperature deviation. Thus, in the case with absence of any internal heat generation, the governing equation is:

$$\frac{\partial\theta(z,t)}{\partial t} = \alpha\frac{\partial^2\theta(z,t)}{\partial z^2} \tag{8.2}$$

Substitution Eq. (8.1) in Eq. (8.2) yields

$$\frac{\partial T(z,t)}{\partial t} = \alpha\frac{\partial^2 T(z,t)}{\partial z^2} \tag{8.3}$$

This way, any factor that might affect thermal exchanges at the air-ground interface was removed. Thus the subsequent analysis will lead to the best estimate of ground temperature deviation (Harris and Chapman 1997).

Another important assumption is that the ground surface temperature has linearly changed over time τ (i.e. it was assumed that the warming began τ years ago), see Eq. (8.5) (Lachenbruch and Marshall 1986; Winterberg 1991). Indeed, assuming nonlinear form of surface warming leads to a more complex solution.

Reference Depth

Because the ground has a relatively low thermal diffusivity, changes in temperature at the Earth's surface propagate slowly into the ground and diminish with depth. The reference depth (hereafter referred to as H) is defined as the depth beyond which, for practical purposes, the effect of warming at surface becomes difficult to observe. This depth depends on both the time elapsed since the start of warming at the surface and thermal diffusivity of the ground. In the current study, Laplace transformation technique was used to give an acceptable approximation to the reference depth, see Eq. (8.4):

$$H = \pi \cdot \sqrt{\alpha \cdot t} \tag{8.4}$$

Note that the temperature change at the depth given by above equation is less than 2.6 % of the surface temperature change. The measurements made by Lachenbruch and Marshall (1986), Goto (2010), Harris and Chapman (1997), Williams and Gold (1976) show no change in ground temperature beyond the depth given by Eq. (8.4). It is good to mention that, Lachenbruch and Marshall defined this depth as the depth beneath where the changes in the ground temperature becomes <5 % of the surface temperature change. Consequently, the reference depth in their study is $H = 2.07\sqrt{\alpha.t}$. However, in this study Eq. (8.4) will be used, and it can be shown that it is more accurate. For alternative approximations it is recommended to consult Chap. 9 given by Özişik (1980).

This way, the boundary conditions for Eq. (8.3):

$$T(0,t) = \frac{\delta}{\tau} \cdot t = \varepsilon \cdot t \tag{8.5}$$

$$T(H,t) = 0 \tag{8.6}$$

While the initial condition is

$$T(z,0) = 0 \tag{8.7}$$

where τ is the time during which the surface temperature increased by δ °C; $\varepsilon = \delta/\tau$ is the warming rate; and H is the reference depth (Eq. 8.4).

Results

To obtain the equation that gives the change in ground temperature due to linear increase of surface air temperature, Eq. (8.3) that is subject to the boundary conditions and initial condition are given by Eqs. (8.5), (8.6) and (8.7), respectively. Using Fourier's theory technique for solving partial deferential equation, the temperature change in ground temperature is:

$$\begin{aligned}
&\textit{if } z \leq \pi\sqrt{\alpha \cdot t} \textit{ then}\\
&T(z,t) = \frac{\varepsilon}{\alpha}\left(\frac{z^2}{2} - \frac{z^3}{6\pi\sqrt{\alpha \cdot t}} - \frac{\pi\sqrt{\alpha \cdot t} \cdot z}{3}\right) + \varepsilon \cdot t \cdot \left(1 - \frac{z}{\pi\sqrt{\alpha \cdot t}} + \frac{2}{\pi}\sum_{n=1}^{\infty} \frac{e^{-n^2} \cdot \sin\left(\frac{n \cdot z}{\sqrt{\alpha \cdot t}}\right)}{n^3}\right)\\
&\textit{if } z > \pi\sqrt{\alpha \cdot t} \textit{ then}\\
&T(z,t) = 0
\end{aligned} \tag{8.8}$$

This equation represents the ground temperature deviation at depth z and time t elapsed since the start of linear warming at the surface.

Figure 8.2 illustrates some of the partial sums (containing only a finite number of terms) for $T(z,t = \tau)$, Eq. (8.8), where terms up to $n = 4$ were considered. The same ground thermal properties assumed by (Lachenbruch and Marshall 1986) were used in the current example, i.e. thermal conductivity λ and thermal diffusivity α are 2.1 W/m K, and 1.10–6 m^2/s, respectively. The ground surface temperature was linearly increased to $\delta = 2.5$ °C over time $\tau = 130$ years. For other assumptions, the curve can be easily adjusted using Eq. (8.8).

To examine the accuracy of the derived equation, Eq. (21) given by (Winterberg 1991) and Eq. (10) given by (Lachenbruch and Marshall 1986) (for $n = 2$) were also illustrated in Fig. 8.2 using the same assumptions made above. As it is shown, the

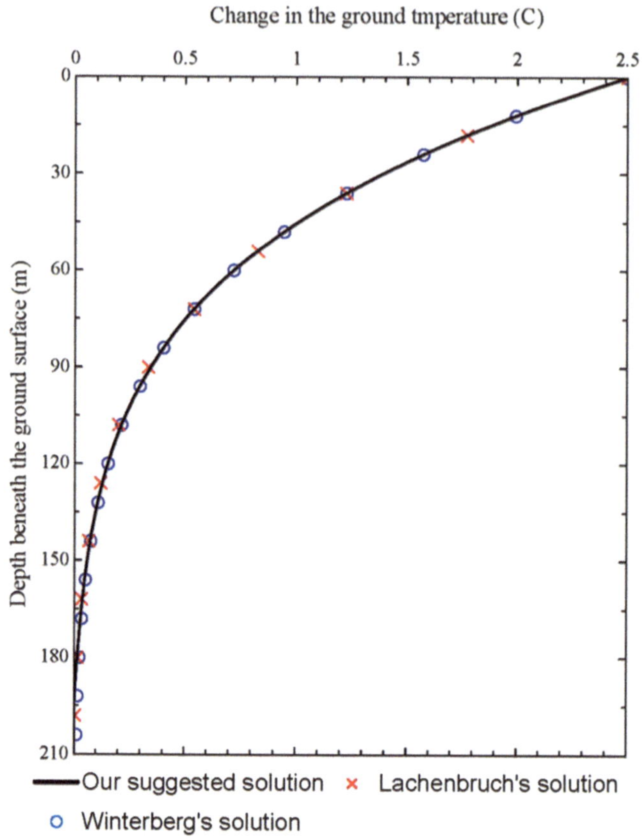

Fig. 8.2 Comparison between our suggested equation and equations derived by Lachenbruch and Marshall (1986) and Winterberg (1991) of the response of the ground to recent warming of the mean annual surface temperature of 2.5 °C started in 1880. The following assumptions were made: Ground thermal conductivity $\lambda = 2.1$ W/m K, thermal diffusivity $\alpha = 1.10\text{–}6$ m^2/s and warming rate of ambient air $\varepsilon = 0.0192$ °C/year

suggested equation (i.e., Eq. 8.8 in the current study) is consistent with the solutions proposed in other studies using other techniques.

Application of Derived Equation

The total heat absorption Q as a result of increased air temperature is about the same for all forms of warming curve at the surface (linear, step, accelerating as the square of time and so forth) (Lachenbruch and Marshall 1986). The knowledge of this heat is of significant importance in analyzing global warming's sources (Nordell 2003). The heat retained by a column of earth surface due to increasing the air can now be calculated by integrating the result of Eq. (8.8) and volumetric heat capacity of the ground during the warming event (i.e., t)

$$Q(t) = \int_{0}^{H} C \cdot T(z,t)dz = K \cdot C \cdot \varepsilon \cdot t \cdot \sqrt{\alpha \cdot t} \tag{8.9}$$

where K is a constant (K = 0.74727) given by Eq. (8.10); C is volumetric heat capacity of the ground that is in the range of 2 MJ/m^3 K.

$$K = \frac{\pi}{2} - \frac{\pi^3}{24} + \frac{4}{\pi}\sum_{n=0}^{\infty}\frac{e^{-(2n+1)^2}}{(2n+1)^4} \approx \frac{\pi}{2} - \frac{\pi^3}{24} + \frac{4}{\pi \cdot e} \tag{8.10}$$

Thus, for the same warming rate and thermal properties assumed above, the total heat absorption by 1 m^2 of ground is 251 MJ/m^2 (about 0.06 W/m^2).

Calculating the mean change in ground temperature over a particular depth, say L, (ΔTm(L,t)) is of great importance for ground source heat pump system design (Kharseh et al. 2011; Kharseh and Altorkmany 2012; Fontaine et al. 2011) . This mean change ΔTm(L,t) is given by integrating Eq. (8.8) from zero to the concerned depth (L) and dividing the result by L as follows:

$$\begin{aligned}
&if\, L \le \pi\sqrt{\alpha \cdot t} \\
&\Delta T_m(L,t) = \frac{\varepsilon}{6\alpha}\left(L^2 - \frac{L^3}{4\pi \cdot \sqrt{\alpha \cdot t}} - \pi\sqrt{\alpha \cdot t} \cdot L\right) + \\
&\quad + \varepsilon \cdot t \cdot \left(1 - \frac{L}{2\pi \cdot \sqrt{\alpha \cdot t}} - \frac{4 \cdot \sqrt{\alpha \cdot t}}{\pi \cdot L}\sum_{n=1}^{\infty}\frac{e^{-n^2} \cdot \sin^2\left(\frac{n \cdot L}{2 \cdot \sqrt{\alpha \cdot t}}\right)}{n^4}\right) \\
&if\, L > \pi\sqrt{\alpha \cdot t} \\
&\Delta T_m(L,t) = \frac{\varepsilon \cdot \pi \cdot t \cdot \sqrt{\alpha \cdot t}}{2L}\left(1 - \frac{\pi^2}{12} + \frac{8}{\pi^2}\sum_{n=0}^{\infty}\frac{e^{-(2n+1)^2}}{(2n+1)^4}\right)
\end{aligned} \tag{8.11}$$

Conclusions

The aim of this study was to find an equation relating ground temperature change with global warming. To achieve this goal, a heat conduction equation was solved using Fourier's theory. The derived equation in the current study and the equations established in other studies are in excellent agreement. The suggested equation is more user-friendly than other equations that are published in previous publications. Because of its simplicity, the suggested equation was used to state the mean change in ground temperature and the heat retained by a column of earth surface due to the increasing surface air temperature. These equations are expected to be useful in situations where the ground temperature is of importance; e.g. design of underground thermal energy storage systems, ground source heat pump, studies of microbiological systems in the ground, plant growth, frost heave, in addition to the understanding of the effect of global warming. Although the study is based on large uncertainties (a fact typical of this subject), the derived equation provides insight into time scales of ground temperature anomalies.

Acknowledgements This work was made possible by an NPRP 7-725-2-270 a grant from the Qatar National Research Fund (a member of The Qatar Foundation). The statements made herein are solely the responsibility of the authors.

Nomenclature

C	Volumetric heat capacity (J/m^3 k)
H	Reference depth (m)
L	A certain depth that is of interest to calculate the ΔT_{ave} (m)
Q(t)	Heat retained by a column of earth surface due to the climatic changes (J/m^2)
t	Time elapsed since start-up warming (s)
T(z,t)	Ground temperature anomaly (°C)
T_s	Air temperature at the surface level (°C)
z	Depth below ground surface (m)
$\alpha = \lambda/C$	Thermal diffusivity (m^2/s)
γ	Temperature gradient due to geothermal heat flux (°C/m)
δ	Temperature change at the ground surface (°C)
$\Delta T_m(L,t)$	Average change of ground temperature over a certain depth L (°C)
ε	Warming rate of air (assumed to be constant over time) (°C/s)
$\theta(z,t)$	The transient ground temperature at depth z and time t
λ	Thermal conductivity (W/m K)

References

Chisholm, T. J., & Chapman, D. S. (1992). Climate change inferred from analysis of borehole temperatures—An example from western Utah. *Journal of Geophysical Research-Solid Earth, 97*, 14155–14175.

Deming, D. (1995). Climatic warming in North America: Analysis of borehole temperatures. *Science, 268*, 1576–1577.

EPA. (2013). Retrieved January 15, 2013, from www.epa.gov/hiri/index.htm.

Ferguson, G. (2006). Transient lateral heat flow due to land-use changes. *Earth and Planetary Science Letters, 242*, 217–222.

Fontaine, P., Marcotte, D., Pasquier, P., & Thibodeau, D. (2011). Modeling of horizontal geoexchange systems for building heating and permafrost stabilization. *Geothermics, 40*, 211–220.

Goto, S. (2010). Reconstruction of the 500-year ground surface temperature history of northern Awaji Island, southwest Japan, using a layered thermal property model. *Physics of the Earth and Planetary Interiors, 183*, 435–446.
Guillou-Frottier, L., Mareschal, J., & Musset, J. (1998). Ground surface temperature history in central Canada inferred from 10 selected borehole temperature profiles. *Journal of Geophysical Research, 103*, 7385–7397.
Hansen, J., & Lebedeff, S. (1987). Global trends of measured surface air-temperature. *Journal of Geophysical Research: Atmospheres, 92*, 13345–13372.
Harris, R. N., & Chapman, D. S. (1997). Borehole temperatures and a baseline for 20th-century global warming estimates. *Science, 275*, 1618–1621.
Kharseh, M., & Altorkmany, L. (2012). How global warming and building envelope will change buildings energy use in central Europe. *Applied Energy, 97*, 999–1004.
Kharseh, M., Altorkmany, L., & Nordell, B. (2011). Global warming's impact on the performance of GSHP. *Renewable Energy, 36*, 1485–1491.
Kolokotroni, M. (2009). Heating and cooling degree day prediction within the London urban heat island area. *Building Services Engineering Research & Technology, 30*, 183–202.
Lachenbruch, A. H., & Marshall, B. V. (1986). Changing climate: Geothermal evidence from permafrost in the Alaskan arctic. *Science, 234*, 689–696.
NCDC and NOAA. (2010). Retrieved February 2013 from www.ncdc.noaa.gov/oa/climate/research/anomalies/index.html.
Nitoiu, D., & Beltrami, H. (2005). Subsurface thermal effects of land use changes. *Journal of Geophysical Research, 110*, F01005.
Nordell, B. (2003). Thermal pollution causes global warming. *Global and Planetary Change, 38*, 305–312.
Özişik, M. N. (1980). Heat conduction. Wiley, New York, 0-471-05481-X.
Peterson, T. C., Baringer, M. O., Thorne, P. W., Menne, M. J., Kennedy, J. J., Christy, J., et al. (2009). *Bulletin of the American Meteorological Society, 90*, S13.
Williams, G. P., & Gold, L. W. (1976). Canadian Building Digest.
Winterberg, F. (1991). Detection of global warming through spatial temperature-variations below the earth surface. *Kerntechnik, 56*, 143–146.
Yoshikawa, K., Bolton, W. R., Romanovsky, V. E., Fukuda, M., & Hinzman, L. D. (2002). Impacts of wildfire on the permafrost in the boreal forests of Interior Alaska. *Journal of Geophysical Research-Atmospheres, 108*, FFR 4-1–FFR 4-14.

Chapter 9
Specific Case: Regional Estimates of Global Climate Change: A Dynamical Downscaling Approach to Southeast Europe

Rafaella-Eleni P. Sotiropoulou, Efthimios Tagaris, Andreas Sotiropoulos, Ioannis Spanos, Panagiotis Milonas, and Antonios Michaelakis

Introduction

Climate change is a major environmental problem that will affect future weather and will modify the meteorological data, locally. General Circulation Models (GCMs) estimate that the global mean surface temperature change for the near future (i.e., for the period 2046–2065 relative to 1986–2005) will likely be in the range of 0.4–2.6 °C while globally-averaged precipitation increases with global mean surface temperature at about 1–3 % °C^{-1} are projected. However, the minimum and the maximum estimated changes cover a wide range of values for all geographical regions (IPCC 2013). For example, under the RCP4.5 experiments, annual temperatures for northern, central, and southern Europe are estimated to be modified in the range of −0.5 °C to 3.8 °C, 0.4 °C to 3.2 °C and 0.7 °C to 3.1 °C, respectively, while annual precipitation rates are estimated to be modified in the

R.-E.P. Sotiropoulou (✉)
Department of Mechanical Engineering, University of Western Macedonia, 50100 Kozani, Greece

Environmental Research Laboratory, NCSR Demokritos, 15310 Athens, Greece
e-mail: rsotiropoulou@uowm.gr

E. Tagaris
Environmental Research Laboratory, NCSR Demokritos, 15310 Athens, Greece
e-mail: tagaris@ipta.demokritos.gr

A. Sotiropoulos • I. Spanos
Environmental Engineering Consultancy, Terra Nova Ltd., 11527 Athens, Greece
e-mail: sotiropoulos@terranova.gr; spanos@terranova.gr

P. Milonas • A. Michaelakis
Department of Entomology and Agricultural Zoology, Benaki Phytopathological Institute, 14561 Kifisia, Greece
e-mail: p.milonas@bpi.gr; a.michaelakis@bpi.gr

P. Grammelis (ed.), *Energy, Transportation and Global Warming*, Green Energy and Technology, DOI 10.1007/978-3-319-30127-3_9

range of −5 % to 17 %, −6 % to 9 %, and −14 % to 3 %, respectively for the period 2046–2065 relative to 1986–2005. However, the outputs from the GCM are relatively coarse for applications to regional scales. To increase their spatial resolution, the coarse resolution data of the GCM are used as initial and boundary conditions by a regional climate model (RCM) (dynamical downscaling). The purpose of doing this is to add-up more detail from local topography, coastline and land use/land cover to obtain high resolution data, with the greatest possible accuracy for the area of interest.

During the last years there is an increasing interest related to the climate change over Mediterranean as it is one of the most vulnerable regions of the globe given that large climate shifts have been observed in the past (Luterbacher et al. 2006). As a result, Mediterranean has been identified as one of the most prominent "Hot-Spots" in future climate change projections (Giorgi 2006). It is located in a transition zone between the arid climate of northern Africa and the wet climate of central Europe and even a minor change in large scale climatic factors might impose large impacts on the climatic conditions of different Mediterranean areas.

The objective of this study is to estimate the changes in temperature and precipitation over southeast Europe in the near future at a very fine resolution since the topography and the coastlines of the region suggest a fine scale spatial variability of the climatic conditions (Gao and Giorgi 2008).

Methods

General Circulation Model

The NASA GISS GCM ModelE (Goddard Institute for Space Studies General Circulation Model) (Schmidt et al. 2006) is used to simulate current and future climate. The GISS model has been chosen for simulations since it is a state-of-the-art and widely used global climate model. The version of the NASA GISS GCM ModelE used here has horizontal resolution $2° \times 2.5°$ latitude by longitude and 20 vertical layers (from surface to 0.1 hPa). The model accounts for both the seasonal and diurnal solar cycles in its temperature calculations. It simulates the emissions, transport, chemical transformation and deposition of several chemical tracers. Sea surface temperatures (SST) are calculated using model-derived surface energy fluxes and specified ocean heat transports. In the present study the IPCC-A1B emissions scenario (IPCC 2000) is used. This scenario describes alternative directions of technological change in the energy system suggesting balance across all sources, where balanced is defined as not relying too heavily on one particular energy source, on the assumption that similar improvement rates apply to all energy supply and end-use technologies. The simulations cover the period from 1880 to 2061. The beginning of the simulations is chosen prior to significant anthropogenic forcing and climate change, but sufficiently recent to contain observations that allow extensive evaluation of the model. The main perturbation to climate over the historic period (up to 2008) is the change in atmospheric composition: primarily the increasing concentration of greenhouse gases and aerosols. Greenhouse gas concentrations up to 2008 are prescribed using ice-core measurements (Schmidt et al. 2011). For the period 2009–2061 the GHG levels are supplied from the IPCC A1B emissions scenario (IPCC 2000).

Regional Climate Downscaling

The outputs from the GCM are relatively coarse (i.e., $2° \times 2.5°$) for applications to regional and local scales. The need for regional climate projections in a finer grid size is assessed, here, using the Weather Research and Forecasting (WRF version 3.4.1) model to dynamically downscale GCM simulations. WRF is a next generation limited area, non hydrostatic, terrain following model designed to serve both research and operational applications (http://www.wrf-model.org/index.php). The domain covers the south—southeast Europe in 273×161 horizontal grids of $9\ \text{km} \times 9\ \text{km}$, with 28 vertical layers (Fig. 9.1).

For the set of physics options we used the following modules: the WRF Single-Moment 3-class scheme for microphysics, the Rapid radiative transfer model (RRTM) for long-wave radiation, the Dudhia scheme for short-wave radiation, the Monin-Obukhov with Carslon-Boland viscous sub-layer for surface layer options, the five-layer thermal diffusion for land surface options, the Yonsei University scheme for the planetary boundary layer and the Kain-Fritsch scheme for the cumulus cloud option. Due to the time needed for the downscaling procedure meteorological conditions are simulated for five current (i.e., 2008–2012) and five future (i.e., 2058–2062) years.

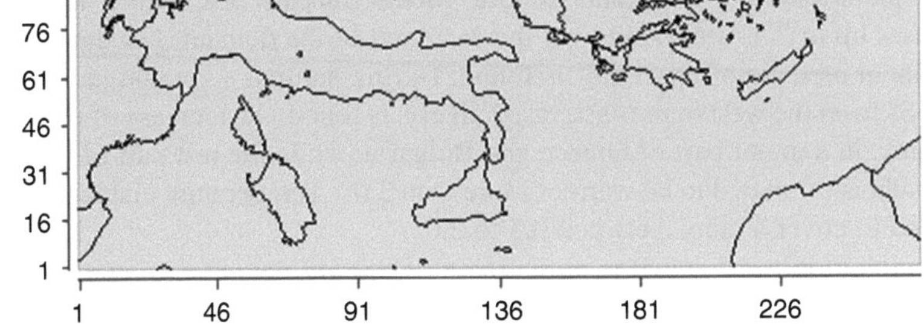

Fig. 9.1 Modeling domain

Results and Discussion

The temperature trend suggests an overall increase in the future, as a result of the reduction of the low temperatures and the increase of the high temperatures for all seasons (Table 9.1). A reduction in the number of days for which the daily average temperature is below 10 °C for both winter and spring is estimated, with a consequent increase in the number of days with daily average temperature above 10 °C. These changes suggest milder **winters** and warmer springs in the future. During summer and autumn, a reduction is estimated in the number of days for which the daily average temperature is below 20 °C, leading to an increase in the number of days above 20 °C. The number of days with daily average temperature above 30 °C is estimated to be ten times higher in the future summers and eight times higher in the future springs. Daily average temperatures above 30 °C start to appear in the future autumns too. What is interesting to point out here is that, the increase in the number of days characterized by high temperatures during future springs and autumns suggests an extension of the future summer period beyond the traditional summer months.

Analysis of the spatial distribution plots suggests that annual average temperature is estimated to be higher in the future all over the domain (Fig. 9.2), while increases in the range of 1.0–1.5° cover the major part of it. A maximum increase of up to 2.0° is estimated in the eastern and southeastern parts of the domain, while a minimum increase in the range of 0.5–1.0° is estimated in the western and southwestern parts of the domain, as well as in the eastern Mediterranean Sea. During winter, an increase up to 1.5° is estimated in the Tyrrhenian Sea, most of Italy, central Mediterranean Sea, Adriatic Sea, Ionian Sea and the western part of the Balkan Peninsula; a smaller increase of up to 1.0° is estimated in north Balkan Peninsula, north Italy and the Ligurian and Balearic Seas. Central Greece, the Aegean Sea and the eastern Mediterranean face temperature increases in the order of 0.5°. The maximum increase is simulated in the greater Venice area (i.e., up to 2.0°). During spring, an increase of up to 1.0° is estimated in the major part of the domain. A higher increase (i.e., up to 1.5°) is found over the east land of the domain, central Mediterranean, north Italy and at the central and north Greece while a milder increase (i.e., up to 0.5°) is found locally over Mediterranean Sea. During summer, an increase of up to 2.0° is estimated in Balkan Peninsula, most of Italy, southeastern France and central Mediterranean Sea, while temperature increases up to 2.5° are found over the east land of the domain. For the rest of the domain an increase of up to 1.5° is found. During autumn a gradual increase from 0.5 to 3° from the west to the east, respectively, is found. An increase of up to 2.5° is estimated in a major part of Greece and Bulgaria, while the rest part of the Balkan Peninsula is estimated to be warmer more than 2.0°. Temperature change over Italy is found to cover a range between 0.5 to 2.0°.

Precipitation trend suggests an increase of the dry days (i.e., precipitation rate less than 1 mm/day) and a decrease of the days with precipitation rate in the range of 1–10 mm/day for future winter, spring and autumn (Table 9.2) over the domain.

Table 9.1 Number of days per temperature range of the daily average temperature for current and future seasons (%)

	Winter		Spring		Summer		Autumn	
	Current	Future	Current	Future	Current	Future	Current	Future
Temperature ranges	Number of days (%)							
T < 0 °C	18.427	15.67	4.169	2.909	0.038	0.014	5.721	1.874
0 °C < T < 10 °C	39.019	38.216	27.011	24.512	1.820	1.289	23.847	23.715
10 °C < T < 20 °C	42.547	45.984	66.997	68.567	36.372	23.448	46.590	44.958
20 °C < T < 30 °C	0.007	0.130	1.817	3.966	61.732	74.818	23.842	29.431
T > 30 °C	0.000	0.000	0.006	0.046	0.039	0.431	0.000	0.022

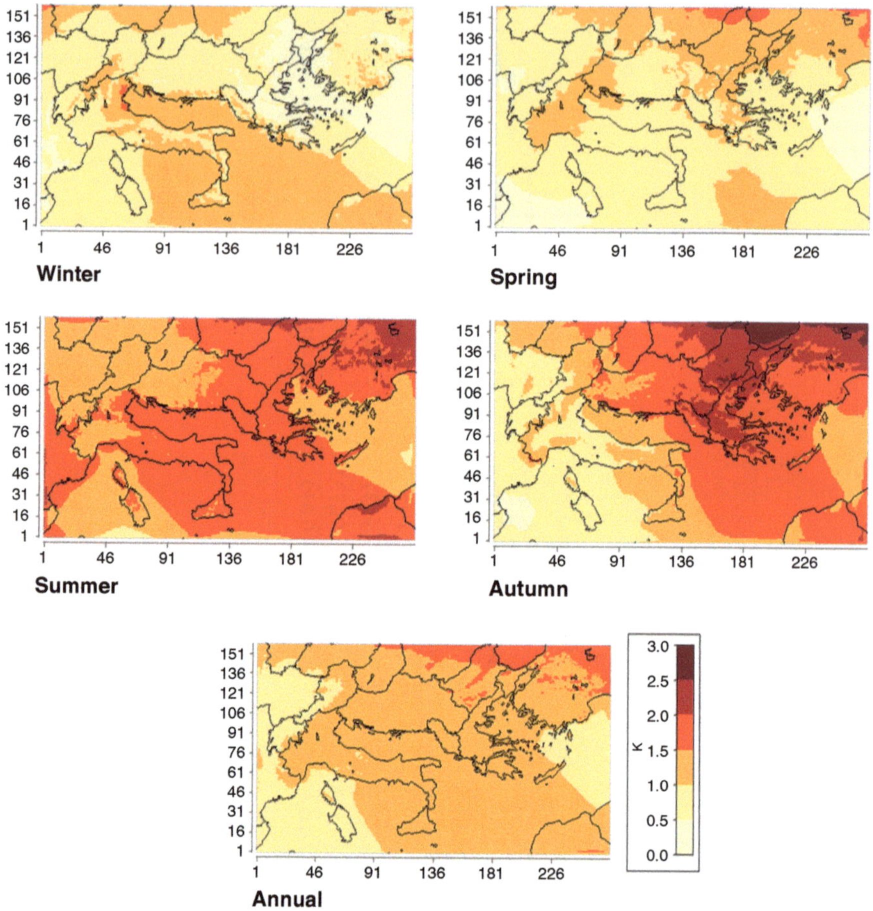

Fig. 9.2 Seasonal and annual temperature change between future and current years

The changes are much higher during winter compared to spring and autumn. More days with precipitation rates in the range of 10–200 mm/day are estimated for future winters and less for future springs. Fewer days with precipitation rates in the range of 10 mm/day up to 100 mm/day are estimated for future autumns, while a small increase is found for precipitation rates in the range of 100–200 mm/day. Future summer is estimated to be slightly wetter than the present; however around 90 % of the days are dry. Extreme precipitation rates (i.e., greater than 200 mm/day) are more frequent during all future seasons; especially for future winter and spring, a doubling in the number of days characterised by extreme rates is fund.

Analysis of the spatial distribution plots, suggests that precipitation change is very location dependent, presenting a mixed trend. Given that, the fine resolution

Table 9.2 Number of days per precipitation range of the daily average precipitation for current and future seasons (%)

	Winter		Spring		Summer		Autumn	
	Current	Future	Current	Future	Current	Future	Current	Future
Precipitation ranges	Number of days (%)							
R < 1 mm/day	72.239	74.009	83.633	84.233	89.643	89.061	77.587	78.478
1 mm/day < R < 10 mm/day	21.924	20.040	12.757	12.267	6.904	7.200	16.060	15.399
10 mm/day < R < 100 mm/day	5.771	5.858	3.521	3.406	3.155	3.361	6.029	5.788
100 mm/day < R < 200 mm/day	0.061	0.083	0.075	0.069	0.229	0.290	0.264	0.266
R > 200 mm/day	0.005	0.011	0.014	0.025	0.069	0.088	0.060	0.070

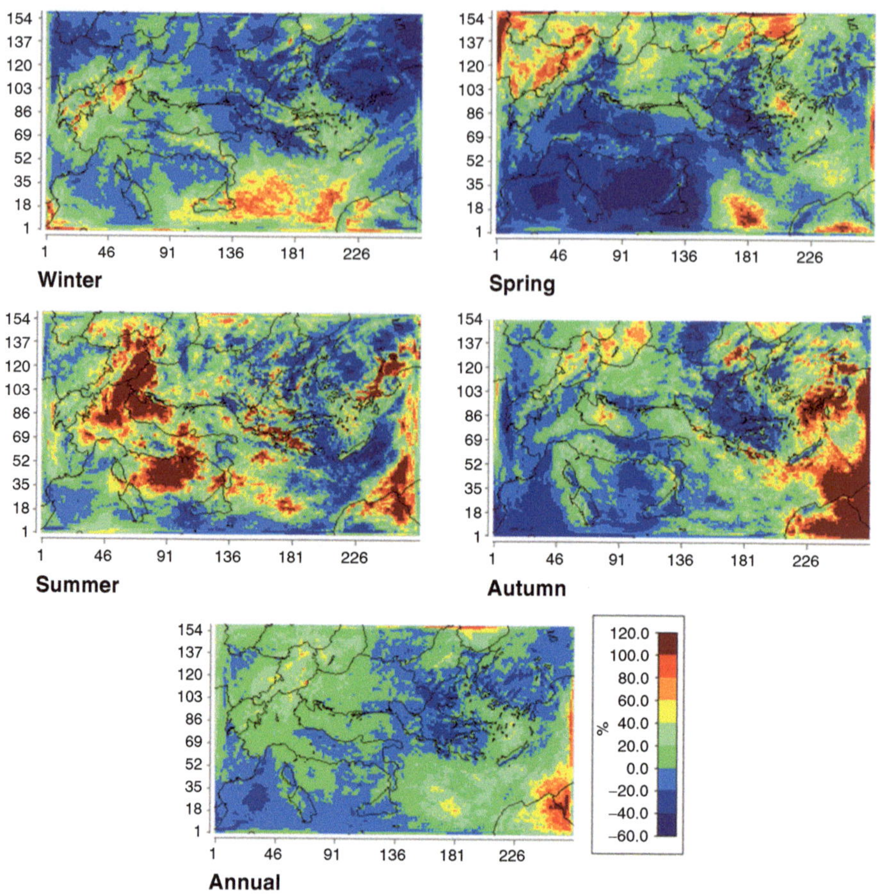

Fig. 9.3 Seasonal and annual precipitation change between future and current years

grid size used in our analysis (i.e., 9 km) provides a detailed insight of the localised precipitation changes (e.g., over islands) happening in the future (Fig. 9.3).

Annual precipitation is estimated to be lower up to 40 % in the major part of the southeast land (i.e., Balkan Peninsula and Turkey) and the west part of the domain, while an increase of up to 60 % is found over central Mediterranean, northern Italy and the central European countries belonging to the domain. A higher increase is simulated at the south of the domain. Seasonally, various changes take place. During winter a decrease of more than 20 % is estimated over the land at the southeast and the north parts of the domain. A decrease is also found over the Balkan Peninsula, except for central Greece. Precipitation is estimated to increase over most of Italy. High precipitation increase (i.e., more than 100 %) is found over northern Italy and Sicily. Precipitation increase is dominant over sea also (i.e., central Mediterranean and Aegean Sea), however, a decrease is found at the east and the west Mediterranean regions of the domain. During spring, an increase at the northern and a decrease at the southern parts of the land of the domain dominate. The major parts of Italy and Greece are found to face less precipitation (up to 60 %) in the future, with some local exceptions. Reduction is also found above sea except for parts of the central Mediterranean and the Aegean Sea. During summer, higher precipitation rates (i.e., more than 100 %) are found mainly at northern Italy, Tyrrhenian Sea, Adriatic Sea, western Greece and at the south—southeast borders of the domain, while lower precipitation rates up to 60 % are found over sea, locally. During autumn, a decrease is found for the west part of the Mediterranean Sea and an increase for the southern and southeast regions of the domain. An increase is also dominant for the central European countries of our domain. Precipitation is projected to be lower in the future over south and north Italy in contrast to the central part of the country. A decrease of up to 60 % is found over Greece, except for the southwest part of the country.

A direct comparison of our results with the results of other researchers (e.g., Christensen et al., 2012; Dubrovsky et al., 2014; García-Ruiz et al., 2011; Gao and Giorgi, 2008) focusing in the area of our interest is not possible due to the different spatiotemporal scales and emissions projection used. However, the following conclusions can be extracted: All the models point to a warmer future. Domain's temperature is projected to increase for all seasons. All models project a homogenous increase which depends on the season. There is less agreement among the models about future precipitation change with differences in regions for increasing, decreasing, or similar-to-current precipitation rates. The general trend is a reduction at the south—southeast part of the domain which turns to an increase at the north with a transition zone between them. Due to the largest grid size used in the other studies, precipitation presents a more homogenous change compared to our results, where seasonal trends are available in a very fine scale providing detailed information for the continental regions and the islands.

Conclusions

Future temperature is estimated to be higher over southeast Europe. The increase depends on both the season and the location, with higher increases during spring and autumn at the south and southeast of the domain. More days with high temperatures and fewer days with low temperatures are estimated in the future, suggesting an extended future warmer period. Precipitation change is very location dependent, presenting a mixed trend for all seasons. A large part of the land at the southeast is estimated to have less annual precipitation in the future, however, seasonal analysis suggests a wetter summer and an increase in the extreme precipitation rates during all future seasons.

Acknowledgement This work was supported by the EU co-funded LIFE-CONOPS project through grand agreement LIFE12 ENV/GR/000466.

References

Christensen, O. B., Goodess, C. M., & Ciscar, J.-C. (2012). Methodological framework of the PESETA project on the impacts of climate change in Europe. *Climatic Change, 112*, 7–28.

Dubrovsky, M., Hayes, M., Duce, P., Trnka, M., Svoboda, M., & Zara, P. (2014). Multi-GCM projections of future drought and climate variability indicators for the Mediterranean region. *Regional Environmental Change, 14*, 1907–1919.

Gao, X., & Giorgi, F. (2008). Increased aridity in the Mediterranean region under greenhouse gas forcing estimated from high resolution simulations with a regional climate model. *Global and Planetary Change, 62*, 195–209.

García-Ruiz, J. M., López-Moreno, J. I., Vicente-Serrano, S. M., Lasanta–Martínez, T., & Beguería, S. (2011). Mediterranean water resources in a global change scenario. *Earth-Science Reviews, 105*, 121–139.

Giorgi, F. (2006). Climate change hot-spots. *Geophysical Research Letters, 33*, L08707.

Intergovernmental Panel on Climate Change (IPCC). (2000). In N. Nakicenovic, & R. Swart (Eds.), *Emissions scenarios*. Cambridge: Cambridge University Press.

Intergovernmental Panel on Climate Change (IPCC). 2013. In T. F. Stocker, D. Qin, G.-K. Plattner, M. Tignor, S. K. Allen, J. Boschung, et al. (Eds.), *Climate change 2013: The physical science basis. Contribution of working group I to the fifth assessment report of the intergovernmental panel on climate change*. Cambridge/New York, NY: Cambridge University Press.

Luterbacher, J., Xoplaki, E., Casty, C., Wanner, H., Pauling, A., Küttel, M., and 43 co-authors, 2006. *Mediterranean climate variability over the last centuries: a review*. Edited by Lionello, P., Malanotte-Rizzoli, P., Boscolo, R., The Mediterannean Climate: An overview of the main characteristics and issues, Amsterdam, Elsevier, pp. 27–148

Schmidt, G. A., Jungclaus, J. H., Ammann, C. M., Bard, E., Braconnot, P., Crowley, T. J., et al. (2011). Climate forcing reconstructions for use in PMIP simulations of the last millennium (v1.0). *Geoscientific Model Development, 4*, 33–45.

Schmidt, G. A., Ruedy, R., Hansen, J. E., Aleinov, I., Bell, N., Bauer, M., et al. (2006). Present day atmospheric simulations using GISS ModelE: Comparison to in-situ, satellite and reanalysis data. *Journal of Climate, 19*, 153–192.

Part III
Global Warming and Climate Change: General Issues and Challenges

Climate Change Observatory

Worldwide climate change observations, specific cases

Chapter 10
Drought Conditions in Turkey Between 2004 and 2013 Via Drought Indices Derived from Remotely Sensed Data

Nazila Molavizadeh, Elif Sertel, and Hande Demirel

Introduction

Drought is one of the major hazards that has a significant adverse effect on the socio-economy, agriculture, and ecosystem. According to the United Nations Convention to Combat Desertification, drought is defined as "the naturally occurring phenomenon that exists when precipitation has been significantly below normal recorded levels, causing serious hydrological imbalances that adversely affect land resource production systems" (UN 2014). Evaporation (affected by temperature and wind), soil types and their ability to store water, the depth and presence of ground water supplies and vegetation are among the most important parameters that influences occurrence of drought. Each drought year is unique in its climatic characteristics and impacts, because drought is related to the timing and the effectiveness of the precipitation. Therefore, it is impossible to make a definition of drought that can be universally accepted (Li and Xiao 1992; Wilhite 1993). Drought can be described by three characteristics: spatial coverage, duration and

N. Molavizadeh (✉)
Satellite Communication & Remote Sensing Program, Department of Advanced Technology, Informatics Institute, Istanbul Technical University, Maslak, 34469 Istanbul, Turkey
e-mail: nazila.itu@gmail.com

E. Sertel
Geomatics Engineering Department, Istanbul Technical University, Maslak, 34469 Istanbul, Turkey

Center for Satellite Communications and Remote Sensing, Istanbul Technical University, Ground Receiving Station, Maslak, 34469 Istanbul, Turkey
e-mail: sertele@itu.edu.tr

H. Demirel
Geomatics Engineering Department, Istanbul Technical University, Maslak, 34469 Istanbul, Turkey
e-mail: demirelha@itu.edu.tr

P. Grammelis (ed.), *Energy, Transportation and Global Warming*, Green Energy and Technology, DOI 10.1007/978-3-319-30127-3_10

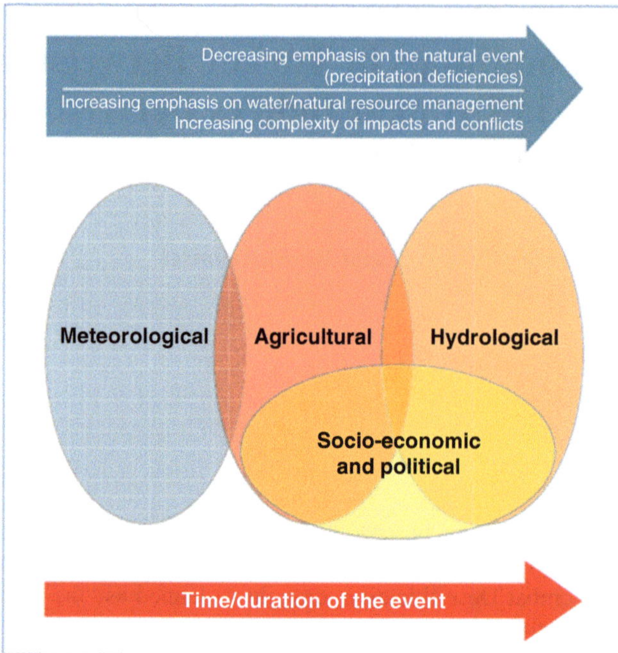

Fig. 10.1 Interrelationships between meteorological, agricultural, hydrological and socioeconomic drought (NDMC 2014)

intensity (Wilhite and Glantz 1985). Briefly, according to (Heim 2002; Keyantash and Dracup 2002), meteorological drought is a precipitation deficit, agricultural drought is a total soil moisture deficit, hydrological drought is a shortage of stream flow, and socioeconomic drought is associated with the shortage of any economic goods affected by the drought process. The relationship between meteorological, agricultural and hydrological drought is illustrated in Fig. 10.1 (NDMC 2014).

According to climate change scenarios such adverse impacts will gain speed in the near future especially for certain locations of the world, including Turkey and its neighborhood. Hence, monitoring its severity has a vital importance. This requires understanding historical droughts in the region as well as impacts of droughts during their occurrences. However, monitoring drought presents various challenges such as inadequate data collection networks with respect to density of stations, high cost of data and insufficient data sharing, difficulties in forecasting, insufficiencies of drought indices for detecting the early beginning and end of the drought, data integration problems such as soil parameters and socioeconomic indicators, mitigation and response programs needs to be regional, therefore regional assessments are required and data sharing and dissemination is limited (WMO 2006). Several of these challenges could be eased by means of spatial information technologies. Remote sensing based drought indices are becoming more popular, since it is possible to determine spatio-temporal distribution of droughts. Remote sensing satellites could monitor large areas rapidly, accurately, periodically, economically and have been widely used to examine drought impacts.

In order to analyze the possibilities of remote sensing based drought indices, three different indices namely Temperature Condition Index (TCI), Vegetation Condition Index (VCI) and Vegetation Health Index (VHI) were generated to determine drought conditions in Turkey for the last decade. MODIS-derived Normalized Difference Vegetation Index (NDVI) and Land Surface Temperature (LST) values were used to derive these indexes. NDVI values could be used to identify healthy and unhealthy vegetation, where LST values derived from thermal bands indicates the variations in temperature. Since, LST and NDVI are negatively correlated; drought areas and periods could be investigated for the research area. The paper is set out as follows: Section "Introduction", introduces the problem and the aim of the study. Section "Data and Methodology" covers case study area, the data and methodology used. In Section "Results and Conclusion", the achieved results, problems encountered, major findings and conclusions from the study is presented.

Data and Methodology

The development of remote sensing, spatial drought monitoring and assessment has become possible in the last decades (Kanellou et al. 2008). Remote Sensing data presents significant advantages and is the integral part of monitoring drought, especially for the spatial and temporal evolution. A number of satellite drought-monitoring indices are developed based on the Advanced Very High Resolution Radiometer (AVHRR) of National Oceanic and Atmospheric Administration (NOAA) polar-orbiting satellite series, MODIS (Moderate Resolution Imaging Spectroradiometer) and others. Historically, the most commonly used remote sensing tool for large-area drought monitoring has been the daily orbiting National Oceanic and Atmospheric Administration (NOAA) Advanced Very High Resolution Radiometer (AVHRR), partially because there is now a sufficiently long time series to let for the identification of anomalies to compare to "normal" conditions. MODIS is superior to AVHRR because it provides higher spatial and spectral resolutions, improved atmospheric corrections and more precise geolocation. When compare the MODIS and AVHRR data sets such as NDVIAVHRR with NDVIMODIS atmospherically corrected MODIS NDVI generally displays a higher dynamic range than atmospherically corrected AVHRR NDVI. This is attributed to the small bandwidth of MODIS (Huete et al. 2002; IWMI 2014). Within this study, The Moderate-resolution Imaging Spectroradiometer (MODIS) data is used. MODIS sensor was launched into Earth orbit by NASA in 1999 on board the Terra (EOS AM) Satellite, and in 2002 on board the Aqua (EOS PM) satellite. The MODIS has 36 spectral bands ranging in wavelength from 0.4 μm to 14.4 μm and at varying spatial resolutions (two bands at 250 m, five bands at 500 m and 29 bands at 1 km). Together the instruments image the entire Earth every 1–2 days.

They are designed to supply measurements in large-scale global dynamics, including changes in Earth's cloud cover, radiation budget and processes occurring in the oceans, on land, and in the lower atmosphere. More information on MODIS data could be obtained from NASA web-page (NASA 2014). Products of one of the well known satellite system, MODIS, have been used for numerous research applications including mapping deforestation, identifying desertification and crop yield estimation, natural and non-natural disasters—such as floods and droughts—in many areas due to the its high temporal resolution and coverage. Terra MODIS monthly composite NDVI images of 1 km resolution and Terra MODIS 8-day composite LST of 1 km resolution were used in this research for the period 2004–2013.

The three remote sensing based drought indices, Vegetation Condition Index (VCI), Temperature Condition Index (TCI) and Vegetation Health Index (VHI) were generated and analyzed to determine drought conditions in Turkey for the last decade.

Vegetation Condition Index (VCI): Kogan (1987) proposed a vegetation condition index based on the relative NDVI change with respect to minimum historical NDVI value. VCI is pixel based normalization of NDVI values. The basic assumption is that drought conditions will weaken the growth of vegetation resulting in lower NDVI values in multiyear NDVI values (Du et al. 2013; Kogan 1995). VCI is a good indicator of drought stress and its values scaled between 0 and 1. In case of very dry month, vegetation conditions will be weakened and the value will close to zero whereas a VCI value of 0.5 reflects fair vegetation conditions (Du et al. 2013). It is defined as following:

$$VCI = \frac{NDVI - NDVI_{min}}{NDVI_{max} - NDVI_{min}} \tag{10.1}$$

where NDVI, $NDVI_{max}$, and $NDVI_{min}$ the smoothed monthly NDVI, multi-year maximum NDVI and multi-year minimum NDVI, respectively, for each grid cell.

Temperature Condition Index (TCI): is a remote sensing based thermal stress indicator that is used to quantify temperature component of droughts. There is a higher LST in the drought year than the same month of normal years and this index assumes that drought event will decrease soil moisture and cause land surface thermal stress (Du et al. 2013; Kogan 1995). TCI is used to find out thermal stressed regions by using the maximum/minimum temperature in a given time series. Since drought is expected to decrease soil moisture and increase LST values, higher LST values are signal for drought conditions and lower LST values indicates favorable conditions, opposite to the NDVI (Du et al. 2013). Thus, the TCI formula was modified as the following expression 10.2.

$$TCI = \frac{LST_{max} - LST}{LST_{max} - LST_{min}} \tag{10.2}$$

where LST, LST_{max} and LST_{min} are the values of LST, maximum LST and minimum LST of each pixel respectively in the same month during the study period of January 2004–December 2013. The is a good indicator to determine drought and to find out its beginning time, intensity, duration and dynamics. The VCI could be successfully used to define both prolonged and short-term droughts either global or localized (Kogan 1995).

Vegetation Health Index (VHI): is calculated from an empirical formula by giving equal weights to Vegetation Condition Index (VCI) and the Temperature Condition Index (TCI) to reflect the impacts of both temperature and vegetation components. The equation of VHI is below:

$$VHI = 0.5\,(VCI + TCI) \tag{10.3}$$

The TCI indicates areas that are hotter than usual and the VCI's normalized NDVI anomalies identify areas where vegetation is more or less dense than usual. VHI reflects both vegetation cover and temperature anomalies and also it has been widely applied for the early drought warning, monitoring of crop yield and production, and assessment of irrigated areas and extreme wetness (Karnieli et al. 2006) The classification of indices for drought conditions are presented in Table 10.1 (Kogan 2001).

Results and Conclusion

Turkey is located in Anatolia and the Balkans, bordering the Black Sea, between Bulgaria and Georgia, and bordering the Aegean Sea and the Mediterranean Sea, between Greece and Syria. The total land area is about 783,562 km^2, of which 756,816 km^2 are in West Asia (Anatolia) and 23,764 km^2 are in Southeastern Europe (Thrace). Climatic conditions of Turkey vary significantly from one region to another, due to topography and the geo-location of the country. While the coastal regions have milder climates, extreme hot summers with limited rain fall and cold winters are generally experienced at the inner Anatolian plateau (Sensoy et al. 2008). The Aegean and Mediterranean coasts have cool, rainy winters and hot, moderately dry summers. Annual precipitation in those areas varies from 580 to 1300 mm, depending on location. The Black Sea coast receives the greatest amount of rainfall. The eastern part of that receives 2200 mm annually and is the only region of Turkey that receives rainfall throughout the year. Climate types of Turkey are presented in Fig. 10.2 (Atalay and Efe 2012).

In order to automate the process of generating indices and data handling, codes were written in Interactive Data Language programming language and all indices

Table 10.1 Classification of VCI, TCI and VHI drought conditions (Kogan 2001)

VCI, TCI, VHI values * 100	<10	10–20	20–30	30–40	40–60	>60
Drought conditions	Extreme	Severe	Moderate	Mild	Normal	Wet

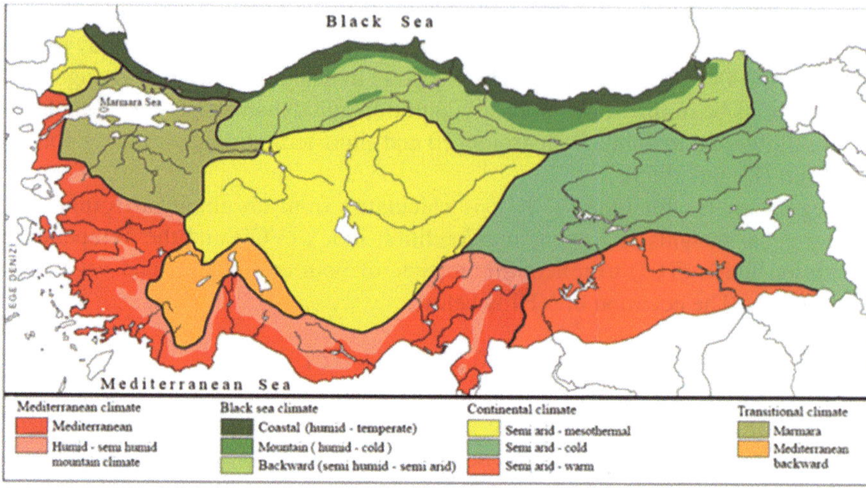

Fig. 10.2 Climate types of Turkey (Atalay and Efe 2012)

were generated automatically. In the last 10 years, Turkey observes two drought periods, the first one start from 2006. This drought period, is observed from November and December of 2006 until December of 2008. This long period between 2007 and 2008 clearly can be observed by analyzing the VCI and VHI maps.

Drought is a slow occurring process starts with precipitation deficits, then causing soil moisture deficits resulted in higher Land Surface Temperature (LST) values. Finally, vegetation growth is affected because of water deficits and temperature increases (Du et al. 2013). In this research LST and NDVI data were used to monitor LST and NDVI anomalies during 10 year period using TCI and VCI indexes. In order to represent both soil moisture deficit and vegetation growth, VHI was calculated using TCI and VCI. VHI values between 40–60 % represent normal conditions whereas VHI values lower than 0.4 represents drought higher than 0.6 represents wet climatic conditions.

TCI is based on LST values and could be used to identify regions that are hotter or colder than usual values. For drought affected regions, higher temperature values resulted in lower TCI values than normal conditions. TCI values lower than 30 % might give the signal of extreme, severe and/or moderate drought conditions. In drought conditions, land surface will have thermal stress and TCI could be used to determine temperature-related drought conditions (Du et al. 2013; Kogan 1995).

VCI is based on NDVI anomaly values and could be used to identify areas where vegetation is more or less than usual values. In normal/usual conditions VCI presents same seasonal pattern and presents higher values in May-September period and lower values in October-April period. Figure 10.3 shows an example of VCI for 2004 a normal year in terms of climatic conditions. As can be seen from the figure VCI values are comparatively higher between May and September. Vegetation response in the May–September period is related to the very wet period from November–April (months where most of the geographic regions receive most of

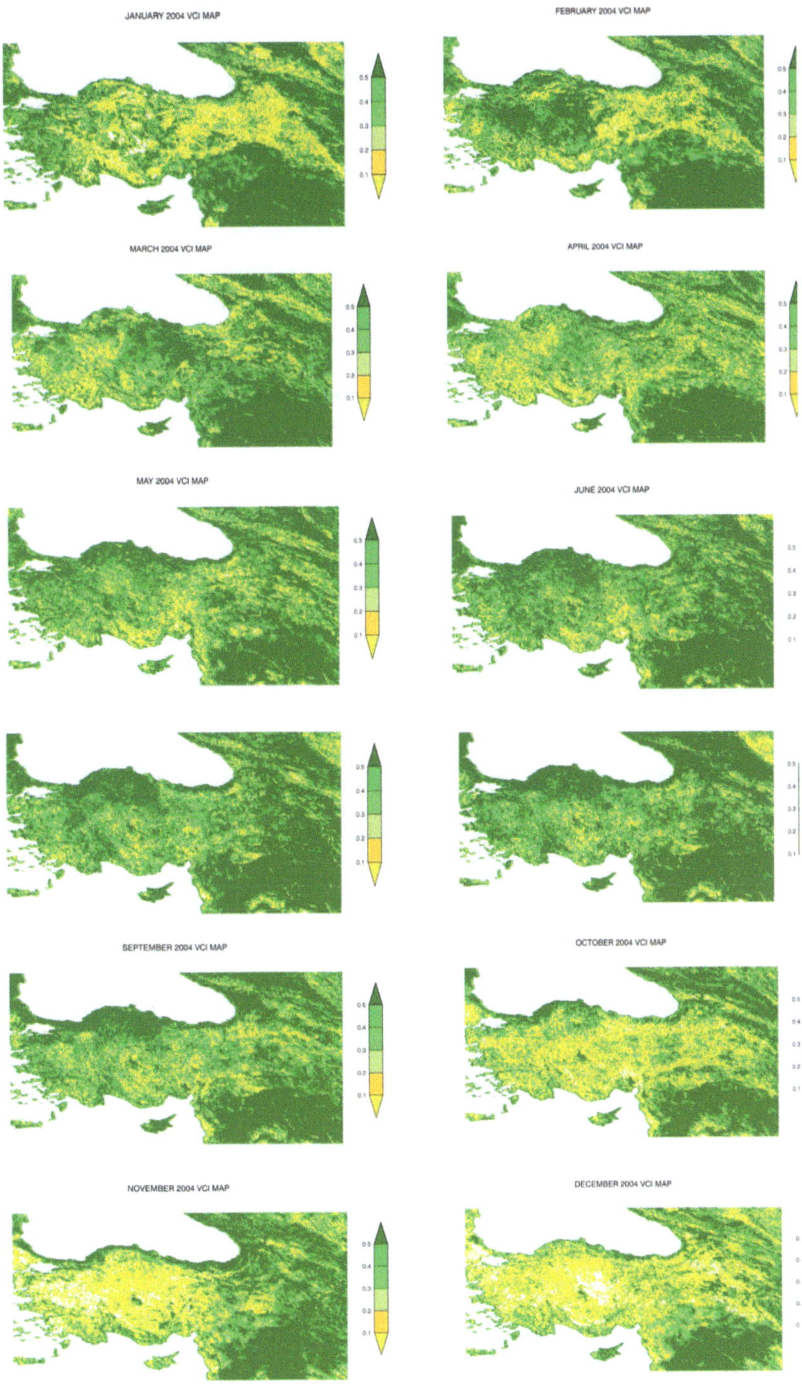

Fig. 10.3 VCI maps for the year 2004

the precipitation). For drought occurring years such as 2007, VCI presents lower values in May–September period as well unlike the normal conditions mentioned.

Turkey has different climatic regions as shown in Fig. 10.2. Black Sea Region having coastal (humid-temperate) climate has vegetation throughout the year resulting in higher VCI values as can be seen in Fig. 10.3. This region is highly vegetated with mostly evergreen forests causing higher NDVI values. Considering Turkey boundaries, during the winter months, most of the country has lower NDVI values due to the fact that there are not planted agricultural areas and deciduous plants.

Analysis of TCI, VCI and VHI maps illustrated that 2007, 2008 and 2013 were drought impacted years in Turkey. In order to consider for temperature and vegetation growth, VHI maps were extensively analyzed. Figure 10.4 illustrates the comparison of a normal (2004) and drought (2007) years. After comparing 12 months of each year, it was found that impact of drought is more detectable between July and September. As can be seen from Fig. 10.4, during the drought year 2007 most of the country has VHI values lower than 0.2 indicating the excess or very excess drought conditions.

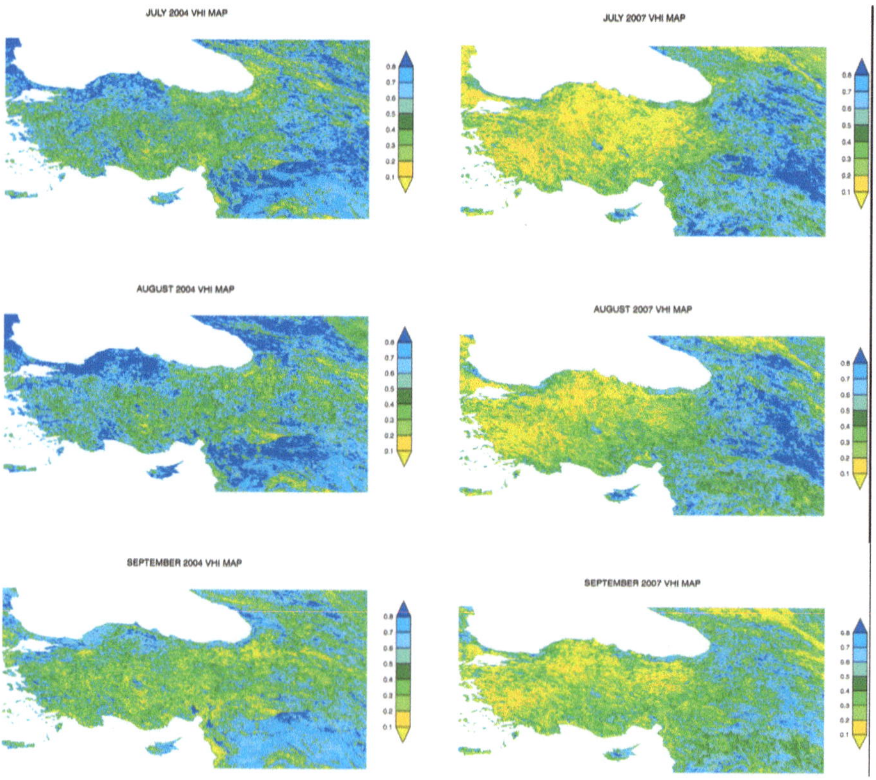

Fig. 10.4 Comparison of VHI maps between 2004–2007 for July, August and September

Droughts can lead significant environmental, social, and economic consequences by impacting agriculture, ecology and socio-economy. Therefore, it is important to monitor droughts and their impacts in local, regional and global scales. Meteorological data and drought indices derived from these data could be used to monitor drought conditions however their spatial and temporal availability are limited. To this end, remotely sensed data derived indices could be an important asset to monitor drought conditions globally and timely.

References

Atalay, I., & Efe, R. (2012). *Ecology of Scots pine (Pinus sylvestris var. sylvestris) forests and their dividing into regions in terms of seed transfer*. Ministry of Environment and Forestry Publ. 45, Ankara. ISBN 975-605-4610-11-2.

Du, H. Y., Shen, Y. Z., & Huang, Z. J. (2013). Function of the wheat TaSIP gene in enhancing drought and salt tolerance in transgenic Arabidopsis and rice. *Plant Molecular Biology, 81*, 417–429.

Heim, R. (2002). A review of twentieth-century drought indices used in the United States. *Bulletin of the American Meteorological Society, 83*, 1149–1165.

Huete, A., Didan, K., Miura, T., Rodriguez, E. P., Gao, X., & Ferreira, L. G. (2002). Overview of the radiometric and biophysical performance of the MODIS vegetation indices. *Remote Sensing of Environment, 83*, 195–213.

IWMI. Retrieved February, 6, 2014, from http://www.iwmi.cgiar.org/.

Kanellou, E., Domenikiotis, C., Tsiros, E., & Dalezios, N. R. (2008). Satellite-based drought estimation in Thessaly. *European Water Publications, 23*(24), 111–122.

Karnieli, A., Bayasgalan, M., Bayarjargal, Y., Agam, N., Khudulmur, S., & Tucker, C. J. (2006). Comments on the use of the Vegetation Health Index over Mongolia. *International Journal of Remote Sensing, 27*(10), 2017–2024.

Keyantash, J., & Dracup, J. (2002). The quantification of drought: An evaluation of drought indices. *Bulletin of the American Meteorological Society, 83*, 1167–1180.

Kogan, F. N. (1987). Vegetation index for areal analysis of crop conditions. In *Proceedings of 18th conference on agricultural and forest meteorology, September 15–18* (pp. 103–106). W. Lafayette, Indiana: AMS.

Kogan, F. N. (1995). Droughts of the late 1980s in the United States as derived from NOAA polar-orbiting satellite data. *Bulletin of the American Meteorological Society, 76*(5), 655–668.

Kogan, F. N. (2001). Operational space technology for global vegetation assessment. *Bulletin of the American Meteorological Society, 82*(9), 1949–1964.

Li, S.-X., & Xiao, L. (1992). Distribution and management of dryland in the People's Republic of China. *Advances in Soil Science, 18*, 148–278.

NASA. Retrieved February 6, 2014, from http://modis.gsfc.nasa.gov/.

NDMC. Retrieved February 6, 2014, from http://drought.unl.edu/.

Sensoy, S., Demircan, M., Ulupınar, U., & Balta, İ. (2008). Türkiye İklimi, DMİ. http://www.dmi.gov.tr/iklim/iklim.aspx.

UN. Retrieved February 6, 2014, from http://www.unccd.int/en/Pages/default.aspx.

Wilhite, D. A. (1993). *Drought assessment, management, and planning: Theory and case study*. Boston: Kluwer.

Wilhite, D. A., & Glantz, M. H. (1985). Understanding the drought phenomenon: The role of definitions. *Water International, 10*, 111–120.

World Meteorological Organization. (2006). *Drought monitoring and early warning: Concepts, progress and future challenges*. WMO-No. 1006. ISBN 92-63-11006-9.

Chapter 11
Carbon Foot Print of a Passanger Aircraft Engine at Landing and Take-Off Cycle

Yasin Şöhret and T. Hikmet Karakoç

Introduction

In last decade, rapid civilization forces us to develop more eco-friendly systems with respect to the environmental issues. Mankind population growth on the earth and utilization of the sources callously leads to consider sustainable development. On the other hand, environmental impact reduction is as important as sustainability at the present time. Especially damage and impact of the energy industry on the environment become more of an issue for the future of the mankind. In this case, transportation industry, which is an ever-growing and consuming energy sources more and more, should be evaluated in terms of environmental issues (Winter 2014; Lee et al. 2009). Under consideration of the aviation transportation growth in Turkey, environmental impact of the aircrafts plays a key role for understanding slice of this industry within the contribution of Turkey to the global warming and climate change. As mentioned in a previous study (Ekici et al. 2013), number of aircrafts in aviation fleet enormously increases day by day.

Many researcher studied on exhaust emission related issues of the aviation with concern of environmental issues. Beck et al. (1992) presented the impact of the aircraft emissions on atmosphere's tropospheric layer by a two-dimensional model. Authors especially focused on ozone formation as a result of aircraft exhaust emissions in the paper. Role of the nitrogen oxides within the global warming was highlighted by authors. In another text, international airports in Korea was

Y. Şöhret (✉)
Aircraft Technology Program, Keciborlu Vocational School,
Suleyman Demirel University, 32700 Isparta, Turkey
e-mail: ysohret@gmail.com

T.H. Karakoç
Faculty of Aeronautics and Astronautics, Anadolu University, 26470 Eskisehir, Turkey
e-mail: hkarakoc@anadolu.edu.tr

P. Grammelis (ed.), *Energy, Transportation and Global Warming*, Green Energy and Technology, DOI 10.1007/978-3-319-30127-3_11

handled and dependence among the air pollution and the greenhouse gases emitted from commercial aircrafts was investigated. Fan et al. (2012) calculated fuel consumption and aircraft emissions from domestic flights for China in 2010. As a result of their study, an airline emitting excessive emissions is selected in China. Hu et al. (2009) evaluated air pollutants near an airport depending on real-time measurements. Aircraft engine exhaust emissions, exhaust gases emitted from other vehicles in the airport were discussed in the paper. Yilmaz and Ilbas (2012), presented the emission parameters of various aircraft engine in service. Consequently, the necessity of alternative fuel utilization in aircraft engines is emphasized by the authors. Aircraft engine emissions over the course of landing and take-off cycle are investigated by Ekici et al. (2013) for the busiest airports in Turkey. In another study, it is aimed to develop a model for nitrogen oxide emissions of a turbofan engine for determining environmental impact of aircraft engine emissions on the basis of the operational and meteorological conditions. In another text (Naugle and Fox 2014), contribution of the emitted gases from aircrafts to air pollution was discussed. Also, aircraft emissions caused health issues were highlighted by the authors. Exhaust emissions of an aircraft engine which is fed with both an alternative fuel and kerosene were evaluated by Santoni et al. (2011). Especially unburned hydrocarbon and nitrogen oxide emissions were given in details by authors. It is reported that, unburned hydrocarbon concentrations in engine exhaust were lower than ambient concentrations at higher thrust levels. Mazaheri et al. (2011) evaluated exhaust emissions of large aircrafts at an airport. Authors presented emission rates depending on aircrafts operating in the airport. Additionally, particulate matter emissions were discussed in details by researchers. Synylo and Duchêne (2014) modelled nitrogen oxide emissions of a turbofan engine to evaluate the impact of operational and meteorological conditions on nitrogen oxide formation. Additionally the environmental sustainability of the engine was discussed in the paper.

In the present paper, a turbofan engine of a commercial aircraft in service is investigated. Carbon footprint of the examined engine is intended to present with the aid of ICAO aircraft engine emissions inventory. The main goal and differences of the current paper from previous studies may be listed as follows:

- Introducing the correlation between emission index and carbon footprint.
- Revealing the carbon footprint of an aircraft engine for the first time.,
- Presenting a different perspective to understand environmental impact of the aviation.

Turbofan Engine and Evaluation Methodology

In this study, a mixed turbofan type aircraft engine is evaluated. The engine which is named as JT8D is operated on many aircraft such as Boeing 727 and 737 series, McDonnell Douglas DC-9 and MD-80. The JT8D serves the industry by providing

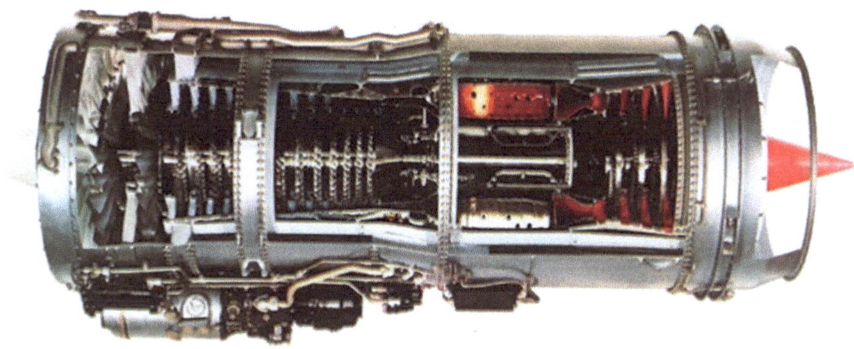

Fig. 11.1 The sectional view of the JT8D engine (Pratt & Whitney 2014)

power to more than 2400 aircraft. The sectional view of the engine is shown in Fig. 11.1 (Pratt & Whitney 2014). For evaluation of the engine, data is provided from the ICAO aircraft engine emissions inventory (EASA 2014).

The ICAO aircraft engine emissions inventory (EASA 2014) contains emission data for the ICAO landing and take-off flight envelope. A typical flight envelope is demonstrated in Fig. 11.2. ICAO landing and take-off flight cycle is the part of this envelope which is under the 3000 feet altitude. In the course of emission measurement, engine was operated under the relevant flight settings. That is to say, engine was operated at 100 %, 85 %, 30 % loads for 0.7, 2.2, 4.0 min at take-off, climb-out and approach flight phases respectively (ICAO 1993).

Carbon footprint is the total sets of greenhouse gas emissions caused by an organization, event, product or person and a useful parameter to understand contribution to the climate change. The calculation of the carbon footprint is performed by finding carbon dioxide equivalent for 100-year time horizon of the greenhouse gases emitted by an organization, event, product or person. In this framework, carbon footprint of the examined turbofan engine can be found as following (Schimel et al. 1996; Wright et al. 2011):

$$C = \sum (CO_2e)_i = \sum EI_i m_f GWP(100)_i \tag{11.1}$$

Here, C notates carbon footprint whereas CO_2e, EI, m_f, GWP(100) and i represents carbon dioxide equivalent, emission index, fuel mass flow rate, global warming potential for 100-year time horizon and ith greenhouse gas emitted from the engine. Carbon dioxide equivalent of any gas means global warming potential of the gas for 100-year time horizon. Emission index indicates emitted gas amount per consumed fuel amount within the combustion chamber. In Table 11.1, global warming potentials of emitted gases for 100-year time horizon are given.

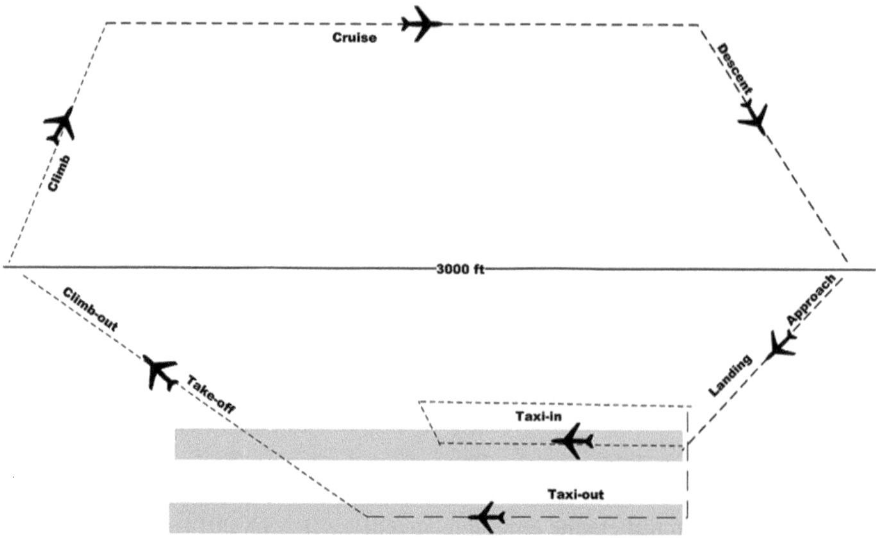

Fig. 11.2 A typical flight envelope for a commercial aircraft (adopted from ICAO 1993)

Table 11.1 Global warming potential values of greenhouse gases emitted from an aircraft engine for 100-year time horizon (Altuntas 2014; Schimel et al. 1996)

Greenhouse gas	GWP(100)
CO_2	1
CO	1
NO_x	310
CH_4	21

Table 11.2 Exhaust gas emissions of the JT8D engine at landing and take-off flight cycle (EASA 2014)

Flight phase	EI_{CO}	EI_{NOx}	EI_{UHC}
Take-off	0.74	19.20	0.69
Climb-out	1.00	15.23	0.79
Approach	8.54	6.10	1.96
Idle	31.00	3.30	0.00
Total	41.28	43.83	3.44

Results and Discussion

In this study, carbon footprint of emitted greenhouse gases from JT8D engine which is still operated on many commercial passenger aircraft. During a typical landing and take-off flight cycle, exhaust gas emissions of the engine are given in Table 11.2.

Calculations based upon data given in Table 11.2, are performed with the aid of Eq. (11.1). Herein, unburned hydrocarbon (UHC) emission was assumed to be

Table 11.3 Carbon footprint equivalents of emitted greenhouse gases from JT8D engine at landing and take-off flight cycle

Flight phase	C(CO)	C(NO_x)	C(UHC)
Take-off	0.921	7410.240	18.040
Climb-out	0.997	4707.136	16.540
Approach	3.023	669.414	14.570
Idle	4.557	150.381	0
Total	9.498	12937.170	49.150

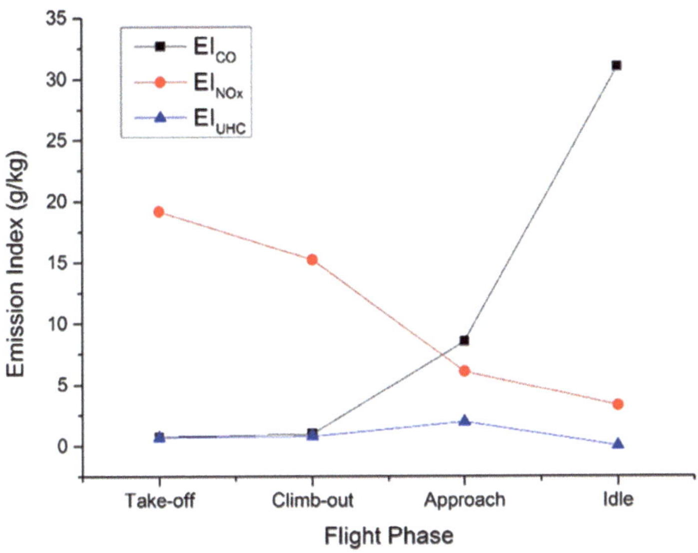

Fig. 11.3 Emission index variation of JT8D engine greenhouse gases with flight phases

methane (CH_4). As a result of the study, obtained carbon footprint equivalent of each emission gas was summarized in Table 11.3 on the basis of flight phases.

For better comprehension of the situation, Figs. 11.3 and 11.4 are plotted. As indicated in Fig. 11.3, unburned hydrocarbon emission index of the engine is approximately same at take-off and climb-out phases of flight. However, emitted unburned hydrocarbon per consumed fuel increases at approach phase and dramatically decreases at idle. Additionally, reverse proportion among carbon monoxide and nitrogen oxide emission index is clarified in the graph. So, it can be asserted that, while amount of emitted carbon monoxide per consumed fuel raises, nitrogen oxide emission descents over through the landing and take-off flight cycle.

Figure 11.4 reveals the carbon footprint of the engine during landing and take-off flight cycle in simple terms. Herein, carbon footprint values of carbon monoxide and unburned hydrocarbon are approximate to zero in comparison with nitrogen oxide emission. However, carbon footprint of emitted nitrogen oxide decreases from 7410.24 g to 150.381 g. So, impact of nitrogen oxides emitted from aircraft engines on global warming is clearly proven.

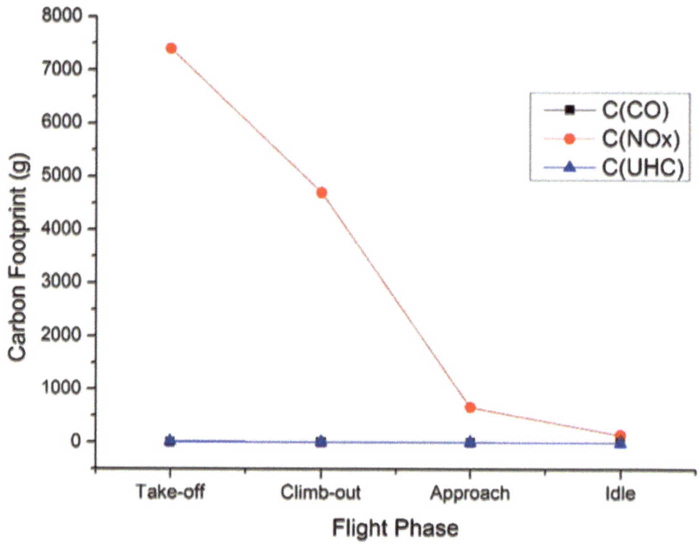

Fig. 11.4 Carbon footprint variation of JT8D engine greenhouse gases with flight phases

Conclusions

As a result of the study, following statements are concluded by authors:

- With regard to emission measurement, emitted greenhouse gases from the engine are concern. On the other, in terms of carbon footprint calculation carbon monoxide and unburned hydrocarbons can be disregarded in future studies. But it is not possible to assert for nitrogen oxide emission.
- Conducting similar study for various engine types can be beneficial as a comparison. So that, contribution of aircraft engine exhaust emissions to global warming and carbon cycle can be revealed in details.
- If the flights all around the world are considered, obtained results can be more meaningful. However, from this point of view, impact of aircraft transportation on the environment may be more comprehensible.

In a future study, it is planned to cover that issue from perspective of aviation fleet and comparison of different aircraft engine types.

Acknowledgements Authors gladly thank to Anadolu and Suleyman Demirel Universities of Turkey.

Nomenclature

C	Carbon footprint
CO_2e	Carbon dioxide equivalent
EI	Emission index (g kg^{-1})
GWP	Global warming potential
m	Mass flow rate (kg s^{-1})

Subscripts

f	Fuel
i	ith greenhouse gas emitted from the engine

References

Altuntas, O. (2014). Calculation of domestic flight-caused global warming potential from aircraft emissions in Turkish airports. *International Journal of Global Warming, 1*, 367–379.

Beck, J. P., Reeves, C. E., De Leeuw, F. A., & Penkett, S. A. (1992). The effect of aircraft emissions on tropospheric ozone in the northern hemisphere. Atmospheric Environment. Part A. *General Topics, 26(1)*, 17–29.

EASA. (2014). *ICAO Aircraft engine emissions databank*. Retrieved December 15, 2014, from http://easa.europa.eu/document-library/icao-aircraft-engine-emissions-databank#1.

Ekici, S., Yalin, G., Altuntas, O., & Karakoc, T. H. (2013). Calculation of HC, CO and NO_x from civil aviation in Turkey in 2012. *International Journal of Environment and Pollution, 53*, 232–244.

Fan, W., Sun, Y., Zhu, T., & Wen, Y. (2012). Emissions of HC, CO, NO_x, CO_2, and SO_2 from civil aviation in China in 2010. *Atmospheric Environment, 56*, 52–57.

Hu, S., Fruin, S., Kozawa, K., Mara, S., Winer, A., & Paulson, S. (2009). Aircraft emission impacts in a neighborhood adjacent to a general aviation airport in southern California. *Environmental Science & Technology, 43*, 8039–8045.

ICAO. (1993). Annex 16. In *Aircraft engine emissions* (2nd ed.).

Lee, D. S., Fahey, D. W., Forster, P. M., Newton, P. J., Wit, R. C. N., Lim, L. L., et al. (2009). Aviation and global climate change in 21st century. *Atmospheric Environment, 43*, 3520–3537.

Mazaheri, M., Johnson, G. R., & Morawska, L. (2011). An inventory of particle and gaseous emissions from large aircraft thrust engine operations at an airport. *Atmospheric Environment, 45*, 3500–3507.

Naugle, D., & Fox, D. (2014). Aircraft and air pollution. *Environmental Science and Technology, 15*, 391–395.

Pratt & Whitney, JT8D engine. Retrieved December 15, 2014, from http://www.pw.utc.com/JT8D_Engine.

Santoni, G., Lee, B., Wood, E., Herndon, S., Miake-Lye, R., Wofsy, S., et al. (2011). Aircraft emissions of methane and nitrous oxide during the alternative aviation fuel experiment. *Environmental Science & Technology, 45*, 7075–7082.

Schimel, D., Alves, D., Enting, I., Heimann, M., Joos, F., Raynaud, D., et al. (1996). Radiative forcing of climate change. In J. T. Houghton, L. G. M. Filho, B. A. Callander, N. Harris, A. Kattenbergan, & K. Maskell (Eds.), *Climate change 1995: The science of climate change*. Cambridge, England: Intergovernmental Panel on Climate Change.

Synylo, K., & Duchêne, N. (2014). NO_x emission model of turbofan engine. *International Journal of Sustainable Aviation, 1*, 72–84.
Winter, R. A. (2014). Innovation and the dynamics of global warming. *Journal of Environmental Economics and Management, 68*, 124–140.
Wright, L. A., Kemp, S., & Williams, I. (2011). 'Carbon footprinting': towards a universally accepted definition. *Carbon Management, 2*, 61–72.
Yilmaz, I., & Ilbas, M. (2012). Investigation of pollutant emissions in aircraft gas turbine engines. *Journal of the Faculty of Engineering and Architecture of Gazi University, 27*, 343–351.

Chapter 12
Greenhouse Gas Emissions Trends from Waste in Greece

L. Kallinikos, I. Sebos, A. Progiou, P. Eleni, I. Katsavou, K. Mangouta, and I. Ziomas

Introduction

United Nations Framework Convention on Climate Change (henceforth the Convention) was adopted on 9 May 1992 and was opened for signature in Rio de Janeiro in June 1992 in response to the emerging evidence that climate change could have a major global impact. Greece signed the Convention in Rio and ratified it in 1994.

The Conference of the Parties (COP) in 2007, by its decision 1/CP.13 (the Bali Action Plan) launched a comprehensive process to enable the full, effective and sustained implementation of the Convention through long-term cooperative action, now, up to and beyond 2012, to be conducted under a subsidiary body of the Convention, the Ad Hoc Working Group on Long-term Cooperative Action under the Convention (AWG-LCA).

The Protocol entered into force on 16 February 2005, after its ratification from 141 Parties (with the exception of USA and Australia) including developed countries with a contribution of more than 55 % to global CO_2 emissions in 1990.

With respect to the EU target under the Kyoto Protocol (i.e. reduction of emissions by 8 % for the period 2008–2012), EU has stated that this will be achieved jointly by EU Member-States under the provisions of Article 4 of the Protocol. The Burden-Sharing agreement between all Member States was finalized during the Environment Council in June 1998 and entered into force with Decision 2002/358/EC concerning the approval, on behalf of the European Community, of the Kyoto Protocol. According to this agreement, Greece is committed to limit its

L. Kallinikos • I. Sebos (✉) • A. Progiou • P. Eleni • I. Katsavou • K. Mangouta • I. Ziomas
Department of Chemical Engineering, National Technical University of Athens, Zografou Campus, 9 Heroon Polytechniou Street, 15780 Zografou, Greece
e-mail: leokalls@central.ntua.gr; isebos@mail.ntua.gr; ap@axonenviro.gr; peleni@central.ntua.gr; katsavou@central.ntua.gr; kon_maggouta@yahoo.gr; ziomas@chemeng.ntua.gr

P. Grammelis (ed.), *Energy, Transportation and Global Warming*, Green Energy and Technology, DOI 10.1007/978-3-319-30127-3_12

GHG emissions increase for the period 2008–2012 to +25 % compared to base year emissions.

For the period 2013–2020 EU adopted the climate and energy package which is a set of binding legislation aiming in ensuring of the European Union meets its ambitious climate and energy targets for 2020. These targets, known as the “20-20-20” targets, set three key objectives for 2020 (http://ec.europa.eu/clima/policies/ets/index_en.htm):

- A 20 % reduction in EU greenhouse gas emissions from 1990 levels.
- Raising the share of EU energy consumption produced from renewable resources to 20 %.
- A 20 % improvement in the EU’s energy efficiency.

Finally, EU leaders agreed on 23 October 2014 the domestic 2030 greenhouse gas reduction target of at least 40 % compared to 1990 together with the other main building blocks of the 2030 policy framework for climate and energy, as proposed by the European Commission in January 2014. This 2030 policy framework aims to make the European Union’s economy and energy system more competitive, secure and sustainable and also sets a target of at least 27 % for renewable energy and energy savings by 2030.

Regarding Greek contribution in the first commitment period, 2008–2012, the country ratified the Kyoto Protocol in 2002 (Law 3017/2002) and adopted a National Programme for achieving its commitment by a decision of the Council of Ministers (DCM5/2003).

Annual inventories of greenhouse and other gases emissions form an essential element of each national environmental policy-making process. They can be used to derive information on emissions trends, with reference to a pre-selected base year, and can assist in monitoring the progress of existing abatement measures for the reduction of greenhouse gases emissions and the fulfillment of the KP target (MRPEE 2014b).

According to UNFCCC structure for the first commitment period, 2008–2012, annual inventories regarding emissions sources could be categorized to the following sectors (IPCC 1997, 2002):

Energy sector
Industrial processes
Solvents and other products
Agriculture
Waste
Land Use, Land Use Change and Forestry (LULUCF)

Waste sector consists one of the sectors as they are categorized by IPCC Guidelines and it is recognized that the waste management industry plays an increasing role in climate change mitigation (Ragossnig and Hilger 2008).

The definition of ‘waste management’ adopted here covers four basic divisions (UN 1998; Bogner et al. 2008; Braschel and Posch 2013):

- Solid waste disposal on land
- Wastewater handling
- Waste incineration
- Other (Biological treatment)

Solid Waste Disposal on Land

Solid waste disposal on land is responsible for methane emissions. Methane is emitted during the anaerobic decomposition of organic waste disposed in various solid waste disposal sites (SWDS). The main characteristic of this process is that organic waste decomposes at a diminishing rate over time and takes many years to decompose completely. Moreover, other factors such as the type of waste disposed, the characteristics of the disposal sites and the climate conditions, affect the decomposition rate.

Carbon dioxide emissions occur during the flaring of biogas released from the decomposition of waste. However, these emissions should not be included in the total GHG emissions of the sector as they are of biogenic origin. Recovery and flaring of biogas constitute a waste management practice in the major managed SWDS.

Methane emissions from solid waste disposal on land consist of emissions from municipal solid waste disposal on sites, emissions from sewage sludge (generated during municipal wastewater handling) landfilled and emissions from industrial solid waste and construction and demolition solid waste disposal in managed and unmanaged sites (MRPEE 2014b; IPCC 1997, 2002).

The estimation of methane emissions from solid waste disposal on land could be based on default emission factors or more detailed methodologies. IPCC Guidelines provide three accounting methods; Tier 1, Tier 2 and Tier 3. Tier 1 provides IPCC default data to countries with little data available (Braschel and Posch 2013). Tier 2 includes first-order decay (FOD) equations with some default parameters defined by the IPCC, while Tier 3 is based on FOD equations with nationally developed parameters, such as the degradable organic content of the waste (Gentil et al. 2009). The method is applied separately for the managed and unmanaged waste disposal, taking account of the different conditions in those sites and the detailed information available regarding the opening and closure years of the operation of the managed sites (MRPEE 2014b; IPCC 1997, 2002, 2006).

Wastewater Handling

Domestic and industrial wastewater handling under anaerobic conditions produces CH_4 while under aerobic treatment they result in sewage sludge production, the disposal of which produces also CH_4. Moreover, from both sub sectors N_2O

emissions are released (MRPEE 2014b; IPCC 1997, 2002). For the estimation of the emissions from waste water handling default methodologies are provided by IPCC (1997), IPCC (2002) and IPCC (2006).

Waste Incineration

Carbon dioxide, Methane and Nitrous oxide emissions produced from the incineration of waste such as clinical waste, biogenic agricultural residues produced in slaughterhouses and from the incineration of small amounts of industrial chemical waste are being estimated (MRPEE 2014b; IPCC 1997, 2002). For the estimation of the emissions from this sector default methodologies are provided by IPCC (1997), IPCC (2002) and IPCC (2006).

Emissions Trend

Total Emissions

Information regarding the methodologies utilized on the estimation of emissions from the different sectors and the data collection are provided by the annual emissions inventory submission of Greece (MRPEE 2014b).

According to these estimations in 2012, GHG emissions (without LULUCF) amounted to 110.78 Mt CO_2 eq showing an increase of 3.63 % compared to base year emissions and of 5.86 % compared to 1990 levels.

Carbon dioxide emissions accounted for 81.67 % of total GHG emissions in 2012 (without LULUCF) and increased by approximately 9.11 % from 1990. Methane emissions accounted for 8.57 % of total GHG emissions in 2012 and decreased by 8.77 % from 1990, while nitrous oxide emissions accounted for 6.15 % of the total GHG emissions in 2012 and decreased by 33.39 % from 1990. Finally, f-gases emissions (from production and consumption) that accounted for 3.95 % of total GHG emissions in 2012, and 41.16 % of the IP sector, has an average increase of 33.19 % from 1995 (base year for F-gases).

In Fig. 12.1, the evolution of the emissions for the period 1990–2012 are provided. In the same figure the evolution of the Gross Domestic Product (GDP) is provided. As it is observed GHG emissions follow the trend of GDP. Over the last 5 years Greek economy faced its most-severe crisis since it experienced a significant recession having recorded a cumulative decline in real terms exceeding 20 % (by the end of 2012 compared to GDP at the end of 2007).

The repercussions from the international financial crisis are unavoidable felt also in Greece especially through the negative impact in the two significant exporting sectors (tourism and ship transportation) but at a large extent economic downturn

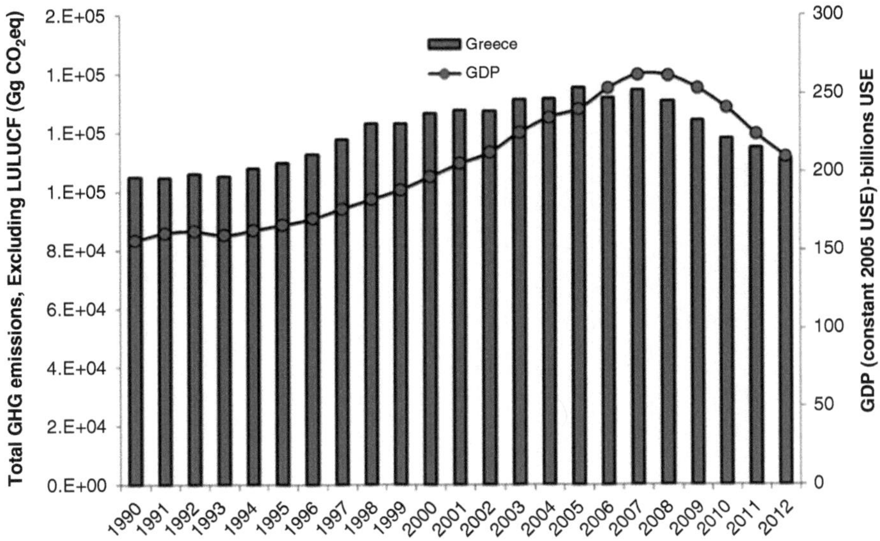

Fig. 12.1 Total GHG emissions of Greece and GDP

relates also to the diminishing growth potential of the country since no significant changes have occurred in the domestic production model towards innovative or high value added activities.

Before this 5 years period, Greek growth performance was impressive. The annual rate of increase of the GDP during the period 2000–2004 was approximately 4.5 % due to the financial market liberalization coupled with membership in the monetary union, which led to substantial increase in credit expansion and reduction in borrowing costs, the stimulus given by the Olympic Games hosted in Athens in 2004 and the Community Structural Funds. Contrary to expectations of a post-Olympics slump, the economy continued to grow briskly in 2005–2007 period. GDP increased by an average growth rate of 3.7 % and thus Greece enjoyed one of the highest growth rates in the EU and the Eurozone.

Regarding EU-15 and E-28 related information are provided by the annual emissions inventory submission of European Commission (European Environmental Agency 2014b). In 2012 total GHG emissions in the EU-15, without LULUCF, were 15.1 % (644 million tonnes CO_2 equivalents) below 1990. Between 2011 and 2012 emissions decreased by 0.8 % (31 Mt of CO_2 equivalents) (http://unfccc.int/).

Total GHG emissions, without LULUCF, in the EU-28 decreased by 19.2 % between 1990 and 2012 (−1082 Mt of CO_2 equivalent). Between 2011 and 2012, emissions decreased by 1.3 % (59 Mt CO_2 equivalent) (Fig. 12.2).

In the same figure the evolution of the GDP is provided. As it is observed GDP of the European Union, for both EU-15 and EU-28, decreases only for 2009, while GHG emissions decreases for the period 2005–2012 as a result of the mitigation measures adopted by its members.

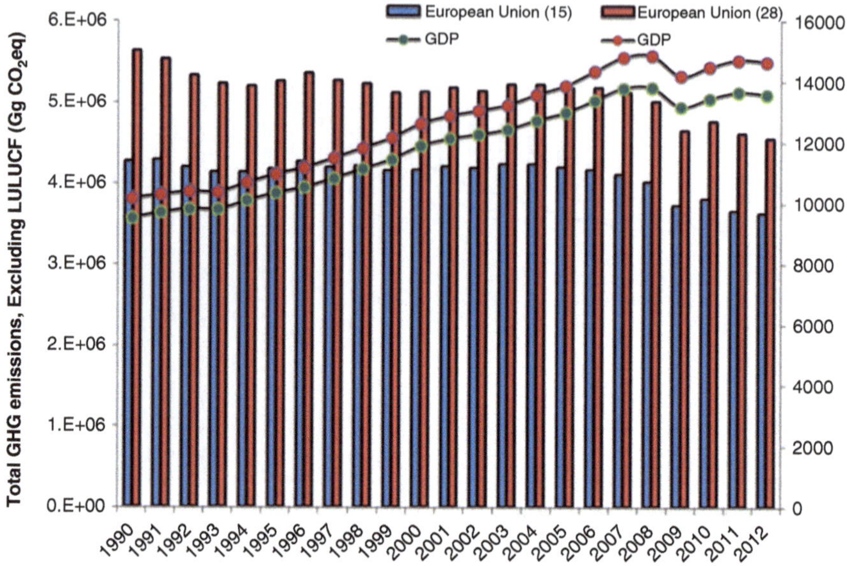

Fig. 12.2 Total GHG emissions for European Union (EU15 and EU28) versus GDP

Waste Sector

In general, waste sector is the fourth largest sector in Greece, contributing 4.3 % of the total greenhouse gas emissions sector in 2012 (excluding Land Use, Land-Use Change and Forestry, LULUCF), while GHG emissions from waste decreased by 19.6 % in 2012 compared to 1990 levels, based on the latest official GHG inventory for the country. Similarly, the waste sector is the fourth largest sector in the EU-28 with approximately 3.1 % contribution on the total GHG emissions for 2012, while the total decrease over the period 1990–2012 is about 31.5 % (Fig. 12.3).

Information regarding the methodologies utilized on the estimation of emissions from the waste sector and the data collection are provided by the annual emissions inventory submission of Greece (MRPEE 2014b).

Specifically, the major Waste category since 1999 for Greece is the solid waste disposal on land with its contribution increasing from 37.8 % in 1990 to 67.7 % in 2012. On the contrary, GHG emissions from wastewater handling was the major source up to 1999, presenting a declining trend with an average annual rate of −2.69 % for the period 1990–2012. Emissions from the incineration of waste present a negligible contribution on the total GHG from the waste sector. Similarly, GHG emissions from waste disposal on land accounts for 72.3 % of the total waste greenhouse gas emissions in the EU-28 for the 2012.

Emissions from the Waste Sector (4.27 % of the total emissions, without LULUCF), decreased by approximately 19.58 % from 1990. Living standards improvement resulted in an increase of the generated waste and thus of emissions. However, the increase of recycling along with the exploitation of the biogas

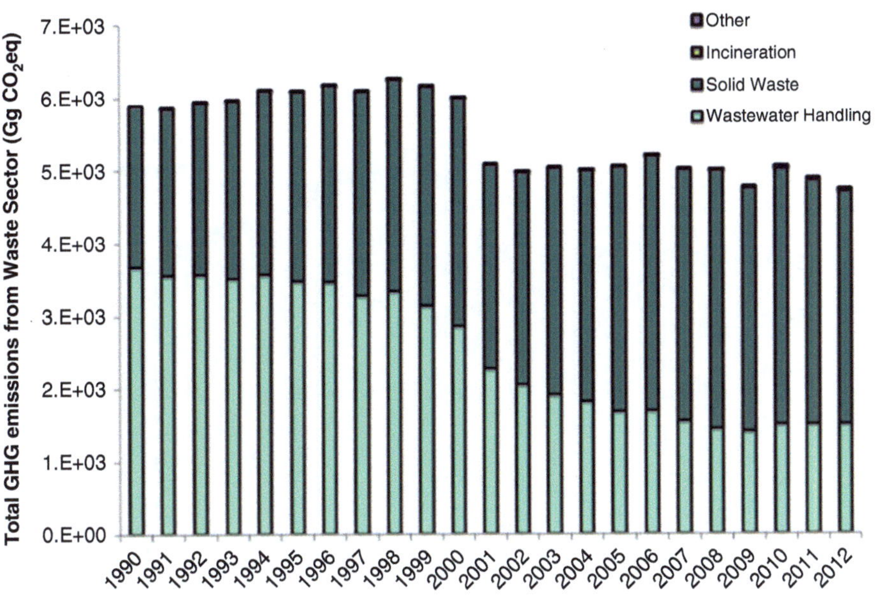

Fig. 12.3 GHG emissions from Waste Sector for Greece

produced limits the increase of methane emissions. At the same time, emissions from wastewater handling have considerably decreased, due to the continuous increase of the population served by aerobic wastewater handling facilities.

Waste sector is the fourth largest sector in the EU-15, after energy, agriculture and industrial processes, contributing 3 % to total GHG emissions (European Environmental Agency 2014b). Total emissions from Waste have been decreasing by 40 % from 171 Tg in 1990 to 102 Tg in 2012 (Fig. 12.4 for EU-15, Fig. 12.5 for EU-28). In 2012, emissions decreased by 3.3 % compared to 2011.

As it shown in this figure, CH_4 emissions from Managed Waste Disposal on Land had the greatest decrease of all waste-related emissions, but still account for 66 % of waste-related GHG emissions in the EU-15 in 2012.

Solid Waste Disposal

Given the fact that solid waste sector is the main source for both Greece and EU, an analysis on it is presented in this paragraph. As it is observed in the Fig. 12.2, Greenhouse gas emissions from solid waste disposal on land present an increasing trend, with some fluctuations due to variations on the CH_4 recovered each year. This is the result of the increasing trend observed on the municipal solid waste disposal on land for the period 1990–2012, Fig. 12.6. While the solid waste generation increase rate is significantly higher than this of disposal on solid waste disposal

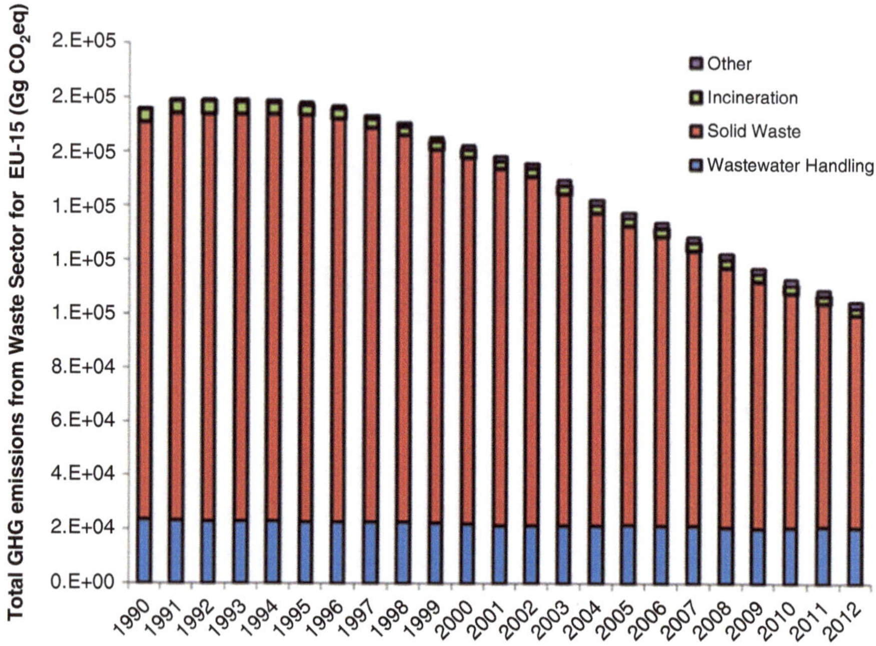

Fig. 12.4 GHG emissions from Waste Sector for European Union (EU15)

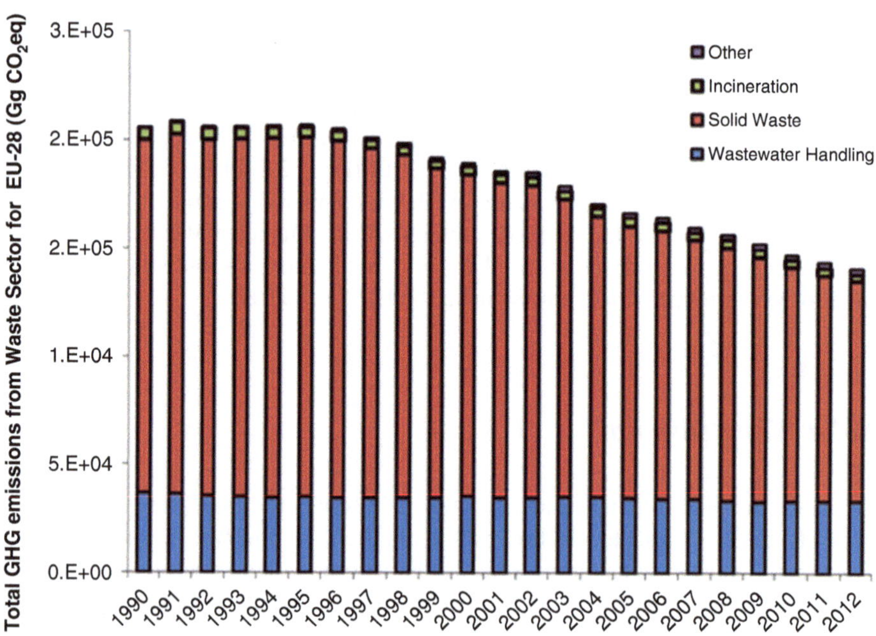

Fig. 12.5 GHG emissions from Waste Sector for European Union (EU28)

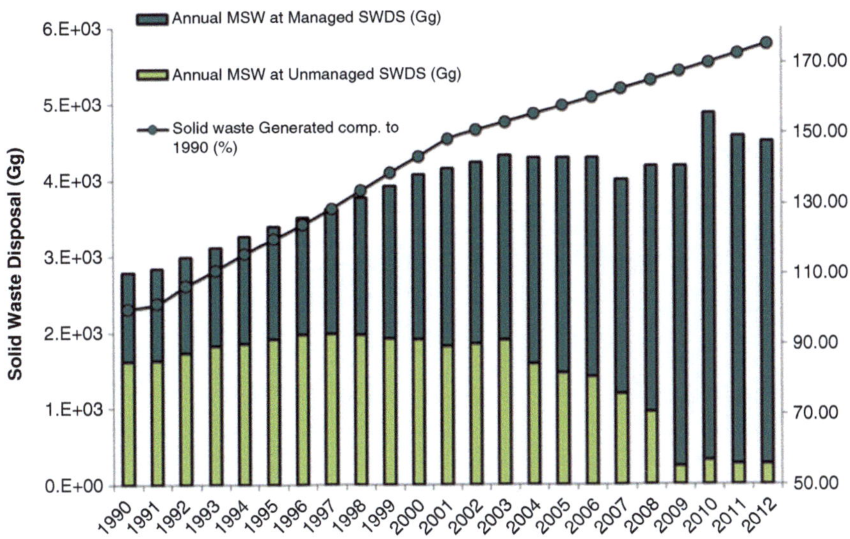

Fig. 12.6 Annual municipal solid waste disposal and solid waste generated compared to 1990 of Greece

sites, the increasing trend on the treatment of solid waste, recycling and composting, is not able to decrease the amounts of solid waste disposal on sites. For this reason, emissions of the sector are showing an increasing trend. In contrast, both for EU-15 and EU-28 the solid waste amount treatment increases for all the period of 1990–2012 resulting in a decreasing trend on the amounts of solid waste disposal on sites and the emissions (Figs. 12.7 and 12.8).

Projections

For the projection of the evolution of the emissions for the period 2013–2030 the impact of the implemented, adopted and planned policies and measures for the mitigation of the emissions are utilized (MRPEE 2014a). This policy is fully harmonized with the EU policy for the mitigation of GHG emissions from waste and is mainly described by the regulation for the Municipal Waste disposal on land (Landfill Dir. 1999/31/EC), the sewage sludge use in agricultures (Dir. 86/278/EEC), the Paper/Cardboard recycling (Dir. on Packaging and Packaging Waste, 94/62/EC) and the Bio-wastes in the Waste Framework Dir. (art. 22, 2008/98/EC).

With Decision 50910/2727 (December 2003), the National Law 4042/2012 (OJG 24 A) and the transposition of Directive 2008/98/EC into national legislation the measures, the terms and the processes for the rational management of waste in national and regional level have been specified.

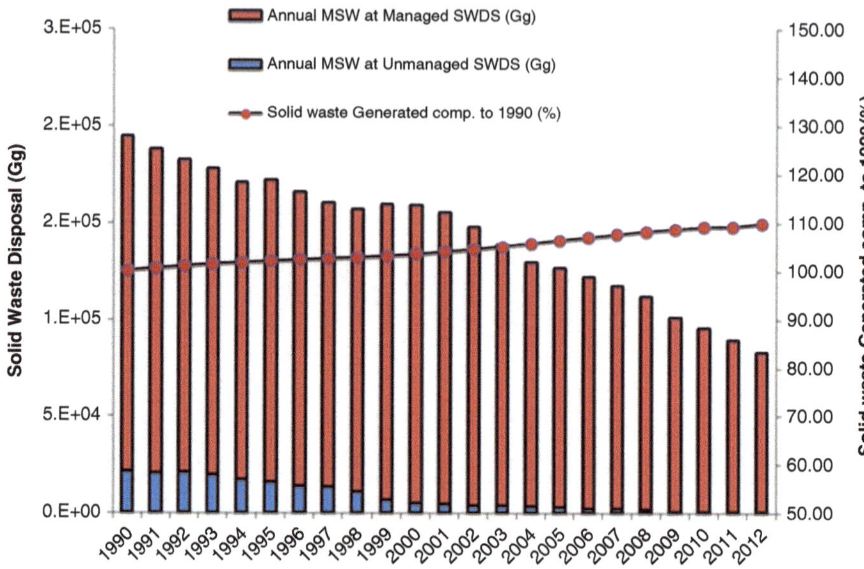

Fig. 12.7 Annual municipal solid waste disposal and solid waste generated compared to 1990 of EU-15

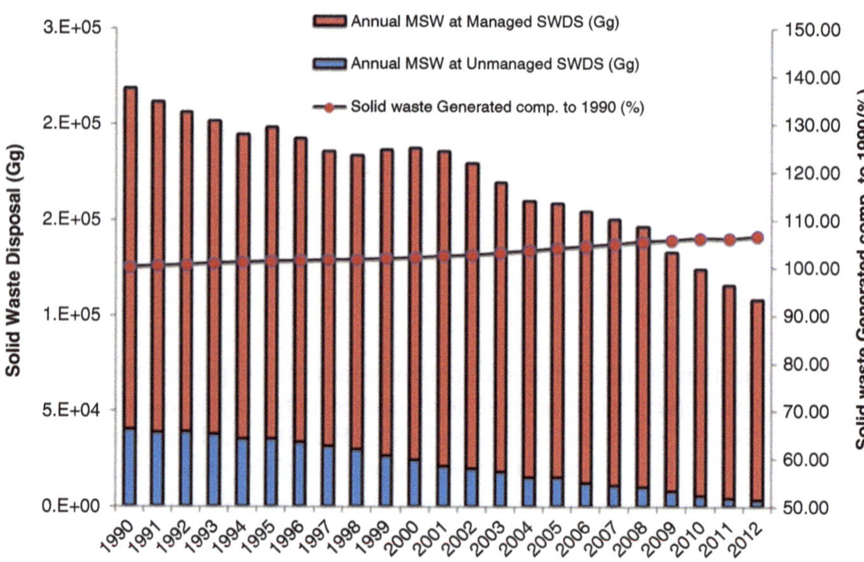

Fig. 12.8 Annual municipal solid waste disposal and solid waste generated compared to 1990 of EU-28

Fundamental objectives are the elimination of unmanaged solid waste disposal sites, the coverage of all urban and rural areas of the country with modern installations for final disposal and the promotion of measures for the prevention and reduction of produced waste, as well as the exploitation of materials with maximization of recycling and recovery of products and energy. The necessity to reduce the quantities of biodegradable wastes landfill through the installation of treatment facilities, Joint Ministerial Decision 29407/3508 in agreement with Directive 1999/31/EC, is acknowledged.

The reduction of biodegradable waste landfill is enhanced by the directive for Packaging and Packaging Waste (94/62/EC) Paper/Cardboard recycling.

Concerning bio-wastes treatment the legislative framework is based in the Waste Framework Dir. (art. 22, 2008/98/EC). According to this, Article 41, until 2015 separately collected bio-waste shall be minimum 5 % of the total produced bio-waste and until 2020, separately collected bio-waste shall be minimum 10 % of the total produced bio-waste.

The flaring of landfill gas in all managed sites for urban centres with population more than 100,000 is partially an integrated measure. Already, the managed disposal sites serving the population of the largest cities of Greece are equipped with systems for the collection or for the flaring of biogas.

As regards wastewater, a collection network with its corresponding wastewater treatment plants has already been developed during the last 5-years, covering the needs of 70 % of the population in 2001 and the 91 % in 2011, in compliance with the Directive 91/271/EEC concerning the collection, treatment and discharge of the urban wastewater. In the Psyttalia wastewater treatment plant that serves approximately four millions of Attica population, a part the sludge produced is treated under anaerobic conditions resulting in the production of biogas. The biogas produced covers the energy needs of the wastewater treatment facilities, while the surplus is flared.

The quantities of the solid waste end out at disposal sites were estimated on the bases of historical data as well as on the implementation of adopted policies and measures taking into consideration the deflection from the targets due to economic recession. The composition of the solid waste landfill at disposal sites was estimated taking into account the composition of MSW generated and the amounts of waste recycling and compost.

The obtained results for Greece are presented in Fig. 12.9 and for EU in Fig. 12.10 (European Environmental Agency 2014a).

As it is shown, in 2020 emissions from waste sector of Greece are estimated to be decreased by 22 % compared to 1990 and in 2030 28 % compared to 1990. At the same time, emissions from waste for EU are estimated to be decreased in 2020 49 % (EU-15) and 44 % (EU-28) compared to 1990 and in 2030 63 % (EU-15) and 50 % (EU-28) compared to 1990. Therefore, it is estimated that if the total targets for the Greenhouse Gas emissions are equal to the targets for the waste sector (20 % in 2020 compared to 1990 and 40 % in 2030 compared to 1990) EU-15 and EU-28 have already achieved with the measures that already have adopted/planned, while for Greece additional measures are needed for 2030.

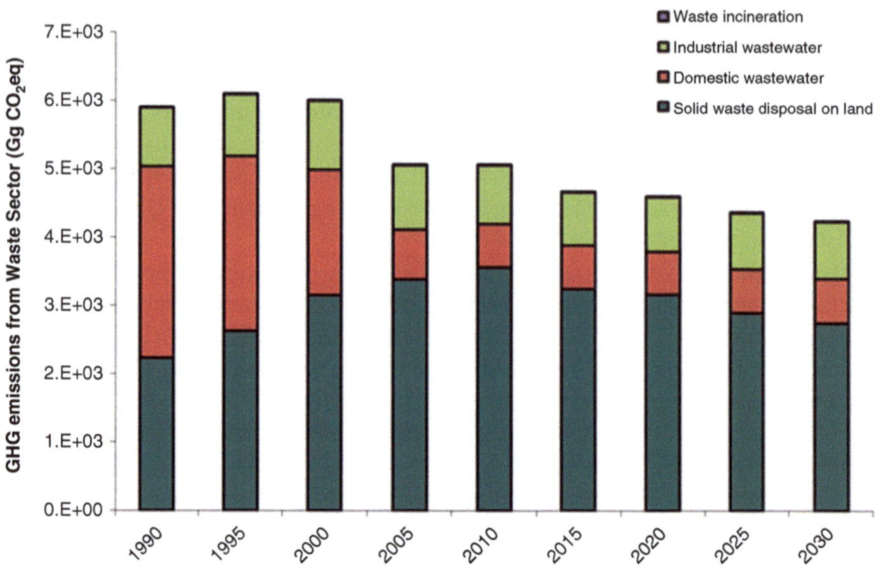

Fig. 12.9 Emissions from waste sector projections for Greece

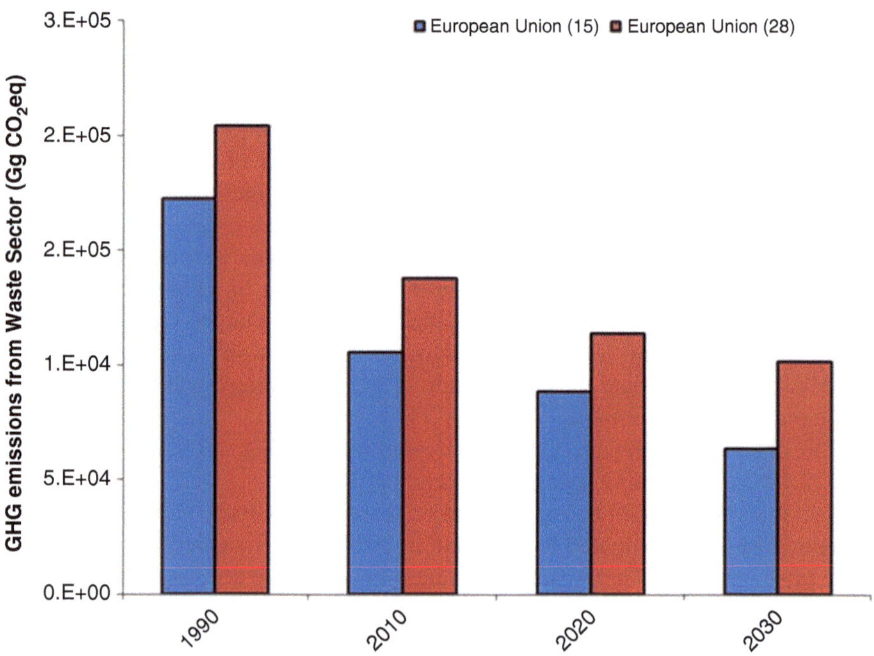

Fig. 12.10 Emissions from waste sector projections for EU-15 and EU-28

Conclusions

Based on the analysis presented in the current work the following conclusions derive:

- Total GHG emissions of Greece follow the trend of GDP. They increase up to 2007 and they decrease in the period 2008–2012 during economic recession.
- Total GHG emissions of EU trend decreases for all the period 1990–2012.
- Waste sector is the fourth largest sector in Greece and it has shown a decreased trend from 2000 mainly due to declining trend of emissions from wastewater handling subsector while GHG emissions from waste sector for EU decrease for all the period 1990–2012.
- If the total targets for the Greenhouse Gas emissions are equal to the targets for the waste sector, EU-15 and EU-28 has already achieved their targets with the measures that already have adopted/planned.
- Greece seems to need some additional measures for 2030.
- It must be mentioned that specific targets for waste sector depends on the disaggregation of total GHG emission targets taking into consideration EU-ETS and non EU-ETS specific targets and the ability for emission decrease of the sectors like F-gases and Agriculture sector.

References

Bogner, J., Pipatti, R., & Hashimoto, S. (2008). Mitigation of global greenhouse gas emissions from waste: Conclusions and strategies from the Intergovernmental Panel on Climate Change (IPCC) Fourth Assessment Report. Working Group III (Mitigation). *Waste Management & Research, 26*, 11–32.

Braschel, N., & Posch, A. (2013). A review of system boundaries of GHG emission inventories in waste management. *Journal of Cleaner Production, 44*, 30–38.

European Environmental Agency. (2014a). *Sixth National Communication and first biennial report from the European Union under the UN framework convention on climate change (UNFCCC), technical report.*

European Environmental Agency. (2014b). *Annual European Union greenhouse gas inventory 1990–2012 and inventory report 2014 Submission to the UNFCCC.*

Gentil, E., Christensen, T. H., & Aoustin, E. (2009). GHG accounting and waste management. *Waste Management and Research, 27*, 696–706.

Intergovernmental Panel on Climate Change (IPCC). (1997). *Revised 1996 IPCC guidelines for national greenhouse gas inventories—Greenhouse gas inventory reference manual* (Vol. 3). Bracknell: IPCC/OECD/IEA, UK Meteorological Office.

Intergovernmental Panel on Climate Change (IPCC). (2002). *Good practice guidance for land use, land use change and forestry*. Japan: IPCC National Greenhouse Gas Inventories Programme, Institute for Global Environmental Strategies.

Intergovernmental Panel on Climate Change (IPCC). (2006). *IPCC guidelines for national GHG inventories*. Japan: IPCC National Greenhouse Gas Inventories Programme, Institute for Global Environmental Strategies.

Ministry of Reconstruction of Production Environment & Energy (MRPEE) (former: Ministry of Environment, Energy and Climate Change). (2014a). *Sixth National Communication and 1st biennial report under the United Nations framework convention on climate change.*

Ministry of Reconstruction of Production Environment & Energy (MRPEE) (former: Ministry of Environment, Energy and Climate Change). (2014b). *Emissions inventory annual inventory submission of Greece under the convention and the Kyoto protocol for greenhouse and other gases for the years 1990–2012.*

Ragossnig, A., & Hilger, H. (2008). Editorial: Waste management: Stepping up to the climate change challenge. *Waste Management & Research, 26*, 3–4.

United Nations (UN). (1998). *Kyoto-protocol to the United Nations Framework Convention on Climate Change.* http://ec.europa.eu/clima/policies/ets/index_en.htm on March 31, 2015. http://unfccc.int/, Accessed on March 31, 2015

Part IV
Global Warming and Climate Change: General Issues and Challenges

Climate Prediction Tools

Development and Application of Climate Prediction Tools

Chapter 13
Development of Models for the Estimation of Global Solar Radiation Over Selected Stations in India

M. Maroof Khan, M. Jamil Ahmad, and Basharat Jamil

Introduction

Solar radiation, passing through the atmosphere, reaching the earth on a horizontal surface can be classified into two components: beam radiation and diffuse radiation. Beam radiation is the solar radiation propagating along the line joining the receiving surface and the center of the sun. It is also referred to as direct radiation. Diffuse radiation on the other hand, is the solar radiation scattered by aerosols, dust and molecules in the atmosphere, it does not have a unique direction. The total radiation is the sum of the beam and diffuses radiation and is sometimes referred to as the global radiation. When the amount of diffuse radiation reaching the earth's surface is less than or equal to 25 % of global radiation, the sky is termed as clear sky (Duffie and Beckman 1991).

In many applications of solar energy, the solar irradiance incident on the surface of the earth at the location of interest is an important input parameter. The temporal and spatial fluctuations of such irradiance necessitate a method to predict them.

M.M. Khan
Mechanical Engineering Section, University Polytechnic, Aligarh Muslim University, Aligarh, 202002 Uttar Pradesh, India
e-mail: maroof.khan08@gmail.com

M.J. Ahmad
Department of Mechanical Engineering, Zakir Husain College of Engineering and Technology, Aligarh Muslim University, Aligarh, 202002 Uttar Pradesh, India
e-mail: mjahmad.me@amu.ac.in

B. Jamil (✉)
Heat Transfer and Solar Energy Laboratory, Department of Mechanical Engineering, Zakir Husain College of Engineering and Technology, Aligarh Muslim University, Aligarh, 202002 Uttar Pradesh, India
e-mail: basharatjamil@zhcet.ac.in

P. Grammelis (ed.), *Energy, Transportation and Global Warming*, Green Energy and Technology, DOI 10.1007/978-3-319-30127-3_13

The systematic variation of solar irradiance outside the earth's atmosphere makes it possible to introduce many models for such prediction (Munroe 1980).

Knowledge of global solar irradiance at a site is essential for the proper design and assessment of solar energy conversion systems. Some of the systems such as concentrating systems require information on direct beam component whereas in the case of tilted plain surfaces the diffuse component of solar irradiance is also important for the computation of system performance (Khogali et al. 1983). However, at locations on the Earth's surface, the solar radiation is also a function of variables such as the nature and extent of cloud cover, the aerosol and water vapor content of atmosphere, etc. Good prediction of the actual value of solar irradiance for a given location requires, in principle, long-term, average meteorological data, which are still scarce for many developing countries like India (Leung 1980; Ezekwe and Ezeilo 1981; Khogali 1983). It is, therefore, not always possible to predict the actual value of solar irradiance for a location of interest.

There are several formulae available in literature on solar radiation modelling that relate global radiation to climatological parameters such as sunshine hours, relative humidity, maximum temperature, and average temperature. The first such correlation proposed for estimating the monthly average daily global irradiation is due Angstrom (1924). The original Angstrom-type regression equation relates monthly average daily radiation to clear day radiation at the location in question and average fraction of possible sunshine hours:

$$H/H_c = a + b\left(\frac{S}{S_o}\right) \tag{13.1}$$

A basic difficulty with Eq. (13.1) lies in the ambiguity of the terms S/S_o and H_o. Page (1961) modified the method to base it on extraterrestrial radiation on a horizontal surface rather than on clear sky day radiation.

$$H/H_0 = a' + b'\left(\frac{S}{S_o}\right) \tag{13.2}$$

Inspite of having complication in Hc calculations, better results were obtained using Hc instead of H_0 (Toğrul 1999). The major objective of this article to investigate usability of clear sky radiation to predict and express the average measured values of solar irradiance on a horizontal surface by using various regression analyses for selected locations in India.

Clear Sky Solar Radiation

Hottel (1976) has presented a method for estimating the beam radiation transmitted through clear atmospheres which takes into account zenith angle and altitude for a standard atmosphere and four climate types. The atmospheric transmittance for beam radiation is given in the form:

Table 13.1 Correction factors for different climates

Climate type	r_0	r_1	r_k
Tropical	0.95	0.98	1.02
Midlatitude summer	0.97	0.99	1.02
Subarctic summer	0.99	0.99	1.01
Midlatitude winter	1.03	1.01	1.00

$$\tau_b = a_o + a_1 \exp(-k \cos \theta_z) \tag{13.3}$$

The constant and k for the standard atmosphere with 23 km visibility are found from a_o*, a_1* and K* which are given for altitudes less than 2.5 km by

$$a_o^* = 0.4237 - 0.00821(6 - A)^2 \tag{13.4}$$

$$a_1^* = 0.5055 + 0.00595(6.5 - A)^2 \tag{13.5}$$

$$K^* = 0.2711 + 0.01858(2.5 - A)^2 \tag{13.6}$$

where, A is the altitude of the observer in kilometers.

The correction factors are applied to a_o*, a_1* and *K** to allow for changes in climate types.

The correction factors $r_o = a_o/a_o^*$, $r_1 = a_1/a_1^*$ and $r_k = k/K^*$ are given in Table 13.1.

Thus, the transmittance of this standard atmosphere for beam radiation can be determined for any zenith angle and any altitude up to 2.5 km. The clear sky beam radiation (G_{cb}, W/m^2) is than

$$G_{cb} = G_{on}\tau_b \tag{13.7}$$

where, G_{on} is the extraterrestrial radiation, measured on the plane normal to the radiation on the nth day of the year and given in the following form (W/m^2).

$$G_{on} = G_{sc}\left(1 + 0.33 \cos\left(\frac{360n}{365}\right)\right) \tag{13.8}$$

where, G_{sc} is the solar constant equal to 1367 W/m^2.

The clear sky horizontal beam radiation is

$$G_{cb} = G_{on}\tau_b \cos \theta_z \tag{13.9}$$

It is also necessary to estimate the clear sky diffuse radiation on a horizontal surface to get the total radiation as suggested by Liu and Jordan (1960). They developed an empirical relationship between the transmission coefficients for beam and diffuse radiation for clear days.

$$\tau_d = 0.271 - 0.294\tau_b \tag{13.10}$$

where τ_d is the ratio of diffuse radiation to the extraterrestrial radiation on the horizontal plane. The clear sky diffuse radiation G_{cd} (W/m^2).

$$G_{cd} = G_{on}\tau_d \cos\theta_z \tag{13.11}$$

Thus, the clear sky global solar radiation is given by

$$G_c = G_{cb} + G_{cd} \tag{13.12}$$

Meteorological Data

In the present study, the monthly average global solar radiation, have been calculated, using sunshine hour data on horizontal surfaces. Six models have been developed which include the effect of latitude and altitude of a location. These six stations have been selected for different geographical locations covering most part of India (Table 13.2).

The measured values of the monthly average global solar radiation (G) and the S/S_o ratio for six locations are obtained from Chandel et al. (2005) and are provided in Table 13.3. G_c values calculated by Hottle's model for six cities are given in Table 13.4.

Development of Models

The following equations were obtained by investigating the relation between G/G_c and S/S_o by different regression analysis. The scatter of monthly mean values between G/G_c and S/S_o are given in Fig. 13.1.

Table 13.2 Geographical location of six Indian cities (Chandel et al. 2005)

Station (State)	Longitude (°E)	Latitude (°N)	Altitude (m)
Palampur (Himachal Pradesh)	76.30	32.60	1270
Amritsar (Punjab)	74.87	31.63	234
Kodaikanal (Tamil Nadu)	77.47	10.23	2339
Nandi Hill (Karnatka)	77.68	13.37	1479
New Delhi (Delhi)	77.20	28.58	216
Shillong (Meghalaya)	91.88	25.57	1600

Table 13.3 Measured values of monthly average global solar radiation (G) and S/S_o for six cities (Chandel et al. 2005)

	Palampur		Amritsar		Kodaikanal		Nandi Hills		Delhi		Shillong	
Month	G	S/S_o	G	S/S_o	G	S/S_o	G	S/S_o	G	S/S_o	G	S/S_o
January	2.80	0.57	3.10	0.66	6.28	0.62	6.37	0.86	3.99	0.73	4.01	0.52
February	3.37	0.55	3.99	0.69	6.71	0.67	6.26	0.81	5.00	0.81	5.08	0.65
March	4.31	0.58	4.88	0.63	6.87	0.63	7.49	0.84	6.14	0.69	5.63	0.68
April	5.24	0.65	6.26	0.73	6.43	0.55	7.33	0.80	6.94	0.68	5.70	0.62
May	6.15	0.69	6.23	0.65	5.84	0.41	7.29	0.62	7.29	0.60	5.23	0.53
June	5.33	0.60	6.60	0.74	5.28	0.27	6.27	0.33	6.54	0.43	4.12	0.33
July	3.92	0.29	6.32	0.63	4.54	0.20	4.92	0.29	5.33	0.43	4.26	0.39
August	4.34	0.33	5.30	0.65	4.70	0.32	5.17	0.32	5.05	0.43	4.21	0.35
September	3.94	0.53	5.45	0.71	4.80	0.33	4.78	0.36	5.60	0.57	3.87	0.41
October	5.08	0.85	4.47	0.76	4.44	0.30	5.66	0.55	5.36	0.78	4.02	0.54
November	3.61	0.82	3.71	0.78	4.74	0.36	3.47	0.39	4.52	0.87	3.94	0.55
December	2.76	0.66	3.01	0.74	5.22	0.53	4.38	0.57	3.84	0.78	4.15	0.57

Table 13.4 Calculated G_c values for each city (using Hottle's Model) in $kW/m^2/day$

Months	Palampur	Amritsar	Kodaikanal	Nandi Hill	Delhi	Shillong
January	3.6401	3.3702	6.6410	6.0672	3.7228	4.6388
February	4.6831	4.3351	7.2561	6.7572	4.6764	5.6139
March	5.9138	5.4802	7.7983	7.4148	5.7955	6.7156
April	6.9983	6.4898	8.0260	7.7831	6.7738	7.6232
May	7.5867	7.0321	7.9153	7.7856	7.2861	8.0549
June	7.7606	7.1892	7.7717	7.6932	7.4271	8.1534
July	7.6540	7.0924	7.8044	7.7038	7.3364	8.0782
August	7.2133	6.6886	7.9330	7.7408	6.9578	7.7667
September	6.3071	5.8471	7.8448	7.5193	6.1472	7.0345
October	5.0445	4.6710	7.3853	6.9265	5.0007	5.9287
November	3.8828	3.5941	6.7671	6.2146	3.9439	4.8615
December	3.3433	3.0965	6.4305	5.8408	3.4459	4.3497

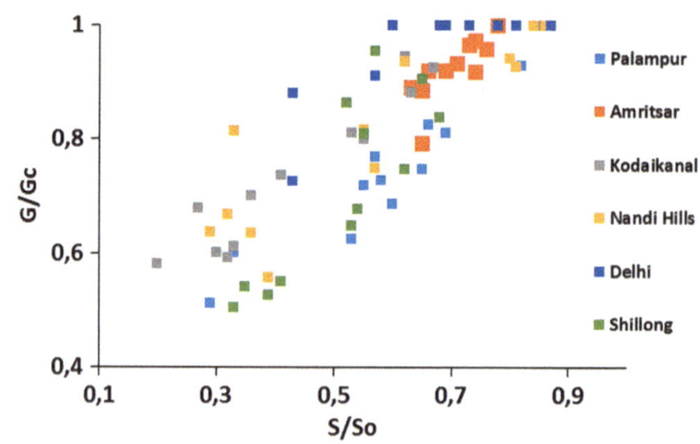

Fig. 13.1 G/G_c vs S/S_o for six cities in India

$$\frac{G}{G_c} = 0.7604\left(\frac{S}{S_o}\right) + 0.3750 \quad \text{(linear)} \tag{13.13}$$

$$\frac{G}{G_c} = 0.0163\left(\frac{S}{S_o}\right)^2 - 0.7424\left(\frac{S}{S_o}\right) + 0.3794 \quad \text{(quadratic)} \tag{13.14}$$

$$\frac{G}{G_c} = -3.2664\left(\frac{S}{S_o}\right)^3 + 5.5108\left(\frac{S}{S_o}\right)^2 - 2.1434\left(\frac{S}{S_o}\right) + 0.8428 \quad \text{(cubic)} \tag{13.15}$$

$$\frac{G}{G_c} = 1.0742\left(\frac{S}{S_o}\right)^{0.4933} \quad \text{(power)} \tag{13.16}$$

$$\frac{G}{G_c} = 0.3761\ln\left(\frac{S}{S_o}\right) + 1.0401 \quad \text{(logarithmic)} \tag{13.17}$$

$$\frac{G}{G_c} = 0.4511\exp^{0.989\left(\frac{S}{S_o}\right)} \quad \text{(exponential)} \tag{13.18}$$

Various statistical methods are available in solar energy literature, which deal with the assessment and comparison of solar radiation estimation models (Stone 1993; Toğrul 1999; Toğrul and Onat 1999; Walpole and Mayers 1989; Şahin 2007). Presently, statistical tests, root mean square error (RMSE), mean bias error (MBE) and t-statistics are utilized to evaluate the accuracy of the correlations developed.

Methods of Comparison

Statistical tests, root mean square error (RMSE) and mean bias error (MBE) were used to evaluate the accuracy of the correlations developed. Also, t-statistics was applied to the developed models to illustrate the statistically significant results.

Root Mean Square Error (RMSE)

The root mean square error is defined as

$$RMSE = \left(\frac{1}{N}\sum_{i=1}^{N}\left(G_{i,pre} - G_{i,meas}\right)^2\right)^{\frac{1}{2}} \tag{13.19}$$

where, $G_{i,pre}$ is ith predicted value, $G_{i,meas}$ is the ith measured value, and N is the total number of observations. The RMSE is always positive, a zero value is ideal. This test provides information on short-terms performance of the correlation by arranging a term by term comparison of the actual deviation between the calculated value and the measured value. The smaller the value, the better the model's performance, however, a few large errors in the sum can produce a significant increase in RMSE.

Mean Bias Error (MBE)

The mean bias error is defined as

$$MBE = \frac{1}{N}\sum_{i=1}^{N}\left(G_{i,pre} - G_{i,meas}\right) \tag{13.20}$$

This test provides information on the long term performance. A low value is desired. Ideally a zero value of MBE should be obtained. A positive value gives the average amount of over-estimation in the calculated value and vice versa. A drawback of this test is that over estimation of an individual observation will cancel under estimation in a separate observation.

It is obvious that each test by itself may not be an adequate indicator of a model's performance. It is possible to have a large RMSE value and at the same time a small MBE (a large scatter about the line of perfect estimation). On the other hand, it is also possible to have a relatively small RMSE and a relatively large MBE (a consistently small over or under estimation).

However, these statistical indicators generally provide a reasonable procedure to compare models, they do not objectively indicate whether a model's estimates are statistically significant, i.e. not significantly different from their measured counterparts. Therefore, an additional statistical indicator, the t-statistic is used. This statistical indicator allows models to be compared and at the same time indicate whether or not a model's estimates are statistically significant at a particular confidence level. It was seen that the t-statistic used in addition to the RMSE and MBE gave more reliable and explanatory results (Walpole and Mayers 1989).

t-Statistics

$$t = \left[\frac{(N-1)MBE^2}{RMSE^2 - MBE^2}\right]^{\frac{1}{2}} \tag{13.21}$$

The smaller the value of t, the better is the model's performance. To determine whether a model's estimates are statistically significant, one simply has to determine a critical t-value obtainable from standard statistical tables, at a particular confidence level, i.e. $t_{\alpha/2}$ at an α-level of significance and $(N\text{-}1)$ degrees of freedom. For the model's estimates to be judged statistically significant at the $(1\text{-}\alpha)$ confidence level, the calculated t-value must be less than the critical t-value.

Results and Discussion

As observed from Table 13.5, good results were not seen in the short term (RMSE) but relatively good results were observed in the long term performance (MBE). Equation (13.17) has the best result among the equations developed.

When investigations were compared among all the equations the best MBE value was seen in Eq. (13.17). The best RMSE value was obtained with Eq. (13.15) which is a cubic equation.

Table 13.5 RMSE and MBE for the whole country (India)

Equation number	RMSE	MBE
13.13	0.5726	−0.0222
13.14	0.5723	−0.0227
13.15	***0.5662***	−0.0233
13.16	0.5869	−0.0496
13.17	0.6096	***−0.0192***
13.18	0.5711	−0.0511

Also, it is observed that considering country as a whole and each city individually, the performance of developed equations is different. Therefore, the MBE and RMSE values of the developed equations separately for each city were calculated. The results of this statistical comparison are given in Table 13.6.

At the first view, it is seen that the MBE values of Table 13.4 are higher than tabulated in Table 13.5 RMSE and MBE for the whole country (India). Each equation developed for the city was compared with the equations in its group and the results obtained were setup in order below (Table 13.7).

Although the models give overall good results for the whole country (Table 13.5) but the errors were higher for individual cities (Table 13.6). These tables did not include adequate information about on usability of the developed equation. In view of this, t-statistics is applied to the developed models to investigate the usability of each model. The critical t-values are shown in Table 13.8, t-values higher than the critical t-values show that the equation has no statistical significance.

Equations which have t-values lower than the critical t-value exhibited good and logical results. In case of Palampur, Eqs. (13.16) and (13.17) are significant. In case of Amritsar and Kodaikan almost all of the equations showed good and logical results. In case of Nandi Hills and Delhi Eqs. (13.13), (13.14) and (13.18) are significant. In case of Shillong, Eqs. (13.15) and (13.18) are significant.

Conclusions

It was observed that the clear sky solar radiation and sunshine hour can be used to estimate the global radiation in India. It was further observed that the cubic equation gave the lowest RMSE error, and the lowest MBE error was obtained by the logarithmic equation developed for the whole year (Table 13.5). It is also observed that the performance of the equations is different for all the stations taking together and for the cities individually. Cubic and logarithmic models gave the best results among all the developed equations for the country as a whole. While for individual cities cubic, logarithmic and exponentials yield better results.

Finally these results clearly indicate that reliance on the RMSE and MBE used separately can lead to a wrong decision in selecting the best model suited from the candidate models and that the use of the RMSE and MBE in isolation is not an adequate indicator of model performance. Therefore, the t-statistics should be used in conjunction with these two indicators to evaluate a model's performance in a better way.

Table 13.6 The RMSE and MBE values of the equation developed for each city

	Palampur		Amritsar		Kodaikanal		Nandi Hills		Delhi		Shillong	
Equation	RMSE	MBE	RMSE	MBE	RMSE	MBE	RMSE	MBE	RMSE	MBE	RMSE	MBE
13.13	0.5992	0.4945	0.2162	−0.0699	0.3796	−0.2552	0.5645	−0.1028	0.7373	−0.6424	0.7448	0.4428
13.14	0.5977	0.4938	0.2164	−0.0711	0.379	−0.2547	0.5643	−0.1016	0.7385	−0.6431	0.744	0.4408
13.15	0.6213	0.1173	**0.2137**	**−0.0067**	**0.3537**	**−0.2075**	**0.5634**	−0.1548	0.7543	−0.6709	**0.6931**	0.4031
13.16	0.5988	0.1115	0.2301	−0.1023	0.433	−0.2941	0.5757	−0.1673	0.7527	−0.666	0.7549	0.4594
13.17	0.6495	0.1194	0.2275	−0.0729	0.4984	−0.2906	0.5864	−0.1608	**0.7171**	**−0.6277**	0.806	0.5307
13.18	**0.5515**	**0.1091**	0.2273	−0.1011	0.3998	−0.2826	0.5823	**−0.0897**	0.7765	−0.6686	0.7064	**0.3729**

Table 13.7 Equations with significant results for selected stations

Station	Equation number
Palampur	13.18
Amritsar	13.15
Kodaikanal	13.15
Nandi Hills	13.15 and 13.18
Delhi	13.17
Shillong	13.15 and 13.18

Table 13.8 Critical *t*-values and the results of *t*-statistics analyses for each city

Equation	Palampur	Amritsar	Kodaikanal	Nandi Hills	Delhi	Shillong
13.13	4.8466	**1.1332**	**3.012**	**0.6143**	**5.8881**	2.4523
13.14	4.8633	**1.1538**	**3.0099**	**0.6071**	**5.875**	2.4392
13.15	4.4236	**0.104**	**2.4026**	0.9478	5.4541	**2.3712**
13.16	**4.2603**	1.6462	**3.0693**	1.0073	6.2982	2.5436
13.17	**4.1172**	**1.1219**	**2.3803**	0.9457	6.0041	2.9015
13.18	5.1024	1.6471	3.3142	**0.5171**	**5.6157**	**2.0614**
Critical *t*	**4.2658**	**1.2702**	**3.1036**	**0.7687**	**5.979**	**2.4175**

Nomenclature

A	Altitude (km)
a, *b*, *a'*, *b'*	Empirical constants
G_c	Clear sky global solar radiation (W/m^2)
G_{cb}	Clear sky beam radiation (W/m^2)
G_{cd}	Clear sky diffuse radiation (W/m^2)
G_{on}	Extraterrestrial radiation (W/m^2)
G_{sc}	Solar constant (=1367 (W/m^2))
H	Monthly mean daily global radiation on a horizontal surface (MJ/m^2)
H_c	Clear sky monthly mean daily global radiation on a horizontal surface (MJ/m^2)
H_o	Monthly mean daily extraterrestrial radiation (MJ/m^2)
n	Day of the year
N	Number of observations
S	Monthly average daily hours of bright sunshine
S_o	Monthly average of maximum possible daily hours of bright sunshine (i.e. day length of average day of the month)

Greek Symbols

α	Level of significance
θ_z	Zenith angle (°)
τ_b	Atmospheric transmittance for beam radiation

τ_d Atmospheric transmittance for diffuse radiation
ω The sunset hour angle (°)
ϕ Latitude (°)
δ Declination angle (°)

References

Angstrom, A. (1924). Solar and terrestrial radiation. Report to the international commission for solar research on actinometric investigations of solar and atmospheric radiation. *Quarterly Journal of the Royal Meteorological Society, 50*(210), 121–126.

Chandel, S. S., Aggarwal, R. K., & Pandey, A. N. (2005). New correlation to estimate global solar radiation on horizontal surfaces using sunshine hour and temperature data for indian sites. *Journal of Solar Energy Engineering, 127*, 417–420.

Duffie, J. A., & Beckman, W. A. (1991). *Solar engineering of thermal processes*. New York: Wiley.

Ezekwe, G. I., & Ezeilo, C. C. O. (1981). Measured solar radiation in a Nigerian environment compared with predicted data. *Solar Energy, 26*(2), 181–186.

Hottel, H. C. (1976). A simple model for estimating the transmittance of direct solar radiation through clear atmosphere. *Solar Energy, 18*, 129–134.

Khogali, A. (1983). Solar radiation over Sudan-Comparison measured and predicted data. *Solar Energy, 31*(1), 45–53.

Khogali, A., Ramadan, M. R. I., Ali, Z. E. H., & Fattah, Y. A. (1983). Global and diffuse solar irradiation in Yemen (Y.A.R.). *Solar Energy, 31*(1), 55–62.

Leung, C. T. (1980). The fluctuation of solar irradiation in Hong Kong. *Solar Energy, 25*(6), 485–494.

Liu, B. Y. H., & Jordan, R. C. (1960). The interrelationship and characteristics distribution of direct, diffuse and total solar radiation. *Solar Energy, 4*(3), 1–19.

Munroe, M. M. (1980). Estimation of totals of irradiance on a horizontal surface from UK average meteorological data. *Solar Energy, 24*, 235–238.

Page, J. K. (1961). The estimate of monthly mean values of daily total short wave radiation on vertical and inclined surfaces from sunshine records for latitudes 40°N–40°S. In *Proceedings of U.N. conference on new sources of energy*, Rome.

Şahin, A. D. (2007). A new formulation for solar radiation and sunshine duration estimation. *International Journal of Energy Research, 31*(2), 109–118.

Stone, R. J. (1993). Improved statistical procedure for the evaluation of solar radiation estimation models. *Solar Energy, 51*, 289–291.

Toğrul, I. T. (1999). Estimation of clear sky radiation in Elazig. *Chimica Acta Turcici, 27*(1), 25–30.

Toğrul, I. T., & Onat, E. (1999). A study for estimating solar radiation in Elazığ using geographical and meteorological data. *Energy Conversion and Management, 40*, 1577–1584.

Walpole, R. E., & Mayers, R. H. (1989). *Probability and statistics for engineers and scientists* (4th ed.). New York: Macmillan.

Chapter 14
Effect of Gap Between Absorber Plate and Condenser Cover on the Performance of a Solar Still

Basharat Jamil and Naiem Akhtar

Introduction

Availability of fresh water for domestic and industrial use is an important issue facing the human world today. Lack of potable sources of water in developing countries is even severe. Rise is pollution levels of the local surface water resources have worsened the conditions leading to fatal water borne diseases and consequently an increase in human mortality rate.

Although several methods to purify water are available in industry, but most of them involve high energy requirements and are technologically complicated, thus, require sophisticated personnel for operation. Therefore, it is required that simple yet effective methods of water purification/desalination must be developed to reduce dependence on conventional energy resources.

Solar energy is a capable prospect for utilization in solar-energy conversion systems. This can be exploited to heat and evaporate brackish or polluted water which can later be condensed to form purified water suitable for drinking. Distillation is one of the important methods of making clean and potable water from brackish or polluted water using the abundant energy supply from the sun. Formation of clouds and rains in the atmosphere of earth is an example of desalination cycle using solar energy to produce huge amounts of pure water from the sea.

B. Jamil (✉) • N. Akhtar
Heat Transfer & Solar Energy Laboratory, Department of Mechanical Engineering, Zakir Husain College of Engineering and Technology, Aligarh Muslim University, Aligarh, 202002 Uttar Pradesh, India
e-mail: basharatjamil@zhcet.ac.in; akhtar_n@rediffmail.com

P. Grammelis (ed.), *Energy, Transportation and Global Warming*, Green Energy and Technology, DOI 10.1007/978-3-319-30127-3_14

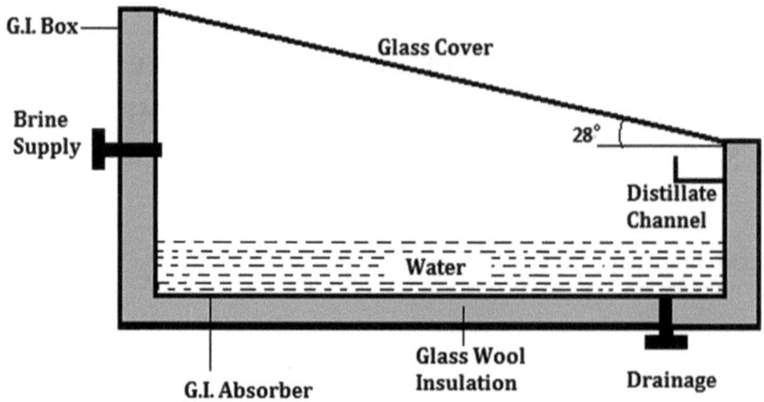

Fig. 14.1 Schematic diagram of conventional single slope solar still

Conventional Solar Still

A solar still is a promising device for the desalination of brackish water into potable form. The schematic diagram of conventional single slope solar still is shown in Fig. 14.1. A solar still imitates the process of desalination through which the earth's atmosphere forms rain. Water is heated through incoming solar radiation in a closed chamber with an inclined transparent glass cover to form vapours. The vapours travel through the chamber height to reach the inclined condensing cover and lose heat to form liquid water. The condensed liquid water trickles on the condensing cover and falls in the collection trough. The condensed liquid produced is usually highly purified and free of microorganisms and other harmful bacteria. This is because only heated water evaporates leaving behind the salts and other dissolved impurities. Also, temperatures within the solar still reaches sufficiently high values to kill the microorganisms and bacteria present in polluted water.

Although the construction, operation, and maintenance of a solar still is quite simple but the main limitation of a simple solar still is its low output (yield) and efficiency of the order of 2 lt/m^2/day and 20–30 % respectively (Tiwari and Tiwari 2006). A large number of research activities around the world address the performance improvement and evaluation of solar stills.

Heat Transfer in a Solar Still

Heat transfer in a conventional solar still can be categorized in two parts: internal heat transfer and external heat transfer (Tiwari and Tiwari 2007). Internal heat transfer in a solar still consists of three modes viz; convection, radiation and evaporation heat transfers.

The incoming solar radiation heats the blackened absorber of the solar still. Heat is transferred from the heated absorber plate to the water in basin and from vapors to inner surface of glass cover by convection. Radiative heat exchange occurs between water interface and glass surface. Evaporative heat transfer occurs due to mass transfer of vapors from the basin to inner surface of glass cover. While the external heat transfer refers to heat exchange between the solar still and the surrounding atmosphere. This consists of convection and radiation heat transfers. Heat is transferred through convection between the glass cover and the ambient; and bottom and side insulations and ambient. Radiative heat exchange occurs between glass cover and sky. Heat transfer interactions of a solar still and the relationships to model these mathematically are presented below.

Factors Affecting Performance of Solar Still

Meteorological parameters (such as solar intensity, ambient temperature and wind velocity etc.), design features (such as slope of condensing cover, different cover materials, various absorber material and their thickness and aspect ratio of absorber, thickness of insulation etc.) and operational parameters (such as input water temperature, effect of water depth, orientation of still etc.) have been extensively studied and presented in literature for different designs under local operating conditions of environment for various locations (Refalo et al. 2014). However, meteorological parameters cannot be controlled. On the other hand design features and operational parameters can be optimized for better yield from the solar still. Some important studies related to factors that affect the performance of a solar still are discussed below.

Manokar et al. (2014) presented an extensive review of factors affecting the rate of evaporation and condensation on passive solar still. They reported that the rate of evaporation is dependent on basin construction materials, depth of water in basin, absorption rate of basin water, absorption rate of still basin, volumetric heat capacity of the basin, inlet temperature of water and temperature of top surface water. They also reported that the condensation rate is affected by glass cover temperature and wind speed. As the wind velocity increases the cover temperature decreases and hence an increase in the overall yield is observed (Dimri et al. 2008). In other word, convective heat transfer increases with increase in wind velocity.

A cover inclination equal to the latitude angle of the site help receive sun rays close to normal throughout the year was reported by Murugavel et al. (2008). Studies conducted on the effect of water depth in stills have shown that the highest outputs and efficiencies occur at lower depths (Nafey et al. 2000).

Performance Gap

It was observed from thorough literature review that although several solar still designs are analysed in literature but the effect of distance (or gap) between the absorber plate and the condensing cover has never been properly addressed. Only a single study concerned with 2-D simulation study performed by Rahbar and Esfahani (2013) reported that when the specific height of the solar still decreases, the distance between the glass-cover and water-surface also decreases. This enhances the convective heat transfer and, therefore, the condensation rate of water-vapor increases. This is an important parameter since the distance between the absorber plate and the condensing cover directly affect the amount of solar radiation incident on the absorber plate. Also, the formation of water vapours in the cavity is greatly affected. The distance vapours travels from the heated water surface in absorber basin to condensing cover determines the yield of the solar still. Hence, this distance (or gap) has to be experimentally analysed for optimum solar still design.

Experimental Setup and Procedure

Fabrication of Solar Still

In this work, a conventional single slope solar still is designed and analysed for desalination of brackish water. Solar still was fabricated using locally available materials. Solar still is made out of Galvanised Iron (GI) sheet of 22 gauge. The area of the absorber plate is 1 m^2, with an aspect ratio of 2:1. Glass wool insulation of 1 in. is provided around all the sides and bottom of the solar still to avoid heat losses. The basin was painted black to maximize the absorption of solar radiation in the solar still. Ordinary glass of 5 mm thickness forms the condensing cover with the inclination of 28° (approximately equal to latitude of Aligarh City, 27.88°). Aluminium frame is used to support the glass cover over the solar still cavity and silicon sealant is used to fix the glass to the aluminium frame. An aluminium trough is fixed along the lower edge of the condenser cover to collect the condensate trickling from the inclined glass cover. Finally, a flexible rubber tube transfers the condensate (distillate) to the collection bottle. Every possible effort is made to make the solar still leak proof. The experimental setup was aligned in a way so that the glass cover faces the exact south direction to receive maximum solar radiation. Figure 14.2 shows the clear view of the experimental setup (conventional solar still) and Fig. 14.3 gives the dimensional details of the absorber/water basin.

Fig. 14.2 Actual view of experimental setup (conventional solar still)

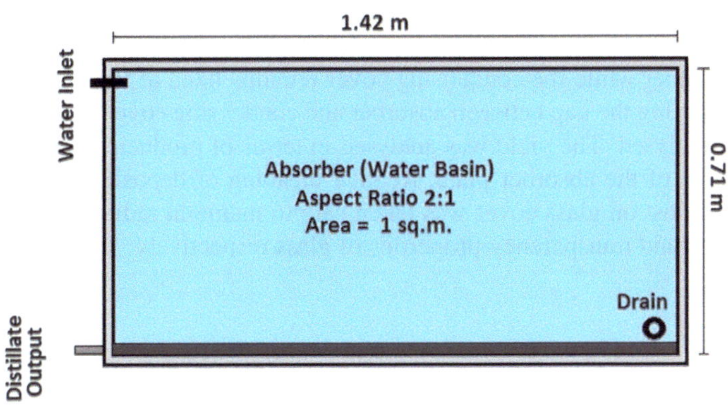

Fig. 14.3 Details of absorber basin

Instrumentation

- Temperatures were measured using T-type (copper-constantan) thermocouples. Thermocouples were used to measure temperatures of absorber, water, air-vapour mixture, glass inner and outer surface and ambient air. The output of the thermocouples was read on a digital multimeter.
- Global solar radiation was measured using Kipp and Zonen Pyranometer (CMP-11) and diffused solar radiation was measured using CMP-11 with CM-121B shading ring. Difference of the two provides the beam radiation. Global and diffused solar radiation values were stored in LOGBOX SD datalogger.

- Wind speed was observed using Precision AM804 CPM/CMM vane-type digital anemometer.
- A borosil measuring flask with a least count of 1 ml was used to measure the yield of distilled water from the solar still.

Observations

Experiments were conducted at Department of Mechanical Engineering, Aligarh Muslim University, Aligarh (27.88°N, 78.08°E), Uttar Pradesh, India. The experiments have been conducted during the month of March 2015 on clear days under similar conditions of environment.

Experiments were firstly conducted to investigate the depth of water. Solar radiation, wind velocity and atmospheric temperatures were measured. The temperatures, solar radiation intensity, wind speed and the distillate yield were observed from morning 8 a.m. to evening 5 p.m.

A separate absorber plate was also fabricated for the solar still so that it can be adjusted to effectively change the distance between the absorber and condenser cover. The separate absorber can be raised from the bottom to change the height of vapour chamber while the condensing cover remains fixed at top of the solar still. Effect of varying the gap between absorber and condensing cover of the solar still is presently analysed. The yield was analysed in terms of productivity of fresh water per unit area of the absorber plate. Regular cleaning of deposits in the basin and cleaning of dust on glass cover was taken care to maintain radiation absorption in the absorber and transparency properties of glass respectively.

Results and Discussion

The first set of experiments was performed to verify the depth of water for maximum water distillation in solar still. Figure 14.4 shows the measured temperature of basin or absorber, water, air-vapours mixture and average glass temperatures for different depths of water.

It is observed that the water temperature rises rapidly in case of lower depths and gradually at higher depths. Also the highest temperature attained shows a phase shift in case of higher depths.

Figure 14.5 shows the hourly and cumulative yield for different depths of water. The variation of metrological parameters viz., solar radiation (global, diffused and direct), ambient temperature and wind speed on the day of experiments are also depicted.

It can be observed that the meteorological parameters were similar on the days of experiment. The effect of increasing the depth of water in the basin has been

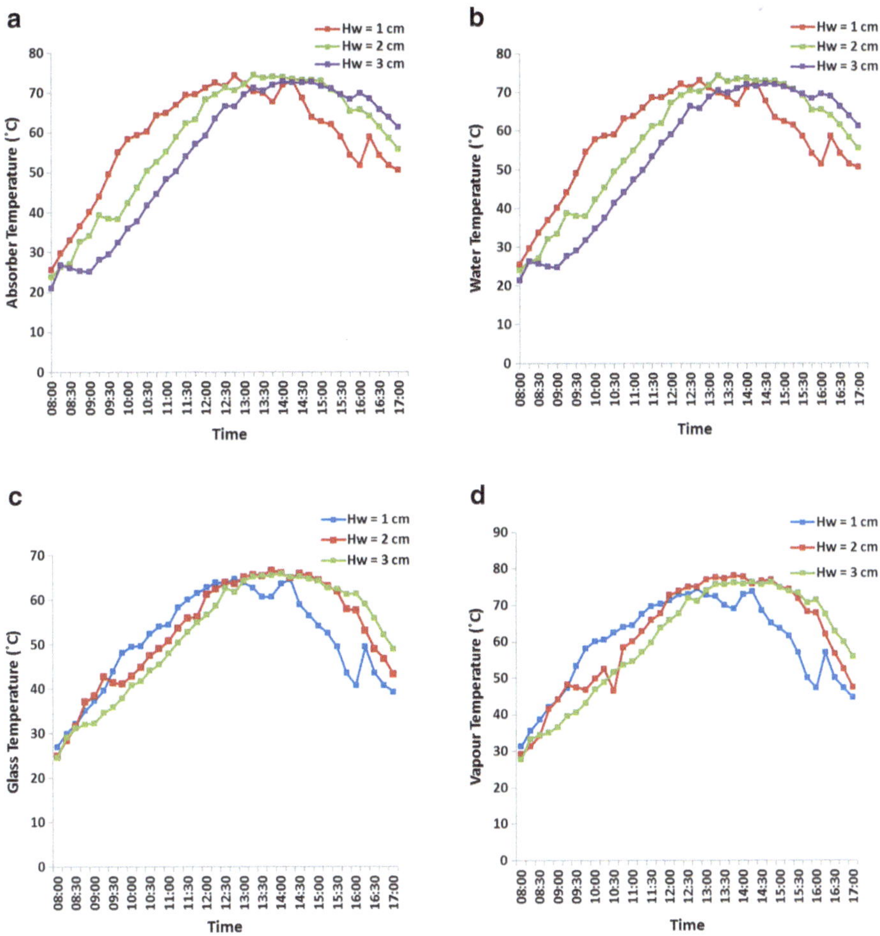

Fig. 14.4 Comparison of temperature variation during the day for different depths of water (**a**) absorber or basin (**b**) water (**c**) vapour and (**d**) glass

analysed. It can be observed that increasing the depth of water in the basin reduces the productivity (or yield) of the solar still.

Figure 14.6a shows the comparison of cumulative distillate yield for three different depths of water ($H_w = 1$ cm, 2 cm and 3 cm). The observed cumulative yield during the day is 1260 ml, 1045 ml and 635 ml respectively. In other words, twice the depth of water reduces the yield by 17.06 % and thrice the depth of water reduces the yield by 49.60 %. This can be attributed to the higher amount of energy required for heating and further evaporating a larger mass of water in the basin. In other words, total specific energy requirement of the basin water increases. Thus, it is verified that lower level of water should be kept in basin for better yield.

The distillation process continue even after the experiments were stopped (at 5 p.m.) due to available thermal energy within the solar still. The distilled water collected between evening 5 p.m. to 7 a.m. the next morning (known as "overnight productivity") was also observed. It was found that increasing the depth of water in the basin increases the overnight productivity (Fig. 14.6b). But the amount of overnight distillate is small in comparison to daytime distillate. Therefore, the total yield is still greater at lower depth of water. The observed values of overnight distilled water were 30 ml, 55 ml, 96 ml for (H_w = 1 cm, 2 cm and 3 cm) respectively.

The second set of experiments was performed to investigate the effect of distance between absorber plate and condenser cover on the performance of the solar still. For this purpose a separate absorber was fabricated in the form of water tray of approximately the same dimensions as the original absorber of the solar still. This secondary absorber or water tray was then raised from the bottom with the help of 1 in. thick polystyrene sheets as shown in Fig. 14.7. This effectively changes the

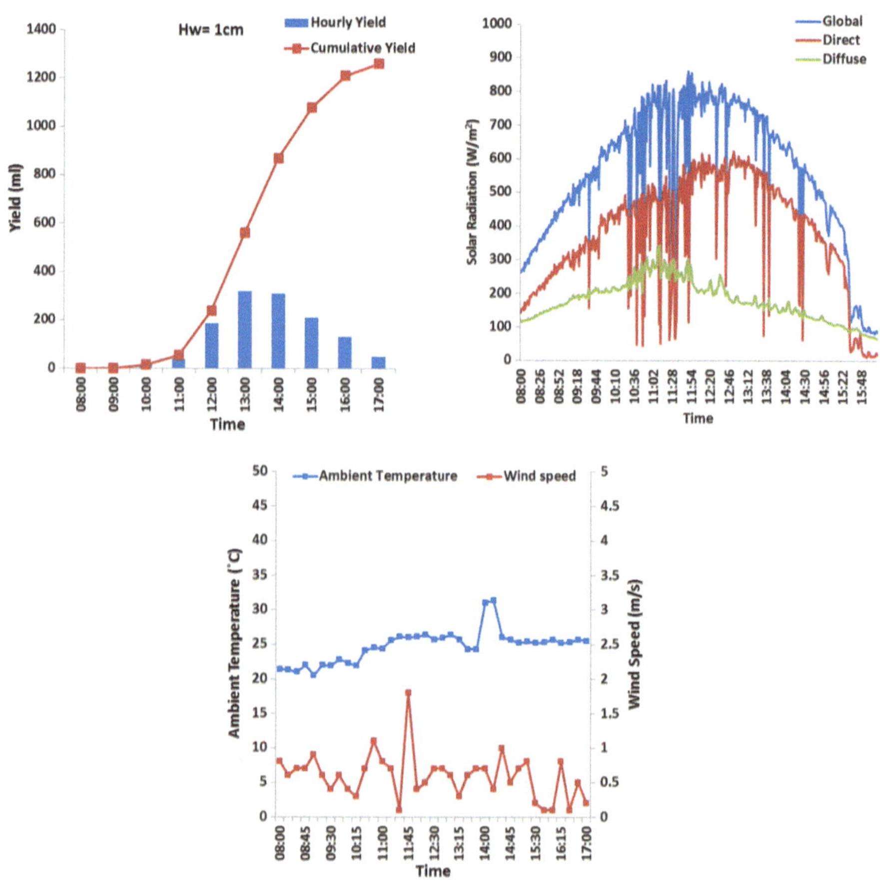

Fig. 14.5 Yield (hourly and cumulative) of distilled water for different depths of water in the basin along with observed solar radiation, ambient temperature and wind speed

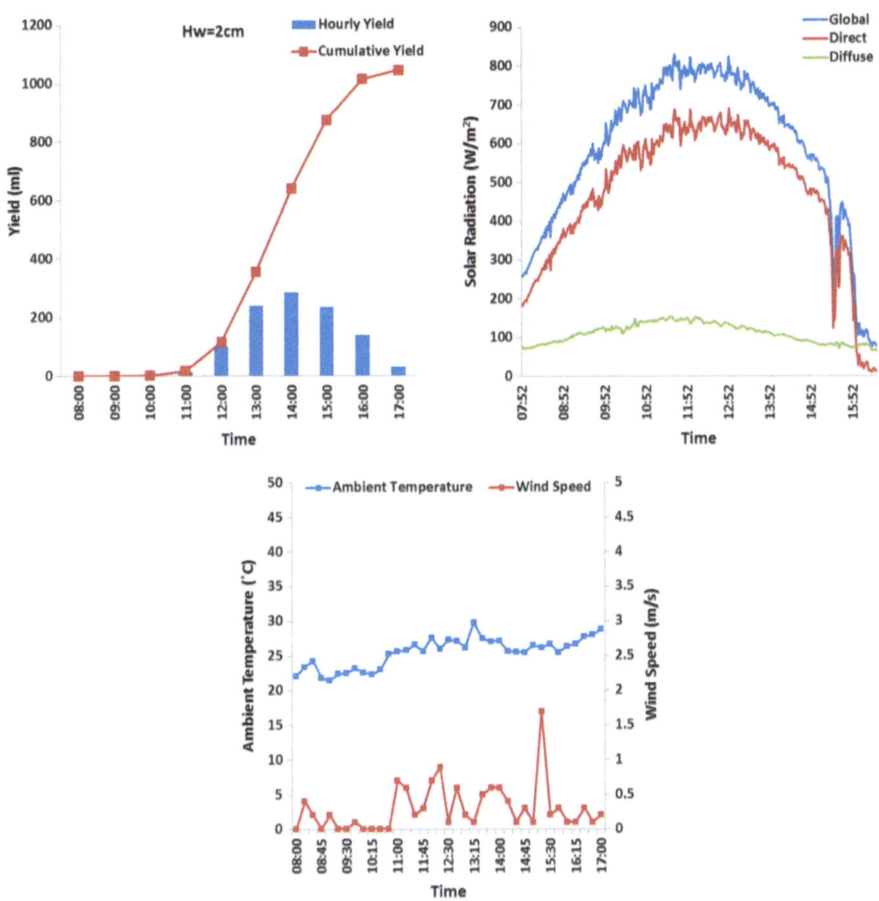

Fig. 14.5 (continued)

average distance between the absorber and condensing glass cover (H), thereby reducing the height of vapour chamber.

Experiments were performed with different gaps as shown in figure above with each time raising the absorber (or water tray) from the base by 1 in. Thus, H1 (=*15 in.*) forms the first experiment with the largest distance between absorber and condenser cover and H4 (=*12 in.*) forms the last experiment of the set with smallest distance between the absorber and condenser. For each experiment 1 cm depth of water was maintained in the water tray throughout the experiment. Figure 14.8 shows the measured absorber, water, air-vapour mixture and glass temperatures for different gaps between the absorber and condenser.

It can be observed that at the largest gap (H1), the observed temperatures are the lowest while at the smallest gap (H4) the observed temperatures are the highest.

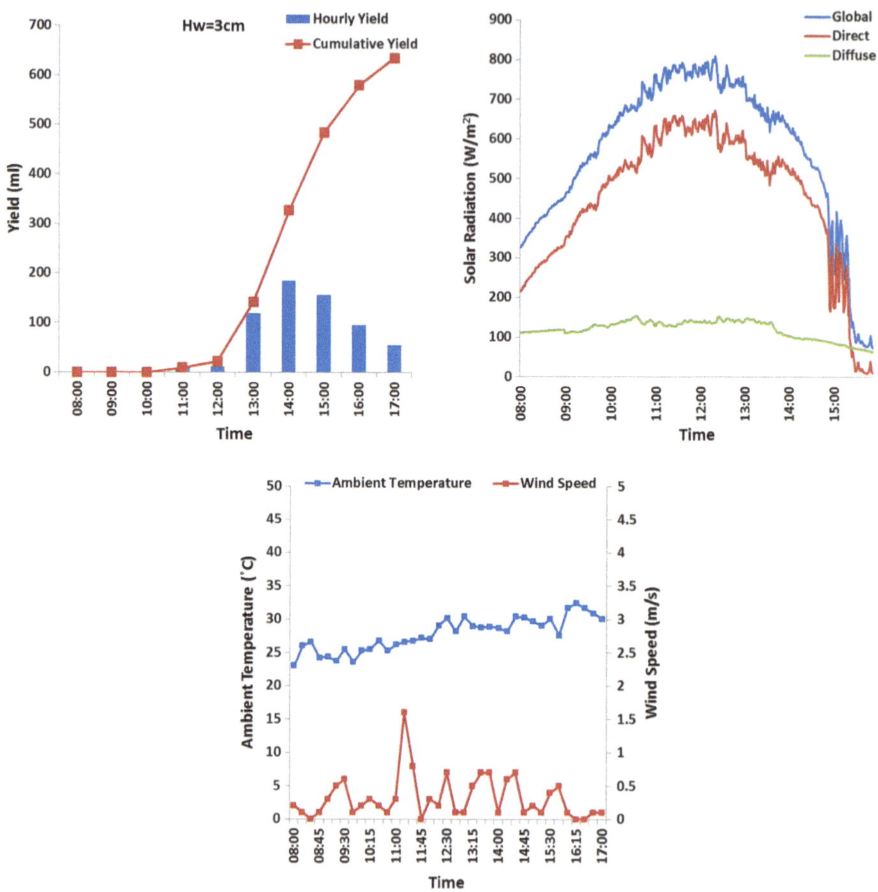

Fig. 14.5 (continued)

Also, all the temperature curves show a similar trend with peak temperatures between 12 p.m. to 1 p.m.

Figure 14.9a shows the observed hourly distillate yield for various gaps and Fig. 14.9b compares the cumulative productivity of the solar still with different gaps. It can clearly be observed that reducing the gap improves the distillate yield significantly. This can be attributed to higher rate of convective heat transfer. As the heated water surface and comparatively cooler glass surface comes closer the rate of distillation increases. Therefore an increase in productivity is observed on reducing the gap.

Observed yield at H1, H2, H3 and H4 are 1260 ml, 1964 ml, 2312 ml and 2695 ml respectively. Therefore, an increase of 55.87 % (for H2), 83.49 % (for H3) and 113.88 % (for H4) is observed in comparison to the base case (H1).

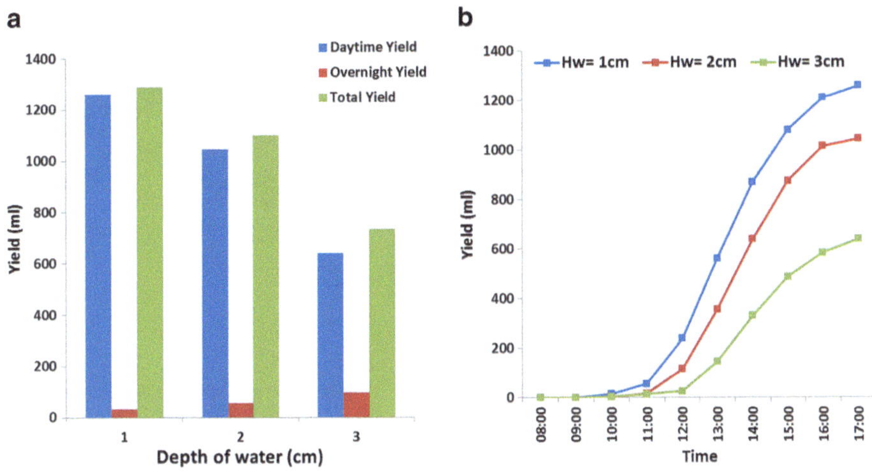

Fig. 14.6 Comparison of (**a**) hourly cumulative yields (**b**) overall daytime, overnight and total yields at different depths of water

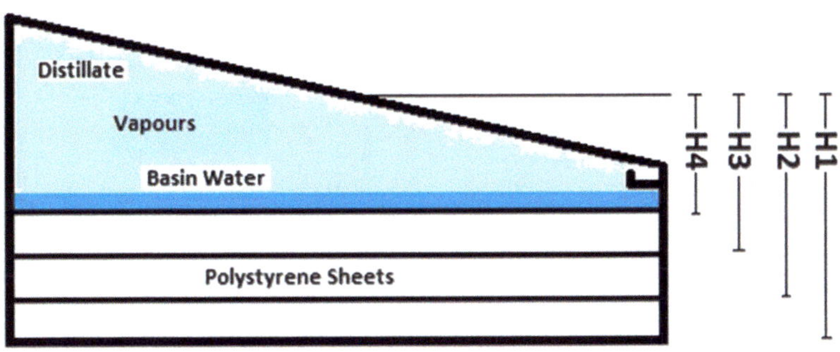

Fig. 14.7 Gap between absorber and condenser

Conclusions

A conventional single slope solar still was designed, fabricated and analysed for desalination of brackish water. Initially, the effect of water depth was investigated. It was found that increasing the depth of water in the basin of solar still decreases the daytime productivity or yield. Also it was observed that the overnight productivity of the solar still increases with increase in water depth. These results verified the investigations of previous researchers. Next, the effect of reducing the gap

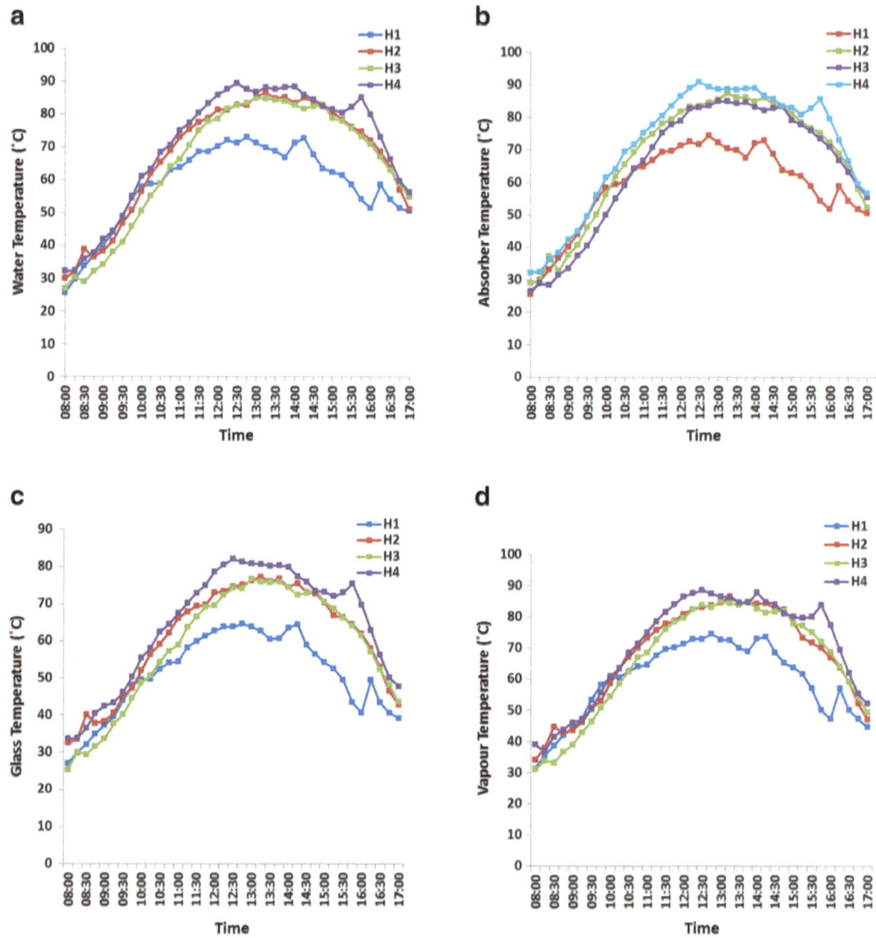

Fig. 14.8 Temperatures of (**a**) absorber or basin (**b**) water mass (**c**) air-vapour mixture (**d**) glass at different gaps between absorber and condenser

between absorber and condenser cover was experimentally evaluated. It was found that reducing the gap increases the daily productivity. Reducing the gap increases the convective heat transfer between the water surface and condenser cover. This distance or gap is important because it directly affects the overall yield of the solar still. It is believed that the results obtained under this study are valuable for future solar still designs and optimization of heat and mass transfer processes involved.

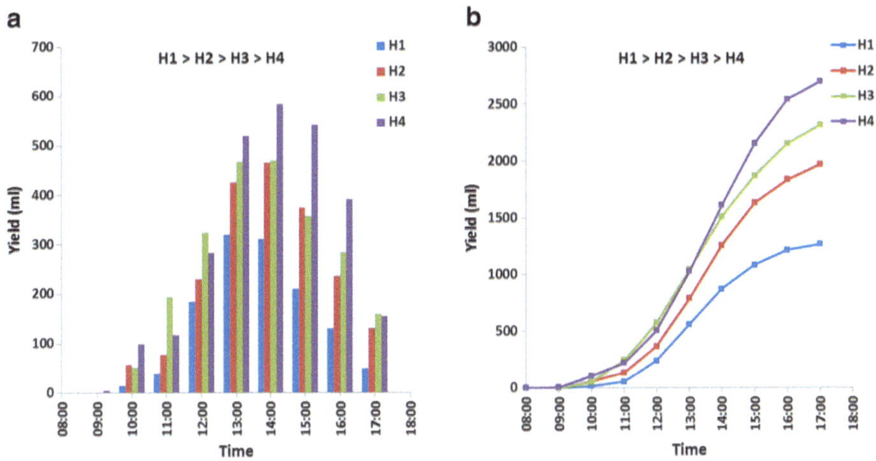

Fig. 14.9 (**a**) Hourly yields for different gaps (H) between absorber and condenser cover. (**b**) Corresponding cumulative productivity

References

Dimri, V., Sarkar, B., Singh, U., & Tiwari, G. (2008). Effect of condensing cover material on yield of an active solar still: An experimental validation. *Desalination, 227*, 178–189.

Manokar, A. M., Murugavel, K. K., & Esakkimuthu, G. (2014). Different parameters affecting the rate of evaporation and condensation on passive solar still-A review. *Renewable and Sustainable Energy Reviews, 38*, 309–322.

Murugavel, K., Chockalingam, K. K. S., & Srithar, K. (2008). Progresses in improving the effectiveness of the single basin passive solar still. *Desalination, 220*, 677–686.

Nafey, A. S., Abdelkader, M., Abdelmotalip, A., & Mabrouk, A. (2000). Parameters affecting solar still productivity. *Energy Conversion and Management, 41*, 1797–1809.

Rahbar, N., & Esfahani, V. (2013). Productivity estimation of a single-slope solar still: Theoretical and numerical analysis. *Energy, 49*, 289–297.

Refalo, P., Ghirlando, R., & Abela, S. (2014). The effect of climatic parameters on the heat transfer mechanisms in a solar distillation still. *Heat Transfer Engineering, 35*(16-17), 1473–1481.

Tiwari, A., & Tiwari, G. (2006). Effect of water depths on heat and mass transfer in a passive solar still: In summer climatic condition. *Desalination, 195*, 78–94.

Tiwari, A., & Tiwari, G. (2007). Thermal modeling based on solar fraction and experimental study of the annual and seasonal performance of a single slope passive solar still: The effect of water depths. *Desalination, 207*, 184–204.

Chapter 15
Solar Assisted Organic Rankine Cycle for Power Generation: A Comparative Analysis for Natural Working Fluids

Önder Kizilkan, Sandro Nižetić, and Gamze Yildirim

Introduction

The steam Rankine cycle is one of the most important ways to transform on large scale thermal energy into power. Because of its good properties, water is the most suitable working fluid for high temperature applications and large centralized systems. Seeking for small and medium scale power plants, the problems encountered with water can be partially mitigated by selecting an appropriate fluid. Organic compounds characterized by higher molecular mass and lower critical temperature than water have been proposed in so called Organic Rankine Cycles (ORC) (Tchanche et al. 2011). ORC, as a method of low grade heat utilization can rise the energy utilization by conversion of heat into electric energy (He et al. 2012). There are lots of low temperature practices in which the ORC can be used such as solar thermal, biomass geothermal oceanic, waste heat from power plants, combined heat and power, waste heat from industrial processes, etc. (Peris et al. 2015).

The development of solar assisted power plants is becoming more crucial because of effects of fossil fuels. For this reason, there is a need to improve existing technologies integrated with solar energy. These systems offer better advantages

Ö. Kizilkan • G. Yildirim (✉)
Department of Energy Systems Engineering, Faculty of Technology,
Süleyman Demirel University, 32200 Isparta, Turkey
e-mail: onderkizilkan@sdu.edu.tr; gazmeyildirim@sdu.edu.tr

S. Nižetić
Department of Thermodynamics and Heat Engines, Faculty of Electrical and Mechanical Engineering and Naval Architecture,
University of Split, R. Boskovica 32, 21000 Split, Croatia
e-mail: snizetic@fesb.hr

P. Grammelis (ed.), *Energy, Transportation and Global Warming*, Green Energy and Technology, DOI 10.1007/978-3-319-30127-3_15

when compared traditional energy sources. Parabolic trough solar collector (PTSC) technology is considered the most established solar thermal technology for power generation. Therefore, this technology has been chosen for this study (Al-Sulaiman 2013).

Organic Rankine Cycle (ORC) as a promising energy transformation technology has been the focus of many investigators (Kerme and Orfi 2014). Over the last decades, significant researches have been carried out in the field of ORC systems. Dai et al. (2009) studied ORC for low grade waste heat recovery using various fluids. They examined thermodynamic parameters of the ORC for every working fluid. They concluded that the cycle with R236EA had the highest exergy performance. He et al. (2012) recommend a theoretical model to check out optimal evaporation temperature of subcritical ORC. In a different study, organic Rankine cycle and supercritical Rankine cycle for the transformation of low-grade heat into electrical power investigated by Chen et al. (2010). They analysed 35 different fluids for two cycles and analysed the effects of the selected fluid characteristics on the cycle performance. Wang et al. (2013) conducted a regenerative ORC to make use of the solar energy over a low temperature series. They used flat-plate solar collectors to collect the solar radiation for their low costs. Lee et al. (2012) investigated the impact of changing the proportion of the cooling water to the condenser on the performance of the ORC. They used R245fa in their analysis. Shengjun et al. (2011) studied performance comparison of the fluids in subcritical and transcritical ORC power cycle in 80–100 °C binary geothermal power system. They conducted the analyses with a program in MATLAB for thermal efficiency, exergy efficiency, recovery efficiency, heat exchanger area and levelized energy cost. Thermodynamic optimization of ORCs for power generation combined heating and power from different average heat source profiles researched by Maraver et al. (2014). Vélez et al. (2011) investigated the use of a low temperature heat source for power generation a carbondioxide transcritical power cycle theoretically. They reported that the efficiency have been increased by the use of an additional internal heat exchanger. Franchini et al. (2013) carried out simulations to estimate the performance of a solar Rankine Cycle and an integrated solar combined cycle. They combined the system with two different solar field configurations based on parabolic trough and power tower systems.

In this study, solar energy driven ORC is investigated thermodynamically for different natural refrigerants. For collecting the solar energy, PTSCs are used because of their good advantages. For the analyses of ORC, eight natural working fluids and a HFC type fluid are considered including, R170 (ethane), R1270 (propylene), R600 (butane), R600a (isobutene), R717 (ammonia), R744 (carbondioxide), R161 (fluoroethane), R218 (octofluoropropane), and a HFC type fluid, R134a. Analyses are made to examine the system performances energetically and exergeticaly. An attempt is also made to evaluate the exergy destruction rates in order to determine how to improve the process. Furthermore, a comprehensive performance assessment of the integrated system is conducted through parametric analysis to investigate the effects of changing operating conditions on the system efficiencies.

Natural Refrigerants and System Description

Natural refrigerants provide alternatives to a number of HCFC, CFC and HFC type refrigerants in addition to their zero ozone depletion potential (ODP) and low or no global warming potential (GWP) (Bolaji and Huan 2013). They are naturally occurring substances, such as hydrocarbons, ammonia, carbon dioxide, water and air. These substances can be used as cooling agents (heat transfer medium) in refrigerators and air conditioners (Refrigerants naturally 2015).

The general properties of the previously mentioned working fluids are given in Table 15.1 (ASHRAE 2004; Restrepo et al. 2008; Calm and Hourahan 2011). In the table, critical properties of the fluids are obtained using EES software (F-Chart 2015). As seen from the table, the selected natural refrigerants have zero ODP and relatively very small GWP values excluding R218 and R134a. Additionally, their atmospheric life times are relatively short. In spite of this, it must be noted that some natural refrigerants such as R170, R1270, R600, R600a and R161 are in A3 safety group which means they are highly flammable. Besides, R744, R218, R134a are belonged to A1 group that means they are non-flammable. The last fluid, R717, is in the lower flammability group with higher toxicity.

The ORC consists of four compounds: a turbine, an evaporator, a condenser and a pump. Required heat energy for the evaporator of the ORC is supplied by means of PTSCs. Therminol-VP1 is for the PTSC system as the heat transfer fluid (HTF), for its good heat transfer properties and good temperature control (Therminol 2014). Because of its good properties, it is used in many high temperature applications driven by PTSC such as power plants (Kumar and Reddy 2009; Vogel et al. 2014; Cheng et al. 2012; Al-Sulaiman 2014). The system operates as follows: The liquid organic working fluid from condenser is compressed by the pump and fed back to the evaporator, where it is heated by the useful heat delivered by the solar PTSCs, and becomes superheated vapor. The superheated vapor then enters to the turbine and expands to a low pressure. Subsequently, the turbine exhaust is intensified to liquid in the condenser by extracting heat to the environment and the cycle completes when the fluid is compressed by the pump. The schematic representation of the system and the T–s diagrams of the case study are given in Figs. 15.1 and 15.2 respectively.

In Fig. 15.2, there are two T–s diagrams, where the first one is for subcritical and the second one is transcritical cycle. In subcritical cycle, the system properties are under the critical values and in transcritical cycle, the heat absorption process takes place above the critical values where heat rejection process takes place below the critical values. For the second case, the gas heater placed instead of evaporator.

Table 15.1 Properties of the selected working fluids

ASHRAE number	Molecular formula	Critical pressure (kPa)	Critical temperature (°C)	Safety group	ODP (relative to R11)	GWP (relative to CO_2)	Atmospheric life time (year)
R170	CH_3CH_3	4872.2	32.172	A3	0	20	0.21
R1270	$CH_3CH{=}CH_2$	4664.6	92.420	A3	0	3	0.001
R600	$CH_3CH_2CH_2CH_3$	3796.0	151.975	A3	0	20	0.018
R600a	$CH(CH_3)_3$	3640.0	134.667	A3	0	20	0.019
R717	NH_3	11,333	132.25	B2	0	0	0.25
R744	CO_2	7377.3	30.98	A1	0	1	120
R161	CH_3CH_2F	5.09	102.20	A3	0	12	0.18
R218	$CF_3CF_2CF_3$	2.6	71.95	A1	0	8830	2600
R134a	CH_2FCF_3	4059.0	101.030	A1	0.000015	1410	14

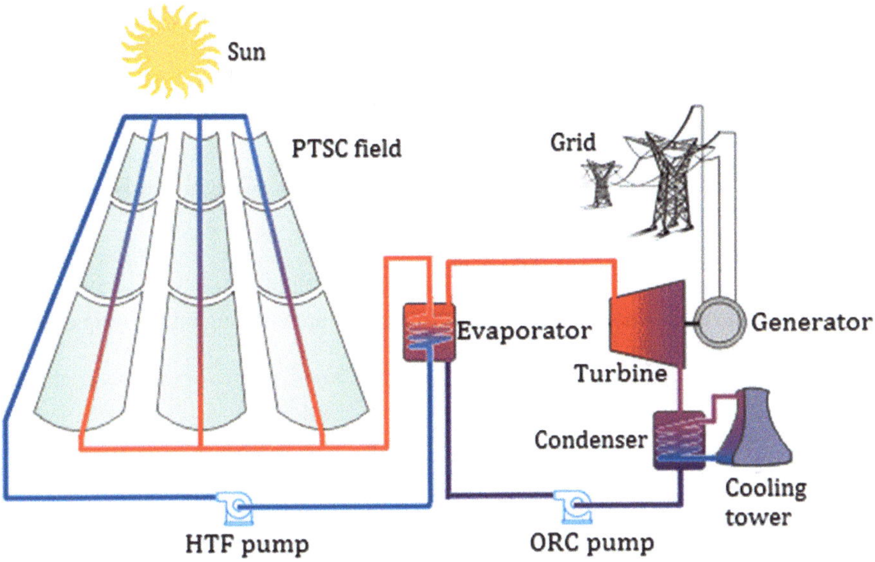

Fig. 15.1 Schematic representation of solar assisted ORC system (modified from Volker 2015)

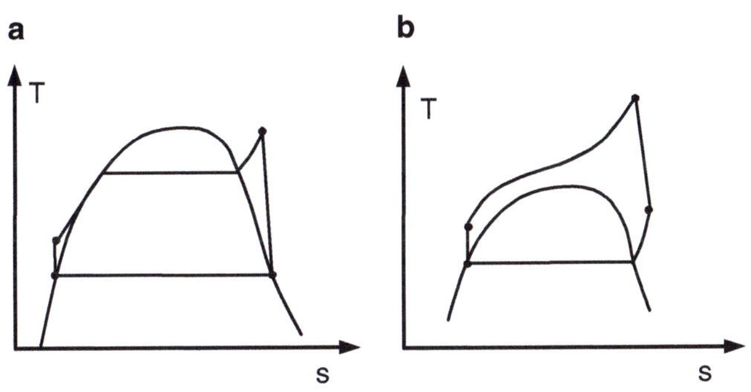

Fig. 15.2 T–s diagrams for ORC (**a**) subcritical (**b**) transcritical

Thermodynamic Analysis

The performance of the solar driven ORC is mathematically modelled using mass, energy and exergy balance equations. Some assumptions are made in order to find the work and heat interactions, the rate of exergy destructions, and the energy and exergy efficiencies. These are as follows:

- All the processes in the system are steady state and steady flow.
- The changes in potential and kinetic energies are neglected.
- The turbine and pumps are adiabatic.
- The heat transfer to/from ambient and pressure drops in the pipes are neglected.

- The natural fluid at pump inlet is saturated liquid.
- The dead state pressure and temperature are taken to be $P_0 = 101.325$ kPa and $T_0 = 25$ °C.
- The pinch point temperature of the evaporator and condenser/gas heater are taken as 10 °C (Al-Sulaiman 2014).

The general mass balance equation can be written as (Cengel and Boles 2006):

$$\sum \dot{m}_{in} = \sum \dot{m}_{out} \tag{15.1}$$

where m is the mass flow rate, subscripts in and out are entering and exiting streams to and from the system, respectively.

The general energy balance

$$\sum \dot{E}_{in} = \sum \dot{E}_{out} \tag{15.2}$$

For steady-flow processes the general energy balance can be written in more detail as:

$$\dot{Q} + \sum \dot{m}_{in} h_{in} = \dot{W} + \sum \dot{m}_{out} h_{out} \tag{15.3}$$

In above equations, E_{in} is the ratio of net energy transfer to the system, E_{out} is the ratio of net energy transfer from the system, Q is the ratio of net heat, W is the ratio of net work, and h is the specific enthalpy.

The rate of useful energy delivered by solar collector is (Tiwari 2003; Kalogirou 2009)

$$\dot{Q}_u = F_R[SA_a - A_r U_L (T_i - T_0)] \tag{15.4}$$

$$\dot{Q}_u = \dot{m} c_p (T_0 - T_i) \tag{15.5}$$

where F_R is the heat removal factor, S is the solar irradiation, A_a is the aperture area, A_r is the receiver area, U_L is the overall heat loss coefficient of solar collector, subscripts i, o and 0 are inlet, outlet and dead state, respectively.

The heat removal factor, F_R can be calculated from (Kalogirou 2009)

$$F_R = \frac{\dot{m} C_p}{A_r U_L}\left[1 - \exp\left(\frac{-A_r U_L F'}{\dot{m} C_p}\right)\right] \tag{15.6}$$

where F' is the collector efficiency factor and given by

$$F' = \frac{U_0}{U_L} \tag{15.7}$$

In Eq. (15.7), U_O is the overall heat transfer coefficient. Based on PTSC properties, U_O and U_L can be determined from (Kalogirou 2009):

$$U_0 = \left[\frac{1}{U_L} + \frac{D_o}{h_f D_i} + \left(\frac{D_o \ln\left(\frac{D_o}{D_i}\right)}{2k}\right)\right]^{-1} \tag{15.8}$$

$$U_L = \left[\frac{A_r}{(h_{c,c-a} + h_{r,c-a})A_g} + \frac{1}{h_{r,r-c}}\right]^{-1} \tag{15.9}$$

In above equations, $h_{c,c\text{-}a}$ is the convection heat loss coefficient between ambient and the cover, $h_{r,c\text{-}a}$ is the radiation heat transfer coefficient for the glass cover to the ambient, $h_{r,r\text{-}c}$ is the radiation heat transfer coefficient between the receiver tube and the glass cover and h_f is the heat transfer coefficient of fluid inside the tube. Additionally, D is the tube diameter and A_g is the glass cover area. The definitions of heat transfer coefficients mentioned above can be found in reference Kalogirou (2009) in more detail.

The general exergy balance equation can be defined as (Dincer and Rosen 2007):

$$\sum \dot{E}x_{in} = \sum \dot{E}x_{out} + \sum \dot{E}x_{dest} \tag{15.10}$$

The exergy balance equation can be expressed more explicitly as:

$$\dot{E}x_Q - \dot{E}x_W - \sum \dot{m}_{in} e_{in} - \sum \dot{m}_{out} e_{out} + T_0 \dot{S}_{gen} \tag{15.11}$$

where, Ex_Q and Ex_W terms are the exergies of heat and work, e is the specific exergy, T_O is the dead state temperature and S_{gen} is the rate of entropy generation. The exergy terms in Eq. (15.11) are described below (Dincer and Rosen 2007):

$$\dot{E}x_{dest} = T_0 \dot{S}_{gen} \tag{15.12}$$

$$\dot{E}x_Q = \dot{Q}\left(\frac{T - T_0}{T}\right) \tag{15.13}$$

$$\dot{E}x_W = \dot{W} \tag{15.14}$$

The specific exergy can be expressed as (Cengel and Boles 2006; Bejan 1997):

$$e = (h - h_0) - T_0(s - s_0) \tag{15.15}$$

where, s is entropy, and subscript 0 stands for dead state properties.

The exergy of the solar radiation in terms of reference and sun's temperature is given by Petela (2005):

$$\dot{E}x_{solar} = SA_a\left(1 + \frac{1}{3}\left(\frac{T_0}{T_{sun}}\right)^4 - \frac{4}{3}\left(\frac{T_0}{T_{sun}}\right)\right) \tag{15.16}$$

where T_{sun} is the temperature of sun's surface and assumed as 5739 K (Tiwari 2003).

The energy and exergy efficiencies are described below (Dincer and Rosen 2007):

$$\eta = \frac{\dot{W}_{ORC}}{\dot{Q}_E} \tag{15.17}$$

$$\eta_{ex} = \frac{\dot{E}x_{out}}{\dot{E}x_{in}} = 1 - \frac{\dot{E}x_{dest}}{\dot{E}x_{in}} \tag{15.18}$$

Results and Discussion

For the thermodynamic analyses of the solar driven ORC, the system parameters for baseline conditions are given in Table 15.2. System characteristics of solar assisted ORC Table 15.2.

Using the general balance equations given in previous section, the analyses were made for the base line conditions first. According to data given in Table 15.2, the turbine power generations for different working fluids are given in Fig. 15.3.

Table 15.2 System characteristics of solar assisted ORC

PTSC system (Al-Sulaiman 2014; Kalogirou 2009)	Pipe receiver inner diameter	0.04 m
	Pipe receiver outer diameter	0.05 m
	Glass cover diameter	0.09 m
	Total length of PTSC	50 m
	Mass flow rate of HTF	0.32 kg/s
	Receiver emissivity	0.92
	Glass cover emissivity	0.87
	Temperature of the sun	5739 K
	Absorbed solar radiation	500 W/m^2
	Wind velocity	5 m/s
ORC	Turbine isentropic efficiency	0.85
	Pump isentropic efficiency	0.90
	Evaporator temperature[a]	50 °C
	Condenser temperature	27 °C
	Turbine inlet temperature	150 °C

[a]Turbine inlet pressure of ORC is determined from saturation pressure corresponding to evaporator temperature excluding R170 and R744, since the critical properties of these fluids are below from the specified values

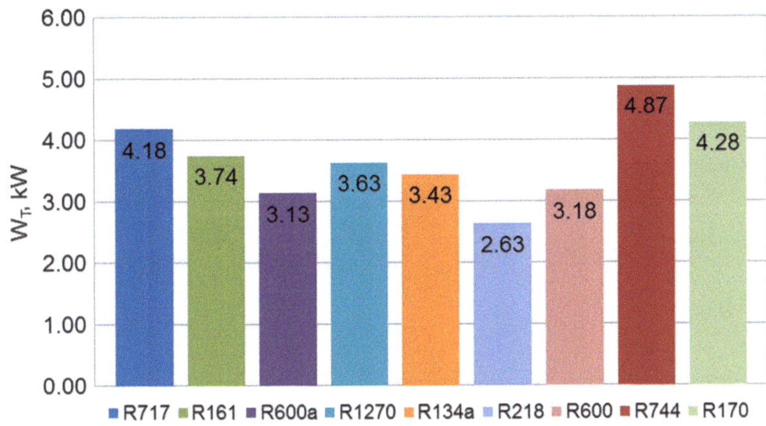

Fig. 15.3 Comparison of net power generation for different natural refrigerants

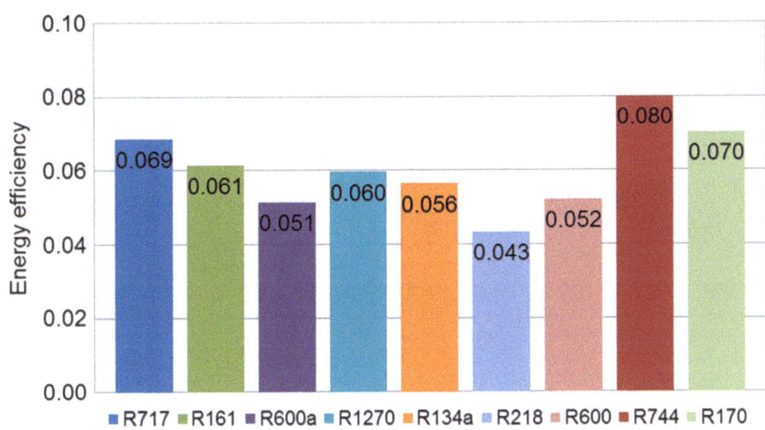

Fig. 15.4 Comparison of energy efficiencies of ORC for different natural refrigerants

As seen from the figure, ORC working with R744 has the highest power generation capacity and followed by R170, R717 and R161.

Figures 15.4 and 15.5 show the energy and exergy efficiencies of ORC for different working fluids according to the energy and exergy analysis. As seen from Fig. 15.4, the best energy performance is obtained using R744 with an energy efficiency of 8 %, whereas the lowest energy efficiency belongs to R218 with an efficiency rate of 4.3 %. The exergy efficiency results show that the highest efficiency is found to be 7.1 % with R744 and followed by R170 and R717.

A comparison of the calculated exergy destruction rates of the solar assisted ORC is given in Fig. 15.6. As seen from the figure, R218 has the highest exergy destruction rate and followed by R600a, R600 and R134a. The lowest exergy destruction rate is found to be 75.33 kW using R717.

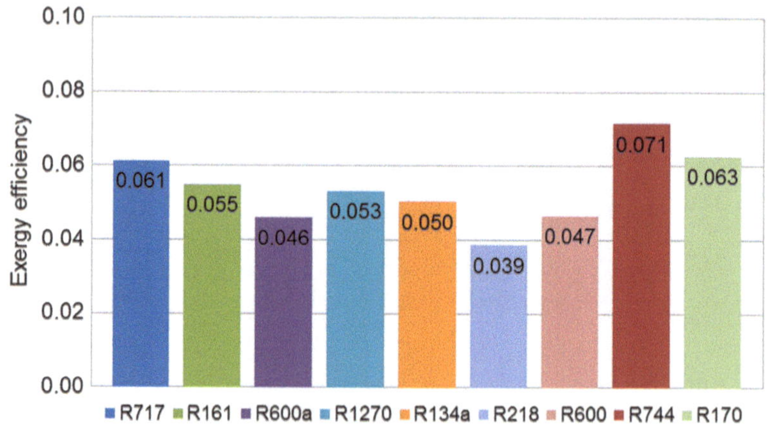

Fig. 15.5 Comparison of exergy efficiencies of ORC for different natural refrigerants

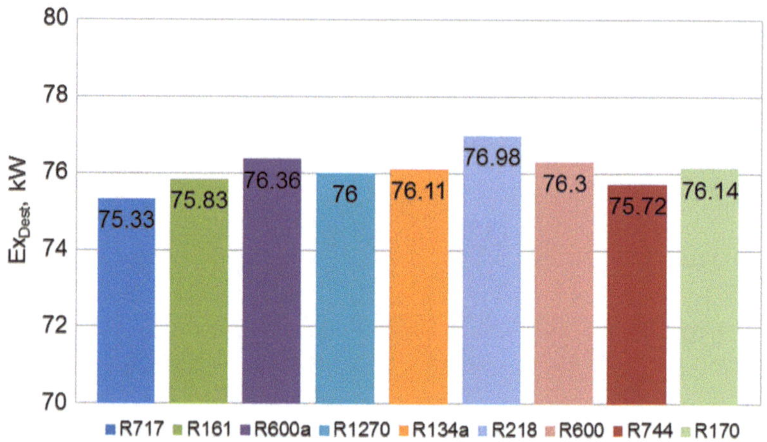

Fig. 15.6 Comparison of exergy destruction rates of ORC for different natural refrigerants

In Fig. 15.7, exergy destruction in the heat transfer process between the HTF and the working fluid in the evaporator is given. As mentioned earlier, Therminol-VP1 was used as HTF and its properties were kept constant during the calculations. Therefore, the differences between exergy destruction of evaporator is mainly depends on the working fluid properties. From the figure, it can be seen that the lowest exergy destruction rate in the evaporator is found to be 16.55 for R170, followed by R218 and R744 where evaporator of R717 has the highest exergy destruction with 19.41 kW.

Through parametric analyses, a comprehensive performance assessment of the integrated system was conducted to investigate the effects of varying operating conditions on the system efficiencies. For the parametric analyses, the variable parameters were selected to be solar radiation intensity, turbine inlet temperature,

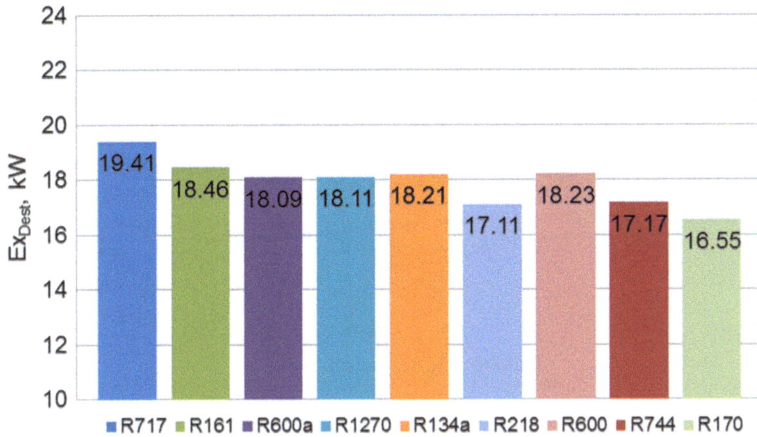

Fig. 15.7 Comparison of exergy destruction rates of evaporator for different natural refrigerants

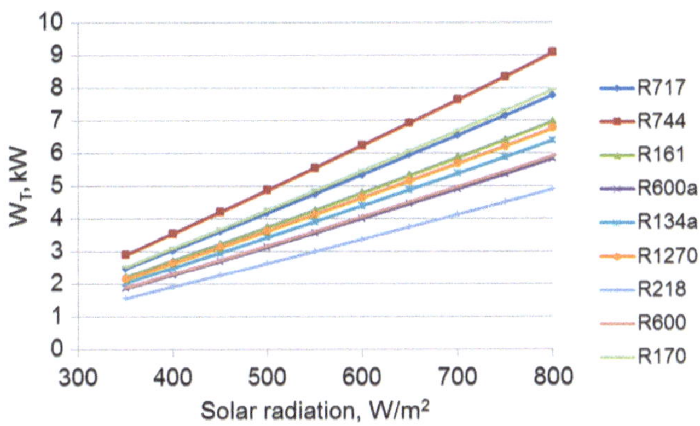

Fig. 15.8 Variation of net power generation with solar radiation

and normalized pressure at the turbine inlet and condenser temperature. Figure 15.8 shows the variation of solar radiation intensity with turbine power generation. As seen from the figure, turbine power generation increases with increase of solar radiation from 350 to 800 W/m^2.

Turbine inlet temperature also affects the cycle performance characteristics. To determine the variation of turbine inlet temperature on energy and exergy efficiencies, it was varied between 70–150 °C (Figs. 15.9 and 15.10). As seen from the figures, with the increase of turbine inlet temperature, energy and exergy efficiencies increase for R717. For the working fluids R744, R170, R134a, R600, R600a and R218, the efficiencies decrease whereas they are not affected so much for R161 and R1210 with the temperature.

Figure 15.11 shows the variation of turbine inlet temperature with the total exergy destruction rate of the system. As seen from the figure, the trend is contrary

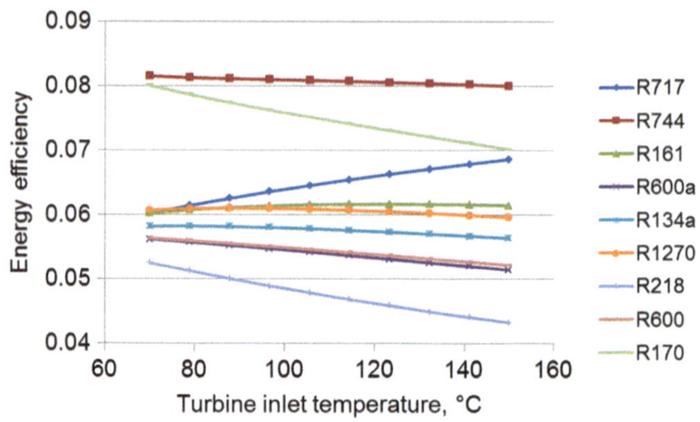

Fig. 15.9 Variation of turbine inlet temperature with energy efficiency

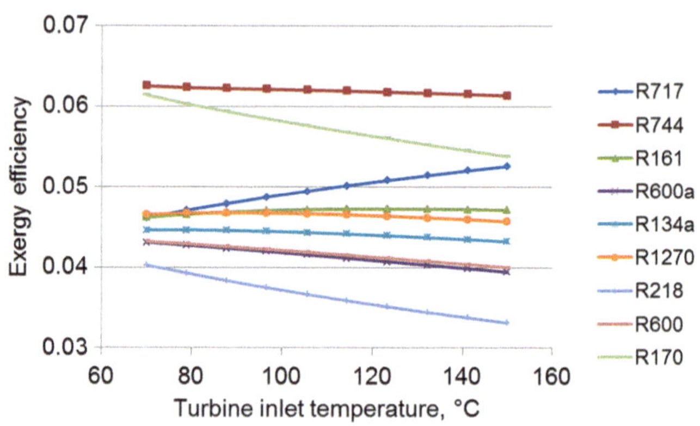

Fig. 15.10 Variation of turbine inlet temperature with exergy efficiency

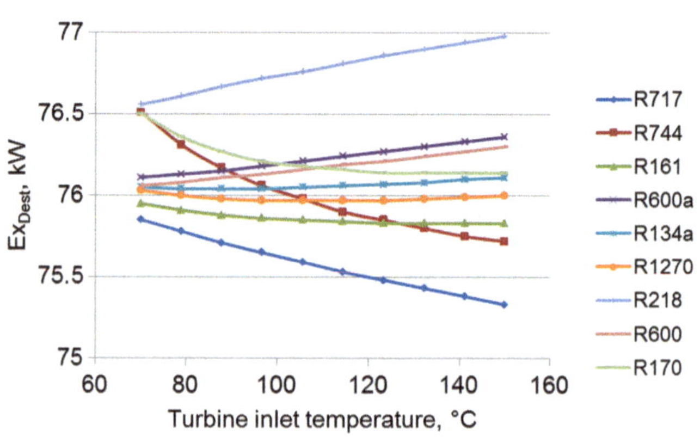

Fig. 15.11 Variation of turbine inlet temperature with exergy destruction rate

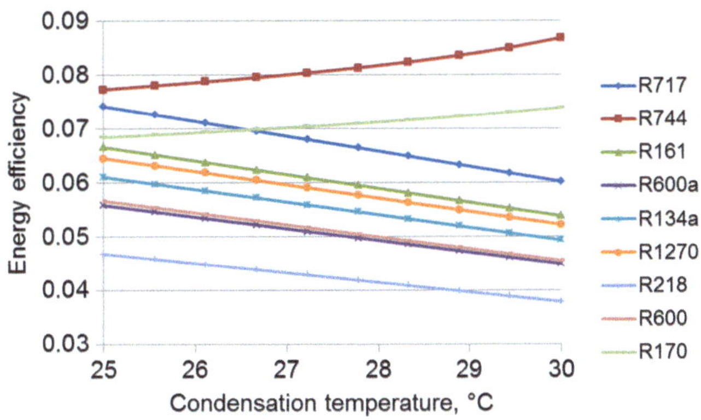

Fig. 15.12 Variation of condensation temperature with energy efficiency

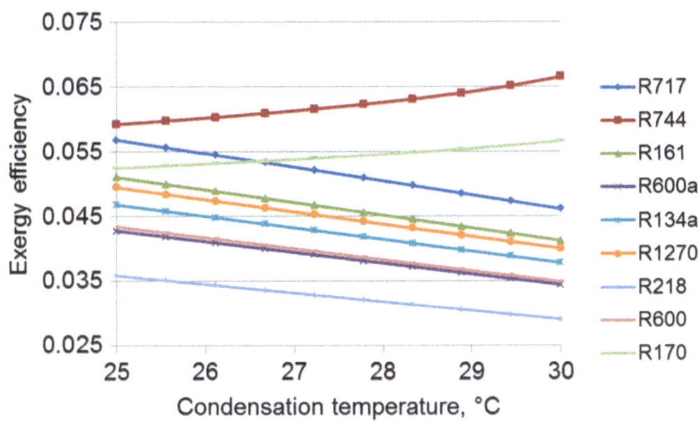

Fig. 15.13 Variation of condensation temperature with exergy efficiency

to energy and exergy efficiency variation. The total exergy destruction rate decreases with increasing the turbine inlet temperature for R1270, R717, R744, R170, it increases for R600a, R134a, R600, R218 and R161.

The condensation temperature is also one of the important parameter for assessment of the system performance. To determine its effect on system performance, it was changed from 25 °C to 30 °C and energy, exergy efficiencies and exergy destruction rates were calculated (Figs. 15.12, 15.13 and 15.14). The results are showed that, the energy and exergetic efficiencies decrease with the increase of the condenser temperature for all working fluids except for R744 and R170. This result is interesting since these two fluids are transcritical fluids. On contrary to this, the total exergy destruction rate increase with the temperature, while for R744 and R170 it decreases.

As declared in Table 15.2 at the beginning of this section, working pressures of solar assisted ORC were determined according to the corresponding saturation

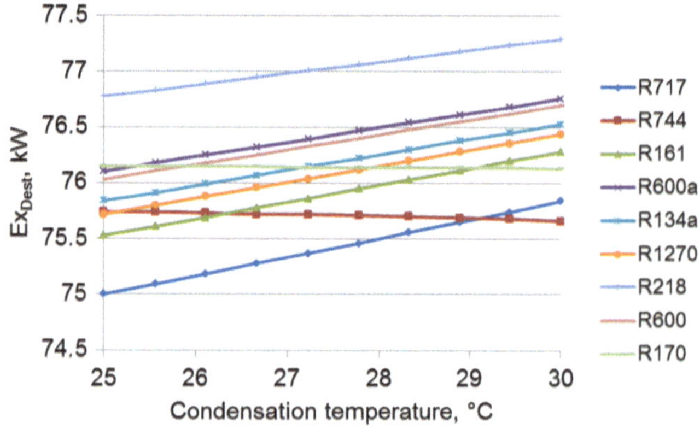

Fig. 15.14 Variation of condensation temperature with exergy destruction rate

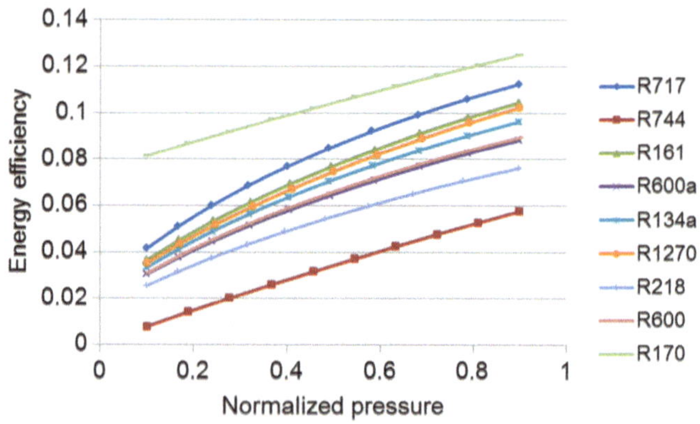

Fig. 15.15 Variation of normalized turbine inlet pressure with energy efficiency

temperatures of evaporator for subcritical working fluids. For transcritical fluids, the pressures were determined from the literature. Since the pressure values of each working fluid differs from each other, they were normalized and the figures were plotted using normalized pressure ranges for better understanding of the results.

Figures 15.15 and 15.16 show the variation of energy and exergy efficiencies with normalized turbine inlet pressure. It is very clear from the figures that with the increase of turbine inlet pressure, efficiencies increase, too.

The effect of normalized turbine inlet pressure on the exergy destruction is given in Fig. 15.17. From the figure, exergy destruction rates decrease with the turbine inlet pressure since the exergy destruction rate is inversely proportional to the energy and exergy efficiencies.

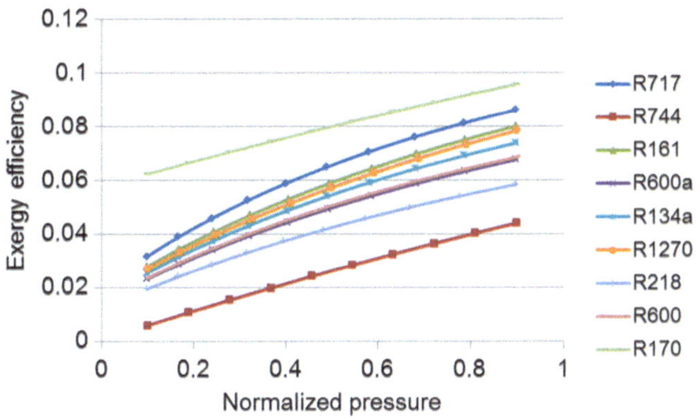

Fig. 15.16 Variation of normalized turbine inlet pressure with exergy efficiency

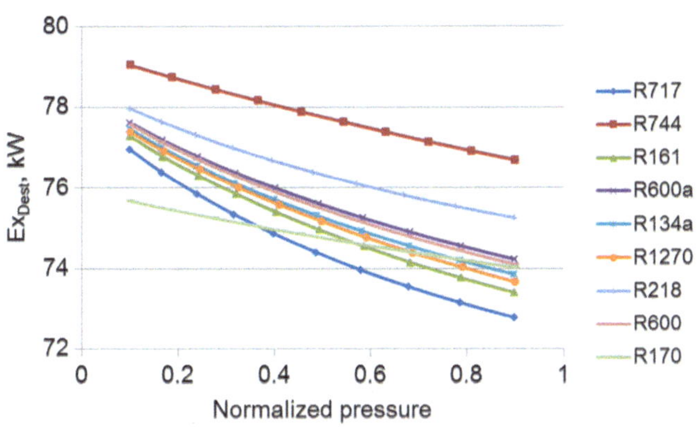

Fig. 15.17 Variation of normalized turbine inlet pressure with exergy destruction rate

Conclusions

A comparative analysis of solar assisted Organic Rankine Cycle for power generation using natural working fluids was investigated. The analyses were carried out for environmentally friendly subcritical and transcritical natural working fluids. The heat energy demand of the ORC was supplied using PTSCs working with Therminol-VP1 as the heat transfer fluid. From the results it was observed that the best cycle performance was obtained using R744 with a power generation rate of 4.87 kW and followed by R170, R717 and R161. Energy analysis results showed that the best cycle had an energy efficiency of 8 % using R744 as working fluid. The exergy efficiency of the same cycle was found to be 7.1 %. The highest total exergy destruction rate was found to be 76.98 kW for R218. Additionally, the effects of turbine inlet temperature, turbine inlet pressure and condensation temperature on system performance were analysed for different working fluids. This study points out that natural refrigerants are compatible with ORC power generation systems and more detailed experimental studies should be carried out for green energy production assisted by solar energy.

Nomenclature

A_a	Aperture area, m^2
A_g	Glass cover area, m^2
A_r	Receiver area, m^2
Cp	Specific heat, kJ/kg K
D	Tube diameter, m
e	Specific exergy, kJ/kg
$\dot{E}$	Energy, kW
$\dot{E}x$	Exergy, kW
F'	Collector efficiency factor
F_R	Heat removal factor
h	Specific enthalpy, kJ/kg
$h_{c,c\text{-}a}$	Convection heat loss coefficient between ambient and the cover, kW/m^2K
$h_{r,c\text{-}a}$	Radiation heat transfer coefficient for the glass cover to the ambient, kW/m^2K
h_f	Heat transfer coefficient of fluid inside the tube, kW/m^2K
$h_{r,r\text{-}c}$	Radiation heat transfer coefficient between the receiver tube and the glass cover, kW/m^2K
$\dot{m}$	Mass flow rate, kg/s
$\dot{Q}$	Heat, kW
s	Specific entropy, kJ/kg K
$\dot{S}$	Entropy, kW/K
S	Solar irradiation, kW/m^2
T	Temperature, °C or K
U_L	Overall heat loss coefficient, kW/m^2 K
$\dot{W}$	Work, kW
U_0	Overall heat transfer coefficient, kW/m^2 K

Greek Letters

η	Efficiency

Subscripts

dest	Destruction
gen	Generation

i	Inlet
o	Outlet
u	Useful
0	Dead state

References

Al-Sulaiman, F. A. (2013). Energy and sizing analyses of parabolic trough solar collector integrated with steam and binary vapor cycles. *Energy, 58*, 561–570.

Al-Sulaiman, F. A. (2014). Exergy analysis of parabolic trough solar collectors integrated with combined steam and organic Rankine cycles. *Energy Conversion and Management, 77*, 441–449.

ASHRAE. (2004). *Designation and safety classification of refrigerants*. Atlanta, GA: ANSI/ ASHRAE Standard 34-2001.

Bejan, A. (1997). *Advanced engineering thermodynamics*. New York: John Wiley and Sons.

Bolaji, B. O., & Huan, Z. (2013). Ozone depletion and global warming: Case for the use of natural refrigerant. *Renewable and Sustainable Energy Reviews, 18*, 49–54.

Calm, J. M., & Hourahan, G. C. (2011). Physical, safety, and environmental data summary for current and alternative refrigerants. In *Proceedings of the 23rd international congress of refrigeration*, Prague, Czech Republic, August 21–26, 2011.

Cengel, Y. A., & Boles, M. A. (2006). *Thermodynamics: An engineering approach* (5th ed.). New York, USA: McGraw-Hill.

Chen, H., Goswami, D. Y., & Stefanakos, E. K. (2010). A review of thermodynamic cycles and working fluids for the conversion of low-grade heat. *Renewable and Sustainable Energy Reviews, 14*, 3059–3067.

Cheng, Z. D., He, Y. L., Cui, F. Q., Xu, R. J., & Tao, Y. B. (2012). Numerical simulation of a parabolic trough solar collector with nonuniform solar flux conditions by coupling FVM and MCRT method. *Solar Energy, 86*, 1770–1784.

Dai, Y., Wang, J., & Gao, L. (2009). Parametric optimization and comparative study of organic Rankine cycle (ORC) for low grade waste heat recovery. *Energy Conversion and Management, 50*, 576–582.

Dincer, I., & Rosen, M. A. (2007). *Exergy: Energy, environment and sustainable development* (1st ed.). Amsterdam: Elsevier Science.

F-Chart, F-Chart Software. Retrieved February 1, 2015, from http://www.fchart.com/.

Franchini, G., Perdichizzi, A., Ravelli, S., & Barigozzi, G. (2013). A comparative study between parabolic trough and solar tower technologies in Solar Rankine Cycle and Integrated Solar Combined Cycle plants. *Solar Energy, 98*, 302–314.

He, C., Liu, C., Gao, H., Xie, H., Li, Y., Wu, S., et al. (2012). The optimal evaporation temperature and working fluids for subcritical organic Rankine cycle. *Energy, 38*, 136–143.

Kalogirou, S. A. (2009). *Solar energy engineering: processes and systems* (1st ed.). Oxford: Academic Press.

Kerme, E. D., & Orfi, J. (2014). Exergy- based thermodynamic analysis of solar driven organic Rankine cycle. *Journal of Thermal Engineering, 33*, 192–202.

Kumar, K. R., & Reddy, K. S. (2009). Thermal analysis of solar parabolic trough with porous disc receiver. *Applied Energy, 86*, 1804–1812.

Lee, Y. R., Kuo, C. R., & Wang, C. C. (2012). Transient response of a 50 kW organic Rankine cycle system. *Energy, 48*, 532–538.

Maraver, D., Royo, J., Lemort, V., & Quoilin, S. (2014). Systematic optimization of subcritical and transcritical organic Rankine cycles (ORCs) constrained by technical parameters in multiple applications. *Applied Energy, 117*, 11–29.

Peris, B., Navarro-Esbri, J., Moles, F., Collado, R., & Mota-Babiloni, A. (2015). Performance evaluation of an organic Rankine cycle for power applications from low grade heat sources. *Applied Thermal Engineering, 75*, 763–769.

Petela, R. (2005). Exergy analysis of the solar cylindrical-parabolic cooker. *Solar Energy, 79*, 221–233.

Refrigerants naturally, C/O Heat International. Retrieved February 2, 2015, from http://www.refrigerantsnaturally.com/.

Restrepo, G., Weckert, M., Brüggemann, R., Gerstmann, S., & Frank, H. (2008). Ranking of refrigerants. *Environmental Science & Technology, 42*, 2925–2930.

Shengjun, Z., Huaixin, W., & Tao, G. (2011). Performance comparison and optimization of subcritical ORC and transcritical power cycle system for low-temperature geothermal power generation. *Applied Energy, 88*, 2740–2754.

Tchanche, B. F., Lambrinos, G., Frangoudakis, A., & Papadakis, G. (2011). Low-grade heat conversion into power using organic Rankine cycles—A review of various applications. *Renewable and Sustainable Energy Reviews, 15*, 3963–3979.

Therminol, Heat transfer fluids by Eastman. Therminol VP-1. Retrieved December 10, 2014, from http://www.therminol.com/products/Therminol-VP1.

Tiwari, G. N. (2003). *Solar energy: Fundamentals, design, modelling and applications.* Pangbourne: Alpha Science International Ltd.. 525 p.

Vélez, F., Segovia, J., Chejne, F., Antolin, G., Quijano, A., & Martín, M. C. (2011). Low temperature heat source for power generation: Exhaustive analysis of a carbondioxide transcritical power cycle. *Energy, 36*, 5497–5507.

Vogel, T., Oeljeklaus, G., Görner, K., Dersch, J., & Polklas, T. (2014). Hybridization of parabolic trough power plants with natural gas. *Energy Procedia, 49*, 1238–1247.

Volker. Retrieved January 14, 2015, from http://www.volker-quaschning.de/articles/fundamentals2/index_e.php.

Wang, M., Wang, J., Zhao, Y., Zhao, P., & Dai, Y. (2013). Thermodynamic analysis and optimization of solar-driven regenerative ORC based on flat-plate solar collectors. *Applied Thermal Engineering, 50*, 816–825.

Chapter 16
Rainfall Trend Analysis in the Region of Curitiba Using Regional Climate Model Scenarios

Robinson Ploszai and Miriam Rita Moro Mine

Introduction

The climate change studies in hydrological time series have a great scientific and practical importance in water management. In general, the water resources systems are designed assuming that the hydrological data series (streamflow or precipitation) are statistically stationary (Alexandre et al. 2010).

The operational management of water resources is predicted assuming the hydrologic stationarity. According to Matalas (1997), the designers are uncertain about how the global warming can be converted in a hydrologic non-stationarity and how this non-stationarity could be incorporated into water resources management.

In the context of climatic changes in hydrological time series, it is not possible to assume whether these changes are related to anthropogenic causes (e.g., land uses changes or greenhouse gas emissions) or to natural climate variability, which is a consequence from multi annual or secular cycles in the Earth's climate (Tucci 2002).

Applying a probabilistic method in a large hydrological time series (e.g., 100 years or more with the minimum gaps or incorrect data), it is possible to identify changes in the time series, even if these changes are cyclical or related to trends. These changes do not distinguish the natural effects from anthropogenic effects.

To detect trends in time series, several studies have been performed worldwide (Coscarelli and Caloiero 2012; Li et al. 2011; Sayemuzzaman and Jha 2014; Oguntunde et al. 2011; Wang et al. 2008; Liang et al. 2011; Stipp et al. 2013; Fisch and Folhes 2006; Romano et al. 2011; Meschiatti et al. 2012; among others).

R. Ploszai (✉) • M.R.M. Mine
Federal University of Parana, PO Box 19011, 81531-990 Curitiba, PR, Brazil
e-mail: ploszai@ufpr.br; mrmine.dhs@ufpr.br

P. Grammelis (ed.), *Energy, Transportation and Global Warming*, Green Energy and Technology, DOI 10.1007/978-3-319-30127-3_16

According to these authors, the most efficient methods in trends detection are the Mann Kendall and Seasonal Mann Kendall tests and linear regression.

Statistical methods in both time series: the measured monthly precipitation and the rainfalls generated by Regional Climate Models (RCM) were applied. This paper verifies whether the RCM simulations can reproduce the historical period data and analyses precipitation trends through 2100 in the region of Curitiba to develop tools that will allow safety measure criteria to be taken for water resources management.

Methods

According to Kundzewicz and Robson (2000), before performing any statistical analysis, such as the application of a test or a method, it is necessary to do the Exploratory Data Analysis (EDA) on the preliminary data. These authors state that an analysis trend study is incomplete if the EDA is not used.

The EDA consists in using graphics and other heuristic methods to explore, understand and represent the information. Several methods are used to evaluate trends in time series, such as histograms and your relation with the normal probability distribution, temporal and autocorrelation graphs, normality tests, and serial independence tests and analysis of the smoothing curves (Kundzewicz and Robson 2000). Linear regression is a powerful tool to evaluate the existence of a secular trend in hydrological variables. The linear regression method in this case is represented by Eq. (16.1)

$$y_t = b_1 \times t + b_0 + \epsilon_t \tag{16.1}$$

where t is a time counter ($t = 1, 2, \ldots, n$); y_t is the hydrological variable (precipitation) on time t; b_1 is the angular coefficient of the regression's line; b_0 is the linear coefficient of the regression's line; and ϵ assumes the errors of the regression. After assuming some hypothesis and that the error ϵ is normally distributed, the parameters b_1 and b_0 are estimated by Ordinary Least Squares, solving the Gauss's normal equations (Johnson 1984).

Adopting the residuals of the linear regression as normally distributed, this hypothesis can be tested using the t-test, well described in Sharma et al. (2000). According to Alemaw and Chaoka (2002), another important tool to detect possible trends in hydrological time series is the Rescaled Adjusted Partial Sums (RAPS) method. According to these authors, the sum of a RAPS represented in a function Y (t) is defined by Eq. (16.2).

$$X_k = \sum_{t=1}^{k} \frac{Y(t) - Y}{S_Y} \tag{16.2}$$

where the value of $k = 1, 2, \ldots, n$. The variable X_k is the RAPS at limit k; Y(t) is the value of the variable at time t; Y is the mean of the sample values; S_Y is the sample's standard deviation; and k is the counter limit of the actual sum. The significance of RAPS analysis and EDA can be tested through the application of statistical tests, such as a t-test and a Fisher-Snedecor test for annual analysis and the Mann Whitney test and rho's Spearman coefficient test for seasonal analysis. These tests are fully described in Tozzi (2014).

Other tools were used (statistical indexes), such as the Nash-Sutcliffe (NS) coefficient, the index Root-Mean-square-Error (RMSE), and the determination coefficient (R^2). All methods are well described by Moriasi et al. (2007), Hyndman and Koehler (2006), and Naghettini and Pinto (2007), respectively. To evaluate the significance of the time series, the mean, median, variance, standard deviation, kurtosis excess, and asymmetry of the series under a confidence interval of 95 % were all estimated.

A common and practical issue in analysing hydrological time series consists in finding statistical techniques to verify the occurrence of some variation along the period (Buishand 1984). In order to do that, some statistical methods, such as the Mann Kendall test and Seasonal Mann Kendall test, were applied to analyse trends in hydrological time series.

The Mann Kendall and Seasonal Mann Kendall tests are largely used to verify homogeneity in hydrological time series. According to Back (2001), Barros et al. (2011), Buchir (2013), and Müller et al. (1998), the Mann Kendall test application is the most powerful in analysing climatic changes from climatological time series because it allows the beginning of the trend can be detected.

The Mann Kendall test consists in comparing each value in a chronological time series with the residuals, always ordered on the time. Hence, the number of times that the result terms are lower than the analysed terms is counted. The statistics of the MK test (S_{MK}) are given by Eq. (16.3).

$$S_{MK} = \sum_{j=1}^{n} \sum_{i=j+1}^{n} signal(R_j - R_i) \tag{16.3}$$

where n is the number of observations; R_i is the previous observation; R_j is the latest observation; i is the position of R_i at the previous time; j is the position of R_j at the latest time. The signal is defined through conditions, as represented by Eq. (16.4)

$$signal(R_j - R_i) = \begin{cases} 1, \ if\ R_j - R_i > 0 \\ 0, \ if\ R_j - R_i = 0 \\ -1, \ if\ R_j - R_i < 0 \end{cases} \tag{16.4}$$

It is possible to apply the Mann Kendall test to investigate trends on time series, only if the series has shown serial independence. Therefore, the series observations are tested to see if they are independent and identically distributed, in other words, to check the hypothesis:

H_0: The series observations are independent and identically distributed; H_1: The series observations have some trends along the period.

It is possible to show under a null hypothesis that S_MK is normally distributed with null mean, according to Eq. (16.5). The statistic is calculated by Eq. (16.6)

$$E(S_{MK}) = 0 \; and \; VAR(S_{MK}) = \frac{n \times (n-1) \times (2n+5)}{18} \tag{16.5}$$

$$MK = \begin{cases} \dfrac{S_{MK}-1}{\sqrt{VAR(S_{MK})}}, \; if \; S_{MK} > 0 \\ 0, \; if \; S_{MK} = 0 \\ \dfrac{S_{MK}+1}{\sqrt{VAR(S_{MK})}}, \; if \; S_{MK} < 0 \end{cases} \tag{16.6}$$

Even for values lower than 30, it is possible to use the statistics MK = Z to do the test. In a bilateral test, the null hypothesis, H_0, is rejected given a 95 % significance interval α, in the case of the quartile values $Z_{\alpha/2}$, pursuing a normal standard distribution, or $MK \leq Z_{\alpha/2}$.

The Seasonal Mann Kendall was proposed by Hirsch et al. (1982) and later was improved by Hirsch and Slack (1984). It has been created due to the difficulties in facing seasonality's climate events. This test is an adaptation of the Mann Kendall's test. It is an alternative to control the difficulties imposed by the high observational seasonal variabilities (Buchir 2013). According to Fisch and Folhes (2006), this test is largely recommended by the World Meteorological Organization (WMO) in detecting trends or abrupt changes in climate time series.

According to Lettenmaier et al. (1994) and Blain (2011), in the case of the monthly series, the MKS can be calculated, organizing the data into a matrix. The columns are filled with the series values, and the factor T_j is initially estimated for each month. Then, the statistics are calculated using the same assumptions: the null hypothesis H_0, is rejected for a significance interval α equal to 95 %. This method is fully described in Stipp et al. (2013).

Study Area and Data

The region of this work is the Curitiba city, Parana State, Brazil. The pluviometric station is located at Centro Politécnico, in the High Iguassu River basin. This station was chosen because of its location and because it is considered as a key station in south Brazil. The station has a good length of the observational time series, in other words, up to 100 years of extension (1889–2013).

Curitiba is the capital and the main city of the Parana state. It is considered the eighth largest city of Brazil with more than 1.75 million people living in the city (IBGE 2013). According to INMET (2013) and Fill et al. (1999), Curitiba has a subtropical wet climate, with average temperatures between 11 °C and 23 °C, in the winter and summer, respectively. These temperatures can oscillate according to each epoch of the year. The annual average precipitation is around 1400 mm in Curitiba; and the average of rainy days is 164 days per year. The relative moisture of the capital is around 80.7 %.

The pluviometric station Curitiba is situated in the Guabirotuba neighbourhood of Curitiba. The station's coordinates are 25°27′S and 49°14′E, and the average altitude is around 924 m. This station is shown in Fig. 16.1.

The rainfall records and the RCM's outputs are used in this work. The historical records of the Curitiba station are used to compare and verify if the RCM's outputs can represent the rainfall records in Curitiba, in the period from 1961 to 2008. The observational data were collected at the Hydrological Information System (Hidroweb), held by ANA (National Waters Agency, in Portuguese).

The observed data and the precipitation scenarios were split into two groups: (1) Historical period is the rainfall observed at the Curitiba hydrological station and the precipitation scenarios generated by three RCM (ETA, PROMES and RCA1); (2) Future period is only the precipitation generated by the RCM. Based on the project, CLARIS LPB was identified as the most representative models for the region, the ETA, PROMES and RCA1 (Mine et al. 2009).

For the preliminary analysis, the total of annual and monthly precipitation and the maximum and minimum values of the monthly rainfalls of RCM's outputs and observed records were used. After the most representative model for the Curitiba region was identified, the annual and seasonal total rainfall in each period were analysed.

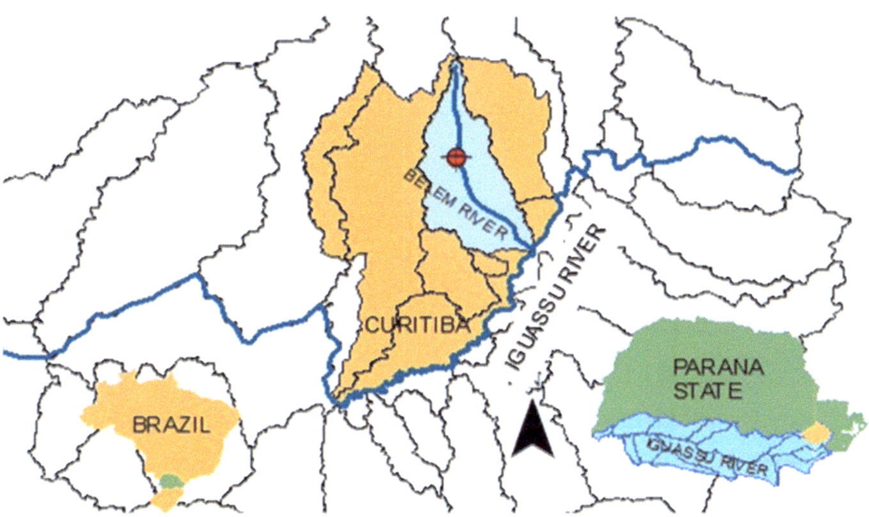

Fig. 16.1 Location of the main rivers, capital and the pluviometric station Curitiba

Preliminary Analysis

Both the historical time series, as the series generated by the RCM, were analysed in order to identify the most representative scenario for the region of Curitiba. First of all, the frequency bias correction method was applied as presented in Saurral (2010). After the bias correction was performed, the gaps were set up both in historical time series, as in RCM's outputs. There were found 75 months and 24 months of gaps in the historical time series and ETA series, respectively. The years with gaps were disregarded because the historical time series were 1491 months long.

Using the frequency polygons it was observed that the ETA model generates better-simulated series, after the bias correction. The series generated by the RCA1 model were not as satisfactory as the PROMES series, both shown in Fig. 16.2a. The rainfall permanence curves are presented in Fig. 16.2b. The periods to construct the curves were the same for the observed and each simulated series (historical period).

Figure 16.2b shows the permanence curves for all precipitations (simulated and observed). From the curves' analysis, it is possible to conclude that the PROMES model does not represent the historical observed series. The analysis of the scenario generated by the RCA1 model has shown a satisfactory adjustment (observed and simulated data curves), but the ETA model is the most representative.

To finalize the analysis of the precipitation scenarios obtained by RCM simulations, the descriptive statistics of monthly precipitation series were computed. Table **16.1** indicates the ETA scenario as the most representative for the region, when comparing to observed rainfall (OBS).

Regarding the scenario generated by the ETA model, both the mean and median between the two samples were so close. The other indicators such as standard deviation, kurtosis excess, and the asymmetry coefficient resulted in close values

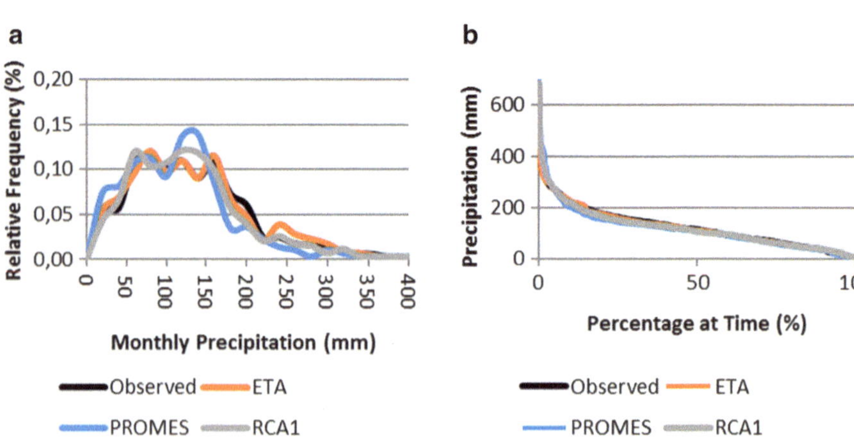

Fig. 16.2 (**a**, **b**) Frequency polygons and permanence curves for all scenarios and models

Table 16.1 Statistical indexes of monthly precipitation during the historical period from 1961 to 2008

Statistics	OBS	ETA	PROMES	RCA1
N (months of precipitation)	564	576	576	576
Mean	122.48	123.36	114.07	121.18
Median	114.35	114.09	103.36	107.56
Standard-deviation	76.29	76.62	87.28	82.52
Kurtosis excess	1.48	1.24	11.83	8.29
Asymmetry	1.00	0.91	2.51	2.02
Minimum	2.00	0.25	1.09	5.71
Maximum	473.80	498.71	833.36	685.61
Nash-Sutcliffe coefficient	–	−0.51	−0.77	−0.67
RMSE	–	93.78	103.78	98.41
R^2	–	0.06	0.05	0.06

between the observed rainfall (OBS) and the ETA outputs. The ETA simulations were not efficient in representing the minimum rainfall records but were efficient in representing the maximum records of each month (Table 16.1).

From all the previous analysis, the ETA model has shown the best results, resulting in its being adopted as the most representative for the region for the next analysis.

Annual Analysis

The precipitation scenario generated by the ETA model in this section was split into two precipitation scenarios: ETA01, representing the historical period, and ETA02, representing the future period. The observed rainfalls at the pluviometric station called Curitiba were represented by OBS.

The common period in the scenarios ETA01 and OBS is limited to December 2008. The initial point for both series is centred in January 1961. There are gaps in the OBS and ETA series. Therefore, the series were limited in 2008. The ETA's gaps correspond to a period of about 24 months: January 2009 to December 2010. The future period represented by ETA02 corresponds to the interval: January 2012 to November 2099. The last month obtained by ETA simulations is November 2099.

In order to precede an initial analysis of the annual precipitations, the descriptive statistics of the three precipitation scenarios: OBS, ETA01 and ETA02, were calculated, as follow the results in Table 16.2. Table 16.2 shows that the ETA01 and OBS series have almost the same mean values as well as almost the same medians and the standard-deviation indexes.

Comparing the statistics between these two scenarios, OBS and ETA01, it is clearly showed that both minimum and maximum values of ETA01 have higher

Table 16.2 Descriptive statistics of the annual precipitation

Statistics	OBS	ETA01	ETA02
N (months of precipitation)	43	48	88
Mean	1467.19	1480.30	1789.49
Median	1473.70	1434.95	1780.11
Standard-deviation	294.24	287.97	370.52
Kurtosis excess	−0.17	1.93	−0.38
Asymmetry	−0.10	1.05	0.22
Minimum	765.50	994.99	950.49
Maximum	2071.20	2497.80	2603.94

values than those represented by scenario OBS, according to Table 16.2. The ETA02 scenario showed an increase of three statistical indexes: mean, median, and standard deviation. It is possible to conclude that there is an increasing trend in precipitation in the region.

Below are presented the temporal graphs of mean and median, with a variation of about 10 years. Following that are the graphs using the linear regressions for all scenarios (Fig. 16.3). As shown in Fig. 16.3a, the historical series has shown an increasing trend throughout its length, according to the linear regression line, which is slightly positive from the series beginning (1889) to the end (2008). This trend can be visualized through the angular coefficient of the equation generated by the linear regression's line (positive signal).

The smoothing curves (mean and median variation around 10 years) have shown a slightly increasing trend in the end of the hydrological series. Figure 16.3b shows that, in the ETA01 scenario, the trend at the period decreases for the techniques mentioned in the legend. The angular coefficient of the linear regression line has shown a decreasing trend (negative signal).

Figure 16.3c depicts an increasing trend on scenario ETA02. Analysing the smoothing functions and linear regression technique shows an increasing trend in rainfall. The smoothing functions, such as mean and median variation, have shown a slightly positive trend.

The test of the linear regression coefficient for the precipitation series of the three scenarios (OBS, ETA01, and ETA02) was applied in order to identify the trend significance. As described in Sharma et al. (2000), this test was applied and compared to the critical values obtained from t-test tables for the pre-specified freedom degrees.

The null hypothesis to a given significance level at $\alpha = 5\%$ is $H_0 : b = 0$, and b is the angular coefficient of the linear regression line. The test results (Table 16.3) showed a trend both in ETA01 and in ETA02 scenarios. From the initial analysis of the angular coefficient and the test applied to this coefficient, it was shown that there is an increasing trend (positive) in ETA02, when in ETA01 there is a decreasing trend (negative).

The results of the coefficient test (Table 16.3) have not shown significant trends at observational series (OBS). As follows, the graphs using the RAPS method were

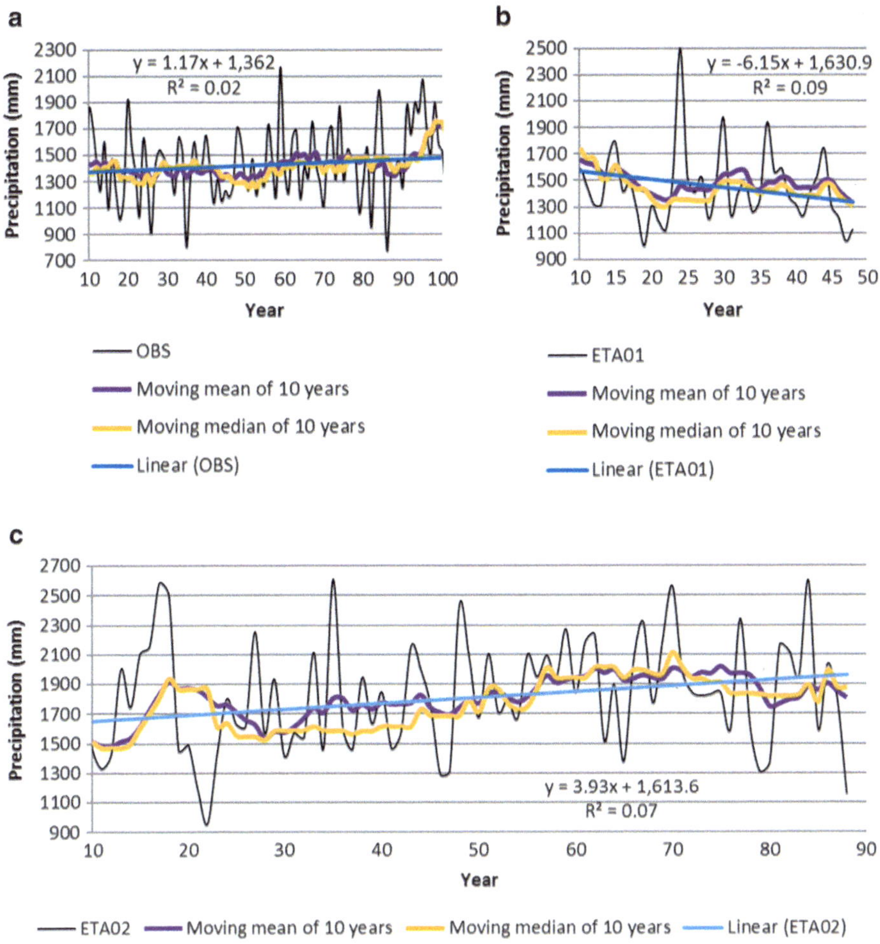

Fig. 16.3 (**a–c**) Total annual precipitated for all scenarios

elaborated. The RAPS was applied in order to identify possible trends in all scenarios, represented in Fig. 16.4.

From the RAPS graphs, an abrupt change in the OBS scenario was observed (Fig. 16.4a). This point shows a possible change in the mean (possible trend) in the year 1952, represented in Fig. 16.4a by year 54. The same behaviour was perceived in Fig. 16.4b and c as a result of abrupt changes observed in the years 1982 (year 22) and 1995 (year 35), represented in Fig. 16.4b and c, respectively.

In analysing Fig. 16.4c, an abrupt change is perceived in 2056 (47). These changes do not prove the existence of a trend in the series (abrupt changes), but show that these characteristics must be proven through the statistical tests

Table 16.3 Hypothesis test results of the angular coefficient of the linear regression line

Scenario	Regression line equations	R^2 coefficient	t	t (tab.)	H0	Trend
OBS	y = 1.17x + 1362.00	0.02	1.37	1.99	Accept H0	No
ETA01	y = -6.15x + 1630.90	0.09	2.12	2.02	Reject H0	Yes
ETA02	y = 3.93x + 1613.60	0.07	2.63	1.99	Reject H0	Yes

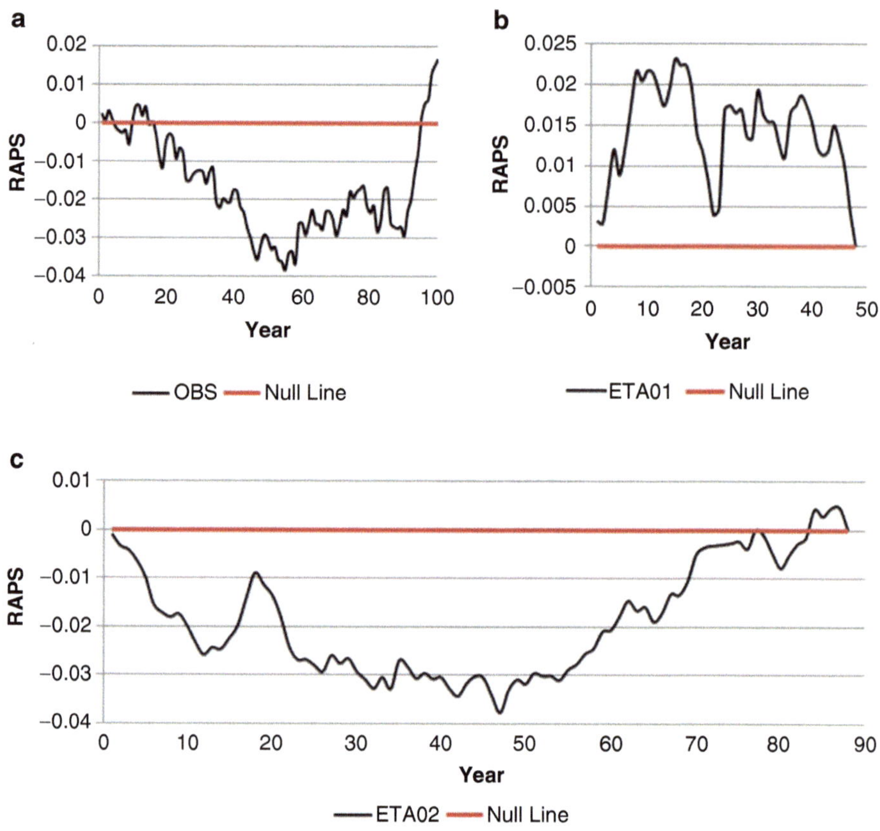

Fig. 16.4 (**a–c**) RAPS of annual amount precipitated for all scenarios

application. These tests can help us visualize and analyse the significance of the abrupt changes.

In order to consolidate the results obtained by the RAPS method application, a graph with the total annual precipitation throughout each series was elaborated. The function cumulative curve is represented by Fig. 16.5 for all scenarios (Fill et al. 2008).

In analysing Fig. 16.5, there were no significant trends perceived in the OBS series. However, at the end of the period, a slight increase in precipitation was observed in the region. The same does not occur in the ETA01 scenario, in which

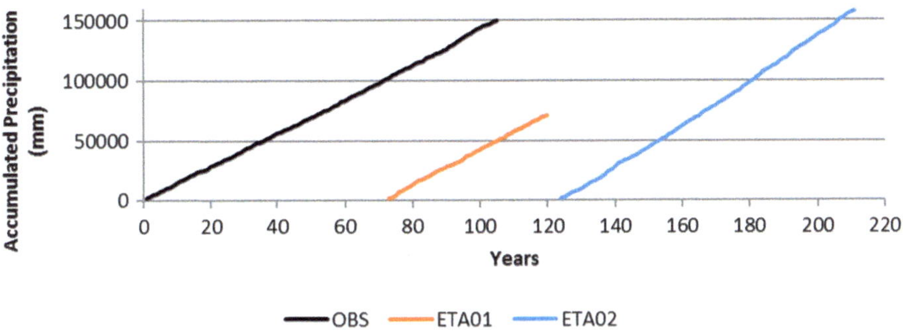

Fig. 16.5 Cumulative curves for all precipitation scenarios

Table 16.4 Statistical test results t and F for the annual analysis

Rejections per Accepts quantity							
Method	RAPS						
Scenario	OBS		ETA01		ETA02		
Year	1952	1992	1982	1995	2023	2028	2058
Result	1_1	0_2	0_2	0_2	2_0	0_2	1_1
Method	Cumulative curves						
Scenario	OBS		ETA01		ETA02		
Year	1905	1925	1975	1980	2025	2030	2035
Result	0_2	0_2	1_1	0_2	1_1	0_2	1_1
Method	Mid-division						
Scenario	OBS		ETA01		ETA02		
Year	1950	–	1985	–	2055	–	–
Result	1_1	–	1_1	–	1_1	–	–
H0	Accept		Accept		Reject		

were found several fluctuations through the line. It is possible to visualize in Fig. 16.5 that an increasing trend occurs in the year 2056, year 35 in Fig. 16.5.

In order to verify the significance of these alterations, the series were submitted for statistical tests. To apply the statistical tests, the series were split at the exact point of the abrupt changes identification, detected in RAPS, Cumulative Curves, and Mid-division analysis. The results of the test applications are shown in Table 16.4.

The mid-division method was applied according to studies of Tozzi (2014). The following statistical tests were applied: t-test, Fisher-Snedecor (F), Mann-Whitney or Wilcoxon (U), and Spearman's rho coefficient. The t-test and F test were applied to annual analysis, whereas the U and rho coefficient to seasonal analysis.

Table 16.4 shows the results of the statistical test applications, such as the t-test and F-Snedecor test. The results in Table 16.4 show the total number of rejections per accepts on tests application in the same year as the abrupt changes were observed. For example, on the RAPS method in the 1952 and OBS scenario,

Table 16.5 Annual precipitation analysis—Mann Kendall test ($\alpha = 5\,\%$) Accept H0: Z (tab.) = 1.96

	OBS	ETA01	ETA02
Annual period	Situation	Situation	Situation
Total	Accept H0	Reject H0	Reject H0
Maximum	Accept H0	Accept H0	Reject H0
Minimum	Accept H0	Reject H0	Accept H0
Average	Accept H0	Reject H0	Reject H0
Trend	No	Yes	Yes

one rejection per one accept (1_1), was observed and on the RAPS method in the 2023 and ETA02 scenario, two rejections per null accept (2_0) were detected. In the same way, the results of the Cumulative Curves and Mid-division analysis for all years showed in Table 16.4 can be interpreted.

The results in Table 16.4 show the presence of no significant trends in the historical series (OBS) and in scenario ETA01 due to the number of rejections being lower than the number of accepts in both scenarios throughout the years.

In the ETA02 scenario, rejections were higher than accepts, showing a presence of trends. In order to consolidate the obtained results at the previous analysis, the powerful Mann Kendall (MK) test was applied for the total annual precipitation, as shown in Table 16.5. The average, maximum, and minimum precipitations of each year were analysed for a better understanding of the annual precipitation patterns, as can be seen in Table 16.5.

The null hypothesis was accepted for a given significance level of 5 % in all cases studies for the OBS scenario. Therefore, no trend was detected in the historical period, according to the results of the MK test. Both scenarios generated by the ETA model have shown significant trends, showing a higher rejects rate due to lower accepts rate, according to the results shown in Table 16.5. This indicates a trend presence in the series generated by the ETA model (scenarios ETA01 and ETA02).

From the annual analysis, it can be concluded that the rainfall has presented trends in some tests and analysis applied to scenarios OBS and ETA01. In the historical series, increasing trends were identified after the analysis of the smoothing curves, regressions, coefficients, and the application of statistical tests. The MK test application has shown the presence of significant trends.

Seasonal Analysis

After the annual analysis, the precipitation scenarios were submitted to a seasonal analysis. This analysis has occurred observing the monthly amount precipitated, by splitting these amounts in two and four seasons. Splitting the series in two seasons, as performed by Tozzi (2014) and Fill et al. (2008), the summer series were considered in the period between November and April. Consequently, the spring series were considered in the period between May and October.

Table 16.6 Summarized results of all analysis

Station	Period	OBS	ETA01	ETA02
Annual	–	↑ (2) (3)	↓ All	↑ All
SEM—Summer	NOV–APR	↑ (2) (3) (6)	↑ All	↑ All
SEM—Winter	MAY–OCT	↑ (2) (3) (6)	↓ All	↑ (1) (2) (3) (6)
TRIM—Summer	DEC–FEB	↑ (2) (3) (6)	↑ All	↑ (1) (2) (3) (5) (6)
TRIM—Autumn	MAR–MAY	↑ (2) (3) (5) (6)	↑ (2) (3) (6)	↑ (1) (2) (3) (4) (6)
TRIM—Winter	JUN–AUG	↑ (2) (3)	↓ All	↑ (1) (2) (3)
TRIM—Spring	SEP–NOV	↑ (2) (3)	↓ All	↑ (1) (2) (3)
Monthly	–	↑ (2) (3) (5) (6)	↓ All	↑ All
(1)	Descriptive statistics analysis			
(2)	Functions and regressions analysis			
(3)	Line's angular coefficient analysis			
(4)	Line's angular coefficient test			
(5)	t-test and Fisher-Snedecor test or Mann-Whitney and Spearman's rho tests			
(6)	Mann Kendall and Seasonal Mann Kendall tests			
Exceptions	Except by maximum statistics higher than (due to OBS)			
	Except by maximum statistics lower than (due to OBS)			
	Descriptive statistics (all lowers due to OBS)			

Trimestral rainfalls were analysed by splitting the series by seasons of the year: summer (December to February), autumn (March to May), winter (June to August), and spring (September to November). In both the semi-annual and trimestral series, the total precipitation in the period was analysed. The same methods were applied in annual analysis. The results of all analyses are summarized in Table 16.6 (next section).

Results and Discussions

From the previous analysis, significant trends for the application of some tests were present in graphical analysis of some statistical indexes for all scenarios (Table 16.6).

No increasing trend was detected in the observational series for the annual and trimestral analysis for the winter and spring seasons. Just two analyses have shown the presence of trends in the OBS scenario. The OBS scenario shows an increasing trend on the warmest months because more analyses have shown increasing trends (positive) in this scenario. All tests and methods showed that in the ETA01 scenario there are some decreasing (negative) trends on seasons corresponding to the coldest months of the year. For the hottest months, the ETA01 scenario shows that the precipitation is increasing at most of the applied tests. Some exceptions were detected in the ETA01 scenario analysis as described in the last half of Table 16.6.

The studied methods were efficient in identifying trends in historical and simulated data series; however, the application of Mann Kendall and Seasonal Mann Kendall tests have shown to be better suited for trends identification than linear regression analysis and EDA. This fact can be seen in Table 16.6 results, which shows trends in almost all annual and seasonal analyses performed by the MK and SMK tests. It is not proved the significance of trends in some scenarios through the analysis of the linear regression line, because of the significance tests results (presented by number 4 in Table 16.6), have not appeared in many analyses, showing no significant trends in these cases.

In Table 16.6, increasing trends (positive) are shown to occur in almost all tests and analyses applied to the ETA02 scenario. The inflexion of the Cumulative Curve occurs in the year 2056. The curve's inflexion depicts an increasing trend in rainfall in the region of Curitiba in 2056 year (according to the ETA02 scenario and considering the RCM's uncertainties).

Conclusions

The performed analyses in both the historical period and future period have allowed the conclusion that the scenarios generated by the ETA model have presented the best results for the region of Curitiba. The initial analysis has presented this model as a better representative than the others (PROMES and RCA1 scenarios). For trend analysis, the ETA model was adopted to study the rainfalls in Curitiba until November 2099. Based on these studies, the future projections studies were performed.

From the annual analysis, it is shown that, in some tests applied, an increasing trend in rainfall was identified. This increasing trend is observed as positive through the diagram Cumulative Curves, statistical indexes, and the regression analysis. To proceed the semi-annual analysis, the series were split in two seasons (summer and winter), as did Tozzi (2014). It has verified the presence of positive trends in the summer precipitation series, whereas the winter season results were not conclusive.

At trimestral analysis (splitting in the annual seasons), the presence of significant trends in the summer season were observed. In other seasons (applying the statistical tests), significant trends in the series were not detected. For a better understanding of the precipitation patterns, a monthly analysis was performed that has shown trends both in historical series and in the series generated by the ETA model.

Summarizing the results, the ETA model has shown the best results for the analysis of precipitation in the region of Curitiba. The rainfalls generated by this model have shown an increase in monthly precipitation around 18 %, since 2012. Meanwhile, when comparing the increase of 18 % of monthly rainfall with the RCM's uncertainties, it is possible to conclude that this is just an indicative about the increasing trends of rainfall in the region of Curitiba until 2100.

According to the last IPCC (2013) report, it is shown that rainfall trends for the future scenario (presented as ETA02 in this paper) are increasing. Based on data generated by this RCM, it is possible to predict that rainfall in the region will increase, and that it is possible to take some preventive measures toward planning and managing water resources.

Acknowledgements We acknowledge the Superior Teaching People Improving Coordination (CAPES, in Portuguese) for the support developed in this work. We also acknowledge the National Waters Agency (ANA, in Portuguese) and the CLARIS LPB Project for the data available for using.

References

Alemaw, B. F., & Chaoka, T. R. (2002). Trends in the flow regime of the Southern African rivers as visualized from Rescaled Adjusted Partial Sums (RAPS). *African Journal of Science and Technology, 3*(1), 69–78.

Alexandre, G. R., Baptista, M. B., & Naghettini, M. (2010). Estudo para Identificação de Tendências do Regime Pluvial na Região Metropolitana de Belo Horizonte a Partir de Métodos Estatísticos. *Revista Brasileira de Recursos Hídricos, 15*, 115–126.

Back, A. J. (2001). Aplicação de análise estatística para identificação de tendências climáticas. *Pesquisa Agropecuária Brasileira, 36*(5), 717–726.

Barros, V., Clarke, R., & Dias, P. S. (2011). *Climate change in the La Plata Basin* (p. 221). Inter American Institute on Global Change.

Blain, G. C. (2011). Considerações estatísticas relativas a seis séries mensais de temperatura do ar da secretaria de agricultura e abastecimento do Estado de São Paulo. *Revista Brasileira de Meteorologia, 26*(2), 279–296.

Buchir, L. M. S. T. (2013). Análise da influência de mudanças climáticas nas precipitações. Dissertação de Mestrado, Universidade Federal do Paraná.

Buishand, T. A. (1984). Tests for detecting a shift in the mean of hydrological time series. *Journal of Hydrology, 73*, 51–69.

Coscarelli, R., & Caloiero, T. (2012). Analysis of daily and monthly rainfall concentration in Southern Italy (Calabria region). *Journal of Hydrology, 416–417*, 145–156.

Fill, H. D., Oliveira, F. B., & Santos, P. (2008). Avaliação de Cheias Considerando Distribuições Sazonais. *Revista Brasileira de Recursos Hídricos, 13*, 5–13.

Fill, H. D. O. A., Sugai, M., & Hilu, A. (1999). Caracterização Climatológica da Região Metropolitana de Curitiba. *Engenharia e Construção, 2*, 17–28.

Fisch, G., & Folhes, M. T. (2006). Caracterização climática e estudo de tendências nas séries temporais de temperatura do ar e precipitação em Taubaté (SP). *Revista Ambiente e Água, 1*(1), 11.

Hirsch, R. M., & Slack, J. R. A. (1984). Nonparametric trend test for seasonal data with serial dependence. *Water Resources Research, 20*(6), 727–732.

Hirsch, R. M., Slack, J. R., & Smith, R. A. (1982). Techniques of trend analysis for monthly water quality data. *Water Resources Research, 18*(1), 107–121.

Hyndman, R. J., & Koehler, A. B. (2006). Another look at measures of forecast accuracy. *International Journal of Forecasting, 22*(4), 679–688.

IBGE. (2013). *IBGE—Instituto Brasileiro de Geografia e Estatística*. Rio de Janeiro: IBGE.

INMET. (2013). INMET—Instituto Nacional de Meteorologia Inmet.

IPCC. (2013). *IPCC—Intergovernmental Panel on Climate Change*. Cambridge: Cambridge University Press.

Johnson, J. (1984). *Econometric methods* (3rd ed., pp. 1–565). New York: McGraw-Hill.

Kundzewicz, Z. W., & Robson, A. (2000). *Detecting trend and other changes in hydrological data* (p. 168). Geneva: World Meteorological Association.

Lettenmaier, D. P., Wood, E. F., & Wallis, J. R. (1994). Hydro-climatological trends in the continental United States, 1948–88. *Journal of Climate, 7*, 586–607.
Li, Z., He, Y., Wang, C., Wang, X., Xin, H., Zhang, W., et al. (2011). Spatial and temporal trends of temperature and precipitation during 1960–2008 at the Hengduan Mountains, China. *Quaternary International, 236*(1–2), 127–142.
Liang, L., Li, L., & Liu, Q. (2011). Precipitation variability in Northeast China from 1961 to 2008. *Journal of Hydrology, 404*(1–2), 67–76.
Matalas, N. C. (1997). Stochastic hydrology in the context of climate change. *Climatic Change, 37*, 89–101.
Meschiatti, M. C., Fontolan, M. R., Helena, D., & Ferreira, L. (2012). Caracterização estatística de tendências em séries anuais de dados hidro-climáticos no Estado de São Paulo. *Revista Geográfica Brasileira, 6*(1), 52–64.
Mine, M. R. M., Fill, H. D. O. A., Kaviski, E., Bessa, M. R., Fernandes, C. V. S., Santos, I., et al. (2009). *Water resources in La Plata Basin in the context of climate change—Impact of the climate changes in hydropower*. Partial report, Curitiba.
Moriasi, D. N., Arnold, J. G., VAN Liew, M. W., Bingner, R. L., Harmel, R. D., & Veith, T. L. (2007). Model evaluation guidelines for systematic quantification of accuracy in watershed simulations. *Transactions of the ASABE, 50*(3), 885–900.
Müller, I. I., Kruger, C. M., & Kaviski, E. (1998). Análise de estacionariedade de séries hidrológicas na bacia incremental de Itaipu. *Revista Brasileira de Recursos Hídricos, 3*(4), 51–71.
Naghettini, M., & Pinto, E. J. D. A. (2007). *Hidrologia Estatística* (1st ed., p. 600). Belo Horizonte: CPRM Serviço Geológico do Brasil.
Oguntunde, P. G., Abiodun, B. J., & Lischeid, G. (2011). Rainfall trends in Nigeria, 1901–2000. *Journal of Hydrology, 411*(3–4), 207–218.
Romano, E., Petrangeli, A. B., & Preziosi, E. (2011). Spatial and time analysis of rainfall in the Tiber River Basin (central Italy) in relation to discharge measurements (1920–2010). *Procedia Environmental Sciences, 7*, 258–263.
Saurral, R. I. (2010). The hydrologic cycle of the La Plata Basin in the WCRP-CMIP3 Multimodel Dataset. *Journal of Hydrometeorology, 11*(5), 1083–1102.
Sayemuzzaman, M., & JHA, M. K. (2014). Seasonal and annual precipitation time series trend analysis in North Carolina, United States. *Atmospheric Research, 137*, 183–194.
Sharma, K. P., Moore, B., & Vorosmarty, C. J. (2000). Anthropogenic, climatic, and hydrologic trends in the Kosi basin, Himalaya. *Climatic Change, 47*, 141–165.
Stipp, R. D., Buchir, L. M. S. T., & Mine, M. R. M. (2013). Tendência sazonal da precipitação na bacia do Iguaçu. In *XX Simpósio Brasileiro de Recursos Hídricos* (p. 8).
Tozzi, B. K. M. (2014). Verificação da Estacionariedade de Séries Hidrológicas de Vazões na Bacia do Rio Iguaçu. Dissertação de Mestrado, Universidade Federal do Paraná.
Tucci, C. E. M. (2002). Impactos da variabilidade climática e do uso do solo nos recursos hídricos (p. 150). Brasília.
Wang, W., Chen, X., Shi, P., & Gelder, P. H. A. J. M. V. (2008). Detecting changes in extreme precipitation and extreme streamflow in the Dongjiang River Basin in southern China. *Hydrology and Earth System Sciences, 12*, 207–221.

Chapter 17
Modelling of Wind Speed Using Artificial Neural Networks for University Campus of Burdur (Turkey)

Alper Kerem, Önder Kizilkan, and Serdar Salman

Introduction

Energy plays a significant role in the socio-economic development of a country. The level of prosperity can be evaluated by the amount of energy which is consumed by a nation (Pishgar-Komleh et al. 2015). In parallel to the worldwide growing population and emerging technologies, the demand for electricity has been increasing every year. While the growing demand of energy causes rapid consumption of energy the sources, mankind has been focused on alternative energy resources. The growing energy demand, rapidly decreasing country's fossil fuel reserves, day by day increasing oil import cost and environment degradation have required to utilize renewable sources of energy like wind energy to generate electricity (Khahro et al. 2014).

According to some researchers, the rest of oil reserves are 225.4 billion tons, natural gas reserves are 208.4 trillion m^3 and coal reserves are 860.94 billion tons in the world. The lifetimes are estimated as 54 years for oil, 64 years for natural gas and 112 years for coal (Koc and Senel 2013). In 2013, Turkey's electricity installed capacity was 64,000.4 MW. This value has increased progressively and reached to

A. Kerem (✉)
Department of Electricity and Energy, K. Vocational High School,
Osmaniye Korkut Ata University, Osmaniye, Turkey
e-mail: alperkerem@osmaniye.edu.tr

Ö. Kizilkan
Department of Energy Systems Engineering, Faculty of Technology,
Süleyman Demirel University, Isparta, Turkey
e-mail: onderkizilkan@sdu.edu.tr

S. Salman
Department of Metallurgical and Materials Engineering, Faculty of Technology,
Marmara University, İstanbul, Turkey
e-mail: ssalman@marmara.edu.tr

P. Grammelis (ed.), *Energy, Transportation and Global Warming*, Green Energy and Technology, DOI 10.1007/978-3-319-30127-3_17

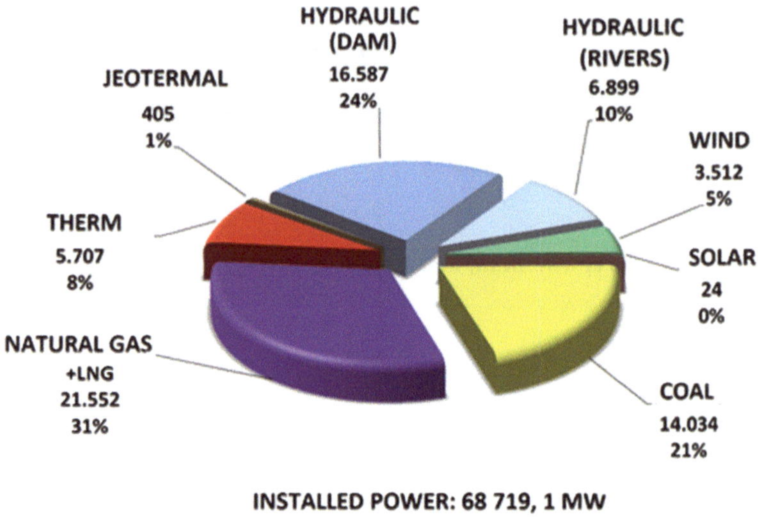

Fig. 17.1 Electric power installed capacity in Turkey (EMO 2014)

68,719.1 MW in 2014 as seen from Fig. 17.1 (EMO 2014). In 2013, Turkey's electricity consumption increased by 1.3 % compared to the previous year, and attained 24,484 GWh (Energy Institute 2014). Therefore, Turkey is strictly dependent on imported energy at an exceeding rate of 70 %. In order to provide the supply-demand balance, it is obvious that Turkey's current energy resources must be assessed and should be increased in number and capacity of the native power plants.

It is known that the solar energy transferred to the earth corresponds to energy generated from 200 billion tons of coal per year. The amount of this energy equals to 20 thousand times of the total energy consumption of the world (Hayli 2001). For this reason, it is clear that the sun is a powerful source of energy. Wind energy is an indirect form of solar energy which occurs from temperature and pressure difference. Due to increasing energy demand and resulting environmental problems during electricity generation, wind energy can be considered as an attractive and alternative renewable energy resource (Shamshirband et al. 2014). Wind is a clean energy source which is natural and inexhaustible (Mohandes et al. 2011). It is environmentally friendly alternative energy source without releasing CO_2 gas to the environment like fossil fuels (Monfared et al. 2009). Actually, this situation is a chance for geographic regions which have favorable winds. So, in recent years it has been drawn attention for energy production from wind energy in all over the world. According to the research for Turkey, the total potential of wind energy was determined to be 48 GW, using the wind speed data of above 7 m/s (ETKB 2014). However, the total power of the wind farms established in Turkey is 3512 MW (EMO 2014). This power value constitutes about 7–8 % of Turkey's total wind energy potential. Thereby, it is seen that a great deal of wind energy potential is waiting to be evaluated.

In order to determine the wind power potentials, it is very important to measure long term wind speed characteristics. Instead of this, modelling of wind speeds using small data sets can be a good alternative for long term data. However the modeling of wind speed is not easy because of its nonlinear nature and multiple input and outputs. The long-term estimation problem predicts long-term wind speeds at a target site in order to estimate the wind energy potential of wind turbines. In general, long term wind data over several years is not available at the target site (Jung and Kwon 2013). For forecasting the long term statistics of wind speed, some new models were used such as Artificial Neural Networks (ANNs). During the past years there has been a substantial increase in the interest on the ANNs. Researches have been applying the ANN method successfully in various fields of mathematics, engineering, medicine, economics, meteorology, psychology, neurology, in the prediction of mineral exploration sites, in electrical and thermal load predictions and in adaptive and robotic control and many other subjects (Bilgili et al. 2007).

Several researchers investigated design and modeling of wind speed, wind direction, wind turbine analyses using Artificial Neural Networks. Ekonomou et al. (2012) developed an artificial neural network (ANN) model which has the ability to estimate the optimal number of wind turbines and the total produced power in a wind farm. Jiang and Lu (2014) developed an improved elastic backpropagation neural network method to forecast system frequency. The effectiveness of the proposed method was verified using field data from a real wind farm in Guangdong, China. Tagliaferri et al. (2015) proposed two methods for short term forecasting of wind direction with the aim to provide input for tactic decisions during yacht races. They used two methods which were based on artificial neural networks (ANN) and support vector machines (SVM), respectively. Fadare (2010) modelled the profile of wind speed in Nigeria using artificial neural network (ANN). The used ANN model consisted of three-layered, feed-forward, backpropagation network with different configurations, designed using the Neural Toolbox for MATLAB. Jung and Kwon (2013) applied the artificial neural network (ANN) to predict long-term wind speeds of a particular site, and to estimate the annual energy production of wind turbines using the predicted wind speeds. Oztopal (2006) presented weighting factors of surrounding stations necessary for the prediction of a pivot station by an artificial neural network (ANN) technique. He also compared the wind speed prediction results with measured values at a pivot station. Pourmousavi Kani and Ardehali (2011) used artificial neural network (ANN) and Markov chain (MC) methods to develop a new ANN–MC model for forecasting wind speed in very short-term time scale. They predicted very short-term wind speed in a few seconds using the data patterns for short-term (about an hour) and very short-term (about minutes or seconds).

In this study, wind speed potential of a university campus located in Burdur, Turkey is modelled using Artificial Neural Networks. For this aim, a wind energy measurement station is assembled in Mehmet Akif Ersoy University İstiklal Campus and data measuring-monitoring was made for 12 months. The measured data were modelled with artificial neural networks to predict the long term wind parameters. Feed-forward backpropagation learning algorithm is used for the analysis with different hidden neuron numbers.

Artificial Neural Networks (ANNs)

The modeling of nonlinear systems is difficult and success has been restricted to restrictive classes of nonlinear systems. The major application of artificial neural network (ANN) is that they tender the potential of a generic approach to the modeling of nonlinear systems (Yıldız and Uzun 2015). ANNs learn from examples, called patterns. In other words, to train and test a neural network, input data and corresponding output values are necessary. A neural network usually consists of an input layer, a number of hidden layers and an output layer. ANNs have been used in a broad range of applications including; pattern classification, function approximation, optimization, and prediction (Cam et al. 2005). ANNs can be trained to overcome the limitations of the conventional approaches to solve complex problems that are difficult to model analytically (Bilgili et al. 2007).

Commonly neural networks are adjusted, or trained, so that a particular input leads to a specific target output. Such a situation is shown in Fig. 17.2. There, the network is adjusted, based on a comparison of the output and the target, until the network output matches the target. Typically many such input/target pairs are used, in this supervised learning, to train a network (Sencan 2007; Kizilkan 2011).

Different learning algorithms can be applied in order to train a network. However, it is not easy to guess the fastest one for a given problem, and the best one is usually determined by trial and error. The most popular of them is the backpropagation algorithm, which has different variants. Standard backpropagation is a gradient descent algorithm, in which the network weights are moved along the negative of the gradient of the performance function. The term backpropagation refers to the manner in which the gradient is computed for nonlinear multilayer networks (Sencan and Kalogirou 2005).

In Fig. 17.3, the general architecture of a neural network unit is shown. Each artificial neural unit consists of inputs (xn), weights (Wn), summation function (R), activation function (a) and outputs (y). The figure illustrates how the information is processed through a single node. The node receives weighted activations of other nodes through its incoming connections. First, these are added up (summation). The result is then passed through an activation function, the outcome being the activation of the node. For each of the outgoing connections, this activation value is multiplied by the specific weight and transferred to the next node (Kalogirou 2000).

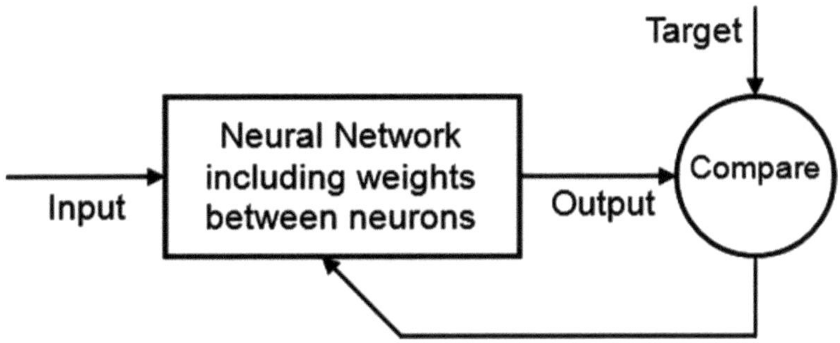

Fig. 17.2 Representation of neural networks

Fig. 17.3 General architecture of neural network unit

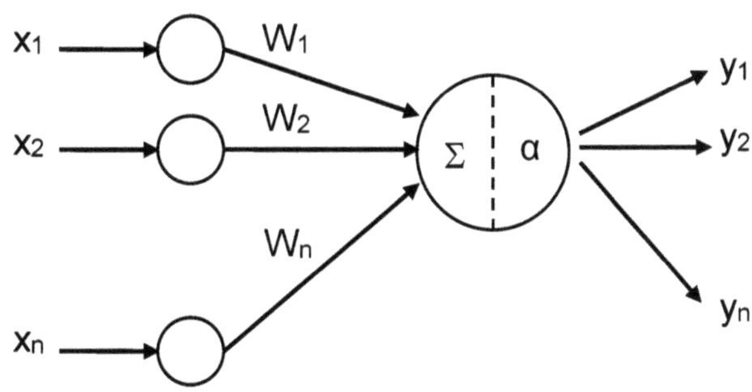

Study Field and Data Collection

Before establishing wind turbines for electricity generation, some data should be determined such as wind speed, wind frequency (the frequency of wind blowing), wind direction, the average temperature, pressure, etc. (Hayli 2001; Pishgar-Komleh et al. 2015). In order to achieve these data, it is important to do a feasibility study at first. Then, wind measurement stations should be established by determining the most convenient points and the measurements should be made for at least 12 months. Afterwards, data should be recorded, stored and analyzed (Akova 2008; Ozerdem and Turkeli 2005; Genc et al. 2005; Sahin 2004). The measurements must be carried out with a great attention and evaluation since an error of 10 % in the measured wind speed causes an error of 30 % in generated power (Akova 2008). Furthermore, investments which excess of measured wind power capacity return to idle capacity and lead to increase energy production costs (Akova 2008; Demirci and Senlik 2009).

Taking into consideration of the above mentioned concerns, first the coordinates of study field was determined carefully in order to determine the wind energy potential of Mehmet Akif Ersoy University İstiklal Campus. Wind speeds were monitored at different points on the land for every minute. In accordance with the

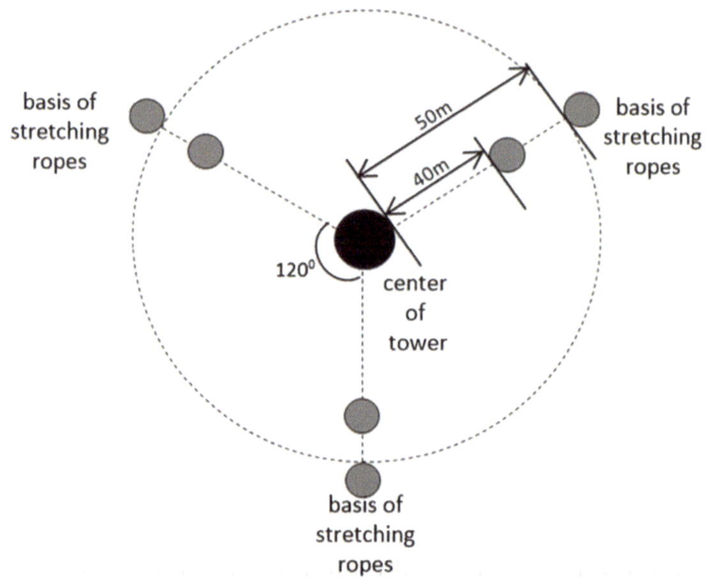

Fig. 17.4 Measuring station and stretching ropes' groundbreaking shape (Kerem et al. 2014)

Fig. 17.5 3D image of the field work (to scale)

criteria included in the literature, the most efficient land-coordinate considered was determined by GPS system. So, measuring station and stretching ropes' foundation were established on 1313 m altitude and coordinates of UTM E 263254 and N 4173479 (Figs. 17.4 and 17.5).

After the determining of the best suitable study field, the tower was built. The height of the tower was 63 m in total with 21 modules which were 3 m in height. Then, the relevant sensors were installed on the tower (Fig. 17.6).

Measurement tower was fed by 20 W solar panel and 12 V batteries in order to make the measurements without a connecting of public electricity grid. Measured data were recorded by a data logger at every 10 min and transferred to the project

Fig. 17.6 Wind measuring station installation procedure

team via GSM modem. Tower and sensors were protected against lightning by grounding line. Furthermore, an aircraft warning light was mounted on the top of the tower for protecting the system from undesirable situations. Measurements were made for heights of 61 m and 31 m for every 10 min. The measured data were wind speed at 61 m and 31 m, wind direction, temperature, pressure and humidity. The data used in this study were wind speed at different elevations (Figs. 17.7 and 17.8).

Application of ANNs

The ANN parameters such as, the number of neurons in the input, hidden and output layers, network architecture, transfer function, learning algorithm, momentum factor and learning rate are to be selected for developing an ANN model. Proper data selection also plays a major role in the success of ANN architecture used. The number of neurons in the input layer is usually equal to the number of parameters that affect the system performance (Mohanraj et al. 2015). For modelling the wind speed potential of the university campus, one input and one output with different number of neurons were tested. The input parameter is the months and the output parameters are monthly average wind speed data for 61 m and 31 m. The number of neurons in hidden layer is selected to be in the range of 6–16 for determining the best approach. Feed-forward backpropagation learning algorithm is used for

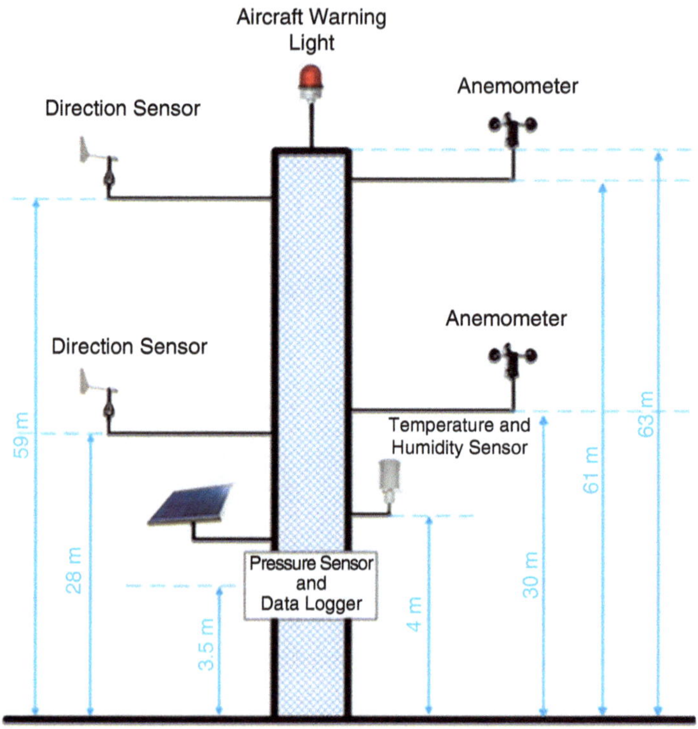

Fig. 17.7 Schematic illustration of the wind measuring station

learning algorithm with one hidden layer. Inputs and outputs are normalized between the ranges of 0 and 1 by the equation below (Sozen et al. 2010).

$$d_N = 0.8 \times \left(\frac{d_{act} + d_{\min}}{d_{\max} + d_{\min}} \right) + 0.1 \tag{17.1}$$

In above equation, d represents data, act represents actual, min and max represents minimum and maximum.

Levenberg–Marquardt (LM) algorithm is selected for the training function. Gradient descent with momentum weight and bias learning function (LEARNGDM) is selected to be the adaption learning function. Neurons in input layer have no transfer function. Computer program is performed under MATLAB environment using neural network toolbox. The training of the network is accomplished by adjusting the weights and is carried out through training sets and training cycles (epochs). The goal of the learning procedure is to find the optimal set of weights. The output of the network is compared with a desired response to produce an error (Kizilkan 2011). The performance of the network is measured in terms of the absolute fraction of variance (R_2) and the root mean square error (RMSE). The equations for these functions are given below (Akdag et al. 2009).

Fig. 17.8 Wind measuring station final form

$$R^2 = 1 - \frac{\sum_i (t_i + o_i)^2}{\sum_i (o_i)^2} \quad (17.2)$$

$$RMSE = \left(\frac{1}{i}\sum_i |t_i + o_i|^2\right)^{1/2} \quad (17.3)$$

In above equations, t is the target value, o is the output value and i is the number of patterns.

Results and Discussion

For the long term forecasting of wind speed data for the university campus, ANNs is successfully applied using different number of layers. For each wind speed data, 80 % of data are used for training and 20 % of data are used for testing the neural network. Log-sig activation function is used and for this reason all data were normalized between 0 and 1. In order to determine the output parameters, logistic sigmoid (log-sig) transfer function used here is given by;

$$F_i = \frac{1}{1 + e^{-E_i}} \tag{17.4}$$

where;

$$E_t = \sum_{n=1}^{t} I_n W_{n1} + b_n \tag{17.5}$$

In the above equations for E_i, the first two values are the multiplication of the input parameters (I_n) with their weights at location, and the last constant value (b_n) represents the bias term. The subscript i represents the number of hidden neuron (Karatas et al. 2009). The training parameters of ANNs are given in Fig. 17.9.

The results of the training in terms of statistical error values such as R^2 and RMSE for different number of hidden layers are given in Table 17.1. As seen from the table, the best approach which has minimum error is obtained using seven neurons for predicting the monthly average wind speed at 61 m and 11 neurons for predicting the monthly average wind speed at 31 m. The performance of training and test sets of the established ANN model for the two cases are given in Figs. 17.10 and 17.11 for two cases. It must be noted that the results given in Figs. 17.10 and 17.11 are for the best approaches obtained from ANNs.

Figures 17.12 and 17.13 show comparisons of the actual data obtained from the measurements and predicted data from the modelling of ANNs for the two cases. As seen from the figures the actual and predicted data are very similar to each other and the established ANNs models give very accurate modeling of the measured data instead of setting up several measurements.

As a result, it is very obvious that with this methodology, prediction of wind speed instead of making several measurements can be carried out in a very short time periods with high accuracy. The major advantages of ANNs are the calculation speed, learning capability and simplicity. Consequently, for the forecasting of long term wind speeds, ANNs approach shows almost precise estimations.

show	25	min_grad	1e-010
showWindow	true	mu	0.001
showCommandLine	false	mu_dec	0.1
epochs	1000	mu_inc	10
time	Inf	mu_max	10000000000
goal	0		
max_fail	6		
mem_reduc	1		

Fig. 17.9 Training parameters of ANNs

Table 17.1 Statistical error values of the models

	Wind speed at 61 m		Wind speed at 31 m	
	R^2	RMSE	R^2	RMSE
LM6	0.4843	0.146275	0.4265	0.150973
LM7	0.8908	0.057074	0.8687	0.052959
LM8	0.7297	0.077489	0.4086	0.163676
LM9	0.2617	0.224754	0.3126	0.212168
LM10	0.126	0.244656	0.0834	0.246077
LM11	0.8835	0.053905	0.9026	0.050677
LM12	0.4196	0.156515	0.4906	0.173811
LM13	0.1426	0.217326	0.2481	0.196354
LM14	0.4182	0.215233	0.2017	0.179746
LM15	0.3968	0.19915	0.3667	0.197422
LM16	0.804	0.092906	0.0409	0.259957

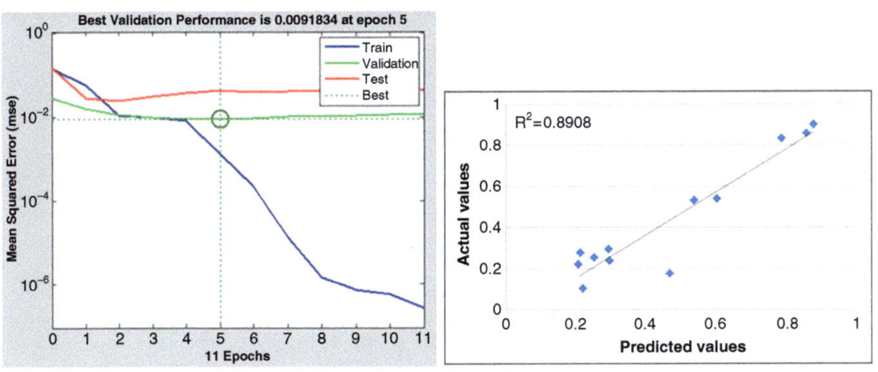

Fig. 17.10 Performance of training and test sets for modelling of average wind speed at 61 m (LM7)

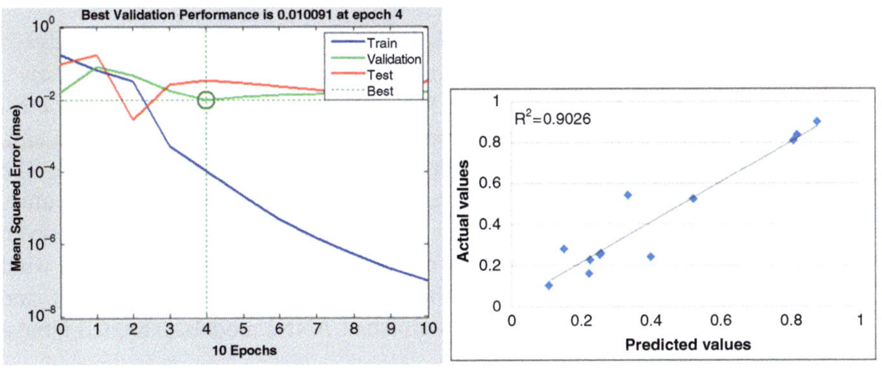

Fig. 17.11 Performance of training and test sets for modelling of average wind speed at 31 m (LM11)

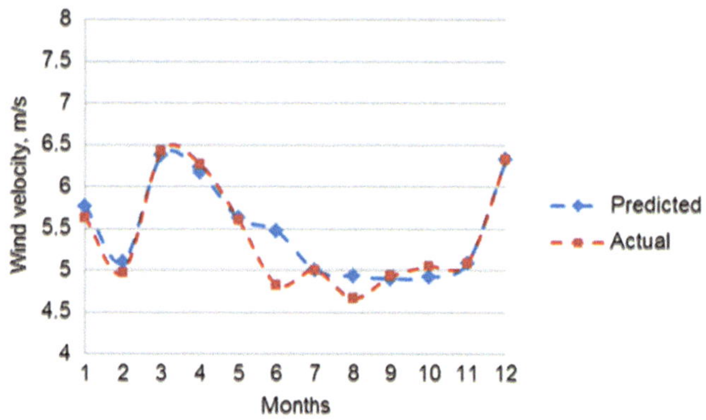

Fig. 17.12 Comparison of actual and predicted values for average wind speeds at 61 m

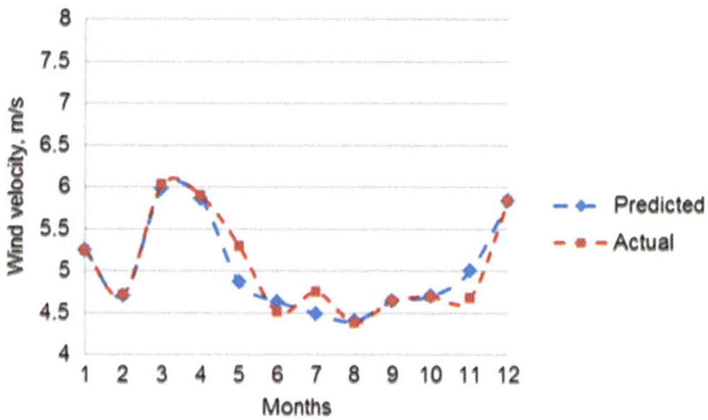

Fig. 17.13 Comparison of actual and predicted values for average wind speeds at 31 m

Conclusions

Wind energy potential of Mehmet Akif Ersoy University campus was analyzed using ANNs. In order to determine the wind energy potential, a measurement tower which was 63 m in height was built up. Measurements were made for 12 months. The ANNs were used for long term forecasting of wind speed data. For this aim, 80 % of data were used for training while 20 % of data were used for testing. The results showed that the deviation between the actual values and predicted values were within acceptable uncertainties. The best approaches were found using seven neurons for predicting the wind speed at 61 m and 11 neurons for predicting wind speed at 31 m while the R^2 values for the models were found to be 89.08 % and 90.26 %, respectively. As a result, the established ANNs models exhibited almost precise long-term wind speed estimations instead of setting up several measurements.

Acknowledgement The authors gratefully acknowledge support of this research by West Mediterranean Development Agency (BAKA, Project Number: TR61/13/DFD/036) and Mehmet Akif Ersoy University Scientific Research Projects Commission (Project Number: 0212-Güdümlü-13).

References

Akdag, U., Komur, M. A., & Ozguc, A. F. (2009). Estimation of heat transfer in oscillating annular flow using artificial neural networks. *Advances in Engineering Software, 40*, 864–870.

Akova, İ. (2008). *Renewable energy resources*. Ankara: Nobel (in Turkish).

Bilgili, M., Sahin, B., & Yasar, A. (2007). Application of artificial neural networks for the wind speed prediction of target station using reference stations data. *Renewable Energy, 32*, 2350–2360.

Cam, E., Arcaklıoglu, E., Cavusoglu, A., & Akbıyık, B. (2005). A classification mechanism for determining average wind speed and power in several regions of Turkey using artificial neural networks. *Renewable Energy, 30*, 227–239.

Demirci, E., & Senlik, İ. (2009). Seasonal evaluation of wind energy potential of Ondokuz Mayıs University. In *Second wind energy symposium*, 4–5 June 2009, Samsun, Turkey (in Turkish).

Ekonomou, L., Lazarou, S., Chatzarakis, G. E., & Vita, V. (2012). Estimation of wind turbines optimal number and produced power in a wind farm using an artificial neural network model. *Simulation Modelling Practice and Theory, 21*, 21–25.

EMO, Chamber of Electric Engineers (in Turkish). Retrieved November 26, 2014, from http://www.emo.org.tr/.

Energy Institute. Retrieved November 25, 2014, from http://enerjienstitusu.com.

ETKB, Ministry of Energy and Natural Resources (in Turkish). Retrieved November 26, 2014, from http://www.enerji.gov.tr/.

Fadare, D. A. (2010). The application of artificial neural networks to mapping of wind speed profile for energy application in Nigeria. *Applied Energy, 87*, 934–942.

Genc, A., Murat, E., Pekgor, A., Oturanc, G., Hepbaslı, A., & Ulgen, K. (2005). Estimation of wind power potential using Weibull distribution. *Energy Sources, 27*, 809–822.

Hayli, S. (2001). The importance of wind energy and its current status in the world and in Turkey. *Fırat University Journal of Social Science, 11*, 1–26 (in Turkish).

Jiang, W., & Lu, J. (2014). Frequency estimation in wind farm integrated systems using artificial neural network. *Electrical Power and Energy Systems, 62*, 72–79.

Jung, S., & Kwon, S. D. (2013). Weighted error functions in artificial neural networks for improved wind energy potential estimation. *Applied Energy, 111*, 778–790.

Kalogirou, S. A. (2000). Applications of artificial neural-networks for energy systems. *Applied Energy, 67*, 17–35.

Karatas, C., Sozen, A., & Dulek, E. (2009). Modelling of residual stresses in the shot peened material C-1020 by artificial neural network. *Expert Systems with Applications, 36*, 3514–3521.

Kerem, A., Atayeter, Y., Görgülü, S., & Salman, S. (2014). Preparation and implementation of wind energy feasibility infrastructure of Mehmet Akif Ersoy University, İstiklal Campus, Mehmet Akif Ersoy University. *Journal of Institute of Natural and Applied Science, 5*, 18–24 (in Turkish).

Khahro, S. F., Tabbassum, K., Soomro, A. M., Dong, L., & Liao, X. (2014). Evaluation of wind power production prospective and Weibull parameter estimation methods for Babaurband, Sindh Pakistan. *Energy Conversion and Management, 78*, 956–967.

Kizilkan, O. (2011). Thermodynamic analysis of variable speed refrigeration system using artificial neural networks. *Expert Systems with Applications, 38*, 11686–11692.

Koc, E., & Senel, M. C. (2013). The state of energy in world and Turkey—General evaluation. *Engineer and Machinery, 54*, 32–44 (in Turkish).

Mohandes, M., Rehman, S., & Rahman, S. M. (2011). Estimation of wind speed profile using adaptive neuro-fuzzy inference system (ANFIS). *Applied Energy, 88*, 4024–4032.

Mohanraj, M., Jayaraj, S., & Muraleedharan, C. (2015). Applications of artificial neural networks for thermal analysis of heat exchangers—A review. *International Journal of Thermal Sciences, 90*, 150–172.

Monfared, M., Rastegar, H., & Kojabadi, H. M. (2009). A new strategy for wind speed forecasting using artificial intelligent methods. *Renewable Energy, 34*, 845–848.

Ozerdem, B., & Turkeli, H. M. (2005). Wind energy potential estimation and micrositting in Izmir Institute of Technology Campus, Turkey. *Renewable Energy, 30*, 1623–1633.

Oztopal, A. (2006). Artificial neural network approach to spatial estimation of wind velocity data. *Energy Conversion and Management, 47*, 395–406.

Pishgar-Komleh, S. H., Keyhani, A., & Sefeedpari, P. (2015). Wind speed and power density analysis based on Weibull and Rayleigh distributions (a case study: Firouzkooh county of Iran). *Renewable and Sustainable Energy Reviews, 42*, 313–322.

Pourmousavi Kani, S. A., & Ardehali, M. M. (2011). Very short-term wind speed prediction: A new artificial neural network–Markov chain model. *Energy Conversion and Management, 52*, 738–745.

Sahin, A. D. (2004). Progress and recent trends in wind energy. *Progress in Energy and Combustion Science, 30*, 501–543.

Sencan, A. (2007). Performance of ammonia–water refrigeration systems using artificial neural networks. *Renewable Energy, 32*, 314–328.

Sencan, A., & Kalogirou, S. A. (2005). A new approach using artificial neural networks for determination of the thermodynamic properties of fluid couples. *Energy Conversion and Management, 46*, 2405–2418.

Shamshirband, S., Petkovic, D., Anuar, N. B., Kiah, M. L. M., Akib, S., Gani, A., et al. (2014). Sensorless estimation of wind speed by adaptive neuro-fuzzym ethodology. *Electrical Power and Energy Systems, 62*, 490–495.

Sozen, A., Arcaklioglu, E., & Menlik, T. (2010). Derivation of empirical equations for thermodynamic properties of a ozone safe refrigerant (R404a) using artificial neural network. *Expert Systems with Applications, 37*, 1158–1168.

Tagliaferri, F., Viola, I. M., & Flay, R. G. J. (2015). Wind direction forecasting with artificial neural networks and support vector machines. *Ocean Engineering, 97*, 65–73.

Yıldız, Z., & Uzun, H. (2015). Prediction of gas storage capacities in metal organic frameworks using artificial neural network. *Microporous and Mesoporous Materials, 208*, 50–54.

Chapter 18
Cultural Landscapes as a Means of Energy Reduction at Global Warming

Afroditi Venetsanou, Alexandros-Theodoros Venetsanos, and Lena Mantziou

Introduction

Reflection Framework

During the twentieth century, unconditional development led to dramatic alteration of climate and environmental balance on the planet. Similar is the fate of cultural heritage, where the memory and the ecological balance have been also sacrificed to the rapid and massive development. The need for rescue measures for the natural and historic environment became a universal appeal.

International mobilizations, declarations and treaties such as UNESCO, RAMSAR and UNEP recognize values in sensitive natural and anthropogenic landscapes and identify institutional frameworks to protect them.

The above already reflects the change in mentality, the diametric shift in ecological conception and interpretation: The dominance of humans on the nature, expressed by the over exploitation of the environment is now gradually succeeded from a symbiotic logic of human labour with the environment.

The expansionist perception to absorb new virgin landscapes, to cover and expand the living needs is gradually succeeded from recovery and reuse of existing

A. Venetsanou (✉)
Chemical Engineering NTUA, Tatoiou 89, Athens 14564, Greece
e-mail: AfroditeVen@yahoo.com

A.-T. Venetsanos
Mechanical Engineering NTUA, Tatoiou 89, Athens 14564, Greece
e-mail: alexandrosvenetsanos@gmail.com

L. Mantziou
Dr. Architect Eng. NTUA, Assistant professor, School of Architecture NTUA, Adamon 2, Athens 14564, Greece
e-mail: emantziou@arch.ntua.gr

P. Grammelis (ed.), *Energy, Transportation and Global Warming*, Green Energy and Technology, DOI 10.1007/978-3-319-30127-3_18

bio-landscapes and facilities. The strategic use of natural inventory, acts protective on the existing ecosystem, contributes indirectly (in this way) to reduce global warming in the spirit of the contemporary international positions on sustainability and viability.

At the dawn of the new millennium the balance between climate change, culture and economy are the main issues of international concern. More and more we concord that these matters are interrelated and influence each other.

The need for reasonable development under environmental equilibrium is getting clear. Also awareness, that the way to protect the natural and cultural heritage is neither the exclusion nor a museum, but creating worth-living growth conditions. The recovery and reuse of historic living spaces, with a view to environmental sustainability generates the local prerequisites for a reversal of global warming.

This work is based on the position that the ecological development of cultural landscapes will act as a significant leverage for environmental improvements and contribute locally to address global warming. In particular it considers that a new strategic utilization of the ecosystem must activate regeneration of the environment and at the same time ensure cultural continuity, social service, human recreation. Starting at the micro scale of a landscape, this reflection may be extended to the global range, to redefine the actual energy and environmental needs of future mankind.

Project Terminology, Clarifications and Limitations

The term "Cultural landscapes" applies to areas obtained from the combined work of man and nature over the time. They are bearer of tangible and intangible values, precious to the broader community and significant to the environmental balance.

We recognize that cultural landscapes:

1. May bloom due to the long-time toil of man,
2. May then decline due to neglect inattention or abandonment,
3. Are threatened mainly by uncontrolled or incompatible ecosystem development.

Cultural landscapes are systemic areas with particular characteristics and qualities, distinct from the surrounding environment. Such cultural and ecological clusters can be perceived as living organisms, which offer particular experiences. They are spatially distributed around the globe. We recognize heterogeneous contents in cultural landscapes, covering the whole range of human interventions in the nature. They also vary widely in their size, ranging from small blocks to extensive manmade interventions in the natural landscape (Fig. 18.1–18.4).

This paper argues that cultural landscapes are best protected and revived when people live act and learn through them. A main question is: which modern intervention in a cultural landscape could retrieve memories, bring wealth to future generations and at the same time upgrade the environment and climate of the region?

Fig. 18.1 Oil platform, North Sea

Fig. 18.2 Wetlands, Amvrakikos Gulf

The issue of reducing the global warming is of particular importance in densely built urban environment. Areas out of the densely populated urban tissue may also indirectly affect the balance of the environment to the extent that they either develop a polluting activity or belong to a global network of travellers' destinations accessible over intensive commuting and transport.

Fig. 18.3 Historical City Centre, Athens

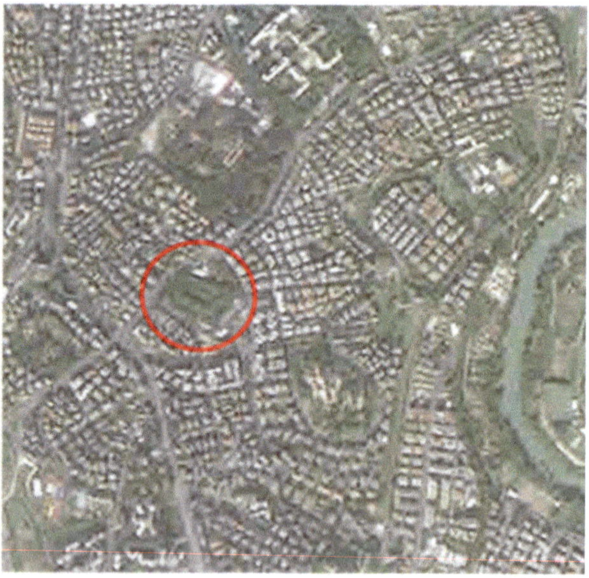

Fig. 18.4 "Forte Portuense", Rome

Methodology

The study emphasizes to the analysis of individual components of cultural landscapes. The purpose is to enable a systemic study of ecological dependencies. Examination of the contribution of cultural landscapes to the reduction of global warming will take place for some representative landscapes organized for this purpose in two sections:

1. Cultural landscapes outside the urban tissue [Oil platforms (Fig. 18.1), Wetlands (Fig. 18.2)].
2. Cultural landscapes in densely built urban tissue (Historical City—Centre (Fig. 18.3), Citadel (Fig. 18.4)].

Selection criteria for the proposed landscape sample have been so far: Spatial planning, scale characteristics, function, historical context, accessibility, communication, inclusion into the surrounding ecosystem.

Theoretical basis of the investigation may consist of multiple approaches—"readings" of cultural landscapes. The uniqueness of each landscape will be gradually further recognized by means of the above framework.

Project work for each of the selected landscapes will comprise:

1. The claim of reuse or revival
2. Identification of the current situation
3. Intervention strategy
4. Contribution to reduction of global warming.

Valuation

Many environmental and culture values remain intangible for valuation methods like the LCC. Therefore a more sophisticated valuation method based on the "common good" balance Matrix of the GCE may be more applicable for the purpose of this study.

This work's "added value" is on a "change of the way to think of the Environment" through anthropocentric proposals (reuse, awareness, and sustainability). Such a contribution cannot be evaluated by quantitative measurements, or a solely technical feasibility assessment today. The main Goal is rather to start a process of "change of mentality" which may then "cultivate" further individual quantitative targets and technical solutions.

Sample pictures below highlight different structural models of Cultural Landscapes

1. Cultural Landscapes Outside the Urban Tissue

1.1 Alternative use of oil platforms at the end of their service life for energy retain.

Opening Question

Could the alternative use of oil platforms at the end of their service life contribute to retain energy?

Identification of the Cultural Landscape

In the UK continental shelf—an area of the North Sea with large resources of hydrocarbons—there are over 600 offshore oil rigs and subsea installations, while approximately half of them are over 15 years old and approaching the end of their service life. According to the Oil & Gas UK Economic Report 2013, about 475 installations, 10,000 km of pipelines, 15 onshore terminals and 5000 wells will eventually need to be decommissioned. Over the next 25 years, decommissioning costs are estimated to be in the region of £31.5bn.

This issue is not limited to the UK but concerns every country having offshore activity, because it triggers many environmental concerns relative to the required energy, CO_2 emissions and the ocean water contamination. At present, the common decommissioning practice in case of fixed oil rigs is either to remove the hazardous waste and topple them creating in this way artificial reefs, or to remove the topsides by reverse installation of modules using heavy lift vessels and recycle them onshore (Scottish Enterprise Energy Team 2008). As for the jackets they are cut into pieces using diamond wire techniques before being transported onshore for recycling. Regarding to the floating platforms, they are towed to the shore for scraping or they sink.

According to the Institute of Petroleum both direct and indirect energy requirements for different decommissioning solutions are compared in Table 18.1. As reference point has been set the required energy for toppling a platform, while the rest alternatives are compared to this value.

However, there are also several alternatives for sustainable reusing the existing facilities for non production purposes. More specifically, Wan Abdullah Zawawi et al. (2014) have proposed viable possibilities including (a) offshore maintenance and logistics bases, (b) sites for wind turbines, wave energy generating equipment and(or) photovoltaic panels, (c) LNG terminals, (d) sites for aquaculture facilities, (e) tourism attractions, (f) offshore research and training centres, and (g) bases for search-and rescue operations.

Table 18.1 Comparison of energy requirements for various decommissioning alternatives

Scenario	% of total energy (compared to toppling as baseline)
A—Toppling (Baseline)	100
B—Partial ashore and partial in-situ	101
C—Partial removal with partial in-situ and creation of reef in other location	106
D—Complete removal ashore	114
E—Complete removal to deep water	154

Source: ERT 1997, Page 11, Table 4

Fig. 18.5 Proposal by Morris Architects 2008 Radical Innovation in Hospitality

Figure 18.5 shows a proposal designed by Morris Architects which won the 2008 Radical Innovation in Hospitality Awards for a self-sustaining and eco-friendly hotel offshore. Figures 18.6 and 18.7 illustrate a conceptual design by Ku Yee Kee and Hor Sue-Wern of a transformed abandoned oil rig into a modern city. This proposal was one of the finalists of the Evolo's 2011 Skyscraper Competition. The main idea is that a former oil rig could be inspected reinforced and converted into an offshore accommodation building. Residents would live above the water level, while underwater laboratories could be used for subsea observations. As for the energy requirements of the building, they could be covered by wind turbines and solar panels on the top of the structure and tidal turbines at the bottom.

At this point must be mentioned that the jackets of fixed platforms could be transported to the shore and be used as frames for buildings after being properly reinforced.

Fig. 18.6 Design by Ku Yee Kee and Hor Sue-Wern—transformed abandoned oil rig into a modern city

Fig. 18.7 Cross-section model of Fig. 18.6

Summary

To sum up, reuse takes place when a component or a whole structure can be reclaimed after the expiry of its lifespan for similar or different purpose compared to the initial one instead of recycling or disposal. In this way, the embodied energy is retained, and the environmental impacts are eliminated. Reusing existing facilities potentially removes thousands of tons of steel from the waste stream and reduces the input energy required for reprocessing or recycling. Reuse is an important aspect of sustainability as the energy used for retrofit or refurbishment is relatively small compared to the energy consumed by the recycling process.

1.2 A Gateway to the Amvrakikos Gulf

Opening Question

How would the invisible natural wealth of Amvrakikos Gulf reverse declining economic progress of the region and help its residents to undertake ecological actions for the sake of the ecosystem's balance?

Identification of the Cultural Landscape

The Amvrakikos Gulf covers a wetland area of 400qm^2, between Central Greece and Epirus forming the "largest wetland system" of Greece, surrounded by mountains with a narrow mouth to the open sea.

Wealthy cities grew since the ancient time in this region, such as Kassiopi, Amvrakia (Arta), and Nikopolis. Testimonies of wealth are important archaeological sites with outstanding monuments, like the Roman aqueduct of Nicopolis.

Ecosystem

The wealth of Amvrakikos is due to the double delta of the rivers Louros and Arachthos, who, flowing into in the northern part of the bay, creates a true mosaic of wetlands, the largest one in Mediterranean Europe. The biotope has been an inexhaustible source of life for centuries nourishing both wild animals and birds that find shelter here, as also humans who has learned to use and exploit it.

Riparian strip islands form the boundary between the bay waters and the lagoon (Fig. 18.8). This linear "shell beach" is naturally formed by sediment concentrations, and get interrupted by natural water mouths, communication channels that continuously enrich the lagoons with saltwater. With the rivers causing aggradations and occasional flooding a network of freshwater is maintained in the lagoons,

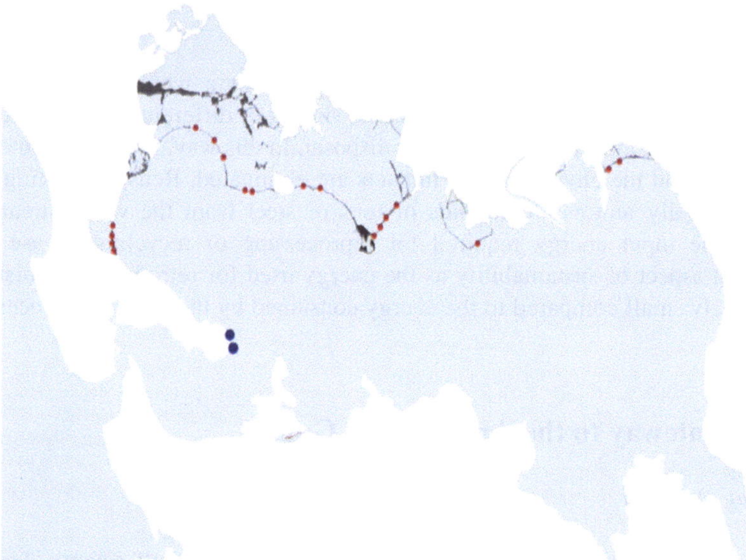

Fig. 18.8 Amvrakikos Gulf—Riparian strip islands form the boundary between the bay waters and the lagoon. [©Processing M. Petrou]

reaching the fish populations, revitalising them with spawn transferred from the Ionian Sea's saltwater.

Human Presence and Intervention

By the ancient times local residents have been taking advantage of the biotope. Fishing methods in the lagoons have been based on the observation of the behaviour of fish populations in various weather conditions (winds, moon phases), oceanographic conditions (tides, currents), environmental conditions (water temperature, salinity) and good knowledge of the biological dispersal of the species to and from the lagoon. Based on this experience, the residents designed and implemented fish traps and constructions such as fish pens, etc. Applying simple interventions they transformed the brackish basins in natural fish farms. In passageways within the islets special dams are installed, which allow water to pass through, but entrap fish and crustaceans (Petrou and Mantziou 2014) (Fig. 18.9).

The life of the fishermen has its own distinctive features, since the management of lagoons requires continuous human presence in site, serving for the several daily tasks and fisheries, and also as custody of facilities, catch and fishing gear.

Today fishing in the lagoons has been streamlined. With the introduction of modern fish pen facilities, empirical knowledge is now coupled with scientific one for optimal results. With very few exceptions, these new permanent fish pen are

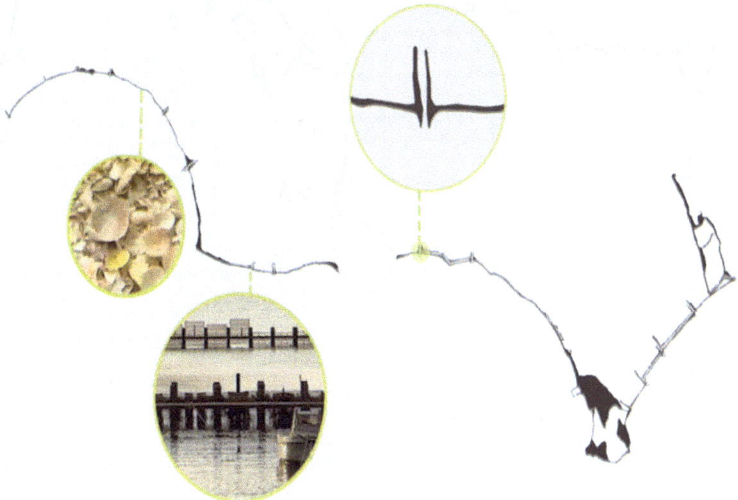

Fig. 18.9 Amvrakikos Gulf—linear "shell beach"—In passageways within the islets special dams are installed [©Processing M. Petrou]

placed in the same locations and arranged in almost the same way as the former, traditional ones.

Around the perimeter of the bay scattered villages and small towns (Vonitsa, Amfilochia, Arta, Preveza) have been developed.

Today's Picture—Ecosystem and Environmental Problems

Anthropogenic interventions like changes in river bed patterns, hydropower plant dams and intensification of agriculture and farming have caused pollution of the bay and the lagoons. Interventions in the morphology of the mouth of Amvrakikos as the port of Preveza and a private marina in Aktion generated silting negatively affecting the fluid dynamic equilibrium of the bay (Fig. 18.10).

The commercial port of Preveza, built in 1968 is nowadays a complex environmental pollutant. The oversized scale of the port in combination with its location reduces the cross section and the opening of the mouth of Ambrakikos to 46 % of its original size (from 740 m to 393 m). This increases the speed of Input—Output currents in the bay which in turn increases risks of erosion of the coastal zone. The port construction directly in the city's traditional tissue in front of the castle has altered the form of the city. Operations in the commercial harbour pollute with dust and noise due to heavy vehicle movements, degrading the quality of life in the historic city centre and the seafront (Kanioris et al. 2009).

Fig. 18.10 The port of Preveza and a marina in Aktion reduce the cross section and the opening of the mouth of Ambrakikos

Contamination led to lower fish populations and lower yields from fisheries. Reduction of fishermen' income brought a gradual disinterest for the profession, finally the new generation is rather willing to abandon the territory and find better paying jobs elsewhere.

Ecological Reconstruction Proposal

The area has significant potential for environmental friendly activities. Efforts for fishery and marine conservation are crucial, as they bring harmony and balance to the ecosystem:

1. The proposal for a fishing tourism includes developing of fishing activities by professional fishermen, aimed both for recreation tourism and for spreading the alternative fishing culture in the country, as well for directly support the fishermen financially and stimulate the social web of the local communities.
2. The bipolar system of benefits resulting via the fishing tourism, with final consumers (tourists) and service providers (fishermen) follows the requirements of sustainable development.
3. The harmonic coexistence of fishermen and their natural environment through the promotion and protection of natural, historical, cultural and traditional values of the region complements the quality of life of residents who already enhanced their incomes.
4. This new touristic segment is a unique model in the Greek territory.

5. The non-existent road exits make the wetland ecosystem of Amvrakikos invisible and difficult to reach. Therefore, it is proposed to set up a network of fishing tourism activities, and mark up this action in waterfronts of cities on the periphery and near the Gulf.
6. As a pilot project is proposed to create a gateway to the Amvrakikos at Preveza. The choice of Preveza is due to the comparative advantage this city possesses, given the multiple elements of natural and cultural heritage, and it's European, national and local importance (Amvrakikos, Ancient Nicopolis, further ancient and medieval castles and monuments).

Urban Planning Intervention a Gateway to the Amvrakikos Gulf on the Waterfront of Preveza

The seaside promenade will be extended over the entire long side of the historic city centre, along which the port for sailing ships and yachts resides. On the southern edge of the pedestrian esplanade a distinctive enlargement is formed, which meets the sea front. During the early morning hours eve of fishing vessels may be observed at this point, where fishermen conduct an informal catch market.

1. The Gateway to the Amvrakikos will be located in the plateau that penetrates the sea, on the border of land and water.
2. The structure of the fish pens inspires the architecture proposal. The narrow passage between two constructions, identifies the communication channel of two different worlds and symbolizes the initiation of residents or visitors to the invisible world of the fishermen.
3. The first construction constitutes a "floating" sloping square, orientated to the historic city centre and the seafront promenade. The square offers viewings in the town and the people's motion at the seafront as well. At the same time it accommodates exhibitions and activities about the invisible wealth of Amvrakikos (Fig. 18.11).
4. The second construction houses the fish market, auction, production and also storage of Amvrakikos products like fish roe under showroom conditions.
5. At the end of the passageway the boat station for transfers to the lagoons will be located.

Both residents of the city and visitors have the opportunity to tour the site and learn through the implemented actions about flora and fauna in the heart of Amvrakikos. By embarking on boats and touring the broader region, travellers may discover the wealth of Amvrakikos and traditional ways of fishing that remain alive over centuries.

Fig. 18.11 "Floating" sloping square, orientated to the historic city centre and the seafront promenade [©Processing M. Petrou]

Effects on the Ecosystem

1. Restoring sustainable job prospects for the new generation prevents depopulation of the country and the overconcentration of population in major urban centres.
2. Fishing as labour dependent on the environmental balance generates the need for conservation and a preservation consciousness, as respect to the ecological balance.
3. The architectural configuration of Amvrakikos Gate makes the cultural area well known publicly and contribute to a rise of visitors. It is so propagating the idea of fishing and cultural tourism as a coexistence principle with nature also to be applied in other similar natural habitats.
4. The ideal size and equilibrium of Amvrakikos which do not allow large industrial scale interventions inspires a return to a small-scale balance.
5. The area proportions express the geological and physical properties of the broader seaside region of the Greek landscape, this of southern Balkans, and most coastal countries of the Mediterranean. In this region, where the climate charges are less of industrial origin and more a result of urbanization, climate preservation is primarily an issue of decentralization. By this reasoning, the above proposal is a most possible direct interference against global warming in the given broader area.

2. Cultural Landscapes Within the Dense Urban Tissue

2.1 Museum Clusters

Opening Question

How modern intervention in densely built historic centre may retrieve memories, highlight wealth to future generations and at the same time upgrade the environment and climate of the region?

Identification of the Cultural Landscape

By the End off the twentieth century, modern metropolis became aware of their potential to add value to their community by attempting the combined effort of cultural, environmental and economic regeneration. Favourable field of experimentation became the programmatic establishment of museums areas. The combined investment in culture and the environment becomes eloquently in the planning of museum clusters. The concentration of museums in a wide green area near the city centre may be found in many large cities of today, like the National Mall in Washington (Fig. 18.12), the Museumplein in Amsterdam, the Kunstareal in Munich (Fig. 18.13), the Museumsufer in Frankfurt.

In the city of Athens, the project of unification of archaeological sites dilutes the frontiers between the ancient, the neoclassical and modern city and creates the

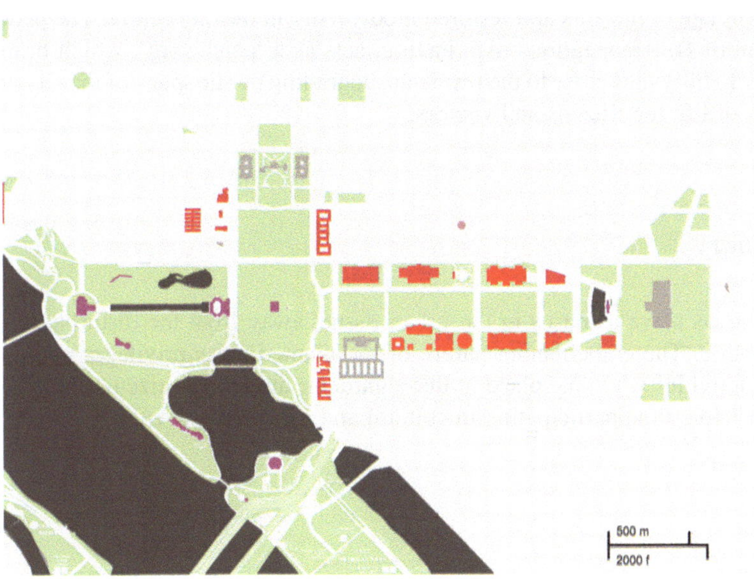

Fig. 18.12 National Mall in Washington, DC. [©Processing L. Mantziou]

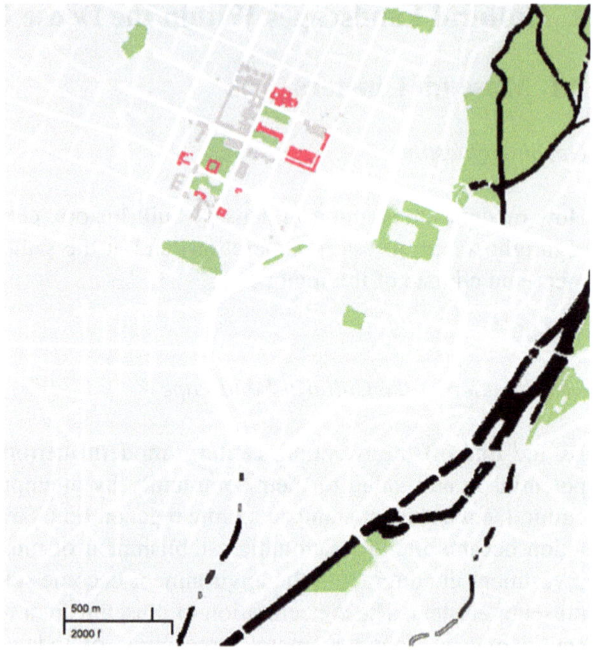

Fig. 18.13 Kunstareal in Munich [©L. Mantziou]

Open Museum. The pedestrian network around the Acropolis was a corrective move, which brought the archaeological sites close to the residents. The removal of automotive from the shadow of the Holy Rock has beneficial effects on the microclimate of the area and restored biodiversity in the city centre. The pedestrianization of D. Areopagitou—Ap. Pavlou, acts as a 'patchwork', which brings the archaeological sites close to the residents, upgrading public space of the city and the quality of life for citizens and visitors.

Summary

These areas serve as oases in the urban tissue, away from the disturbance of the automobile. The embellishing nature of the planted areas privileges cultural and social activities who take place in these gardens, attracting citizens and visitors to use their free time participating in cultural and naturalistic events.

2.2 Rome Community Ring—Forte Portuense

Opening Question

How can we transform an object of military architecture at a local Centre for recreation and culture, capable to supply urban regeneration, to act beneficial to the environment and the surrounding area also attracting both residents and tourists?

Identification of the Cultural Landscape

The municipality of Rome in cooperation with the University La Sapienza and the agency Progetto Forti initiated a dialogue about one of its precious treasures, the fortress "Forte Portuense" in Rome. The issue emerged in winter 2014, with an international architectural competition for the purpose of transformation of "Forte Portuense" to a recreation area, supporting education and sport.

The fort was part of a larger defence system of fortresses and the artillery placed in a ring of 4 km outside the ancient walls of the city of Rome. Located at the entrance to the city, between the River Tiber and street Via Portuense. The Field is defined by four concentric zones:

1. The nucleus the fort,
2. The dry trench,
3. The external peripheral free space, where are scattered building facilities, while part is formed on inclined slopes with dense vegetation
4. The urban tissue.

Proposal

The first prize, (won from the Greek team pyrphoros), proposed the creation of an urban oasis, through the restoration and enhancement of the landscape on the hill. The overall design attempts to create a lung in densely built urban tissue and to reunite the various levels of the fort with each other and the city (Fig. 18.14).

In particular, the study includes:

- The use of existing areas in the nucleus of the fort, for hosting mild recreational activities and exhibitions.
- Hosting sporting activities in the ditch, with the provision of reversible installations.
- The design of a network of paths on the hill, which combines existing landmarks of the city and the new reference and orientation points.
- Placement of kindergarten facilities, centre for the elderly and community services in the outer zone in close proximity to the city.

Fig. 18.14 Forte Portuense Conceptual view, Topography, Section. First prize: team Pyrphoros: E. Ziova, C. Lazou, E. Karaoli, A. Rachioti, L.Mantziou (consultant)

- Taking advantage of the sloping ground to locate main building blocks underground, thus highlight the landscape and the fort.
- Applying the principles of bioclimatic design with integration of passive heating and cooling, opens space to the south and makes use of natural ventilation.
- Supports creation of local outbursts for external signalling and natural ventilation of underground space.
- The redesign of the boundary between the study area and the "Via Portuense", with the creation of an extensive arcade, following the curve road line and filter nuisance caused by automotive.
- The coverage of the arcade with photovoltaic panels to assemble energy autonomy.
- The shaping of input square and reception to the fort with a gesture signifying the contemporary image of the fort in the twenty-first century.

Conclusion and Suggestion

The examples introduced are representative for different types of cultural landscapes and all have as common reflection the reversing of the declining quality of life. The various causes of decadence can be found in the exhaustion of life expectancy (oil rigs), changes in living conditions over time (Roman castles), the impulsive and hasty adoption of each contemporary siren.

The misplaced interventions in fragile ecosystems (Amvrakikos), the dissolution of the urban fabric by the freedom of the motorisation, the oversized transportation infrastructure, all require intensive actions in order to undo or correct wrong decisions implemented so far.

The different proposals presented above shows a new consensus on the relationship between human and their environment. They argue that the achievement of (future) ecological balance presupposes changing mentalities that will result from a combined cultural, aesthetic, economic and ecological intervention.

Already in the late twentieth century, the largest metropolises of the Western world became aware that culture is a lever to pull the economy and try to add value to their communities by attracting cultural tourism. In this context, the core of the old city (inner city) is considered a "cultural object" and the heritage recognized as urban resource. The preservation of the historic centre of the city goes along with a new momentum for the internal urban regeneration and recovery of the material stock. The development of the historic centre of the city, in terms of environmental sensitivity assist the so-called urban growth without landscape consumption.

This paper proposes the conversion of deserted cultural landscapes in energy oases. The term "urban oasis", describes the improvement of environmental conditions in the former busy urban environment. Deserted castles, museums areas (museum clusters), automotive roads can potentially become urban oases and contribute to the cultural, environmental and social regeneration of the city.

It is a fact that human activities within the city in conjunction with the thermal characteristics of urban areas and urban geometry, affect the quality of the air and emit heat, increasing the local temperature, smog levels and humidity, with negative effects on the carbon footprint of urban buildings and the thermal comfort of the residents. Energy upgrading of building stock, due to a marked reduction of energy that can offer, is a requirement in modern societies. Till now, the energy upgrade focused on interventions on individual buildings. The possibility to develop green areas in densely built urban tissue is limited. Considering that free spaces are expensive, another type of strategies are being sought to address the urban heat island.

The new concept redefines the scope of energy upgrade, expands the study limits beyond the enclosed space and captures the building unity—surroundings. The building is not considered in isolation but as a whole with its surroundings. The new approach seeks thermal discharge of the microenvironment. The strategy of urban oases utilizes the stock material, serves the perception of preservation of the city centre as a "cultural object", goes along with a new momentum for the internal urban regeneration and enhances biodiversity in the city.

Both within cities as also outside urban tissue, a defensive strategy is often followed, which intervenes to reduce the symptom and is exhausted in quantitative measurements. But there are significant hidden factors which are not currently quantifiable—traceable and they affect global warming.

One of them is considered the material stock of abandoned structures, which multiply in an uncontrolled rate and strongly raises the question of their maintenance. Dismantling and removal is often not feasible due to the excessive cost and energy consumption. Intangible values, like cultural memories or embodied energy, suggest the strategy of reuse as a source of hope—a most efficient environmental option. The energy upgrading of existing cultural landscapes will be founded on the interdisciplinarity and will demonstrate synergy—in involvement of different factors and viewpoints—in a collective global climate regeneration effort.

References

Ambrakikos Gulf. Retrieved March 21, 2016, from https://www.wettbewerbe-aktuell.de/de/contents/5125/Rome+Community+Ring+RCR.html and https://www.pinterest.com/yacpinterest/rome-community-ring-awarded-proposals/

"Common Good Matrix", Economy for the Common Good and the CG Matrix. Retrieved March 21, 2016, from http://balance.ecogood.org/matrix-4-1-en/ecg-matrix-en.pdf

Consolidation of archaeological sites in Athens (in Greek language). Retrieved March 31, 2015, from http://www.astynet.gr/static.php?c=5

Davey, A. (2005). Oil & Gas UK Economic Report, 2013. Retrieved March 21, 2016, from http://oilandgasuk.co.uk/wp-content/uploads/2015/07/Adam-Davey-Oil-Gas-UK.pdf

European Landscape Convention. Retrieved April 24, 2015, from http://conventions.coe.int/Treaty/en/Treaties/Html/176.htm.

Galani, D. Pedestrianization D. Areopagitou—Ap. Pavlou (in Greek language). Retrieved March 31, 2015 from http://www.archaiologia.gr/wp-content/uploads/2011/07/70-15.pdf.

GCE. Retrieved March 21, 2016, from https://www.ecogood.org/en/ecg-balance-sheet/what-common-good-balance-sheet

Kanioris, S., Lahana, E., & Logothetis, G. (2009 December). *Port of Preveza. Prospects for development in connection with the operation of the traditional city center—implementation proposals* (in Greek language). Preveza: TEE/Department Epirus. Retrieved April 23, 2015, from http://issuu.com/pamepreveza/docs/.

Kunstareal. Retrieved March 31, 2015, from http://projekt.kunstareal.de/projekt/images/Publikationen/KAM%20Workshop%20Intro.pdf.

LCC. Economical valuation for sustainable construction using the Life Cycle Costing (EU). http://ec.europa.eu/enterprise/sectors/construction/files/compet/life_cycle_costing/common_methodology_en.pdf

Museumplein 1891–2020. Retrieved March 31, 2015, from http://www.tentoonstellingmuseumplein.nl.

National Mall. Retrieved March 31, 2015, from https://web.archive.org/web/20140304025556, http://www.savethemall.org/mall/resource.

Petrou, M., & Mantziou, L. (2014). *The Invisible Amvrakikos* (in Greek language). Dissertation. School of Architecture NTUA, Athens.

RAMSAR. Retrieved April 24, 2015, from http://www.ramsar.org/, http://www.ramsar.org/wetland/greece.

Rome Community Ring—Forte Portuense. Retrieved March 21, 2016, from https://www.wettbewerbe-aktuell.de/de/contents/5125/Rome+Community+Ring+RCR.html and https://www.pinterest.com/yacpinterest/rome-community-ring-awarded-proposals/

Scottish Enterprise Energy Team. (2008). The decommissioning market report.

The Institute of Petroleum. (2002, February). *Guidelines for the calculation of estimates of energy use and gaseous emissions in the decommissioning of offshore structures.*

UNEP (Environment Protection Global Issues and Greece Protected Areas).

UNESCO. Retrieved March 21, 2016, from http://whc.unesco.org/en/culturallandscape/

Wan Abdullah Zawawi, N. A., Liew, M. S., & Na K. L. (2014). *Decommissioning of offshore platform: A sustainable framework.* 2014 MSSA Malaysian Structural Steel Association.

WWF on Wetlands. Retrieved March 21, 2016, from http://wwf.panda.org/about_our_earth/about_freshwater/intro/value/, Amvrakikos Rodia Wetlands Center. Retrieved March 21, 2016, from http://www.rodiawetlands.gr/

Part V
Energy Technologies and Their Effect on Global Warming

Solar Energy

Investigation of solar still performance and novel combination of solar energy with methane reforming reactors and organic Rankine cycle for power generation.

Chapter 19
Solar-Driven Continuous Methane Reforming Reactor

M. Lange, J. Lapp, R. Rieping, L. de Oliveira, M. Roeb, and C. Sattler

Introduction

The world energy demand is covered mainly by the use of fossil fuels. Apart from the direct use of fossil fuels, however, the use of synthetic (mainly liquid) fuels has been attracting a lot of interest lately. Currently, synthetic fuels are produced at a commercial scale from carbonaceous sources with low energy content, which are upgraded at the expense of additional energy. This additional energy is also obtained from the combustion of fossil fuels. In order to make the latter procedure more attractive and environmentally friendlier, these processes could be combined with a renewable energy source, such as solar energy.

Although there is an increase in the use of alternative energy vectors, e.g. renewable electricity, the reliance on fossil fuels is still extensive. The main advantage of liquid fuels when compared for example to electricity is their prompt availability upon demand regardless of temporal or location-related restrictions.

The preservation of economic stability and growth creates the most significant motive for industries to start pursuing environmentally friendlier energy vectors. Existing technologies are further investigated and improved in order to minimize CO_2 production and renewable energy installations have started to appear with higher frequency. An important step forward would be to employ solar energy, an inexhaustible energy source, in order to produce synthetic fuels.

As a consequence, in the last decade an increased research activity could be noted regarding the exploitation of the solar potential to produce Hydrogen or synthesis gas (hydrogen + carbon monoxide) through the dissociation of water and carbon dioxide (Ermanoski et al. 2014; Roeb et al. 2008, 2012; Muhich

M. Lange (✉) • J. Lapp • R. Rieping • L. de Oliveira • M. Roeb • C. Sattler
German Aerospace Center (DLR), Institute of Solar Research, Linder Höhe,
51147 Köln, Germany
e-mail: matthias.lange@dlr.de

P. Grammelis (ed.), *Energy, Transportation and Global Warming*, Green Energy and Technology, DOI 10.1007/978-3-319-30127-3_19

et al. 2013). On one hand solar water-splitting is a totally renewable process with zero greenhouse gas emissions, while on the other hand solar carbon dioxide splitting utilizes CO_2—a material which is currently characterized as an undesirable waste—as a reactant to finally produce valuable solar fuels. However, these processes have not yet reached the essential maturity for large-scale industrial implementation.

A process that is considered to have one of the highest short term potential in terms of technology and economics is solar reforming of methane (Agrafiotis et al. 2014). With such an approach longer term carbon-neutral or carbon-free energy solutions, such as water and carbon dioxide splitting with the aid of solar energy for the production of solar fuels, can be given the required time to be developed. Further, the feedstock natural gas may be replaced in the future by biogas sources (Bereketidou and Goula 2012).

In this paper we propose a new reactor concept which acts as a solar receiver-reactor and heat exchanger at the same time. This multi-use approach enables continuous methane reforming and even night time operation is possible.

Concept

Several solar reactors for methane reforming have been developed and studied previously (Wörner and Tamme 1998; Agrafiotis et al. 2014). The novelty of the proposed concept is its flexibility regarding the solar radiation. The key to this flexibility lies in the receiver geometry. The core of the receiver consists of a monolithic honeycomb structure with two different and separated sets of channels, see Fig. 19.1.

For daytime operation, concentrated solar radiation is focused on the front face of the monolith, heating it up to 900 °C. In one set of channels, the solar heat is used to run the endothermic reforming reaction. Steam reforming (19.1) and dry reforming (19.2) or a mixture of both is possible (Rostrup-Nielsen 2002):

$$CH_4 + H_2O \rightarrow 3H_2 + CO \quad \Delta H^\circ\, 298K = +206\, kj/mol \qquad (19.1)$$

$$CH_4 + CO_2 \rightarrow 2H_2 + 2CO \quad \Delta H^\circ\, 298K = \; + 247\, kj/mol \qquad (19.2)$$

In the other set of channels, heat is taken up by a heat transfer medium, which we propose to be air. This heat transfer medium transfers the solar heat to a thermal storage close to the reactor.

During nighttime operation, the air flow is reversed in order to transfer the heat from the storage to the receiver. With this heat, the reforming reaction in the first set of channels can be powered even without direct solar input.

A process flow diagram of the concept is shown in Fig. 19.2.

This new concept brings along the potential to reduce costs, because by continuous operation all the downstream components can be reduced in size and thus

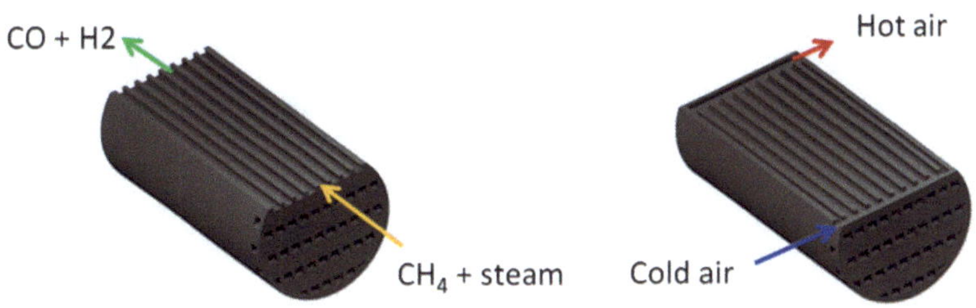

Fig. 19.1 Concept drawing of novel receiver reactor monolithic structure

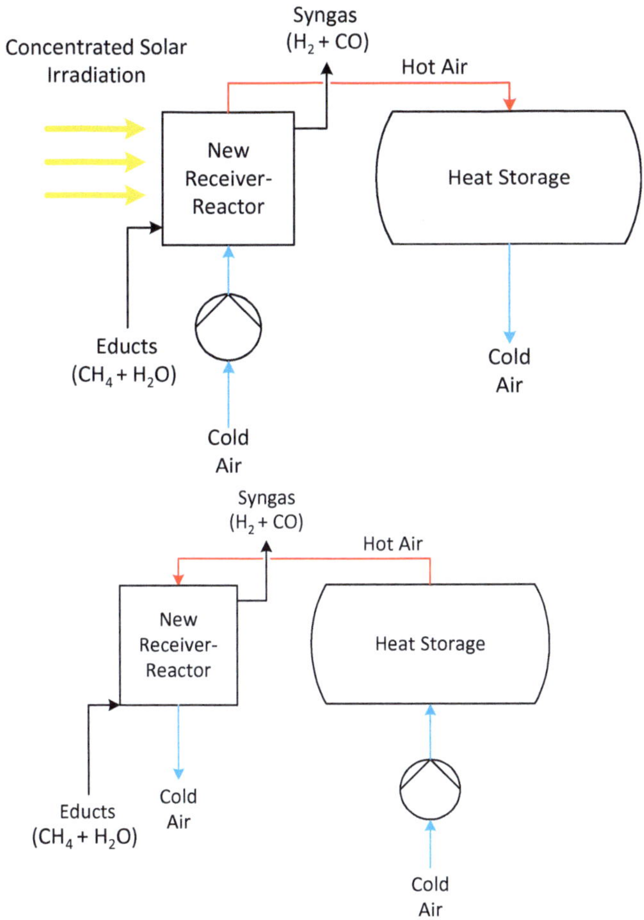

Fig. 19.2 Process scheme (*top*: on-sun operation mode; *bottom*: off-sun operation mode)

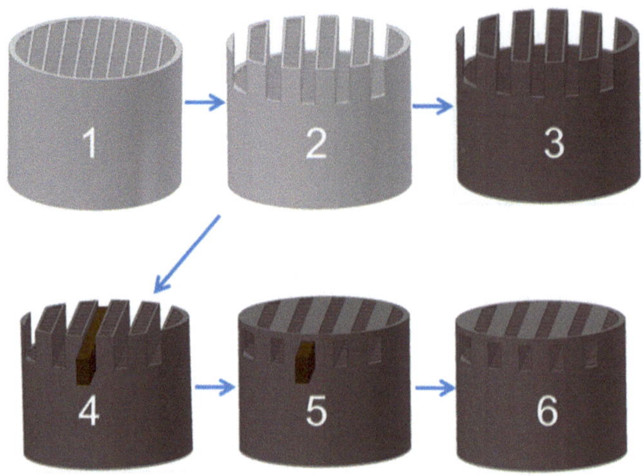

Fig. 19.3 Production method of custom honeycomb (only upper half shown here, lower half is symmetric). (*1*) Extruded RSiC monolith, (*2*) cutting of slots, (*3*) infiltration to SiSiC, (*4*) catalyst coating and preparation for closing of slots, (*5*) closing of slots, (*6*) burning and removal of stick for closing of slots

costs. Additionally, less start-up and shut-down cycles will be necessary which is beneficial from an operational as well as from a durability point of view.

Production of Monolith

As the monolith concept is novel, no standard production method is readily available to create such a geometry. Therefore, a custom-made honeycomb was produced. The production method, which is illustrated in Fig. 19.3, is explained in the following.

First, a common honeycomb made from recrystallized silicon carbide (RSiC) was acquired. Such pieces are extruded mainly for the automotive industry (catalytic converters, diesel particulate filters). As the channel walls of the monolith need to be very gas tight in order to avoid crossover flow between the two channel types, the porous RSiC needs to be infiltrated with further silicon. Only then, the channel walls are gas tight. However, the Si-infiltrated silicon carbide (SiSiC) is extremely hard to machine. Therefore, before infiltration, the slots which are necessary for the side inlet channels were cut into the RSiC piece. Then, infiltration was carried out. After that, the monolith walls were coated with a Rhodium and Al_2O_3 catalyst which is necessary for the reforming reaction. In a final step, the slots which had been cut into the front and back of the monolith were closed with ceramic glue and SiC-sticks.

Some pictures showing the real production steps are shown in Fig. 19.4.

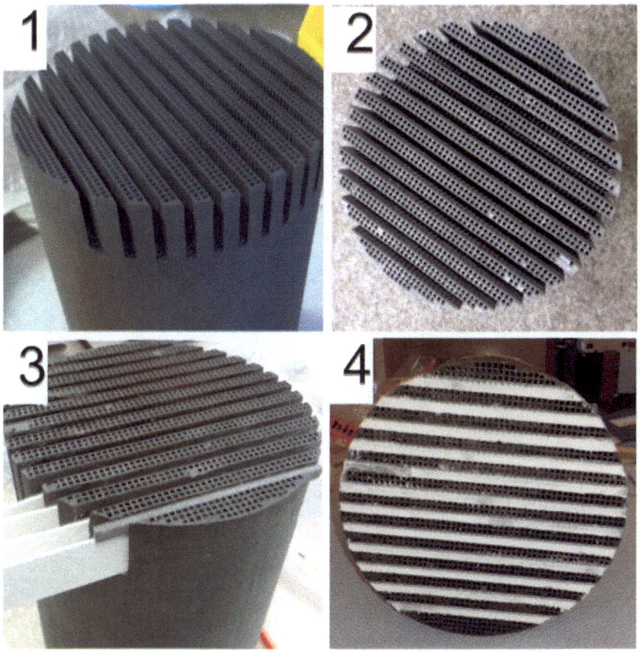

Fig. 19.4 Steps of producing the monolith. (*1*) Slots in RSiC block, (*2*) SiSiC block, (*3*) closing of slots, (*4*) final geometry

Experimental Set-up

The concept is currently being experimentally evaluated at the high flux solar simulator of DLR, Cologne. The reactor during assembly is displayed in Fig. 19.5.

The solar simulator has a maximum power input of 20 kW. The simulator consists of 10 xenon lamps which have a spectrum similar to the solar spectrum. This solar-like concentrated radiation will be used to simulate the daytime operation. Half of the input energy will be used for the reforming reaction and the other half will be taken up by the air stream.

A scheme of the experimental set-up is shown in Fig. 19.6.

This first test campaign focuses on only thermal characterization of the concept. This way, a more sound characterization of the thermal behaviour of the system is possible, because no heat of reaction distorts the thermal result. Further, valuable information can be collected to decide how the reactor should be operated.

In the experimental setup, two gas heaters are included. The first one, which is located upstream of the reactor in the reactants stream, is employed in order to simulate heat recovery from the product gas stream. In a future plant, such heat recovery would most likely be implemented. As real heat recovery with a heat exchanger would reduce the flexibility of the set-up and thus also the scientific output of the experiments, we decided to replace a real physical heat recovery by

Fig. 19.5 *Left*: assembly of monolith in reactor shell, *right*: final design of reactor with glass window for solar input

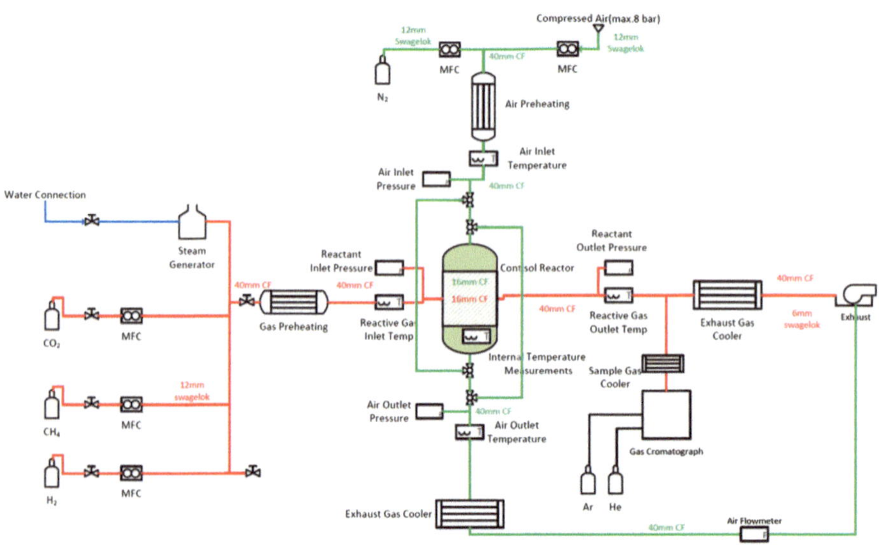

Fig. 19.6 Experimental set-up for continuous methane reforming

this first gas heater. The second gas heater is placed in the air stream before entering the reactor. This heater enables to simulate nighttime operation. Again, we apply this electrical heater instead of a heat storage in order to stay as flexible as possible, in order to increase the scientific output and also to save expenses.

For the experimental campaign, the methodology called Design of Experiments is applied. This method is based on statistical evaluation of the results. During the experiments, not one factor at a time (OFAT) will be changed, but several parameters (like mass flows, inlet temperatures and radiation) will be changed at once. By statistical analysis, the results have a more general character than results obtained with the OFAT method, because interactions between parameters can be assessed.

Conclusion

A novel receiver reactor concept for continuous operation of solar thermochemical processes has been proposed. The core of the concept is a honeycomb monolith with two different sets of channels. One set of channels is used for a solar-powered reaction, the other set of channels acts as a heat exchanger to take up heat during the day and bring it back into the reactor at night.

As the monolith is not a standard part, it was custom made. The production process was described. Further, the monolith is integrated into a reactor, which in turn is integrated into an experimental set-up in a high flux solar simulator. In this experimental set-up, it is possible to simulate daytime as well as night time operation.

Acknowledgement The authors thank the German Ministry of Education and Research for funding this work within the project ContiSol (Contract Number 03SF0468). We also greatly acknowledge the work of our colleagues and project partners Dr. Souzana Lorentzou and Christodoulos Lekkos from Aerosol & Particle Technology Laboratory, CERTH-CPERI.

References

Agrafiotis, C., von Storch, H., Roeb, M., Sattler, C. (2014). Solar thermal reforming of methane feedstocks for hydrogen and syngas production—A review. *Renewable and Sustainable Energy Reviews, 29*, 656–682.

Bereketidou, O. A., & Goula, M. A. (2012). Biogas reforming for syngas production over nickel supported on ceria–alumina catalysts. *Catalysis Today, 195*(1), 93–100.

Ermanoski, I., McDaniel, A. H. (2014). Solar hydrogen production with a metal oxide based thermochemical cycle. *DOE Annual Merit Review*: 25.

Muhich, C. L., Evanko, B. W., Weston, K. C., Lichty, P., Liang, X., Martinek, J. (2013). Efficient generation of H2 by splitting water with an isothermal redox cycle. *Science, 341*(6145), 540–542.

Roeb, M., Neises, M., Monnerie, N., Call, F., Simon, H., Sattler, C. (2012). Materials-related aspects of thermochemical water and carbon dioxide splitting: A review. *Materials, 5*(11), 2015–2054.

Roeb, M., Neises, M., Säck, J.-P., Rietbrock, P., Monnerie, N., Dersch, J. (2008). Operational strategy of a two-step thermochemical process for solar hydrogen production. *International Journal of Hydrogen Energy, 34*(10), 4537–4545.

Rostrup-Nielsen, J. R. (2002). Syngas in perspective. *Catalysis Today, 71*(3–4), 243–247.

Wörner, A., & Tamme, R. (1998). CO_2 reforming of methane in a solar driven volumetric receiver-reactor. *Catalysis Today, 46*, 165–174.

Chapter 20
Specific Applications/Examples: Use of Solar Energy in Fishing: Community Initiatives

J. Vincent Jain and Satish Babu

Research Background

Any developmental project or research project without considering the climate change factor will be incomplete. Indeed, 'climate change has become the defining generational challenge for the international community.' It is considered one of the biggest threats to our world today. The aftermath of today's global condition, if not given due attention locally and globally, would harshly affect the young generation and the generations to come and they won't find the earth a better place to live. The significant rise in global temperature and human activities has worsened the situation. Energy saving, conservation of water and minimization of emission of CO_2 help to balance the Global Warming.

Introduction

Fishery is an area where an alternative energy technology, products and its services are needed to minimize the fossil fuel dependency which could save large amount of energy, water and reduce carbon emission. Fishing is considered as the most energy-intensive food production method in the world. In India, thousands of large- and medium-size fishing boats—from trawlers and seiners to liners and

J.V. Jain (✉)
Association of Deep Sea Going Artisanal Fishermen, Shark Street, Thoothoor P.O., Kanyakumari District, Tamil Nadu, India
e-mail: jain1.siffs@gmail.com

S. Babu
ICFOSS, Technopark, Trivandrum, Kerala, India
e-mail: sbabu@ieee.org

P. Grammelis (ed.), *Energy, Transportation and Global Warming*, Green Energy and Technology, DOI 10.1007/978-3-319-30127-3_20

gillnetters—depend primarily on one fossil fuel, diesel or kerosene. Large volumes of energy and water are used to produce inputs for pre harvest and post harvest activities. If we are able to use an alternate energy technology, products and its services properly we could save large amount energy, water and avoid unnecessary emission of CO_2. There is scope to reduce CO_2. by setting emission norms and improving fuel efficiency of marine fishing boats.

Problem

An artisanal deep sea fisherman takes 15–20 days for a fishing voyage. A few others take 25–30 days and more. During the voyage they have to keep the engine running the engine for charging of the batteries. The engine keeps running even they are not fishing. Some fishermen keep running the engine to prevent restarting trouble. About 20–25 % of fuel is used for the continuous operation of the engine. The signal lights, navigational and safety equipments are operating with the support of batteries in the boat. Many times fishermen have met with accidents and have been unable to use the navigational and safety equipments because of engine failure.

Community Initiatives

Though tapping solar energy for land-based applications is very popular, its use in fisheries sector is uncommon. In India, the idea of using solar energy in fishing boats was first initiated in the year 2010 by the Association of Deep Sea Going Artisanal fishermen (ADSGAF), a fishermen organization based in Thoothoor, Kanniyakumari District, Tamil Nadu, India.

The ADSGAF was established in 1992. It has 340 full-time members who are engaged in long lining and gill-netting using small mechanized boats (15–21 m in length) within the Indian Exclusive Economic Zone (EEZ). There are around 3400 fish workers and their dependents associated with the organization. There are 600 deep sea going fishing vessels in Thoothoor area that they fish all over the Indian EEZ, which covers an area of 2.02 million sq. km. During 2009, the members of ADSGAF raised the issue of increasing searching time for fish coupled with rising fuel cost leading to decline in their income. The members also expressed concerns about increasing cost of running the engine during idle time for lighting and signalling requirements. It was found that fuel cost is accounting for about 70 % of the total operational costs.

Faced with this challenge and also considering the carbon footprints generated by this fishery, the ADSGAF decided to experiment with the use of solar energy in fishing vessels in 2010. Initially, the ADSGAF made a prototype boat (3 feet on scale) equipped with solar panels and associated batteries etc and demonstrated its operation and technical feasibility under laboratory conditions. In August 2012, the ADSGAF further demonstrated the pilot-scale initiative in Chennai with the involvement of the Bay of Bengal Programme, an Inter-Governmental Organisation (BOBP-IGO) working in Bangladesh, India, Maldives and Sri Lanka.

Why Solar Energy

Solar energy is the cleanest, most abundant and renewable energy source available. Today's technology allows us to capture this power in several ways giving the public and commercial entities flexible ways to employ both the heat and light of the sun. Solar energy can be produced on a distributed basis, called distributed generation, with equipment located on rooftops or on ground-mounted fixtures close to where the energy is used. Large-scale concentrating solar power systems can also produce energy at a central power plant.

Techno-Economic Assessments and Pilot Trials

The ADSGAF since then carried out a couple of techno-economic assessments and pilot trials to measure the efficacy and viability of their venture in different categories of fishing boats. It is found that while at sea the engine is run for several hours at partial load to provide power for auxiliary activities such as lighting, communication, navigation, signalling and on-board entertainment. Therefore, if these services can be run by using alternative power sources there could be a significant saving of fuel and this will improve the viability of the operation as well as reduce the carbon footprints. From the pilot- scale trials, it is seen that switching off the engine for 3 h can save 30 l of fuel a day and 7500 l per year. The potential reduction of carbon dioxide emissions per year if the entire fleet switches off the engine for three hours a day is estimated to be 117,000,00 kg (11,700 tonnes) (refer Table 20.1). Therefore, solar power seems to be a viable alternative to reduce fuel cost in fishing operations.

The size of the boat (Fig. 20.1) is 15.5 m length and 4.5 m width which undertake a voyage of about 15 days. It operates long lines and gill nets. Four solar panels of 1 kW (250 W each) are fitted at the top of the wheel house of the boat. Presently there are six such boats are in operation. The solar energy accumulated during the day is able to light the bulbs/tube lights/focus lights and also charge/operate the GPS, fish finder and the two VHF sets. Additionally, when required during night time, the transistor batteries can also be charged.

Mr. Sunil who is the skipper of the vessel for the past 10 years told that he is satisfied with the light while fishing and he never find trouble in starting and re-starting the engine during the fishing voyage. Mr. Jose, a crew member said, "Usually light in the boat will not be bright and need utmost care for removing

Table 20.1 The economics of using solar energy in fishing boats calculated from the Thoothoor experience

Sl. No.	Parameter	Units	Conversion
1.	Number of fishing boats in Thoothoor	600	
2.	Fuel required to run engine for one hour	10 l	
3.	Extra running of engine (mainly during night time)	03 h	
4.	Number of fishing days in one year	250	
5.	Cost of fuel- diesel per litre	INR. 60	USD. 1
By switching off the engine for 3 h and running it on solar power:			
6.	Saving of fuel per year for 1 boat × 30 l × 250 fishing days	7500 l	
7.	*Saving of fuel per year for 600 boats × 30 l × 250 fishing days*	*45,00,000 l*	
8.	Potential cash savings per year per boat (7500 × INR. 60)	INR. 4,50,000	USD. 7500
9.	*Potential cash savings per year for 600 boats × INR.4,50,000*	*INR. 27,00,000,00*	*USD. 45,000,00*
10.	Approximate cost for 1kw solar power plant including modification of wheel house and other fitting charges	INR. 2,50,000	USD. 4167
11.	*Approximate cost for installing 1kw solar power plant each in 600 boats including modification of wheel house and other fitting charges INR. 2,50,000 × 600*	*INR. 15,00,000,00*	*USD. 25,000,00*
12.	Net saving to a boat per year	INR. 2,00,000	USD. 3333
13.	*Net saving to 600 boats*	*INR. 12,00,000,00*	*USD. 19,998,00*
Environmental benefits by reducing Carbon Emission:			
14.	Emission of CO_2 per litre of diesel consumption	2.60 kg	
15.	Potential reduction in emission of CO_2 (1 boat × 7500 l × 2.60 kg)	19,500 kg	19.5 ton
16.	*Potential reduction in emission of CO_2 (600 boats × 7500 l × 2.60 kg)*	*117,000,00 kg*	*11,700 ton*

the fish from the hooks and line. But solar energy fitted lights are bright which enabled us to handle the fish and hooks and line without much stress. It also saved considerable amount of time." Mr. Salin, another crew said that they had full lighting in their boat. The other fishing boats and ships passed over by them gave a strange look because of the unusual brightness in their boat. He added that usually they are stingy or miser in using lights during night because of the recharging problem. But with the solar energy, they are encouraged to use the lights for the whole nights. Mr. Sunil said that they could save approximately 320 l of diesel in this voyage. Cost wise, they could save an amount of INR. 19,200 (320 l × INR. 60).

Fig. 20.1 Solar power equipped boat (funded by National Bank for Agriculture and Rural Development (NABARD))

Solar Power for Multi-Day Fishing Boats

This initiative was based on the successful trial and subsequent operation of solar energy system in the deep sea fishing boats. Generally, fishermen carry charged battery in boats for lights and for the operation of navigational and safety equipments. Recharging will be a major problem for smaller boats. They need to run the engine exclusively for recharging because while fishing, they do not run the engine. Running the engine for recharging is highly expensive and affects the income. The major expenditure in each fishing voyage is fuel. The department of fisheries, Government of Tamil Nadu is supplying 250 l of kerosene for one month at the cost of INR. 25 per litres. But, it is sufficient for only one voyage. The duration per voyage is 5–6 days. They buy from the black market for their requirements and the cost per litre varies from INR. 40 to INR. 45. It is expected that the solar energy system will help them to save fuel cost and will be a boon to the community (Fig. 20.2).

Fig. 20.2 Solar power equipped multi-day fishing boat

Solar Lanterns for Catamaran Fishermen and Fish Vending Women

The solar lantern uses solar energy and helps to reduce carbon emission and free from the risk of fire. The emission of carbon in 1 litter kerosene is 3 kg. Solar Lanterns are environmental friendly and can be used even in remote villages where there is no electricity. Even in the urban areas people prefer a solar lantern as an alternative during power cuts because of its simple mechanism. Photovoltaic cells are solid-state semiconductor devices that convert Solar Energy into Electricity (direct current). These Solar cells are connected in series or in parallel to obtain desired output and are called solar modules. Solar lantern is a simple application of solar photovoltaic technology, which has found good operations in rural regions where the power supply is irregular. ADSGAF initiated to encourage the fish vending women and Catamaran fishermen to use solar lanterns for their occupational operations.

Fish Vending Women use Candles

Generally fish vending women use candles or kerosene lamps at the fish markets. These lights are not safe and not bright enough to use in the market. Poor quality

Fig. 20.3 Fish vending women use candles

and irregular lighting blights lives. The light and the smoke from kerosene lamps not only strains eyes but also release harmful carbon and also there is risk of fire (Figs. 20.3 and 20.4).

Catamaran Fishermen Use Solar Lantern

Catamarans,[1] those engage in fishing either at day or at night which operate generally one man manually. During night they use kerosene lamps or gas lights for handling fishing gears, overseeing the fishing accessories etc. The kerosene lights don't have brightness unlike the gas lights. The kerosene lights are not water proof (Fig. 20.5; Table 20.2).

Cost Benefit Analysis

No of working days in a year are 250 (Tables 20.3, 20.4, and 20.5).

Many types and designs of solar lanterns are available with a price range. They are ideal for not only fish vendors or fishers but also good for farmers and other vendors, and children to read and write. Solar lights can also be used at home as light after the sunset. They are highly durable and having longer life than other lanterns. Solar lanterns would be cheaper than kerosene in the long run but the short

[1] Catamaran is a fishing craft, made of logs of *Albizzia falcataria* timber lashed together with rope.

Fig. 20.4 Fish vending women use solar lantern

Fig. 20.5 Solar lantern fitted catamaran

Table 20.2 Gas lights, kerosene lamp and candle VS solar lanterns

Sl. N	Problems identified in using of Gas lights, Kerosene lamp and candle	Merits of the use of solar lanterns
1	Harmful effects—breathing of smoke	There is no harmful effect to health
2	The brightness of candle/kerosene lamp is not sufficient	The solar lanterns are brighter than candle/kerosene lights
3	Can't keep the candle/kerosene lamp when it is windy	Can keep the solar lanterns when it is windy
4	Need utmost care in using	Easy to handle
5	Risk of fire	No risk
6	Non availability of gas/kerosene on time	Can charge at all sunny days
7	Need to pay for candle, kerosene, and gas filling charge every day	One time investment
8	Gas lights are water proof but corrosion	No corrosion

Table 20.3 In the case of a fish vending woman who use candle

	Particulars	INR	USD
1	Candle expenditure per day	20	00.33
2	The total cost is (20 × 250 days)	5000	83.00
3	The cost of 1 solar lantern is	1800	30.00
4	Total saving in an year	3200	53.00

Table 20.4 In the case of a Catamaran fisherman who use Kerosene

	Particulars	INR	USD
1	Kerosene expenditure per day	60	01.00
2	Total cost in an year (60 × 250)	15,000	250.00
3	The cost of a solar lantern is	2800	46.70
4	Total savings in an year	12,200	203.30

Table 20.5 In the case of a Catamaran fisherman who use gas

	Particulars	INR	USD
1	Gas filling expenditure per day	70	01.17
2	Total cost in an year (70 × 250)	17,500	291.70
3	The cost of a solar lantern is	2800	46.70
4	Total savings in an year	14,700	245.00

run cost is higher. There are no authorized gas filling stations and it is always a problem to the fishermen for filling gas for every fishing voyage. The fishermen may not be in hurry or may not be in worry in gas filling of every fishing voyage, if they have solar lanterns. Solar lanterns are safe and easy to handle and no need to pay recharging fee. We could eliminate unauthorized gas filing stations run by some people in the villages.

Solar Powered Refrigerated Trucks for the Transportation of Fresh Fish

The dispersed and distant living population requires proper means for food preservation and distribution without its spoilage. The existing food preservation and transportation technology has developed sufficiently to preserve the wide variety of foods for a considerable long time without losing the freshness. In the case of fish, a highly perishable commodity needs best preservation and transportation to avoid spoilage and to maintain freshness. This is significantly enhancing the costs—both capital and operational. The ADSGAF has initiated a means of reducing the operational expenses of refrigerated transport using solar power (Fig. 20.6).

Features

- Fuel not required for operation of the system
- Alternate system is arranged, if solar system fails
- Better reliability, simple and versatile
- Easy to operate
- Money and fuel saving
- Back up for night operation (Table 20.6).

Fig. 20.6 Solar powered refrigerated truck (funded by National Fisheries Development Board, Government of India)

Table 20.6 Description of solar powered refrigerated truck

Sl.No	Particulars	INR	USD
1	Cost of Vehicle Mahindra Mxi Truck Plus	5,23,000	8717
2	Cost of body building with PUF insulation	2,10,000	3500
3	Cost of refrigeration Refrigerant-R134a	2,70,000	4500
4	Cost of solar system (1.5 KW) Solar panels 250 W × 6	1,50,000	2500
5	Total	11,53,000	19,217

The Larger Picture

India has a fleet size of 1,94,490 crafts comprising 72,559 (37.3 %), mechanized crafts, 71,313 (36.7 %) motorized crafts and 50,618 (26 %) non-mechanized crafts.[2] It indicates that 1,43,872 (74 %) of the fishing crafts depends on fossil fuel either on diesel or kerosene. If 1 l of fuel is saved per day, the daily fuel saving will amount to 1,43,872 l (0.15 million) and reduction of carbon emission will amount to a total of (1,43,872 × 2.6) 3,74,067 kg (374 ton) day. Assuming that the Indian fishing fleet fishes for 250 days in a year, the total fuel saving will amount to 35.97 million litres and the carbon reduction will amount to 9,35,167,50 kg (93,516 ton) (3,74,067 × 250 fishing days). The cash saving in a day will be INR. 8,63,232,0 (USD. 14,387,2) and in a year will be INR. 2,15,808,00,00 (USD. 35,968,00).

Ongoing Programme

Based on the immediate and future needs in the fisheries, the community established a SIGHT[3] under IEEE[4] called Artisanal Deep Sea Fishers SIGHT. The main objective of the SIGHT is to develop ideas, concepts, projects and products with the help of existing scientific know-how, managerial efficacy and engineering expertise and along with the traditional knowledge of fishermen. Presently the ADSGAF is on the process of introducing solar powered OBM[5] fitted boats, solar powered small ice-making units and tricycles for fish vending women and also establishing solar power charging stations.

---

[2] As per the 2010 National Marine Fisheries Census conducted by the CMFRI, Kochi.

[3] Special Interest Group on Humanitarian Technology.

[4] Institute of Electrical and Electronic Engineers.

[5] Out Board Motor.

Suggestions

The initial investment cost of the solar power projects is high. People do not realize the long term benefits. Hence it is necessary to organize intensive awareness programmes to motivate the public to undertake investments in such projects. Further the people have to be familiarised with the handling of solar power equipments. There is need for conducting regular training sessions for handling solar equipments. It is necessary to establish after sale services and also make available solar power parts locally which are lacking today and which is one of the major reasons for lack of interest using solar power systems. It is necessary to increase the number of quality technicians in this field. It is also necessary to do more study, experimentation and need sufficient funding to bring out right combination of goods and services in the fisheries.

Conclusion

From the energy consumption, sustainability and cost perspectives, there is an urgent need for initiatives that reduce fuel consumption, substituting carbon-based fuels with renewable alternatives. The initiatives taken by the community that promote reducing the operating costs minimize dependency of fossil fuel and minimize the carbon emission. Thus use and application of solar energy will not only be saving fuel, money, time but also help to minimize the carbon emission which is one of the major problems for global warming. This will lead to green revolution in the fisheries sector.

Acknowledgment The authors acknowledge the contributions of Dr. Y.S. Yadava, Director and Mr. Prajdeep Mukherjee, Policy Analyst BOBP-IGO, Chennai, India, Mr. V. Vivekanandan, Director, FishMARC, Trivandrum, India, Mr. Sebastian Mathew, Executive Secretary, International Collective in Support of Fishworkers, Chennai, India, Dr. E. Vivekanandan, Emeritus Scientist, CMFRI, Chennai, India, Mr. Sadhish Kumar, Project Head, Jagath Jothi Solar Energy Pvt Ltd, Chennai, India, Mr. Vivek Dwivedi, Managing Director, BRIJ EXIM, Pune, India, Dr. Chinnappan Gasper, Professor, Centre for Development Studies, Trivandrum, India and Dr. Teresa Shobana, Professor, St. Jude's College, Thoothoor, Tamil Nadu, India in the preparation of this paper. Special thanks to BOBP-IGO for providing advice and support to the initiatives taken by ADSGAF.

References

Department of Economic and Social Affairs, World Economic and Social Survey 2008: Overcoming Economic Insecurity, E/2008/50/Rev.1, United Nations, New York, 2008.

Wilson, J. D. K. (1999). Fuel and financial savings for operators of small fishing vessels. *FAO Fisheries Technical Papers, 383*, 46.

Vivekanandan, E., Singh, V. V., & Kizhakudan, J. K. (2013). Carbon footprint by marine fishing boats of India. CMFRI, India. *Current Science, 105*(3), 10 August 2013.

Chapter 21
Diagnostic of Sensors for Induction Machine Powered by Photovoltaic Generator Based on Fuzzy Logic Techniques

A. Amrane, A. Larabi, and A. Hamzaoui

Introduction

Many industrial applications require fault tolerance and continuity of service (Benbouzid et al. 2007). This is due to the growing need to improve the availability of systems. Therefore, strategies are defined in the early stages of design, to facilitate fault detection, localization and reconfiguration of the order. For this, several recent works (Green et al. 2003; Abolhassani and Toliyat 2009) deal with fault tolerant control of electric drives.

In this article, we focus our study on the induction machine (IM), conventionally designed for constant speed applications, has become, due to its simple, robustness structure and the evolution of electronics power and control of the control vector machine most commonly used for variable speed drives. This machine has the advantage of being more robust and less expensive, with equal power, as other machines. However, it has drawn backs. This allowed the opening of various lines

A. Amrane (✉)
Laboratory of Systems Electric and Industrial (LSEI), Faculty of Electronics and Informatics (FEI), University of Science and Technology (USTHB), BP 32, ElAlia, Bab-Ezzouar, 16111 Algiers, Algeria

National School of Technology, National Road N 5 Z.I., Rouiba, 16013 Algiers, Algeria
e-mail: amraneahmed@gmail.com

A. Larabi
Laboratory of Systems Electric and Industrial (LSEI), Faculty of Electronics and Informatics (FEI), University of Science and Technology (USTHB), BP 32, ElAlia, Bab-Ezzouar, 16111 Algiers, Algeria
e-mail: larabiabdelkader@yahoo.fr

A. Hamzaoui
CReSTIC of Troyes University of Reims Champagne Ardennes Troyes, Reims, France
e-mail: abdelaziz.hamzaoui@univ-reims.fr

P. Grammelis (ed.), *Energy, Transportation and Global Warming*, Green Energy and Technology, DOI 10.1007/978-3-319-30127-3_21

of research on its control and power (Shoudao et al. 2004; Montanari et al. 2000; Ouhrouche 2002).

We introduced in this work, an approach widely used for the detection and isolation of faults based on the method of estimator (Chen and Patton 1999; Holtz 1996), the main components of a tolerant control strategy to defects (CTD) sensor. It is divided into three stages: detection, isolation and reconfiguration.

A novel estimator schemes based on fuzzy logic algorithm for the speed estimation where derived using Lyapunov's stability theorem (Ouhrouche 2002; Amrane and Larabi 2012; Amrane et al. 2014). Several strategies have been proposed for the estimation in the vectorial induction motor drives (Ouhrouche 2002; Amrane and Larabi 2012; Amrane et al. 2013). Among these techniques, model reference adaptive systems (MRAS) schemes are the most common strategies employed due to their relative simplicity and low computational effort (Amrane and Larabi 2012; Amrane et al. 2013).

Can be seen, the majority of estimation schemes described in the literature for MRAS observer employ a PI controller to generate the desired value. However, due to the continuous variation in the machine parameters and the operating conditions, in addition to the nonlinearities present in the inverter, the PI controllers may not be able to provide the required performance. Not much attention has been devoted to study other types of estimation scheme (Amrane and Larabi 2012; Amrane et al. 2013; Larabi et al. 2013).

In this article, the performances of the diagnostic of sensors, the detection of faults and the fuzzy estimation of speed for the induction machine are analyzed by digital simulation. To achieve our objective, this paper is organized as follows. Section "Introduction" presents the control of the asynchronous machine; the rest of the paper is organized as follows. Sections "Control of Asynchronous Machines" and "Sensorless Control of the Induction Machine" present the control of the IM and the MRAS algorithm using to estimate the speed. Section "The Fuzzy Logic Adaptation" depicts fuzzy logic algorithm for the estimation. Section "Modelisation and Control of Generator Photovoltaique" provides and discusses the simulation results and Section "Simulation Results" sketches some conclusions.

Control of Asynchronous Machines

Control Scheme of an Asynchronous Machine

Figure 21.1 shows a simplified block diagram illustrating the essential of our control system. This system consists of a dc power source, a dc link filter, a voltage source inverter, an induction motor, and same circuit of control system. The dc power source converts the constant-frequency ac power to dc power by a three-phase, full wave diode bridge rectifier; the dc voltage is smothered by a smoothing capacitor dc link filter and then applied to a three-phase bridge inverter witch converts dc power to variable voltage variable frequency ac power supply to the motor. The two control parameters required are frequency and voltage, the frequency command also generates the voltage command through a volts/hertz ratio.

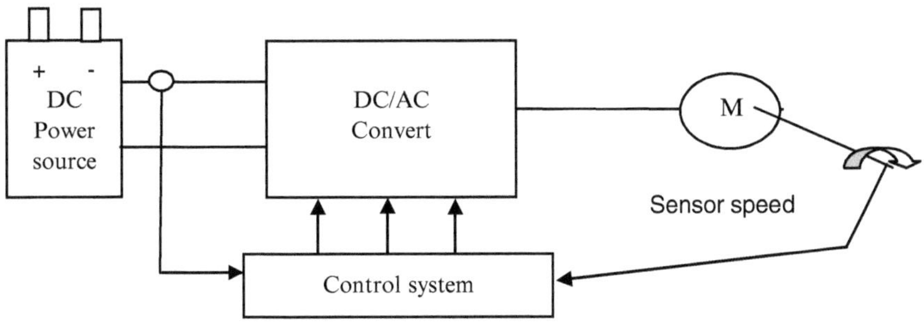

Machine powered by
photovoltaic generator

Fig. 21.1 Block diagram of control system

Induction Machine Model

The considered system is an induction machine with a squirrel cage of three phases. The state representation of the mathematical model of this machine can be represented according to usual d-q axes, as follows (Ouhrouche 2002; Amrane and Larabi 2012; Amrane et al. 2013, 2014; Ho and Yeh 2010)

$$
\begin{aligned}
\frac{d\,ids}{dt} &= 1/(\sigma^{*}Ls)\left[\text{-}Rsm^{*}\,ids + \omega s^{*}\sigma^{*}Ls^{*}\,iqs + \frac{Lm}{(Lr^{*}Tr)}{}^{*}\varphi dr\right]\\
\frac{d\,iqs}{dt} &= 1/(\sigma^{*}Ls)\left[\text{-}Rsm^{*}\,iqs\text{-}\omega s^{*}\sigma^{*}Ls^{*}\,ids\text{-}\frac{Lm}{(Lr^{*}Tr)}{}^{*}\varphi qr\right]\\
\frac{d\,\varphi dr}{dt} &= \frac{Lm}{Tr}{}^{*}ids\text{-}\frac{1}{Tr}{}^{*}\varphi dr + (\omega s - \omega r)^{*}\varphi qr\\
\frac{d\,\varphi qr}{dt} &= \frac{Lm}{Tr}{}^{*}iqs\text{-}\frac{1}{Tr}{}^{*}\varphi qr - (\omega s - \omega r)^{*}\varphi dr\\
Cem &= \frac{P^{*}Lm}{Tr}(iqs^{*}\varphi dr\text{-}\varphi qr^{*}ids)\\
\frac{d\,\Omega r}{dt} &= \frac{1}{Tr}(Cem\text{-}Cr\text{-}Kf^{*}\Omega r)
\end{aligned}
\tag{21.1}
$$

Indirect Field-Oriented Control of an Induction Machine

This technique consists in assimilating the behaviour of asynchronous machine to the DC machine; this is for applications requiring raised dynamic performances (Montanari et al. 2000; Amrane et al. 2013; Larabi et al. 2013; Louri et al. 2013). The aim of such a control technique of the torque and a choice of the (d,q) reference, in order to obtain a decoupling between the torque and the field. From equation system (21.1); the torque control is made on the components of current iq and is. The electromagnetic torque depends only on component iq. It is a maximum for a given current if we impose is = 0. Consequently the obtained torque is then proportional to the current of machine power supply as in case of separately excited DC machine. The model given by Eqs. (21.1) is expressed as follows:

$$\begin{aligned}
&Vds = \sigma^*Ls^*\frac{d\,ids}{dt}\ \text{-}1/(\sigma^*Ls)\left[\text{-}Rsm^*\ ids + \omega s^*\sigma^*Ls^*\ iqs + \frac{Lm}{(Lr^*Tr)}\ {}^*\phi dr\right]\\
&Vqs = \sigma^*Ls^*\frac{d\,iqs}{dt}\ \text{-}1/(\sigma^*Ls)\left[\text{-}Rsm^*\ iqs\text{-}\omega s^*\sigma^*Ls^*\ ids\text{-}\frac{Lm}{(Lr^*Tr)}\ {}^*\phi qr\text{-}\omega r^*\frac{Lsr}{Lr}\ {}^*\phi dr\right]\\
&\frac{d\,\phi dr}{dt} = \frac{Lm}{Tr}\ {}^*ids\text{-}\frac{1}{Tr}\ {}^*\phi dr\\
&\omega s = \frac{Lm}{Tr} * \frac{iqs}{\phi dr} + \omega r\\
&Cem = \frac{P^*Lm}{Tr}(iqs^*\phi dr)\\
&\frac{d\,\Omega r}{dt} = \frac{1}{Tr}(Cem\text{-}Cr\text{-}Kf^*\Omega r)
\end{aligned} \tag{21.2}$$

It is noted that rotor flux depends only on the satatoric current is and iq, that the electromagnetic couple depends only on the quadratic current iq.

Sensorless Control of the Induction Machine

The Structure of Model Reference Adaptive System MRAS

The aim of this technique of control is to replace the speed sensor by a speed estimator. Into our study, we introduced a speed estimator of type MRAS (Model Reference Adaptive System). The latest replace the mechanical sensor without changing the dynamics of our machine.

The principle of the method MRAS speed estimation rests on the comparison of the sizes obtained in two different ways. One model of such a method is the voltage one (or stator equation) and the other is current model (or rotor equation); because the voltage method doesn't include rotor speed then it does not depend explicitly of speed (model of reference) and the other includes rotor speed (adjustable adaptive model). Figure 21.2 illustrates the derived MRAS scheme for speed adaptation.

Reference Model

One uses the equations of the currents (21.2) of the IM, the current expressed in the reference fixed to the stator. The two equations of the model of reference become:

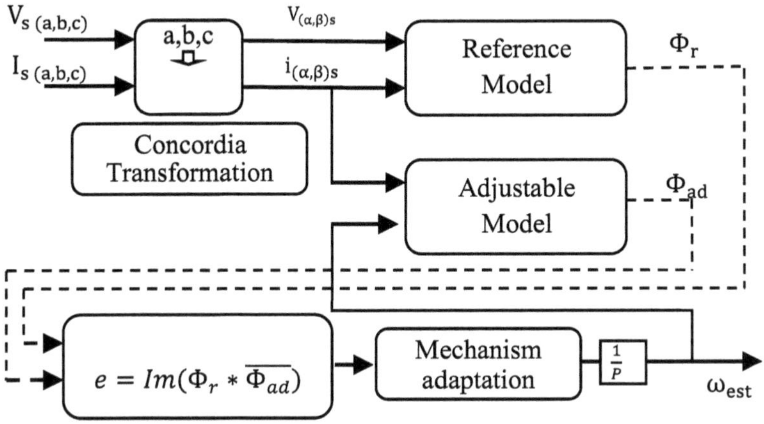

Fig. 21.2 Structure of MRAS estimation of speed

$$\begin{aligned}\frac{d\,\varphi\alpha r}{dt} &= \frac{Lr}{Lsr}\left[V\alpha s\text{-}\sigma^*Ls^*\ \frac{di\alpha s}{dt}\text{-}Rs\ ^*i\alpha s\right]\\ \frac{d\varphi\beta r}{dt} &= \frac{Lr}{Lsr}\left[V\beta s\text{-}\sigma^*Ls^*\ \frac{di\beta s}{dt}\text{-}Rs\ ^*i\beta s\right]\end{aligned} \tag{21.3}$$

Adjustable Model

To establish the adaptive model, we expressed the current in the reference fixed to the stator, the fluxes expressed with the rotor sizes in a reference αβ are:

$$\begin{aligned}\frac{d\,\varphi\alpha ad}{dt} &= -\frac{1}{Tr}\varphi\alpha ad + \frac{Lsr}{Tr}{}^*i\alpha s\ \text{-P}\ ^*\Omega est^*\varphi\beta ad\\ \frac{d\,\varphi\beta ad}{dt} &= -\frac{1}{Tr}\varphi\alpha ad + \frac{Lsr}{Tr}{}^*i\beta s\ \text{-P}\ ^*\Omega est^*\varphi\alpha ad\end{aligned} \tag{21.4}$$

Adaptation Mechanism

The entry of an adaptive mechanism is activated by the error between the reference field and adaptive field. By carrying out the difference between the reference model and the adjustable model, we obtain the following system of equations which govern the adaptive mechanism.

$$\begin{bmatrix}\frac{de\alpha}{dt}\\ \frac{de\beta}{dt}\end{bmatrix} = \begin{bmatrix}-\frac{1}{Lr} & -\omega\\ \omega & -\frac{1}{Lr}\end{bmatrix} + (\omega - \omega ad)\begin{bmatrix}\varphi\alpha ad\\ \varphi\beta ad\end{bmatrix} \tag{21.5}$$

The adaptation law chosen to ensure the convergence of ωad towards ω is:

$$\omega ad = Tp^{*}\delta e + Ti\int_{0}^{t}\delta e^{*}dx \tag{21.6}$$

The adaptive mechanism has an integral proportional form:

$$\omega est = {}^{1}/_{P}{}^{*}\left(Tp^{*}e + Ti\int e^{*}dt\right) \tag{21.7}$$

where Tp and Ti are positive gain.*

The Fuzzy Logic Adaptation

Control by Fuzzy Logic

Fuzzy logic technique makes it possible to control nonlinear systems and complicated models (Amrane et al. 2013, 2014; Larabi et al. 2013). In fact, the calculation of the parameters of the system is not necessary to carry out this control (Amrane et al. 2013, 2014; Larabi et al. 2013).

On opposite of the adaptation of the traditional techniques; the fuzzy logic does not treat a mathematical relations well defined, but uses inferences with a several rules, being based on variables linguistic.

These inferences are treated by operators suitable for fuzzy logic (Amrane et al. 2013, 2014; Larabi et al. 2013). Figure 21.3 shows the structure of a fuzzy regulator with two input (X1 and X2) and one output (Xr).

So, we can note that the calculation of the control is carried out starting from three fundamental stages: an interface of fuzzification; a mechanism of inference (rules); and an interface of defuzzification.

The Fuzzification

The entries and exits are defined of the fuzzy are defined by membership functions with 7, 5 or 3 sets. The various sets are characterized by standard designations (Amrane and Larabi 2012; Amrane et al. 2013; Larabi et al. 2013):

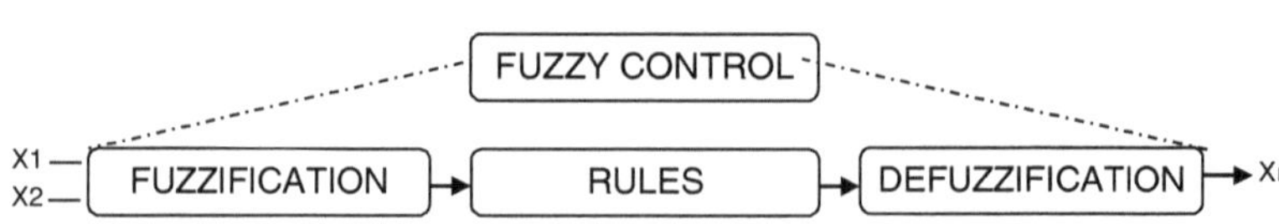

Fig. 21.3 Functional diagram of the fuzzy control

Negative Big NB, Negative Medium NM, and Negative Small NS, Zero Z, Positive Small PS, Positive Medium PM, Positive Big PB.

Inference Mechanism

It is well known that the realizations of the matrix of the rules are deduced by experiment, the experiment of the human operators and rests on the analysis of the system.

This analysis must take into an account the trajectory which one wants to give to the system (Amrane and Larabi 2012; Amrane et al. 2013; Larabi et al. 2013).

The Defuzzification

By this stage; the return to the sets of real exit will be made. It is a question of calculating, from the degrees of membership of all the sets variable of output, the coordinate which corresponds with the value of this exit. Various methods are used (Amrane and Larabi 2012; Amrane et al. 2013; Larabi et al. 2013).

The Fuzzy Logic Adaptation Mechanism Principle

The structure of the proposed PI controller used in the adaptation mechanism will be replaced by a fuzzy logic controller as shown in Fig. 21.4.

In our work, we adopted seven sets for the two variables of input (En and dEn), and a same number of sets for the variable of output dUn. The rules of the controller can be presented in a matrix with seven sets known as matrix of inference shown in Table 21.1.

In this article, the memberships function are chosen as the triangular type and trapezoidal, and the method of reasoning is considered as the max-min method, the

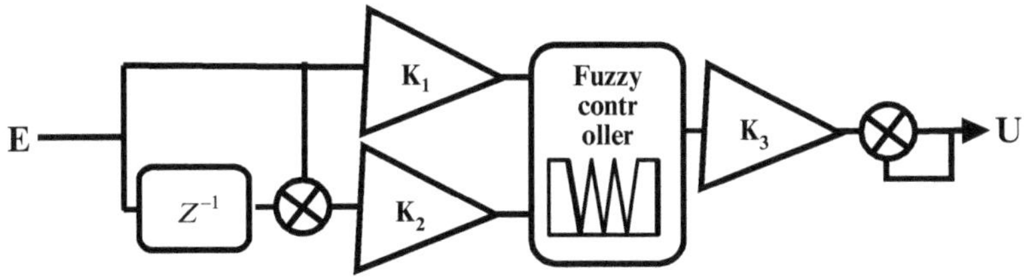

Fig. 21.4 Synoptic diagram of a fuzzy adaptation

Table 21.1 The fuzzy rules (a regulator with five sets)

		E_n				
dU_n		NG	NP	EZ	PP	PG
dE_n	NG	NG	NG	NP	NP	EZ
	NP	NG	NP	NP	EZ	PP
	EZ	NP	NP	EZ	PP	PP
	PP	NP	EZ	PP	PP	PG
	PG	EZ	PP	PP	PG	PG

defuzzification stage is done based on the gravity centre method, as it is frequently quoted in the literature and because they requiring less time computing will be also adopted in our work (Amrane et al. 2013, 2014; Larabi et al. 2013).

The Basic Rules of the Fuzzy Controller

The first step in the design of fuzzy controller is to generate fuzzy rules based on the knowledge of the expert. According to the expert, three situations can be distinguished for the constant time estimation, above, around and below the reference desired.

In our control, in order to determine the rules table that generates a command that will be presented later Based on Fig. 21.5.

Modelisation and Control of Generator Photovoltaique

To profit from solar energy and to use it to power the IM, it is necessary to understand the operation of this energy and optimize its use. In this way, we are interested here, to describe the basic concepts of solar energy and the production of electricity by the photovoltaic effect.

The connection of a generator PV to a load requires the adaptation of the generator so that it functions with its maximum power, and then to provide the maximum of energy delivered by the generator to the load, it is permanently necessary to track the corresponding that called Maximum Power Point (MPP). So, to provide this we try here to adapt the photovoltaic generator (PVG) to the asynchronous machine. This adaptation between the source and the load is carried out by the insertion of a converter DC-DC (chopper) controlled by a tracking mechanism of Maximum Power Point. This method is known as MPPT (Ho and Yeh 2010; Agrebi et al. 2006).

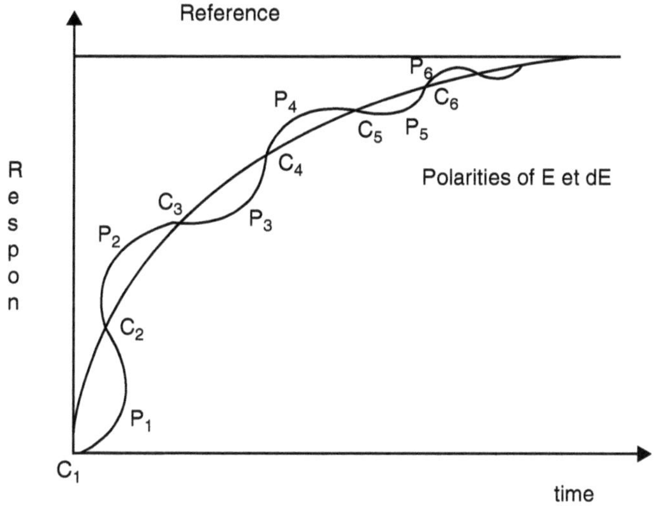

Fig. 21.5 Temporal analysis of the rules of the regulator

Model of a Generator Photovoltaic

The mathematical model of the physical process in (Ho and Yeh 2010; Agrebi et al. 2006; Peng and Fukao 1994) a PV cell are given in figure (Fig. 21.6), known under the name "two-diode model", it is largely used in the literature. It is most widespread because it gives the results closest to those obtained by a real GPV.

So, to obtain the wished values of the tensions and currents, the modules are connected in series, parallel or parallel-series. The later association "parallel-series" enables us to obtain the values of the tension and the desired current. The characteristic equation of the system is given by the following relation:

$$I = N_P \times I_{Ph} - N_P \times I_{s1}\left[exp\left[\frac{q[N_P V + (N_s \times I \times R_{se})]}{N_s N_P A_1 KT}\right] - 1\right] - N_P \times I_{s2}\left[exp\left[\frac{q[N_P V + (N_s \times I \times R_{se})]}{N_s N_P A_2 KT}\right] - 1\right] - \frac{[N_P V + (N_s \times I \times R_{se})]}{N_s N_P R_{Pe}} \tag{21.8}$$

With:
I and V: the current and the voltage of the photovoltaic cell.
q: the constant elementary charge.
$I_{ph:}$ the photocurrent produced.
K: the Boltzmann constant.
I_{S1} and I_{S2}: the saturation currents of the diodes.
A_1 and A_2 factors of purity of diode ($A_1 = 1$ and $A_2 = 2$).

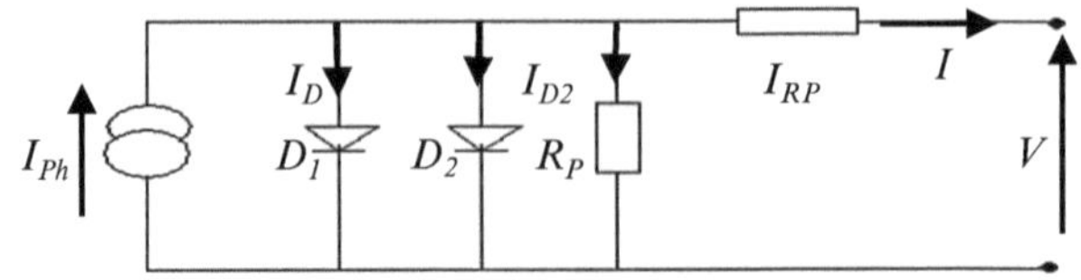

Fig. 21.6 Model of solar cell

$$I_{Ph} = I_{Ph}\big|_{(T=298)} \times \left[1 + (T - 298) \times 5 \times 10^{-4}\right]$$
$$\text{with} \quad I_{Ph}\big|_{(T=298)} = 3.25\ \text{A}$$

R_S and R_P: respectively resistance series and parallel resistance.
T: the absolute temperature °K.

$$I_{S1} = K_1 T^3 e^{\frac{-Eg}{KT}} \quad \text{and} \quad I_{S2} = K_2 T^{\frac{5}{2}} e^{\frac{-Eg}{KT}}$$

E_g: The energy band of the semi-conductor
$K_1 = 1.2\ A/Cm^2\ K^3$ and $K_2 = 2.9\ 10^{5/2}\ A/Cm^2\ K^{5/2}$

GPV Control

To convert solar energy with the best manner possible, it is essential to work in the zone of maximum power of generator. In other words, they need to track the MPP in order to maximize the power delivered to load by ordering converter DC/DC. There are different control methods of tracking of the point of maximum power, based rest on the generator power-tension characteristic (Ho and Yeh 2010; Agrebi et al. 2006; Peng and Fukao 1994).

The approach of artificial fuzzy logic intelligence is implemented to improve the performances of control and to track the point of maximum power, by the modelling and the simulation of a controller containing fuzzy logic (FLC) (Fig. 21.7).

The fuzzy controller proposed has two variables of entries, the error E and the variation of error CE, are expressed at a rate of sampling K. They are defined as follows:

$$E(k) = \frac{P(k) - P(k-1)}{V(k) - V(k-1)} \tag{21.9}$$
$$CE(k) = P(k) - P(k-1)$$

With: P (k) ET V (k): are respectively the power and the tension of PVG.

Once the two sizes of entry is calculated and converted into fuzzy linguistics variable, the exit which is the variation of the cyclic report dα, as mentioned in table (Table 21.2), intended to control a chopper back-booster. The control rules must be indicated so that the entry variable E must be always zero.

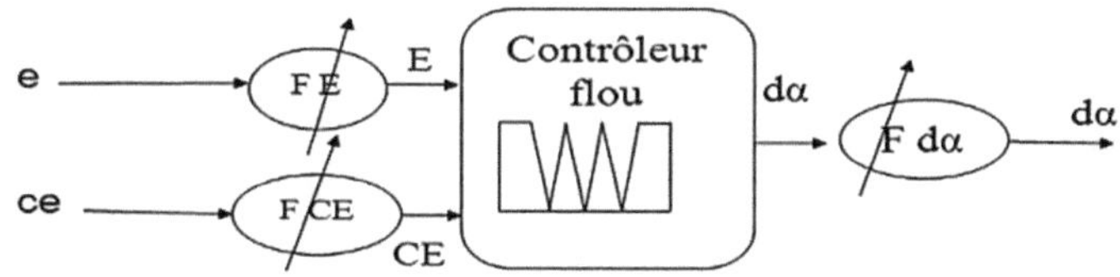

Fig. 21.7 Structure of the fuzzy controller for the tracking of the PPM

Table 21.2 The fuzzy rules for algorithm MPPT (a regulator with five sets)

		E_n				
dα		NL	NS	Z	PS	PL
dE_n	NL	Z	Z	PS	NS	NL
	NS	Z	Z	Z	NS	NL
	Z	PL	PS	Z	NS	NL
	PS	PL	PS	Z	Z	Z
	PL	PL	PS	NS	Z	Z

The synoptic diagram of the chain PV: Photovoltaic generator—Converters—Induction Machine—Battery is given by Fig. 21.8.

Simulation Results

To diagnose the speed sensor, a MRAS estimator is used as an observer. Then, an algorithm for selecting is used to enable the detection of defects and perform the selection between the measured signals and the estimated signals.

So, to validate the performances of the MRAS, the proposed simulation of the dynamic behaviour of the machine has been done using the MATLAB/SIMULINK and that for the following conditions.

With, the value of the torque is fixed to zero and the field is fixed to 1 Wb. To highlight the influence of speed variations and uncertainties, particularly those of the control process, we gave reference speed of 100 rad/s (Fig. 21.9).

Afterwards, constant field and a reference speed $\Omega ref = 100$ rad/s, the field rotor field is fixed to 1 Wb, the initial values of the torque assumed by the machine is zero, and between 1 s and 2.5 s, one will apply a nominal torque load (10 N.m) (Fig. 21.10).

Finally, we will use our estimator to diagnose faults sensor for speed exceeds the rated speed to a value of the torque is fixed to zero, the field is fixed to 1 Wb and the initial values of the speed assumed by the machine is zero, and between 1 s and 2.5 s, we will apply variations (Fig. 21.11).

According to the simulation results we can say that the MRAS technique of detection provides an effective solution to the problem of diagnostic. We can say that our objective here has been successfully achieved.

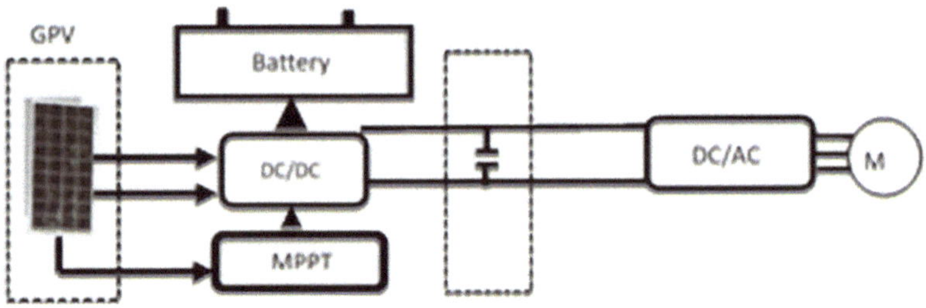

Fig. 21.8 Total structure of the photovoltaic chain

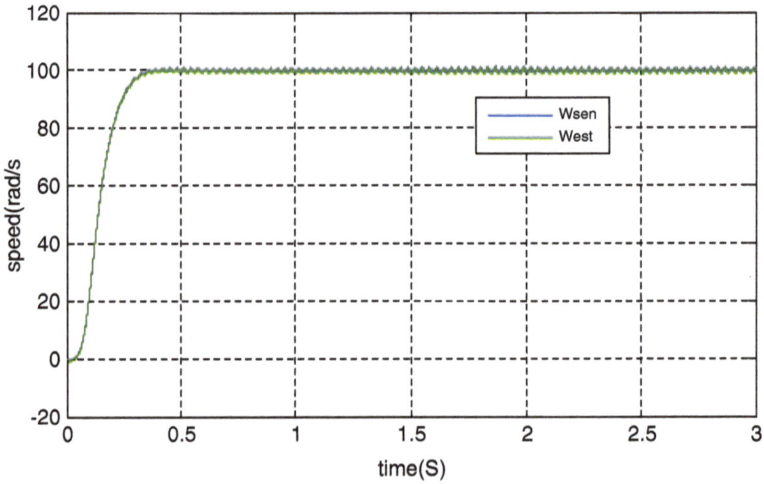

Fig. 21.9 Estimation of speed in the case of an indirect field oriented of the induction machine, speed reference (Wref), estimated speed (West) and speed with sensor (Wavec). With the torque is fixed to zero and the field is fixed to 1 Wb and the speed fixed to 100 rad/s

The results presented on the previous figures, show that the dynamics of the flux magnitude are presented can highlight the decoupling role of the flux controller where the flux tracks its nominal value of 1.1 Wb for all speed ranges.

The MRAS observer has an advantage of avoiding integration of stator voltages, which is of particular interest for hardware implementation. The proposed MRAS strategy has been incorporated within the classical IM control system. Numerical simulations are performed and shown good results for tricking the reference. Implementation of the proposed algorithm can be used for the diagnostic issues suitable for IM applications requiring high-performance.

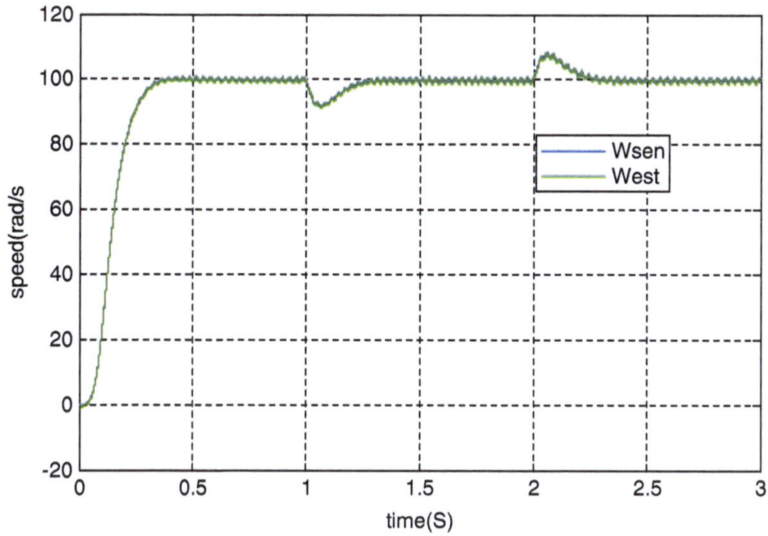

Fig. 21.10 Estimation of speed in the case of an indirect field oriented of the induction machine, speed reference (Wref), estimated speed (West) and speed with sensor (Wavec). With the torque is fixed to 10 N.m and the field is fixed to 1 Wb and the speed fixed to 100 rad/s

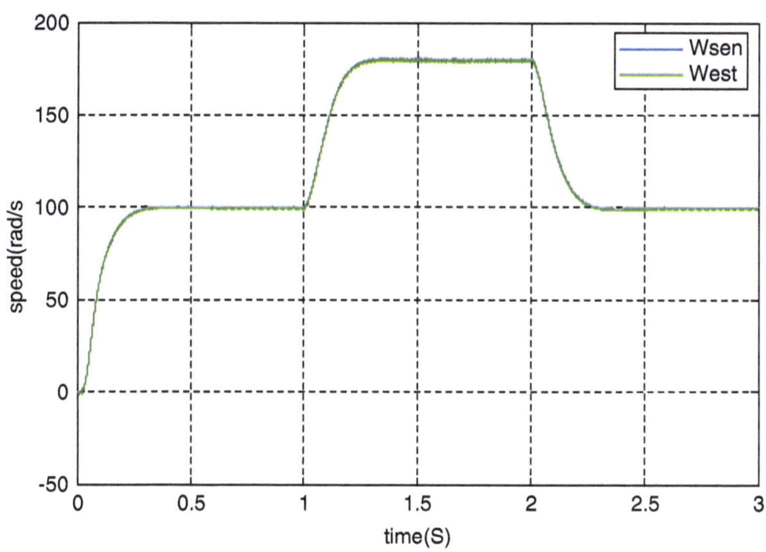

Fig. 21.11 Estimation of speed in the case of an indirect field oriented of the induction machine, speed reference (Wref), estimated speed (West) and speed with sensor (Wavec). With the torque is fixed to zero and the field is fixed to 1 Wb and between 1 s and 2.5 s, we will apply variations of speed

Conclusion

The work that we presented contributes to the analysis and the synthesis of a robust diagnostic applied to the induction machine. The MRAS is employed for the detection of the faults of sensor speed. The use of the fuzzy logic is a powerful tool in realization of the robust and reliable diagnostic.

We have proposed a method for detection faults of sensor speed using the MRAS and the fuzzy logic algorithm, to ensure a good diagnostic of induction machines. The method proposed in this paper is applicable to a large category of induction motor drives with a gradually varying load torque. The tests of robustness show clearly that the performances of the diagnostic in the presence of estimator, type MRAS, for the tracking the references is always fast. The validity of this method is checked by several tests.

The results obtained show that the model suggested for the MRAS adapts perfectly to all the diagnostic of the IM.

The next step of this work is the integration of a real induction motor and estimator MRAS for testing the diagnostic is required for the practical case.

Annexes

The parameters of the induction machine cage used are shown below:

Rated power:	1.5 kw
Nominal voltage:	220/380 V
Rated power factor:	0.8.Rated
Speed:	1420 rev/min
Nominal frequency:	50 Hz
Stator resistance:	4.85 Ω
Rotor resistance:	3.805 Ω
Stator cyclic inductance:	0.274 H
Cyclic inductance of Rotor:	0.274 H
Cyclic mutual inductance:	0.258 H
Number of pole pairs:	2
Moment of Inertia:	0.031 Nm-s^2/rad
Friction:	0.008 Nms/r

Nomenclature

Ls Stator inductance [H]
Lr Rotor inductance [H]
Lm Mutual magnetizing inductance

Lsr	Mutual Inductance between the stator and rotor [H]
Kf	Friction coefficient [N.s/rad]
J	Total inertia [kg.m^2]
P	Number of pole pairs
ωs	Synchronous Pulsation [rad/s]
ωr	Electrical angular Pulsation [rad/s]
Cem	Electromagnetic torque [N.m]
Cr	Resistive torque [N.m]
Tr	Rotor time constant [s]

References

Abolhassani, M. T., & Toliyat, H. A. (2009, May). Fault tolerant permanent magnet motor drives for electric vehicles. In *IEEE international on electric machines and drives conference, 2009. IEMDC '09* (pp. 1146–1152, 3–6).

Agrebi, Y., Triki, M., Koubaa, Y., & Boussak, M. (2006). Commande avec et sans capteur mécanique d'une machine asynchrone. In *7ème conférence internationale des sciences et techniques de l'automatique STA'2006*, 17–19 décembre, Hammamet Tunisie (pp. 573–584).

Amrane, A., & Larabi, A. (2012). Speed control sensorless of an induction machine powered by photovoltaic generator. In *10th global conference on sustainable manufacturing (GCSM'12)*, 31 October–02 November 2012, Istanbul, Turkey.

Amrane, A., Larabi, A., & Hamzaoui, A. (2014). Robust sensorless control of an induction machine based on a fuzzy MRAS. In *International conference on electrical and electronics engineering (ICEEE 2014)*, du 21–23 April, Antalya, Turquie.

Amrane, A., Louri, M., Larabi, A., & Hamzaoui, A. (2013). A fuzzy model reference adaptive system control for induction. In *3rd international conference on systems and control*, October 29–31, Algiers, Algeria.

Benbouzid, M. E. H., Diallo, D., & Zeraoulia, M. (2007, March). Advanced fault-tolerant control of induction-motor drives for EV/HEV traction applications: from conventional to modern and intelligent control techniques. *IEEE Transactions on Vehicular Technology, 56*(2), 519–528.

Chen, J., & Patton, R. (1999). *Robust model-based fault diagnosis for dynamic systems*. Boston: Kluwer.

Green, S., Atkinson, D. J., Jack, A. G., Mecrow, B. C., & King, A. (2003). Sensorless operation of a fault tolerant PM drive. *IEEE Proceedings Electric Power Applications, 150*(2), 117–125.

Ho, T.-J., & Yeh, L.-Y. (2010). Design of a hybrid PID plus fuzzy controller for speed control of induction motors. In *5th IEEE conference on industrial electronics and applications* (pp. 1352–1357), October 27–30. Gyeonggi-do, Korea: KINTEX.

Holtz, J. (1996). Methods for speed sensorless control of AC drives. In *Sensorless control of AC motor drives*. Piscataway, NJ: IEEE Press.

Larabi, A., Amrane, A. & Boucherit, M. S. (2013). Robust speed sensorless fuzzy control of the induction machine drive with adaptation of the rotor time constant supplied by photovoltaic solar energy. In *International conference on power engineering, energy and electrical drives, POWERENG-2013*, 13–17 May, Istanbul, Turkey.

Louri, M., Amrane, A., & Barazane, L. (2013). Comparison between the performances of variables structures control and the theory of synergetic on applied to the squirrel motor drives. In *3rd international conference on systems and control*, October 29–31, Algiers, Algeria.

Montanari, M., Peresada, S., Tilli, A., & Tonielli, A. (2000). Speed sensorless control of induction motor based on indirect field-orientation. In *35th annual meeting, IEEE industrial application society, Rome* (pp. 1858–1865).

Ouhrouche, M. (2002). Estimation of speed, rotor flux and rotor resistance in cage induction motor using the EKF algorithm. *International Journal of Power and Energy Systems, 22*, 103–109.
Peng, F., & Fukao, T. (1994). Robust speed identification for speed-sensorless vector control of induction motors. *IEEE Transactions on Industry Applications, 30*(Sept./Oct.), 1234–1240.
Shoudao, H., Yaonan, W., Jian, G., Jiantao, L., & Sihai, Q. (2004). The vector control based on MRAS speed sensorless induction machine drive, intelligent control and automation. In *WCICA, 5th world congress* (pp. 4550–4553).

Chapter 22
Air Conditioning Based on Hydroxides with Solar Driving for Low GHG Emissions

R.J. Romero, S. Silva-Sotelo, Y.R. Galindo-Luna, C.V. Valdéz-Morales, J. Ibarra-Bahena, A. Hdz-Jasso, and A. Rodríguez-Martínez

Introduction

There is a large amount of energy transformed in several ways. Ranging from the clean use for the environment such as solar energy is the energy that is taken directly from the sun, which with various processes is transformed into mechanical energy and then transformed to power. Other clean energy to the environment is nuclear energy, which is generated by reactions usually radioactive chemical elements, this type of energy promises much but may be dangerous at the same time; if that is not under controlled handled conditions. Some people do not consider it as clean source of energy production due to highly toxic waste than generates it.

Another clean energy is hydropower, which involves harnessing the potential energy accumulated in the water to generate electricity, is a classic form of energy. About 20 % of the electricity used in the world comes from this source.

In the world today most of the energy used (60–70 %) is generated by burning fossil fuels, which have been generated over the years many waste from the combustion of these have caused a negative impact to the environment to be the main pollutants of this, the problem with the use of fossil fuels to generate energy is not renewable, is that fewer and fewer of these fuels, and to use excessive release of GHG and another contaminants.

According to International Energy Agency Moreover we have the domestic air conditioners which take a very important role in the big cities and they generate about 20–30 % of the total to the urban emissions atmosphere.

R.J. Romero (✉) • S. Silva-Sotelo • Y.R. Galindo-Luna • C.V. Valdéz-Morales
J. Ibarra-Bahena • A. Hdz-Jasso • A. Rodríguez-Martínez
Centro de Investigación en Ingeniería y Ciencias Aplicadas, Universidad Autónoma del Estado de Morelos, Av. Universidad 1001, Chamilpa 62209, Morelos, Mexico
e-mail: rosenberg@uaem.mx

P. Grammelis (ed.), *Energy, Transportation and Global Warming*, Green Energy and Technology, DOI 10.1007/978-3-319-30127-3_22

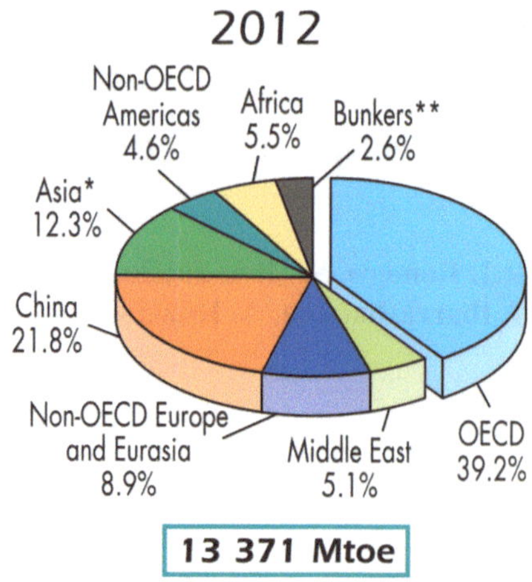

Fig. 22.1 Regional total primary energy supplies at 2012 (© iea.org)

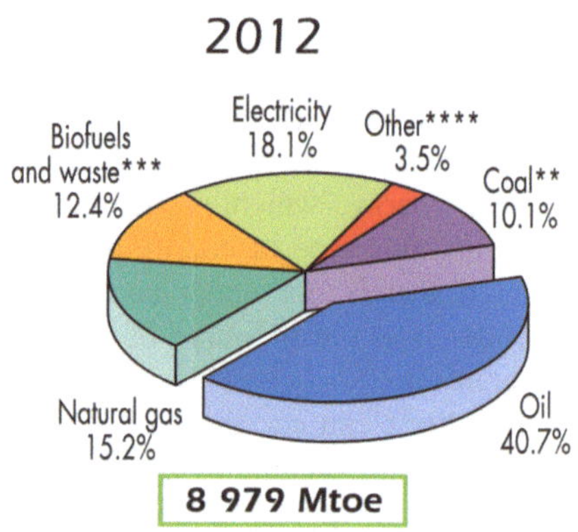

Fig. 22.2 World Final consumption at 2012 (© iea.org)

OECD countries supplies almost 40 % of primary energy in the world, but consumption in the OECD are large as the population number. The OECD consumption for 2012 was 3582 Mtoe from 8980 World's total consumption. That is practically the 40 % of the total. OECD renewable energies for 2012 were geothermal 32 Mtoe, hydropower 119 Mtoe and Solar + Tide + Wind energies 48 Mtoe. These renewable balances are 5.5 % of the OECD consumption. Biggest energy consumption is based on oil. See Figs. 22.1, 22.2, and 22.3 taken from "2014 Key World energy statistics" by International Energy Agency.

Solar Air Conditioning Proposal

Solar air conditioning is an old dream for technician from worldwide. A proposal for solar absorption air conditioning for building space cooling is due to the coincidence between the availability of maximum solar irradiance and the

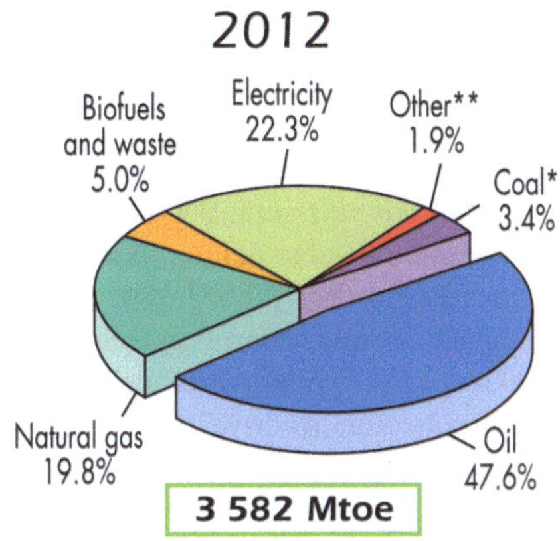

Fig. 22.3 OECD Final consumption at 2012

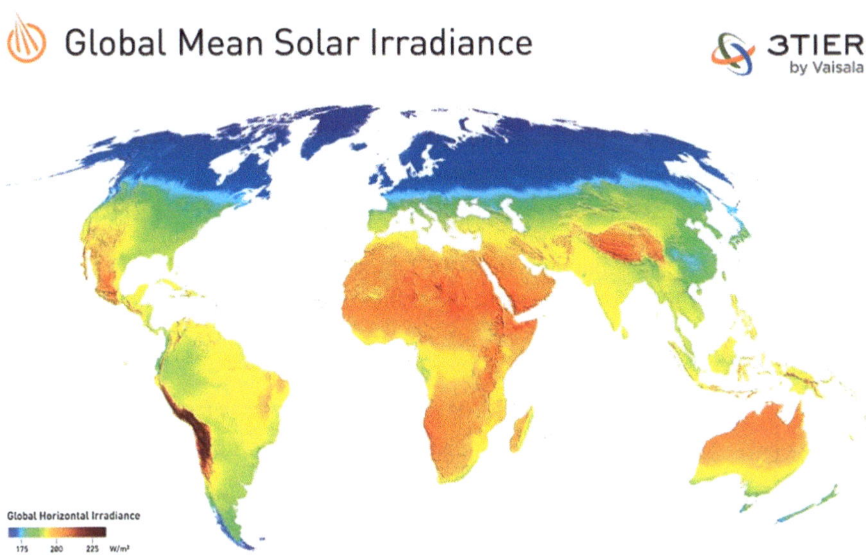

Fig. 22.4 Global solar irradiance map (© 3tier.com)

maximum load for cooling, particularly in office buildings. According to average solar irradiance, the solar potential is available to provide the energy for the air conditioning. An earlier patent for this concept was realized by on 1977 (Kumm 1979).

Figure 22.4 shows the global potential for installation of solar systems. Solar air conditioning systems were studied from 1957 as a recent review reports (Siddiqui and Said 2015). However there are some limitations for the implementation of the thermodynamic cycle.

Description of the Hydroxide Based Solar System

Conventional air conditioning requires mechanical vapor compression by fossil fuel driven. Electrical power energy into the compressor unit causes a pressure increase in a gaseous refrigerant (typically an HCFC) and then condenses this at higher temperature level to deliver the exchanged energy al lower temperature level in the evaporator; as Fig. 22.5 shows.

Space cooling absorption technology from 14 to 22 °C requires heat supply from 100 to 150 °C while hydroxide blend is used as working pair (Romero et al. 2000). Solar collectors to drive an air conditioning system must be concentrators, cylindrical parabolic concentrator and evacuated tube collectors to heat at generator temperature requirements. Additionally, photovoltaic cells must be installed to drive the centrifugal pumps.

A *HBAC* system basically consists of an evaporator, a condenser, a generator, an absorber and an economizer as shown in Fig. 22.6. A quantity of waste heat Q_{GE} is added at a relatively high temperature T_{GE} to the generator to vaporize the working fluid from the solution. The vaporized working fluid goes to the condenser, where it is condensed delivering an amount of heat Q_{CO} at an intermediate temperature T_{CO}. The liquid leaving the condenser passes through a valve to the evaporator where it is evaporated by means of a quantity of heat Q_{EV} at a low temperature T_{EV} producing the cooling effect. After this, the vaporized working fluid goes to the absorber where it is absorbed by the solution coming from the generator, delivering an amount of heat Q_{AB} at an intermediate temperature T_{AB}. Finally, the solution leaving the absorber is pre-heated in the economizer and returns to the generator starting the cycle again.

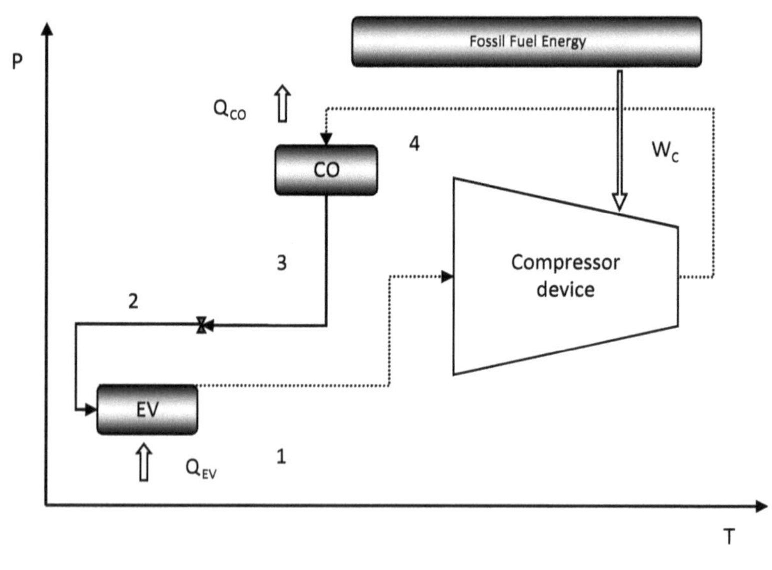

Fig. 22.5 Conventional air conditioning system

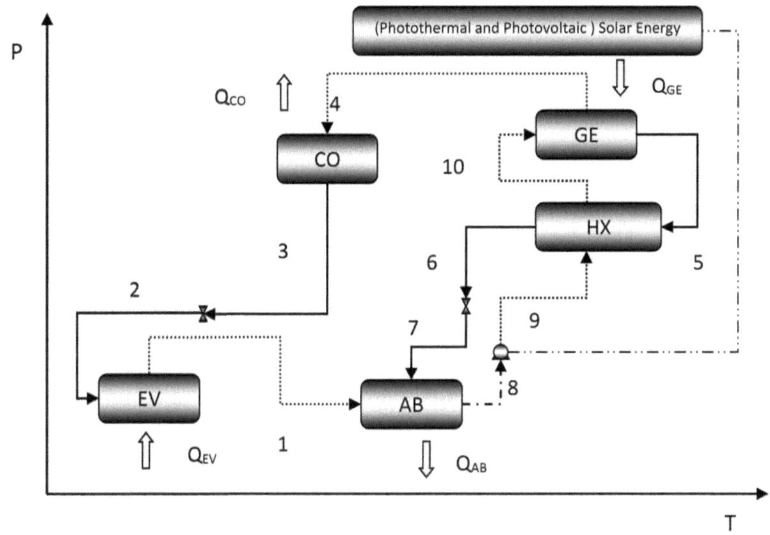

Fig. 22.6 Proposal hydroxide based thermodynamic cycle for solar air conditioning system

Heat Transfer Effectiveness

Heat transfer effectiveness is a function of three thermal parameters (George and Srinivasa Murthy 1989):

$$\eta = \frac{X_{AB} - X_{GE}}{X_{AB\_EQ} - X_{GE}} \tag{22.1}$$

Economizer effectiveness has been analyzed into refined thermodynamic cycle for some authors (Romero et al. 2000; Balghouthi et al. 2005; Fabrizio et al. 2014; Siddiqui and Said 2015). The effect for the economizer causes an internal energy recover from absorber to the generator.

Results

From mass and energy balances, the power into the heat transfer components is calculated as follow:

$$Q_{EV} = M_{EV}(H_{GE,V} - H_{CO,S}) \tag{22.2}$$

$$Q_{GE} = M_{GE,V}\, H_{GE,V} + M_{GE,S}\, H_{GE,S} - M_{AB,S}\, H_{VA,S} \tag{22.3}$$

$$Q_{CO} = M_{CO}\left(H_{CO,S} - H_{GE,V}\right) \tag{22.4}$$

$$Q_{AB} = M_{CO}H_{EV,V} + M_{GE,S}H_{GE,S} - M_{AB,S}H_{AB,S} \tag{22.5}$$

$$COP = \frac{Q_{EV}}{Q_{solar}} \tag{22.6}$$

COP is a dimensionless parameter that indicates the ratio between useful energy and inlet energy. This parameter allows the comparison between air conditioning devices. For this case, the effectiveness of the components were calculated and the values for each are $\eta_{EV} = 0.9$, $\eta_{GE} = 0.8$, $\eta_{AB} = 0.75$, $\eta_{CO} = 0.7$ and $\eta_{HX} = 0.7$.

This mass and energy balance was computed with the following algorithm (Fig. 22.7).

The energy for drive the centrifugal pump into the cycle is assumed that comes from photovoltaic panels to avoid fossil fuel derived energy. That power was calculated as 2 % of the air conditioning power. That value is 350 W at 12 VCD.

Figure 22.8 show the HBAC's COP as function of solar energy temperature for constant temperature in absorber and condenser component. The air conditioning keeps an insulated space at 20 °C. An optimal ideal condition for the hydroxide based air conditioning system is noted in this figure. This optimal condition requires solar energy at 125 °C for ambient temperature at 40 °C and the COP is 0.7 than indicates the CPC energy is 25 kW for a 17.5 kW of air cooling load.

Figure 22.9 show the modified HBAC's COP as function of solar energy temperature as in previous figure. The modified COP was computed with the shown component effectiveness. The behavior for both lines in Figs. 22.8 and 22.9 are the same. But there is a variation in the value for the optimal thermodynamic conditions. Solar energy is required at 120 °C for ambient temperature at 40 °C but the calculated COP is 0.5 that means the CPC energy is 31 kW for 15.5 kW of air cooling at same room.

The CPC area for this solar energy panel is 20 % larger than the ideal cycle, and the air conditioning load is diminished 11 % anyway. The non-ideal calculation leads to solar engineers to precise areas to fossil fuel free air conditioning systems.

Unfortunately at 2010 the air conditionings powers were higher than 25 GW as shown Fig. 22.10, and the associated CO^2 emission are higher as well. Solar air conditioning systems as the proposal and others may reduce the CO_2 impact to 2050 (iea.org 2011). OECD countries dominate the market for space and water heating, but not for cooling. China leads the world for annual installed capacity of solar thermal systems and air conditioners devices.

Conclusion

Hydroxide based solar air conditioning may be a proposal for CO_2 reduction at 2050. The entire system drives with solar energy from CPC and photovoltaic panels. The effectiveness addition for calculation of air conditioning leads to

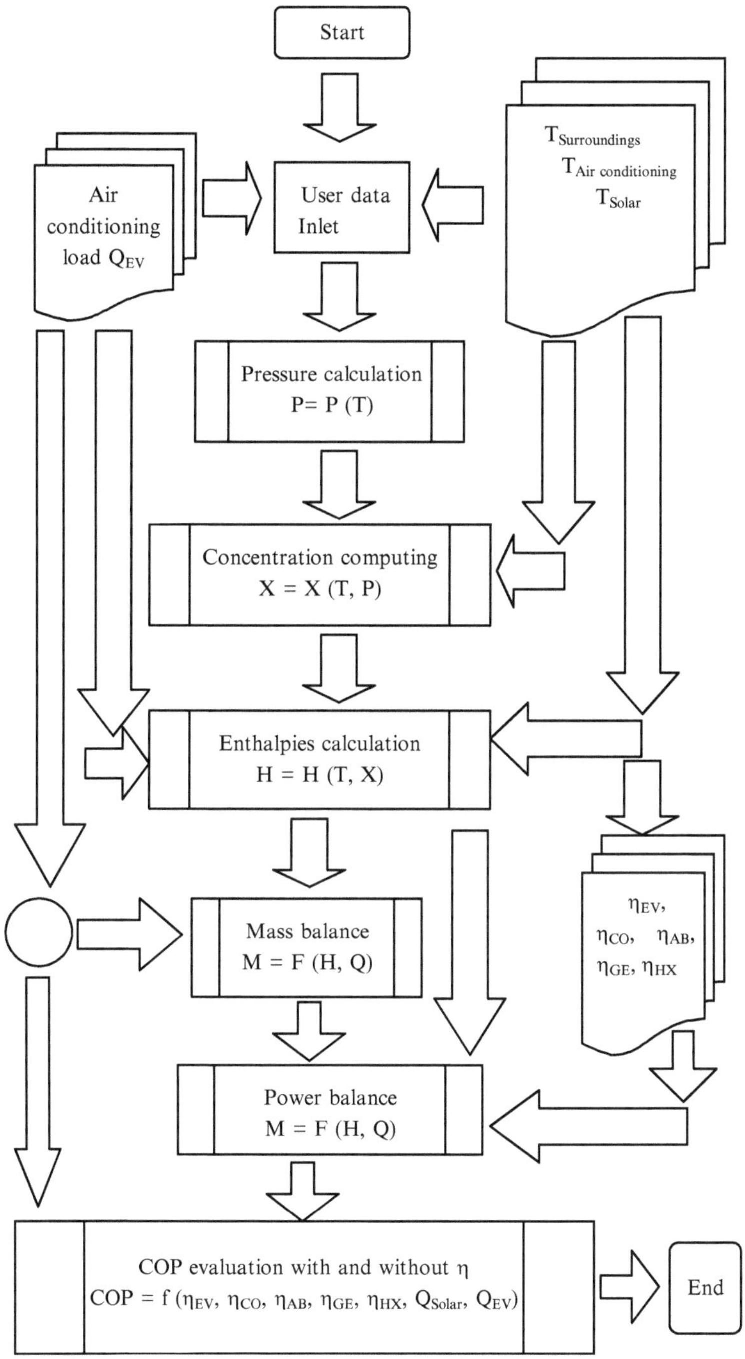

Fig. 22.7 Algorithm for calculation of powers and COP for proposal system

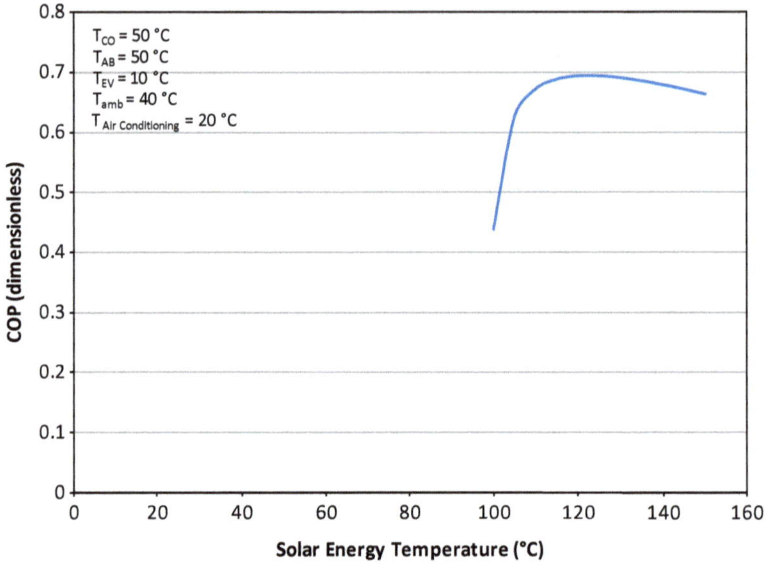

Fig. 22.8 HBAC's COP as function of solar energy temperature for ideal cycle

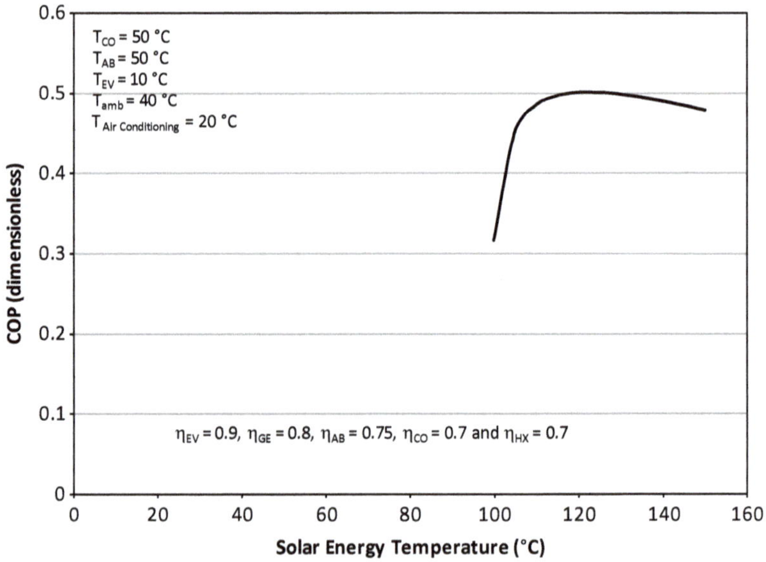

Fig. 22.9 Modified HBAC's COP as function of solar energy temperature

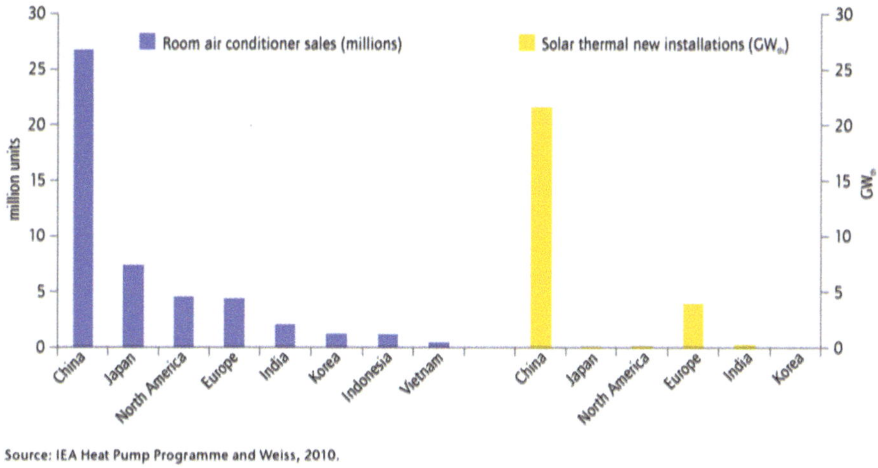

Fig. 22.10 Sales and solar thermal installations for air conditioning at 2010

lower COP values. For two selected scenarios the air conditioning load for 17 kW requires solar energy of 25 kW in an ideal thermodynamic cycle. For a non-ideal scenario the hydroxides-based air conditioning load with same flow is just 15.5 kW and solar energy of 31 kW. Due all energy in the HBAC is solar, there is not GHG compared with conventional compression air conditioning.

Acknowledgement To CEMIE-Sol P09 and CB-167434 by CONACYT—Mexico.

Nomenclature

H	Enthalpy, kJ/kg
M	Flow, kg/s
P	Pressure, Pa
Q	Power, W
T	Temperature, °C
W	Mechanical power, W
X	Hidroxide concentration, kg/kg

Greek Letters

η	Effectiveness, dimensionless

Subscripts

AB	Absorber
CO	Condenser

EV	Evaporator
GE	Generator
HX	Economizer
S	Liquid solution
Solar	Solar
V	Vapour

References

Balghouthi, M., Hachemi Chahbani, M., & Guizani, A. (2005). Solar powered air conditioning as a solution to reduce environmental pollution in Tunisia. *Desalination, 185*, 105–110.

Fabrizio, E., Seguro, F., & Filippi, M. (2014). Integrated HVAC and DHW production systems for zero energy buildings. *Renewable and Sustainable Energy Reviews, 40*, 515–541.

George, J. M., & Srinivasa Murthy, S. (1989). Influence of absorber effectiveness on performance of vapour absorption heat transformers. *International Journal of Energy Research, 13*, 629–638.

International Energy Agency. (2011). Technology roadmap energy-efficient buildings: Heating and cooling equipment. https://www.iea.org/publications/freepublications/publication/buildings_roadmap.pdf.

Kumm, E. L. (1979). U.S. Patent No. 4,151,721. Washington, DC: U.S. Patent and Trademark Office.

Retrieved March 26, 2015 from http://www.3tier.com/en/support/resource-maps/.

Retrieved March 26, 2015, from http://www.iea.org.

Romero, R. J., Rivera, W., & Best, R. (2000). Comparison of the theoretical performance of a solar air conditioning system operating with water/lithium bromide and an aqueous ternary hydroxide. *Solar Energy Materials and Solar Cells, 63*, 387–399.

Siddiqui, M. U., & Said, S. A. M. (2015). A review of solar powered absorption systems. *Renewable and Sustainable Energy Reviews, 42*, 93–115.

Chapter 23
Life Cycle Analysis as a Decision Criterion for the Implementation of Solar Photovoltaic Panels in as Northeast Brazil Hospital

Monica Carvalho, Danielle Bandeira de Mello Delgado, and Ricardo Chacartegui

Introduction

The environmental performance of products, services and processes has become one of the key issues in today's world, and it is important to examine ways in which negative effects on the environment are assessed. One of the reasons for undertaking a Life Cycle Assessment (LCA) study is that there are growing concerns about a variety of environmental issues expressed by public opinion, political bodies and industry. Environmental concern may be related to the long-term resource base of human societies or may be more health related or it may be a concern for the natural environment as such (Baumann and Tillmann 2004). LCA is one of the tools that can be used for such a purpose.

LCA can be defined as the "compilation and evaluation of the inputs, outputs and potential environmental impacts of a product system throughout its life cycle" (Guinee et al. 2001). The objective of LCA is to describe and evaluate the overall environmental impacts of a certain action by analyzing all stages of the entire process from raw materials supply, production, transport and energy generation to recycling and disposal stages—following actual use, in other words, "from the cradle to the grave" (WEC 2004). The environmental burden covers all types of impacts upon the environment, including extraction of different types of resources, emission of hazardous substances and different types of land use. The term

M. Carvalho (✉) • D.B. de Mello Delgado
Department of Renewable Energy Engineering, Center of Alternative and Renewable Energy, Federal University of Paraíba, Caixa Postal 5115, Cidade Universitária, João Pessoa 58051-900, Paraíba, Brazil
e-mail: monica@cear.ufpb.br; danielle.delgado@cear.ufpb.br

R. Chacartegui
Energy Engineering Department, University of Seville, Camino de los descubrimientos, s/n, Seville 41092, Spain
e-mail: ricardoch@us.es

P. Grammelis (ed.), *Energy, Transportation and Global Warming*, Green Energy and Technology, DOI 10.1007/978-3-319-30127-3_23

'product' is taken in its broadest sense—including physical goods as well as services; it includes goods and services at both operational and strategic levels (Guinee et al. 2001).

Energy use and its associated primary energy consumption and emissions are important contributors to the life cycle impacts of many products and processes, as well as to greenhouse gas inventories of entities, products, and countries. LCA can be used to calculate the environmental loads (e.g., emissions of greenhouse gases) associated with the consumption of units of energy (electricity, heat, coolth, for instance). In this way, alternative systems may be ranked according to their respective emissions of greenhouse gases. The advantage of analyzing full life cycles as opposed to using only emission factors is that renewable and nuclear systems with no direct (stack) emissions can be compared with systems based on fossil fuel combustion (WEC 2004).

There are actually two innovations presented herein: (1) utilization of environmental data in the objective function of a mathematical model for the optimization of an energy supply system, and (2) application of the methodology to a hospital located in Northeast Brazil. Regarding the state of LCA in Brazil, most of the studies have been undertaken in the South and Southeast regions—therefore the application to the Northeast region could help disseminate good practice. The energy supply system is optimized through the solution of an optimization model based on Mixed Integer Linear Programming (MILP). The optimization encompasses the configuration of the system simultaneously with the operational strategy, providing an energy supply system with minimum environmental impacts. These environmental impacts are associated with the production of each piece of equipment installed plus the consumption of energy resources (natural gas, diesel, biomass, etc.) to satisfy the energy demands of the consumer center.

Polygeneration System

The study case is a university hospital, with 420 beds, located in the city of João Pessoa (Northeast Brazil). The energy demands considered were electricity (lighting and equipment), hot water (internal consumption of the hospital), steam (laundry and sterilization) and coolth (a/c for comfort). Hospitals are good study cases for optimization because they present regular, stable energy demands. The study considered an entire operational year, subdivided into two representative days per month (weekday and holiday/weekend), with 24 hourly periods each.

Real electricity data was provided by the hospital. Coolth, hot water and steam demands were disaggregated through the utilization of climate data, degree-days method, occupation and energy audit data (Erbs et al. 1983; Climaticus 4.2, 2005; Nepote et al. 2009; Araújo 2004). The hospital presents the following annual energy demands: 2791 MWh of electricity, 1947 MWh of hot water, 138 MWh of steam, and 2309 MWh of coolth. Table 23.1 shows the energy demands for the hospital, throughout the year.

Table 23.1 Energy demands of the hospital

Month representative day	n_d (days/year)	Electricity Total MWh/day	Hot Water Total MWh/day	Steam Total MWh/day	Coolth Total MWh/day
Jan w[a]	20	8538	3543	0.465	5961
Jan h[b]	11	6315	3543	0.233	4360
Feb w	19	9696	6436	0.465	6472
Feb h	9	6247	6436	0.233	4290
Mar w	20	9861	6436	0.465	8013
Mar h	11	7474	6436	0.233	5443
Apr w	20	9949	6436	0.465	7071
Apr h	10	7348	6436	0.233	4737
May w	20	8901	6436	0.465	7895
May h	11	6383	6436	0.233	5534
Jun w	19	7489	5032	0.465	7768
Jun h	11	6832	5032	0.233	5259
Jul w	20	6568	3881	0.465	5186
Jul h	11	6045	3881	0.233	3760
Aug w	20	5954	3881	0.465	7343
Aug h	11	4681	3881	0.233	4941
Sep w	21	6686	3881	0.465	7483
Sep h	9	4755	3881	0.233	5046
Oct w	20	8041	5284	0.465	8834
Oct h	11	5969	5284	0.233	6057
Nov w	20	9309	6436	0.465	8260
Nov h	10	6022	6436	0.233	5627
Dec w	20	1036	6436	0.465	5912
Dec h	11	7389	6436	0.233	3879
Σ		MWh/year	MWh/year	MWh/year	MWh/year
Year	365	2791	1947	138	2309

[a]w = weekday
[b]h = holiday/weekend

One of the major steps in Mixed Integer Programming is the definition of a superstructure that represents all options available for equipment and energy resources. Figure 23.1 depicts the superstructure of the energy supply system for the hospital located in João Pessoa, following the scheme of Carvalho and Millar (2012).

The available utilities were electricity (EE), diesel (DI), biomass (BM), Natural gas (NG), steam (VA, 180 °C), hot water (HW, 90 °C), cooling water (CW, $t_0 + 5$ °C), ambient air (t_0°C), and chilled water (CO, 5 °C). A positive node denotes supply/production of energy and a negative node denotes consumption. The horizontal lines represent the energy forms considered at the site, into which the equipment (vertical lines) are connected to consume or produce energy.

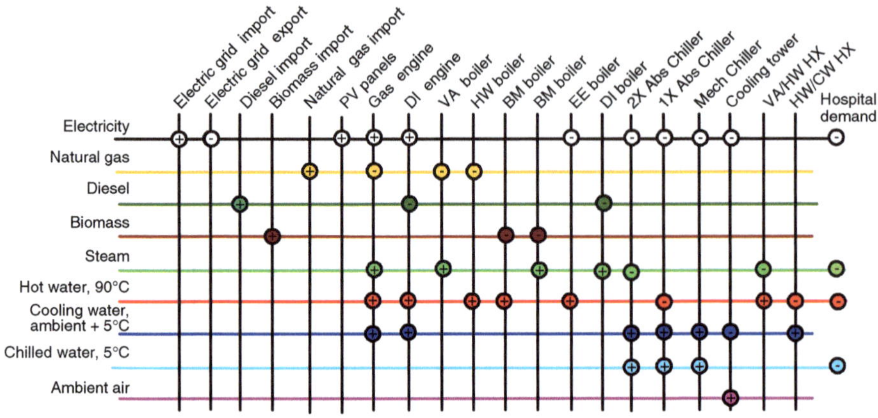

Fig. 23.1 Superstructure illustrating all potential conversion pathways considered

Table 23.2 shows the technical and economic characteristics of the equipment that belong to the superstructure (Fig. 23.1). The rows indicate the technologies available, and the columns indicate the energy resources. The coefficient in bold indicates the flow that defines the productive capacity of the equipment, and positive and negative coefficients, indicate, respectively, that an energy flow is either produced or consumed. At the time of writing, 1 US$ = 2.50 R$.

The PV system is constituted by PV panels and inverters (electricity storage will not be considered). Information was obtained from consultation to PV panel (Kyocera 2014) and inverter (Santermo 2014) manufacturers. The cost of the system was US$ 880/panel (including the cost for the panel, inverter, installation materials, transportation and assembly). The area of each panel is 1.64 m^2. The maintenance costs are US$ 10/m^2 per year. As the geographic latitude of the municipality of João Pessoa is 07°S, the inclination angle of the panels to be installed in the hospital must be 10°. Historic hourly radiation data (W/m^2) were available from the CLIMATICUS database (2005).

Electricity was charged with an hour-seasonal tariff, presenting differentiated tariffs for peak and off-peak periods: US$51.60/MWh for off-peak periods, and between 18 h and 21 h the peak value was US$79.60/MWh. Steam demand was considered constant during the period in which the sterilization central was open, during 6 h and 20 h, in addition to the steam demand of a hot counter in the restaurant, during lunch and supper times. The hot water demand presents two contributions: laundry (operating between 8 h and 18 h), and the internal use of the hospital (no interruptions). The tariff for natural gas (US$48/MWh—PBGAS 2013) does not include hourly or seasonal differentiation, as is the case of diesel (US $59.60/MWh). Due to the location of João Pessoa, the biomass considered herein originates from sugar cane bagasse, and the final price was U$ 13.20/MWh (considering 15.40 MJ/kg dry matter and 0.787 dry matter/kg).

A lifetime of 15 years was considered for the system, with an interest rate of 10 % y^{-1}, obtaining a capital recovery factor of 0.13 yy^{-1}. The system was

Table 23.2 Matrix of technical production coefficients and technology data (Carvalho et al. 2014)

			Technical production coefficients							Equipment		
	NG	BM	DI	VA	HW	CW	AA	CO	EE	Cost CINV (10^3 US$)	Cost O&M (US$/MWh)	Nom. Power PNOM (MW)
Gas engine	−2.63				1.10	0.45			**1**	185.00	6.00	0.41
Diesel engine			−2.66		0.80	0.50			**1**	90.80	6.00	0.36
Steam boil (BM)		−1.40		**1**						20.40	3.20	0.25
Steam boil (NG)	−1.18			**1**						21.29	0.89	0.30
Steam boil (EE)				**1**					−1.15	0.40	0.89	0.15
HX (VA-HW)				−1.10	**1**					3.56	0.89	0.40
Hot water boil (BM)		−1.25			**1**					25.00	3.20	0.17
Hot water boil (NG)	−1.22				**1**					19.72	0.89	0.30
Hot water boil (EE)					**1**				−1.11	11.28	0.89	0.15
HX (HW-CW)					−1.10	**1**				2.96	0.89	0.40
Abs chil (2x)				−0.77		1.77		**1**	−0.01	186.08	4.00	0.46
Abs chil (1x)					−1.32	2.32		**1**	−0.01	215.88	4.00	0.49
Mechanical chiller						1.21		**1**	−0.21	58.00	1.77	0.28
Cooling Tower						−1.00	**1**		−0.02	11.28	4.00	1.00

designed to interact with the electric grid also through the export of autogenerated electricity, considering the credit compensation scheme set out by BRASIL (2012). In this energy compensation system, surplus generated electricity can be fed into the distribution grid and registered as energy credits, which can be consumed in up to 36 months. This special regime for electricity exports applies to photovoltaic- and natural gas-produced electricity. Finally, the problem to be solved consists of two simultaneous tasks: selection of the combination of technologies (type and power installed) that are capable of meeting the energy demands of the hospital, and establishment of the operation mode for all equipment installed, for each time interval defined.

Life Cycle Analysis

It takes considerable effort to conduct an LCA study, exploring large industrial systems, collecting and analyzing a great amount of environmental information (Baumann and Tillmann 2004). A brief overview of the basic LCA process is provided for context when interpreting the findings. A framework for LCA has been standardized by the International Organization for Standardization (ISO) in the ISO 14040 series (ISO 14040 2006; ISO 14044 2006). This LCA framework consists of the following elements (Carvalho et al. 2011): (1) Goal and Scope definition, which specifies the goal and intended use of the LCA and delineates the assessment (system boundaries, function and flow, required data quality, technology and assessment parameters); (2) Life Cycle Inventory analysis (LCI), which includes the collection of data on inputs and outputs for all processes in the product system; (3) Life Cycle Impact Assessment (LCIA), which translates inventory data on inputs and outputs into indicators about the product system's potential impacts on the environment, human health, and availability of natural resources; and (4) Interpretation, the phase where the results of the LCI and LCIA are interpreted according to the goal of the study and where sensitivity and uncertainty analysis are performed to qualify the results and conclusions.

Because of the potential impacts of global warming and its associated climate change, greenhouse gases (GHG) were chosen to represent the environmental loads. The more general category name climate change replaced the formerly used greenhouse effect and global warming (Udo de Haes 2002). Because the average tropospheric lifetime of all greenhouse gases, even of the relatively short-lived ones (CH_4, about 10 years), exceeds the tropospheric mixing time (about 1 year), it is not important where the emissions occur. Climate change is therefore a truly global impact category (Udo de Haes 2002).

Although the geographic scale of GHG emissions is global, special care must be taken at the time of defining the LCA processes and data utilized, as the results will vary greatly if, for example, different production methods and transportation are used. This will result in a great difference in environmental loads if natural gas is consumed in Spain (originating from Algeria, liquefied, transported in Liquefied

Natural Gas carriers, and regasified before being distributed via pipelines) or Canada (originating from the province of Alberta, transported and distributed in pipelines).

The impact evaluation method chosen to represent the environmental loads was the 2013 version of IPCC, which expresses environmental impacts in kg CO_2-eq after atmospheric emissions are quantified and multiplied by a conversion factor (IPCC 2014). The reader is directed to IPCC (2014) for a more complete description of the method. Software SimaPro v. 8.0.4 (PréConsultants 2015) was utilized for the quantification of environmental impacts, and the database utilized was EcoInvent version 3 (EcoInvent 2014).

Equipment

For each item present in the superstructure (Fig. 23.1), the environmental emissions were calculated as the sum of the partial emissions for each stage (construction, operation and maintenance, and dismantlement and disposal). Data on the material composition of each piece of equipment were obtained after exhaustive and extensive consultation of catalogues and manufacturers. Table 23.3 shows the main material composition and CO_2-eq emissions associated (COE).

The best available disposal scenario was considered for each material (recycling), and all waste flows (lubricating oil, wastewater, Li-Br, etc.) received adequate treatment before disposal. The exception was the gas engine, which in its specific case considered reuse of materials by the manufacturer, resulting in the impressive low emissions shown in Table 23.3. All recycling options also considered transportation to the recycling site, which significantly contributed to the final emissions, as a distance of 2200 km was considered (the recycling site is located in São Paulo, Southeast Brazil).

Energy Utilities

Special care was taken to correctly adapt each existing process that would be applicable to the optimization study herein presented. For the electricity mix, a 2014 average was considered (ONS 2014) for the state of Paraíba: hydro 48.04 %, thermoelectrical (oil) 44.58 %, and eolic 7.38 %. Considering these percentages, the final emissions associated with the consumption of 1 kWh from the electric grid were 0.580 kg CO_2-eq. This figure is very different from the overall Brazilian electricity mix, which presents a much lower carbon content.

For the sugar cane bagasse, the emissions were 0.147 kg CO_2-eq/kWh. These emissions already take into account the drying of the bagasse. For natural gas, the emissions were 0.256 kg CO_2-eq/kWh consumed, and for diesel, 0.319 kg CO_2-eq/ kWh consumed. Please note that these emissions refer to the consumption of one

Table 23.3 Main material composition of technologies and corresponding environmental emissions

	Main material composition	COE kg CO_2-eq
Gas engine	Cast iron 1300 kg, Oil 540 kg, Low-alloyed steel 225 kg, PVC 5 kg, Steel sheet 2 m^2	$3.53\ 10^2$
Diesel engine	Cast iron 3500 kg, Oil 1292 kg, Low-alloyed steel 300 kg, Rubber 50 kg, Copper 105 kg	$6.98\ 10^6$
Steam boil (BM)	Chromium steel 1980 kg, Copper 50 kg, PVC 10 kg	$2.73\ 10^3$
Steam boil (NG)	Rock wool 115 kg, Al sheet 14 m^2, Stainless steel 980 kg, low-alloyed steel 490 kg	$2.23\ 10^6$
Steam boil (EE)	Rock wool 115 kg, Al sheet 14 m^2, Stainless steel 980 kg, low-alloyed steel 490 kg	$2.23\ 10^6$
HX (VA-HW)	Fe-Ni-Cr alloy 94 kg, Epoxy resin 0.5 kg, Titanium dioxide 3 kg	$1.50\ 10^3$
Hot water boil (BM)	Chromium steel 1980 kg, Copper 50 kg, PVC 10 kg	$2.73\ 10^3$
Hot water boil (NG)	Rock wool 115 kg, Al sheet 14 m^2, Stainless steel 980 kg, low-alloyed steel 490 kg	$2.23\ 10^6$
Hot water boil (EE)	Rock wool 115 kg, Al sheet 14 m^2, Stainless steel 980 kg, low-alloyed steel 490 kg	$2.23\ 10^6$
HX (HW-CW)	Titanium dioxide 2 kg, Fe-Ni-Cr alloy 75 kg, Epoxy resin 0.5 kg	$1.47\ 10^3$
Absorption chil (2x)	Li-Br 1600 kg, Stainless steel 7338 kg, Copper 30 kg, PVC 10 kg	$2.47\ 10^5$
Absorption chil (1x)	Li-Br 2200 kg, Stainless steel 7418 kg, Copper 30 kg, PVC 10 kg	$3.25\ 10^5$
Mechanical chiller	Polyurethane 30 kg, Stainless steel 2000 kg, Copper 30 kg, PVC 10 kg	$1.39\ 10^4$
Cooling Tower	Glass fiber 2622 kg, Cast iron 200 kg, Copper 10 kg, PVC 10 kg, Steel sheet 1 m^2	$9.71\ 10^3$
PV panel	PV panel 1.64 m^2, Inverter 1, Copper 1.5 kg, PVC 0.125 kg, Steel sheet 0.005 m^2	$7.80\ 10^3$

energy unit of the utility; however, the optimization model will consider adequate conversion (Table 23.2) to different energy forms. For example, the diesel generator will produce electricity at 0.848 kg CO_2-eq/kWh, considering its electrical efficiency (1/2.66 = 37.6 %).

The system boundaries were defined as in Fig. 23.2, where the selfgenerated electricity exported to the electric grid is evaluated at the same “environmental impact” as the electricity imported from the grid, introducing the concept of avoided emissions as the emissions avoided elsewhere by the production of electricity by the PV panels (avoiding the purchase of electricity from the grid).

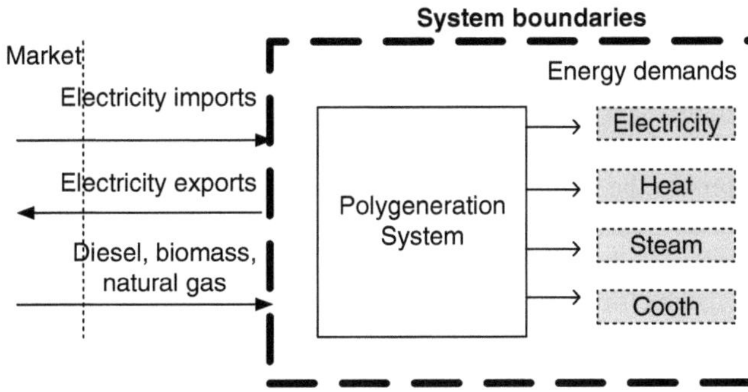

Fig. 23.2 Boundaries of the polygeneration system

Optimization Model

A combinatorial optimization method, based on mathematical programming (MILP), determines the optimal design and operation for the energy supply system. The optimization model was implemented in LINGO 14.0 (2015), an optimization tool that utilizes the combination of branch and bound and simplex methods in its solution algorithm. The environmental objective function was the minimization of the total annual impact (CO_{tot}):

$$\text{Min } CO_{tot} = CO_{fix} + CO_{ope} \tag{23.1}$$

Where CO_{fix} referred to the impact associated with the installation of equipment, and CO_{ope} referred to the consumption of energy utilities to satisfy the energy demands of the hospital. Then:

$$CO_{fix} = fam_e \bullet \Sigma_i[NEI(i) \bullet COE(i) + PV] \tag{23.2}$$

where NEI(i) and COE(i) are, respectively, number of pieces of equipment installed for technology i and the CO_2 emissions for the production of each piece of equipment. PV refers to the PV panels. The environmental amortization factor fam_e allocates the global emissions throughout the lifetime of the system, and was 0.10 y^{-1}.

Considering that the year was divided into *d* representative days, which were in turn subdivided into *h* hours, (d,h) represented the *h*th hour of the *d*th representative day. The annual operation impact (CO_{ope}), associated with the operation of the system was expressed by:

$$CO_{ope} = \Sigma_d \Sigma_h \begin{bmatrix} CO_{ng} \bullet F_{ng}(d,h) + CO_{ee} \bullet E_p(d,h) - CO_{ee} \bullet E_s(d,h) \\ + CO_{bm} \bullet F_{bm}(d,h) + CO_{di} \bullet F_{di}(d,h) \end{bmatrix} \quad (23.3)$$

where CO refers to the emissions associated with the consumption of the different utilities: ng (natural gas), ee (electricity), bm (biomass), and di (diesel). F refers to the amount imported for each fuel, E_p refers to electricity imported from the grid and E_s refers to electricity exported to the grid (and therefore counted as avoided emissions). Equations (23.1)–(23.3) can be changed to consider economic costs. Capacity limits, production restrictions, and balance equations were enforced in the optimization model.

Capacity limits: For each period (d,h) and for each technology i

$$POE(i,d,h) \leq NEI(i) \cdot P_{nom}(i) \quad (23.4)$$

where POE(i,d,h) is the energy production of technology i in the period (d,h), and P_{nom} is the nominal power of the equipment.

Production restrictions: For each period (d,h) and for cogeneration modules, the production of energy was restricted to:

$$POE(i,d,h) = NES(i,d,h) \bullet P_{nom}(i) \quad \text{with} NOP(i,d,h) \in \{0, 1, \ldots, NEI(i)\} \quad (23.5)$$

where NES(i,d,h) is the number of operational equipment of technology i in the period (d,h). For each technology i, For each utility j,

$$UI(i,j,d,h) = PC(i,j) \bullet POE(i,d,h) \quad (23.6)$$

where UI(i,j,d,h) is the energy flow of utility j interchanged with technology i in the period (d,h) and PC(i,j) is the absolute value of the production coefficient (Table 23.2).

Balance equations: For each period (d,h) and for each utility j:

$$Prod(j,d,h) - Cons(j,d,h) + Imp(j,d,h) - Exp(j,d,h) - Waste(j,d,h) - Demand(j,d,h) = 0 \quad (23.7)$$

$$Prod(j,d,h) = \Sigma_i X(i,j,d,h) \cdot YTUP(i,j) \quad \text{with} YTUP(i,j) \in \{0,1\} \quad (23.8)$$

$$Cons(j,d,h) = \Sigma_i X(i,j,d,h) \bullet YTUC(i,j) \quad \text{with} YTUC(i,j) \in \{0,1\} \quad (23.9)$$

$$Imp(j,d,h) \leq YUP(j) \bullet (Cons(j,d,h) + D(j,d,h)) \quad \text{with} YUI(j) \in \{0,1\} \quad (23.10)$$

$$Exp(j,d,h) \leq YUS(j) \bullet Prod(j,d,h) \quad \text{with } YUE(j) \in \{0,1\} \quad (23.11)$$

$$Waste(j,d,h) \leq YUW(j) \bullet Prod(j,d,h) \quad \text{with } YUW(j) \in \{0,1\} \quad (23.12)$$

$$\text{Demand}(j,d,h) \leq \text{YUD}(j) \bullet (\text{Prod}(j,d,h) + \text{P}(j,d,h))\text{with}\text{YUD}(j) \in \{0,1\} \tag{23.13}$$

YTUP(i,j) was 1 when technology i produced utility j. YTUC(i,j) was 1 when technology i consumed utility j. Production (Prod), and Consumption (Cons) corresponded to internal utility flows. Imports(Imp), Exports (Exp), Waste and Demand correspond to the utility flows interchanged between the energy system and the market environment. Binary variables YUI(j), YUE(j), YUW(j) and YUD (j) indicated, respectively, the possibility of such interchanges.

Results and Discussion

The main criterion chosen to carry out the optimization of the energy supply system was the total annual emissions (kg CO_2-eq/year). Table 23.4 presents the results for the optimizations carried out, considering that there are 200 PV panels available for installation.

Table 23.4 presents three optimal solutions: economic and environmental optimals, and a reference solution. The reference solution reflects the conventional way of installing energy supply systems: restrictions were introduced within the optimization model to prevent PV panels and biomass-driven equipment to be installed, while also excluding the possibility of installing cogeneration modules and absorption chillers.

The reference solution installed hot water and steam boilers that operated on natural gas, as well as one electrical boiler. All the electricity necessary to supply the hospital was purchased directly from the grid. The cooling demand was met by mechanical chillers. This solution presented the worst environmental emissions, almost 40 % higher than those of the environmental optimal.

When optimizing the system applying the economic criteria, the system took advantage of the lower purchase cost of biomass and imported 2663 MWh/year for this energy source to operate boilers. For the economic optimal, steam-hot water heat exchangers, and exclusively mechanical chillers were installed. Electricity was used to drive the mechanical chillers and auxiliary machinery in this system.

A very different configuration was obtained for the environmental optimal (Fig. 23.3): a natural gas engine with heat recovery was installed for the production of electricity. In Fig. 23.3, the installed equipment is shown in bold, while available but not installed equipment are shown in gray. Only biomass boilers were installed along with both types of heat exchangers (steam-hot water and hot water-refrigeration water). Electricity was supplied to users by operating the gas engine cogeneration module, by the production of the PV panels, and by purchasing a small quantity from an outside electric power company. The environmental optimal suggested that environmental benefits are realized through electricity exports. Trigeneration was installed in this case, as the cold water for space cooling was

Table 23.4 Minimum annual cost for the polygeneration system with 200 PV panels installed

	Economic Optimal		Reference system[a]		CO_2-eq optimal	
Composition of system	Number (Installed Power)		Number (Installed Power)		Number (Installed Power)	
Gas engine	0 (0 MW)		– (0 MW)		1 (0.41 MW)	
Diesel engine	0 (0 MW)		– (0 MW)		0 (0 MW)	
Steam boiler (NG)	0 (0 MW)		1 (0.300 MW)		0 (0 MW)	
Steam boiler (BM)	1 (0.250 MW)		0 (0 MW)		1 (0.25 MW)	
Steam boiler (EE)	0 (0 MW)		0 (0 MW)		0 (0 MW)	
Heat exchanger (VA-HW)	1 (0.400 MW)		1 (0.400 MW)		1 (0.40 MW)	
Hot water boiler (NG)	0 (0 MW)		1 (0.300 MW)		0 (0 MW)	
Hot water boiler (BM)	2 (0.340 MW)		0 (0 MW)		1 (0.17 MW)	
Hot water boiler (EE)	1 (0.150 MW)		1 (0.150 MW)		0 (0 MW)	
Heat exchanger (HW-CW)	0 (0 MW)		0 (0 MW)		1 (0.40 MW)	
Double-effect absorption chiller	0 (0 MW)		0 (0 MW)		0 (0 MW)	
Single-effect absorption chiller	0 (0 MW)		0 (0 MW)		1 (0.49 MW)	
Mechanical chiller	3 (0.810 MW)		3 (0.810 MW)		3 (0.81 MW)	
Cooling tower	1 (1.000 MW)		1 (1.000 MW)		2 (2.00 MW)	
Photovoltaic panels	200 units		–		200 units	
Electricity imports	3244 MWh/year		4281 MWh/year		297 MWh/year	
Electricity produced by PV panels	129 MWh/year		–		129 MWh/year	
Electricity credits	–		–		760 MWh/year	
Natural gas imports	–		1528 MWh/year		9554 MWh/year	
Diesel imports	–		–		–	
Biomass imports	2663 MWh/year		–		281 MWh/year	
Initial investment in equipment	US$	587,089	US$	374,670	US$	1,022,892
Annual cost of electricity imports	US$/year	180,489	US$/year	233,979	US$/year	15,342
Annual credit with electricity exports	US$/year	–	US$/year	–	US$/year	43,488
Annual cost of natural gas imports	US$/year	–	US$/year	73,335	US$/year	458,576
Annual cost of diesel imports	US$/year	–	US$/year	–	US$/year	–
Annual cost of biomass imports	US$/year	35,150	US$/year	–	US$/year	3708
Operation and Maintenance costs	US$/year	21,813	US$/year	16,716	US$/year	53,519
Annual cost of equipment[b]	US$/year	76,322	US$/year	48,707	US$/year	132,976
TOTAL annual cost	US$/year	331,774	US$/year	372,758	US$/year	620,633

(continued)

Table 23.4 (continued)

	Economic Optimal		Reference system[a]		CO_2-eq optimal	
Composition of system	Number (Installed Power)		Number (Installed Power)		Number (Installed Power)	
CO_2 emissions for equipment only	kg CO_2-eq/year	3,851,100	kg CO_2-eq/year	2,291,100	kg CO_2-eq/year	1,954,903
CO_2-eq emissions for electricity imports	kg CO_2-eq/year	1,881,711	kg CO_2-eq/year	2,482,917	kg CO_2-eq/year	172,222
CO_2-eq emissions for electricity exports	kg CO_2-eq/year	–	kg CO_2-eq/year	–	kg CO_2-eq/year	441,039
CO_2-eq emissions for natural gas imports	kg CO_2-eq/year	–	kg CO_2-eq/year	391,230	kg CO_2-eq/year	2,445,736
CO_2-eq emissions for diesel imports	kg CO_2-eq/year	–	kg CO_2-eq/year	–	kg CO_2-eq/year	–
CO_2-eq emissions for biomass imports	kg CO_2-eq/year	391,448	kg CO_2-eq/year	–	kg CO_2-eq/year	41,296
CO_2-eq emissions for equipment[b]	kg CO_2-eq/year	385,110	kg CO_2-eq/year	674,291	kg CO_2-eq/year	195,490
TOTAL annual CO_2-eq emissions	kg CO_2-eq/year	2,658,269	kg CO_2-eq/year	3,548,438	kg CO_2-eq/year	2,557,232

[a]Carvalho et al. (2014)
[b]Lifetime of system 15 years, interest rate 10 %/year

supplied by the single effect absorption chiller and mechanical chillers. Surplus not-consumed cogenerated heat was disposed of through heat exchangers. This result was not totally unexpected as trigeneration systems have long been seen as good alternatives to reduce emissions; however, the potential to reduce emissions depends also on the electricity mix supplied to the system.

Distributed energy resource systems have been recognized as an environmentally-friendly in comparison with conventional energy systems, as mentioned by Ren et al. (2010); these systems can employ combined heat and power systems, PV systems, etc. and help address global environmental problems. Saner et al. (2014) optimized the energy supply for a Swiss building considering greenhouse gas emissions and particulate matter formation, with a drastic shift from fossil fuel-dominated system to a biomass-dominated system. You (2012) and Carvalho et al. (2011) have obtained divergent solutions when optimizing from economic and environmental perspectives, corroborating the results obtained herein.

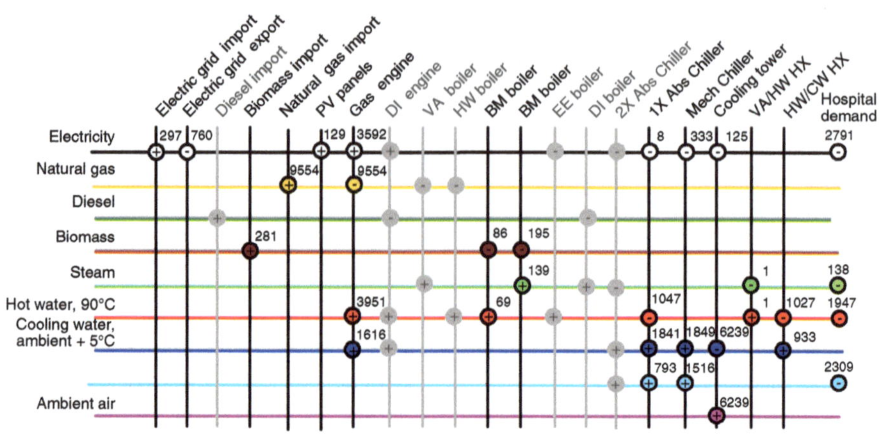

Fig. 23.3 Optimal environmental system annual flows (MWh/year)

Conclusions

A progressive change has been observed in recent years: consumers, and society in general, are starting to evolve towards a more environmentally-friendly conscience. This conscience, along with restrictions on the emissions of environmental impacts, have in turn changed how optimal solutions are obtained. Initially optimal solutions were based exclusively on economic aspects. Nowadays, environmental aspects are beginning to be introduced in optimizations, either as a main objective or as part of the objective (which could include energy efficiency or economic terms).

This manuscript explained the incorporation of environmental information in an optimization model for the energy supply of a hospital located in Northeast Brazil. The Life Cycle Assessment methodology was applied for the quantification of environmental loads, expressed by the emissions of CO_2-eq (IPCC methodology). Environmental data was introduced into the linear programming model, which determined the equipment to be installed and how to operate each technology, in order to satisfy the energy demands of the hospital. The optimal solution also provided all energy, economic, and environmental flows.

The environmental optimal solution suggested that trigeneration is the best option to minimize the environmental loads produced by the energy supply system. Natural gas was imported from the grid to drive the gas engine, which produced hot water and electricity to meet the energy requirements of the hospital. PV panels produced electricity, and biomass was imported for hot water and steam boilers. A single-effect absorption chiller utilized part of the cogenerated heat to produce coolth. Mechanical chillers helped with the production of coolth also. This optimal solution took advantage of the electricity exports to achieve higher rates of avoided emissions, and realize environmental benefits to lower the overall emissions of the system.

The optimization carried out herein was based on the emission on greenhouse gases (IPCC 2014). Future works could include the consideration of a different, perhaps more global, impact assessment method for the calculation of environmental impacts. This would likely provide a different result and help verify the effect of utilizing a more wider perspective in the optimization. Also, sensitivity analyses could verify whether the configuration varies when different electricity mixes are applied (different regions present different mixes). The high emissions associated with the purchase of electricity could have a considerable effect on the configuration of the system, as the municipality of João Pessoa counts with the largest thermal plant in the world (oil-based engines).

Acknowledgments The authors wish to acknowledge the support of the National Council for Scientific and Technological Development (CNPq), through project n° 475879/2013-9.

References

Araújo, M. M. D. (2004). Methodological contribution for the exergetic diagnosis of thermal and electric systems—Study case at the Lauro Wanderley University Hospital. M.Sc. dissertation, Department of Mechanical Engineering, Federal University of Paraíba, João Pessoa, Brazil (in Portuguese).

Baumann, H., & Tillmann, A. M. (2004). *The hitch hiker's guide to LCA. An orientation in life cycle assessment methodology and application.* Gazelle Book Services.

Brasil, A. (2012). Resolution N° 482, of April, 17, 2012. Establishes the general conditions for the access of distributed micro- and mini-generation to the electricity distribution system, electricity compensation system, and other matters. Retrieved August 04, 2014, from http://www.aneel.gov.br/cedoc/ren2012482.pdf (in Portuguese).

Carvalho, M., & Millar, D. L. (2012). Concept development of optimal mine site energy supply. *Energies, 5*, 4726–4745.

Carvalho, M., Romero, A., & Millar, D. L. (2014). Biomass and thermal solar energy in the synthesis and optimization of a polygeneration system. IN *Proceedings of the V Brazilian solar energy conference*, Recife.

Carvalho, M., Serra, L. M., & Lozano, M. A. (2011). Optimal synthesis of trigeneration systems subject to environmental constraints. *Energy, 36*, 3779–3790.

CLIMATICUS 4.2. Climatic database. Project strategies for 58 Brazilian cities. INMET1961-1990 database. Beta version. Faculty of Architecture and Urbanization, University of São Paulo, Technology department, Laboratory of environmental comfort and energy efficiency, 2005. Retrieved November 29, 2013, from www.usp.br/fau/pesquisa/laboratorios/labaut/conforto/Climaticus_4_2.xls (in Portuguese).

EcoInvent database. (2015). Retrieved March 25, 2015, from http://www.ecoinvent.ch.

Erbs, D. G., Klein, S. A., & Beckman, W. A. (1983). Estimation of degree-days and ambient temperature bin data from monthly-average temperatures. *ASHARE Journal, 25*, 60–65.

Guinee, J. B. (Ed.). (2001). Life cycle assessment—An operational guide to the ISO standards. http://media.leidenuniv.nl/legacy/new-dutch-lca-guide-part-1.pdf.

IPCC. (2014). 2013 revised supplementary methods and good practice guidance arising from the Kyoto protocol. Retrieved March 25, 2015, from http://www.ipcc-nggip.iges.or.jp/public/kpsg/.

ISO 14040. (2006). *Environmental management—Life cycle assessment—Principles and framework.* Geneva: International Organization for Standardization (ISO).

ISO 14044. (2006). *Environmental management—Life cycle assessment—Requirements and guidelines*. Geneva: International Organization for Standardization (ISO).

Kyocera. Kyocera Brasil. Retrieved July 30, 2014, from http://www.kyocera.com.br/.

Lingo 14.0. Lindo systems. Retrieved March 27, 2015, from http://www.lindo.com.

Nepote, M. H. A., Monteiro, I. U., & Hardy, E. (2009). Association between operational indices and occupation rates of a general surgery center. Latino-Am. Enfermagem [online] Journal 17 (in Portuguese).

ONS—Brazilian National Electric System Operator. (2014). *Daily operation bulletin*. Retrieved March 27, 2015, from http://www.ons.org.br/resultados_operacao/ipdo_new.aspx.

PBGÁS—Companhia Paraibana de Gás. Tariffs. Retrieved December 24, 2013, from http://www.pbgas.com.br/?page_id=1477 (in Portuguese).

PréConsultants. (2013). SimaPro software. Retrieved March 27, 2013, from http://www.simapro.nl.

Ren, H., Zhou, W., Nakagami, K., Gao, W., & Wu, W. (2010). Multi-objective optimization for the operation of distributed energy systems considering economic and environmental aspects. *Applied Energy, 87*, 3642–3651.

Saner, D., Vadenbo, C., Steubing, B., & Hellweg, S. (2014). Regionalized LCA-based optimization of building energy supply: Method and case study for a Swiss municipality. *Environmental Science & Technology, 48*, 7651–7659.

Santerno. Carraro group webpage. Monophasic solar inverters. 2014. Available at: <http://www.santerno.com/br/produtos/energia-solar/inversores-solares-monofasicos.html>. Accessed on 10 mar 2016 [In portuguese]

Udo de Haes, H. A. (Ed.). (2002). *Life-cycle impact assessment: Striving towards best practice*. Published by the Society of Environmental Toxicology and Chemistry (SETAC). Pensacola, USA.

WEC—World Energy Council. (2004). *Comparison of energy systems using life cycle assessment*. http://www.worldenergy.org/documents/lca2.pdf.

You, F. (2012). Multiobjective optimization of energy-environmental systems. In *2012 CAPD annual meeting*, Carnegie Mellon University, Pittsburgh, PA.

Part VI
Energy Technologies and Their Effect on Global Warming

Green Buildings Technology

Available technologies and numerical investigation tools for buildings performance optimization

Chapter 24
Energy Conservation Through Sunrays Reflecting Coating on Buildings

Imre Benkö

Introduction

Sun reflecting coating materials are used in ambient temperature applications (under 80 °C) such as walls, roofs, heat shields, etc. where thermal radiation is a dominant mechanism of heat transfer. Thus there is a real need for data concerning the radiative properties of such materials.

The determination of the emissivity of coating materials is not straightforward mater. Severe practical difficulties arise in connection with the measurement of test specimen surface temperature using thermocouples embedded just below the surface. It is extremely difficult to achieve uniformity of temperature over the surface of a coated specimen operating at sunshine and it is probably for this reason that the calorimetric method of emissivity measurement has not found favor in the case of such materials and attention has instead been concentrated on radiometric techniques as IR-imagery, too. But in the case of practical approach, the use of IR-imagery gives a direct temperature difference between the coated and uncoated surfaces resulting a lower emissivity of the coating.

Method of Examination

The measuring method consists of two successive steps. In the first step IR-imaging equipment, which viewed the speciment at right angle to the surface of the sample, received both the radiation emitted by the speciment and the sun radiation reflected

I. Benkö (✉)
Faculty of Mechanical Engineering, Budapest University of Technology and Economics, H-1112 Budapest, Cirmos u.1., Hungary
e-mail: ibenko@freestart.hu

P. Grammelis (ed.), *Energy, Transportation and Global Warming*, Green Energy and Technology, DOI 10.1007/978-3-319-30127-3_24

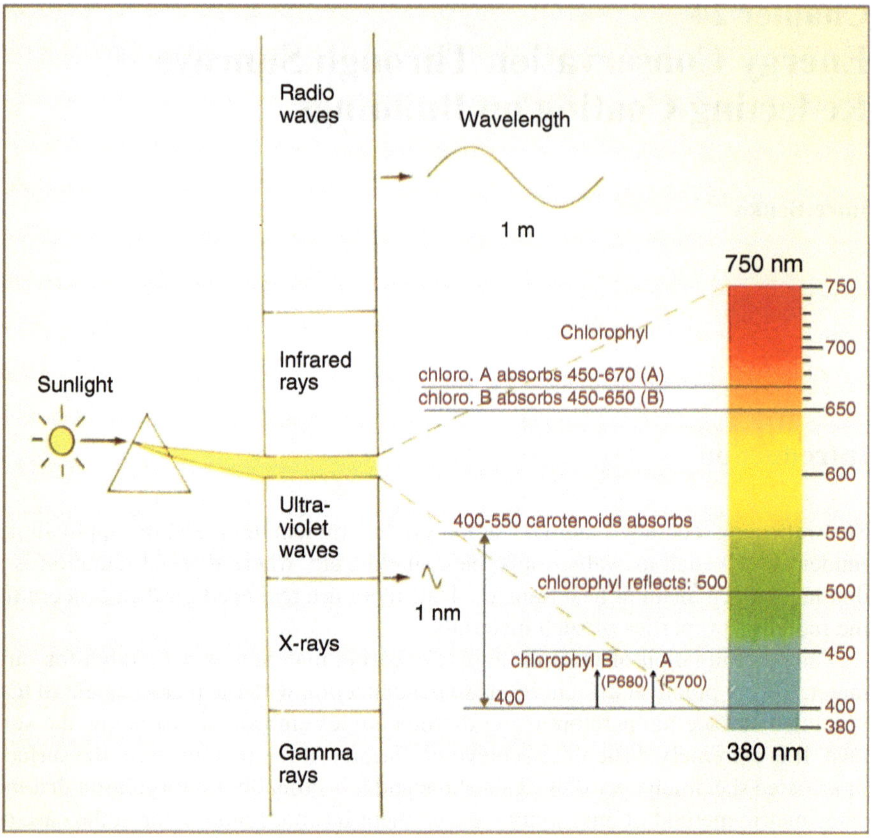

Fig. 24.1 Visible light and electromagnetic spectrum of pigments and absorption spectra

from it. In the second step, after getting to the thermal steady state, shielded the sample from the sunshine, the IR-imaging equipment received only the radiation emitted by the nonuniform heated coated and uncoated sides of the specimen (that is the 'own radiation').

The special feature of the arrangement used in this case was a heat shield mounted on rotating arm. The shield consisted of a plate backed by insulating material. The shield was designed to fit over the surface of the specimen without touching it, held in position by a spring.

The arrangement was designed so that when the specimen reached thermal equilibrium on the sunshine, the shield could then be quickly swung on the rotating arm, thus shielding the specimen from the sunshine and allowing only the own radiant an convective heat transfer to take place. Measurement was terminated after about 4–6 s (Figs. 24.1 and 24.2).

In the case of each measurements it was possible to vary the angle of view of the measuring equipment with respect to the normal to the surface of the speciment (see Fig. 24.3). A well defined distribution of emissivity as a function of the quality of

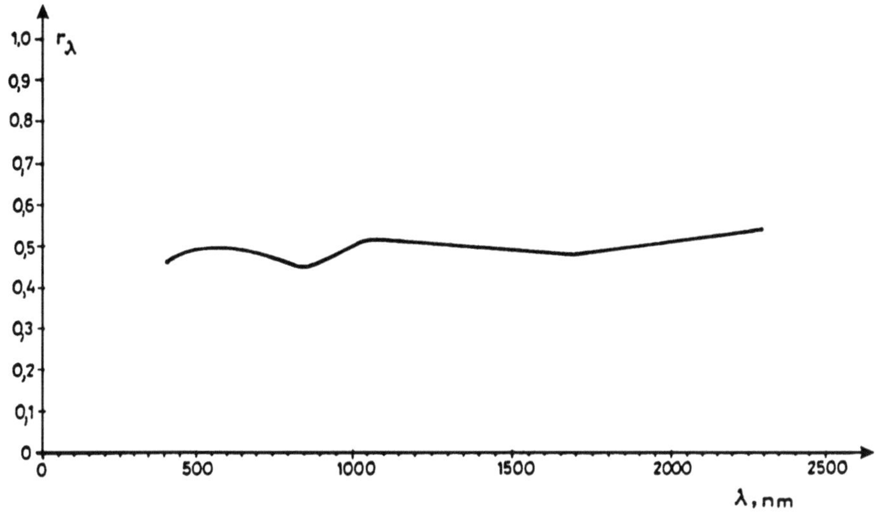

Fig. 24.2 Spectral values of the reflectivity of AL-CHR coating

the specimen has been also presented for the coated and uncoated surfaces (see Figs. 24.5 and 24.6). The average value of the emissivity for these surfaces can be calculated by histographical method.

Theoretical Basis of the Direct Comparative Method

The IR-radiation of different samples was examined under practical conditions, in direct sunshine. The half of the each sample was coated by selective coating and the other half was uncoated (Fig. 24.4). Infrared radiation emitted by the surfaces was measured by AGEMA thermal imaging equipment with a special method.

As indicated before, the arrangement for the 'direct' method of emissivity measurement consists of a relatively small heated specimen having a surface temperature Tw within a large environment which is at uniform temperature Te . The specimen is assumed to be opaque to thermal radiation.

The sunshine radiation views a given target area on the specimen surface in the normal direction or at some angle θ to the normal. In the second step of the examination, the solar radiation absorbed by the specimen is transferred to the environment by emission and convective heat transfer.

Thus

$$q_s^{\bullet}(\theta)\,.\varepsilon(\theta,T) = \varepsilon(\theta,T).\sigma.\left(T_w{}^4 - T_c{}^4\right) + 2.\alpha(T).\,(T_w - T_c) \qquad (24.1)$$

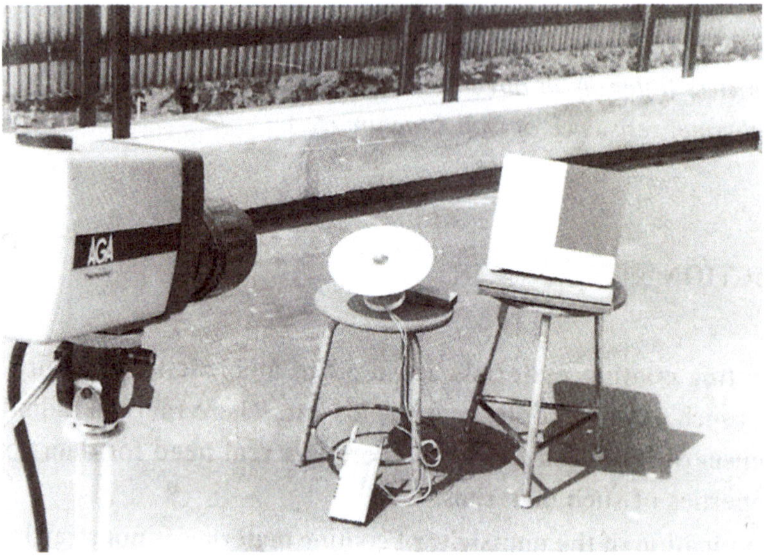

Fig. 24.3 Experimental arrangement for steady state and shielded measurements of emissivity

Fig. 24.4 A sample (Pace_04) of a slate roof

Here σ is the Stefan-Boltzmann constant, α is the coefficient of heat transfer in the heated specimen surface and $q^{\bullet}_{s}(\Theta)$ intensity of direct sunshine radiation in direction Θ.

From Eq. (24.1) it follows that

$$\varepsilon(\theta, T). = 2.\alpha(T).\ (T_w - T_c)/\left[\ q_s^{\bullet}(\theta) + \sigma.\ \left(Tc^4 - T_w{}^4\right)\right] \quad (24.2)$$

Noting that according to Kirchhoff's Low applied to total values of absorbtivity a (θ,T) and emissivity

$$a\ (\theta, T)\ = \varepsilon(\theta, T) \quad (24.3)$$

Comparison of Surfaces with Different Emissivities

The surface emissivity of the coating was compared with that of specimen material. In Figs. 24.5 and 24.6 on the thermogram are two quadrangles marked for which temperature distribution was computed. The temperature difference between the area coated with SRC-coating and the uncoated surface shows that the SRC-coating directly decreases the IR-emissivity of the material. Numerical values for the temperature represent the so called' effective black body temperature (at thermal imaging equipment's $\varepsilon = 1$).

Equation between temperature and emissivity ratios comes from the next equality

$$\sigma.T_1{}^4.\varepsilon_1 = \sigma.\ T_2{}^4.\varepsilon_2 \quad (24.4)$$

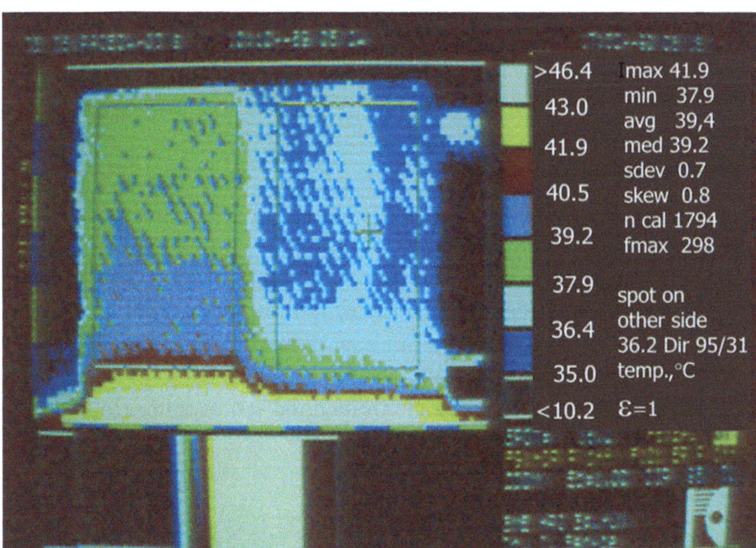

Fig. 24.5 IV-thermogram of a sample of Fig. 24.4 on direct sunshine

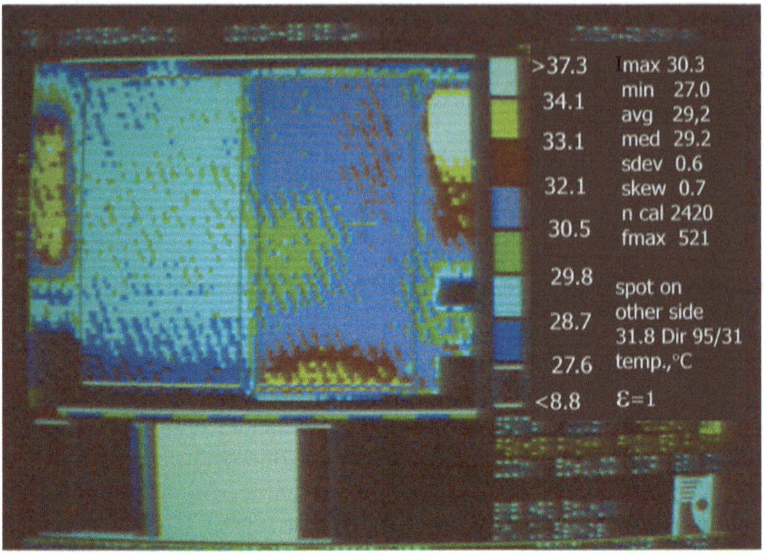

Fig. 24.6 IV-thermogram of a sample of Fig. 24.4 shielded from sunshine

So from Eq. (24.4), it follows that

$$\varepsilon_1/\varepsilon_2 = (T_2/T_1)^4. \tag{24.5}$$

Presentation of Results

The results are presented in the form of:

(a) A plot of spectral reflectivity of SRC-coating as a function of wavelength of the radiation (Fig. 24.2).
(b) IR-thermograms (Figs. 24.5 and 24.6) and table (Tables 24.1 and 24.2) of sample for presenting the thermal effects of different material/coating couples.
(c) Four plots of equilibrium temperature and intensity of 'heat gain' of different wall structures versus emissivity of surfaces, intensity of direct sunshine, coefficient of heat transfer and environmental temperature (Figs. 24.7, 24.8, 24.9, and 24.10).

Table 24.1 Technical data of AL-CHR (ALUMA-CHRON) elastomeric roof/wall coating

Wet material
Metallic aluminium: 6 %
Pigments: 2 %
Nonvolatile liquid elastomer: 23 %
Dry to complete: 3 days
Solids by volume: 48 %
Flashpoint of solvent: 40 °C
Application temperature: 5–49 °C
Wet film thickness: 1 L/sq m, 1 mm
Cured coating
Elongation at 25 °C: 400 %
Elongation at 0 °C:200 %
Return after elongation: >95 %
U.V. reflection, white: 90 %
Light gray: 70 %
Colors: <40 %
U.V. resistance: excellent
I.R. emissivity:
At low temperature, 2 ÷ 6 μm: 0.5 ÷ 0.7
I.R. reflectance:
At sunshine: 0.4 ÷ 2.4 μm: 0.45 ÷ 0.48
Dry film thickness: 1 L/sq m: 388 μm
Dry film weight: 1 L/sq m: 372 g/sq m
Re-coat interval: normally 10 years or more

Table 24.2 Collection of histographical data of IR-thermograms (Figs. 24.5 and 24.6) for comparison of a specimen (Fig. 24.4) on direct (s) sunshine and shielded (t) conditions: slate roof/SRC_04

On the right: slate roof	Temperature, °C	
On the left (Pace_04) temperature values, °C	On sunshine (s)	Shielded (t)
Spot on the other side	36.2	31.8
Dir x/y	95/31	95/31
max	41.9	30.3
min	37.9	27.0
avg	39.4	29.2
med	39.2	29.2
sdev	0.7	0.6
skew	0.8	0.7
Ncal, pixel	1794	2420
Fmax, pixels	248	521

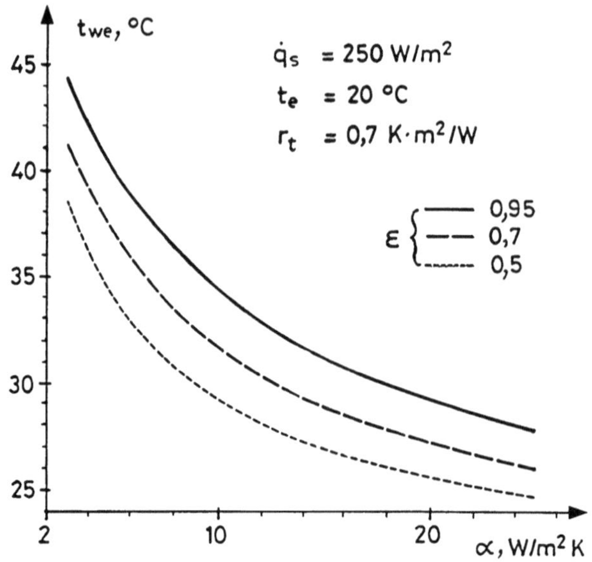

Fig. 24.7 Equilibrium temperatures of a wall structure (rt denotes the thermal resistance of wall structure) against coefficient of heat transfer

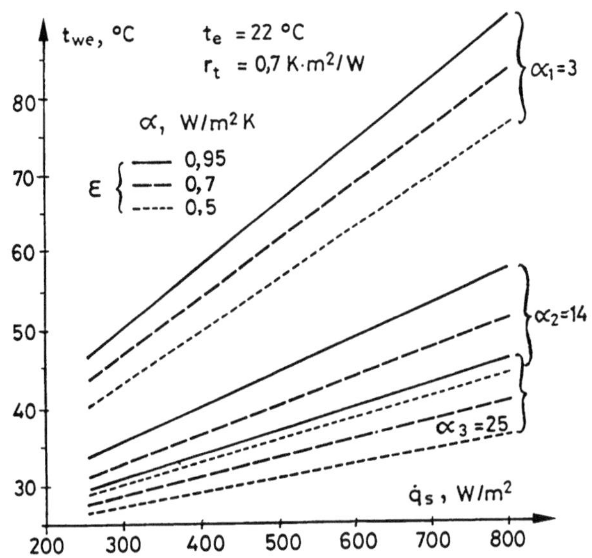

Fig. 24.8 Equilibrium temperatures of a wall structure (rt = 0.7 K m²/W) versus intensity of direct sunshine radiation

Applications in the Thermotechnique of Buildings

The factors affecting the infrared radiation of building surfaces and low temperature industrial equipment have not been widely enough studied. As far as building construction is concerned, the infrared radiation emissivity of walls has been

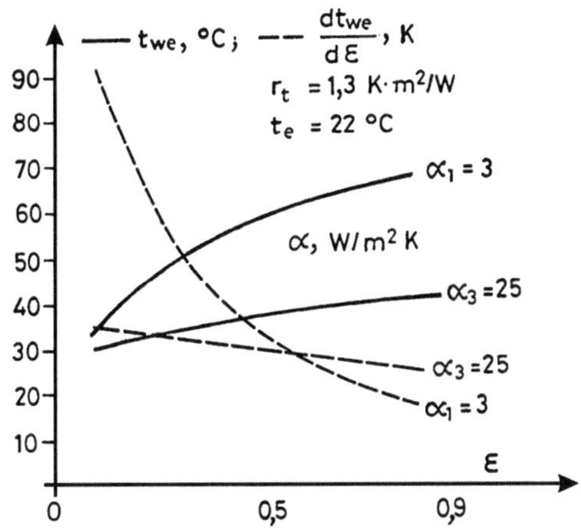

Fig. 24.9 Values and differential ratios of equilibrium temperatures of wall structure vs emissivity

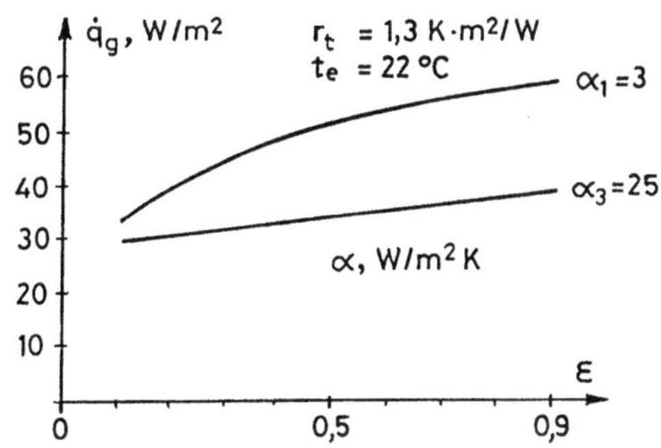

Fig. 24.10 Intensity of 'heat gain' of wall structure vs emissivity

regarded as an unchangeable factor basically dependent upon the temperature and quality of their material.

The research was based on different elastometric roof coatings (e.g. SRC-coatings), commercialized on name 'ALUMA-CHRON' (Data Specs of PACE Products International, Inc.), which are available also in Hungary (MULTITREND Ltd.). The special coating is liquid-applied and cures the surface. It is applied to forming single, tough, seem-free rubberized roof/wall membrane. From thermal point of view the coating is a double function membrane: it saves energy years-round, as it lowers interior temperatures in summer and its low emissivity slows heat loss in winter.

ALUMA-CHRON's Properties

Materials of buildings have generally good heat radiation properties, but from energy management and building physics point of view this fact is not always convenient. Some coating materials, e.g. ALUMA-CRON (further on: AL-CHR) have good properties: their emissivity is considerably lower that of the material onto which they are applied. This paper gives an account only for the infrared emissivity decreasing effect of SRC-coating.

Performance Characteristics

AL-CHR is an elastomeric roof/wall coating available in different colours. It is liquid-applied and it perfectly conforms to shapes. It is an economical, practical alternative to total roof replacement. This rubberized coating provides excellent resistance to both chemical and weather deterioration. Extensive elongation and nearly 100 % recovery allows the coating to stretch to meet changing weather requirements as the roof expands and contracts.

AL-CHR remains flexible. It will not become rigid, crack or peel even at temperatures of −40 °C and it keeps in heat stability under +82 °C.

Basic Uses of AL-CHR Coating

One-coat AL-CHR waterproofs, rustproofs, seal leaks, saves energy, deadens sound of metal roofs and stretches to accommodate all sudden temperature changes without breaking.

It restores and protects aged metal, asphalt shingles, sealed and unglazed tile and felt (with stabilized asphalt) roofs.

It can recoat vinyl, chlorinated rubber, and epoxy coatings.

It was originally designed to fill the need for a coloured roof and tank membrane that could handle unusual contours and steep slopes. It proved so effective that it now is used to coat all types of roofs, tanks, silos, and stucco or masonry sidewall where appearance, whether resistance and corrosion control are essential.

Role of Pigment in Sun's Rays Utilization

Coloured pigments in living cells trap light. Light and pigments are vital to important biological processes. There is an essential difference between chlorophyll molecules and the other pigments (e.g. in coat of animals): chlorophyll participates in photosynthesis as well as giving a green leaf its colour.

Visible light and the electromagnetic spectrum. We are constantly bathed by electromagnetic energy, from radio waves to infrared rays (heat), X-rays, and gamma rays. Each category has a range of energies measured in oscillating waves of specific lengths. Only the small portion of the entire electromagnetic spectrum with wavelengths in the range 380–750 nm is visible to us as white light that can be broken into coloured light, each colour with different wavelengths. The shorter the wavelength, the more energetic the wave, the longer the wavelength, the less energetic

Pigments and Absorption Spectra

The colours we see in a given object depend on which wavelengths of light the pigments in that object absorb and which are reflected back to our because all wavelengths are reflected. Black pigment in a coat absorbs light throughout the whole colour spectrum and reflect nothing. Thus, it looks black to us. A leaf contains chlorophyll-'a', which absorbs most strongly around 450 and 670 nm, and chlorophyll-'b', which absorbs most strongly around 490 and 650 nm; green light in the 500 nm range is reflected, and thus the leaf looks green. The peelings of fruit often contain carotenoids, which absorb, wavelengths from 400 to about 550 nm and reflect back red, yellow and orange light. Because a leaf often contains both kinds of chlorophyll plus carotenoid, its pigments can absorb and use for photosynthesis most of the wavelengths of visible light that strike it. However plants, algae and other autrophs use less than one percent of the energy of sunlight in photosynthesis.

In the techniques, the spectral reflectivity of a special (Al-CHR) coating in function of wavelength of the radiation (between 200–600 nm) is approx. $r = 0.5–0.7$.

The content of this coating is 22 % pigments, 6 % metallic aluminium and 23 % non-volatile liquid elastomer and lowers exterior temperature of walls of dwelling in summer by about 5–10 K, so decreasing incoming heat of structures by some 33–38 %. It slows infrared radiation heat loss by some 35–40 % in winter (Data Specs of PACE Products International, Inc.).

Results

The sunshine radiation of different samples was examined under practical conditions, in direct sunshine (Benko and Kiss 1977; Benko 1981; Benko 2008). Infrared radiation was measured by AGEMA thermal imaging equipment with a special method. The temperature field of specimen's surface was measured by thermocouples, too, in some control points (Benko 1992). The thermal imaging equipment shows the non-uniformity of the temperature field of the specimen surface.

Conclusions

SRC coating lowers exterior temperature of wall in summer by about 5–10 °C, so decreasing incoming heat of structure by some 33–38 %. It slows infrared radiated heat loss by some 35–40 % in winter (Benko 1992).

References

Benko, I., & Kiss, L. I. (1977). The struggle of mankind to ensure a comfortable thermal environment. In *The River Valley as a focus of interdisciplinary research; proceedings of the international conference to Commemorate Maupertuis' expedition to the River Tornio, Northern Finland*, 1736–1737, Oulu, Finland, 1977, p. 8.

Benko, I., Kiss, L. I. (1981). Effects of isolation and integration on the development of systems. In *Proceedings of the international conference: The Archipelago as a focus for interdisciplinary research*, Turku, Finland, 1978 (pp. 49–68). Karachi: Hamdan Foundation Pakistan.

Benko, I. (1981). Possibilities of utilising renewable energy sources (Vol. 25, No. 1). Periodica Polytechnica, Mech. Eng., Budapest, Hungary, pp. 67–86.

Benko, I. (1992). Examination of low emissivity coatings by infrared .imagery Advanced Infrared Technology and Applications. Firenze: Grafiche Troya Publ., 1992, pp. 185–197 and 1993, pp. 255–270.

Benko, I. (2003). The role of exergy and entropy in thermal engineering applications. In *The 1st international exergy, energy and environment symposium (IEEES-1)* (p. 81), Izmir, Turkey.

Benko, I. (2008). Thermal behaviour and entropy of living organisms through examples. In *Proceedings of the 2nd WSEAS/IASME international conference on energy planning, energy saving, environmental education (EPESE'08)* (pp. 19–24), Corfu, Greece, October 26–28, 2008.

COEC. (1981). *The solar handbook*. Denver, CO: COEC.

Data Specs of PACE Products International, Inc.

MULTITREND Ltd., H-1031 Budapest, Zahony u. 13., Hungary.

Postlethwait, J. H., & Hopson, J. L. (1989). *The Nature of Life*. New York: McGraw-Hill.

Rostvik, H. N. (1992). *The sunshine revolution*. Stavanger, Norway: SUNLAB Publ.

Chapter 25
Comparison of Heating and Cooling Loads of a Typical Building with TRNSYS and eQUEST

E. Bellos, C. Tzivanidis, A. Kouvari, and K.A. Antonopoulos

Introduction

Building energy consumption is a very important parameter in the design especially in our days with the increasing rate in the electricity cost. More specifically, the energy consumption has increased about 50 % the last years (Development Ministry 2009) in residential houses. In Europe a typical house with an average household occupation of 2.5 persons consumes 14.2 kWh daily (http://www.mpoweruk.com/electricity_demand.htm) and the yearly electricity consumptions is about 60 TWh in Greece (http://en.wikipedia.org/wiki/List_of_countries_by_electricity_consumption). It is obvious that energy consumption is a very important issue which influences on our society and the new life-style. Building energy consumption for heating and cooling is a great proportion of the total energy used. So, the knowledge of this energy is able to create more efficient houses and systems in order to optimize them. More specifically, by calculating with accuracy the heating and cooling loads of a house, an optimum HVAC is able to be selected which leads to lower electricity and fuel consumption.

It is important to find ways to lower the energy consumption of a building. For this reason a wide variety of building energy simulation programs have been developed, enhanced and are in use. So building designers today can use those programs to optimize the design to reduce the heating, cooling loads and electrical loads of the building. Some of the programs that are used worldwide on the market either purchase or as free software are:

E. Bellos (✉) • C. Tzivanidis • A. Kouvari • K.A. Antonopoulos
Department of Thermal Engineering, National Technical University of Athens, Zografou, Heroon Polytechniou 9, 15780 Athens, Greece
e-mail: bellose@central.ntua.gr

P. Grammelis (ed.), *Energy, Transportation and Global Warming*, Green Energy and Technology, DOI 10.1007/978-3-319-30127-3_25

DOE-2, Equest, EnergyPlus, ESP-r, ECOTECT, DeST, TRNSYS, BLAST, TRACE, HEED, Hap, Energy Express, BSim, Ener-Win, DOE-2.1E (Han et al. 2014; Crwley et al. 2008).

Some studies of the building performance with simulation tools are presented below. Xing et al. (2014) used eQuest to simulate the energy performance of an existing four-star hotel in Tianjin, China. Terziotti et al. modelled a seasonal solar thermal energy building with TRNSYS 16 (Terziotti et al. 2011). The combination of ESP-r and TRNSYS is applied on a case study of a low-energy building serviced by a solar-thermal and PV/t system (Beausoleil-Morrison et al. 2012).

Also many comparisons between these programs have been made by many researchers. TRNSYS and Energy Plus were compared for a single module with 100 m^2 of floor area for two Italian climatic conditions and the results had a difference of approximately 10 % (Gasparella and Pernigotto 2012). A comparison between the energy analysis model established for building by eQuest and the actual energy consumption (IPMVP method D), for an actual office building in Taiwan was made and the final results had errors from 0.5 % to 27 % (Kea et al. 2013). Zhu et al. (2013) made a comparison for loads with EnergyPlus, DeST and DOE-2.1E. Finally, Brahme et al. (2009) compared the three more known programs which are Equest, Energy Plus TRNSYS. According this comparison, eQuest is the easiest tool to use but it has a limited number of choices in simulated systems, EnergyPlus is able to model many net zero energy technologies with an easy way but the lack of a friendly interface is an important drawback and TRNSYS has the greater variety in simulations but learning this program has larger difficulties.

In our study, a comparison between eQuest and TRNSYS is made in order to predict how close the results of these programs are. A typical building is analyzed and parametric analyses made by changing the main parameters that influences on heating and cooling loads.

Simulation Data

In order to make a realistic comparison the same building was developed in the two programs and the yearly loads was calculated for the same case studies. Typical building parameters have been selected in order to make general results. The analysis is made for a building in Athens and the meteorological data was taken from data bases.

Examined Building

The characteristics of the analyzed building are given in following tables (Tables 25.1–25.3).

Table 25.1 Building parameters

Parameter	Value	Parameter	Value
Area	100 m^2	Specific gains	14 W/m^2 K
Height	3 m	Occupants density	0.1 $person/m^2$
East glass	double-2 m^2	Specific light	10 W/m^2
West glass	double-2 m^2	Infiltration rate	1.12 changes per hour
South glass	double-4 m^2	Rotation	0° (walls in four directions)
Shading coefficient	70 %	Insulation conductivity	0.035 W/m K

Table 25.2 U for structural components

Program	External walls U (W m^2 K)	Roof U (W m^2 K)	Ground U (W m^2 K)
eQuest	0.39	0.37	0.30
TRNSYS	0.38	0.39	0.29

Table 25.3 Layers of structural components

Wall construction	Thickness (cm)	Roof construction	Thickness (cm)	Ground construction	Thickness (cm)
ComBrick	10.15	Conc LW 30 lb	5.09	LightSoil, Damp	0.31
Polystyrene	7.62	Blt-UpRoof	0.95	Conc HW 140 lb	3.05
ComBrick	10.15	Polystyrene	7.62	Polystyrene	12.19
–		Conc HW 140 lb	15.24	–	

After these parameters it is important to refer the structural components and their properties. Table 25.2 gives the u-value of the external walls, of the roof and of the ground.

The values are close to each other but there are differences because every simulation tool uses different materials. Table 25.3 gives the analytical structure of all the components in eQuest.

In TRNSYS similar structures was selected in order to have similar u-values and thermal masses. Now it is important to state other differences between theses simulation tools. Another difference is in lighting strategy. TRNSYS lighting system is depended from outside the solar irradiation level but the eQuest has an indoor system. There are two photo sensors per zone which each one controls 50 % of lights. Each photo sensor is located 0.76 m height above the floor and 33 % of zone depth, which is 3.3 m from East and West wall. Also, it is important to state that the exact location of every window in the respect wall is influences the results in the eQuest and for the reason the windows are located in the center of the wall in every case.

Finally it is important to say that TRNSYS calculated the theoretical loads of the building when eQuest calculates the electrical consumption of the heat pumps that operates in the house. By setting the coefficient of performance equal to 1, the electrical consumption is the same as the heating and the cooling load respectively. So, TRNSYS calculates the loads with a direct way and the loads can be calculated from eQuest by an indirect way. This is may lead to differences in the final results.

A last main difference of the programs is that TRNSYS uses the S.I. system but the eQuest uses English Units.

Results

In this section the results of the simulations are presented and the comparison of the energy tools is made. First of all the results of the main case are given and after parametric analyses are presented.

Main Case Comparison

The yearly cooling and heating loads for the analyzed building were calculated analytically and are presented in Fig. 25.1.

It is shown that the heating loads are close to each other for all months, except January where TRNSYS gives greater loads. It is also important to say that eQuest does not give heating loads for October and May. This can be explained because the accuracy of eQuest is lower than TRNSYS, and the very low loads are taken as zero (Fig. 25.2).

It is obvious that the results of the programs are similar with low differences. The months April to July, TRNSYS gives greater loads but on the August and on the September eQuest calculated greater cooling loads (Table 25.4).

The above table presents the yearly cooling and heating loads which calculated from the two programs. It is shown that TRNSYS gives greater loads of about 5 %. This difference is acceptable and can be explained by the small differences between the programs. These differences are related to the way that the two programs calculate the loads and in the used materials. TRNSYS calculates the exact load in every case in order to keep the temperature in the desire level while Equest uses standard equipment which covers the loads. Moreover, eQuest uses a library for the building materials but TRNSYS gives the choice of selecting the material properties in every case.

Parametric Study

In this section parametric diagrams are given in order to compare the two programs for different cases. The parameters that changes are the following:

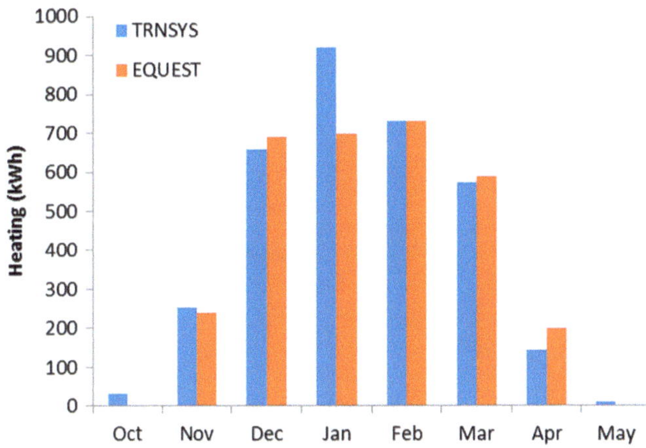

Fig. 25.1 Monthly heating loads of the building

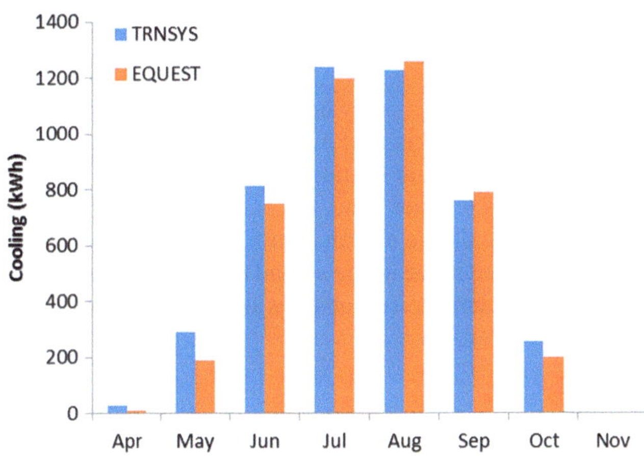

Fig. 25.2 Monthly cooling loads of the building

Table 25.4 Comparison of loads

Programs	TRNSYS	eQuest	Difference
Heating	3318	3150	5.06 %
Cooling	4627	4400	4.91 %

- Infiltration rate
- Building orientation
- Insulation thickness
- Windows area

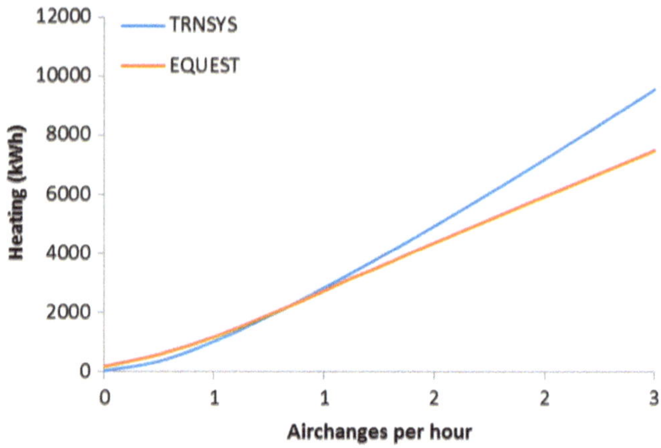

Fig. 25.3 Heating loads for different infiltration rate

Infiltration Rate

The infiltration rate of a building is very important and influences on the loads a lot. So the parametric analysis for this is of a high importance (Figs. 25.3 and 25.4).

From Fig. 25.3 it is shown that heating loads has an increasing rate for greater values of infiltration. On the other hand, the cooling load is lower in the cases of greater infiltration rate. A simple explanation for these results is that the summer heat from windows is stored in the inside air and the more air changes help fresh and colder air to get into the house. It is obvious that the two programs give similar results with low differences.

Building Orientation

The orientation of the building is able to influence on the loads because of the existence on the windows. For example, a south window let greater amount of solar irradiation to through into the house compared to a north window. Figures 25.5 and 25.6 present the heating and cooling loads for a full reverse of the building, from $-180°$ to $+180°$.

Low differences between the programs are observed from Figs. 25.5 and 25.6. The curves are not symmetrical to the vertical line to 0°, because the East and the West irradiation is not the same. The minimum heating load is achieved when the south wall azimuth is 0° because of the large window is located in this wall and in this case it lets the solar irradiation to through into the house with the greatest way. On the other hand, the cooling load in this case is the maximum, because of the high irradiation leads to a great cooling load.

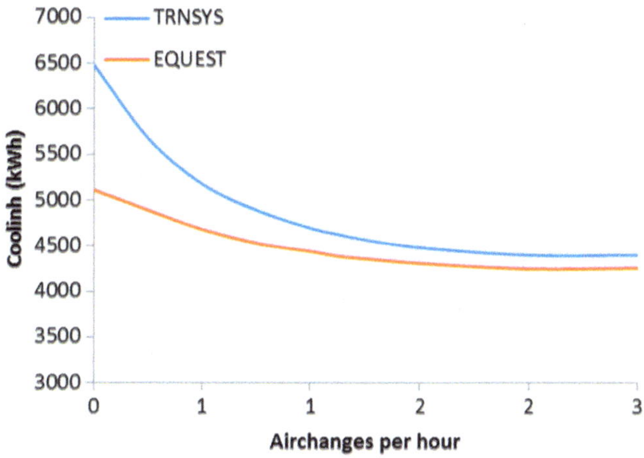

Fig. 25.4 Cooling loads for different infiltration rate

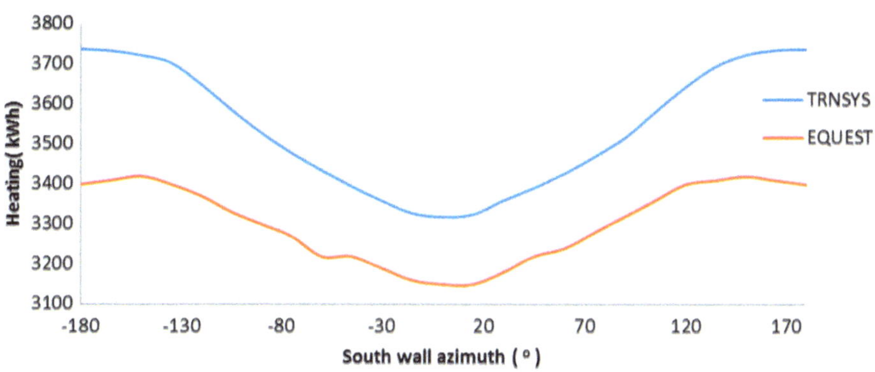

Fig. 25.5 Heating loads for different building orientation

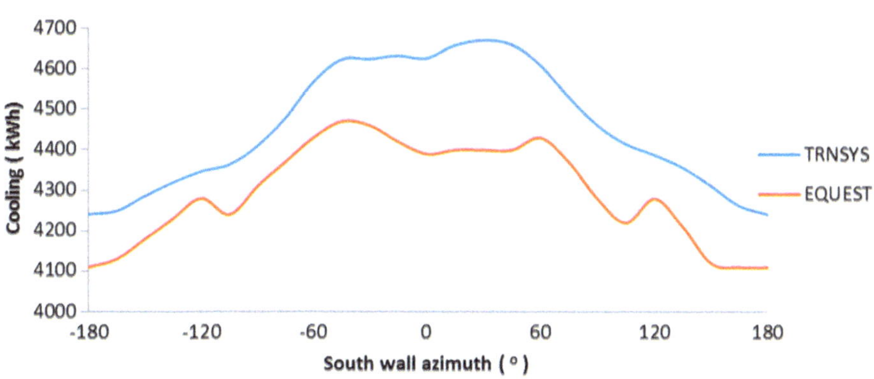

Fig. 25.6 Cooling loads for different building orientation

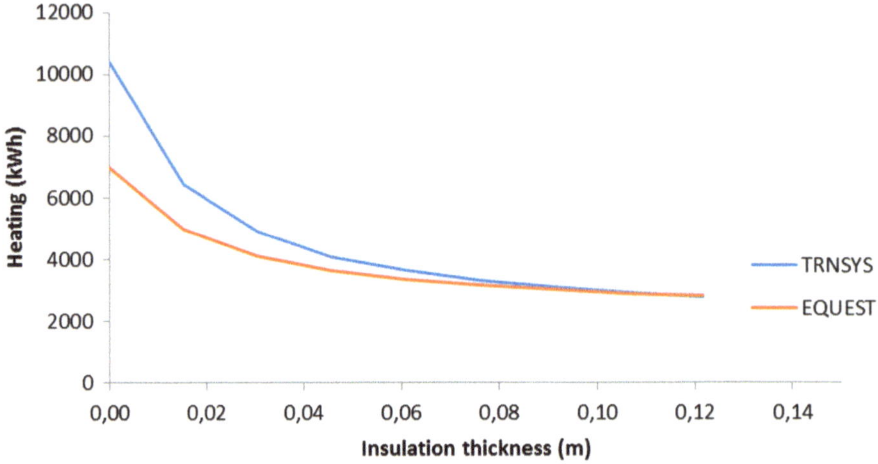

Fig. 25.7 Heating loads for different insulation layer thickness

Insulation Thickness

The existence of insulation is the most important parameter for the load reduction and especially for the heating load. The thickness of this layer is an optimization issue and it is analyzed in the following schemes (Fig. 25.7).

Figure 25.8 shows that the higher thickness of insulation lead to lower heating energy consumption, but after 8 cm, the influence on the loads is very low. TRNSYS and eQuest give similar results according the curves and while the thickness is greater, the curves come very close to each other. On the other hand, the cooling energy consumption has a strange behavior. More specifically, the eQuest simulation proves that the greater insulation is beneficial for the building load, by decreasing them when TRNSYS shows that the insulation layer increases these. The values for cooling does not differs a lot as values, but the derivative of the one is negative and of the other positive. A possible explanation is that the two programs make the energy balance of the building with different way.

Windows Area

The existence of windows in a building changes a lot the heating and the cooling loads. The windows let the solar energy through into the house, which means that the indoor air is getting warmer. Simultaneously, the U-value of the windows is greater than the wall, fact that leads to high losses in the night. The following diagrams assume the building performance for different values of total windows area. It is important to say that the 50 % of the total windows area is located to the south wall, the 25 % to the east and the other 25 % to the west (Figs. 25.9 and 25.10).

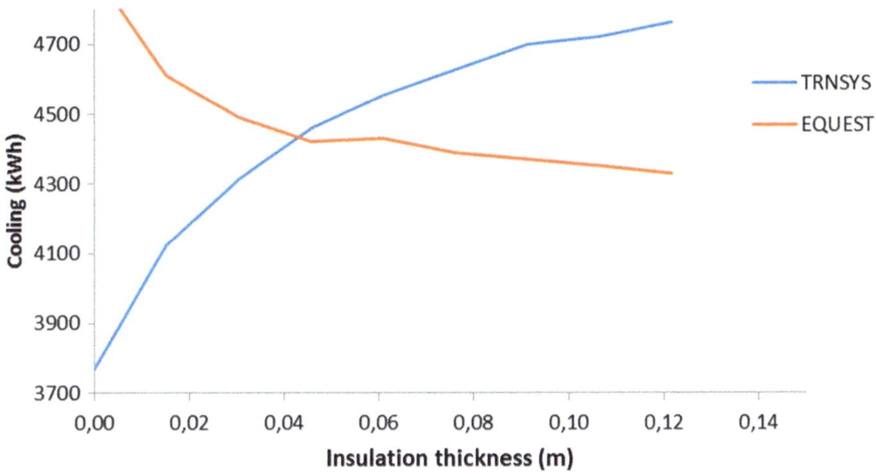

Fig. 25.8 Cooling loads for different insulation layer thickness

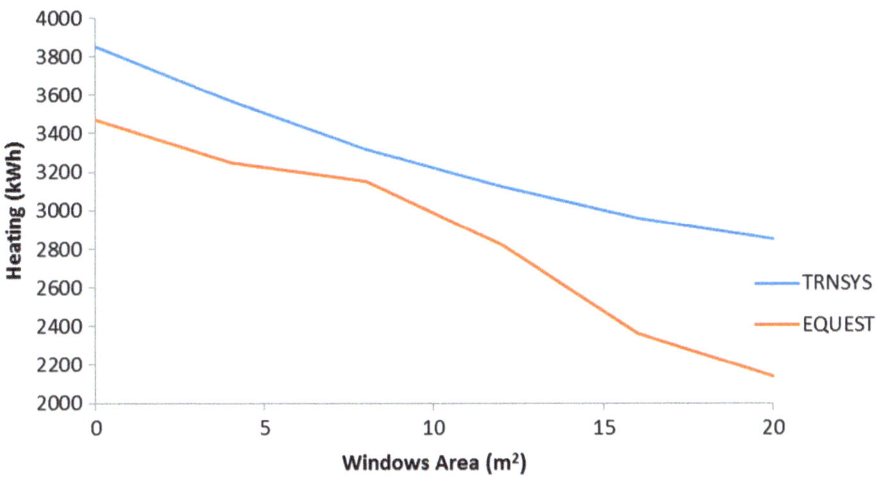

Fig. 25.9 Heating loads for different Windows Area

The above diagrams show that the two programs give similar result, with the heating load to be lower for a higher windows area and the cooling load has the reverse behavior. The climate of Athens explains these results because of the high summer irradiation level in the summer increases the cooling loads when greater windows are put in the building.

Table 25.5 gives the men differences between the two energy tools for all study cases.

Table 25.5 shows all the results by giving the mean differences between programs for the all cases. The main case, which is about the firstly building, which

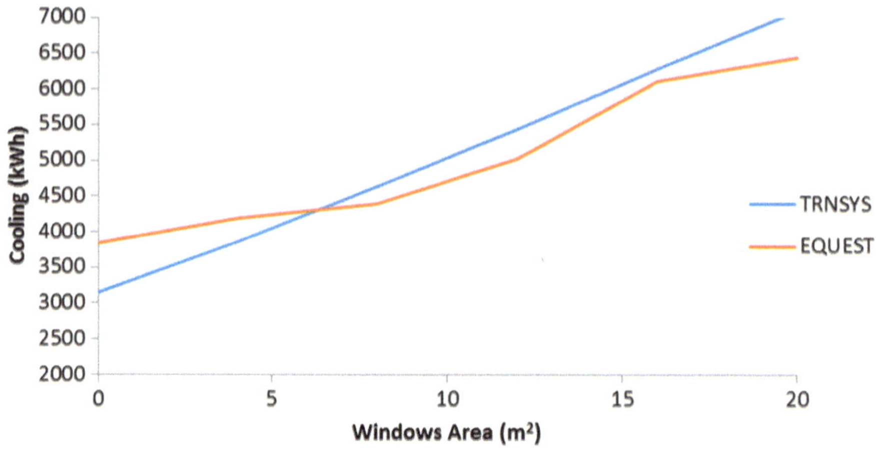

Fig. 25.10 Cooling loads for different Windows Area

Table 25.5 Final comparison for all the studied cases

Case	Heating difference (%)	Heating difference (%)
Main case	5.06	4.91
Infiltration rate	11.00	13.36
Orientation	6.54	3.65
Insulation thickness	15.89	8.76
Windows area	12.61	9.24
Mean value of all the cases	10.22	7.98

examined, gives differences of about 5 % in cooling and heating loads which is acceptable. The difference in the orientation change is in the same proportion; while in the other cases differences of 10–15 % are observed. Finally the mean difference of the heating load is about 10 % and in the cooling 8 %, assuming all the cases of our analysis.

Conclusions

In this analysis two very known simulation tools for energy calculations in buildings are compared in order to validate the one from the other. The differences were low which shows that the results are similar in the majority of the cases. More specifically:

- It is important to state that the way that the loads are calculated are different because of TRNSYS calculates the theoretical load every moment in order to keep the desired temperature while eQUEST uses a heat pump system to satisfy the preferable inside temperature.

- The results of this analysis are easily generated because in every parameter typical and usual values have been selected which match with the real constructions.
- In the main case, the difference of the loads in the two programs is very low, about 5 %.
- In other parametric cases, greater differences of about 10–15 % are observed.
- It is important to state that the heating loads have greater differences between the two programs compared to the differences in cooling loads with 10 % and 8 % respectively.
- In all cases, the derivative of the curves is the same except from Fig. 25.8, which shows the cooling energy consumption for different thickness of the insulation layer. TRNSYS gives greater cooling load for greater thickness, while eQuest has the gives results.
- The main conclusion of this comparison is that the two programs can be validated by this analysis and the small difference can be explained by the different way that the programs cover the heating and cooling loads. So the two programs use different approximations in order to predict the loads of the building.
- Finally it is important to state that TRNSYS is able to calculate the building loads with greater accuracy because the material properties are able to be fully determined by the user and calculate the exact value of the loads. So, to make a more realistic approach for the Greek environmental conditions, TRNSYS is the suitable choice.

References

Beausoleil-Morrison, I., Kummert, M., Macdonald, F., Jost, R., McDowell, T., & Ferguson, A. (2012). *Demonstration of the new ESP-r and TRNSYS co-simulator for modeling solar buildings, Energy Procedia, 30, 505–514.*

Brahme, R., O'Neill, Z., Sisson, W., & Otto, K. (2009). Using existing whole building energy tools for designing net-zero energy buildings—Challenges and workarounds. In *Eleventh international IBPSA conference*, Glasgow, Scotland.

Crwley, D. B., Hand, J. W., Kummert, M., & Griffith, B. T. (2008). *Contrasting the capabilities of building performance simulation programs, Building and Environment, 43(4), 661–673.*

Development Ministry. (2009). *The Greek energy system.*

Gasparella, A., & Pernigotto, G. (2012). *Extensive comparative analysis of two building energy simulation codes for southern Europe climates: Heating and cooling energy needs and peak loads calculation in TRNSYS and EnergyPlus, International High Performance Buildings Conference, 3556, Purdue.*

Han, Y., Liu, X., & Chang, L. (2014). Comparison of software for building energy simulation. *Journal of Chemical and Pharmaceutical Research, 6*(3), 467–471.

http://en.wikipedia.org/wiki/List_of_countries_by_electricity_consumption

http://www.mpoweruk.com/electricity_demand.htm.

Kea, M.-T., Yeh, C.-H., & Jian, J.-T. (2013). *Analysis of building energy consumption parameters and energy savings measurement and verification by applying eQUEST software, Energy and Buildings, 61, 100–107.*

Terziotti, L. T., Sweet, M. L., & McLeskey Jr., J. T. (2011). *Modeling seasonal solar thermal energy storage in a large urban residential building using TRNSYS 16, Energy and Buildings, 45, 28–31.*

Xing, J., Ren, P., & Ling, J. (2014). *Analysis of energy efficiency retrofit scheme for hotel buildings usingeQuest software: A case study from Tianjin, China.* www.elsevier.com/locate/enbuild.

Zhu, D., Hong, T., Yan, D., & Wang, C. (2013). A detailed loads comparison of three building energy modeling programs: EnergyPlus, DeST and DOE 2.1E. Building Simulation.

Chapter 26
Optimum Insulation Thickness for Cooling Applications Through Exergy Analysis and Environmental Methods

Beyza Nur Daldal, İbrahim Sarıoğlu, Gülcan Özel Erol, Emin Açıkkalp, and Hasan Yamık

Introduction

In Turkey, heating and cooling applications constitute 80 % of the total energy consumption in the buildings. Also, the building sector is the second largest energy consumer thus it is an important area for improving of energy savings and efficiency (Dombaycı 2009).

Energy is basic for humankind and it one of the most important parts of the sustainable development, however, environmental problems becomes a really serious problem, while energy needs increases. Until recently, energy conversion systems has been only designed and improved by using the economic and efficiency parameters. However, environmental impact analysis has gain importance with the increasing environmental problems such as global warming, natural resource depletion and climate changes in recent years. It is shown that some new design and improvement parameters that provide us evaluation for environmental impact are an obligation for the energy conversion systems. Some countries have already obligated the environmental impact assessment with laws (Dasdemir 2011).

In the literature, there are many study about optimum insulation thickness for cooling and heating applications (Gürel and Dasdemir 2011; Dagıdır 2011;Arslan et. all 2010; Kurekci 2016). They are used energy, exergy and economic analysis for determining optimum insulation thickness for different cooling applications. In this study, it presented a novel method that combined environmental impact and the exergy analysis to determine the optimum insulation thickness is performed. Calculations are carried out for two insulation materials and wall types.

B.N. Daldal (✉) • İ. Sarıoğlu • G. Özel Erol • E. Açıkkalp • H. Yamık
Department of Mechanical and Manufacturing Engineering, Faculty of Engineering, Bilecik S.E. University, Bilecik, Turkey
e-mail: bnurdaldal@gmail.com; saribrahim2401@gmail.com; gulcan.ozel@bilecik.edu.tr; eacikkalp@gmail.com; hasan.yamik@bilecik.edu.tr

P. Grammelis (ed.), *Energy, Transportation and Global Warming*, Green Energy and Technology, DOI 10.1007/978-3-319-30127-3_26

Modeling and Analysis

In this study, exergy analysis is combined with the environmental impact analysis for determining the optimum insulation thickness for the cooling applications. Ambient temperature is assumed as 313 K and the indoor temperature is assumed as 295 K. Electricity is used as fuel for the cooling system (air conditioner). Optimum insulation thickness is investigated for two different wall and two different insulation materials. The investigated walls are illustrated in Fig. 26.1. All calculations are conducted for the unit wall area and annual process (Table 26.1).Please check the renumbering of figures/tables/equations and their citations, and correct if necessary. equation 26.4 have been deleted and the other equatins have revised.

Annual cooling loss in unit area (kJ/m^2-year) from the wall is calculated as follow (Cengel 1998):

$$q = 0.024(CDD)U \tag{26.1}$$

where CDD is cooling degree-days (°C.day), and U is heat transfer coefficient (W/m K). Annual energy need (kJ/m^2-year) for the unit wall area is calculated as following (Cengel and Boles 2006):

$$E = \frac{0.024(CDD)U}{COP} \tag{26.2}$$

Annual exergy loss (kJ/m^2-year) is defined as (Cengel and Boles 2006):

$$Ex_{loss} = \frac{0.024(CDD)U}{COP}\left(1 - \frac{T_o}{T_i}\right) \tag{26.3}$$

where T_i is the indoor temperature (K) and T_o is the ambient temperature (K). COP is the coefficient of the performance for the cooling system. Heat transfer coefficient (W/m K) for no-insulation and insulation wall conditions are shown in Eqs. (26.4) and (26.5) respectively (Cengel 1998).

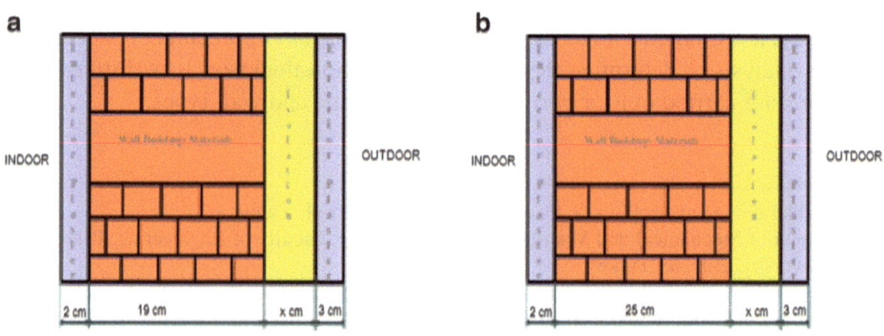

Fig. 26.1 Schematic of the investigated system for wall I (**a**) and wall II (**b**)

Table 26.1 Properties of walls (TS 825) used in this study

Wall types	Thickness (m)	Thermal conductivity (W/m K)	Overall heat transfer coefficient (W/m^2 K)	Heat resistance (m^2 K/W)
Wall I			1.572	0.636
External plaster	0.03	1.40		
Internal plaster	0.02	0.87		
Horizontal coring brick	0.19	0.45		
Wall II			1.772	0.564
External plaster	0.03	1.60		
Internal plaster	0.02	1		
Gasconcrete	0.25	0.16		

$$U_{nins} = \frac{1}{R_i + R_w + R_o} = \frac{1}{R} \tag{26.4}$$

$$\mathrm{U_{ins}} = \frac{1}{\mathrm{R_i + R_w + R_o + R_{ins}}} = \frac{1}{\mathrm{R + x/k}} \tag{26.5}$$

Environmental impact function of the system (mPts/m^2-year) is;

$$B_T = (b_f.Ex_{loss}) + (b_{ins}.\rho.x) \tag{26.6}$$

where b_f is the environmental impact of the electricity (mPts/kWh), b_{ins} is the environmental impact of the insulation materials (mPts/ kg), ρ is the density of the insulation materials and x is the insulation thickness. CO_2 emission is given in the environmental impact of the electricity. The net saving of the environmental impact (mPts/m2-year) is;

$$S = \left((b_f.Ex_{loss}) + (b_{ins}.\rho.x)\right)_{ins} - (b_f.Ex_{loss})_{nins} \tag{26.7}$$

The net saving of the exergy loss (kJ/m^2-year) is;

$$SEx = (Ex_{loss})_{nins} - (Ex_{loss})_{ins} \tag{26.8}$$

The optimum insulation thickness is calculated by taking the derivative of B_T and set equal to zero. Data used in calculations can be seen in Table 26.2.

Table 26.2 Parameters used in calculations

Parameter	Unit	Value
Heat conduction coefficient (Izocam 2013)	W/m K	
Rockwool		0.04
Glasswool		0.032
Environmental impact point	mpts/kg	
Rockwool (Eco-indicator 99 2014)		4.2
Glasswool (Eco-indicator 99 2014)		2.1
Electricity (Eco-indicator 95 2014)		27
Mean temperature for heating period	°C	22
Heating degree day (Dombaycı 2009)	°C-days	661
COP (Dasdemir 2011)		2.4
Density of insulation material	kg/m^3	
Rockwool (Dombaycı 2009)		105
Glasswool (Dombaycı 2009)		45

Results and Discussion

In this study, environmental impact analysis is combined with exergy analysis for determining the optimum insulation thickness of a building wall. The rockwool and glasswool are used as insulation material for the analysis. Wall I and wall II are the investigated systems which are the most commonly used wall arrangements in Turkey.

In Fig. 26.2a and b, variations of the total environmental impact and the net savings of the environmental impact with insulation thickness for the glasswool are shown for wall I and II. As appears in the graphics, initially the environmental impact decreases dramatically and achieves a minimum point called the optimum point. After the optimum point, it starts to increase in small steps. When glasswool is used as the insulation material for wall 1 and 2, optimum points are calculated as 0.052 m and 0.054 m respectively. While the environmental impact decreases until the optimum point, the net savings of the environmental impact increases continuously. At optimum points for wall I and II, environmental impact points are determined as 11.67 and 11.88 $mPts/m^2$-year respectively.

In Fig. 26.3a and b, change of the exergy loss and the net exergy savings according to the insulation thickness from the wall 1 and 2 is illustrated. Exergy losses decreases with insulation thickness. At optimum points exery loss decreases by 60.8 % and 74.95 % compared with the no insulation conditions. In Fig. 26.4a and b, the variations of the total environmental effect and the net savings of the environmental impact with insulation thickness for the rockwool are shown for wall I and II. When the rockwool is used as the insulation material for wall I and II, optimum points are calculated as 0.01 m and 0.015 m respectively. At these points, the environmental impact of the system reaches its minimum. Rockwool has relatively lower optimum points than the glasswool.

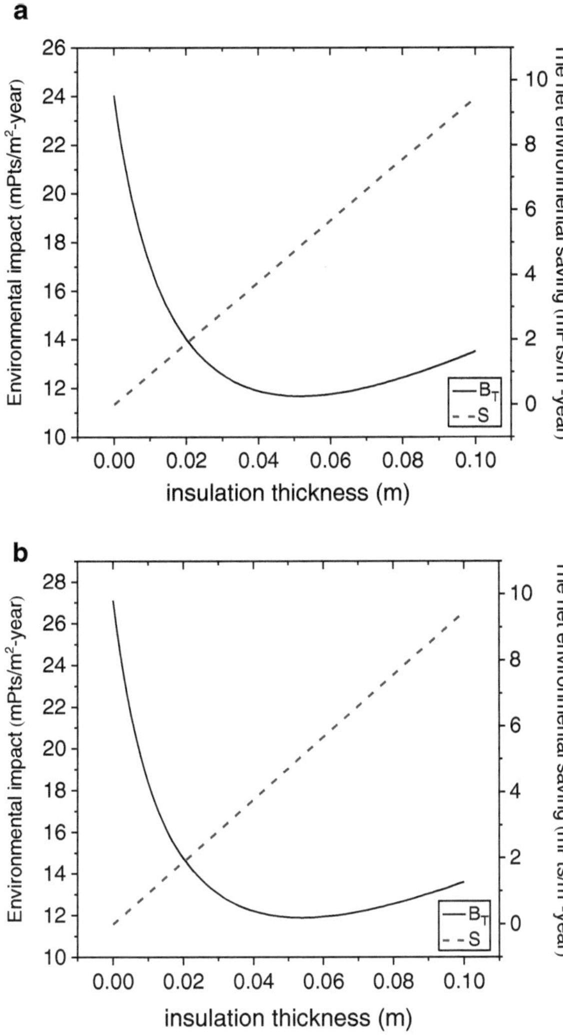

Fig. 26.2 Change of the environmental impact and the net saving with insulation thickness for glasswool, (**a**) wall I, (**b**) wall II

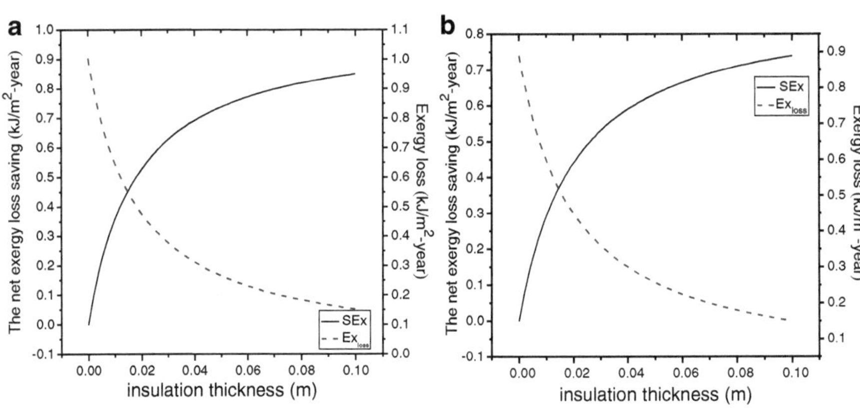

Fig. 26.3 Change of the exergy loss and the net exergy loss saving with insulation thickness for glasswool, (**a**) wall I, (**b**) wall II

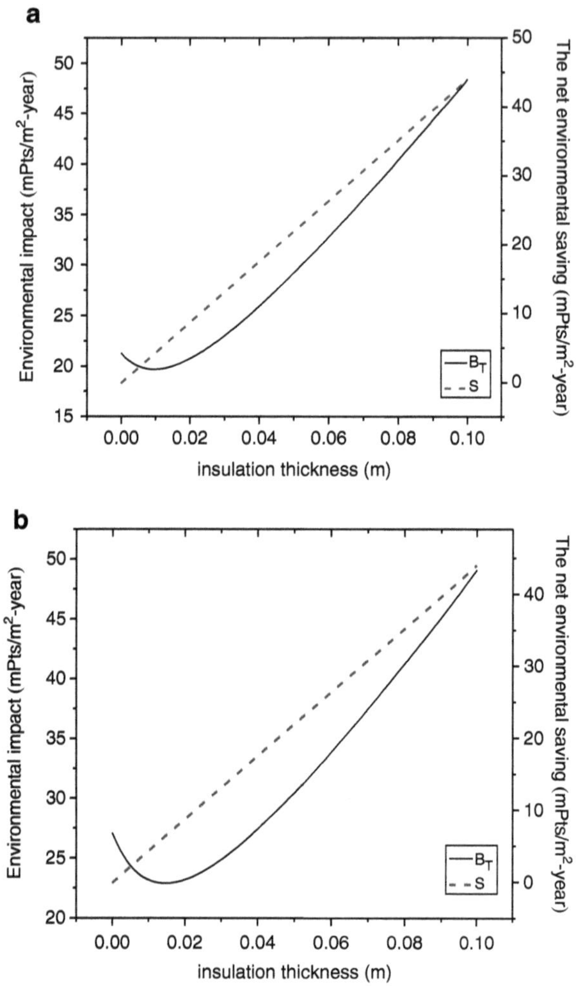

Fig. 26.4 Change of the environmental impact and the net saving with insulation thickness for rockwool, (**a**) wall I, (**b**) wall II

In Fig. 26.5a and b, the change of the exergy loss and the net exergy saving of the optimum insulation thickness are presented for wall I and II. Using the rockwool as insulation material it cause to decrease the exergy saving from the wall I and II by 17 % and 40 % compared with the no insulation conditions

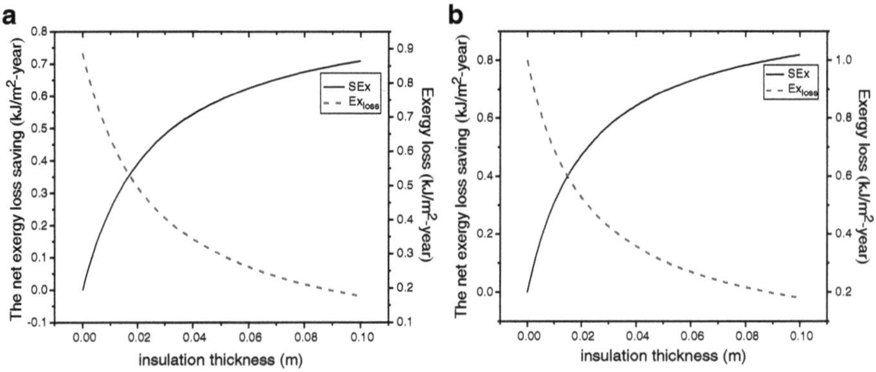

Fig. 26.5 Change of the exergy loss and the net exergy loss saving with insulation thickness for rockwool, (**a**) wall I, (**b**) wall II

Conclusions

In this study, exergy based environmental impact analysis is used to determine the optimum insulation thickness for cooling applications. As mentioned above, global warming has gained attention in recent decades and buildings have an important effect on energy consumption and environmental issues. Some important results obtained from the analysis can be listed as follows:

- Environmental impacts decrease according to the optimum insulation thickness.
- For the rockwool, optimum points are as 0.052 m and 0.054 m for wall I and II respectively.
- For the glasswool, optimum points are as 0.01 m and 0.015 m for wall I and II respectively.
- At optimum points exergy savings increase considerably compared with the no insulation conditions.

We get more efficient, less energy consumption and more environmental in the optimum insulation thickness.

According to these results, the environmental effects should be investigated as well as the economic effects to provide sustainable development.

Nomenclature

b	Environmental impact point, mPts/kg-mPts/kWh
B	Environmental impact rate associated with exergy, mPts
CDD	Cooling degree-days, °C.day
COP	Coefficient of performance for the cooling system
E	Annual energy need for unit wall area, kJ/m^2-year

El	Annual electricity consumption for unit wall area, kJ/m^{2}year
Ex	Annual heat exergy, kJ/m^2-year
k	Thermal conductivity, W/m K
q	Annual heat loss for unit wall area, kJ/m^2-year
R	Heat resistance, K/W
S	Annual net saving of environmental impact, mPts/m^2-year
SEx	Annual net saving of the exergy loss, kJ/m^2-year
T	Temperature, °C or K
U	Heat transfer coefficient, W/K
x	Insulation thickness, m

Subscripts

o	Ambient
f	Fuel
i	Indoor
ins	Insulation
loss	Loss
nins	No-insulation
T	Total
w	composition

Greek Letter

ρ	Density, kg/m^3

References

Arslan, O., Ozgur, M. A., Yıldızay H. D., Kose, R., (2010). Fuel Effects on Optimum Insulation Thickness: An Exergitic Approach, Energy Sources, Part A, 32, 128–147.

Cengel, Y. A. (1998). *Heat transfer: A practical approach* (1st ed.). Hightstown: McGraw-Hill.

Cengel, Y. A., & Boles, M. A. (2006). *Thermodynamics an engineering approach* (5th ed.). Hightstown: McGraw-Hill.

Dagıdır, C., (2011). Determination of Optimum Insulation Thickness by Considering Solar Radiation in Buildings of Warm Climate Regions, M. Sc Thesis, Graduate School of Applied and Natural Sciences, Süleyman Demirel University, Isparta, TURKEY.

Dasdemir, A. (2011). Heating and cooling applications energy saving of optimum thickness and the effect of the flue gas emission M. Sc Thesis, Graduate School of Applied and Natural Sciences, Karabuk University, Karabuk, TURKEY.

Dombaycı, Ö. A. (2009). Degree-days maps of Turkey for various base temperatures. *Energy, 34*, 1807–1812.

Goedkoop, M., & Spriensma, R. (2014, March). The Eco-indicator 99: A damage oriented method for life cycle impact assessment. Methodology report, 2000, Amersfoort, Netherlands. http://www.pre.nl.

Gürel, A. E., Dasdemir, A. (2011). Economical and Environmental Effects of Thermal Insulation Thickness in Four Different Climatic Regions of Turkey, International Journal of Renewable Energy Research, Vol.1 , No.1, 1–10.

Kurekci, N.A., (2016). Determination of optimum insulation thickness for building walls by using heating and cooling degree-day values of all Turkey's provincial centers, Energy and Buildings, 118, 197–213.

Retrieved March 2013 from http://www.izocam.com.tr/tr-tr/urunler/yalitim-malzemeleri.aspx.

The Eco-indicator 95. (2014, March). http://www.pre-sustainability.com/eco-indicator-95-final-report, http://www.pre.nl.

Turkish Standards Institution (TSE). (1998). *TS 825: Thermal insulation in buildings.* Ankara, Turkey: TSE Publications.

Chapter 27
Novel Tungsten Bronze Nanoparticles for Shielding Near Infrared Ray and Decreasing CO_2 Emission

Tsugio Sato, Chong-shen Guo, and Shu Yin

Introduction

Recently, much attention has been attracted for the smart windows to reduce the energy consumption for air conditioning, and, thereby, to decrease the carbon dioxide emission by shielding the heat-ray from the sunlight into the room in summer, and from the room to the outside in winter for automobiles, buildings, etc. In case of the smart window, the high transparency of visible light and high shielding ability of NIR light are required. In order to cut off the heat-rays, the shielding of the near infrared light by the electronic plasma is used (Kreibig and Vollmer 1995). Namely, the electronic plasma frequency (Wp) of the conductive materials is expressed by Eq. (27.1)

$$W_p{}^2 = 4\pi ne^2/m \tag{27.1}$$

where n is free electron density and m is electronic mass. It is known that the electromagnetic wave with the lower frequency than plasma frequency is not able to pass the plasma. Therefore, it is possible to shield the near infrared rays by controlling the free electron density of the conductive materials. In addition, it is well known that the scattering of lights by the particles greatly changes depending on the particle size, and nano particles less than 20 nm in diameter are quite transparent, since the Rayleigh scattering intensity (s) is proportional to the particle size (d) to the six power as shown by Eq. (27.2).

T. Sato (✉) • C.-s. Guo • S. Yin
Institute of Multidisciplinary Research for Advanced Materials, Tohoku University, 2-1-1 Katahira, Aoba-ku, Sendai 980-8577, Japan
e-mail: tsusato@tagen.tohoku.ac.jp

P. Grammelis (ed.), *Energy, Transportation and Global Warming*, Green Energy and Technology, DOI 10.1007/978-3-319-30127-3_27

$$s = Kd^6/\lambda^4 \quad (27.2)$$

Nowadays, the thin films of various nanosized conductive compounds such as silver (Kreibig and Vollmer 1995), indium tin oxides (ITO) (Yang et al. 2009), etc. are widely used as solar filters (Okada et al. 2003; Hammarberg and Roos 2003; Adachi and Miratsu 2010; Takeda et al. 1998), however, the performances of the these materials are not satisfactory. We focused the attention on the hexagonal tungsten bronze compounds with the chemical formula of M_xWO_3 consisting of the mixed valence of W^{6+} and W^{5+}, where M is alkali metal ion which locates in the space of the WO_3 layer. The free electron density of tungsten bronze can be controlled by controlling the amounts of M^+. There are lots of tungsten bronze compounds, but the cesium tungsten bronze shows the most excellent chemical stability because of the adequate ionic size matched with the space size in the WO_3 layer. Takeda and Adachi (2007) prepared cesium tungsten bronze nanoparticles for near infrared light shielding by the solid state reaction under severe reaction conditions, such as hydrogen gas atmosphere above 800 °C, followed by energy consuming milling process. Now, we tried the direct fabrication of the cesium tungsten bronze nanocrystals by solvothermal reaction under mild conditions.

Experimental

In a typical experiment, after dissolving desired amounts of WCl_6 and CsOH in ethanol separately, both solutions were added into the ethanol and acetic acid mixed solution to precipitate amorphous gel precursor, followed by solvothermal reactions at a desired temperature to form Cs_xWO_3 nanocrystals by the dissolution-crystallization mechanism. In order to evaluate NIR shielding performance, the product was dispersed in a collodion-ethanol mixed solution, and then, coated on a quartz glass by a doctor-blade method using an applicator of 12.5 μm film thickness.

Results and Discussion

Figures 27.1 and 27.2 shows the X-ray diffraction patterns of the products obtained in (a) pure ethanol and (b) 90 vol.% ethanol-10 vol.% acetic acid mixture at 240 °C. Both samples consisted of the single phase of the hexagonal cesium tungsten bronze, but the morphology was greatly different. It can be seen that the irregularly agglomerated large particles were formed in pure ethanol. In contrast, the samples synthesized in 10 vol.% acetic acid contained ethanol solution consisted of well-dispersed rod-like nanoparticles with about 15 nm in diameter and 50 nm in length.

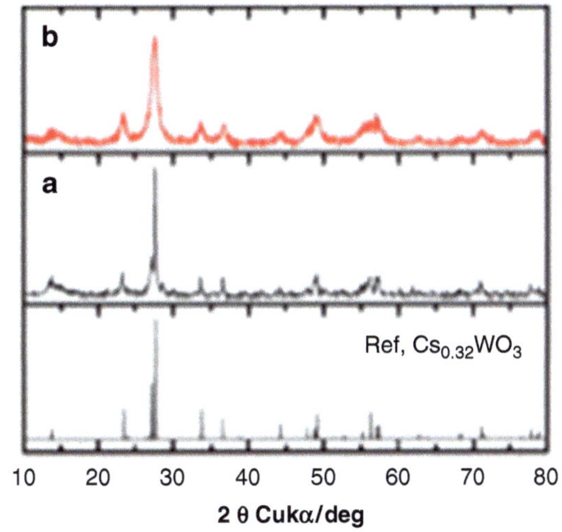

Fig. 27.1 XRD patterns of samples synthesized in (**a**) pure ethanol and (**b**) 90 vol.% ethanol-10 vol.% acetic acid mixed solutions at 240 °C for 20 h (reference: $Cs_{0.32}WO_3$, JCPDS No. 831334)

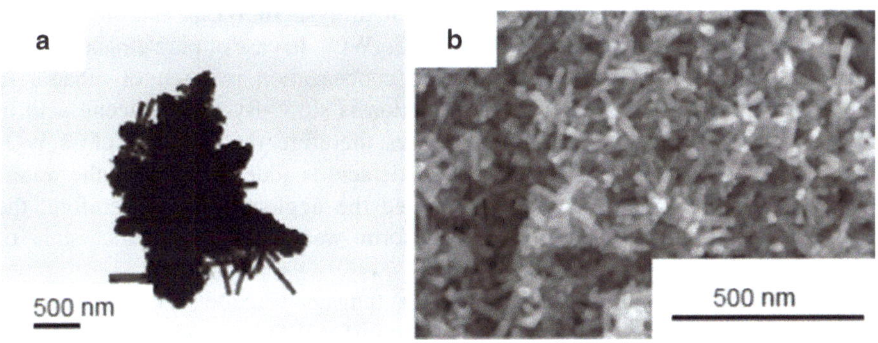

Fig. 27.2 Scanning electron micrographs of samples synthesized in (**a**) pure ethanol and (**b**) 90 vol.% ethanol-10 vol.% acetic acid mixed solutions at 240 °C for 20 h with a nominal Cs/W atomic ratio of 0.5

By increasing the acetic acid content up to 40 vol.%, the size of nanorods could be decreased, but the agglomeration of particles proceeded.

Table 27.1 shows the Cs/W atomic ratio and Cs_xWO_3 yield in ethanol mixed with various amounts of acetic acids. It can be seen that the Cs/W atomic ratio and Cs_xWO_3 yield in pure ethanol were quite low, but Cs/W atomic ratio increased to almost 0.33, which is the ideal value of tungsten bronze structure, and Cs_xWO_3 yield increased to nearly 100 % by adding acetic acid.

In order to confirm the reaction mechanism to promote the formation of tungsten bronze by adding carboxylic acid, the chemical compositions of the reaction solutions after solvothermal reaction were analyzed by NMR. It was found that the peaks corresponding to not only the ethanol and acetic acid which were used as starting solvents, but also ethyl acetate and diethyl ether were determined. These

Table 27.1 Cs/W atomic ratio and Cs_xWO_3 yield in ethanol mixed with various amounts of acetic acids

CH_3COOH content/vol.%	Cs/W atomic ratio	Cs_xWO_3 yield/%
0	0.26	65.1
10	0.31	90.0
20	0.32	96.2
40	0.29	95.5

results indicate that the esterification and etherification shown by Eqs. (27.3) and (27.4) proceeded during the solvothermal reaction, and water generated by these reactions played an important role to form Cs_xWO_3 nanocrystals.

$$CH_3CH_2OH + CH_3COOH = CH_3CH_2OOCCH_3 + H_2O \quad \text{(Esterification)} \tag{27.3}$$

$$2CH_3CH_2OH = CH_3CH_2OCH_2CH_3 + H_2O \quad \text{(Etherification)} \tag{27.4}$$

It is considered that water promotes the hydrolysis of WCl_6, and ethanol promotes the reduction of W^{6+} to W^{5+} to form Cs_xWO_3. In case of pure ethanol, due to the low reaction rate of the dehydration condensation reaction of ethanol to generate water, the rate of Cs_xWO_3 formation is slow. By adding acetic acid in ethanol, the rate of water generation increases, therefore, the formation of Cs_xWO_3 was promoted. When excess amount of acetic acid is added in ethanol, the water-releasing rate becomes too high to proceed the agglomeration. Therefore, the optimum acetic acid content exists to form well dispersed nanocrystals of $Cs_{0.33}WO_3$.

By the XPS analysis, it was confirmed that tungsten trioxide (WO_3) consisted of only W^{6+}, but cesium tungsten bronze crystal (Cs_xWO_3) consisted of the mixed valences of W^{6+} and W^{5+} as shown in Fig. 27.3.

Transmittance spectra of Cs_xWO_3 particles synthesized in (a) pure ethanol (b) 90 vol.% ethanol-10 vol.% acetic acid, (c) 80 vol.% ethanol-20 vol.% acetic acid, (d) 60 vol.% ethanol-40 vol.% acetic acid solutions at 240 °C for 20 h with Cs/W atomic ratio of 0.5 and (e) WO_3 particles are shown in Fig. 27.4. As-expected, tungsten trioxide (WO_3) consisted of only W^{6+} did not show NIR shielding ability, but Cs_xWO_3 consisted of the mixed valences of W^{6+} and W^{5+} showed NIR-shielding ability, where the NIR-shielding ability greatly changed depending on the reaction conditions, indicating that well dispersed nanocrystals of Cs_xWO_3 showed excellent heat ray shielding performance.

In order to confirm the heat-ray shielding performance of Cs_xWO_3, the simulated experiments were carried out by a 50 W Halogen lamp radiating to three boxes sealed by quartz glass, Cs_xWO_3 coated quartz glass and ITO glass. The temperatures in the boxes after 1 h irradiation are listed in Table 27.2. It can be seen that the temperature in the box covered by a quartz glass increased up to 34.2 °C after 1 h irradiation. The temperature increment could be depressed by using an ITO glass,

Fig. 27.3 W_{4f} core-level XPS spectra of Cs_xWO_3 nanorods prepared in 90 vol. % ethanol-20 vol.% acetic acid mixed solution at 240 °C for 20 h with Cs/W atomic ratio of 0.5 and WO_3

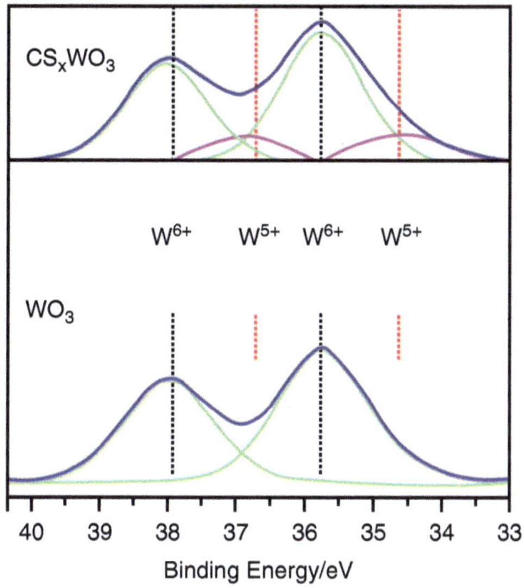

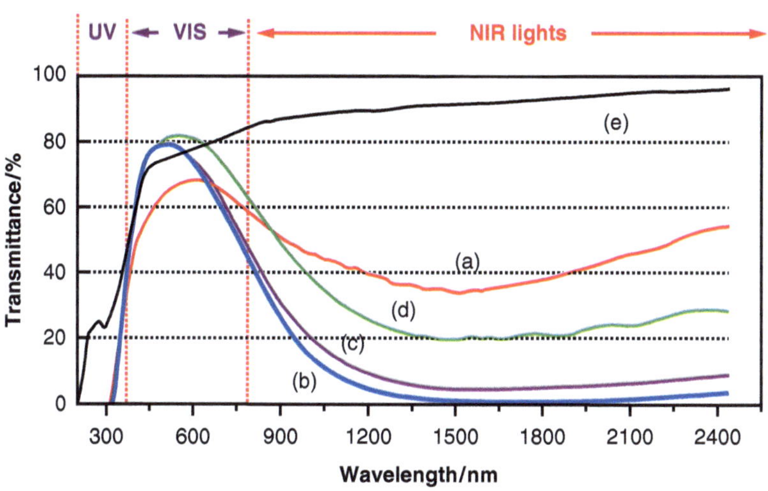

Fig. 27.4 Transmittance spectra of Cs_xWO_3 particles synthesized in (**a**) pure ethanol (**b**) 90 vol.% ethanol-10 vol.% acetic acid, (**c**) 80 vol.% ethanol-20 vol.% acetic acid, (**d**) 60 vol.% ethanol-40 vol.% acetic acid solutions at 240 °C for 20 h with Cs/W atomic ratio of 0.5 and (**e**) WO_3 particles

Table 27.2 Temperatures in the boxes covered by quartz glass, ITO glass and Cs_xWO_3 coated quartz glass after 1 h irradiation by a 50 W halogen lamp

Box	Temperature/°C
Quartz glass	34.2
ITO glass	30.0
Cs_xWO_3 coated quartz glass	24.4
Temperature before irradiation:	20.0

but still caused a remarkable increase in temperature up to 30.0 °C. In contrast, the temperature increment was limited to 24.4 °C by using a $Cs_{0.32}WO_3$ coated quartz glass, indicating the excellent heat-ray shielding performance of the Cs_xWO_3 nanocrystals.

Conclusions

From the present results, following conclusions may be drawn. (1) The Cs_xWO_3 nanocrystals were successfully synthesized by solvothermal process. (2) The well dispersed Cs_xWO_3 nanocrystals showed the excellent NIR shielding ability as well as high transmittance in the visible light range.

References

Adachi, K., & Miratsu, M. (2010). Absorption and scattering of near-infrared light by dispersed lanthanum hexaboride nanoparticles for solar control filters. *Journal of Materials Research, 25*, 510–521.

Hammarberg, E., & Roos, A. (2003). Antireflection treatment of low-emitting glazings for energy efficient windows with high visible transmittance. *Thin Solid Films, 442*, 222–226.

Kreibig, U., & Vollmer, M. (1995). Optical properties of metal clusters. In J. P. Toennies (Ed.), *Springer series in materials science* (Vol. 25). New York: Springer.

Okada, M., Yamada, Y., Jin, P., Tazawa, M., & Yoshimura, K. (2003). Fabrication of multifunctional coating which combines low-e property and visible-light-responsive photocatalytic activity. *Thin Solid Films, 442*, 217–221.

Takeda, H., & Adachi, K. (2007). New infrared absorption of tungsten oxide nanoparticle dispersions. *Journal of American Ceramic Society, 90*, 4059–4061.

Takeda, H., Otsuka, Y., Kuno, H., & Adachi, K. (1998). Coating solution for a heat-ray shielding film and a process for forming a heat-ray shielding film by employing the same. US Patent No. 5,840,364.

Yang, L. L., Ge, D., Wei, H., He, F., & He, X. D. (2009). Morphology and characterization of ITO-Ag-ITO films on fibers by layer-by-layer method. *Applied Surface Science, 255*, 8197–8201.

Chapter 28
Modelling of a Solar Assisted Floor Heating System with TRNSYS

E. Bellos, C. Tzivanidis, A. Prassas, and K.A. Antonopoulos

Introduction

Energy consumption in buildings has a very important role with a proportion of 40 % of the total energy in Europe. Especially energy consumption for heating and cooling is responsible for 70 % of building consumption which is a great percentage (European Environment Agency 2010). More specifically, the specific heating consumption for typical buildings varies from 15 kWh/m^2 year to 100 kWh/m^2 year (Asimakopoulos et al. 2012; Theodoridou et al. 2011). In order to cover this load, many technologies are used and this research field evolves rapidly. Solar energy utilization is very beneficial because of the zero operating cost and is able to cover great loads in countries with high irradiation level as Greece. With a solar potential of 1600 kWh/m^2 year in Athens (Fantidis et al. 2013; Kouremenos et al. 1985), the solar energy exploitation is a feasible solution for covering the building heating loads.

Simultaneously, underfloor heating systems are a heating technology which is used with an increasing rate because provide to occupants more satisfactory indoor micro-climate (Gao et al. 2011). The first steps of this technology were in Germany in 1970s in commercial buildings (David 1984) and in data centers of high heat and fewer people (Sodec and Craig 1990). Nowadays, the 50 % if the new constructed Europe buildings uses this technology (Kilkis et al. 1994) because underfloor heating systems reduces the seasonal heating about 18 % (Ghali 2007) and improve the thermal comfort conditions inside the buildings (Inard et al. 1998).

Many analyses have been done in this area, Alajmi and El-Amer developed an underfloor-air-distribution system and proved that this system saves energy

E. Bellos (✉) • C. Tzivanidis • A. Prassas • K.A. Antonopoulos
Department of Thermal Engineering, National Technical University of Athens, Zografou, Heroon Polytechniou 9, 15780 Athens, Greece
e-mail: bellose@central.ntua.gr

P. Grammelis (ed.), *Energy, Transportation and Global Warming*, Green Energy and Technology, DOI 10.1007/978-3-319-30127-3_28

compared to ceiling-based air distribution (Alajmi and El-Amer 2010). Other studies prove the improvements in thermal comfort conditions by using an underfloor heating system (Myhren and Holmberg 2008; Sekhar and Ching 2002; Hasan et al. 2009). Zhai et al. (2009) studied the design and performance of the solar-powered floor heating system in for a typical building. Badran and Hamdan made a theoretical and an experimental study for under-floor heating system using solar collectors and solar pond. They finally proved that the system with solar collector is 7 % more efficient than the solar pond system (Badran and Hamdan 2004). In other studies, Athienitis made a numerical model to predict the underfloor heating load under high solar gain and conclude that controllers should be used in order the floor temperature does not exceed the comfort limits (Athienitis 1994). Al-Khalayleh et al. in a theoretical investigation of an underfloor heating system coupled with solar collectors proved that solar energy is able to cover the heating load sufficiently (Al-Khalayleh et al. 1999).

In this study the solar energy utilization is analyzed for heating a typical building with an underfloor heating system. The combination of these is suitable for Greek high solar potential and leads to an optimum performance. The simulation is done by commercial software TRNSYS which uses an active layer to simulate the underfloor heating system.

General Information

In this paragraph the building description, the underfloor system modelling and the basic mathematical equations are presented in order to make the next analysis feasible.

Examined Building

The building of this study is a cubic building with double glass windows in south, east and west direction. The characteristics of the building are presented in Table 28.1. Simultaneously the specific loads of the building are presented in the same table.

The structure and the composition of outer walls influence on the thermal behavior of the building. Table 28.2 presents the material properties which have the default TRNSYS values.

The building consists of four outer same walls and a roof without having internal walls. The roof of the building has three layers; the outer is cement with 20 cm thickness, the intermediate is insulation of 8 cm thickness and the inner plaster of 1.5 cm thickness with a U-value equal to 0.439 W/m^2 K. The outer wall is a five layer wall with the following materials: 1.5 cm plaster, 12 cm brick, 6 cm insulation, 12 cm brick and 1.5 cm plaster with a U-value equal to 0.510 W/m^2 K.

Table 28.1 Building parameters

Parameter	Value	Parameter	Value
City	Athens	Specific gains	14 W/m^2 K
Area	100 m^2	Occupants density	0.1 person/m^2
Height	3 m	Specific light	10 W/m^2
East glass	3 m^2	Infiltration rate	1 change per hour
West glass	3 m^2	Thermal capacitance	720 kJ/K
South glass	6 m^2	h_{out}	17.8 W/m^2 K
Shading coefficient	70 %	h_{in}	3.1 W/m^2 K

Table 28.2 Properties of structural materials

Material	k (W/m K)	Cp (kJ/kg K)	ρ (kg/m^3)
Brick	0.89	1.0	1800
Plaster	1.39	1.0	2000
Insulation	0.04	0.8	40
Concrete	7.10	0.8	2400

The ground contains the heating pipes inside the concrete layer and also has insulation layer of 8 cm with U-Value about 0.314 W/m^2 K.

For thermal comfort conditions, the daily temperature should be greater than 22 °C which means that a control system is used to determine the operation of the heating system. The minimum theoretical heating energy to cover this thermal comfort condition is about 5020 kWh which was calculated by TRNSYS.

Active Layer (Underfloor Heating System)

The underfloor heating system is inserted in the ground of the building as an active layer which exchanges data with other system components. The hot water flows inside the tubes giving heating in the house and let the house with a lower temperature. There is a thermostat which controls when there is heating need. Figure 28.1 shows the TRNSYS environment which related with the underfloor heating.

In this window the parameters of the active layer are determined in order to take data for the hot water temperature and mass flow rate. Also, the minimum desired inlet mass flow rate is given which is a very important parameter. This parameter determines the segmentation of the area for the underfloor system. In this figure, the specific mass flow rate of 3 kg/h m^2 separates the area in four segments as it is shown in Fig. 28.2. By changing this parameter, the area segmentation changes in order to improve the system operation.

Table 28.3 gives the main parameters for the tubes of the heating system.

These values are the default TRNSYS values and are realistic according the bibliography for building cases.

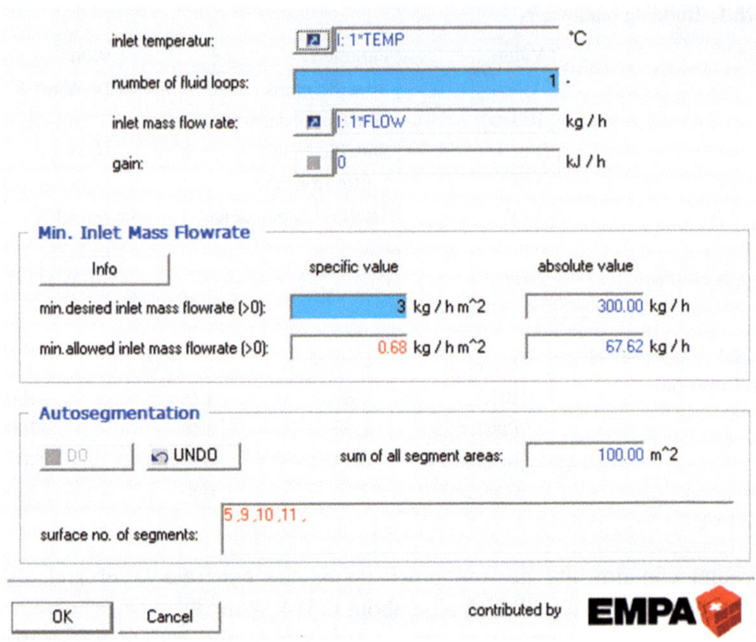

Fig. 28.1 TRNSYS active layer environment

Fig. 28.2 Floor segmentation with four different loops. The *red line* shows the hot water inlet flow and the *blue line* the return flow (Color figure online)

Table 28.3 Properties of piping

Parameter	Value
Pipe wall conductivity	0.35 W/m K
Pipe spacing	0.2 m
Pipe outside diameter	2 cm
Pipe wall thickness	2 mm
Liquid specific thermal capacity	4.19 kJ/kg K

Mathematical Equations

The basic equations which describe the analyzed case are presented and explained in this section. These equations are related with the solar collectors, the system energy balance, the auxiliary heater and the building.

Building

The thermal resistance of the structural components is given from the next equation:

$$R = \sum_{j=1}^{N} \left(\frac{L}{k}\right)_j + \frac{1}{h_{in}} + \frac{1}{h_{out}} \tag{28.1}$$

$$U = \frac{1}{R} \tag{28.2}$$

where N is the number of layers of every structural component.

The heating load of the house is approximately given from the next equation (28.3):

$$Q_{heating} = m_s \cdot Cp \cdot \left(T_{in}^{B} - T_{out}^{B}\right) \tag{28.3}$$

where Cp is the specific heat capacity of water which was taken 4.19 kJ/kg K.

Flat Plate Collectors

The following equations are the basic equations for the useful collector energy and for the collector efficiency.

$$Q_u = \dot{m_c} \cdot Cp \cdot \left(T_{f,o} - T_{f,i}\right) \tag{28.4}$$

$$\eta = \frac{Q_u}{A_c \cdot G_T} \tag{28.5}$$

$$\eta = F_R(\tau\alpha) - F_R U_L \cdot \left(\frac{T_{f,i} - T_{amb}}{G_T}\right) \tag{28.6}$$

Storage Tank

A simple energy balance in the storage tank is given below:

$$Q_{source} = Q_{load} + Q_{loss} \tag{28.7}$$

where source is the collectors, load the heat to the building and loss the radiant losses to the environment.

Auxiliary Heater

The auxiliary heater operation is described from Eqs. (28.8) and (28.9):

$$Q_{A.H.} \leq Q_{A.H.,\max} \tag{28.8}$$

$$Q_{A.H.} = \min\left(Q_{A.H.,\max}, m_s \cdot Cp \cdot (T_{set} - T_l)\right) \tag{28.9}$$

Other System Parameters

The next equation is a simple energy balance in the total system

$$Q_u + Q_{A.H.} - Q_{loss} \approx Q_{heating} \tag{28.10}$$

The following equation gives the solar cover of the system for heating need:

$$f = 1 - \frac{Q_{A.H.}}{Q_{heating}} \tag{28.11}$$

The last equation shows the temperature set value in the mixer and in the auxiliary heater. This is a function of the ambient temperature and gives acceptable results. It was able to use a constant value of 40–45 °C, but this relation leads the system to use warmer water in the colder days which covers better the heating loads.

$$T_{set} = 50 - \frac{T_{amb}}{2} \tag{28.12}$$

System Description

The total system is simulated with TRNSYS which is able to make a dynamic simulation. The months from November to April are simulated because in these months there is heating need according the simulation results. The heating is given from the solar collectors and from an auxiliary heater and the goal is to maximize the solar cover in order reduce the operation cost of this system. Figure 28.3 shows the analyzed system.

The red line shows the hot water which flows in the flat plate collectors and is stored in the tank, while the blue line shows the water stream which flows in the underfloor heating system. More specifically, the water leaves the storage tank and goes to auxiliary heater. After the heater this flow goes into the house and when it leaves the house piping it returns to the storage tank. In the case that the temperature of the water which goes to the house is great, this is mixed with the return of the house in order to reduce the inlet temperature and to protect the piping from destroying. It is important to say that there are different thermostats which control the system and the heater operation in order to reduce the heater energy usage.

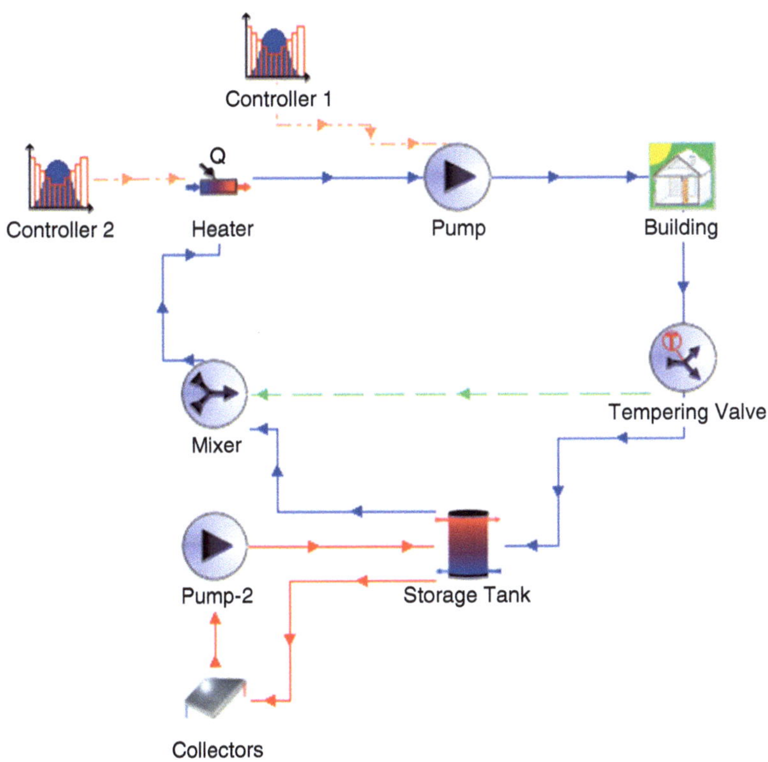

Fig. 28.3 Basic components of the analyzed system

Table 28.4 System parameters

Collector Parameters	Value	Other Parameters	Value
$F_R(\tau\alpha)$	0.8	Heating system water flow rate	1500 kg/h
F_RU_L	3.61	Maximum system temperature	98 °C
Slope angle	45°	Tank losses coefficient	0.7 W/m^2 K
Azimuth	0°	Tank volume	1.5–6 m^3
Operation hours	9:00–17:00	Heater thermostat temperature	20–22 °C
Collector water mass flow rate	2000 kg/h	Heater power	0–12 kW

In this system is important to determine the optimum values for critical parameters in order minimize the auxiliary heater energy for the winter period. The collecting area, the storage tank volume, the maximum heater power, the controller strategy and the area segmentation are the parameters that are examined in this study.

Table 28.4 gives the values of the system parameters. In the case that one parameter is analyzed for different values, these are also given:

Results

In this paragraph the results of the simulation of the system are presented. The diagrams show the system performance for different values of important parameters. By evaluating these data is able to determine the optimum system which leads to a feasible situation. The goal of the optimization is to minimize the auxiliary heater energy by using a realistic collecting area. It is important to state that the storage tank volume influences on the results, so in every case the suitable table is used in order to have an optimal system.

Area Segmentation and Heater Control Strategy

First of all, is important to determine the way that the underfloor system should be separated in parts in order to heat the building efficiently. So, the floor area is either separated in four segments, in two segments and in one segment and the three cases are compared. Simultaneously, the control of the thermostat temperature is analyzed; because of this is a crucial parameter for the heater energy. By reducing this temperature, the heater energy is dramatically diminished. It is important to state that the heater power is selected to be at 6 kW, which is close to maximum heating load of the building (Figs. 28.4, 28.5, and 28.6).

From the above diagrams, it is obvious that the thermostat temperature of heater is very important parameter. By using 21 °C as the critical temperature, the energy is low and the thermal comfort conditions are in preferable limits (Fig. 28.7). If the

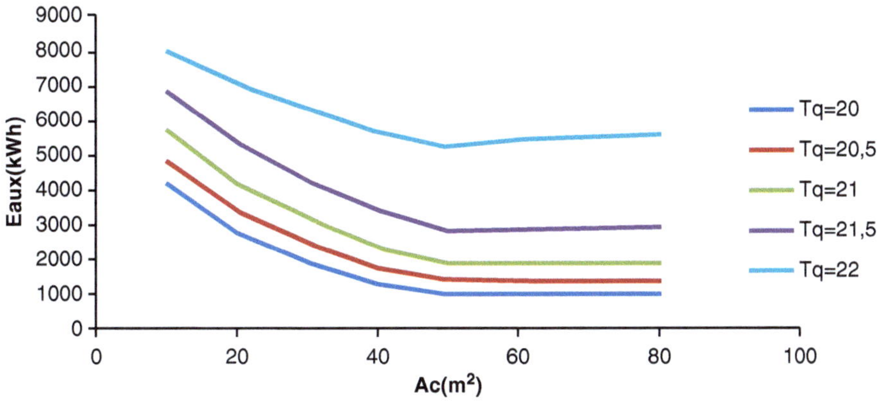

Fig. 28.4 Heater energy for one underfloor loop (area segmentation = 1 part)

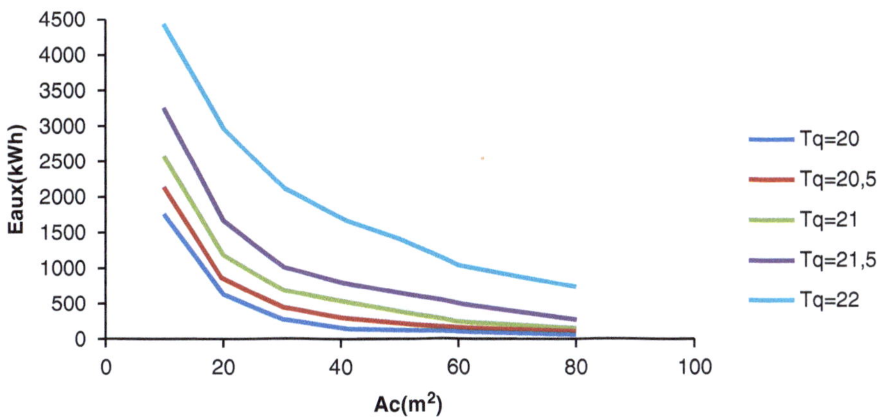

Fig. 28.5 Heater energy for two underfloor loops (area segmentation = 2 parts)

thermostat was set at 22 °C, which is also the set temperature for the underfloor heating system, then the energy is about four times greater which leads to a high energy consumption and is not sustainable.

It is obvious that the temperature is over 22 °C for the greater part of the winter and only a few times is lower than this limit. The next diagram (Fig. 28.8) compares the three heating systems.

It is obvious that the segmentation of four parts gives the better results because the heat transfer from the piping to the indoor air becomes easily. In the following analysis, this case is selected as the optimum, but also the case of the two segments gives similar results. Finally it is important to stat that with only one loop the auxiliary energy is about five times greater which to an unaffordable solution.

The optimum collecting area is 30 m^2 according the Fig. 28.7. More specifically, a greater collecting area means lower heater energy, but after the 30 m^2, the

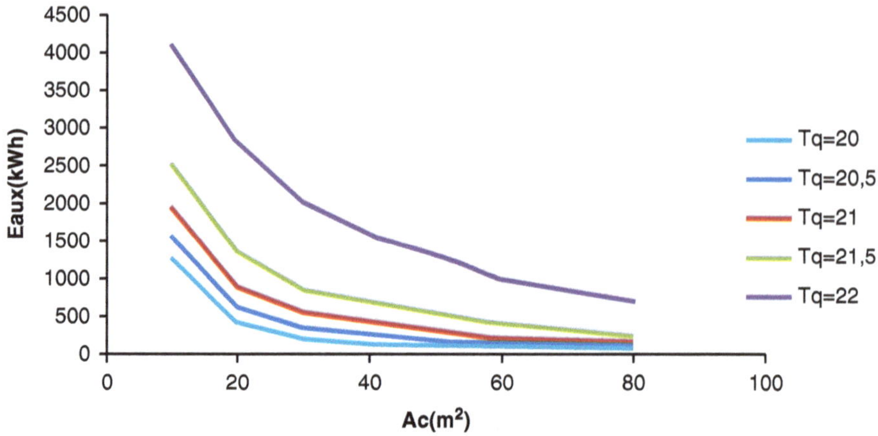

Fig. 28.6 Heater energy for four underfloor loops (area segmentation = 4 parts)

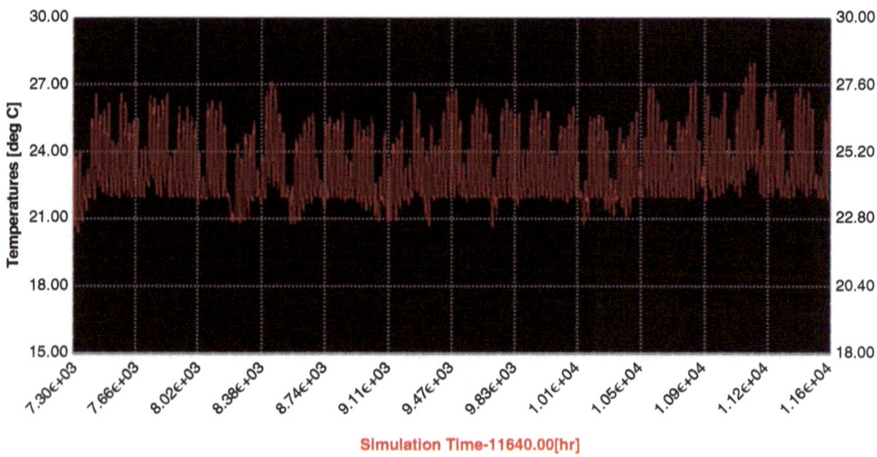

Fig. 28.7 Distribution of the indoor temperature for all over the winter period, from November to April. The thermostat set temperature of the heater is 21 °C and the floor heating system is consisted of four segments

reduction in the heating energy is very low. Table 28.5 gives the optimum tank volume for every collecting area and the heater energy consumption for every case.

Heater Power Analyses

In this section a parametric analysis is presented in order to predict the value of the auxiliary heater power. Figure 28.9 shows how the mean value and the minimum value of the indoor temperature changes with different values of heater power.

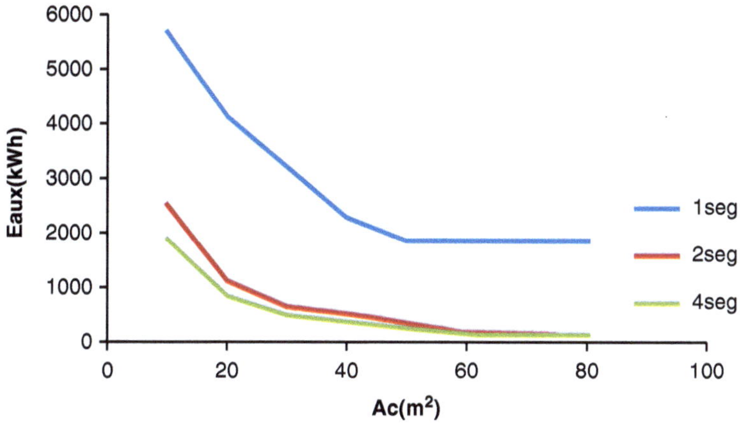

Fig. 28.8 Comparison of the three heating system with 1 (*blue*), 2 (*red*) and 4 (*green*) area segmentation (Color figure online)

Table 28.5 Results of the 4 segmentation heating system

A_c (m^2)	V (m^3)	1seg	2seg	4seg	f (%) for 4seg
10	1.50	5706	2569	1940	61.35
20	2.00	4159	1171	871	82.65
30	**2.50**	**3190**	**689**	**533**	**89.38**
40	2.50	2347	557	404	91.95
50	2.75	1886	410	314	93.75
60	4.00	1879	261	204	95.94
70	5.00	1882	217	182	96.37
80	6.00	1906	193	164	96.73

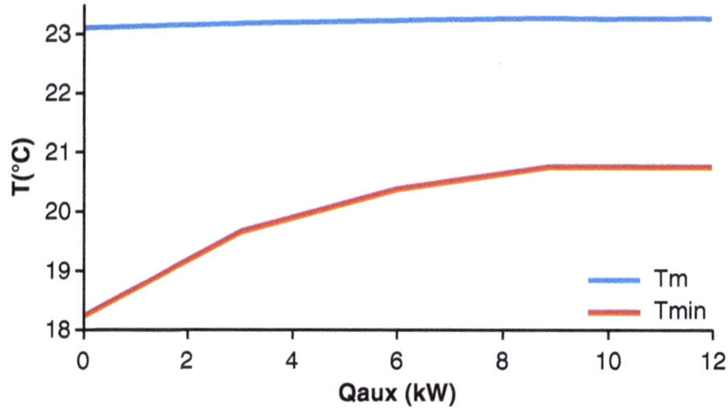

Fig. 28.9 Mean (*blue line*) and minimum (*red line*) indoor temperature for different values of heater power

From this diagram it is obvious that the mean temperature is not changes a lot for different heating power, but the minimum temperature is influenced a lot. More specifically, a heater of 6 or 9 kW is suitable by giving the desired indoor conditions.

Figure 28.10 presents the operation time of the heating system and of the heater. These times are different, because the auxiliary heater operates when there is great need for heating.

The Solar Stand-Alone Underfloor Heating System

In this section the case that the solar collectors give all the energy to the heating system is analyzed. It is very important to determine if this case is feasible because the energy consumption from electricity or fuel is zero (Fig. 28.11).

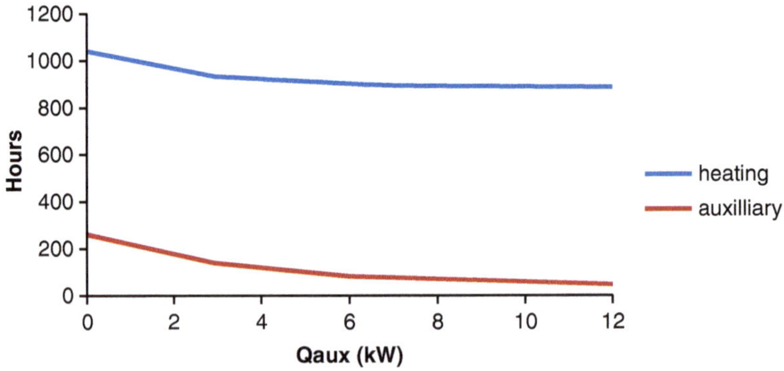

Fig. 28.10 Operation time of the heating system (*blue line*) and of the auxiliary heater (*red line*) (Color figure online)

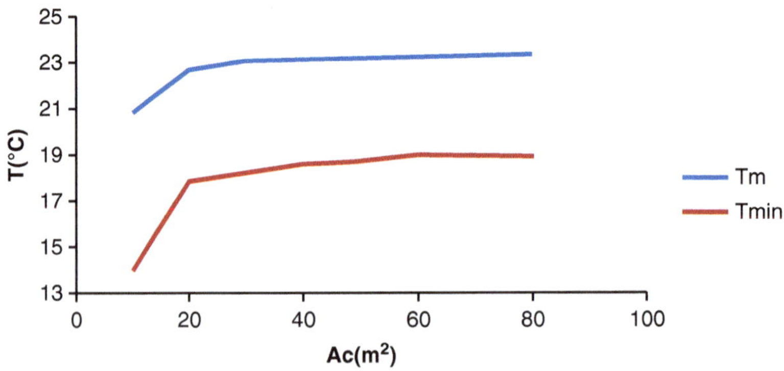

Fig. 28.11 Mean (*blue line*) and minimum (*red line*) indoor temperature for different values of collecting area in the case without heater (Color figure online)

In order to have an acceptable minimum temperature, 50 m^2 of flat plate collectors should be used. In this case the minimum temperature of the winter period is about 18.70 °C. The mean temperature is not influenced a lot after the 20 m^2 fact that shows a low collecting area is able to cover a great part of the heating load.

Conclusions

The main conclusions of this study are the following:

- It is able to heat a building efficiently with an underfloor heating system. By using 30 m^2 flat plate selective collectors is able to cover the 90 % of the heating load. The optimum volume of the storage tank in this case is about 2.5 m^3.
- A for segmentation of the building area gives better results than the other cases. Especially, in the case of one underfloor loop, the energy consumption is five times greater than the four loops case.
- By using two thermostats, one for heating system and one for auxiliary heater, is able the reduction of the energy consumption. More specifically the heating system thermostat is set to 22 °C and the heater to 21 °C which lead to a great indoor temperature profile and to a 75 % auxiliary heating energy reduction. Simultaneously, the operations hours of the heater are about the 10 % of the total operation hours of the underfloor heating system. This shows that the heater operates in moments with high heating need.
- The optimum value for the auxiliary heater power is 6 kW, which is close to the maximum heating load of the building. This means that the heater should be able to cover the total building loads in the moments which solar energy is unsufficient.
- An underfloor heating system is able to operate without auxiliary energy by using more collecting area. More specifically, 50 m^2 are able to keep the thermal comfort condition in high levels inside the house. This leads to zero consumption which is a feasible solution for new buildings.
- The ratio of collecting are to building area is about 1–3 for the case with heater and 1–2 for the case with only solar collectors. For the heater case, other studies (Zhai et al. 2009) give similar results which validate this analysis.

Nomenclature

A_c Collecting area, m^2
C_p Specific heat capacity, J/kg K
E Energy, kWh
f Solar cover
F_R Heat removal factor

G_T Solar titled radiation, W/m^2
h Heat transfer coefficient, W/m^2 K
k Thermal conduction, W/m K
L Layer length, m
m Mass flow rate, kg/s
Q Heat flux, W
R Thermal resistance, mK/W
T Temperature, °C
U Thermal transmittance, W/m^2 K
U_L Collector losses coefficient, W/m^2 K
V Storage tank volume, m^3

Greek Letters

η Collector efficiency
ρ Density, kg/m^3
$(\tau\alpha)$ Transmittance–absorptance product

Subscripts

A.H. Auxiliary heater
amb Ambient
aux Auxiliary
c Collector
f,i Fluid inlet
f,o Fluid outlet
in Indoor
l Load
min Minimum
max Maximum
out Outdoor
q Heater thermostat
s Heating system
u Useful

Superscripts

B Building

References

Alajmi, A., & El-Amer, W. (2010). Saving energy by using underfloor-air-distribution (UFAD) system in commercial buildings. *Energy Conversion and Management, 51*, 1637–1642.
Al-Khalayleh, M. T., et al. (1999). Modelling and simulation of solar-pond floor heating system. *Renewable Energy, 18*, 1–14.

Asimakopoulos, D. A., Santamouris, M., Farrou, I., Laskari, M., Saliari, M., Zanis, G., Giannakidis, G., Tigas, K., Kapsomenakis, J., Douvis, C., Zerefos, S.C., Antonakaki, T., Giannakopoulos, C. (2012). Modelling the energy demand projection of the building sector in Greece in the 21st century. *Energy and Buildings, 49*(June), 488–498.

Athienitis, A. K. (1994). Numerical model of floor heating system. *ASHRAE Transactions, 100*, 1024–1030.

Badran, A. A., & Hamdan, M. A. (2004). Comparative study for under-floor heating using solar collectors or solar ponds. *Applied Energy, 77*, 107–117.

David, J. (1984). Under floor air conditioning. *Journal of the Charted Institution of Building Services, 6*, 29–34.

European Environment Agency. (2010). Consumption and the environment—state and outlook.

Fantidis, J. G., Bandekas, D. V., Potolias, C., & Vordos, N. (2013). Cost of PV electricity—Case study of Greece. *Solar Energy, 91*, 120–130.

Gao, R., Li, A., Zhang, O., & Zhang, H. (2011). Comparison of indoor air temperatures of different under-floor heating pipe layouts. *Energy Conversion and Management, 52*, 1295–1304.

Ghali, K. (2007). Economic viability of underfloor heating system: A case study in Beirut climate. In *International conference on renewable energies & power quality*, Sevilla, Spain.

Hasan, A., Kurnitski, J., & Jokiranta, K. (2009). A combined low temperature water heating system consisting of radiators and floor heating. *Energy and Buildings, 41*, 470–479.

Inard, C., Meslem, A., & Depecker, P. (1998). Energy consumption and thermal comfort in dwelling-cells: A zonal-model approach. *Building and Environment, 33*(5), 279–291.

Kilkis, B. I., Sager, S. S., & Uludag, M. (1994). A simplified model for radiant heating and cooling panels. *Simulation Practice and Theory, 2*, 61–76.

Kouremenos, D. A., Antonopoulos, K. A., & Domazakis, E. S. (1985). Solar radiation correlations for the Athens, Greece, area. *Solar Energy, 35*(3), 259–269.

Myhren, J. A., & Holmberg, S. (2008). Flow patterns and thermal comfort in a room with panel, floor and wall heating. *Energy and Buildings, 40*, 524–536.

Sekhar, S. C., & Ching, C. S. (2002). Indoor air quality and thermal comfort studies of an under-floor air-conditioning system in the tropics. *Energy and Buildings, 34*, 421–444.

Sodec, F., & Craig, R. (1990). The underfloor air supply system—The European experience. *ASHRAE Transactions, 96*(2), 690–695.

Theodoridou, I., Papadopoulos, A. M., & Hegger, M. (2011). A typological classification of the Greek residential building stock. *Energy and Buildings, 43*, 2779–2787.

Zhai, X. Q., Yang, J. R., & Wang, R. Z. (2009). Design and performance of the solar-powered floor heating system in a green building. *Renewable Energy, 34*, 1700–1708.

Part VII
Energy Technologies and Their Effect on Global Warming

Bioenergy–Biofuel Technologies

Novel biofuels

Chapter 29
Biogas Production from Napier Grass at Various Cutting Intervals

Nusara Sinbuathong, Yuwadee Sangsil, and Suriya Sawanon

Introduction

Anaerobic digestion can be applied to convert biodegradable wastes such as plant biomass, energy crops and grass to produce biogas (Seppala et al. 2009). Napier grass (*Pennisetum purpureum*) is known as elephant grass and is a perennial grass with high biomass. Normally it is used for cattle feeds and is cultivated as forage. The grass has the advantage of being familiar to farmers and suitable for harvesting and storing with existing methods and machinery. It grows fast and is highly adaptive, hence is suitable to be grown on marginal lands. Climate, soil fertility, cutting interval, variety and management practices may affect on chemical composition and digestibility of Napier grass. It is interesting to digest the grass in order to produce biogas. However, grass is not a mature feedstock for anaerobic digestion facilities due to its complex structure; lignin and cellulose are resistant to hydrolysis reaction and microbial actions (Nizami et al. 2009). Previous studies about biological pretreatment were mainly focused on use of pure cultures, such as anaerobic bacteria, fungi, and actinomycetes, which were able to degrade lignocellulose (Desvaux et al. 2000; Xu and Goodell 2001). However, pure cultures of

N. Sinbuathong (✉)
Scientific Equipment and Research Division, Kasetsart University Research and Development Institute (KURDI), Kasetsart University, Bangkok 10900, Thailand
e-mail: rdinrs@ku.ac.th

Y. Sangsil
Department of Environmental Engineering, Faculty of Engineering, Kasetsart University, Bangkok 10900, Thailand
e-mail: ysangsil@gmail.com

S. Sawanon
Department of Animal Science, Faculty of Agriculture at Kamphaeng Saen, Kasetsart University, Kamphaeng Saen Campus, Nakhon Pathom 73140, Thailand
e-mail: agrsusa@ku.ac.th

P. Grammelis (ed.), *Energy, Transportation and Global Warming*, Green Energy and Technology, DOI 10.1007/978-3-319-30127-3_29

microorganisms are difficult to maintain in an open system, and often require long pretreatment time. In fact, it has been approved that the complete decomposition of lignocellulose requires the combined actions of various microorganisms (Slater and Lovatt 1984). Several studies have demonstrated the efficiency of constructed microbial consortia for the rapid degradation of lignocellulose (Haruta et al. 2002; Wongwilaiwalin et al. 2010)

Cows rely on rumen microorganisms to convert feed components into useable source of energy and protein. Rumen microorganisms ferment carbohydrates to make volatiles fatty acids and grasses. The rate of digestion of feeds depends on the quality and composition of feed. It is affected by the number and type of microorganism, pH in the rumen and nutrients. These factors limit the growth of the microorganisms.

Hence, two phase anaerobic digestion with mixed microorganisms may enhance the efficiency of digestion. In this study, a lab scale, two-stage anaerobic digestion system was used for digesting Napier grass for biogas production. In the first stage, the acidogenic condition provides acidic surroundings to digest the recalcitrant structure and enhances digestibility. The aim of the current study is to digest Napier grass of different cutting intervals by using mixed ruminal microorganisms from cow rumen in semi-continuous two-stage reactor in order to minimize inhibition of the hydrolysis phase by low pH and accumulation of inhibitors. The test was performed under optimum hydraulic retention time (HRT). Annual CH_4 yield per hectare obtained from Napier grass at various cutting interval was investigated.

Materials and Method

Two-Stage Anaerobic Operation

Two separate bioreactors were designed specifically for the acidogenic and methanogenic anaerobic digestion. Four set of bioreactors was constructed with a PET plastic reactors with a working volume of 4 L and 5 L for acidogenic and methanogenic stage respectively (Fig. 29.1). The reactor was connected to a gas collection system, which was based on water displacement by the exiting gases (Fig. 29.1).

Fresh Napier grass (Pakchong 1) of approximately 30 kg. was cut at various cutting intervals (30, 45, 60 and 90 days) and then bagged and stored in a cold room. The grass was chopped to a size of 1 cm and ground to a size of approximately 2 mm. The stems and leaves were blended thoroughly into a harmonious whole. The ground Napier grass was stored in a refrigerator at 4 °C before use. Without pretreatment and supplement with nutrient addition, the grass was prepared with tap

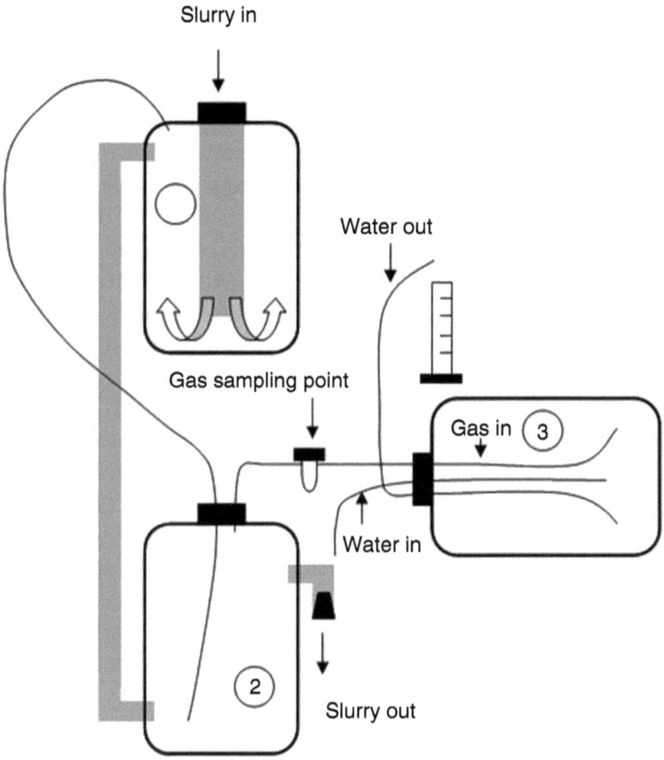

Fig. 29.1 Two-stage anaerobic reactor with gas collection system

water at a ratio of 1:5 by weight to create a slurry. Napier grass was used as a sole carbon source. Four systems were developed by varying the cutting intervals of the Napier grass in the acidogenic and methanogenic reactors. A slurry of grass at various cutting intervals at a 1:5 grass-water ratio was added to the acidogenic reactor at a rate of 0.5 L/day for 8 days. This acidogenic reactor was operated for 2 days in batch mode before feeding once daily with the slurry semicontinuously in an upflow mode at a feed rate of 200 mL/day in each set of reactors, giving rise to HRT of 20 days for acidogenic and 25 days for methanogenic reactor stage. Mixed ruminal microorganisms of approximately 10 g mixed liquor volatile suspended solid/L were used as inoculum. The reactors were operated at ambient temperature of 30 + 1 °C. pH was adjusted to be 7.5 for all methanogenic reactors at the start-up period. The reactors functioned without pH control during an operation period. In this study, two-stage anaerobic digestion of Napier grass was investigated by controlling the HRT of the two digesters. The start-up and operating period were approximately 55 days. The study was carried out using steady-state conditions.

Analysis of Materials

Napier grass (Pakchong 1) used in the test was cultivated in well fertilized field at Khamphaeng Saen, Nakorn Pathom Province, Thailand. The grass at various cutting intervals was measured for height and analysed for major (N, P, K) and minor elements (K, Ca, Mg, Cu, Fe). These elements are nutrients in the digestion.

Nitrogen (N) was determined by the Kjeldahl methods by digesting samples to convert organic N to NH_4^+-N and determining NH_4^+-N in the digest. Total phosphorous (P) was determined by digesting samples with sulphuric acid and analysed by the Vanadomolybdophosphoric Acid Colourimetric method. Methane (CH_4) was determined by employing a gas chromatograph (Model GC 14B, Shimadzu, Japan) equipped with a thermal conductivity detector (TCD) and a molecular sieve 13X column. The temperatures of the injector, column and TCD detector were 60, 40 and 70 °C, respectively. Argon was used as the carrier gas. The slurry fed to reactors and the digested slurry from the methanogenic reactor were analysed for Chemical Oxygen Demand (COD), Total Volatile Solids (TVS), Total Solids (TS) and pH according to the procedures of the Standard Methods (APHA and AWWA 2005). Organic degradation efficiency (in terms of COD and TVS) and CH_4 production from the system obtained from the Napier grass at various cutting intervals were used as indicators of reactor performance. All experiments were conducted in duplicate and the results are calculated using the mean of the experimental values.

Results

The grass yield from the field is as shown in Fig. 29.2. A slurry of Napier grass at a 1:5 grass-water ratio was analysed for COD, TS, TVS also for the major and minor biological nutrients (Table 29.1). The composition and quantity of Napier grass varied on the cutting intervals (Table 29.1), thus impacts on the biogas production.

During sample preparation, it was noticeable that the higher age grass (higher cutting interval) had more stems and had fewer leaves than the lower age. Thus, the proportion of leaf to stem of the higher age grass was less comparing with that of the lower age grass. Crowder and Chheda (1995) had also reported this. Napier grass is easy in propagation and management, thus it has high yield. The data from field found that when the cutting interval increased, the grass yield increased (Fig. 29.2). The grass yield obtained from the Napier grass at the cutting interval of 45, 60 and 90 days was double, quadruple and penta times of that from 30 days (Fig. 29.2). On the other hand, at higher cutting intervals, the major and minor elements by dry weight decreased (Table 29.1). This is in accordance with the data reported by Crowder and Chheda 1995. Nutrients are crucial that limit microbial growth in anaerobic digestion. If the nutrients are in short supply, microbial growth is retarded, and so is the digestibility of feed (Moran 2005).

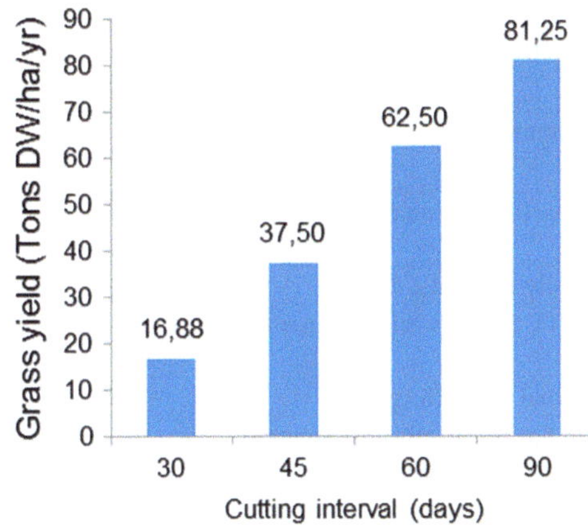

Fig. 29.2 Annual Napier grass yield at various cutting intervals

Table 29.1 Characteristics of the slurry of Napier grass (Napier grass:water = 1:5) fed to reactors at various cutting intervals

Parameters	Napier grass at various cutting interval (days)			
	30	45	60	90
Chemical oxygen demand (COD, g/L)	19.32	24.00	27.32	30.87
Total solids (TS, g/L)	17.73	24.00	26.94	32.30
Total volatile solids (TVS, g/L)	14.43	18.42	20.88	29.18
pH	5.86	7.22	6.92	6.17
Nitrogen (TKN, mg/L as N)	661	343	300	251
Phosphorus (PO_4^{3-}, mg/L as P)	2.74	1.84	2.14	3.11
Calcium (Ca, mg/L)	50.19	32.45	24.11	35.87
Copper (Cu, mg/L)	0.15	0.07	0.07	0.02
Iron (Fe, mg/L)	14.47	2.55	2.99	4.51
Potassium (K, mg/L)	1334	864	972	1017
Magnesium (Mg, mg/L)	34.79	19.79	39.50	35.47
COD:N	100:3.42	100:1.43	100:1.10	100:0.81
COD:N:P	100:3.42:0.014	100:1.43:0.008	100:1.10:0.008	100:0.81:0.010

The grass was fairly high in organic substrate (Table 29.2) and therefore had good potential for biogas generation. The COD content in the slurry of Napier grass (Napier grass:water = 1:5) at the cutting intervals of 30, 45, 60 and 90 days was 19.32, 24.00, 27.32 and 30.87 g/L and TKN of the grass was 661, 343, 300, 251 mg/L as N respectively; thus, the C:N ratio was 100:3.42, 100:1.43, 100:1.10 and 100:0.81 respectively. Microorganisms that decompose organic matter use carbon as a source of energy and nitrogen for building cell structure. The recommended COD:N of organic substrate is 100:2.5 in order to assimilate the microorganisms in

Table 29.2 Summary of two-stage anaerobic digestion of the Napier grass as a sole carbon source at various cutting intervals

Parameters	Napier grass at various cutting intervals (days)			
	30	45	60	90
Reactor volume for each acidogenic, methanogenic reactor (L)	4, 5	4, 5	4, 5	4, 5
Daily feed (mL/day)	200	200	200	200
HRT for each acidogenic, methanogenic reactor (day)	20, 25	20, 25	20, 25	20, 25
Initial MLVSS (g/L)	10.0	10.0	10.0	10.0
Initial pH of substrate solution	5.86	7.22	6.92	6.17
Initial COD (g/L)	19.32	24.00	27.32	30.87
Initial TS (g/L)	17.73	24.00	26.94	32.30
Initial TVS (g/L)	14.43	18.42	20.88	29.18
OLR (kgCOD/m^3.day)	0.97	1.20	1.37	1.54
OLR (kgTVS/m^3.day)	0.72	0.92	1.04	1.46
CH_4 produced at steady state (mL at STP/day)	568	610	561	480
Average CH_4 (%)	41.34	43.51	42.99	39.43
COD at steady state (g/L)	5.32	6.04	6.60	7.69
COD degradation efficiency (%)	72.46	74.82	75.84	75.88
TVS at steady state (g/L)	5.21	6.09	6.62	7.50
TVS degradation efficiency (%)	63.88	66.94	68.30	74.30
CH_4 yield (L at STP/kgCOD added)	147	127	103	78
CH_4 yield (L at STP/kgTS added)	160	127	104	74
CH_4 yield (L at STP/kgTVS added)	197	166	134	82
Napier grass (DW) for pure CH_4 1 m^3 (kg)	6.25	7.87	9.62	13.51
CH_4 yield per area (m^3/ha/year)	2700	4763	6500	6013
Average pH of acidogenic reactor at steady state	5.20	5.24	5.35	5.35
Average pH of methanogenic reactor at steady state	7.25	7.20	7.18	7.10

anaerobic digestion (Anderson et al. 2003). If the COD:N ratio is too high, decomposition slows because the nitrogen is used up and some microorganisms die. On the other hand, if the COD:N ratio is too low, microorganisms make full use of the available carbon and get rid of the excess nitrogen as ammonia. Ammonia caused toxic to microorganisms in the system (Hartmann and Ahring 2006). In this study, the ratio of COD:N:P was 100:3.42:0.014, 100:1.43:0.008, 100:1.10:0.008, 100:0.810:010. These caused differences in the amount of carbon source and nutrients fed to reactors.

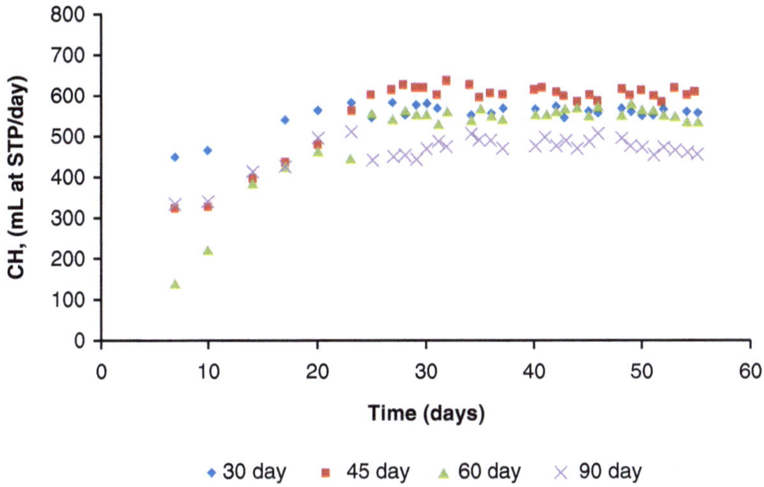

Fig. 29.3 Time course of CH_4 production of Napier grass slurry at various cutting intervals

Biogas Production

At steady-state conditions, total biogas production and the performance of digestion were reported. In this study, the results are reported for 25–55 days of digestion period. The amount of CH_4 production from the reactors are shown in Fig. 29.3 and the average CH_4 content are shown in Table 29.2.

Differences in the amount of CH_4 production (Fig. 29.3) by the Napier grass of each cutting interval were observed. At steady-state condition, the average daily CH_4 produced from the Napier grass at the cutting interval of 45 days was 605 mL at STP/day and that from the cutting interval of 30 and 60 days was comparable and was recorded at 570 mL at STP/day while that from the cutting interval of 90 days was 480 mL at STP/day (Fig. 29.3). However, the key parameter that makes more sense to use to consider the appropriate cutting interval is the annual CH_4 yield per area of cultivation, which is calculated by using the CH_4 gas produced and the Napier grass added to reactors including the mass of the grass yield obtained from field.

Organic Substrate Degradation

The initial COD, TS, TVS and other parameters of the slurry of Napier grass (Napier grass: water = 1:5) are summarised in Table 29.2. The COD and TVS of the slurry of Napier grass at the higher cutting interval was higher than that of the lower (Table 29.2). Ansah et al. (2010) also reported that the higher age grass had higher stem and higher organic matter (OM), resulted in higher COD and TVS than that of the lower age grass. The steady-state COD and TVS obtained from the various cutting intervals are as shown in Table 29.2. The COD and TVS removal efficiency were over 70 % and 60 %, respectively for all reactors that contained a slurry of Napier grass of all cutting intervals (Table 29.2).

CH_4 Yield

CH_4 yield is the amount of CH_4 produced per kilogram of organic substrate added to reactor. For CH_4 production and based on the COD, TVS and TS added to reactors at the steady-state conditions, CH_4 yield obtained at various cutting intervals are shown in Table 29.2, respectively. Differences in the CH_4 yield by the Napier grass of each cutting interval were obviously observed (Table 29.2). The composition and quantity of Napier grass varied on the cutting intervals (Table 29.1) impacts on the biogas production (Fig. 29.3). Hence, the CH_4 yield was influenced by the cutting interval of the grass and the grass yield. Napier grass at the cutting interval of 30 days showed a maximum value of CH_4 yield at 147 L at STP/kg of COD added (197 L at STP/kg of TVS added or 160 L at STP/kg DW of Napier grass added). CH_4 yield obviously dropped when the cutting intervals increased (Table 29.2). For Napier grass of higher cutting interval (Napier grass of 90 days), grass is composed of more stems and it is not a mature feedstock for anaerobic digestion facilities due to its complex structure; lignin and cellulose are resistant to hydrolysis and microbial actions.

The highest CH_4 yield was observed from the Napier grass at the cutting interval of 30 days. For the Napier grass of 30 days, the CH_4 yield was 160 L at STP/kg of dry Napier grass added to the reactor (Table 29.2), which indicated that 6.25 kg of dry Napier grass is needed to produce 1 m^3 of pure CH_4. COD and TVS degradation efficiency was 72 % and 64 %, respectively (Table 29.2). The average pH in the acidogenic and methanogenic reactors was 5.20 and 7.25, respectively (Table 29.2). For the Napier grass at the cutting intervals of 45, 60 and 90 days, the highest CH_4 yield was observed at 127, 104 and 74 L at STP/kg of dry Napier grass added to the reactor, respectively (Table 29.2). The higher the cutting interval, the lower the CH_4 yield obtained. Although the higher age grass had more stems, however, the higher age grass had the more fibre and lignin (lignocelluloses) which are nonbiodegradable organics. Hence, the higher age grass gave less CH_4 yield during digestion than the lower age grass.

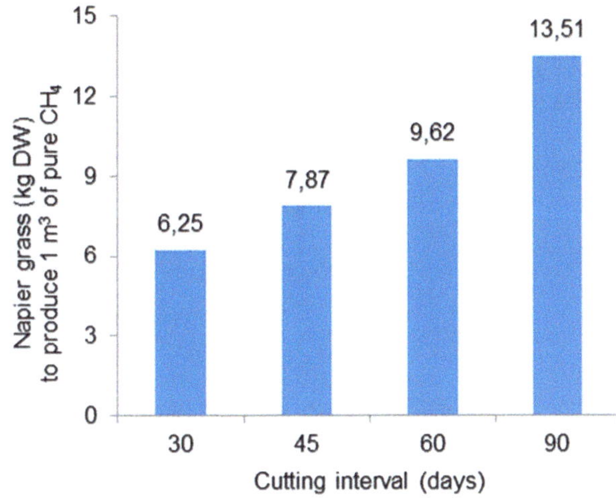

Fig. 29.4 Napier grass at various cutting intervals to produce 1 m^3 of pure CH_4

Annual CH_4 Gas Produced per the Napier Grass Added to Reactors

For the digestion of Napier grass at the cutting intervals of 30, 45, 60, 90 day, the CH_4 yield was 160, 127, 104, 74 L at STP/kg of dry Napier grass added to the reactor (Table 29.2), which indicated that 6.25, 7.87, 9.62, 13.51 kg of dry Napier grass is needed to produce 1 m^3 of pure CH_4 (Fig. 29.4).

However, when considering the Napier grass yield from cultivation in the field (the grass yield at the cutting intervals of 30, 45, 60 and 90 days was 16.88, 37.50, 62.50 and 81.25 tons dry weight/hectare/year as mentioned earlier), it was found from calculation that the CH_4 yield per area of cultivation which was obtained from the Napier grass at the cutting interval of 30, 45, 60 and 90 was 2700, 4763, 6500 and 6013 m^3/hectare/year (Fig. 29.5). The annually highest CH_4 yield per area of cultivation was obtained from the Napier grass at the cutting interval of 60 days (Fig. 29.5).

The result of this study is in accordance with that reported by Ansah et al. (2010). Ansah et al. (2010) reported that the lower age grass gives the higher CH_4 yield than the higher age grass due to its higher crude protein and fibre. From this study, even though the Napier grass of the cutting interval of 30 days gave the highest CH_4 yield but it gave the lowest annual grass yield per area, resulting in least annual CH_4 yield per area.

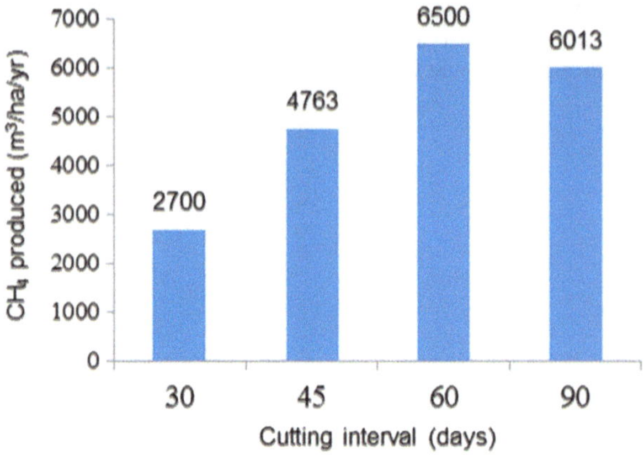

Fig. 29.5 Annual CH_4 produced per hectare from Napier grass at various cutting intervals

Conclusions

Napier grass (Pakchong 1) at the cutting interval of 60 days gave the highest annual CH_4 yield per area of all various cutting intervals. By two-stage anaerobic digestion using the Napier grass as the sole carbon source and mixed ruminal microorganisms from cow as seed source, the CH_4 yield obtained was 104 L at STP/kg of dry Napier grass added to the reactor, which indicated that 9.62 kg of dry Napier grass is needed to produce 1 m^3 of pure CH_4. The grass yield obtained from the field was 62.50 tons dry weight/hectare/year, resulting in the annual CH_4 yield per area obtained was 6500 m^3/hectare/year.

Acknowledgement This research was supported by the Kasetsart University Research and Development Institute (KURDI), Kasetsart University, Bangkok, Thailand.

References

Anderson, G. K., Sallis, P. J., & Uyanik, S. (2003). Anaerobic treatment processes. In D. Mara & N. J. Horan Jr. (Eds.), *Handbook of water and wastewater microbiology* (pp. 391–426). London: Academic.

Ansah, T., Osafo, E. L. K., & Hansen, H. H. (2010). Herbage yield and chemical composition of four varieties of Napier (*Pennisetum purpureum*) grass harvested at three different days after planting. *Agriculture and Biology Journal of North America, 1*, 923–929.

APHA and AWWA (American Public Health Association and American Water Works Association). (2005). *Standard methods for the examination of water and wastewater* (21st ed.). Washington, DC: APHA and AWWA.

Crowder, L. V., & Chheda, H. R. (1995). *Tropical grassland husbandry* (1st ed.). NJ: Wiley.

Desvaux, M., Guedon, E., & Petitdemange, H. (2000). Cellulose catabolism by Clostridium cellulolyticum growing in batch culture on defined medium. *Applied and Environmental Microbiology, 66*, 2461–2470.

Hartmann, H., & Ahring, B. K. (2006). Strategies for the anaerobic digestion of the organic fraction of municipal solid waste: An overview. *Water Science and Technology, 53*, 7–22.

Haruta, S., Cui, Z., Huang, Z., Li, M., Ishii, M., & Igarashi, Y. (2002). Construction of a stable microbial community with high cellulose-degradation ability. *Applied Microbiology and Biotechnology, 59*, 529–534.

Moran, J. (2005). *How the rumen work? Tropical dairy farming: Feeding Management for small dairy farmers in the humid tropics* (pp. 41–49). Collingwood: Land Links Press.

Nizami, A. S., Korres, N. E., & Murphy, J. D. (2009). A review of the integrated process for the production of grass biomethane. *Environmental Science and Technology, 43*, 8496–8508.

Seppala, M., Paavolla, T., Lehtomaki, A., & Rintala, J. (2009). Biogas production from boreal herbaceous grass-specific methane yield and methane yield per hectare. *Bioresource Technology, 100*, 2952–2958.

Slater, J. H., & Lovatt, D. (1984). Biodegradation and the significance of microbial communities. In D. T. Gibson (Ed.), *Microbial degradation of organic compounds* (pp. 439–485). New York: Marcel Dekker.

Wongwilaiwalin, S., Rattanachomsri, U., Laothanachareon, T., Eurwilaichitr, L., Igarashi, Y., & Champreda, V. (2010). Analysis of a thermophilic lignocelluloses degrading microbial consortium and multi-species lignocellulolytic enzyme system. *Enzyme and Microbial Technology, 47*, 283–290.

Xu, G., & Goodell, B. (2001). Mechanisms of wood degradation by brown-rot fungi: chelator-mediated cellulose degradation and binding of iron by cellulose. *Journal of Biotechnology, 87*, 43–57.

Chapter 30
Sustainability Assessment of Fuels Production via Hydrotreating Waste Lipids and Co-processing Waste Lipids with Petroleum Fractions

Stella Bezergianni and Loukia P. Chrysikou

Introduction

Considering the depletion of fossil energy resources and the increased greenhouse gas (GHG) emissions of the transportation sector, biofuels are being extensively explored as renewable and sustainable energy sources (Demirbas 2009). The transportation sector contributes around 15 % to the global GHG emissions, while according to the EU targets by 2020, 10 % of fuels in the transportation sector should be biofuels (Directive 2009/28/EC). In particular, biofuels are being promoted as one of the most substantial substitutes of fossil fuels, accounting their favorable environmental performance, especially if utilizing residual biomass as feedstock (Gnansounou et al. 2009). FAME (Fatty Acid Methyl Esters) biodiesel constitutes the most common biofuel mainly derived from energy crops, whereas residual biomass can be utilized in minor percentages. For instance, waste cooking oil (WCO) generated from frying and cooking with edible oils has been investigated as feedstock for FAME biodiesel production through transesterification and esterification processes. FAME biodiesel is used as a low-blend component in transport fuels (up to 7 % v/v) in Europe. However, FAME biodiesel cultivation processes raised significant issues, such as food-versus fuel debate, land-use alteration rendering the use of bio-based feedstocks imperative, aiming to address global warming concerns, as well (Gnansounou et al. 2009). Catalytic hydroprocessing is an alternative conversion process of lipid feedstocks (e.g. vegetable oils, animal wastes/fats etc.) into bio-based diesel fuels and has been applied in industrial scale for the production of hydrotreated vegetable oils (HVOs). More specifically, HVOs

S. Bezergianni (✉) • L.P. Chrysikou
Chemical Process & Energy Resources Institute (CPERI), Centre for Research & Technology Hellas (CERTH), Thermi Thessaloniki, Greece
e-mail: sbezerg@cperi.certh.gr; loukia@cperi.certh.gr

P. Grammelis (ed.), *Energy, Transportation and Global Warming*, Green Energy and Technology, DOI 10.1007/978-3-319-30127-3_30

are produced from vegetable oils and/or grease by Universal Oil Products (UOP) (product's trade name green diesel) (Kalnes et al. 2009) and from vegetable oils and/or animal fats by NesteOil Corporation (product's trade name NExBTL-acronym for "Next Generation Bio-to-Liquid) (Arvidsson et al. 2011).

WCO has been extensively explored as potential feedstock for renewable diesel production via catalytic hydrotreatment in CPERI/CERTH (Bezergianni et al. 2010a, b, 2011). The renewable diesel is a high quality fuel based on its remarkable properties (Bezergianni and Dimitriadis 2013a). Nevertheless, the high capital cost investment of hydroprocessing units of vegetable oils and/or animal fats supports the exploration of co-hydroprocessing petroleum fractions with oils utilizing existing refining facilities, since catalytic hydroprocessing units are available in all petroleum refineries. Recently, co-hydroprocessing of petroleum fractions with WCO was also examined in CPERI/CERTH, resulting in the production of a new hybrid diesel. In particular, catalytic hydroprocessing of a heavy petroleum fraction (heavy atmospheric gas oil, HAGO) with WCO (at a ratio 95/5 % v/v) was investigated towards the production of a hybrid diesel of increased sustainability (Bezergianni and Dimitriadis 2013b; Bezergianni et al. 2014a, b). This new hybrid diesel was evaluated as a blending component of fossil diesel (Bezergianni et al. 2014b). In this field Garraín et al. (2014) studied the production of a renewable diesel fuel from co-processing soybean vegetable oil (at variant ratios from 9 up to 13.8 % v/v/) with conventional fossil fuel.

However, the environmental performance of the production processes involving the exploitation of existing petroleum refining processes to convert vegetable oils to renewable bio-based diesel fuels, receives also noticeable attention, accounting the global necessity to mitigate the climate change and to promote sustainable technologies (Garraín et al. 2014). Life Cycle Assessment (LCA) is a commonly applied powerful method for the global warming impact assessment of conventional fuels and biofuels. LCA quantifies the environmental impacts of conventional fuels and biofuels throughout their life-cycle stages, proposing concurrently possible ways to reduce them. A noticeable amount of LCA studies related to fuels is available in the literature, in order to identify potential sustainable fuels that could contribute to the transportation sectors' GHG emissions reduction (Pleanjai et al. 2009; Arvidsson et al. 2011; Kochaphum et al. 2012; Bezergianni et al. 2014c; Garraín et al. 2014). Based on the large number of publications analyzing the life-cycle of fuels and biofuels significant controversies arise, even for the same biofuel type and conversion technology, mainly due to variant methodological assumptions, inventory data etc. An accurate assessment of a new biofuel technology requires a comprehensive fuel-cycle analysis, commonly called Well-To-Wheel (WTW) analysis, involving both the Well-To-Tank analysis (WTT), as well as the Tank-To-Wheel analysis (TTW). The WTT approach covers the fuel's production chain while the TTW the vehicle's operation. The majority of the LCA studies separate the analysis in the WTT and TTW counterparts, whilst the WTT is particularly appropriate if the purpose of the study is concerned with the fuel's production process (Malça and Freire 2011).

The goal of this study is to assess the environmental profile of the renewable diesel and the hybrid diesel produced via WCO hydrotreatment and co-hydroprocessing of petroleum fractions with WCO, respectively. In addition, the environmental impacts of the renewable diesel's blend (up to 7 % v/v) with fossil diesel were also investigated. A secondary goal of this study is to compare the environmental performance of all aforementioned fuels with fossil diesel (containing 7 % v/v FAME biodiesel) in order to address the most sustainable process and to verify the advantages of utilizing WCO as feedstock and its prospective integration in the refinery. Aiming to establish a reliable comparison basis of the aforementioned fuels production processes, WTT analysis was conducted addressing their GHG intensity.

Methodology

The scope of the LCA study is to investigate the environmental impacts of bio-based fuels produced via hydrotreating WCO and co-processing WCO with petroleum fractions, and to compare their production processes GHG emissions with fossil diesel, as well. The experimental procedure followed for the WCO catalytic hydrotreatment has been previously described in detail (Bezergianni and Dimitriadis 2013a), utilizing a large pilot plant scale hydroprocessing unit of CPERI/CERTH. Likewise, the production process of the new hybrid diesel has been described in authors' previous work (Bezergianni et al. 2014b), exploiting data collected from the Hellenic Petroleum refinery in Thessaloniki as reported in authors' previous work (Bezergianni et al. 2014b). The experiments were carried out in a small-scale hydroprocessing plant of CPERI/CERTH, while the unit is described extensively by Bezergianni and Dimitriadis (2013a). Apart from the real experimental data collected for the LCA study, results reported by JRC (2014) were also employed for comparison purposes referring to the fossil diesel's and FAME biodiesel's production processes GHG intensity.

The fuels production processes examined in the study are shown in Figs. 30.1, 30.2, 30.3 and 30.4 depicting the system boundaries encompassed for each of the examined fuels. Particularly, Fig. 30.1 presents the production process of the renewable diesel via WCO catalytic hydrotreatment, including WCO collection and transportation to the hydrotreating unit and the hydrotreatment process as well.

The production of the new hybrid diesel from co-hydroprocessing HAGO and WCO in the refinery under consideration is shown in Fig. 30.2, involving the crude oil extraction and transportation to the refinery, the refining of the crude oil for the production of petroleum fractions (HAGO and light atmospheric gas oil-LAGO), the co-processing HAGO with WCO (95/5 % v/v), the hydrotreatment of LAGO and finally the blending with FAME for the production of the new hybrid diesel (the precise blending ratios are reported below). Figure 30.3 presents the production chain of fossil diesel, encompassing the stages of crude oil extraction and transportation to the refinery and the refining process based on previous studies (JRC

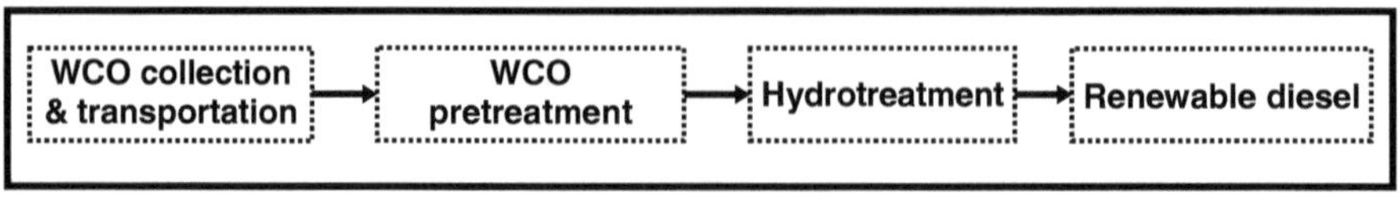

Fig. 30.1 Production chain of the renewable diesel via WCO catalytic hydrotreatment

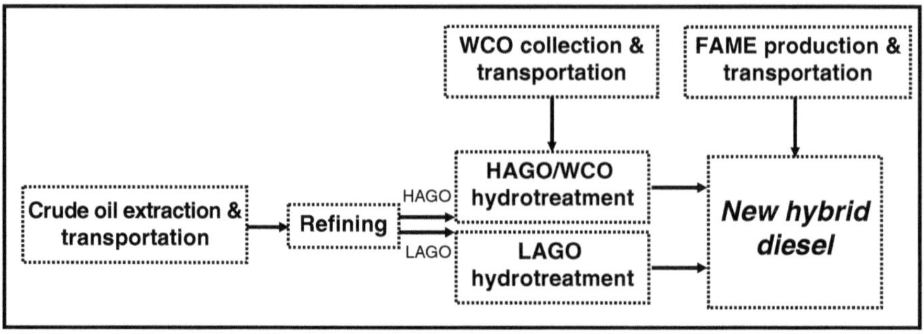

Fig. 30.2 Production chain of the new hybrid diesel via co-hydroprocessing HAGO with WCO

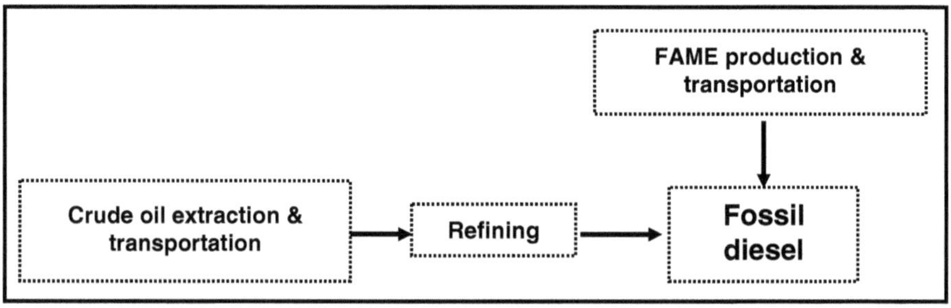

Fig. 30.3 Production chain of the fossil diesel

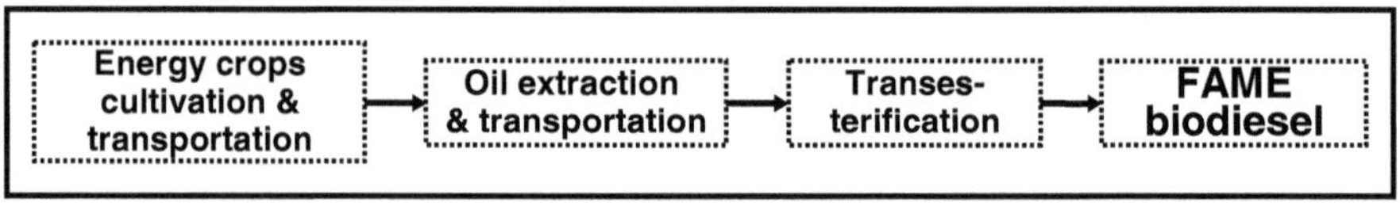

Fig. 30.4 Production chain of the FAME biodiesel

2014). The final stage of the production process involves its blending with FAME biodiesel (~7 % v/v) as previously mentioned in the introduction. The FAME biodiesel production chain considered is shown in Fig. 30.4 containing the energy crops cultivation and the oil extraction from them and finally the transesterification process, taking into consideration the transport processes as well. More details for FAME biodiesel production process can be found on a previous thorough study (JRC 2014).

The functional unit was 1 MJ of the renewable diesel and the hybrid diesel. The inventory data sources employed for each of the investigated fuels were reproduced based on this functional unit. The impact category assessed in this study is the global warming potential (GWP) expressed in CO_2 equivalents (CO_2 eq-), since GHG emissions and energy use are of high interest in bio-based fuels LCA studies. The CO_2 eq-outlines the total effect of GHG emissions including related emissions of CO_2, CH_4 and N_2O. Emissions contributing to acidification and eutrophication are also often considered as impact categories, yet in the present study were not accounted, since WCO does not impose these emissions.

Inventory Data

This section includes the inventory data of the fuels production processes previously described.

Renewable Diesel

Real experimental data were collected from the production of the renewable diesel via WCO catalytic hydrotreatment while data were derived from current literature as well. The assumptions and considerations for the environmental evaluation of the renewable diesel's production process are listed below:

- the impacts associated to the collection and transportation stages of WCO are negligible in comparison to the impacts of the hydrotreatment process, accounting also the short distances within the city limits.
- main energy supply systems considered were the electric power consumption from the grid and steam generation. The inventory data referring to the energy requirements of the hydrotreatment process (electricity and steam) and the associated GHG emissions were adapted from literature data, considering the production of a HVO using vegetable oils and/or animal fats (Nikander 2008).
- the GHG emissions of the H_2 consumption in the hydrotreatment process have been estimated based on data retrieved from the literature (Nikander 2008), produced via natural gas steam reforming.
- construction and demolition phases were eliminated from this study, due to deficiency of related GHG emissions data.

Table 30.1 gathers the main inventory data (inputs and outputs) for the production of renewable diesel from WCO catalytic hydrotreatment, based on experimental data computed by previously evaluations (Nikander 2008).

At this point it should be underlined that the renewable diesel could be possibly employed as a blending component to fossil diesel up to ~7 % v/v, as HVOs. Therefore the environmental profile of this blend was also addressed in order to

Table 30.1 Inventory data for the production of renewable diesel via WCO catalytic hydrotreatment

	Unit	Input	Output
WCO	g/MJ	23	–
H_2	g/MJ	1.268	
Renewable diesel	g/MJ	–	20.41
Gas product[a]	g/MJ	–	1.587
Total liquid product	g/MJ	–	3.648
Energy	MJ/MJ	0.002	–
Steam	MJ/MJ	0.014	–

[a]Including excess H_2

determine the most sustainable fuel in terms of GHG intensity and the results are presented below. Data for the GHG emissions of fossil diesel production process were obtained from literature sources in view of a European context (JRC 2014).

Hybrid Diesel

The key inventory data for the new hybrid fuel production process for the quantification of the GHG emissions (e.g information on crude oil origin, typical catalytic hydrotreatment operation conditions, electric power consumption, fuels used for heat production, main input and output flows used during the production etc.) were collected from the refinery of the Hellenic Petroleum in Thessaloniki. Secondary data were taken from assessments previously undertaken (Nikander 2008; Cetinkaya et al. 2012). For assessing the GHG emissions of the new hybrid diesel profile GEMIS (Global Emission Model for Integrated Systems, Version 4.9) was used. GEMIS is a life-cycle analysis program and database for energy, material, and transport systems providing information for fuels, processes for electricity and heat etc.

The primary considerations and assumptions of the LCA study are the following:

- the impacts associated with the collection and transportation of WCO to the refinery were considered negligible, accounting short distances (within city limits).
- the petroleum refineries are extremely complicated plants with combined and sequenced processes and as a result the refinery process of the crude oil is often handled as a "black box" and not partitioned into sub-processes. In this study, the refining stage is also depicted as a "black box" and the energy consumption and GHG emissions were allocated according to the final products distribution. Figure 30.5 illustrates a generic flow diagram of the refinery studied. Briefly, after crude's oil refining process the treated petroleum fractions, HAGO and LAGO are further treated to hydrotreating units. In the HAGO hydrotreating unit, HAGO with WCO (95/5 % v/v ratio) are co-hydroprocessed for the production of a new hydrotreated hybrid fuel. Then this fuel is mixed with hydrotreated LAGO and with FAME biodiesel for the production of the new

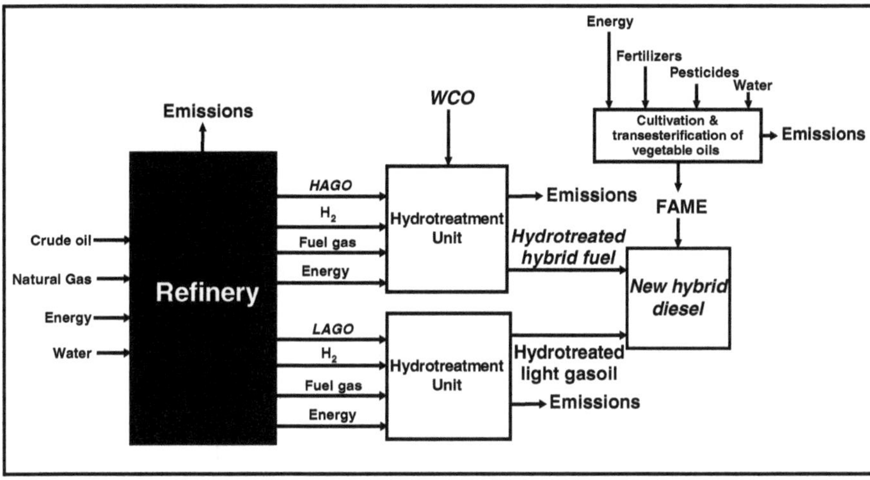

Fig. 30.5 Generic flow diagram of the refinery describing the production process of the new hybrid diesel via co-processing HAGO and WCO

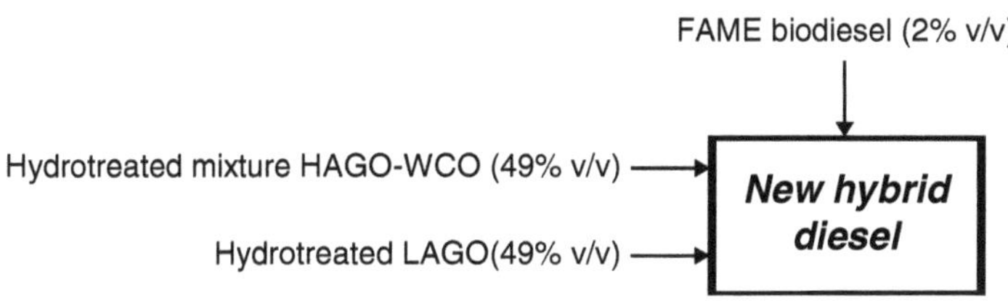

Fig. 30.6 Blending ratios of the components for the production of the new hybrid diesel

hybrid diesel, based on the blending strategy of the refinery of the Hellenic Petroleum in Thessaloniki. A detailed analysis of the blending followed is presented in Fig. 30.6.

- data regarding the GHG emissions of the fossil diesel's and FAME biodiesel's production processes were retrieved form scientific papers (JRC 2014) considering typical refining in the EU. Specifically, the GHG balances of FAME biodiesel refer to farming emissions not taking account of any land use change emissions.
- electricity used in the refinery is assumed to be taken from the local grid, while a small quantity is produced internally by fuels combustion (fuel gas, fuel oil, natural gas etc.). The GHG emissions from the electricity consumption were calculated based on the Greek electricity grid mix. The combustion of carbon containing fuels internally results in the formation of CO_2 (very small amounts of N_2O and CH_4 are also formed). The contribution of these emissions has been taken into account in the refinery's GHG emissions. The electricity power consumption in the hydrotreatment unit of HAGO did not alter due to the addition of WCO.

Table 30.2 Inventory data of the HAGO/WCO (95/5 % v/v) hydrotreating unit for the production of the new hybrid diesel

	Unit	Inputs	Outputs
HAGO	m^3/m^3 hydrotreated hybrid fuel	0.986	–
WCO	m^3/m^3 hydrotreated hybrid fuel	0.052	–
H_2	m^3/m^3 hydrotreated hybrid fuel	70.924	–
Electricity	kWh/m^3 hydrotreated hybrid fuel	15	–
Steam	kg/m^3 hydrotreated hybrid fuel	24	–
Fuel gas	$mmBtu/m^3$ hydrotreated hybrid fuel	0.238	–
Hydrotreated hybrid fuel	m^3/m^3 hydrotreated hybrid fuel	–	1

Table 30.3 Inventory data of the LAGO hydrotreating unit for the production of the new hybrid diesel

	Unit	Inputs	Outputs
LAGO	m^3/m^3 hydrotreated LAGO	1.04	–
H_2	m^3/m^3 hydrotreated LAGO	55	–
Electricity	kWh/m^3 hydrotreated LAGO	15	–
Steam	kg/m^3 hydrotreated LAGO	24	–
Fuel gas	$mmBtu/m^3$ hydrotreated LAGO	0.238	–
Hydrotreated LAGO	m^3/m^3 hydrotreated LAGO	–	1

- in the examined production process of the hybrid diesel the H_2 consumption of the HAGO hydrotreating unit is increased by about 6.25 %, as compared with conventional HAGO hydrotreating, due to the addition of 5 % v/v WCO. This increase is expected, due to the more demanding WCO hydroconversion reactions. The additional H_2 required can be supplied without operating cost increase, since in the examined refinery there is a surplus of H_2 from the natural gas steam reforming unit (Bezergianni et al. 2014b).
- the GHG emissions of H_2 production process from natural gas steam reforming were based on literature data (Cetinkaya et al. 2012).
- the GHG emissions of construction and decommissioning of all process steps as well as the maintenance operations are not accounted, due to lack of detailed related GHG emissions data.

The inventory data of the new hybrid diesel's production process are detailed in Tables 30.2 and 30.3.

Results and Discussion

Environmental Evaluation of the Renewable Diesel and the Hybrid Diesel

The inventory data previously presented were the basis for the calculation of the GWP associated with the investigated fuels production processes. The results are

Table 30.4 GWP emissions of the renewable diesel production process

	Renewable diesel	
Process stage	g CO_2 eq/MJ	% Contribution to production process
WCO collection & transportation	~0	~0
Pretreatment		
Energy	0.957	7.873
Hydrotreatment		
Energy	0.088	0.724
H_2	11.11	91.403
Total	12.15	100

Table 30.5 GWP emissions of the renewable diesel production process

	Hybrid diesel	
Process stage	g CO_2 eq/MJ	% Contribution to production process
Refinery		
HAGO	3.640	39.052
LAGO	2.463	26.454
Hydrotreatment HAGO/WCO		
Energy	0.410	4.398
H_2	0.736	7.896
Fuel gas	0.185	1.985
Hydrotreatment LAGO		
Energy	0.410	4.399
H_2	0.572	6.137
Fuel gas	0.175	1.985
FAME	0.720	7.724
Total	9.321	100

given in Tables 30.4 and 30.5 for the renewable diesel and the hybrid diesel, respectively. As it can be observed, the hybrid diesel emits lower GHG emissions than the renewable diesel, indicative of its optimal environmental characteristics. Since this study aims to compare the environmental impacts of the aforementioned fuels among themselves and also relative blend of the renewable diesel with fossil diesel and fossil diesel, further comparisons results are exhibited in the following subsection.

The contribution of the process stages encompassed in each production chain was accounted so as to identify the major parameters affecting the GHG emissions and is depicted in Fig. 30.7. In the case of the renewable diesel, the H_2 demands of the hydrotreating process arose as the main source of GHG emissions (91.41 %). This high contribution to the GWP was attributed to the increased H_2

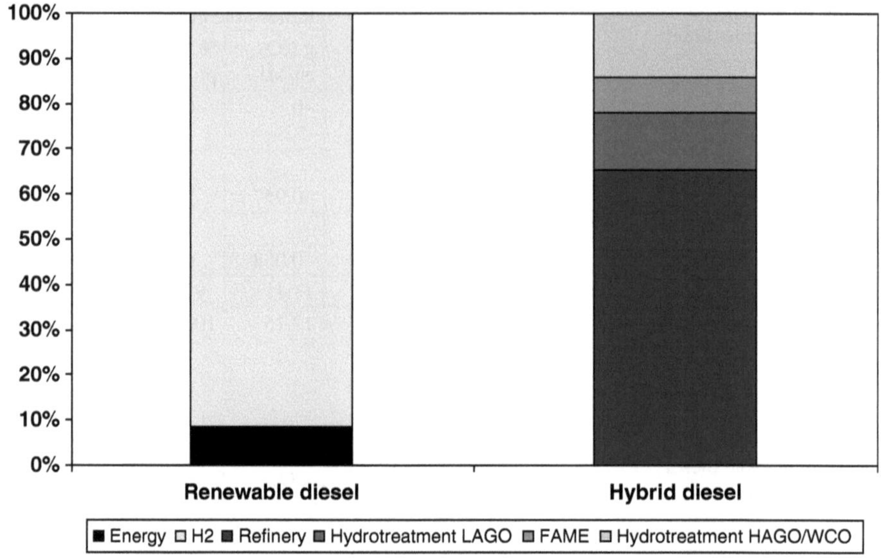

Fig. 30.7 Percentage contributions to the GWP of the renewable diesel and hybrid diesel

consumption required for the conversion of WCO to renewable diesel. Likewise, the contribution of H_2 production has also been identified as the core source of GHG emissions in the production chain of HVOs from vegetable oils (Arvidsson et al. 2011). As expected, the hybrid diesel from co-processing HAGO/WCO mixtures showed a different contribution profile, since the emissions from the refinery were found to be dominating (~65.4 %, encompassing the stage of crude oil extraction and transportation to the refinery), linked to crude oil refining for the production of the petroleum fractions HAGO and LAGO. It should be noted that the H_2 requirements in the co-processing of HAGO/WCO mixtures were also substantial (~7.896 %), while FAME biodiesel was also a noticeable GHG emissions source (~7.724 %).

The GHG emissions of the hybrid diesel examined could be potentially further reduced by energy efficiency improvements steps in the refinery, focusing more on optimizing energy utilization. Improvement actions to minimize the H_2 requirements in the co-processing of HAGO and WCO are also encouraged via incorporating H_2 from renewable energy sources for instance solar energy. Based on literature data the dominant contributor to the GWP of the fossil diesel's production process is the refining process (JRC 2014). Emissions from the cultivation stage represent the highest GHG emissions source in the carbon footprint of a renewable diesel fuel obtained from co-processing soybean vegetable oil with fossil diesel (mainly emissions connected with nitrogen fertilizers) (Garraín et al. 2014). Likewise, cultivation processes were identified as key contributors to the GHG emissions of HVOs production processes (Uusitalo et al. 2014).

Table 30.6 GWP emissions of the examined fuels production processes

Fuel type	gCO_2 eq/MJ
Renewable diesel (via WCO hydrotreating)	12.15
Renewable diesel's blend (7 % v/v) with fossil diesel	14.15
Hybrid diesel (via co-processing WCO with HAGO)	9.321
Fossil diesel (containing 7 % v/v FAME biodiesel)	16.974

Comparison with Other Fuels

This section compares the GHG emissions of the bio-based renewable fuels produced via hydrotreating WCO and co-processing HAGO with WCO among themselves and to relative blends and also to conventional fossil diesel. The comparison includes only the fuels production processes and in the case of fossil diesel and FAME biodiesel literature data were retrieved (JRC 2014).

Table 30.6 contrasts the GHG emissions of the targeted fuels production processes. As it is shown, the GWP of the new hybrid diesel was estimated as 9.321 g CO_2 eq/MJ, presenting a better environmental profile compared to the renewable diesel and to the other fuels. This observation validates that the integration of a bio-based feedstock in the refinery and its co-processing with petroleum fractions, constitutes a favorable environmental production process over stand-alone biomass conversion technologies from GHG intensity viewpoint. Furthermore, the environmental superiority of the renewable diesel is also evident, while its blend (7 % v/v) with fossil diesel showed also a better environmental profile than fossil diesel (containing 7 % v/v FAME biodiesel). As a result, the blend of the renewable diesel with fossil diesel constitutes an alternative to fossil diesel and conventional bio-based counterparts. As regards to the fossil diesel (containing 7 % v/v FAME biodiesel) it entailed the highest GWP.

Furthermore, when comparing the renewable fuels with HVOs originating from rapeseed, palm oil and jatropha (GHG emissions range from 20.97 to 57.57 g CO_2 eq/MJ, Uusitalo et al. 2014) their sustainable environmental profile is further validated. Although, it should be underlined that the comparison of life-cycle studies is frequently ambiguous since a large number of publications present sometimes contradictory results, due to variant methodological approaches and assumptions, inventory data for key input parameters etc. The high variability of the life-cycle results from the various assessments demonstrates the necessity to appoint key issues used in all the studies (Malça and Freire 2011). Nevertheless, the variability in the results presented in this study could be high and subject to various details of all stages included in the investigated production chains.

Overall, the renewable fuels investigated in this study are environmental advantageous than conventional fossil diesel in terms of GHG emissions, identifying the utilization of WCO as a potential favorable bio-based feedstock. The integration of WCO in the refinery for the production of fuels with high bio-content via co-processing technology could be conducive to the global warming mitigation contrasting to stand-alone biomass conversion technologies. These observations could be guiding future actions towards the decarbonization of the transportation sector and guarantee the environmental sustainability of the EU policy at this level.

According to the European legislation the Directive 2009/28/EC establishes sustainability criteria for biofuels, setting minimum GHG saving limits. In detail, the GHG emissions savings should be at least 35 % for any biofuel. This specific minimum limit shall be at least 50 % by the year 2017 and 60 % by 2018 for biofuels and bioliquids. Therefore, the GHG emissions savings of the renewable fuels examined were quantified and compared to those of fossil diesel. GHG emission savings of 89.48, 86.28 and 84.03 % were calculated for the hybrid diesel, the renewable diesel and its blend with fossil diesel, respectively. In all cases the GHG emission savings meet the sustainability criteria of the regulation for the 2017 and 2018 limits. The new hybrid diesel could be assigned as an alternative to fossil diesel and be conducive to significant GHG emissions reductions. These results further validate the environmental performance of renewable fuels derived from residual biomass via the co-processing approach and the catalytic hydrotreatment conversion technology.

Conclusions

LCA is a valuable method applied to evaluate the environmental burdens of processes accounting the inputs and outputs on a life-cycle basis. Renewable diesel via WCO catalytic hydrotreatment has been produced, while the co-processing of petroleum fractions/WCO mixtures has been explored as an alternative approach to integrate residual biomass in the refinery. This new hybrid diesel constitutes a blending component of fossil diesel. The present study aims to provide the environmental characterization of bio-based fuels production systems via hydrotreating WCO and co-processing petroleum fractions with WCO utilizing existing refinery infrastructure. A comparison of these fuels with fossil diesel (containing 7 % v/v FAME biodiesel) and the renewable diesel's blend (7 % v/v) with fossil diesel was also conducted through a WTT analysis. In this respect, the LCA study was performed by utilizing real experimental data and inventory data sources from previous environmental studies in the field as well as GEMIS software. The new hybrid diesel produced via co-processing HAGO/WCO (95/5 % v/v) exhibited the most favorable environmental performance, in terms of GHG emissions. The renewable diesels' production chain GHG emissions are also relatively low, whereas its blend with fossil diesel presents also environmental benefits compared with fossil diesel, enforcing its usage as a blending component. The results also showed that in the case of the renewable diesel the emissions for the hydrotreatment process (H_2 consumption) were the dominant contributors to the GWP of the process. Furthermore, in the case of the hybrid diesel the emissions from the refining processes contributed notably to its GWP, while the emissions of the hydrotreatment unit of HAGO/WCO mixtures were also prominent. Improvements actions targeted primarily to energy efficiency techniques in the refinery and to the incorporation of H_2 originating from RES could probably lead to further GHG emissions from the hybrid diesel production chain. The GHG emission savings calculated for the new hybrid diesel further promotes its optimal environmental characteristics, validating that WCO constitutes a suitable feedstock for bio-based fuels production from GHG emissions perspective. Based on these findings, it could be inferred that the co-processing approach of petroleum fractions with WCO could contribute to the production of sustainable and environmentally efficient fuels. Supplementary to this observation, it may also be underlined that the integration of residual biomass in the refinery via co-processing technology for the production of bio-based fuels compatible with their fossil counterparts is preferable in terms of GHG balance compared to stand-alone biomass conversion technologies. However, future studies should also address thoroughly the issue of WCO availability in Greece in the case the above conversion technologies become commercially viable.

Acknowledgements This work was supported by the EU-Program LIFE+ LIFE08 ENV/GR/000569 and by the project SustainDiesel-09SYN-32-328 which is funded 50 % by the European Commission and the Greek Government.

Nomenclature

FAME	Fatty acid methyl esters
GHG	Greenhouse gas
GWP	Global warming potential
HAGO	Heavy atmospheric gas oil
HVOs	Hydrotreated vegetable oils
LAGO	Light atmospheric gas oil
LCA	Life cycle assessment
WCO	Waste cooking oil
WTT	Well-to-tank
TTW	Tank-to-wheel
WTW	Well-to-wheel

References

Arvidsson, R., Persson, S., Fröling, M., & Svanströmb, M. (2011). Life cycle assessment of hydrotreated vegetable oil from rape, oil palm and Jatropha. *Journal of Cleaner Production, 19*, 129–137.

Bezergianni, S., & Dimitriadis, A. (2013a). Temperature effect on co-hydroprocessing of heavy gas oil-waste cooking oil mixtures for hybrid diesel production. *Fuel, 103*, 579–584.

Bezergianni, S., & Dimitriadis, A. (2013a). Catalytic hydrotreating of waste cooking oil for renewable diesel production. *Proceeding's of the 9th International Colloquium on Fuels Conventional and Future Energy for Automobiles* (pp. 551–557). Germany: Stuttgart. Retrieved Jan 15–17, 2013.

Bezergianni, S., Dimitriadis, A., Kalogianni, A., & Knudsen, K. G. (2011). Towards hydrotreating of waste cooking oil for biodiesel production. Effect of pressure, H_2/Oil ratio and liquid hourly space velocity. *Industrial Engineering Chemistry Research, 50*, 3874–3879.

Bezergianni, S., Dimitriadis, A., Kalogianni, A., & Pilavachi, P. A. (2010a). Hydrotreating of waste cooking oil for biodiesel production. Part I: Effect of temperature on product yields and heteroatom removal. *Bioresource Technology, 101*, 6651–6656.

Bezergianni, S., Dimitriadis, A., Sfetsas, T., & Kalogianni, A. (2010b). Hydrotreating of waste cooking oil for biodiesel production. Part II: Effect of temperature on hydrocarbon composition. *Bioresource Technology, 101*, 7658–60.

Bezergianni, S., Dimitriadis, A., & Chrysikou, L. P. (2014a). Quality and sustainability comparison of one- vs two-step catalytic hydroprocessing of waste cooking oil. *Fuel, 118*, 300–307.

Bezergianni, S., Dimitriadis, A., & Karonis, D. (2014b). Diesel decarbonization via effective catalytic co-hydroprocessing of residual lipids with gas–oil. *Fuel, 136*, 366–373.

Bezergianni, S., Dimitriadis, A., & Meletidis, G. (2014c). Effectiveness of CoMo and NiMo catalysts on co-hydroprocessing of heavy atmospheric gas oil–waste cooking oil mixtures. *Fuel, 125*, 129–136.

Cetinkaya, E., Dincer, I., & Naterer, G. F. (2012). Life cycle assessment of various hydrogen production. *International Journal of Hydrogen Energy, 37*, 2071–2080.

Demirbas, A. (2009). Political, economic and environmental impacts of biofuels: A review. *Applied Energy, Suppl 1*, 108–117.

Garraín, D., Herrera, I., Lechón, Y., & Lago, C. (2014). Well-to-Tank environmental analysis of a renewable diesel fuel from vegetable oil through co-processing in a hydrotreatment unit. *Biomass and Bioenergy, 63*, 239–249.

Gnansounou, E., Dauriat, A., Villegas, J., & Panichelli, L. (2009). Life cycle assessment of biofuels: energy and greenhouse gas balances. *Bioresource Technology, 100*, 4919–4930.

JRC. (2014). Well-to-Tank appendix 2—version 4a. Summary of energy and GHG balance of individual pathways. Well-to-Wheels analysis of future automotive fuels and powertrains in the European context.

Kalnes, T. N., Koers, K. P., Marker, T., & Shonnard, D. R. (2009). A technoeconomic and environmental life cycle comparison of green diesel to biodiesel and syndiesel. *Environmental Progress & Sustainable Energy, 28*, 111–120.

Kochaphum, C., Gheewala, S. H., & Vinitnantharat, S. (2012). Environmental comparison of straight diesel and cracked diesel. *Journal of Cleaner Production, 37*, 142–146.

Malça, J., & Freire, F. (2011). Life-cycle studies of biodiesel in Europe: A review addressing the variability of results and modeling issues. *Renewable and Sustainable Energy Reviews, 15*, 338–351.

Nikander, S. (2008). Greenhouse gas and energy intensity assessment of product chain: case transport biofuel. Master's Thesis. Helsinki University of Technology, Faculty of Engineering and Architecture, Department of Civil and Environmental Engineering (pp. 1–112).

Pleanjai, S., Gheewala, S. H., & Garivait, S. (2009). Greenhouse gas emissions from production and use of used cooking oil methyl ester as transport fuel in Thailand. *Journal of Cleaner Production, 17*, 873–876.

Uusitalo, V., Väisänen, S., Havukainen, J., Havukainen, M., Soukka, R., & Luoranen, M. (2014). Carbon footprint of renewable diesel from palm oil, jatropha oil and rapeseed oil. *Renewable Energy, 69*, 103–113.

Part VIII
Energy Technologies and Their Effect on Global Warming

Hydrogen Energy and Technologies

Chapter 31
Spark-Ignition Engine Fueled with Methane-Hydrogen Blends

G.M. Kosmadakis, F. Moreno, J. Arroyo, M. Muñoz, and C.D. Rakopoulos

Introduction

Apart from engine-related techniques to meet stringent imposed emissions regulations and higher efficiency over a large range of operating conditions and loads, engine researchers have focused their interest on the domain of fuel-related techniques (Rakopoulos and Giakoumis 2009). It is true that considerable attention has been paid in the development of alternative fuel sources, with emphasis on liquid bio-fuels that possess the added advantage of being renewable, such as vegetable oil, bio-diesel, ethanol, butanol, and diethyl ether (Agathou and Kyritsis 2014; Rakopoulos 2012, 2013; Wang et al. 2010). Moreover, many internal combustion engines, usually converted from commercial compression- or spark-ignition engines, have been fueled with various alternative gaseous fuels, such as the most common methane (or natural gas) and liquefied petroleum gas to the most rarely syngas and biogas, for use in power generation and transportation, showing inherently clean nature of combustion with reduced emitted pollutants (Papagiannakis and Zannis 2013; Rakopoulos and Kyritsis 2006; Rakopoulos and Michos 2009). Finally, fuel blends show some interesting features, especially when hydrogen is added to a conventional fuel for combustion enhancement, due to its very high combustion speed (Kosmadakis et al. 2012).

G.M. Kosmadakis (✉) • C.D. Rakopoulos
Internal Combustion Engines Laboratory, Thermal Engineering Department, School of Mechanical Engineering, National Technical University of Athens, 9 Heroon Polytechniou St., Zografou Campus, Athens 15780, Greece
e-mail: gkosmad@central.ntua.gr; cdrakops@central.ntua.gr

F. Moreno • J. Arroyo • M. Muñoz
Laboratory of Engines, Department of Mechanical Engineering, University of Zaragoza, C/María de Luna s/n, Zaragoza 50018, Spain
e-mail: fmoreno@unizar.es; jarroyo@unizar.es; mmunoz@unizar.es

P. Grammelis (ed.), *Energy, Transportation and Global Warming*, Green Energy and Technology, DOI 10.1007/978-3-319-30127-3_31

One promising fuel blend is the methane/hydrogen one. The addition of even a small hydrogen quantity in methane increases the combustion speed, improving engine efficiency, while the fuel flammability limit is extended, allowing the use of leaner mixtures. Moreover, CO_2 and HC emissions are decreased, since the initial amount of carbon in the fuel blend is decreased (Ma et al. 2008). On the other hand, nitrogen oxides (NOx) emissions are increased, because of the higher combustion velocity and burned gas temperature, while the event of knock is likely to occur for high hydrogen percentage (Moreno et al. 2012).

Various experimental facilities have been developed for the investigation of the combustion processes of methane/hydrogen blends, and the performance of SI engines under different operating conditions and strategies (Demuynck et al. 2009; Moreno et al. 2012), as well as for the in-cylinder processes and their exhaust emissions (Akansu et al. 2007). Apart from these experimental studies, there exist many simulation works focusing mainly on the engine performance parameters (Aliramezani, et al. 2013) and to some extent to the nitric oxide (NO) emissions, using the thermal NO production path (Dimopoulos et al. 2007), widely known as the "three-equation Zeldovich mechanism".

The present study concerns the extension of the combustion model of an in-house CFD code developed by the NTUA team. This model has been initially developed for the simulation of the power cycle of hydrogen-fueled SI engines, and validated over many different conditions, including variable ignition timing, compression ratio, large range of equivalence ratios, and EGR (Kosmadakis et al. 2012; Kosmadakis and Rakopoulos 2014a, b; Rakopoulos et al. 2010a, 2011). The extension of this combustion model and of the CFD code mainly concerns the introduction of the carbon chemistry. Three more species have been added (carbon dioxide, carbon monoxide and methane), together with an appropriate laminar flame speed correlation for methane fuel, making the code capable of simulating such combustion processes.

In order to validate the CFD code, appropriate measured data are used from a two-cylinder engine. It has been decided here to focus on the variable hydrogen content, keeping all other parameters constant, whereas it is planned to conduct further investigation at other conditions as well in the future. For the validation of this model, the calculated results regarding engine performance and nitric oxide emissions are compared with the corresponding measured ones. From the analysis it is concluded that the presented combustion model manages to predict adequately the engine performance within the range of examined hydrogen content. As far as the nitric oxide exhaust emissions is concerned, the model manages to capture qualitatively correct the trend as hydrogen is increased, while in absolute values a small difference exists from the measured ones. Then, a detailed investigation of flame propagation and in-cylinder combustion is implemented, revealing the capabilities of the CFD code for detailed combustion analysis and engine research.

Experimental Facilities

The experimental facilities are briefly presented next, while further details can be found in (Moreno et al. 2012).

Engine Modifications

The experimental tests were carried out on a Lombardini LGW 523 MPI spark ignition engine, with two cylinders and total swept volume of 505 cm^3. The engine was properly modified to work with gaseous fuels containing hydrogen. These modifications are briefly mentioned next. First, an additional gaseous fuel supply circuit was mounted in parallel with the original gasoline circuit. Gas injectors are designed and manufactured by Quantum, and prevent from premature failure in dry gas applications. To favour the homogeneity of the air-fuel mixture inside the cylinder, a new gas injection port was designed and installed after the inlet manifold, where the gas injectors are fitted. Both gas injectors are fed by a gas ramp connected to the gaseous fuel supply circuit of the test rig, which is explained next. The ignition system was also modified, adding two BERU coils to ignite two NGK spark plugs with low enough thermal grade to prevent problems like pre-ignition. The original engine management system has been replaced by an AEM 30-1900U Universal Programmable Engine Management System, which controls all working parameters of the engine. The crankshaft angle at which the gaseous fuels are injected has a big influence on performance, and therefore the engine was converted to sequential injection. To this end, the installation of a cam sensor was necessary, which allowed the synchronization between the strokes of the cylinders. Moreover, the original narrow band oxygen sensor was replaced by a wide band universal exhaust gas oxygen sensor (BOSCH LSU 4.2) with better control of the fuel-air equivalence ratio, especially at lean equivalence ratios. Finally, in order to perform an analysis of the combustion behavior, a piezoelectric pressure sensor (Kistler 6053CC60) was mounted in the combustion chamber, in order to acquire and analyze the pressure inside the cylinder.

Experimental Procedures

The engine was coupled to a Tecner E315 dynamometer, which allows testing engines at constant speed or load. Data acquisition system of the test bench recorded the load, speed, temperatures, pressures and fuel and air consumptions. Pollutant emissions were measured using a Signal Instruments exhaust gas analysis installation. The measurement principles are based on non-dispersive infrared for CO and CO_2, flame detector ionization for HC, chemiluminescence for NOx and paramagnetism for O_2. The pressure inside the cylinder was measured with the piezoelectric sensor mentioned above connected to a signal conditioning and data acquisition system from National Instruments, controlled by a specific proprietary software programmed in LabVIEW. The volumetric composition of the different gaseous fuel blends tested is presented in Table 31.1.

The engine was tested at full load and at a wide range of speeds (from 2000 to 4500 rpm), but the results selected in this study to validate the model were those corresponding to 4500 rpm. The use of hydrogen blends allows the extension of the lean operation limits of the methane (Moreno et al. 2012), so an equivalence ratio of 0.8 was selected for all cases. The ignition timing was fixed at 33° CA BTDC, in order to focus on the combustion differences due to the different blend composition. The injection timing was set to 15° CA BTDC during the intake stroke for all the fuel blends. According to engine valve timing, the injection at this moment ensures that all fuel is injected, while the inlet valve is open for all cases.

All relevant measurements have been conducted at wide-open throttle (WOT), in order to reduce the pumping losses. The main specifications of this engine are given in Table 31.2.

Table 31.1 Composition of the gaseous fuel blends tested

	H_2 composition (% by vol.)	CH_4 composition (% by vol.)
Methane	0	100
G1	10	90
G2	30	70
G3	50	50

Table 31.2 Lombardini engine main specifications

Engine model and type		Lombardini engine, two-cylinder, naturally aspirated, four stroke, water-cooled
Bore		72 mm
Stroke		62 mm
Swept volume (per cylinder)		252.5 cm^3
Connecting rod length		107 mm
Compression ratio		10.7
Valves (per cylinder)		2
Engine speed		4500 rpm
Ignition timing		33° CA BTDC
Equivalence ratio		~0.8
Number of piston rings		3
Valve timing events	Inlet valve opening	5° CA BTDC
	Inlet valve closure	59° CA ABDC
	Exhaust valve opening	57° CA BBDC
	Exhaust valve closure	7° CA ATDC

CFD Code

The firing version of the CFD code developed by the NTUA team can simulate three-dimensional curvilinear domains, using the finite volume method in a collocated grid. It has been validated in previous published works (Kosmadakis et al. 2012; Kosmadakis and Rakopoulos 2014a, b; Rakopoulos et al. 2010a, 2011), where it has been revealed that it can adequately simulate the power cycle of hydrogen-fueled spark-ignition engines. In the next subsections the extended combustion model is briefly presented, as applied in the present study and its main differences with the addition of methane fuel are discussed. A detailed presentation can be found in previous publications (Kosmadakis et al. 2012, 2013; Kosmadakis and Rakopoulos 2014a, b; Rakopoulos et al. 2010a, b, 2011).

Combustion Model

The combustion model is briefly described in this section, which has been already developed and applied in previous studies for hydrogen combustion. The main difference here is that methane fuel and carbon chemistry are added in the transport equations. Only its basic features are given, since its detailed presentation is provided in previous publications (Kosmadakis et al. 2012; Rakopoulos et al. 2010a).

For the calculation of the species reaction rates, the characteristic conversion time-scale method has been followed (Abraham et al. 1985). A characteristic time-scale (τ_c) is calculated, as the sum of the laminar conversion time (τ_l) and the turbulent mixing time (τ_t), which determines to what extent the combustible gas content of each computational cell has reached its chemical equilibrium during a specific duration (equal to the computational time step). This method is followed for the calculation of the reaction rates of all chemical species considered here, namely H_2, O_2, N_2, H_2O, H, O, N, OH, CO, CO_2, and CH_4. The reaction rates for the nitric oxide (NO) derived from three different paths (thermal NO, NNH and N_2O paths) have already been discussed in (Kosmadakis and Rakopoulos 2014a, b) and will not be repeated here. Further details concerning the calculation of the reaction rates with the use of the characteristic conversion time-scale method can be found in (Rakopoulos et al. 2010a).

The correlation for the hydrogen laminar flame speed is the one proposed by (Gerke et al. 2010), which includes stretch effects. A similar correlation for methane is used, which is proposed by (Ouimette and Seers 2009), which is derived under a wide range of pressure and temperature similar to the gas properties during spark ignition. Other correlations have been examined as well (applicable over a smaller range though), with similar results. The fuel blend laminar flame speed is calculated using a Le Chatelier rule (Rakopoulos and Michos 2009), which is a rather simple calculation method, but ensures that the most reliable correlations are used for the two fuels (hydrogen and methane), taking into consideration stretch effects. The laminar flame speed is used to track the computational cells, in which the flame has propagated during the ignition phase through the calculation of the flame radius, assuming that the flame propagation velocity is approximately equal to the laminar flame speed during this phase. When the flame reaches a critical radius, turbulent effects become dominant and then a turbulent burning velocity (u_t) is used to track the flame. The expression used for this velocity is the one proposed by Zimont/Lipatnikov and Chomiak (Lipatnikov and Chomiak 1997; Zimont 2000), in which one calibration constant exists (constant "A"), which is appropriately tuned for each fuel blend (Rakopoulos et al. 2010a). This correlation is given in Eq. (31.1).

$$u_t = A\left(u'\right)^{3/4} u_l^{1/2} D_{T,u}^{-1/4} L_t^{1/4} + u_l \tag{31.1}$$

where A is the calibration constant that needs to be tuned, u' the rms turbulent velocity given by $u' = \sqrt{2k/3}$, u_l the stretched laminar flame speed, $D_{T,u}$ the thermal diffusivity of the unburned mixture, and L_t the turbulent integral length scale given by $L_t = 0.37\left(u'\right)^3/\varepsilon$.

Computational Details

Some initial and boundary conditions are provided from the experimental data, such as the pressure at IVC, and the inlet air (humidity in the inlet air is neglected) and hydrogen/methane flow rates. For the rest of the required conditions, reliable estimations are used, such as the gas temperature at IVC (from 380 up to 405 K as the hydrogen content increases), wall temperature at IVC (from 440 up to 450 K as the hydrogen content increases), residual gas mass fraction (around 6–7 %), and turbulent kinetic energy at IVC (constant and equal to 55 m^2/s^2 for all cases).

The constructed computational mesh has (40 × 40 × 30) grid lines along each axis (see Fig. 31.1) and has been tested for grid independence. The addition and removal of mesh layers according to the piston movement is also used, since during compression the height of the cells decreases at the dead volume. The engine is equipped with a small bowl-in-piston, and its geometry has been extrapolated with the highest possible detail from the engine design provided by the manufacturer.

The computational time step used is equal to 1° CA, which is further decreased from the beginning of the combustion period and afterwards becoming equal to 0.4° CA. Such time step is sufficient small, since the engine speed is high and equal to 4500 rpm. Finally, the only calibration constant used in the in-house CFD code, which exists in the expression of the turbulent flame speed (no calibration factors exist in the heat transfer and crevice models), has a different value for each fuel blend, which is however in the range of 0.7 (similar with hydrogen combustion).

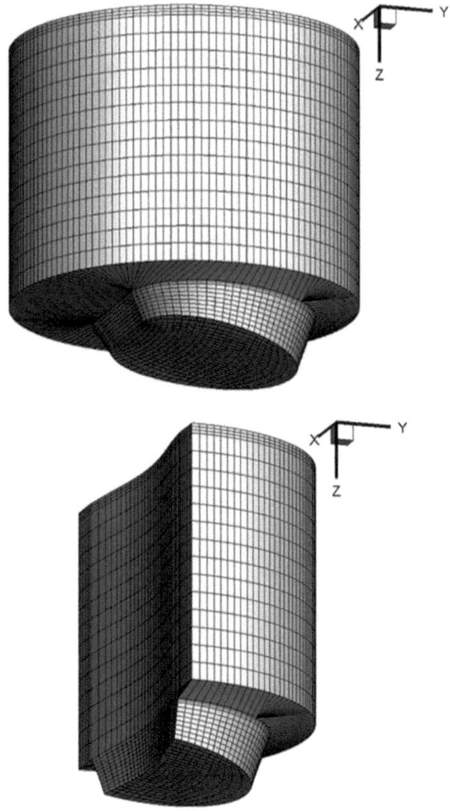

Fig. 31.1 Computational grid at IVC with 40 × 40 × 30 grid lines along the x, y, z axis. The bowl-in-piston is depicted at the bottom

Results and Discussion

Test Cases Considered

The main engine conditions and settings of the test cases examined in the current study have been shown previously in Table 31.2. Focus is given on the effect of hydrogen content in the fuel (four test cases considered).

CFD Code Validation

The validation of the CFD code is presented here, which includes the comparison of the predicted performance and NO emissions with the available measured data for the methane/hydrogen-fueled SI engine.

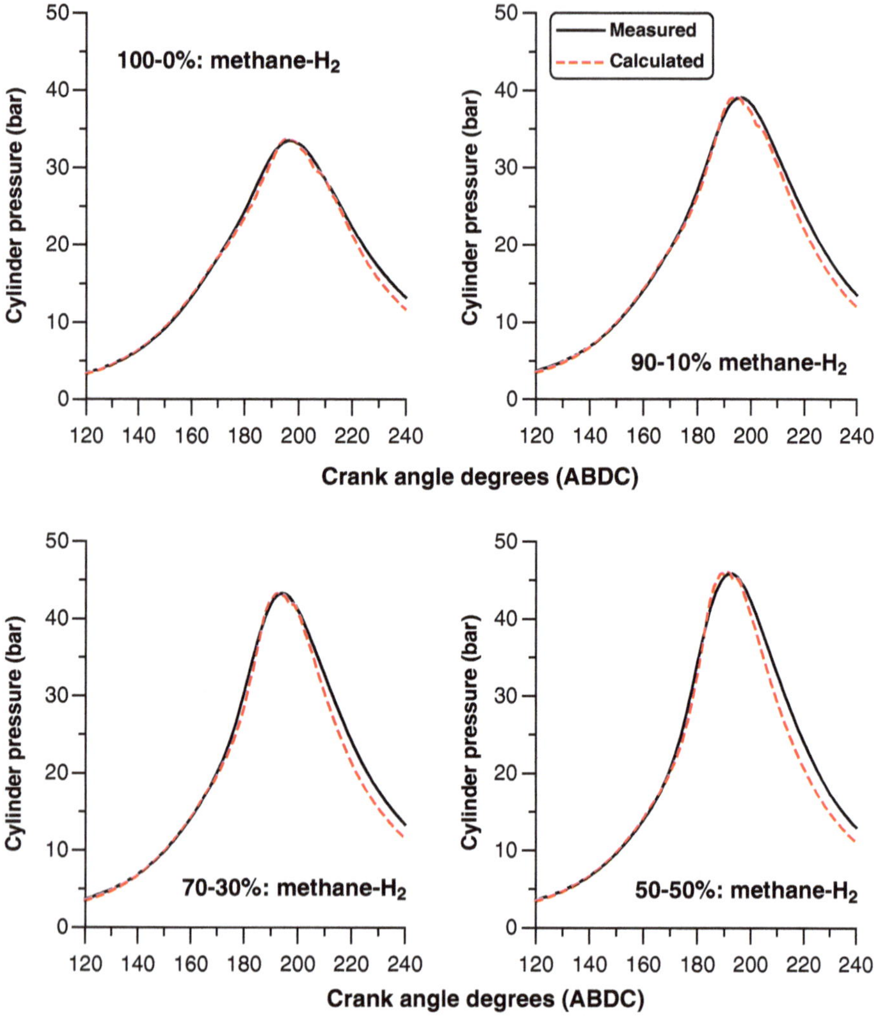

Fig. 31.2 Comparison of calculated pressure traces with the corresponding measured ones for variable hydrogen content

Engine Performance

The CFD code has been applied for the simulation of the four test cases. In Fig. 31.2 is shown the comparison of the predicted with the measured cylinder pressure for variable hydrogen content.

The pressure traces are predicted with high accuracy, since for every case considered the peak pressure and its timing are very well captured, with only some minor discrepancies. These can be partly attributed to the correlation of the laminar flame speeds used in the CFD code, being near to their application range,

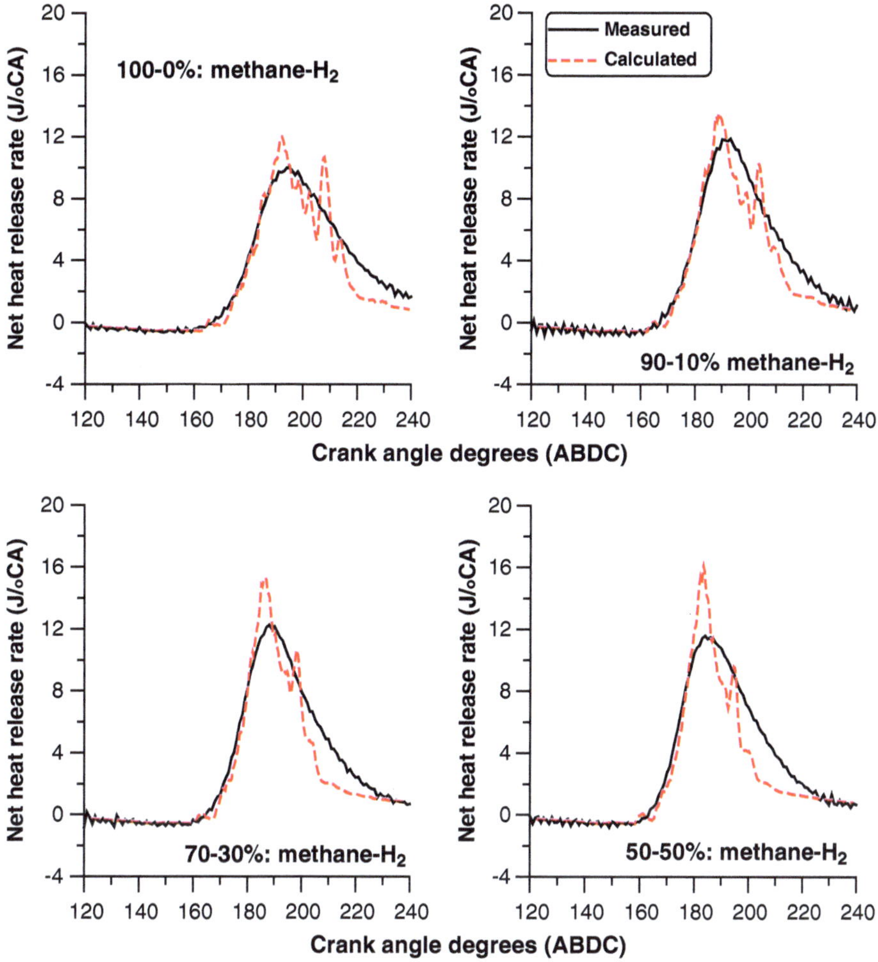

Fig. 31.3 Comparison of calculated net heat release rates with the corresponding measured ones for variable hydrogen content

and to the higher heat loss calculated during the end of combustion. Nevertheless, the cylinder pressure numerical results can be considered reliable, providing a good proof of the reliability of the CFD code developed and of the presented results in the next subsections.

This conclusion is also justified with the good comparison of the net heat release rates, which is shown in Fig. 31.3. The maximum rate is well captured from the CFD code, together with the combustion timing and profile.

Again, the process during the expansion stroke shows some differences, which are under further investigation. Some possible causes can be some differences with the exact bowl geometry, a non-homogeneous gas mixture at IVC, and the uncertainty on the values of the initial gas temperature at IVC and wall temperature.

The local increase of the heat release rates during the expansion stroke at around 200° CA ABDC is due to the rapid combustion of the end gas near the cylinder liner. This is advanced as the hydrogen content is increased, due to the more rapid combustion. According to the measured values, a different trend exists, since this end gas slowly burns during a larger period at the expansion stroke.

Similar comparisons have been made concerning the indicated mean effective pressure (IMEP) and indicated efficiency. These parameters provide a more general insight on the accuracy of the numerical results using the CFD code. It has been shown that both IMEP and indicated efficiency are adequately predicted, since for all cases considered a good match exists with few differences. These exist due to the discrepancies of cylinder pressure during the exhaust stroke.

The main conclusion of this validation process is that the CFD code can predict adequately the performance of such engine with methane/hydrogen fuel blends and its application can be extended for the investigation of various conditions and engine designs.

NO Emissions

After the validation of the performance, the study focuses on the NO exhaust emissions, which are investigated for the test cases of Table 31.1. The calculated NO values are obtained using the CFD code and are compared with the measured values (see Fig. 31.4).

It can be observed that the trend of the NO emissions is adequately captured, as the hydrogen content increases. Concerning the absolute emission values there is some discrepancy, which is almost equal for all cases (around 800 ppm). For high hydrogen content, the relevant variation is maintained lower than around 30–40 %, which can be considered acceptable, since no calibration factor is included in the

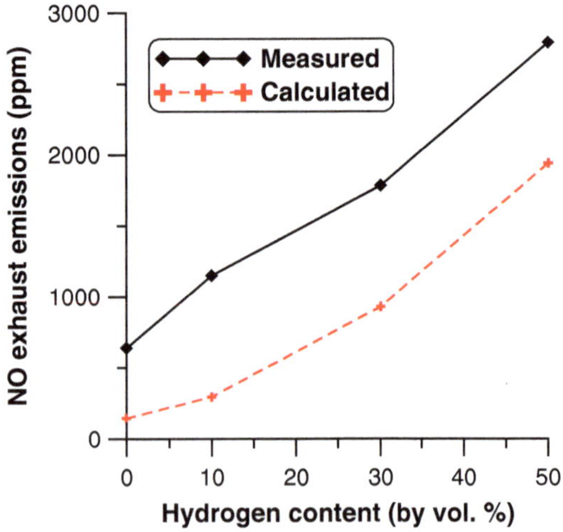

Fig. 31.4 Comparison of calculated NO emissions with the corresponding measured ones for variable hydrogen content

emissions model. Nevertheless, such prediction level is adequate, having in mind all uncertainty factors in the calculation of NO emissions (Kosmadakis et al. 2012; Kosmadakis and Rakopoulos 2014a).

Combustion Analysis

The validation of the extended combustion model revealed that the power cycle of a SI engine with variable hydrogen content can be simulated with detail, while both performance and NO emissions are adequately calculated. Therefore, the CFD code can be then applied for further combustion analysis, by providing some insight on the in-cylinder processes.

As the hydrogen content increases, the combustion speed is higher, increasing the gas temperatures during combustion. This is the main reason that NO emissions are higher, which is one of the main concerns for the use of hydrogen in blends (Moreno et al. 2012). Such indication is provided in Fig. 31.5, where the calculated mean gas temperature is observed. In the same figure are also shown the mean NO reaction rates, which depend a lot on the local gas temperature and the local species concentration.

It is observed that the maximum temperature (around 2400 K) is reached for the 50–50 % blend, which is observed some CA degrees before the 100–0 % blend. Also, as the hydrogen content increases, the NO reaction rates rapidly increase, since there is an exponential dependence from the temperature (Kosmadakis and Rakopoulos 2014a). Moreover, even some small addition of hydrogen in the fuel

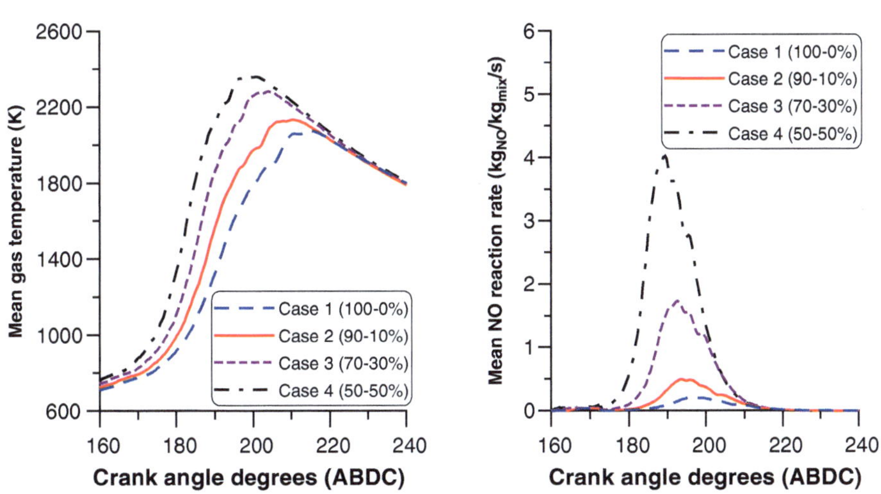

Fig. 31.5 Mean gas temperature and mean NO reaction rates during combustion and early expansion for variable hydrogen content

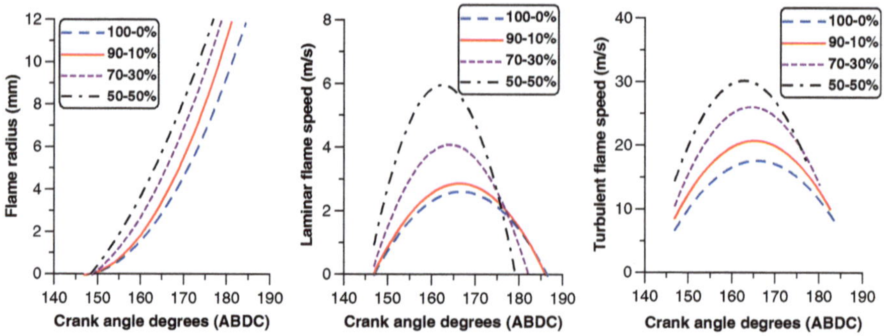

Fig. 31.6 Flame radius, laminar flame speed and turbulent flame speed for variable hydrogen content

blend, as is the case of 90–10 %, increases the maximum mean gas temperature by more than 100 K and almost doubles the mean NO reaction rate.

Some more detailed aspects of flame propagation are presented next. The differences observed are mainly due to the variable hydrogen content, since all other important parameters and conditions are exactly the same for all cases. The initial temperature and pressure at IVC show some small differences, which are however negligible and do not introduce any important changes. In Fig. 31.6 is shown the flame radius, calculated from the computational cells, which are located at an axial direction crossing the spark-plug, and until the flame reaches the piston. In the same figure is also shown the mean laminar flame speed during the early stages of combustion and the turbulent flame speed. It should be stressed that these speeds are mean values averaged over the cells that are inside the flame front and therefore should be only treated as indicative ones.

The spark-timing is constant and equal to 147° CA ABDC. Therefore, for all cases the combustion starts at the same time and it is possible to identify the effect of hydrogen addition on the flame speeds and flame propagation inside the cylinder. It is observed that the flame radius increases more quickly for higher hydrogen content, which is directly related with the mixture laminar flame speed (the hydrogen laminar flame speed is many times higher than the methane one) and turbulent flame speed. The latter mainly depends on the local turbulent properties and the laminar flame speed. These properties are more or less the same for all cases, since a constant value of turbulence has been considered for all cases (engine speed is 4500 rpm). Therefore, the turbulent flame speed depends almost exclusively on the laminar flame speed. For the 50–50 % blend, its maximum value is around twice the maximum value for the pure methane case. In other words, flame propagation is around two times higher, in case hydrogen is added in methane by 50 % by vol. This is verified by the combustion duration angles (0–5 and 5–95 %), but it can be better visualized by the flame images, shown in Fig. 31.7. These images depict how the flame propagates inside the cylinder after the ignition timing,

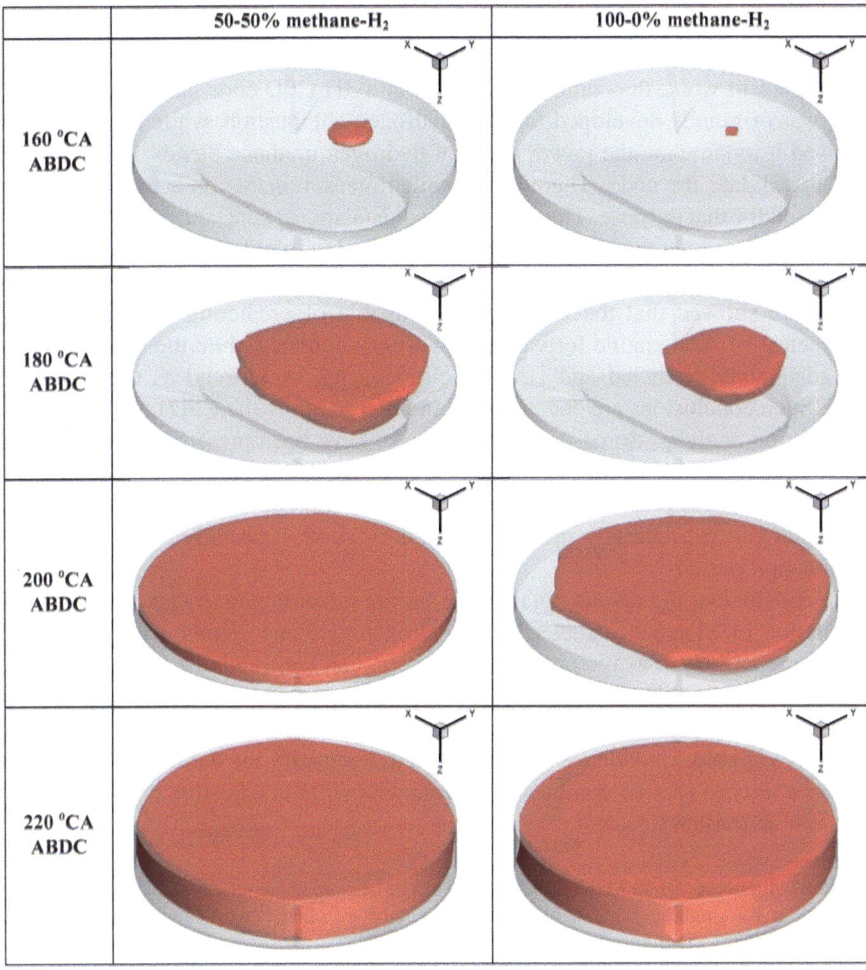

Fig. 31.7 In-cylinder flame images for 50–50 % hydrogen/methane blend and pure methane

noticing the differences for the various hydrogen contents. The spark-plug is offset from the cylinder axis, as one can guess from the location of the ignition kernel.

Owing to the high burning velocity of hydrogen, the flame has significantly expanded at TDC for the 50–50 % blend. Moreover, the flame has covered the whole combustion chamber at around 200° CA ABDC for the 50–50 % blend, and at 220° CA ABDC for the pure methane case. The main differences can be identified during early combustion, where the flame propagates with its laminar flame speed, which is much lower for the pure methane case, as mentioned previously.

Summary and Conclusions

In the present work, the combustion model of a 3D-CFD code has been extended. This was originally developed for pure hydrogen combustion, while this extension allowed it to simulate the power cycle of hydrogen/methane blends. The first step was to validate the code. Therefore, detailed measurements on a SI engine have been used for that purpose. The experimental data are for constant equivalence ratio and load, while only the hydrogen content is varied within the range 0–50 % by vol. The validation has been conducted for both performance and NO emissions. This procedure showed that the CFD code is capable of predicting the performance parameters of such engine for variable hydrogen content, while the NO emissions are adequately predicted and closely following the same trend as the measured ones. Such conclusions are encouraging, in order to apply the CFD code for further combustion analysis. Some first relevant tasks have been presented in the second part of this work, where focus has been given on the flame propagation for variable hydrogen content. It has been shown that the addition of even a small amount of hydrogen can substantially increase the flame speed, especially during the early combustion period.

In conclusion, the results obtained so far are encouraging, by showing that the simulation of SI engines fueled with hydrogen/methane blends using a detailed numerical tool, such as the one presented here, can yield reliable results, with the use of just one calibration constant per blend. However, a more extended study should be conducted in the future, covering a wider range of operating conditions (such as variable equivalence ratio and speed), to justify the validity of the model and its ability to capture combustion phenomena without adjusting the value of the calibration constant.

Acknowledgement Dr. G.M. Kosmadakis (NTUA) wishes to thank the Greek State Scholarships Foundation for granting him an "IKY Fellowship of excellence for postgraduate studies in Greece—Siemens Program, years 2013–2015."

References

Abraham, J., Bracco, F. V., & Reitz, R. D. (1985). Comparison of computed and measured premixed charge engine combustion. *Combustion and Flame, 60*(3), 309–322.

Agathou M. S., & Kyritsis, D. C. (2014). Experimental study of steady, quasi cone-jet electrostatic sprays of bio-butanol for engine applications. *Journal of Energy Engineering 140*(3). doi: 10.1061/(ASCE)EY.1943-7897.0000141, A4014008.

Akansu, S. O., Kahraman, N., & Ceper, B. (2007). Experimental study on a spark ignition engine fuelled by methane-hydrogen mixtures. *International Journal of Hydrogen Energy, 32*(17), 4279–4284.

Aliramezani, M., Chitsaz, I., & Mozafari, A. A. (2013). Thermodynamic modeling of partially stratified charge engine characteristics for hydrogen-methane blends at ultra-lean conditions. *International Journal of Hydrogen Energy, 38*(25), 10640–10647.

Demuynck, J., Raes, N., Zuliani, M., De Paepe, M., Sierens, R., & Verhelst, S. (2009). Local heat flux measurements in a hydrogen and methane spark ignition engine with a thermopile sensor. *International Journal of Hydrogen Energy, 34*(24), 9857–9868.

Dimopoulos, P., Rechsteiner, C., Soltic, P., Laemmle, C., & Boulouchos, K. (2007). Increase of passenger car engine efficiency with low engine-out emissions using hydrogen–natural gas mixtures: A thermodynamic analysis. *International Journal of Hydrogen Energy, 32*(14), 3073–3083.

Gerke, U., Steurs, K., Rebecchi, P., & Boulouchos, K. (2010). Derivation of burning velocities of premixed hydrogen/air flames at engine-relevant conditions using a single-cylinder compression machine with optical access. *International Journal of Hydrogen Energy, 35*(6), 2566–2577.

Kosmadakis, G. M., Pariotis, E. G., & Rakopoulos, C. D. (2013). Heat transfer and crevice flow in a hydrogen-fueled spark-ignition engine: Effect on the engine performance and NO exhaust emissions. *International Journal of Hydrogen Energy, 38*(18), 7477–7489.

Kosmadakis, G. M., & Rakopoulos, C. D. (2014a). Computational fluid dynamics investigation of alternative nitric oxide emission mechanisms in a hydrogen-fueled spark-ignition engine. *International Journal of Hydrogen Energy, 39*(22), 11774–11791.

Kosmadakis, G. M., & Rakopoulos, C. D. (2014b). Computational fluid dynamics study of alternative nitric-oxide emission mechanisms in a spark-ignition engine fueled with hydrogen and operating in a wide range of exhaust gas recirculation rates for load control. *Journal of Energy Engineering* C4014008. DOI: 10.1061/(ASCE)EY.1943-7897.0000229.

Kosmadakis, G. M., Rakopoulos, C. D., Demuynck, J., De Paepe, M., & Verhelst, S. (2012). CFD modeling and experimental study of combustion and nitric oxide emissions in hydrogen-fueled spark-ignition engine operating in a very wide range of EGR rates. *International Journal of Hydrogen Energy, 37*(14), 10917–10934.

Lipatnikov, A. N., & Chomiak, J. (1997). A simple model of unsteady turbulent flame propagation. *Transactions SAE Journal of Engines 106:*2441–2452 [SAE Paper no. 972993].

Ma, F., Liu, H., Wang, Y., Li, Y., Wang, J., & Zhao, S. (2008). Combustion and emission characteristics of a port-injection HCNG engine under various ignition timings. *International Journal of Hydrogen Energy, 33*(2), 816–822.

Moreno, F., Muñoz, M., Arroyo, J., Magén, O., Monné, C., & Suelves, I. (2012). Efficiency and emissions in a vehicle spark ignition engine fueled with hydrogen and methane blends. *International Journal of Hydrogen Energy, 37*(15), 11495–11503.

Ouimette, P., & Seers, P. (2009). Numerical comparison of premixed laminar flame velocity of methane and wood syngas. *Fuel, 88*(3), 528–533.

Papagiannakis, R. G., & Zannis, T. C. (2013). Thermodynamic analysis of combustion and pollutants formation in a wood-gas spark-ignited heavy-duty engine. *International Journal of Hydrogen Energy, 38*(28), 12446–12464.

Rakopoulos, D. C. (2012). Heat release analysis of combustion in heavy-duty turbocharged diesel engine operating on blends of diesel fuel with cottonseed or sunflower oils and their bio-diesel. *Fuel, 96*, 524–534.

Rakopoulos, D. C. (2013). Combustion and emissions of cottonseed oil and its bio-diesel in blends with either n- butanol or diethyl ether in HSDI diesel engine. *Fuel, 105*, 603–613.

Rakopoulos, C. D., & Giakoumis, E. G. (2009). *Diesel engine transient operation—principles of operation and simulation analysis*. London: Springer.

Rakopoulos, C. D., Kosmadakis, G. M., Demuynck, J., De Paepe, M., & Verhelst, S. (2011). A combined experimental and numerical study of thermal processes, performance and nitric oxide emissions in a hydrogen-fueled spark-ignition engine. *International Journal of Hydrogen Energy, 36*(8), 5163–5180.

Rakopoulos, C. D., Kosmadakis, G. M., & Pariotis, E. G. (2010a). Evaluation of a combustion model for the simulation of hydrogen spark-ignition engines using a CFD code. *International Journal of Hydrogen Energy, 35*(22), 12545–12560.

Rakopoulos, C. D., Kosmadakis, G. M., & Pariotis, E. G. (2010b). Critical evaluation of current heat transfer models used in CFD in-cylinder engine simulations and establishment of a comprehensive wall-function formulation. *Applied Energy, 87*(5), 1612–1630.

Rakopoulos, C. D., & Kyritsis, D. C. (2006). Hydrogen enrichment effects on the second law analysis of natural and landfill gas combustion in engine cylinders. *International Journal of Hydrogen Energy, 31*(10), 1384–1393.

Rakopoulos, C. D., & Michos, C. N. (2009). Generation of combustion irreversibilities in a spark ignition engine under biogas-hydrogen mixtures fueling. *International Journal of Hydrogen Energy, 34*(10), 4422–4437.

Wang, S., Ji, C., & Zhang, B. (2010). Effect of hydrogen addition on combustion and emissions performance of a spark-ignited ethanol engine at idle and stoichiometric conditions. *International Journal of Hydrogen Energy, 35*(17), 9205–9213.

Zimont, V. L. (2000). Gas premixed combustion at high turbulence. Turbulent flame closure combustion model. *Experimental Thermal and Fluid Science, 21*(1–3), 179–186.

Chapter 32
Hydrogen Fueled Airplanes, Test Case: Aviation in Libya

Satya P. Bindra, Ali Alwafi, Ashour Saasi, Elbahlul Musa Abogrean, Mohsen Masaud A. Maatugh, and Khaled Khalifa

Introduction

World over CO_2 emission from aviation is on the rise. See Fig. 32.1 Hydrogen fueled airplane offers the most promising zero-emission alternative to conventional aircraft since they have a similar flying range and refueling time. While Hydrogen fueled airplanes require massive investment and government subsidies for fuelling stations, it is noted that hydrogen—the most abundant element in the universe—could be extracted from many different sources and had the advantage of being portable and more easily stored than electricity. The example is the southern Japanese city of Fukuoka that recently completed a fuelling station that uses hydrogen made from sewage. Of course, there are technological hurdles that need to be cleared to make this commercially viable.

Low carbon environmental friendly science and technology for sustainable civil aviation in new Libya is being planned in line with UNCSD Rio+20 Focal Point Libya (Bindra 2012; Bindra et al. 2012) objectives (a) To Make the airport environment friendly and achieve higher level of customer satisfaction; (b) To lay special emphasis on the development of infrastructure for remote and inaccessible areas; (c) To provide airport capacity ahead of demand; (d) To encourage greater efficiency in Airport Operations on ground and in the air; (e) To provide multi-modal linkages for flexibility. The focus regarding the internal infrastructure lies on the vehicles moving at, in, on and around the airport as well as an innovative taxiway and runway constellation using: (1) CAV: Clean Air Vehicles; (2) ITC: Innovative Taxiway Constellation; (3) ERS: Efficient Runway system. The purpose

S.P. Bindra (✉)
UNCSD Rio+20 Focal Point, Sbia, Libya
e-mail: s.p.bindra@gmail.com

A. Alwafi • A. Saasi • E.M. Abogrean • M.M.A. Maatugh • K. Khalifa
Civil Aviation and Meteorology Technical College, Sbia, Libya

P. Grammelis (ed.), *Energy, Transportation and Global Warming*, Green Energy and Technology, DOI 10.1007/978-3-319-30127-3_32

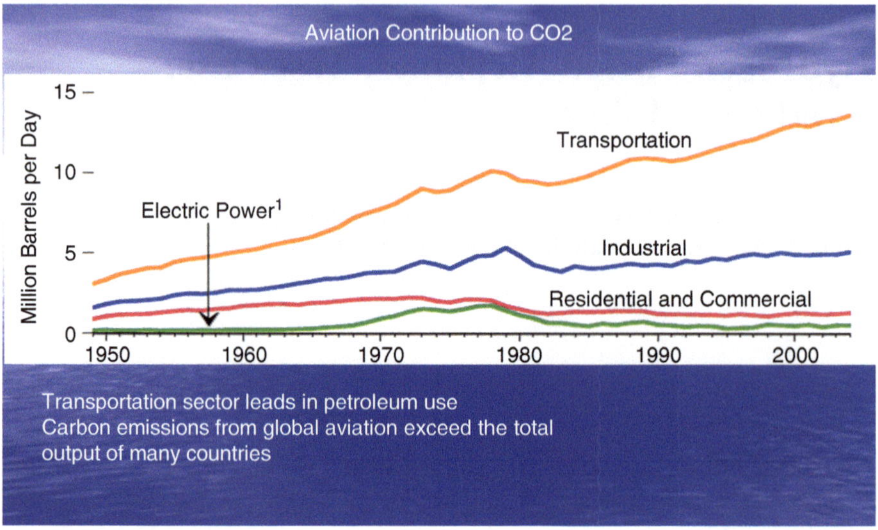

Fig. 32.1 Aviation contribution to CO_2

is to meet and exceed ICAO guidelines issued to its Member States. The idea is to commit the country to help develop solutions for the sustainable future of international civil aviation, in cooperation with industry, sister United Nations agencies and Non-governmental Organizations. It would serve as a catalyst for economic and social development. Therefore, air transport work force needs to be equipped with tools, techniques and technologies that focus on help reduce environmental footprint, in line with the guidance and leadership of Environmental General Authority and ICAO. Its goal is to strike a balance among the three pillars of sustainability—social, economic and environmental so that air transport grows in an environmentally sustainable manner, while continuing to ensure freedom to travel by air.

Policy of Libya is helping establish democratization of air travel by access to sustainable air transport in a guaranteed manner for its future generations. Libyan national UNCSD Rio+20 Focal Point has set clear objectives and provides framework to address the new energy challenges as we transition to a green economy (Bindra 2014; Robertson 2008). This includes use of sustainable alternative fuels as one of the most promising win-win-win solutions for aviation's future. Our Green Airport Concept [4.5] is based on a Clean Air Vehicle Policy. The fleet operating at or near the airport consists of:- Compressed Natural Gas Vehicles (CNG);- Electric Vehicles;- Hybrid-Electric Vehicles;- Green Airport TRAIN powered by hydro electricity.

Furthermore it supports Diesel Vehicles Filters for all transporter and charter bus operators and shuttles. Libya is committed to make available of these fuels through policies that incentivize their development and their deployment. This paper after review of the situation is designed to show that how the civil aviation industry in the country is committed to meet its environmental commitments through its own resources. For this the country policy is devised based on better co-ordination of Libyan national Civil aviation activities at the international level by elimination of duplication in all areas, and to build upon the work of specialized agencies and existing sustainable development frameworks.

An Overview of Civil Aviation and Airports in Libya

Civil aviation in Libya in 1987 was the responsibility of the Secretariat of Communications, which operated all airports, and the Civil Aviation Institute, which trained all personnel. The three international airports in 1985 were located at Tripoli (Al Aziziyah), Benghazi (Benina), and Sabha. Smaller airfields were located at Marsa al Burayqah, Tobruk Ghat, Ghadamis, Al Kufrah, and several other locations. Most civil air personnel went abroad for training. Britain suspended its air traffic control training program for Libyans in 1985, but Pakistan subsequently agreed to train about 78 Libyan air traffic controllers (Bindra 2014).

The national air carrier, Libyan Arab Airlines (LAA), was nationalized in 1973. Possessing only a dozen aircraft in 1977, it grew rapidly, expanding its fleet to 25 aircraft in 1985. The main aircraft in service were manufactured by Boeing, Hawker Siddeley, Caravelle, and Fokker. In 1980, LAA carried 1.17 million passengers. A new carrier called United African Airlines (UAA) was created in 1985 in association with LAA.

Airports in New Libya

Currently air transportation is one of the most pressing issues for the new Libya. The country's infrastructure has suffered from decades of neglect and the transport network is essential to rebuilding the economy. The ports and the airports, in particular, have been neglected and underdeveloped. In a country which imports some 80 % of its food, the freight sector requires urgent attention. Despite some damage to facilities and aircraft, Libya's airports steadily reopened after the revolution. Most recently Sirte airport reopened at the end of October, following extensive repair and maintenance work to address damage sustained during the fighting. Even before the airports were fully-functional, airlines resumed flights to Libya, with the very first being Turkish Airlines, which restarted daily Istanbul to Benghazi flights from 13 September 2011.

Concerns have been raised about safety, not least by the European Union. In accordance with the International Civil Aviation Organisation (ICAO), in March the EU banned Libya's two state-owned carriers, Afriqiyah Airways and Libyan Airlines from flying to Europe, on the grounds of flight safety and airport security. Both carriers were obliged to wet-lease aircraft to maintain services to some European destinations. The EU said when it imposed the ban last April that it would review it this month. The ban is expected to remain in force for some further while. Aviation freight at Libya's 18 main airports faces similar to the ports, in particular limited and obsolete equipment. For smaller-scale cargo, air freight is often more straightforward and cheaper, as well as faster, than sea freight, because the airports do not have the congestion problems of the ports. Urgent cargo, such as medicine, hospital equipment and essential equipment for the oil industry usually comes via air freight. A state monopoly operates at all airports, where only one company, the state-run Shorooq Handling and Ground Services company, can deal with freight deliveries. "The country's airports need to be rebuilt to accommodate big planes," said Mohamed Douas, CEO of Libya Shipping Line "and the services need to be liberalised, because the current state-run services are inadequate, inefficient and totally under-equipped." Recently Civil Aviation Authority of Libya (LCAA) has launched a pioneering program to perform a fundamental country-wide upgrade of the existing Aeronautical Fixed Telecommunication Network (AFTN) and Automatic Identification System (AIS) infrastructure in order to enhance flight safety. LCAA have been effectively operating COMSOFT's highly dependable AIDA-NG and CADAS products since 2004, and have again put their full confidence in the German expert to address Libya's high technical and operational expectations. With the upgrade Libya will join the still exclusive club of Air navigation service provider (ANSPs) processing the aeronautical information data in line with the ICAO Roadmap AIS to AIM. In addition, LCAA will benefit from the fast and highly reliable Aeronautical Message Handling System (AMHS) that already connects many countries in the Arab peninsula and North African Region.

Home to the world's hottest desert, The Sahara, Libya's expansive sandy landscape makes air travel essential in order to facilitate Libya's transportation infrastructure; meanwhile a number of national airports are of growing importance.

COMSOFT's innovative AMHS and AIM solutions will be provided to the central site in capital city Tripoli and countrywide messaging and information management services will also be delivered via remote installations at Metiga, Mistrata, Sahba, Benina.

During the course of the ambitious project LCAA will upgrade their NOTAM, OPMET, Briefing and Flight Plan Management facilities and simultaneously integrate the components in a state-of-the-art AIXM 5.1-based AIM concept (Bindra 2012; Bindra et al. 2012). This includes the central handling of static data and the fully integrated ability to generate electronic AIP and aeronautical charts directly from the central AIXM 5.1 database contents.

Tripoli International Airport is a public airport serving international and domestic passengers from most European and Arab airlines, and handling over three million travelers per year. With the installation of two new terminals they hope to increase this capacity to 20 million passengers annually, therefore, increasing the need for advanced and reliable message handling. Installation of an end around taxiway, which is based on the newest developments regarding safety and efficiency: ITC—Innovative Taxiway Constellation is planned to ensure that:

- Aircraft will be allowed to taxi around the end of the runway instead of waiting to cross them.
- Even if aircraft have to expect a longer taxi, it can be another way of avoiding a lengthy queue at the holding point.
- New taxiway constellation significantly decreases runway crossings thereby increasing safety and efficiency, thereby reducing the probability of an runway incursion.
- Airlines, passengers and the environment are the ultimate benefactors of the new taxiway, because it will safe time, fuel and fumes.

Hydrogen Powered Airplanes

Hydrogen powered aircraft are usually designed with the liquid hydrogen fuel carried inside the fuselage, in order to minimize surface-area and reduce boil-off. The future of low carbon science & technology airplanes are assessed, evaluated & its findings are being implemented & monitored after a fact finding mission. The information was gathered to help assess and create three new concentration curriculums. One of the current green sky initiative related curriculum objectives being developed is the specialty in Aviation Management as the first of its kind in Libya (Bindra et al. 2012). It is based on the meetings with industry & academia leaders representing a vast cross-section of the Libyan aviation industry. This includes the Civil Aviation Authority, Libyan Airlines, Air Libya, ICAO—International Civil Aviation Organization, Airport Company of Libya, local airport managers, and local and regional government agencies. The overall impression is that there is a great need for greener aviation in the higher education system of Libya. This paper examines the needs assessment for aviation fuel curriculum in the higher educational system and also explores the content of this curriculum.

Rationale for Change and Development

Aviation management encompasses a multitude of aviation interest since the industry has delineated itself into various categories such as airline, airport, civil aviation, air traffic control, and piloting (as evident by the number of professional associations). All these represent the planning and managing of a business that brings people together with a specific mission and need to accomplish a specific organizational goal. Aviation helps contribute to growth in many other areas: tourism hinges on airports and the airlines to deliver passengers to resort locations, corporate enterprises depend on civil aviation businesses to maintain and fuel their aircraft and to take care of their flying customers. All these aviation services require a strong regulatory body like the Libyan Civil Aviation Authority (CAA) and a strong airport management system.

The current state of the Libyan Aviation Industry is in turmoil and requires a growing and prospering environment for great strides to modernize and expand. A move by the country to become a North African hub requires participation at the highest forum with regards to aviation matters. There is a need for better trained individuals to work in this dynamic industry. In the Libyan Civil Aviation Authority's Strategic Plan in line with Ministry of planning 2013–2030 3rd generation planning requires, strategic mapping process (Bindra 2012, 2014). This outlines the new strategic objective to attract, develop, and retain human capital. The critical issues outlined in this process include: establishing competitive remuneration structures, favorable working environments, and commitment by management to training development, creation of a management development program, and staff Morale. It is hoped that these issues will assist in accomplishing one of its main priorities that is aviation safety, security, transformation and growth.

UNCSD RIO + 20 Libya Focal Point for Sustainable Development

Rio + 20, like the past Rio Earth Summit & Johannesburg Conferences, did provide an opportunity for the international community to assess achievements and progress since 1992 together (Bindra et al. 2012). It also gave a forum to highlight the unmet objectives and gaps that have hindered implementation of the decisions and recommendations of the two previous summits, with a view to making the improvements needed for the post-2012 global agenda. This agenda acknowledges the legitimate expectations by incorporating economic, social and environmental concerns, in order to better respond to the necessity for growth and progress that can ensure a decent standard of living for citizens in a context of multidimensional international crisis. New Libya's input to the United Nations Conference on Sustainable Development held in Rio de Janeiro in June 20–22, 2012 is part of the both national New Libya rebuilding efforts in line with international community's global effort to promote support and enhance development that is respectful of the environment. It is the outcome of a non-exhaustive analysis of New Libya's socio-economic and environmental achievements as a result of the commitments made at the 1992 Rio Summit, Johannesburg Summit and numerous subsequent events & conferences. It also reflects the concerns that free Libyan society emerging out of the recent crisis and faced with formidable task of rebuilding the nation shares with its neighbour regional and international partners in the Mediterranean & Euro Med region, Maghreb Union, African Union, the League of Arab States and the Group of 77 and BRICS, which it shared once again in 2012.

Libya has initiated a revitalized national environment strategy (NES) and a national plan of action for environment and sustainable development (NPAE-SD) in the context of a participatory approach involving all stakeholders as national actors. The national environmental strategy aims to achieve sustainable economic growth both during and after rebuilding by putting in place public policies to improve health and quality of life, conserve and increase the productivity of natural resources especially its Oil & Gas reserves, reduce economic losses and increase competitiveness and, lastly, protect the environment. It is committed to put into practice through the rebuilding New Libya laws on sustainable development.

A number of initiatives (Bindra et al. 2012; Bindra 2014) have also been created, including the First Libyan Environmental Engineering & Sciences Center (EESC), Libyan Business Council for Sustainable Development (LBCSD), Libyan 350 org, Libyan National Platform for Risk Reduction, Care Libya Brand Foundation (CLBF), Green Sky Initiative, Desert Prosperity Initiative, Sustainable Solid Waste Management Initiative. The national, regional rebuilding schemes, and initiatives provide a policy framework for a wide range of environmental protection and regional rebuilding development after the Arab Spring aimed at sustainable development.

The country is committed to implement many initiatives to foster education and training on sustainable development. Authors as focal point in Libya have set up a network of technical institutions committed to education and training on Eco-Efficiency and sustainable development in the country (Bindra 2014).

Greener Skies Initiative in Libya

The green skies initiative by a team of leading former and serving professors and experts in Libya is based on principles that govern both ecological systems and control economic systems. Airspace design for the Environment considerations are shown below:

It considers that "If you violate ecological principles you might be able to make a fast buck on the short term," "but long-term, you'll have an economic disaster" It is based on UNCSD Rio +20 Future we want that advocates and requires the need to use citizen power backed by intelligent management of the ecosystems and nature-based infrastructure (Bindra et al. 2012). It provides a way forward to think about pragmatic and realistic approaches to the challenges of climate change and environmental sustainability, and the critical role of innovation and greening of industry. Equity, the right to development and the principle of common but differentiated responsibilities are the basis of Libyan approach."

In addition even for external infrastructure there is plan to introduce Inherently Low Emission Vehicle Program (ILEAV). The purpose of the Program is to evaluate airport use of alternative fuel vehicles (AFV) and infrastructure to determine their reliability, performance, and cost-effectiveness in the airport environment (Robertson 2008; Maniaci 2006). The Requirements are:

- Public-use airports in areas that do not meet federal air quality standards designed to protect public health.

Fig. 32.2 H_2 tanks of liquid hydrogen study plane

- Vehicles operating exclusively on six alternative fuels: Compressed Natural Gas (CNG); Liquefied Natural Gas (LNG); Liquefied Petroleum Gas (LPG); electricity, hydrogen, or methanol-85. See Fig. 32.2 below with H2 Tanks of liquid hydrogen study plane.

Being an alternative to traditional jet fuel, hydrogen has a higher energy density per unit *mass* but a lower energy density per unit *volume*, and containing the hydrogen at high pressure requires a heavy container. In aircraft heavy containers are not an option, and therefore regular carbon fibre tanks are often used, which can only sustain a pressure of about 350 bar. This is significantly lower compared to steel hydrogen containers (used in cars and ships) which can sustain 500–700 bar. This limitation decreases the amount of energy that can be spent on the propulsion by about half. Alternatively, as with some rockets, cryogenic liquid hydrogen could be employed. Libya being a country blessed with highest solar radiation intensity supports the recent Si2 as the first sun-powered aircraft able to stay aloft for several days and nights. That is powered by more than 17,000 solar cells built into wings that, at 72 m (236 feet), are almost as long as those of an Airbus A380 superjumbo. This is an innovative design, the light-weight carbon fibre aircraft weighs only 2.3-tonnes, about the same as a family 4 × 4 and less than one percent of the weight of the A380. The propellor craft has four 17.5 horsepower electric motors with rechargeable lithium batteries. It is planned to land 12 times on the trip lasting about 25 days spread over five months.

The longest single leg allows a lone pilot fly non-stop for 5 days across the Pacific Ocean between Nanjing, China and Hawaii, a distance of 8500 km (5270 miles) (Fig. 32.3).

From the Gulf Si2 it did cross the Arabian Sea to India before heading on to Myanmar, China, Hawaii and New York. Landings are also earmarked for the Mid-Western United States and either southern Europe or North Africa, depending on weather conditions.

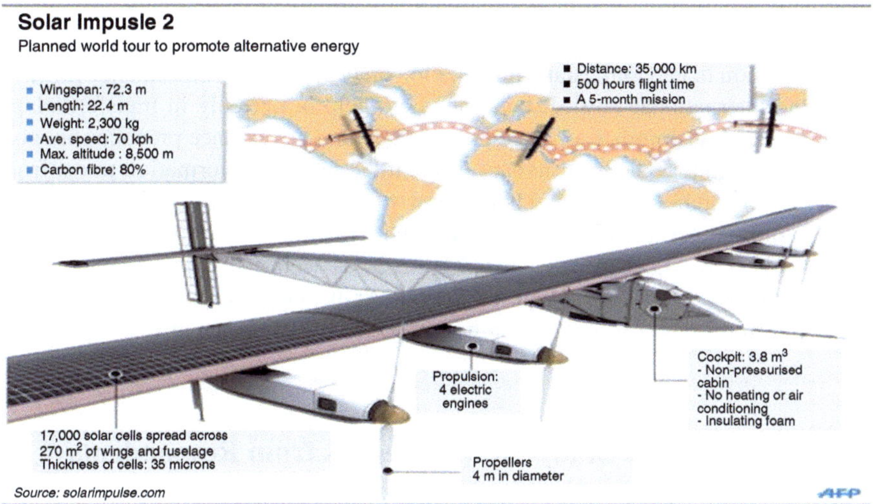

Fig. 32.3 Factfile on the new solar-powered aircraft solar impulse 2

Si2 is the successor to Solar Impulse, a smaller craft that notched up a 26-h flight in 2010, proving its ability to store enough power in the batteries during the day to keep flying at night. It travels at 50–100 km per hour, with the slower speeds at night to prevent the batteries from draining too quickly.

Libyan Strategy

The Libyan strategy includes a project to use information and communications technologies to reform the technical education and scientific research system, which has the potential to become a model for the proper integration of ICTs in cleaner and greener skies using fuels like Hydrogen in the country.

Libya has also helped the country in developing system level improvement, institutional level improvement and individual level improvement based on key principles for GEF-UNDP supported national capacity self assessment (NCSA). These principles are (1) Ensuring national ownership, leadership and policy commitment, (2) Using existing mechanisms and structures whenever possible, (3) Promoting multi-stakeholder participation, (4) Adopting a holistic approach to capacity building, (5) Adopting a long-term approach to capacity building, (6) Considering obligations of the three Rio Conventions; (7) Balance Between Process and Product so that (a) The process should increase stakeholders buy-in and (b) The Product should be of high quality so that government take recommendations/actions as priorities into national budgeting and donor can support it.

The current proposal under the form of civil aviation cleaner and greener airplanes fuels aims at ensuring at relatively short notice, access to and progressive ownership by the Libyans, of the required technical and instrumental capacities. This would also provide opportunities for implementation of practical steps like rapid diffusion of economic data, information on best practices through networking.

At the international level, Libya intends to help in not only in transfer, adapt, adopt, absorb and embrace advanced technology and governance practices but also access to most up to date and unbiased scientific data and information to confront challenges of future. This would help country to play an effective and leading role in promoting international cooperation with organizations of the UN System, among others ICAO, UNEP, UNESCO, FAO, as well as, regional counterparts (Mediterranean, Euro-Med, Maghreb Union, Arab, Sahel and other African) and also other multi- and bilateral funding or assistance providing sources.

Discussion of Case Study Survey Results from Respondents

Based on pilot scale surveys, interviews and meetings with stakeholders it is found that every civil aviation entity has offered advice and input to the potential new airplanes fuels & alternative Fuels See Figs. 32.2 and 32.4 and expressed immediate support of the new program in Libya. All organizations express that they are struggling to find relevant personal with appropriate aviation qualifications to hire for entry level positions. Presently they are spending much time and money to introduce the basic knowledge of aviation to new hires. They both agree there is a

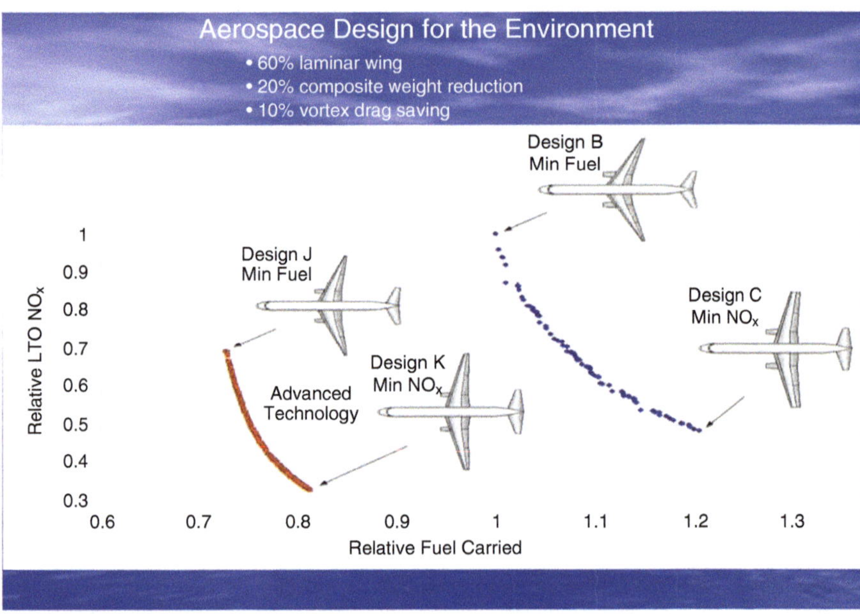

Fig. 32.4 Airspace design for the environment

Fig. 32.5 Jet fuel alternatives

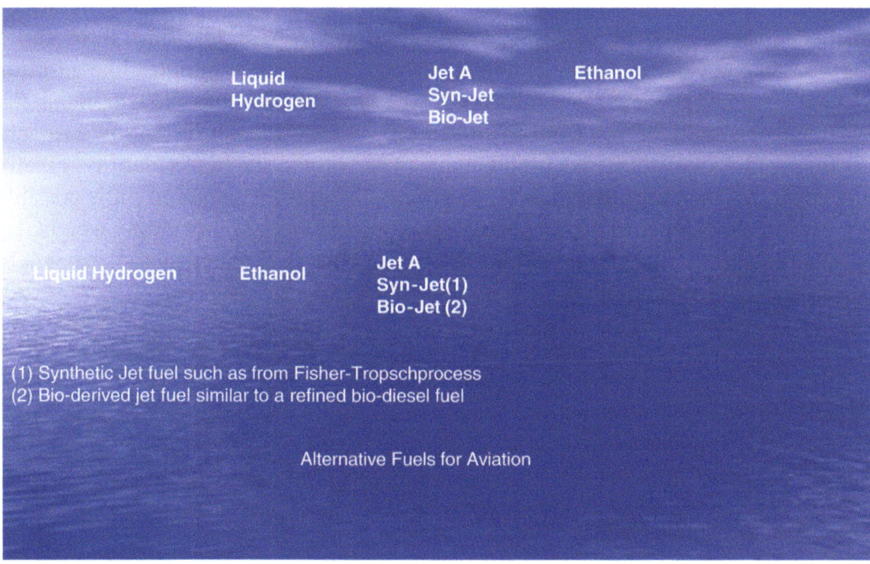

Fig. 32.6 Alternative fuels for aviation

need for greener aviation management curriculum in higher education (Figs. 32.5 and 32.6).

Many respondents suggest that new Libya needs airport and airline managers who are able to communicate with the public, understand their needs and deliver a product to meet these needs within budget. They all feel that professional aviation management skills must be help to coordinate elements of marketing, media, sponsorship, budgets, financial management, public speaking, planning, and complying with regulatory bodies. Most stakeholders say that new Libya needs more of sophisticated IT systems. These systems are required along with skill sets needed for financial management, logistics, reservations, and certification.

Concluding Remarks

Hydrogen-fueled airplanes use hydrogen as a power source. Hydrogen is either burned in jet engine, or internal combustion engine, or used to power a fuel cell to generate electricity to power a propeller. The science & technology of hydrogen fuelled airplanes for aviation industry impacted by low carbon requirement in new Libya is in the process of change. The appraisal of existing situation and needs assessment surveys of industry and government officials in Libya clearly indicate the need for improved hydrogen fuelled airplanes. Survey results indicate that new aviation fuel is feasible alternative similar to Toyota Mirai sedan—powered by hydrogen and emitting nothing but water vapour from its tailpipe. The hydrogen fuelled airplane like Mirai car can travel without refuelling, some three times further than an electric car, and its tank can be filled in a few minutes like gasoline engine vehicles. However, aircrafts need to be configured in such a fashion that the current low energy output from hydrogen propulsion (a result of the low-pressure hydrogen tanks) do not pose a problem due to its volume and high volatility. Finally the paper demonstrates that in order to achieve aviation sustainability it is possible to meet the challenges of $4\times$ larger fuel tanks, higher drag, higher empty weight, H_2 production issues, new infrastructure, affordable cost, high technical risk, passenger acceptance and environmental effects.

References

Bindra, S. P. (2012). Rio+20 Libyan National Report. UNCSD. Brazil: Rio. Retrieved June 2012.

Bindra, S. P. (2014). Civil aviation technical college on low carbon sky related science & technology education in Libya, Istanbul Turkey.

Bindra, S. P., et al. (2012). Change and development in civil aviation. *ICERI 5th International Conference of Education, Research an Innovation 19th–21st of November, 2012*. Madrid, Spain: IATED.

Maniaci, D. C. (2006). Operational performance prediction of a hydrogen-fueled commercial transport. 2006 Symposium.

Robertson, D. (2008). *Boeing tests first hydrogen powered plane*. London: The Times.

Chapter 33
Perspectives of Hydrogen Automotive Applications in Croatia

Ankica Đukić, Ivan Güttler, and Robert Pašičko

Introduction

Global Warming

The warming signal in the global near-surface air and ocean temperature began to emerge from the middle of the 20th century. Up to then, it was masked by the internal variability of the Earths climate system and influences of the events such as changes in the solar energy output and volcanic activities. Globally averaged surface temperature change was estimated to equal 0.85 °C over the period from the 1880 to the 2012. However, the rate of this warming is increasing in recent decades and regionally warming can acquire even higher levels. The changes in the climate system are not limited only to temperature changes. Other changes in the climate system include the melting of the Arctic sea ice, melting of the most of the worlds glaciers, reduction of the land ice over Greenland and Antarctica. Furthermore, the increasing rates of the sea level rise as well as the ocean heat content, various changes in the hydrological system (this can regionally and locally

A. Đukić (✉)
Faculty of Mechanical Engineering and Naval Architecture, University of Zagreb, Ivana Lučića 5, Zagreb 10000, Croatia
e-mail: ankica.djukic@fsb.hr

I. Güttler
Meteorological and Hydrological Service of Croatia, Grič 3, Zagreb 10000, Croatia
e-mail: ivan.guettler@cirus.dhz.hr

R. Pašičko
United Nations Development Programme, Radnička cesta 41, Zagreb 10000, Croatia
e-mail: robert.pasicko@undp.org

P. Grammelis (ed.), *Energy, Transportation and Global Warming*, Green Energy and Technology, DOI 10.1007/978-3-319-30127-3_33

emerge either as a positive or negative change in the precipitation, runoff and soil-moisture signal) and other changes that put pressure to the biosphere and society (IPCC 2014).

Observed and projected temperature and precipitation changes over Croatia were assessed in (Branković et al. 2013) and (MZOIP 2014). Various human activities related to economic activities and population growth are intrinsically related to the energy production and emission of various combustion products. Although there are human related impacts such as changes in the land types and land use practices (this can locally and regionally change the energy, momentum and mass flows), most important impact is related to the changing the chemical composition of the atmosphere, primarily through the emission of the GHG such as CO_2, CH_4, N_2O etc., and secondly through emission of different aerosols. Current GHG emissions are approximately 50 Gt CO_2 eq. per year, and rising.

Therewithal, oceans and biosphere are still net sinks of the additional CO_2, but the amount of CO_2 is accumulating in atmosphere and currently equals approximately 400 ppmv in standard conditions. In nutshell, although different human activities can induce warming and cooling effects, the current net impact on the state of the climate system is substantial warming related to the GHG emissions. IPCC AR5 presented estimates indicating that in order to constrain anthropogenic warming to less than 2 °C (relative to 1861–1880) would require cumulative CO_2 emissions since 1870 to amount below approximately 3000 $GtCO_2$. At the same time, by the year 2011, about 2000 $GtCO_2$ was already emitted, strongly controlling the available time period and the scale of the changes that are needed to be implemented in order to keep the global mean temperature increase below 2 °C (IPCC 2014). Of course, the most of these estimations are also constrained by the various uncertainties. However, independent studies indicate the need for rapid and large-scale upgrade of the existing energy, transport, housing and food production systems.

Transition to Low-Carbon Society and Hydrogen Economy

Concepts of green economy, low emission development, Hydrogen Economy or achieving a 100 % renewable energy systems became increasingly important after 2008 both in the number of published studies, analyses or papers as well as in the attention it gained on an international level (Pašičko 2014). It also became clear that the existing model of economic development needs to change and transition towards different foundations is needed. There are several reasons that can be found and understood or that were mentioned in the literature: (1) Economic crisis that started in 2008 was heavily influenced by the energy crisis (record high prices of oil and gas) and combined and/or alone resulted in other crises such as food crisis, political crisis etc; (2) Increased warnings that society is in danger of transgressing a number of planetary boundaries or ecological limits (Raworth 2012); contrary to economic philosophy that resources come from the market, it is more and more clear they come from nature. Every incident including a deep oil drilling is a reminder that harm to society and environment are increasing as fewer fossil fuels are available; (3) Global warming is becoming more evident while on the other side there is a clear lack of political decision on the global level to tackle this issue; (4) Avoid the trap of 'growing first, cleaning up later' (Atkisson 2012); (5) Growing pressure from people excluded from economic growth (1.3 billion still do not have access to the electricity, 2.6 billion do not have access to the sanitation and 900 million lack access to the clean drinking water); (6) Given the fact that fossil fuels are still heavily subsidized by governments (Kovacevic 2011), in a much higher share than renewable energy or energy efficiency, the present economic model represents a sort of lock-in against low emission development; (7) Energy security issues—with higher and higher energy import dependences, renewable energy and energy efficiency are becoming more and more important.

The term of Green Economy was first coined in 1989 in a pioneering report for the Government of the United Kingdom by a group of leading environmental economists, entitled Blueprint for a Green Economy (Pearce et al. 1989). In 2008 the term was revived in the context of discussions on the policy response to the multiple global crises. The United Nations Environment Programme (UNEP) introduced Green Economy, which it defines as one that 'results in improved human well-being and social equity, while significantly reducing environmental risks and ecological scarcities' (UNEP 2011).

During the financial crisis it inspired several governments to implement significant 'green stimulus' packages as part of their economy recovery efforts. The concept of Green Growth has its origins at the Fifth Ministerial Conference on Environment and Development (MCED) held in March 2005 in Seoul. 52 Governments and other stakeholders from Asia and Pacific agreed to move beyond the sustainable development rhetoric and pursue a path of Green Growth. As with Green Economy, Green Growth attracted significant attention as a way out of economic problems in the aftermath of the 2008 financial crisis. Green Growth is principally a 'top-down' approach, driven by government initiatives. It can also be criticized for being 'green' in relatively limited ways, but it emerged as a serious and very mainstream alternative to Growth as Usual with both environmental and social dimensions.

To define *Sustainable Development*, an international exercise was performed that catalogued, analyzed, and synthesized: written submissions and expert testimony from 'senior government representatives, scientists and experts, research institutes, industrialists, representatives of non-governmental organizations, and the general public' held at public hearings throughout the world, in 1987, the United Nations released the Brundtland Report, that included what is now one of the most widely recognized definitions: 'Sustainable development is development that meets the needs of the present without compromising the ability of the future generations to meet their own needs' (UNWCED). The concept of Low Carbon Development has its roots in the United Nations Framework Convention on Climate Change (UNFCCC) adopted in Rio in 1992. In the context of this convention, Low Carbon Development is now generally expressed using the term Low-Emission

Development Strategies (LEDS). Though no formally agreed definition exists, LEDS are generally used to describe forward-looking national economic development plans or strategies that encompass low-emission and/or climate-resilient economic growth.

Whilst the concepts of Green Economy, Green Growth, Sustainable Development and Low Carbon Development emerged from different sources, through the work of different organizations and with different target audiences, the distinctions among them became blurred and they are now being used almost interchangeably. A main driver behind the development of these concepts was the move towards a more integrated and holistic approach to incorporating environment in policies, economic decision making and planning.

Croatia has recently joined the growing number of countries in the world that have already started to develop a long-term strategy for low-carbon development and thereby firmly embarked on the road to sustainable development. The approach that determines impacts of Hydrogen Economy in the context of low emission development was the core for the methodology used in this work. The design of the low-carbon society requires innovative energy systems considering a trans-disciplinary approach that integrates multi-dimensional elements, related to the social, economic, and environmental aspects (Nakata et al. 2011). Development of energy and transportation sector is complex and increasingly often evaluated as a part of society-biosphere-climate-economy-energy system (Akhtar et al. 2013).

Description of Croatian Power System and Renewable Energy

Croatia is currently importing around 50 % of its energy of which 80 % is oil. Almost all electricity generation capacity is in ownership of Croatian Electrical Utility, HEP Group. From 3993 MW installed, 2056 MW is in hydro, 1589 MW in thermal power plants and 348 in Croatian part of nuclear power plant Krsko. Specificity of Croatian power system is large percentage of installed power in hydro (52 %) that requires reserve capacity during summer period when water level is low. Ever growing electricity consumption is not offset with new generation capacities and electricity import is rapidly rising. That emphasizes need for new generation capacities. Currently, LEDS is in progress—with focus up to 2030 and outlook to 2050.

Hydrogen in Transport

Based on today's global climate changes, the main goal of this work was to research how new technologies application can help in mitigation of GHG. Due to the increasing of the GHG concentrations, this work focuses on climate change analysis, transition of energy, and transportation sector towards low carbon paradigm. As one of the most critical barriers to achieve 100 % renewable energy system is intermittent nature of renewable energy such as solar and wind, demand for energy storage becomes important (Moriarty and Honnery 2009; Najjar 2013). And that's the point where hydrogen as the great energy carrier can be implemented. This is where the focus of this work was positioned, related to the transport sector as the highest energy consumer and polluter (Shafiei et al. 2014). Given the intermittent nature of renewables, energy peaks from wind and solar energy can be transferred to the hydrogen for later use and / or utilize in transport sector (Đukić 2015). This also provides more flexibility in the power system network to offset production peaks to energy for use when needed (Cipriani et al. 2014).

A global trend of population growth leads to the increasing of energy demands, resulting in decreasing of fossil fuels reserves, increasing of CO_2 emission and increasing of the temperature, respectively. Here is where the transfer to the Hydrogen Economy has to be applied. The drivers of Hydrogen Economy in the countries worldwide are energy safety, climate changes, air pollution, and competitiveness in the world market (Andrews and Shabani 2012). Hydrogen technologies and their applications are investigated worldwide, from the aspect of sustainable development, cleaner technologies, energy efficiency and energy savings. A transport sector is the leader of energy consumption and air pollution, the most evident in urban cities. Due to that hydrogen fuel cells powered vehicles present one of the rare technologies capable to switch over internal combustion engines vehicles (Biswas and Thompson 2013). As it is written above importance of involving of hydrogen fuel cells powered vehicles is recognized not just in Europe but all over the world. US Department of Energy predicts demands for hydrogen of 150 Mt until 2040 (Hamad et al. 2014). Thus, countries of Asia such as Japan, China, Korea, India, Malaysia, and Taiwan invest millions dollars to make green transport sector (Lu et al. 2013). For instance, Japan set the goal to have each new vehicle as hydrogen fuel cells powered until 2020. In India the goal is to have 1 million hydrogen fuel cells powered vehicles. A rapid implementation of hydrogen fuel cells powered vehicles in Asia also started because 13 of 20 the most polluted cities are located in China (Pudukudy et al. 2014).

The goal of EU is set to be the first Hydrogen based Economy in the world. In order to realize this plan the list of legislatives has to be set. In other words, EU country states need to reform their statutes to ensure successful transition to the hydrogen technologies (Apak et al. 2012). Nowadays EU Commission applied directive in order to approve hydrogen fuel cells powered vehicles. Furthermore, EU established platform for hydrogen and fuel cells to encourage the development, making strategy, and introduction of standards and legislatives. Aside of legislatives, public-private partnership should be highlighted as a step forward in the financing major projects. These partnerships is going to be applied within several years. During that time period both parties should invest 470 millions euros each. As the result, future EU investment in hydrogen technologies will be around one billions euros.

Croatia as a full member of EU should aim to shift to the new technologies as soon as it can. Meeting this demand the increased investment in hydrogen technologies is required. Croatia is a touristic country with large amount of motorized vehicles during the year with trend of double / triple increasing in summer months. It is important to highlight that hydrogen as a fuel is clean as much as is clean energy and processes used for its production. To satisfy this claim hydrogen production via water electrolysis using renewable energy sources is unquestionable (Đukić and Firak 2011). In Croatia there are around 1.9 millions vehicles. From this number, approximately 1.5 millions belong to the passenger cars. To accomplish hypothesis of clean and environmentally friendly hydrogen, there were performed calculations for the systems that include use of solar and / or wind energy, i.e. photovoltaic cells and / or wind turbines (Nowotny and Veziroglu 2011).

Given results represent directions in the system development based on installed power of solar and wind energy, taking into account demands for hydrogen amounts. Conclusions and recommendations are based on analyzing three different scenarios with hydrogen as the energy storage in Croatia; in what extent can hydrogen be used as the energy carrier to store excessive energy from intermittent renewable energy sources. In order to estimate potential savings in CO_2 emissions, in this study some methodological approaches from (Pascala and Socolow 2004) were applied. These options are related to the improved fuel economy, reduced reliance on the vehicles and the use of the renewable hydrogen in the transportation sector.

Methodology

Following potential changes in the transportation system were analyzed in this study: (1) A fraction of the hydrogen powered vehicles was assumed to be equal 10 %, 50 %, and 75 % by the year of 2030. Starting fraction of the hydrogen fuel cells powered vehicles in Croatia was assumed as 0 %; (2) The kilometers values were assumed to be equal 600, 700 or 800 km / 5 kg of H_2 by the year of 2030 with starting kilometers value assumed to be equal 500 km / 5 kg of H_2 as an average value of 500 km that can be traveled by the current generation hydrogen fuel cells powered vehicles. Total number of scenarios in this set was 9, where all combinations of the fraction of the hydrogen fuel cells powered cars and the average efficiency were evaluated. It was also applied following 3 constants in this set of scenarios: the number of the passenger cars was assumed to be equal 1448229 (DZS 2014) and constant by the year of 2030. Also, average number of kilometers traveled per year was assumed to be equal 11000 km / year and constant in this set of scenarios, while the average gasoline consumption used to estimate potential savings was assumed to be equal 10 km / L.

Although various assumptions in this set of scenarios were subject to the various uncertainties and limitations, they provided simple and realistic framework to evaluate possible hydrogen production and storage needs in Croatia. In order to compare hydrogen scenarios in this work to potential savings in the gasoline consuming cars, next set of scenarios was also evaluated. There were additionally constructed 9 scenarios based on the following options: (1) The average distance was assumed to decrease up to the year of 2030. Starting from 11000 km / year, potential reductions in the gasoline consumptions and reductions in the CO_2 emissions, when distance was reduced to the 9000 km, 7000 km or 5000 km per year, were evaluated; (2) The average distance was assumed to reach 11 km, 15 km or 20 km per liter by the year 2030. In this set of scenarios the same number of passenger cars as in the hydrogen scenarios and the same starting average distance and gasoline consumption were assumed. All changes in the projections were assumed to vary as a linear function of time, and were estimated on the yearly time scale. Scenarios were compared in term of the CO_2 emission saved by the year of 2030 and in the year of 2030. Some other details related to the various conversion factors and comparisons were introduced in the results and discussion sections where needed.

Results and Discussion

Due to the described methodology, two cases were evaluated. First one was related to the hydrogen demands till the end of 2030 and second one was related to the saved gasoline till the end of 2030. Case of hydrogen was consisted of the fraction of the hydrogen fuel cells powered vehicles (included scenarios 1A: 10 %, 1B: 50 % and 1C: 75 %), and average efficiency of the fuel cells, i.e. kilometers per 5 kg of hydrogen (included scenarios 2A: 600, 2B: 700 and 2C: 800). Further discussion of these scenarios was given in the Table 33.1 with conversion factors included.

In the Fig. 33.1 there is given a projected increase of the hydrogen fuel cells powered vehicles in Croatia by the end of 2030 (a) and average kilometers range with 5 kg of hydrogen by the end of 2030.

Due to the average kilometers per 5 kg of hydrogen, Fig. 33.1b presents projected ranges of 600 km, 700 km, and 800 km. However these approximations are related to the expected improving of hydrogen storage that allows to have these ranges. From the Fig. 33.2a it is seen how much hydrogen is needed per year from the 2015 to the 2030. Related to the hydrogen automotive application, Fig. 33.2b brings amounts of saved gasoline. Given results promise CO_2 emissions reducing.

All Scenarios were compared in term of the CO_2 emission saved by the year of 2030. The results are given in the Table 33.2

In the Table 33.3 are given results of calculations for saved gasoline within Scenarios 1A, 1B, and 1C. Accumulated gasoline not spent / L for Scenario 1A is 1.27×10^9 L, for Scenario 1B is 6.37×10^9 L, and for Scenario 1C is 9.56E + 09 L.

Table 33.4 presents data within 6 Scenarios for gasoline of average kilometers per year and average kilometers per litre started from the 2015 to the end of the 2030.

To compare savings in CO_2 emissions and gasoline, further calculations based on current needs of gasoline were performed. Figures 33.3a and 33.3b give view on average kilometers per year and average kilometers per liter of gasoline.

Table 33.1 Hydrogen by the end of the 2030

Starting number of the hydrogen powered cars	0
Starting efficiency of fuel cells (km / 5 kg H_2)	500
Number of passenger cars (assumed constant)	1448299
Average kilometers per year (assumed constant)	11000
Average kilometers per litre of gasoline (km / L) (assumed constant)	10
Fraction of hydrogen fuel cells powered vehicles by the end of 2030	
Scenario 1A	10 %
Scenario 1B	50 %
Scenario 1C	75 %
Average kilometers (km / 5 kg H_2) by the end of 2030	
Scenario 2A	600
Scenario 2B	700
Scenario 2C	800
Conversion factors	
1 gal of gasoline > X kg of C	2.40
1 L of gasoline > X kg of C	0.63
1 kg of C corresponds to X kg of CO_2	3.67
kWh needed in the electrolyzer to get 1 kg of H_2	52.5
Peak power output produced by a windturbine (MWp)	3
Peak power output produced by a photovoltaic array (MWp)	0.01
Capacity factor for the windturbine	30 %
Capacity factor for the photovoltaic cells	15 %

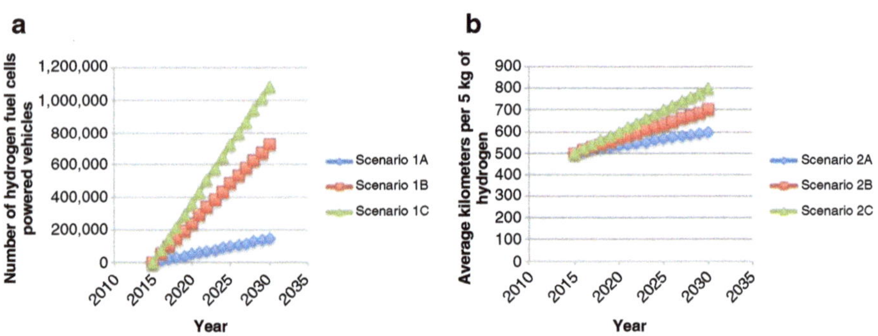

Fig. 33.1 Hydrogen fuel cells powered vehicles: 10 %, 50 %, and 75 % (**a**); Average kilometers per 5 kg of hydrogen (**b**)

Both calculations results were presented in term of accumulated CO_2 not released and accumulated C not spent including 25 % overheads.

In the Fig. 33.4 the combined Scenarios included in the Fig. 33.3 were presented that gives overall data of amount of gasoline needed per year.

Results were grouped into 3 additional Scenarios with group of 3 discussed above. Calculated values of each grouped Scenarios were given in the Table 33.5.

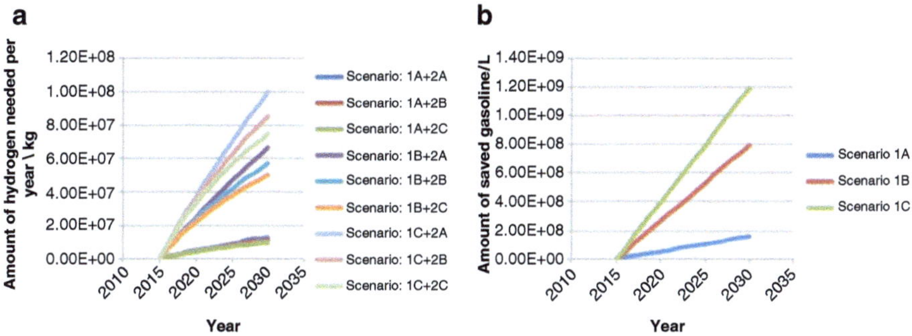

Fig. 33.2 Amount of hydrogen needed per year (**a**); Amount of saved gasoline (**b**)

Table 33.2 Amount of H_2 needed per year (kg)

Scenario	**1A + 2A**	**1A + 2B**	**1A + 2C**
Accumulated H_2 (1000 t)	112.23	100.53	91.22
2030 - demand for H_2 (1000 t)	13.28	11.38	9.96
Electricity needed in 2030 for H_2 production (MWh)	696.994	597.423	522.745
Installed power of the windturbines needed in 2030 (MW)	265.2	227.3	198.9
Installed power in the photovoltaic array needed in 2030 (MW)	530.4	454.6	397.8
Scenario	**1B + 2A**	**1B + 2B**	**1B + 2C**
Accumulated H_2 (1000 t)	561.13	502.64	456.08
2030 - demand for H_2 (1000 t)	66.38	56.90	49.79
Electricity needed in 2030 for H_2 production (MWh)	3484969.47	2987116.69	2613727.10
Installed power of the windturbines needed in 2030 (MW)	1,326.1	1,136.6	920.9
Installed power in the photovoltaic array needed in 2030 (MW)	2,652.2	2,273.2	1,841.8
Scenario	**1C + 2A**	**1C + 2B**	**1C + 2C**
Accumulated H_2 (1000 t)	841.70	753.97	684.13
2030 - demand for H_2 (1000 t)	99.57	85.35	74.68
Electricity needed in 2030 for H_2 production (MWh)	5227454.20	4480675.03	3920590.65
Installed power of the windturbines needed in 2030 (MW)	1989.1	1705	1491.8
Installed power in the photovoltaic array needed in 2030 (MW)	3978.2	3410	2983.7

The main Scenarios studied within this work were based on the hydrogen and gasoline calculations. Results give good directions to the shift to the hydrogen fuel cells powered vehicles. A stronghold is basically in high CO_2 saving as well as savings in fuel cell consumption. Hence, as a very rich with renewables such as solar and wind energy, Croatia has a good potential to follow trends of Hydrogen based Economy what gives tailwind to its transport sector.

Table 33.3 Amount of gasoline saved in Scenarios 1A, 1B, and 1C

Scenario	1A	1B	1C
Accumulated gasoline not spent (L)	1.27E + 09	6.37E + 09	9.56E + 09
Accumulated C not spent (kg)	8.08E + 08	4.01E + 09	6.02E + 09
Accumulated C not spent (kg) + 25 % overheads	1.01E + 09	5.02E + 09	7.53E + 09
Accumulated C not spent (1000 t) + 25 % overheads	1010.06	5018.36	7527.53
Accumulated CO_2 not released (1000 t) + 25 % overheads	3706.94	18417.37	27626.05
2030 - not released CO_2 (1000 t) + 25 % overheads	463.37	2316.84	3475.25

Table 33.4 Gasoline by the end of 2030 (Mt CO_2 = 3.71E + 00 for the 2015)

Number of passenger cars (assumed constant)	1448299
Starting average kilometers per year	11000
Starting average kilometer per liter of gasoline (km / L)	10
Average kilometers per year by the end of the 2030	
Scenario G1A	9000
Scenario G1B	7000
Scenario G1C	5000
Average kilometers per liter of gasoline (km / L) by the end of the 2030	
Scenario G2A	11
Scenario G2B	15
Scenario G2C	20
Conversion factors	
1 gal of gasoline > X kg of C	2.40
1 L of gasoline > X kg of C	0.63
1 kg of C corresponds to X kg of CO_2	3.67

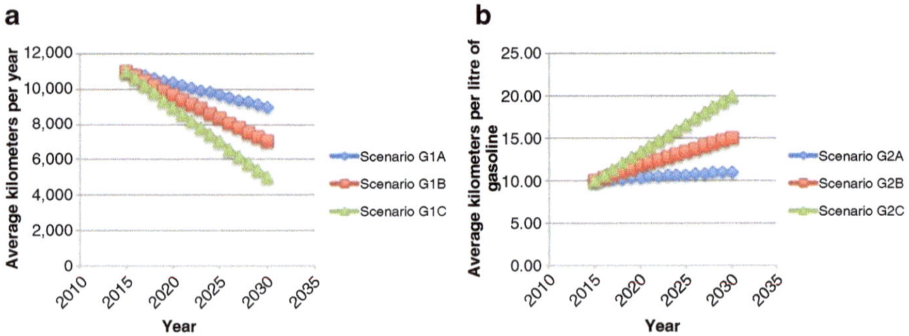

Fig. 33.3 Average kilometers per year: 9000, 7000, and 5000 km (**a**); Average kilometers per liter of gasoline: 11 km, 15 km, and 20 km (**b**)

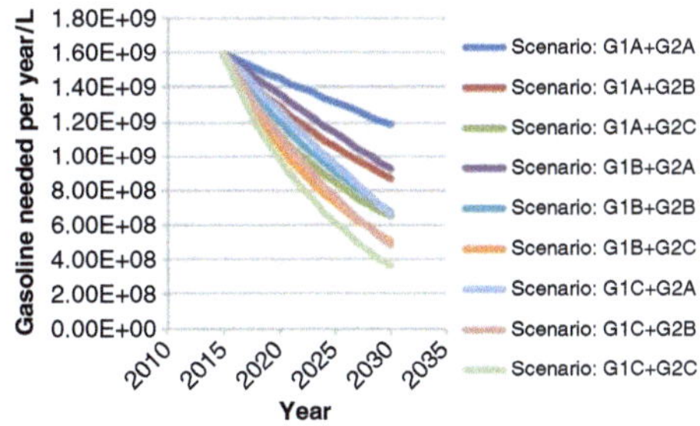

Fig. 33.4 Amount of gasoline needed per year

Table 33.5 Gasoline needed per year (L)

Scenario	**G1A + G2A**	**G1A + G2B**	**G1A + G2C**
Accumulated gasoline not spent (L)	3.36E + 09	6.52E + 09	9.13E + 09
Accumulated C not spent (kg)	2.13E + 09	4.13E + 09	5.79E + 09
Accumulated C not spent (kg) + 25 % overheads	2.66E + 09	5.17E + 09	7.23E + 09
Accumulated C not spent (1000 t) + 25 % overheads	2664.48	5167.12	7234.84
Accumulated CO_2 not released (1000 t) + 25 % overheads	9778.63	18963.33	26551.85
Scenario	**G1B + G2A**	**G1B + G2B**	**G1B + G2C**
Accumulated gasoline not spent (L)	5.53E + 09	8.26E + 09	1.05E + 10
Accumulated C not spent (kg)	3.51E + 09	5.24E + 09	6.68E + 09
Accumulated C not spent (kg) + 25 % overheads	4.38E + 09	6.55E + 09	8.35E + 09
Accumulated C not spent (1000 t) + 25 % overheads	4383.50	6545.02	8347.89
Accumulated CO_2 not released (1000 t) + 25 % overheads	16087.43	24020.21	30636.75
Scenario	**G1C + G2A**	**G1C + G2B**	**G1C + G2C**
Accumulated gasoline not spent (L)	7.70E + 09	1.00E + 10	1.19E + 10
Accumulated C not spent (kg)	4.88E + 09	6.34E + 09	7.57E + 09
Accumulated C not spent (kg) + 25 % overheads	6.10E + 09	7.92E + 09	9.46E + 09
Accumulated C not spent (1000 t) + 25 % overheads	6102.51	7922.92	9460.94
Accumulated CO_2 not released (1000 t) + 25 % overheads	22396.23	29077.10	34721.65

Conclusions and Future Work

Croatia as a touristic country has a great potential for hydrogen automotive application. The number of motorized tourists is rapidly increasing during summer months. This fact opens enough room to the hydrogen fuel cell powered vehicles shift starting with 10 %, 50 %, and 75 % up to the end of the 2030. Due to the EU directives flows of hydrogen powered vehicles should be set for every 300 km. With area of 56594 km^2 (89810 km^2 with associated coastal see) starting point was suggested to be installation of hydrogen refueling stations along the seaside. With seaside of 1777 km (over 6000 km including islands) number of demanded hydrogen refueling stations was calculated to be 6. Hydrogen production technology is water electrolysis using solar energy installed on-site during transition period of infrastructure construction. Obtained results are talking of 27626.05 accumulated CO_2 not released (1000 tones) plus 25 % overheads. Energy needed for fuel cells was calculated as it comes from renewable energy—wind power plants (3 MW each unit, with 30 % capacity factor) and PV systems installed on the roof (10 kW each, 15 % capacity factor). However, methodology used in this work can be easily applied in every country for calculations of demands, taking into account total number of internal combustion engines vehicles. Future work of the authors will be adjusted by changing the time horizon from the year of 2030 to the year of 2050. It will produce two additional sets of Scenarios each including nine of them.

References

Akhtar, M., Wibe, J., Simonovic, S., & MacGee, J. (2013). Integrated assessment model of society-biosphere-climate-economy-energy system. *Environmental Modelling & Software, 49*, 1–21.

Andrews, J., & Shabani, B. (2012). Re-envisioning the role of hydrogen in a sustainable energy economy. *International Journal of Hydrogen Energy, 37*, 1184–1203.

Apak, S., Atay, E., & Tuncer, G. (2012). Renewable hydrogen energy regulations, codes and standards: Challenges faced by an EU candidate country. *International Journal of Hydrogen Energy, 37*, 5481–5497.

Atkisson, A. (2012)). Life beyond growth, annual survey report of the institute for studies in happiness, economy, and society, Japan.

Biswas, W. K., & Thompson, B. C. (2013). Environmental life cycle feasibility assessment of hydrogen as an automotive fuel in Western Australia. *International Journal of Hydrogen Energy, 38*, 246–254.

Branković, Č., Güttler, I., & Gajić-Čapka, M. (2013). Evaluating climate change at the Croatian Adriatic from observations and regional climate models' simulations. *Climate Dynamics, 41*, 2353–2373.

Cipriani, G., Di Dio, V., Genduso, F., La Cascia, D., Liga, R., Miceli, R., et al. (2014). Perspective on hydrogen energy carrier and its automotive applications. *International Journal of Hydrogen Energy, 39*, 8482–8494.

Đukić, A. (2015). Autonomous hydrogen production system. *International Journal of Hydrogen Energy*, In Press, Corrected Proof, http://dx.doi.org/10.1016/j.ijhydene.2015.02.003.

Đukić, A., & Firak, M. (2011). Hydrogen production using alkaline electrolyzer and photovoltaic (PV) module. *International Journal of Hydrogen Energy, 36*, 7799–7806.

DZS, (2014). Registered road vehicles and road traffic accidents 2013. http://www.dzs.hr/Hrv_Eng/publication/2014/05-01-08_01_2014.htm. Retrieved Mar 24, 2014.

Hamad, Y. M., Hamad, T. A., Agil, A. A. A., Bapat, S. G., Bauer, C., Clum, A., et al. (2014). A design for hydrogen production and dispensing for northeastern United States, along with its infrastructural development timeline. *International Journal of Hydrogen Energy, 39*, 9943–9961.

IPCC. (2014). *Climate change 2014: Synthesis report. Contribution of Working Groups I, II and III to the Fifth Assessment Report of the Intergovernmental Panel on Climate Change* [Core Writing Team, R. K. Pachauri & L. A. Meyer (Eds.)]. IPCC, Geneva, Switzerland, 151 pp.
Kovacevic, A. (2011). Fossil fuel subsidies in the Western Balkans, A report for UNDP. p. 18.
Lu, J., Zahedi, A., Yang, C., Wang, M., & Peng, B. (2013). Building the hydrogen economy in China: Drivers, resources and technologies. *Renewable and Sustainable Energy Reviews, 23*, 543–556.
Moriarty, P., & Honnery, D. (2009). Hydrogen's role in an uncertain energy future. *International Journal of Hydrogen Energy, 34*, 31–39.
MZOIP. (2014). *Sixth National Communication and First Biennial Report of the Republic of Croatia under the United Nation Framework Convention on the Climate Change. Republic of Croatia Ministry of Environmental and Nature Protection (MZOIP)*, Zagreb, p. 247.
Najjar, Y. S. H. (2013). Hydrogen safety: The road toward green technology. *International Journal of Hydrogen Energy, 38*, 10716–10728.
Nakata, T., Silva, D., & Rodionov, M. (2011). Application of energy system models for designing a low carbon society. *Progress in Energy and Combustion Science, 37*(4), 462–502.
Nowotny, J., & Veziroglu, T. N. (2011). Impact of hydrogen on the environment. *International Journal of Hydrogen Energy, 36*, 13218–13224.
Pascala, S., & Socolow, R. (2004). Stabilization wedges: Solving the climate problem for the next 50 years with current technologies (review). *Science, 305*, 968–972.
Pašičko, R. (2014). *Optimization of power system operation and development under emission trading scheme*. PhD dissertation: University of Zagreb.
Pearce, D., Markandya, A., & Barbier, E. (1989). *Blueprint for green economy*. U.K.: Earthscan.
Pudukudy, M., Yaakob, Z., Mohammad, M., Narayanan, B., & Sopian, K. (2014). Renewable hydrogen economy in Asia—opportunities and challenges: An overview. *Renewable and Sustainable Energy Reviews, 30*, 743–757.
Raworth, K. (2012). A Safe and Just Space for Humanity. Oxfam Discussion Paper. http://policy-practice.oxfam.org.uk/publications/a-safe-and-just-space-for-humanity-can-we-live-within-the-doughnut-210490.
Shafiei, E., Davidsdottir, B., Leaver, J., Stefansson, H., & Asgeirsson, E. I. (2014). Potential impact of transition to a low-carbon transport system in Iceland. *Energy Policy, 69*, 127–142.
UN WCED. (1987). Our Common Future—The Brundtland Report, Report of the World Commission on Environment and Development.
UNEP. (2011). *Towards a green economy: Pathways to sustainable development and poverty eradication*. France: UNEP.

Part IX
Energy Technologies and Their Effect on Global Warming

Fossil Fuels and Climate Change

Chapter 34
Contribution to Multi-Criteria Evaluation of the Impacts of Air Pollution: Case of Cement Plant (Ain Touta- ALGERIA)

Lylia Bahmed, Samia Daas, Mourad Chebila, and Loubna Khadoudja Aggabou

Introduction

The implementation of a strategy or an Environmental Management Plan (EMP), to minimize air pollution that results from the manufacture of a product, fall within the scope of the EMS policy of most of the cement companies in Algeria, and in accordance with the requirements of the ISO14001 standard (Hariz and Bahmed 2013; Bahmed et al. 2009), in terms of improving their EMS, and depending on the environmental context and legal restrictions that subject the companies regarding the dust releases' limits and their technical capacity to address the problems of air pollution (Hariz 2009).

In this article and after presenting the proposed methodology for the quantitative assessment of the impacts of air pollution of cement by means of a multi-criteria approach, a state of play on the life cycle of the cement product modeled by the SADT method of the process of the cement plant of Ain-Touta, and its environmental impacts will be quantified by the AHP method before concluding on the prospects and expectations from this study.

L. Bahmed (✉) • S. Daas • M. Chebila • L.K. Aggabou
Laboratory of Research in Industrial Prevention- Industrial Health and Safety Institute, University of Batna 2, Batna, Algeria
e-mail: bahmed_lylia@yahoo.fr; samia_hse@yahoo.fr; hsemourad@gmail.com; loubna.aggabou@hotmail.fr

P. Grammelis (ed.), *Energy, Transportation and Global Warming*, Green Energy and Technology, DOI 10.1007/978-3-319-30127-3_34

Method and Material

This study allows the proposition of a first perspective of the different actions required to implement an effective management of the EMS. These objectives constitute the indices that permit the assessment and evaluation of the impacts of air pollution related to the manufacturing of a product.

Proposed Methodology

LCA is a global approach standardized according to the ISO 14040 standard, which quantifies the impacts of a product. LCA encrypts the impacts of each component of the product from the extraction of raw materials to the end of the cycle. The final result is presented according to the specific impacts on climate change, human health, ecosystems and non-renewable resources (ISO 14044 2006).

The phase of the LCA and its standardized methodology and high precision, is an internationally recognized tool for decision support, it allows the reduction of the impacts of the new products on the environment, the identification of what can be improved in the already existing products, avoiding the improvement of an aspect that creates more serious problems to another stage of the product's life and the comparison of the environmental performances of products with the same utility (Fig. 34.1).

In order to evaluate the environmental impacts associated with the manufacture of cement, we have chosen the multi-criteria approach to decision support AHP, taking into account the product life cycle and at the same time the two steps of the LCA method (Bouhidel 2009): evacuation of the environmental impacts and interpretation of the results.

The articulation of these tools in a methodological approach is of great importance especially for the quantitative evaluation of the environmental impacts.

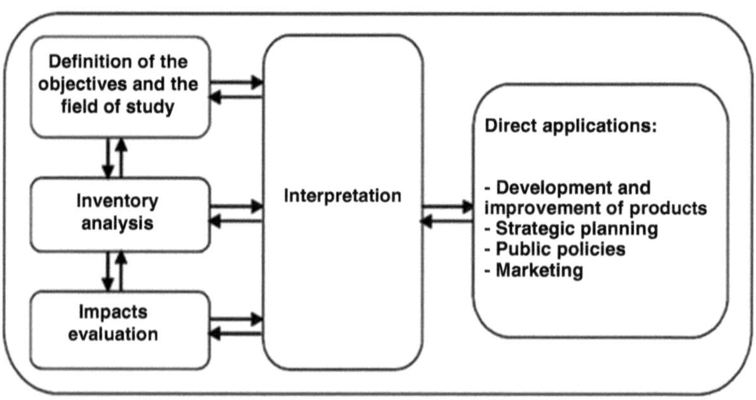

Fig. 34.1 Framework of a LCA according to ISO 14040 (2006)

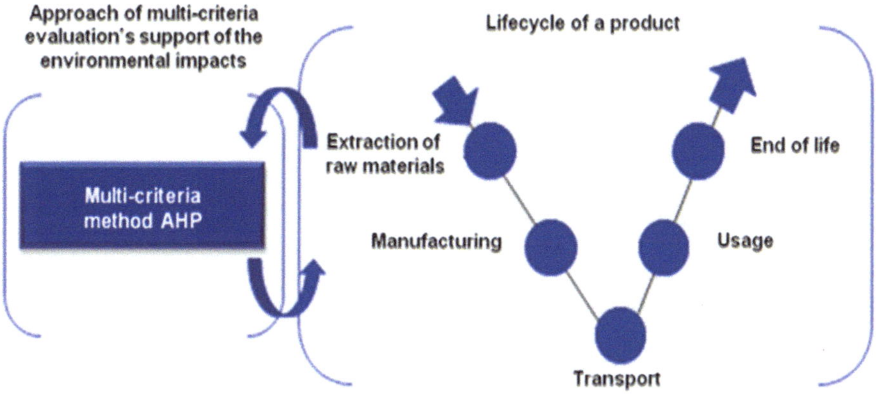

Fig. 34.2 The proposed methodology

Hence, the major stages of the proposed methodology are based on the multi-criteria approach to decision support (Fig. 34.2).

The Analytic Hierarchy Process (AHP) method of T.L. Saaty (1994) was developed in the 70 to facilitate the decision making in the complex situations using a structured approach (Rousval and Bouyssou 2009). The AHP method takes into consideration qualitative and quantitative criteria. This method is simple to use, very flexible, as it is a computerized method by software called Expert Choice.

The AHP method consists of representing a decision problem for evaluating the impacts of air pollution by a hierarchical structure reflecting the interactions between the various phases of the life cycle of cement, then to make paired comparisons of the criteria of the evaluations of the hierarchy, and finally to determine the priorities of the environmental impacts.

Presentation of the AHP Method

The multi-criteria hierarchy method (MHM) or Analytic Hierarchy Process (AHP) of T.L. Saaty (1994) was developed in the 70s to facilitate the decision making in the complex situations using a structured approach (Aull-Hyde et al. 2006). The AHP method allows taking into account qualitative and quantitative criteria. This method is simple to use, very flexible, as it is a computerized method by software called Expert Choice. Actually, several steps compose the AHP method. These steps are sequentially detailed hereinafter.

- **Step 1: Hierarchical decomposition of the complex problem into simple elements.**
  This essential step aims to pose the problem to be solved, to position it in its environment in order to accurately identify its various aspects and features and

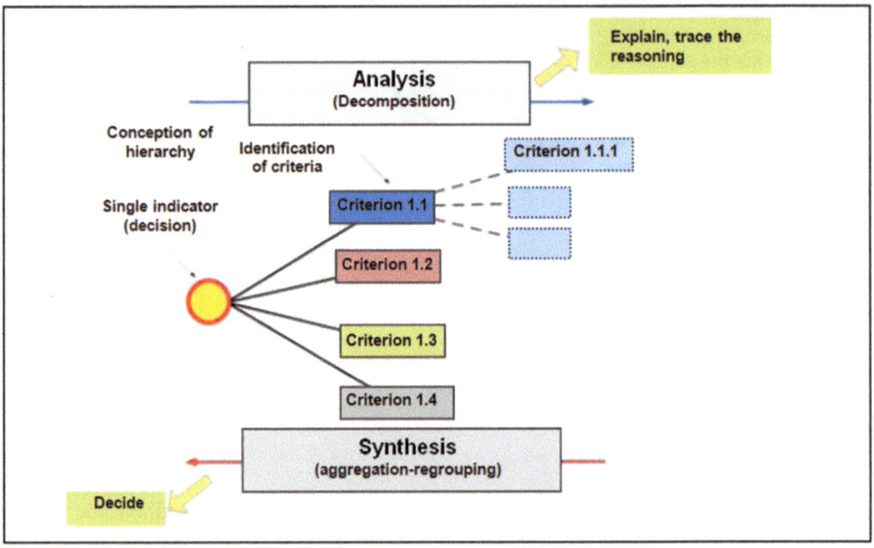

Fig. 34.3 Hierarchical structure of the problem (Aull-Hyde et al. 2006)

also to identify the stakeholders or experts of the decision-making process (Fig. 34.3).

- **Step 2: Comparison of criteria by means of binary comparison matrix.** The preferences of experts reflecting the relative importance between the criteria are expressed by the weights *ajj* resulting from pair-wise comparison matrices between the criteria of the same hierarchical level according to a scale of linguistic ratio. The aggregation and evaluation of weights is done from the roots to the leaves according to a principle that akin to a weighted sum.

  Configure a reciprocal square matrix formed by the evaluations of the weight ratios $(N*N)$ where, N represents the number of the compared elements. In this way, we obtain:

$$a = a_{ij}\text{With}a_{jj} = 1$$

  Let $A = a_{ij}$ to be a pair-wise comparison matrix:

$$A = \begin{pmatrix} a_{11} & a_{12} & a_{1n} \\ a_{21} & a_{22} & a_{2n} \\ a_{n1} & a_{n2} & a_{nn} \end{pmatrix} a_{ij} = \frac{1}{a_{ji}}, \forall_i, j = 1, 2, \ldots.n$$

  The values of *aij* can be determined through the scale of Saaty (Table 34.1.).

Table 34.1 Weighting scale of the AHP method (Saaty and Hu 1998)

Numerical scale	Verbal scale
1	**Importance equals to the two elements**
3	**One element is a little more important than the other**
5	**An element is more important than the other**
7	**An element is much more important than the other**
9	**An element is absolutely more important than the other**
2, 4, 6, 8	**Intermediate values between two judgments used to refine the judgment**

The reciprocal comparison according to (Saaty 1994) can be:

- Favorable and positive, the appreciation will correspond to an integer value.
- Unfavorable and therefore negative, the appreciation will be expressed using a fraction.

- **Step 3: Determine the relative importance of the criteria.**
The determination of the elements' priorities of each matrix is done by solving the eigenvector problem. The relative importance of the different criteria is expressed by the values of the normalized vector that is equal to 1.
So the priorities are determined by solving a system of equations (Saaty 1994):

$$A.w = \lambda_{\max}.w \tag{34.1}$$

Where:
A: is the comparison matrix of dimension $(N{*}N)$.
$w = (w_1, w_2, \dots\dots\dots w_n)$ Is an eigenvector of dimension $N{*}1$ associated with the eigenvalue.
$\lambda_{\max}$: The maximum Eigen value of the matrix.
Therefore,

$$A.w = \lambda_{\max}.w \Rightarrow A.w = \left[\sum_{j=}^{n} a_{ij}.w_j\right], \forall_i = 1, \dots\dots n \tag{34.2}$$

Then, we obtain:

$$\lambda_{\max} = \frac{1}{n}.\sum_{i=1}^{n}\sum_{j=1}^{n} a_{ij}.\frac{w_j}{w_i} \tag{34.3}$$

This step of the appreciations' collect is the most controversial. Indeed, a major criticism pronounced towards AHP method concerns the scale of judgment on the lack of consideration of the uncertainty in the opinions and perceptions of experts (Vincke 1998).

- **Step 4:** Consistency of judgments.

The obtained answers often exhibit some degree of inconsistency. The method does not require that judgments neither be consistent nor transitive, contrariwise the method to define an index of consistency IC (Saaty and Hu 1998).

$$IC = \frac{\lambda_{\max} - 1}{n - 1} \tag{34.4}$$

Where:

$\lambda_{\max}$: is the maximal eigenvalue corresponding to the matrix of pair-wise comparisons.

n: number of the compared elements.

The ratio of coherence can be interpreted as the probability that the matrix is completed randomly. The ratio of coherence is given by the following formula:

$$RC = \frac{IC}{IA} \tag{34.5}$$

Where:

RC: is the ratio of coherence.

IA: is a random index.

IC: is the index of coherence (Table 34.2).

The overall coherence of appreciation is evaluated using the ratio of coherence RC. Still according Saaty, the value of this latter must be at most equal to 10 %.

In the case where this value exceeds 10 %, the appreciations may require some revisions.

- **Step 5: Establish the relative performance of each criterion.**
  This last step aims to assign a performance rating for each action P_{ai}, which will be multiplied by a weighting coefficient w_i. The sum over n actions gives an aggregated performance results expressed in an overall rating (Saaty 1994):

$$P_{ci} = \sum_{i=1}^{n} P_{aji} \times w_i \tag{34.6}$$

Where:

P_{ci}: Rating of performance for the criterion c_j.

P_{aji}: Rating of performance for the action a_i of the criterion c_j.

w_i: Weighting coefficient for the action a_i of the criterion c_j.

Table 34.2 Table of the acceptable coherency's ratios (Saaty 1994)

Size of the matrix (n)	Ratio of the acceptable coherence
3	0.05
4	0.08
5 et +	0.10

Application of the Methodology

Our case study is applied to the life cycle of cement manufacturing of the cement plant Ain-Touta Batna.

The application of the SADT method allows us to model the cement life cycle and determine its major phases and their interactions (Fig. 34.4).

This method facilitates the passage of the environmental impacts' evaluation during the analysis of the life cycle of the cement product, by the application of normative LCA methods and AHP method.

This method facilitates the passage of the environmental impact evaluation in the analysis of the life cycle of the cement product, by the application of normative LCA methods and the AHP method.

The characterization of environmental impacts in the cement plant of Ain Touta results in an approach which consists of developing the hierarchical structure of our multi-criteria system (Fig. 34.5).

Three criteria are determined; the extraction of raw materials, cement manufacturing and transport of cement. Each criterion is defined by several sub-criteria, and each sub-criterion is divided into three alternatives (air pollution, soil pollution and depletion of the natural resources).

The evaluation of the overall environmental impacts in the cement plant of Ain Touta implies an evaluation of the impacts of the extraction, manufacturing, transport of cement, and each criterion has sub criteria. Three types of impacts

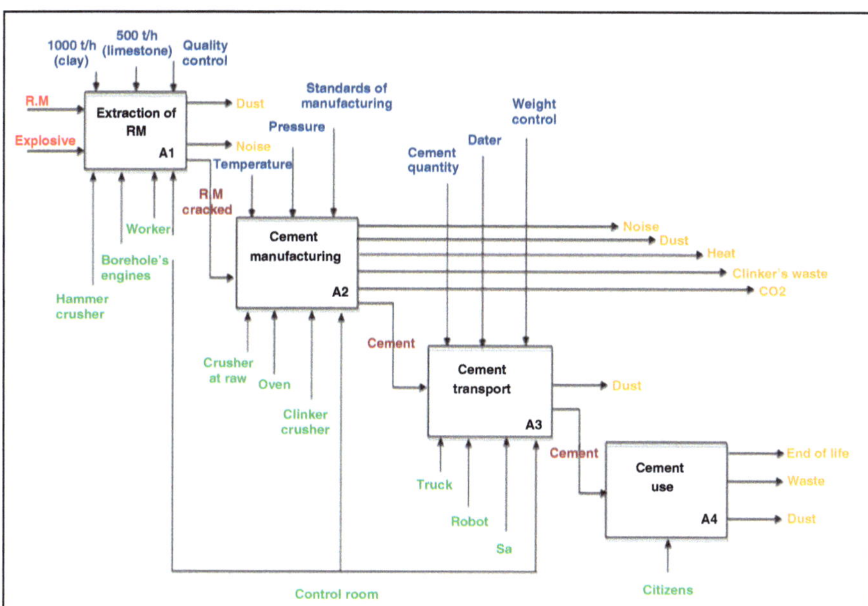

Fig. 34.4 Phase of the life cycle by the method SADT

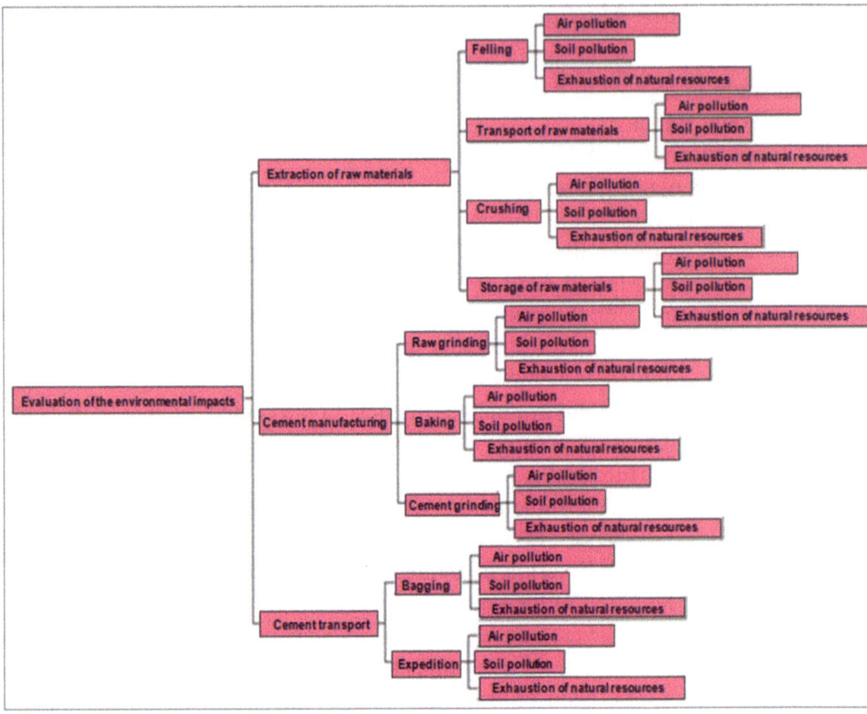

Fig. 34.5 Hierarchical structure for the evaluation of the environmental impacts in the cement plant of Ain touta

(air pollution, soil pollution and depletion of the natural resources) were selected as alternatives.

In order to evaluate the environmental impacts of cement in the cement plant of Ain Touta, a survey was conducted to consolidate the various opinions of the judgment of the services' heads of the different specialties of the cement plant of Ain Touta (SCIMAT 2015).

The table below presents the results of the evaluation of the environmental impacts related to the cement product by means of the decision support multi-criteria AHP method (INERIS, 2009).

Results and Discussion

The main objective of this study is to evaluate the environmental impacts related to the manufacture of cement. Several steps compose the AHP method. These steps will present the results obtained using the Expert Choice software, through the cement life cycle. Therefore, The obtained results using the AHP method (Table 34.3) revealed that the highest pollution levels at the manufacturing stage in which we must take all preventive and corrective measures to reduce this critical

Table 34.3 The obtained results using the AHP method (SCIMAT 2015)

Phase of the cement life cycle	Criteria of the hierarchy	Results
Extraction of raw materials	Felling	32 % Atmospheric pollution 48 % Soil pollution 20 % Depletion of natural resources
	Transport of MP	
	Crushing	
	Storage of MP	
Cement manufacturing	Raw grinding	65 % Atmospheric pollution 23 % Depletion of natural resources 12 % Soil pollution
	Baking	
	Cement grinding	
Transport of cement	Bagging	38 % Soil pollution 55 % Atmospheric pollution 7 % Depletion of natural resources
	Expedition	

and significant rate (65 % atmospheric pollution). Pollution level of other steps (extraction of raw materials and transport of cement) is not negligible, but remains less than that of manufacturing.

Sensitivity analysis is a good tool that allows us to validate the robustness of an environmental decision. To do this, we can change the weight of our criteria to determine the limits of our best choice. This permits us to check the shift thresholds of our results regarding the importance of our criteria. The Expert Choice software skillfully synthesizes the results by graphics of weights of life phases and the associated impacts.

The Fig. 34.5 illustrates hierarchical structure for the evaluation of the environmental impacts in the cement plant of Ain touta, through analysis of the life cycle.

The Fig. 34.6 illustrates the performance graph in the basis of the final eigenvectors of the three impacts per cycle phase, we notice that the alternative of winning solution is either, the "air pollution" before the resources depletion and soil pollution at all stages of the life cycle of the cement except the phase "extraction of RM" where we can see that "transport of cement" has less impact on the environment in the same criterion performances better than the others.

Conclusion

The main objective of this study is the evaluation of the environmental impacts of different life cycle's phases of cement of the cement plant of Ain-Touta Batna. For this, we have applied the LCA method combined with the method of multi-criteria hierarchy AHP which allows the quantification of the different categories of environmental impacts of the cement product, particularly when the software Expert Choice is used (www.expertchoice.com 2015).

The results of the application of LCA and AHP method in this study show that the environment is seriously threatened by the environmental impacts. It is affected

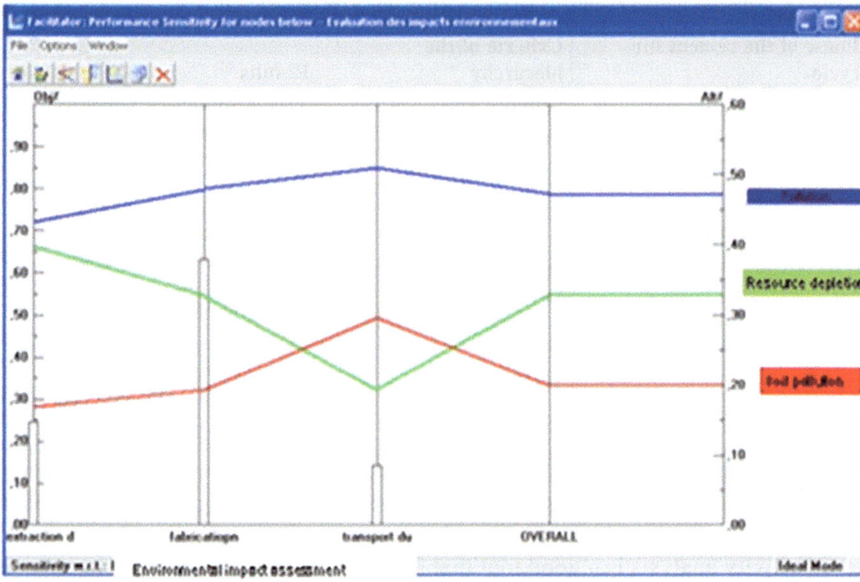

Fig. 34.6 The final eigenvectors of the three impacts per cycle phase

by the impacts of air pollution generated throughout the different stages of the life cycle of the cement such as the phase of cement manufacturing and this in the sub-step of raw grinding of semi finished product, extraction in the sub-step of felling of raw materials and the sub-step of bagging the cement.

Therefore, special attention should be paid to these problems:

From an environmental viewpoint and for the problem of air pollution, the current remedy is to ensure the capture of dust during the cement manufacturing phase to reduce dust levels released into the air, in order to avoid environmental problems.

Despite the fact that many of the dust collection techniques such as the electro-bag filters are implemented, unfortunately they are ineffective. For this reason, it is necessary to proceed preventive maintenance plans, to make the best environmental decision as well as the regulatory compliance.

The regulatory texts play a key role in this area, which will require companies to take serious decisions about the environment. Therefore, to improve the conservation of the environment and to respect the principles of the sustainable development, the strategic environmental planning must be implemented to meet the needs of green products and also the needs of the future generations.

The study has allowed us to expand our knowledge on methods and tools for assessment of environmental impacts related to air pollution from cement plants, based on a multi-criteria evaluation and analysis of the life cycle by method SADT, which has allowed us to develop an original analytical approach on the basis of which we will in future perspective make further quantitative assessments.

References

Aull-Hyde, R., Erdogan, S., & Duke, J. M. (2006). An experiment on the consistency of aggregated comparison matrices in AHP. *European Journal of Operational Research, 171*, 290–295.

Bahmed, L., Djebabra, M., Boubakeur, L., & Boukhalfa, A. (2009). Implementing the ISO 14001 certification—an empirical study of an Ajgerian company in the process of certification. *Journal of Management of Environmental Quality: An International Journal, 20*(2), 155–164. Emerald Group Publishing Limited ISSN 1477-7835.

Bouhidel, M. (2009). Application d'analyse du cycle de vie (ACV) pour un développement durable: cas des cimenteries Algériennes. Mémoire de Magister. Institut d'Hygiène et Sécurité Industrielle, Université Hadj Lakhdar de Batna.

Hariz, S. (2009). Etude critique du système de management environnemental au niveau des entreprises Algériennes, Mémoire de Magister, Institut d'Hygiène et Sécurité Industrielle, Université Hadj Lakhdar de Batna.

Hariz, S., & Bahmed, L. (2013). Assessment of environmental management system performance in the Algerian Companies Certified ISO 14001. *Management of Environmental Quality: An International Journal, 24*(2), 228–243. Emerald Group Publishing Limited ISSN 1477-7835.

INERIS. (2009). Panorama des méthodes d'analyse multicritère comme outils d'aide à la décision.

ISO 14040. (2006). Environmental management: Life-cycle assessment. Principles and framework. International Organization for Standardization.

ISO 14044. (2006). Environmental management: Life-cycle assessment. Requirements and guidelines. International Organization for Standardization.

Rousval, B., & Bouyssou, D. (2009). De l'aide multicritère à la décision à l'aide multicritère à l'évaluation Un cadre et une application aux transports et à l'environnement. Cahier du Lamsade. (p. 293).

Saaty, T. L. (1994). *Décider face à la complexité—Une approche analytique multicritères d'aide à la décision*. Paris: Collection Université-Entreprise, Entreprise Moderne d'Edition. 232p.

Saaty, T. L., & Hu, G. (1998). Ranking by eigenvector versus other method in the analytic hierarchy process. *Applied Mathematics Letters, 11*(4), 121–125.

SCIMAT. (2015). *Technical report*. Algeria: Cemet Plant Ain Touta.

Vincke, P. (1998). *L'aide multicritères à la décision*. Editions de l'université de Bruxelles, Editions Ellipses, 179 p. www.expertchoice.com. Retrieved February 12, 2015.

Chapter 35
Environmental Impact Assessment of Electricity Production, A Case Study of Turkey

Fatih Yılmaz, M. Tolga Balta, Reşat Selbaş, and Doğan Demiral

Introduction

Energy has significant effects in the evolution of economy and technology. Electricity was the major actor during the transition from the industrial to the technological revolution, almost all the financial and economic activities depend directly on it and any development coincides with an increasing demand of electric energy. Recent year's electricity energy majority is produced from fossil fuels. Fossil fuels have historically been the world's primary source of energy and will continue to dominate the world in the next few decades. Combustion of fossil fuels results in carbon dioxide (CO_2). The bigger contributor to global warming from fossil fuel combustion is CO_2. In recent years, the problem of air pollution has reached limits of remarkably dangerousness both for the environment and human health (Nicoletti 2001; Churchill 1997). Greenhouse gases are causing climate change and increasing the world's average temperatures.

Electricity production accounts for 32 % of total global fossil fuel use and around 41 % of total energy-related CO_2 emissions. Improving the efficiency with which electricity is produced is therefore one of the most important ways of

F. Yılmaz (✉)
Department of Energy, Vocational School of Technical Sciences, Aksaray University, Aksaray 68100, Turkey
e-mail: fatiyilmaz7@gmail.com

M.T. Balta • D. Demiral
Department of Mechanical Engineering, Faculty of Engineering, Aksaray University, Aksaray 68100, Turkey

R. Selbaş
Department of Energy Systems Engineering, Faculty of Technology, Suleyman Demirel University, Isparta 32100, Turkey

P. Grammelis (ed.), *Energy, Transportation and Global Warming*, Green Energy and Technology, DOI 10.1007/978-3-319-30127-3_35

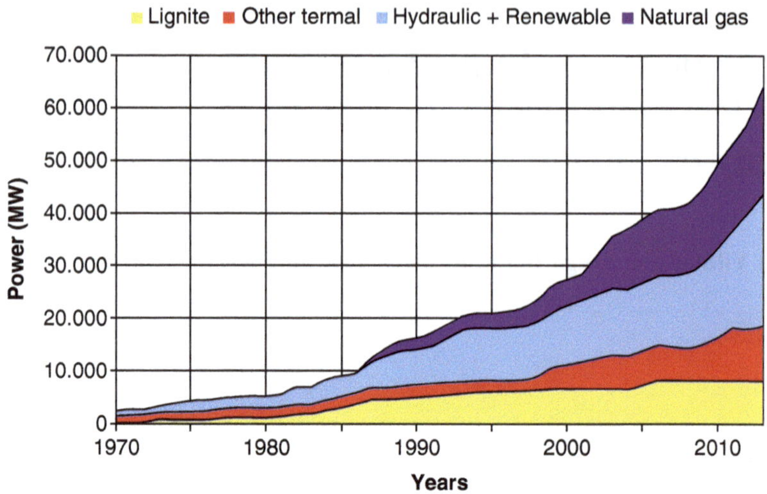

Fig. 35.1 Electricity production in Turkey from 1970 to 2013 (TEİAS 2015)

reducing the world's dependence on fossil fuels, thus helping both to combat climate change and improve energy security (IEA 2007).

Fossil fuel comprises from coal, natural gas, petroleum shale oil and bitumen. They all contain carbon. Coal usage of for a long time to generate heat and electrical energy. But natural gas since 19985 year, it has been usage for heat and electrical energy in Turkey.

Currently, Turkey is turn into utilizing electric power in a clean energy resource because of the coal-based thermal power with the high occurrence of environmental pollution. Figure 35.1 shows the electricity generation mix in Turkey from 1970 to 2010.

An increase in the use of hydropower, renewable energy, and natural gas has been observed over the years in Fig. 35.1.

When examining the literature, several studies are available. Mukhopadhyay and Forssell 2005, stated that empirical investigation of air pollution from fossil fuel combustion and its impact on health in India during 1973–1974 to 1996–1997. The study categorizes the changes in the amount of CO_2, SO_2 and NO_x emissions into four factors: the pollution intensity or eco-efficiency, technology or input-mix, composition of final demand, and the level of final demand. The main factors for these changes were the pollution intensity, technology, and the volume of final demand. A number of studies concentrate on energy and environment (greenhouse gas emissions) by applying input–output structural decomposition analysis. Lin and Chang 1996, used the divisia index approach to examine emissions of CO_2, NO_x and SO_2 from major economic sectors in Taiwan during 1980–1992. They found that economic growth had the greatest impact on the variation of emission intensities during this period, while the influence of fuel mix was limited.

Zhang 2000, analyzed the relationships of fuel mix, energy saving, economic productivity and population expansion to the increase of China's CO_2 emissions during 1980–1997. Paul and Bhattacharya 2004, clear that, paper selected pollution coefficient, energy intensity, structural changes and economic activity as primary input to identify the major factors affecting the energy-related CO_2 emissions from the major economic sectors in India from 1980 to 1996. Their results showed that economic growth was the most important component of CO_2 emissions.

Wier 1998, illustrate that paper explores the anatomy of Danish energy consumption and emissions of carbon dioxide (CO_2), sulphur dioxide (SO_2) and nitrogen oxides (NO_2). Changes in emissions between **1966** and **1988** were investigated using input-output structural decomposition analysis. Hayami et al. (1993), stated that paper estimation of air pollutions and evaluating CO_2 emissions from production activities: using Japan's 1985. The aim of this research is to extend the input-output tables for environmental analysis in the most detailed classification possible. It can also be shown that it plays an important role in evaluating accurately the emissions of CO_2, NO_x and SO_x.

Liaskas et al. (2000), stated that his paper aims at identifying the factors that have influenced changes in the level of industrial CO_2 emissions. By means of an algebraic decomposition method the observed changes are analyzed into four different factors: output level, energy intensity, fuel mix and structural change. The application study refers to the industrial sector of European Union countries. Proops et al. (1996), illustrate that paper examines the UK economy wide, lifecycle implications of eight forms of electricity generations for the emission of three air pollutants, CO_2, SO_2 and NO, The lifecycle of the generating stations is considered in three phases: construction, operation and decommissioning.

Felix and Grewal (2012), stated that environmental assessment of electricity production in Tanzania. They paper presents the environmental assessment of the centralized grid-connected electricity production in Tanzania using a life cycle approach for the years 2000, 2015, 2020, 2026 and 2030, according to the Tanzania Electricity Supply Company Limited, TANESCO's, plans for power system expansion (power system master plan of the year 2009). The results of their study show that in all the chosen impacts categories, based on the absolute annual electricity production, environmental impacts increase significantly with time when comparing the years 2000, 2015, 2020, 2026 and 2030.

Kim 2007, illustrated that evaluation of negative environmental impacts of electricity generation: Neoclassical and institutional approaches. This paper focuses on three dimensions: theoretical and methodological backgrounds; critical review of specific studies: methodologies, results, and limitations; and discussing their results and implications for environmental policy and further research.

In the literature research is understood that in the Turkey almost isn't on this topic. This study focuses on the environmental impacts associated with the atmospheric emissions and other wastes that are produced from fossil fuels from electricity generation. And also a case study is presented for Turkey from 1975 to 2011, according to the TUİK (2015) data.

Materials and Methods

The world energy demand has increased rapidly during the last decades with the increase in the industrialization and population (Balta and Eke, 2011). There has been a continuous increase in the utilization of electricity produced from consumption of fossil fuels, which further causes several problems such as increased CO_2 emissions mainly from fossil fuel combustion products and leads to the global problems. The ongoing growth in fossil fuels consumption suggest that global carbon dioxide emission are still rising (Yilanci et al. 2008) The combustion of fossil fuels for generating electricity is the largest source of CO_2 emissions.

The environmental impacts associated with the electricity generated and supplied in Turkey from 1975 to 2011 were evaluated and compared for comparison purposes. This study mainly focused on the assessment of Turkey's electricity production. Electricity production from fossil fuels such as coal, liquid fuels and natural gas are considered, in this study. Table 35.1 illustrates that between 1975 and 2011 year's electricity production and energy resource in Turkey.

Table 35.1 1975–2011 years electricity production and shares by energy resources in Turkey (TUIK 2015)

Years	Coal (TJ)	Liquids Fuel (TJ)	Natural gas (TJ)	Hyro (TJ)	Renewable energy (TJ)
1975	36.97964	55.30073	0	22.98182	3.936996
1976	38.99764	55.52474	0	32.58676	2.961846
1977	44.05023	70.68396	0	33.37282`	4.07187
1978	50.0567	68.6559	0	36.35603	2.346408
1979	57.97163	58.00001	0	40.0545	2.432376
1980	53.6256	59.93918	0	44.20157	2.5137
1981	55.29219	59.48907	0	49.06492	1.776456
1982	58.06922	61.02287	0	55.17809	0
1983	77.28262	76.31782	0	44.16579	0
1984	91.19911	72.24292	0	52.30134	0.551052
1985	135.1993	72.675	0.447913	46.87466	0
1986	175.055	71.67964	8.833058	46.18864	0.71451
1987	158.8724	56.42766	16.54605	72.49374	0.798354
1988	112.4347	34.01581	21.06956	112.7535	0.864882
1989	182.2025	43.78482	62.33182	69.87304	0.936774
1990	181.7783	40.73699	66.65951	90.0216	1.035774
1991	194.1126	33.99682	82.40829	88.38891	2.168856
1992	221.2185	53.89246	70.95913	103.5169	2.424312
1993	213.8956	53.00891	70.52656	132.1264	2.657088
1994	253.7633	57.05444	90.21792	119.176	2.819592
1995	252.2725	59.28791	108.3781	138.2832	6.209784
1996	274.0563	67.1566	112.3744	157.6334	5.122548
1997	304.9298	73.12737	144.6751	154.7649	7.437312

(continued)

Table 35.1 (continued)

Years	Coal (TJ)	Liquids Fuel (TJ)	Natural gas (TJ)	Hyro (TJ)	Renewable energy (TJ)
1998	320.7426	80.87509	162.7622	164.1802	5.995188
1999	333.2513	82.43253	237.7679	135.0351	6.28776
2000	344.0352	96.12748	302.5081	120.0782	6.745788
2001	345.7163	105.7693	324.4971	93.60894	6.62715
2002	288.8208	110.1945	343.8402	130.929	6.9876
2003	289.7374	93.75347	415.8737	137.3183	5.060916
2004	310.5886	78.85424	407.3371	179.4556	5.425128
2005	389.1803	56.49673	480.1662	153.7854	5.830416
2006	420.4755	45.22095	528.4625	172.2084	6.3468
2007	481.0021	66.82309	621.8396	139.4024	13.79218
2008	519.6567	77.35921	645.4072	129.7235	21.42914
2009	501.4487	49.96953	628.581	140.2548	42.07961
2010	596.6564	23.53593	698.1251	218.7148	78.45309
2011	612.1543	9.828957	711.6268	212.502	112.0846

Table 35.2 GHG emissions values (IPCC 1996; CIPEC 2002; Hepbasli 2010)

Fuel type	CO_2 (t/TJ)	CH_4 (kg/TJ)	NO_x (kg/TJ)
Natural gas	49.68	0.13–1.27	0.62
Liquid Fuels[a]	59.84–74	0.16–121.11	3.11–12.5
Coal	93.8–95	–	–

t ton, *kg* kilogram, *TJ* terajoule
[a]Gasoline, kerosene, jet fuel, LPG, diesel, jet fuel, aircraft

The Table 35.2 is given from combustion resulting factor GHG emission values. It is clear that from the Table 35.2 coal of CO_2 emission is much higher than liquid fuels and natural gas. Therefore, if we want to reduce GHG emissions, we should decrease the usage of coal.

GHG emission values which are given in Table 35.2, used for calculating the CO_2, CH_4 and NO_x emissions in Table 35.3. The average values are used [14-16], while calculating the CO_2, CH_4 and NO_x emissions.

Table 35.3 illustrates that the electricity production emission values from 1975 to 2011 in Turkey. It is obviously seen that CO_2 emissions are the highest energy resource is coal as expectedly.

Total electricity productions 1975–2011 is shown in Table 35.1. Table 35.2 show that between 1975 and 2011 years emission release into atmosphere in Turkey. Coal emission release CH_4 and NO_x values could not be found. As the approached the 2000s years, because of increase in population growth and electricity demand, has increased emissions released into the atmosphere. Table 35.1 and Table 35.2 data's for this paper has been collected from the Turkish Statistical Institute (TUİK) database (TUIK 2015).

Figure 35.2 shows that the electricity production percent of distribution from years 1980 to 2010 in Turkey [13]. In 1980 year, electricity production as percent of

Table 35.3 Emission release values between the years 1975–2011 in Turkey (TUİK, 2015)

Years	Coal			Liquids Fuel			Natural gas		
	CO_2 (t/year)	CH_4 (t/year)	NO_x (t/year)	CO_2 (t/year)	CH_4 (kg/year)	NO_x (kg/year)	CO_2 (t/year)	CH_4 (kg/year)	NO_x (kg/year)
1975	3505.67	–	–	3700.725	3351.224	431.3457	0	0	0
1976	3696.976	–	–	3715.716	3364.799	433.093	0	0	0
1977	4175.962	–	–	4730.171	4283.448	551.3349	0	0	0
1978	4745.376	–	–	4594.453	4160.547	535.516	0	0	0
1979	5495.71	–	–	3881.36	3514.8	452.4	0	0	0
1980	5083.707	–	–	4011.13	3632.314	467.5256	0	0	0
1981	5241.7	–	–	3981.009	3605.038	464.0147	0	0	0
1982	5504.962	–	–	4083.65	3697.986	475.9784	0	0	0
1983	7326.393	–	–	5107.189	4624.86	595.279	0	0	0
1984	8645.675	–	–	4834.496	4377.921	563.4948	0	0	0
1985	12,816.89	–	–	4863.411	4404.105	566.865	22.25232	0.313539	0.277706
1986	16,595.21	–	–	4796.802	4343.786	559.1012	438.8263	6.183141	5.476496
1987	15,061.11	–	–	3776.139	3419.516	440.1358	822.0076	11.58223	10.25855
1988	10,658.81	–	–	2276.338	2061.358	265.3233	1046.736	14.74869	13.06313
1989	17,272.8	–	–	2930.08	2653.36	341.5216	3096.645	43.63227	38.64573
1990	17,232.59	–	–	2726.119	2468.662	317.7485	3311.645	46.66166	41.3289
1991	18,401.88	–	–	2275.067	2060.207	265.1752	4094.044	57.6858	51.09314
1992	20,971.51	–	–	3606.483	3265.883	420.3612	3525.249	49.67139	43.99466
1993	20,277.3	–	–	3547.356	3212.34	413.4695	3503.759	49.36859	43.72646
1994	24,056.76	–	–	3818.083	3457.499	445.0247	4482.026	63.15254	55.93511
1995	23,915.43	–	–	3967.547	3592.848	462.4457	5384.225	75.86468	67.19443
1996	25,980.54	–	–	4494.12	4069.69	523.8215	5582.758	78.66205	69.6721
1997	28,907.34	–	–	4893.684	4431.519	570.3935	7187.457	101.2725	89.69853
1998	30,406.39	–	–	5412.161	4901.03	630.8257	8086.024	113.9335	100.9125

1999	31,592.22	–	–	5516.385	4995.412	642.9738	11,812.31	166.4375	147.4161
2000	32,614.54	–	–	6432.851	5825.325	749.7943	15,028.6	211.7557	187.555
2001	32,773.91	–	–	7078.082	6409.62	825.0006	16,121.01	227.148	201.1882
2002	27,380.21	–	–	7374.213	6677.784	859.5167	17,081.98	240.6882	213.1809
2003	27,467.11	–	–	6273.982	5681.46	731.2771	20,660.6	291.1116	257.8417
2004	29,443.8	–	–	5276.925	4778.567	615.063	20,236.5	285.1359	252.549
2005	36,894.29	–	–	3780.761	3423.702	440.6745	23,854.66	336.1164	297.7031
2006	39,861.08	–	–	3026.186	2740.39	352.7234	26,254.02	369.9237	327.6467
2007	45,599	–	–	4471.801	4049.479	521.2201	30,892.99	435.2877	385.5405
2008	49,263.46	–	–	5176.878	4687.968	603.4018	32,063.83	451.7851	400.1525
2009	47,537.33	–	–	3343.961	3028.154	389.7624	31,227.9	440.0067	389.7202
2010	56,563.03	–	–	1575.024	1426.277	183.5802	34,682.86	488.6876	432.8376
2011	58,032.23	–	–	657.7538	595.6348	76.66586	35,353.62	498.1388	441.2086

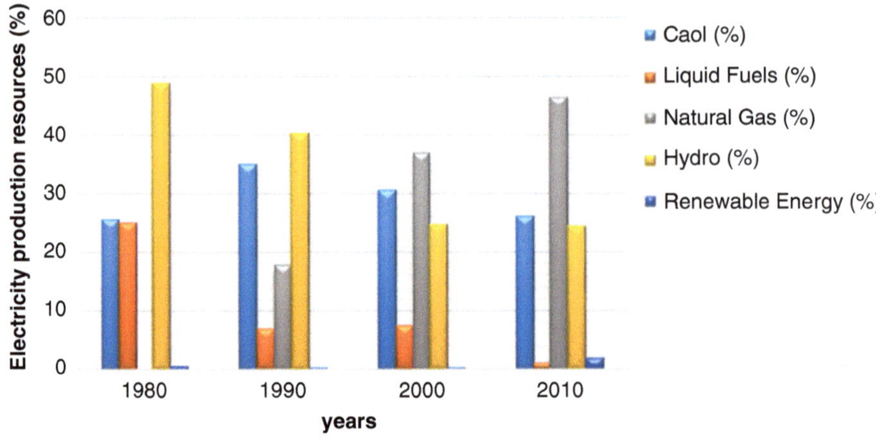

Fig. 35.2 Electricity production resources of Turkey from 1980 to 2010

rate respectively coal 25.6 %, liquid fuels 25.1 %, natural gas 0 %, hydro 48.8 % and renewable energy 0.6 %. Moreover, in 2000 years, while from coal and natural gas increase in for electricity production, hydro and liquids fuels usage of decreased shows that from Fig. 35.1. Approximately, percentage of the electricity production from coal and natural gas at about 72 % in 2010 year. As can be seen in this figure that electricity production from natural gas increases from 1988 in Turkey.

Results and Discussion

In this paper CO_2, CH_4 and NO_x emissions are calculated from 1975 to 2011 for Turkey. The Turkish Statistical Institute (TUİK) database is considered (TUIK 2015) while calculating these emission values.

Figure 35.3 shows that CO_2 emissions from electricity production in Turkey between 1975 and 2011. The CO_2 emission values from liquid fuels did not change too much, between 1975 and 2011. But, an increase is seen for coal and natural gas CO_2 emission values in this figure. As expected that CO_2 emissions values from coal are higher than that of others. CO_2 emissions values from coal and natural gas were observed sharply changes in 2002. CO_2 emissions values from coal decreased while the CO_2 emissions values of natural gas increased at 2002.

Figure 35.3 illustrates the change of electricity production resource in Turkey from 1975 to 2011. Especially, the usage of natural gas subsequently increase after 1998. The usage of liquid fuels decrease because of the commonly usage of the natural gas after 1998. The use of natural gas is better than coal in terms of emissions. In 2001 year was observed that from natural gas electricity production is higher than coal (Fig. 35.4).

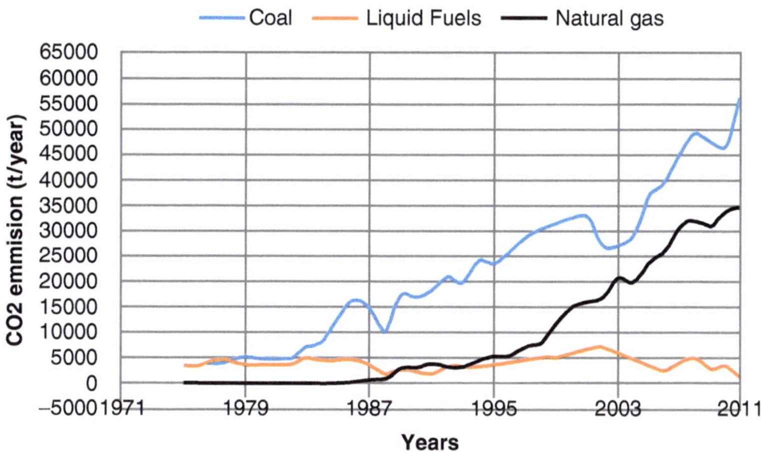

Fig. 35.3 Between 1975 and 2011 from electricity production release CO_2 emission

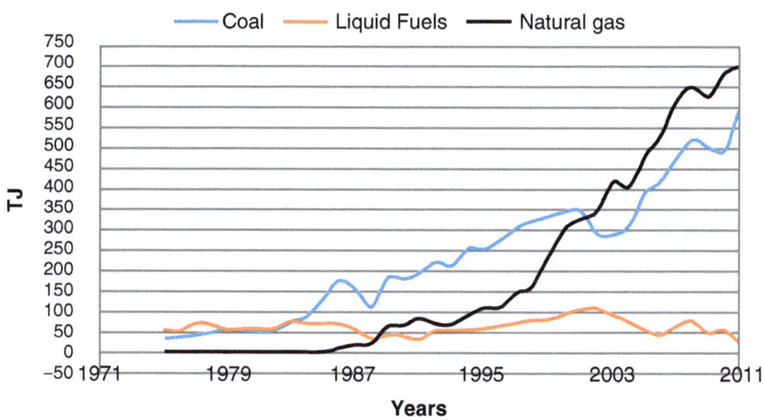

Fig. 35.4 Between 1975 and 2011 electricity production resource (TUİK 2015)

CH_4 emission values from 1975 to 2011 for electricity production resource is given in Fig. 35.5. While resulting from natural gas emissions quite a few, but liquid fuels are higher. Also as can be seen in this figure that, CH_4 emissions from liquid fuels decreases, while CH_4 emissions from natural gas increases after than 2003.

NO_x emission values from 1975 to 2011 for electricity production resource is given in Fig. 35.6. Also as can be seen in this figure that, NO_x emissions from liquid fuels decreases, while NO_x emissions from natural gas increases after than 1988.

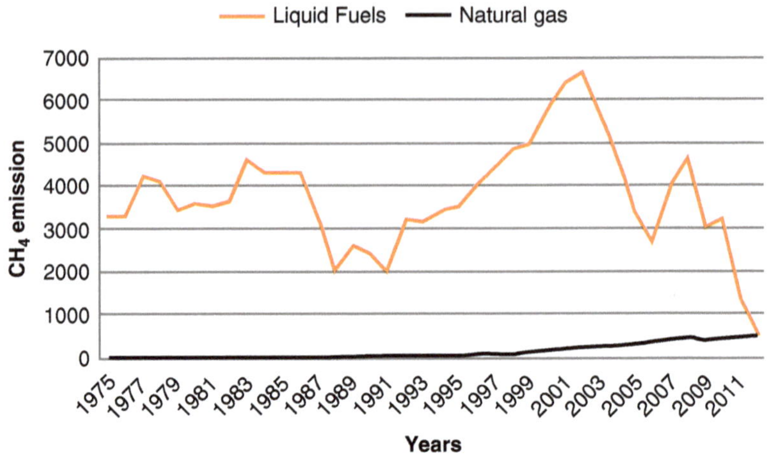

Fig. 35.5 Between 1975 and 2011 from electricity production release CH_4 emission

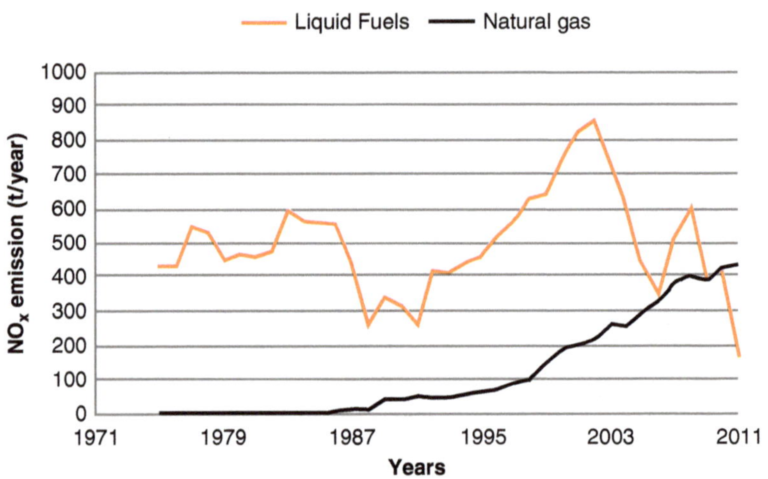

Fig. 35.6 Between 1975 and 2011 from electricity production release NO_x emission

Conclusions

In this study, an environmental impact assessment of electricity production was investigated and compared for CO_2, CH_4 and NO_x emissions in Turkey.

Some concluding remarks from this study can be extracted as follows:

(a) In 2011, the electricity production from coal is 642.15 TJ, and corresponding CO_2 emission is 58032.23 t/year. And 711.62 TJ for natural gas and corresponding CO_2 emission is 498.13 kg/years.

(b) It has been observed a continuous increase in electricity production from coal and natural gas from 1975 to 2011.
(c) It has been observed that most of the CO_2 emissions arise from coal and CH_4 emissions from liquid fuels higher than natural gas.
(d) Until 2009 year, it has been observed that from liquid fuels NO_x emissions higher than that of natural gas, but after this year this situation is reversed.

As can we concluded that, in Turkey the electricity produced mostly from coal and natural gas. The usage of liquid fuels decreased while the usage of the natural gas increased.

Nomenclature

GHG	Greenhouse gas
MW	Megawatt
TJ	Terajoule
t	Ton
kg	Kilogram

References

Balta, M.O., & Eke, F. (2011). Spatial reflection of urban planning in metropolitan areas and urban rent; A case study of Cayyolu, Ankara. European Planning Studies, 19 (10), 1817–1838.

Churchill, S.W. (1997). The reduction of air pollution by improved combustion. *Energy Conversion and Management 38*, 1335–1341.

CIPEC. (2002). Canadian Industry Program for Energy Conversation. "Energy Efficiency Planning and Management Guide" the Office of Energy Efficiency of Natural Resource Canada, Ottawa.

Felix, M., & Grewal, S. H. (2012). Environmental assessment of electricity production in Tanzania. *Energy for Sustainable Development, 16*, 439–447.

Hayami, H., et al. (1993). Estimation of air pollutions and evaluating CO_2 emissions from production activities: using Japan's 1985 input–output tables. *Journal of Applied Input–Output Analysis, 1*(2), 29–44.

Hepbasli, A. (2010). Enerji Verimliliği ve Yönetim Sistemi Yaklaşımlar Ve Uygulamalar. *Schneider Electric Enerji Verimliliği Serisi, 1*, 45–63.

IEA. (2007). *Fossil fuel-fired power generation. Case studies of recently constructed coal- and gas-fired power plants*. Paris: IEA/OECD.

IPCC. (1996). Intergovernmental panel on climate change, climate change 1995; The Science of Climate Change (Türkiye İstatiitik Kurumu, 2015). (p. 120). Cambridge, UK: Cambridge University Press. Tablo 2-9 "Radiative Forcing of Climate Change."

Kim, S. H. (2007). Evaluation of negative environmental impacts of electricity generation: Neoclassical and institutional approaches. *Energy Policy, 35*, 413–423.

Liaskas, K., et al. (2000). Decomposition of industrial CO2 emissions: the case of European Union. *Energy Economics, 22*, 383–394.

Lin, S. J., & Chang, T. C. (1996). Decomposition of SO2, NOx and CO2 emissions from energy use of major economic sectors in Taiwan. *The Energy Journal, 17*, 1–17.

Mukhopadhyay, K., & Forssell, O. (2005). An empirical investigation of air pollution from fossil fuel combustion and its impact on health in India during 1973–1974 to 1996–1997. *Ecological Economics, 55*, 235–250.

Nicoletti, G. (2001). Limpatto ambientale dei combustibili generalmente impiegati peril riscaldamento urbano degli edifici. In: *Proceeding of Italian Conference 56 Congresso Nazionale ATI*, Naples.

Paul, S., & Bhattacharya, R. N. (2004). CO2 emission from energy use in India: A decomposition analysis. *Energy Policy, 32*(5), 585–593.

Proops, J. L. R., et al. (1996). The life time pollution implications of various types of electricity generation: an input–output analysis. *Energy Policy, 24*(3), 229–237.

TEİAŞ. Turkish electricity transmission company. http://www.teias.gov.tr/. Retrieved February 16, 2015.

TUIK. (2015). Turkish Statistical Institute. www.turkstat.gov.tr. Retrieved March 2, 2015.

Wier, M. (1998). Sources of changes in emissions from energy: A structural decomposition analysis. *Economic Systems Research, 10*(2), 99–112.

Yilanci, A., Dincer, I., & Ozturk, H. K. (2008). A review on solar-hydrogen/fuel cell hybrid energy systems for stationary applications. *Progress in Energy and Combustion Science*. doi:10.1016/j.pecs.2008.07.004.

Zhang, Z. (2000). Decoupling China's carbon emissions increase from economic growth: An economic analysis and policy implications. *World Development, 28*(4), 739–752.

Chapter 36
Hybridization of Parabolic Trough Power Plants with Natural Gas Through Integration of Industrial Gas Turbines

Tobias Vogel, Gerd Oeljeklaus, and Klaus Görner

Introduction

The operation of concentrating solar power (CSP) plant concepts without thermal energy storage is completely dependent on the course of the sun. Because of this dependency such power plants have only a limited capability of regulating the electrical grid. The operation temperature is also very closely linked to collector technology and available solar radiation. The input of fossil fuel, however, can be controlled and therefore used according to needs. Through the systematic combination of solar and fossil energy a more flexible power plant is created. Even from financial standpoint hybrid plants offer in mid-term a cost reduction potential and thereby lessen economic inhibitions.

The total electricity consumption will grow until 2035 by roughly 65 %, as predicted in (BP 2014) based on the reference year 2012, whereas the absolute share of natural gas will be approximately constant at 20.5 % (BP 2014). This indicates that natural gas will play in mid-term an important role. In terms of the combined usage of natural gas and concentrated solar thermal energy in one power plant the most common solution is the ISCC (Integrated Solar Combined Cycle). Due to technical reasons the amount of integrable solar energy is strongly limited, for instance the annual solar share is relatively low at 2–8 % (Turchi et al. 2011). An alternative approach is to integrate smaller gas turbines, like industrial or aeroderivative gas turbines, into a CSP plant. A fairer performance ratio and a comparatively lower natural gas usage are achievable, for what reason this idea is increasingly studied in the last years (e.g. Turchi et al. 2011; Vogel et al. 2013; Servert et al. 2014).

T. Vogel (✉) • G. Oeljeklaus • K. Görner
LUAT, University of Duisburg-Essen, Leimkugelstrasse 10, Essen 45141, Germany
e-mail: tobias.vogel@uni-due.de

P. Grammelis (ed.), *Energy, Transportation and Global Warming*, Green Energy and Technology, DOI 10.1007/978-3-319-30127-3_36

For this case study the configuration of the parabolic trough power plant Shams One (cf. Goebel 2010), located Madinat Zayed in Abu Dhabi part of the United Arab Emirates (UAE) has been used. The power plant itself as well as the region fulfills all requirements to make such an analysis worthwhile.

The UAE have the world's seventh largest proved reserves of natural gas (BP 2014). For that reason natural gas covers roughly 63 % of the UAE's primary energy consumption and thus represents their main energy source (BP 2014). In 2011 the installed gross electricity capacity has been divided into 18.5 GW_{el} of combined cycle, 5.4 GW_{el} gas turbine and 2.3 GW_{el} steam turbine (UAE 2012). But in Abu Dhabi itself are still 0.4 GW_{el} of open cycle gas turbine capacity installed, which indicates a potential of combining such gas turbines with Shams One. Due to an expected rising energy consumption and sharper summerly load transients the local grid operator TRANSCO is forecasting an amount of 1.3 GW_{el} installed gross capacity of open cycle gas turbines until 2020 (TRANSCO 2014).

In addition it should be noted, that there is one additional power plant located in Madinat Zayed with the name "Madinat Zayed Power Plant," owned by the Al Mirfa Power Company. It consists out of four open cycle gas turbines with an installed gross capacity of 0.1 GW_{el} (Al Mirfa Power Company 2014). Altogether it can be assumed that at least for this application example from outer perspective a market potential for the developed hybridization configuration exists.

The Site

As site for the calculations Madinat Zayed in the United Arab Emirates was chosen, where also Shams One is located. The weather profile was taken in hourly time resolution from the meteonorm database (version 7.0). This averaged dataset has a yearly total direct normal irradiance (DNI) of 1889 kWh/(m^2a), which corresponds well with the number of 1925 kWh/(m^2a), published in (Goebel 2010), whose value is based on ground data collected during the project development of Shams One. The sorted distribution curve and the interval frequency of the aperture effective DNI (cosine correction included) are shown in Fig. 36.1, whereas for the interval frequency only all hours with an aperture effective DNI greater than 1 Wh/(m^2h) are considered.

Within respect of the cosine correction the yearly available aperture effective DNI is lowered by 8.5 % to 1729 kWh/m^2. An average aperture effective DNI of over 700 Wh/(m^2h) appears only in 204 h, which underpins the medium insolation quality for the analyzed site.

The design point was set to be on 21st March at solar noon, because of the comparatively low aerosol charge of the air in spring compared to the summer month. Referring to (Goebel 2010) the DNI for the design point is 750 W/m^2.

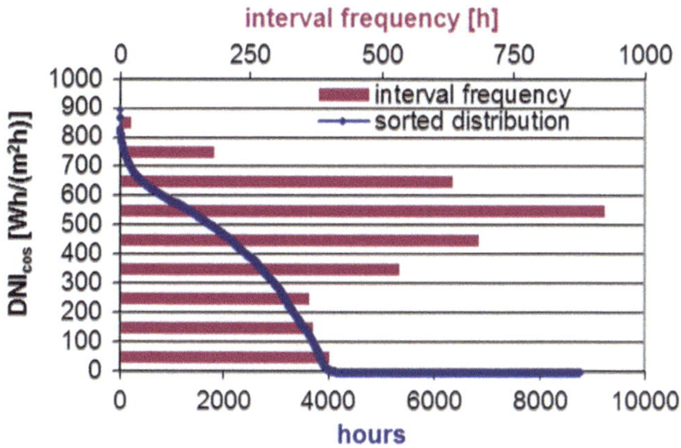

Fig. 36.1 Sorted distribution curve and interval frequency of the aperture effective DNI for the used weather profile from Madinat Zayed

The Base Case Power Plant

The layout of the base case power plant is shown in Fig. 36.2 and the main technical performance data from the design point are given in the diagram in the lower left part of Fig. 36.2. The thermodynamic model can be divided into the solar-heat-transfer-fluid-cycle (solar-HTF-cycle) and the water/steam cycle. The main parameters were primarily taken from publications or otherwise estimated based on typical values and design guidelines.

Solar-HTF-Cycle

The solar-HTF-cycle has the following two main tasks:

- transform the solar irradiation into heat.
- transfer the gained heat to a power generation process.

Therefore the solar-HTF-cycle consists out of the main components solar field, HTF-pump, HTF-heater and the connecting solar steam generator (SSG).

As heat transfer fluid (HTF) the thermal oil Therminol VP1 has been used, similar to Shams One. It is thermally stable up to 400 °C, whereas the related vapor pressure is 10.9 bar (SOLUTIA 2013). For security reasons a HTF-evaporation should be avoided at any time. In order to fulfill this requirement the minimum pressure (position: suction HTF-pump) in the solar-HTF-cycle has been set to 15 bar. The pressure loss over the complete solar field is assumed to be 5.25 bar (Balan 2014) for nominal conditions and in the SSG 2.5 bar, so that the HTF-pump discharge pressure overall amounts to 22.75 bar.

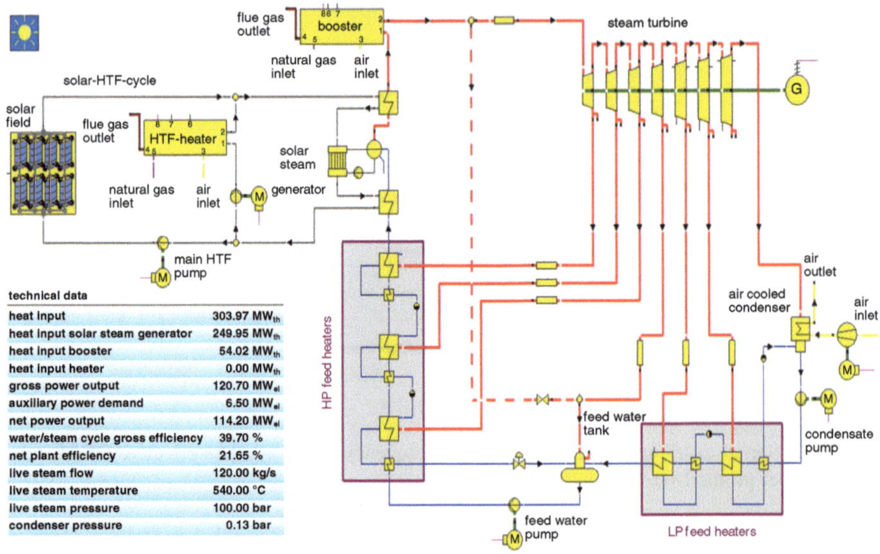

technical data	
heat input	303.97 MW_{th}
heat input solar steam generator	249.95 MW_{th}
heat input booster	54.02 MW_{th}
heat input heater	0.00 MW_{th}
gross power output	120.70 MW_{el}
auxiliary power demand	6.50 MW_{el}
net power output	114.20 MW_{el}
water/steam cycle gross efficiency	39.70 %
net plant efficiency	21.65 %
live steam flow	120.00 kg/s
live steam temperature	540.00 °C
live steam pressure	100.00 bar
condenser pressure	0.13 bar

Fig. 36.2 Process flow diagram of the base case power plant based on the configuration of Shams One

The solar field of Shams One consists out of 768 ASTRØ 150 collectors (Goebel 2010), in a typical 4 collectors per loop arrangement. Due to the fact that the solar library (Pawellek et al. 2009) of the power plant simulation program Ebsilon® Professional includes only a model of the EUROTrough ET 150 collector, this comparable collector type has been used. In order to serve a sufficient thermal heat input even in case of low solar insolation, e.g. due to cloud passing, the solar-HTF-cycle is additionally furnished with a HTF-heater, which is fuelled with natural gas.

At nominal conditions 1061.6 kg/s of HTF are circulating, whereby its temperature is elevated from 296.6 °C (HTF-side outlet of SSG) to 393 °C (HTF-side inlet of SSG) while passing the solar field.

Water/Steam Cycle

The main components in the water/steam cycle are the SSG, the booster, the turbogenerator (steam turbine and generator), the air-cooled condenser, condensate and feedwater pump as well as the low and the high pressure feed heaters.

The solar heat is transferred to the water steam cycle via the SSG. By that heat input the feedwater entering the SSG at 219.46 °C and 111 bar is preheated, evaporated and superheated up to 380 °C at 102.5 bar. Subsequently, the steam is further superheated in a natural-gas-fired booster. At live steam parameters of 540 °C and 100 bar (Reuß 2012) the steam enters a 35-staged, single flow steam turbine (MAN 2014). The 35 turbine stages can be arranged for the thermodynamic

Table 36.1 Efficiency data (isentropic, mechanical and electrical) for the process machinery

Component	η_s (%)	η_{mech} (%)	η_{el} (%)
Electric motor	–	99.80	97.00
Generator	–	–	98.55
Pump	80.00	99.80	–
Blower	86.00	99.80	–
HP steam turbine (1. stage group)	84.00	99.80	–
IP steam turbine (2.-6. stage group)	90.00	99.80	–
LP turbine (7. stage group)	86.00	99.80	–

modelling related to the required extractions for the feedwater preheating into 7 stage groups. Through passing the steam turbine the steam is expanded to the condenser pressure of 0.13 bar (Reuß 2012), while six partial flows are extracted for the feedwater preheating. For the last 3 stage groups the wetness correction for the isentropic efficiency was used because of the present steam wetness. Following, the discharge steam streams through an air-cooled condenser (ACC). Then the fully condensed water is delivered by the condensate pump to the feedwater tank, while passing two low pressure (LP) feed heaters. From the feedwater tank the feedwater is led through the feedwater pump to the SSG, while passing three high pressure (HP) feed heaters. The extraction pressure levels have been chosen in order to enable a temperature step of 28 K in each preheater (regulation cf. Schröder 1966). All assumed machinery efficiency data are given in Table 36.1.

At nominal conditions with a heat input of 301 MW_{th} 120 kg/s of water/steam circulates in the system. The power plant therefore generates 120.7 MW_{el} of gross electricity, which represents a water/steam cycle gross efficiency of 39.45 %. In part load the live steam pressure level is set in general by the steam generator while the pressure entering the first stage of the turbine undergoes the Stodola's law. Because of that coherence for part load operation a modified sliding pressure mode was assumed for the steam generator.

Technical Boundary Conditions

An important basis for the development of the operating logic of a power plant represents the knowledge of the elementary technical boundary conditions, whereby for reasons of presentation a differentiation into the three areas solar field, water/steam-cycle and natural-gas-firing has been made.

For the solar field off-design operation a minimum circulating HTF mass flow of 200 kg/s has been considered, while the outlet temperature is limited to 393 °C. In case of high solar insolation when the heat-entry oversteps the nominal value, the troughs are defocussed automatically until the heat-entry achieves the allowed nominal value.

Regarding the water/steam cycle the SSG and the steam turbine are in general the limiting components, whereby in this work the steam turbine has been used as benchmark. For that reason an industry-related, steam turbine model has been developed, whose technical boundary conditions mainly occur during the off-design operation. The following main technical boundary conditions for the steam turbine with maximum allowable steam wetness (15 %), maximum turbine-exhaust pressure (0.23 bar), smoothed live steam temperature and permissible temperature gradients for heating (5 K/min) and cooling (−2 K/min) have been identified in (Vogel et al. 2013), thus this reference is referred for further explanation. It has to be mentioned, that as specific state point for the steam wetness at the steam turbine outlet the expansion line end point has been used. Furthermore the maximum turbine-exhaust volume flow factor is limited to the factor 1.5 of the nominal value for the volume flow at the turbine-exhaust. Besides, the live steam temperature is smoothed by the booster until 3 O'clock pm, in order to avoid negative temperature gradients during main steam turbine operation time.

According to (Goebel 2010) the amount of natural gas is limited to 600,000 MMBTU/a. From that the booster has an approximated contribution share of 81.8 % and the HTF-heater one of 18.2 % (Goebel and Luque 2012). Due to the fact that the composition of natural gas fluctuates and is not clear defined, in this case study methane was used as reference fuel with a lower heating value of 50,015 kJ/kg. Based on these boundary conditions the heater has in total a potential of 2,301.3 $t_{methane}$/a and the booster one of 10,355.6 $t_{methane}$/a. In addition the booster's min-load is set to 5 % of nominal fuel mass flow.

Operational Logic

The power plant operation can be divided in four main operating states, namely: night, below min-load, normal day and heater operation. During night no insolation is available and the power plant is out of operation. In exceptional situations the grid operator can prescribe power plant operation during night by using the fossil-only mode (Goebel and Luque 2012), but that exception has been neglected for the performed case study. The other three operating states are during the day. If the insolation is too low to provide the minimum amount of circulating HTF-mass flow the power plant is out of operation as well. If the DNI is high enough the power plant is in typical normal day operation. Is the power plant in normal operation and it appears a sudden (or forecasted) short-term insolation drop in the next time step, for example caused by cloud passing, the power plant should still operate in order to avoid start-up and shut-down losses by using the HTF-heater.

The following three identification criteria need to be fulfilled for HTF-heater operation. At first the HTF-heater shall operate only up to 16 o'clock to follow the typical diurnal variation of the insolation. Secondly the insolation drop compared to the previous DNI must be higher than 250 W/(m^2h). As third criteria for HTF-heater operation the DNI must be lower than 400 Wh/(m^{2}h). When HTF-heater operation is identified, the HTF-heater can be operated with its load in different ways. For this case study the HTF-heater operates in order to enable the same load, meaning entering the SSG with the same HTF-mass flow, as in the time step before.

The booster operates in respect to the given technical boundary conditions. When the steam generator achieves a load level close to full load in the previous and the actual hour the booster goes to full load and delivers live steam at 540 °C and 100 bar, as long as none of the named boundary conditions are opposed.

In case of the necessity to reduce the power plant load, e.g. due to the turbine-exhaust pressure limitation, firstly the fossil-firing is reduced to the technical lowest limit, e.g. booster min-load or smoothed live steam temperature. Secondly, the load of the solar field is reduced through defocussing.

The Base Case Power Plant with Front End Gasturbine

Concept Layout

Figure 36.3 shows the process flow diagram of the developed concept with the additional integration of a gas turbine into the base case power plant.

In general there are different ways for the integration of the gas turbine's waste heat possible, for instance the integration into the booster's air flow or directly into the water/steam-cycle. In order to avoid technical changes of the installed booster arrangement the waste heat has been integrated directly through two heat

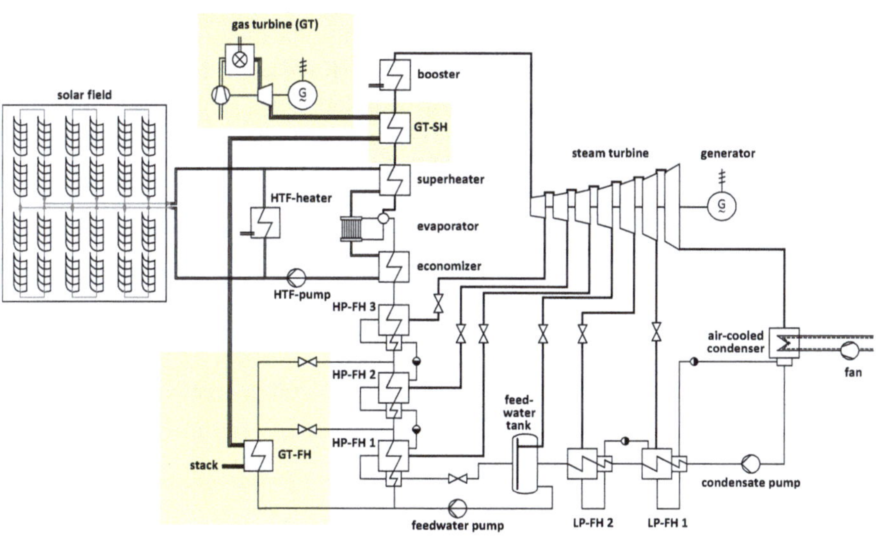

Fig. 36.3 Process flow diagram of the base case power plant added with a front end gas turbine

exchangers into the water/steam cycle. The waste heat at high temperature level is firstly integrated through the gas turbine waste heat recovery superheater (GT-SH), located between the SSG-superheater and the booster, in order to lower the booster's load. The remaining waste heat with a temperature of up to 390 °C can be used part vise further for preheating the feed water. Replacing HP bled off steam is thermodynamically favorable. For that reason secondly the gas turbine waste heat recovery feed heater (GT-FH) is installed in parallel to the HP feed heaters (HP-FH). Thus, the required amount of bled steam is reduced and consequently a higher steam turbine power output reveals. In terms of corrosion by the cause of condensation of exhaust gas components it has to be mentioned, that through the chosen arrangement the stack temperature stays above 100 °C even in low load due to the minimum pressure in the feedwater tank of 1.15 bar. A displacement of the HP-FH 3 was not undertaken in order to secure always the required feed water temperature at the inlet of the economizer.

Gas Turbine Selection

In respect to the developed concept layout the following specification requirements have been used for the gas turbine selection:

- grid frequency: 50 Hz
- gas turbine performance class: 10–20 MW_{el}
- emission reduction technology: dry, resp.no water/steam injection
- exhaust gas temperature: at least 540 °C

Based on the market review, given in (Gas Turbine World 2011), out of the 276 listed gas turbines only four gas turbines fulfill the above mentioned requirements. Namely: THM 1304-14 (MAN), L20A (Kawasaki), H-15 (Hitachi) and SGT-400 (Siemens). In order to cover the complete range of electrical efficiency the THM 1304-14 (31.0 %) and the SGT-400 (34.8 %) have been chosen for the conducted case study. The technical data of these two gas turbines at nominal conditions (cf. ISO 2314-2009) are summed up in Table 36.2.

The operation of gas turbines has a strong dependence on the ambient conditions. For example, a change in the elevation level (resp. ambient pressure) or the

Table 36.2 Technical data for the two gas turbines for generation drive at nominal conditions (GTW-Performance Specs 2011)

Parameter	Unit	GTa)	GTb)
Type		THM 1304-14	SGT-400
Electric power	kW_{el}	12,680	12,900
Heat rate	kJ/kWh	11,610	10,355
Electric efficiency	%	31.0	34.8
Exhaust temperature	°C	545.0	555.0
Exhaust mass flow	kg/s	49.0	39.4
Pressure ratio	–	11.0	16.9

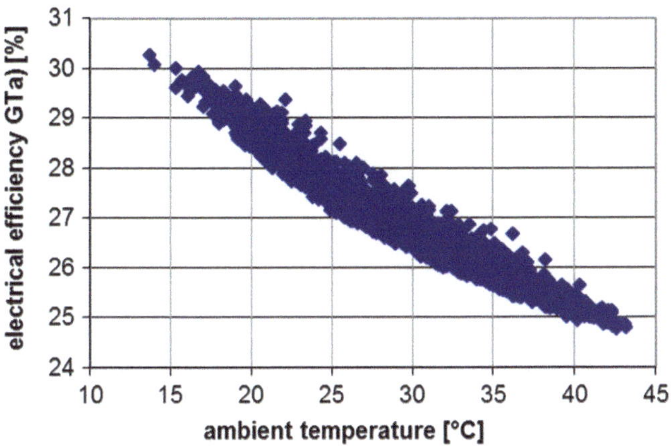

Fig. 36.4 Annual operating behavior of the gas turbine GTa)

ambient temperature have a strong influence on the performance of the gas turbine. For that reason the operating characteristic of the gas turbines need to be taken into account for the annual yield simulation. Therefore two different approaches have been used. For the GTa) a full thermodynamic model has been developed based on the design point data and the instructions given in (Lechner and Seume 2010), which covers the part load and operating behavior. Contrary for the GTb) characteristic curves provided by the manufacturer have been used. It should be noted, that GTb) has a customized design referring to the studied site. For both gas turbines the same site boundary conditions with an ambient pressure at the design-point of 99.9 kPa and pressure losses at gas turbine inlet and gas turbine outlet (due to waste heat recovery) of 1 and 2.5 kPa have been used.

Figure 36.4 shows for GTa) the distribution of the electrical efficiency for each operating point as subject to the ambient temperature.

The operating points of GTa) are arranged in Fig. 36.4 as a passel. That means that even for operating points with an equal ambient temperature different efficiencies occur, due to various ambient pressure and ambient humidity. With decreasing ambient temperature the GT-efficiency is raised. Summarizing, the developed thermodynamic model for GTa) enables a typical operating behavior.

The Solar Only Bench Mark Model

For the closing assessment of the concepts in terms of power generation efficiency from natural gas it is necessary to perform the annual yield simulations as well with a, so called, solar only model. Using the base case power plant and operating it in solar only mode is not feasible, because for example technical limitations like steam turbine wetness could not be fulfilled with the given design of the components. For

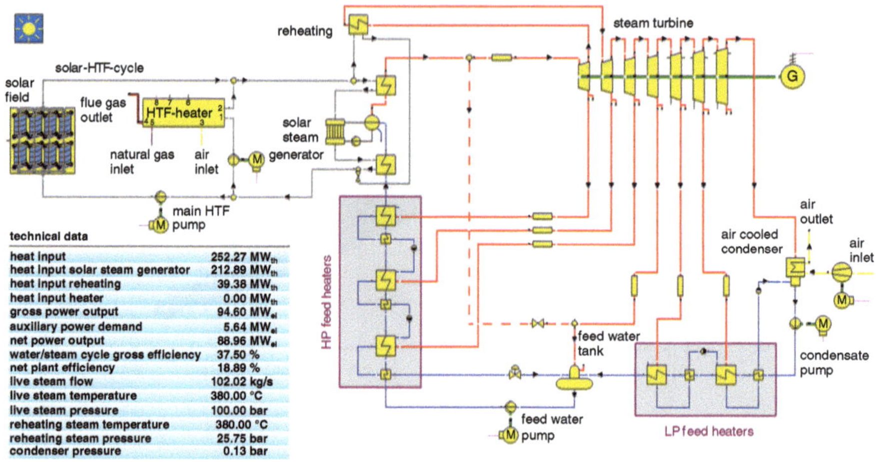

Fig. 36.5 Process flow diagram of the solar only benchmark model

that reason a solar only benchmark model has been developed, whose process flow diagram and technical data for the design-point are given in Fig. 36.5.

In comparison to the base case power plant, shown in Fig. 36.5, the booster has been removed. With the objective to fulfill still the steam turbine wetness limitation the power plant has been additionally equipped with a reheating. Therefore the superheated steam leaving the first turbine stage group at 225.54 °C and 25.75 bar is reheated up to 380 °C. Following the reheated steam is returned to the steam turbine for the further expansion. Through this arrangement the steam turbine-exhaust wetness limitation could be fulfilled, e.g. at the design-point the steam wetness is 11.4 %. The solar field and the nominal HTF-mass flow are still the same as for the base case power plant. But the HTF-mass flow is now split up between the solar steam generator and the reheating. For that reason the transferred heat into the water/steam cycle through the solar steam generator is lowered to 252.27 MW_{th}, while heat input in reheating amounts to 39.38 MW_{th}. As consequence the circulating steam mass flow is lowered to 102.02 kg/s. Using the efficiency data given Table 36.1 altogether a net power output of 88.96 MW_{el} at the design-point could be generated. Whereas the gross efficiency is 37.5 %, which represents a reduction of 2.2 %-points related to the base case power plant.

Results and Discussion

For all in the previous chapters mentioned configurations, namely

- solar only bench mark model: abbr.: solar only
- base case power plant: abbr.: BC

- base case power plant plus GTa) (THM1304-14): abbr.: BC + GTa)
- base case power plant plus GTb) (SGT-400): abbr.: BC + GTb),

annual yield simulations have been performed in hourly time resolution, based on the given information regarding the operational logic and the used exemplary annual weather profile.

In order to enable a fundamental understanding of the power plant model at first the behavior of the base case power plant is analyzed. Therefore initially the daily behavior is discussed based on two example daily curves, before subsequently the attention is given to the annual behavior.

Base Case Power Plant

In Fig. 36.6 the daily variation for process parameters (steam turbine inlet temperature, steam turbine outlet pressure and DNI), shown as curves, and performance values (heat input through SSG, HTF-heater and booster, generated net electricity), shown as bars, are displayed exemplary for (a) 25th March (spring) and (b) 17th July (summer).

Figure 36.6a shows with the 25th March an ideal sunny day in spring. The DNI is very constant and lies during the day at around 600 W/m^2. The power plant starts in solar only mode, wherefore the steam turbine inlet temperature without booster-operation is roughly 380 °C. In the next hour the solar heat input raises analog to the DNI and the daily shape of the incidence correction. Following, the steam mass flow is increasing and so the shape of the process in the T,s-diagram is moving to the left. Assuming a similar expansion in the steam turbine the steam wetness is increasing and the steam turbine wetness limitation moves into focus. As a result, the steam turbine inlet temperature is raised by the booster to 524 °C. Also in the following hour through the SSG a heat input of nearly 250 MWh/h is transferred to the water steam cycle. Thus, the booster goes into full load and delivers steam at

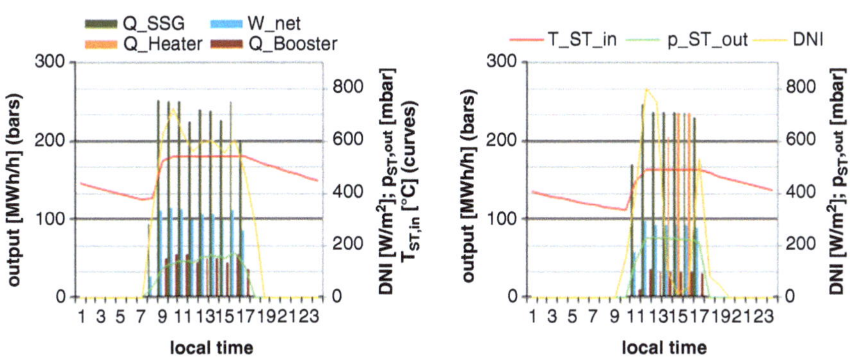

Fig. 36.6 Performance values (bars) and process parameters (curves) for the (**a**) 25th March and (**b**) 17th July

540 °C. Despite the later on slight reduction of the in the SSG transferred heat into the water steam cycle the steam temperature at steam turbine inlet is constant at 540 °C, due to the approach of smoothed live steam temperature. During night the steam turbine is cooling out. In spring the ambient temperature is on a moderate level, wherefore the steam turbine outlet pressure is obvious lower than the limitation of 230 mbar.

A summer day with a short-term insolation drop is shown in Fig. 36.6b with the 17th July. In comparison with Fig. 36.6a it is easy to identify that for that summer day the maximum electrical net output is not reached. This is caused by the hot ambient temperature (at that day roughly 39 °C) and is typical for the chosen design of the power plant with an ACC, whereby a solution could be to overdesign the ACC. The high ambient temperature yields to a large steam turbine outlet pressure with an ACC. For the shown example the maximum steam turbine outlet pressure is reached. For that reason the fossil input is lowered to the technical lowest level, e.g. steam turbine wetness limitation. Therefore the steam turbine inlet temperature is even at high DNI not at 540 °C but rather at 490 °C (12:00), which is typical for the summer period. The drop in the transferred heat through the SSG between 12:00 and 13:00 indicates that supportive the solar field is defocussed in order to lower the power plant load. After 13:00 there is a sudden insolation drop (753 $Wh/(m^2h) \rightarrow$ 111 $Wh/(m^2h)$), which is why the HTF-heater starts its operation and enables a steady transferred heat to the water/steam cycle. In the late afternoon the DNI is raising again, so that the HTF-heater is turned off.

The distribution of all base case power plant operating points with a positive net power output are shown in Figs. 36.7 and 36.8. Based on these two diagrams the functionality of the developed operational logic will be discussed in detail.

In Fig. 36.7 the steam turbine exhaust pressure is plotted against the ambient temperature. Due to the different part load status of the power plant a certain ambient temperature leads to different steam turbine exhaust pressures. The steam turbine exhaust pressure increases with rising power plant load. For that

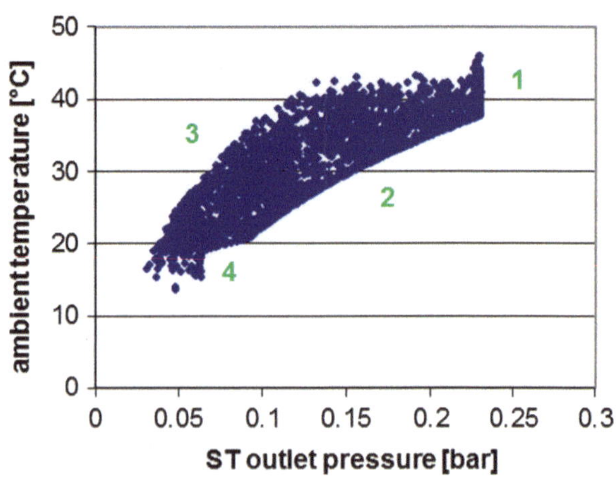

Fig. 36.7 Annual distribution of steam turbine (ST) outlet pressure with respect to the ambient temperature

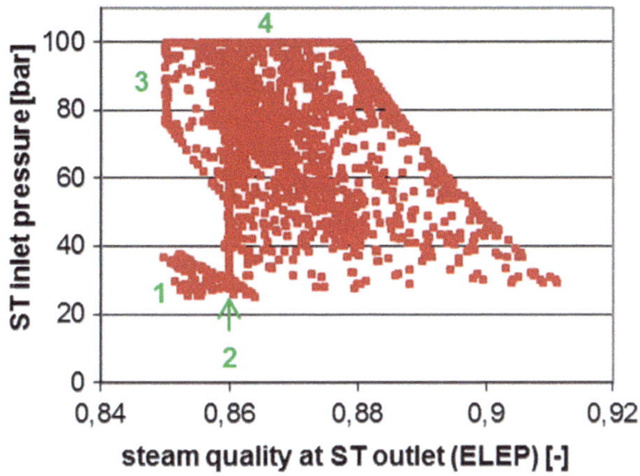

Fig. 36.8 Annual distribution of steam quality at the steam turbine (ST) outlet with respect to the live steam pressure

reason two curves form out, which represent the curve for min load (point 3) and curve for full load (point 2) of the power block as function of the ambient temperature. As already mentioned the maximum tolerable turbine exhaust pressure is 0.23 bar. This limitation is trespassed at high ambient temperatures with concurrent high load status of the power block. Through the operating logic these critical operating points are transferred into an operating point where this limitation is satisfied. Herefrom the perpendicular in the upper right part of the passel in Fig. 36.7 indicated by point 1, is formed out. The volume factor of the volume flow at steam turbine outlet is limited to 1.5. This limitation is most critical at low condenser pressure with simultaneously high part load status and is depicted in Fig. 36.7 by the notch in the passel at point 4.

Figure 36.8 shows the dependency of live steam pressure versus steam quality at steam turbine outlet at ELEP for the operating points. The steam quality describes the share of saturated steam (vapor) in a saturated condensate (liquid) and saturated steam (vapor) mixture. The passel at point 1, with operating points corresponding to a steam turbine inlet pressure of lower than 40 bar and a steam quality at steam turbine outlet of lower than 0.87, located in the lower left part in Fig. 36.8 represents the solar only operating points. In the morning as long as the steam wetness limitation is satisfied the power plants runs in solar only mode. When the load is increasing due to higher solar insolation with respect to the simultaneously raising live steam pressure, caused by the sliding pressure mode of the steam generator, the steam wetness is also increasing. For that reason the booster goes into operation. The minimum allowed steam wetness at ELEP is 15 %, which corresponds with a steam quality of 0.85. Because of the booster's min load of 5 % the set point of the controller regulating the boosters load is set to adapt the booster load in order to enable a steam quality at ELEP of 0.86. For that reason is at a steam quality of 0.86 a perpendicular (point 2). But when the booster is already at full load and delivers a live steam temperature of 540 °C and owing to a relatively low ambient temperature a low condenser pressure arises the booster cannot fulfill the steam wetness limitation. In order to fulfill the requirement the load of the power block must be reduced for example by reducing the solar heat through defocussing. The upper end of the passel at 100 bar (point 4) indicates the power blocks full load. There the steam quality is only depending on the ambient temperature.

In conclusion the annual yield simulation for the base case power plant has shown so far, that due to the chosen operating logic and the process layout nearly all operating points differ from each other. Therefore it is very important to perform the annual yield simulations with a fully thermodynamically model and not by characteristic curves.

Cross Comparison

A cross comparison of the results from the annual yield simulation for the four examined configurations of the case study is given with their main performance parameters in Table 36.3.

As a common constraint the incident solar energy is in all configurations the same, owing to the identical solar field and the uniform site. The yearly operating hours of all configurations are very similar and lay in a range of 3231–3246 h per year. For that reason the theoretical as well as the used solar heat input to the HTF are also in a similar magnitude. Referring to the incident solar energy the amount of

Table 36.3 Comparison of the simulation results of the annual yield analysis for the configurations of the case study

Simulation results	Solar only	BC	BC + GTa)	BC + GTb)	Units
Available DNI	1889.38	1889.38	1889.38	1889.38	$kWh/(m^2\ a)$
Net aperture area	627,786.55	627,786.55	627,786.55	627,786.55	m^2
Incident solar energy	783.72	783.72	783.72	783.72	GWh_{th}/a
Operating hours	3241	3231	3243	3246	h
Theoretical available solar heat input to HTF	682.63	682.27	683.30	683.51	GWh_{th}/a
Used solar heat input to HTF	636.46	635.99	637.26	637.54	GWh_{th}/a
Total gross electricity steam turbine	234.57	282.45	286.84	285.64	GWh_{el}/a
Total net electricity gas turbine	0.00	0.00	26.05	28.98	GWh_{el}/a
Net electricity	220.65	266.55	296.99	298.72	GWh_{el}/a
Supplied energy from natural gas	13.38	126.05	198.29	195.22	GWh_{th}/a
Design-point net plant efficiency	18.89	21.65	22.75	22.66	%
Annual net plant efficiency	18.40	20.31	21.45	21.62	%
Annual solar fraction	97.94	83.46	76.27	76.56	%

used solar heat input to HTF ranges between 81.2 % (solar only) to 81.3 % (BC + GT configurations). The gross electricity of the BC + GT configurations is elevated based on the BC model by 3.2–4.4 GWh_{el}/a owing to the gas turbine's waste heat integration. Comparing the BC + GT configurations, the larger available waste heat amount for GTa) yields to a 1.2 GWh_{el}/a larger steam turbine gross electricity output. Due to the better electrical efficiency of GTb) the benefit of GTa) in terms of steam turbine gross electricity output has been overcompensated so that the overall produced net electricity of GTb) is 1.7 GWh_{el}/a larger than the one for GTa). The net electricity output of the solar only system amounts to 220.65 GWh_{el}/a. The BC offers a 20.8 % higher net electricity output whereby the supplied energy from natural gas is raised by the factor 9.4. Through an additional GT-integration the net electricity output is additional raised by 11.4–12.1 %, while the gas consumption is increased by 50 %. The solar only model has because of the lower live steam temperature the lowest design-point net efficiency of 18.89 %. Following the hybridization the live steam temperature is elevated from 380 to 540 °C which leads for the BC to a 2.76 %-points higher design-point net efficiency. Through the further combination with gas turbines the design-point net efficiency could be raised by additional 1–1.1 %-points. For the solar only model the design point net plant efficiency and the annual net plant efficiency differ with absolute 0.5 %-points not that much, owing to very constant process parameters. For the hybrid configurations the difference between design-point and the annual value is larger and ranges between 1- and 1.3 %-points. This is especially caused by different process parameters between different operating states, e.g. the live steam temperature varies during operation between 380 and 540 °C. In analogy to the supplied energy from natural gas the annual solar fraction decreases beginning with 97.9 % for the solar only model upon further hybridization to 83.5 % (BC) and finally round about 76.3–76.6 % for the BC + GT configurations.

For the energetic comparison between the hybrid configurations the conversion efficiency of fuel into electricity η_{fuel} is introduced, which is defined by Eq. (36.1).

$$\eta_{fuel} = \frac{\Delta P_{net,hybrid-solar}}{\Delta \dot{Q}_{hybrid-solar}} = \frac{P_{net,hybrid} - P_{net,solar}}{\dot{Q}_{hybrid} - \dot{Q}_{solar}} \tag{36.1}$$

Therefore the difference of the produced net electricity of the hybrid power plant and the solar only benchmark model is divided by the difference of the used fuel, whereby also the hybrid case is subtracted from the solar only benchmark model. The resulting η_{fuel} for the hybrid configurations is visualized in Fig. 36.9.

Referring to the solar only bench mark model described in chapter 5 and the results of the performed annual yield simulation the η_{fuel} amounts for the base case power plant to 40.74 %. Through integrating industrial gas turbines the overall conversion efficiency of fuel into electricity could be elevated compared to the base case power plant. The highest increase, with absolute 2.20 %-points could be achieved by integrating GTb) in the BC. But even with the configuration BC + GTa) an enhancement of η_{fuel} could be realized with a plus of 0.55 %-points. Comparing these numbers, while taking the 3.8 %-points better electrical efficiency

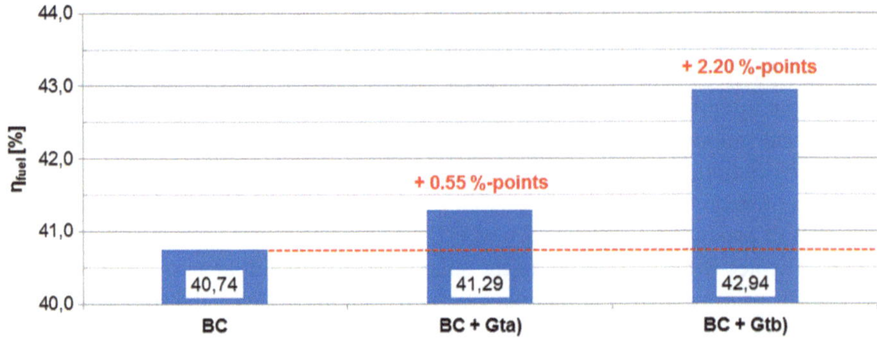

Fig. 36.9 Annual conversion efficiency of fuel into electricity for the three hybrid configurations BC, BC + GTa) and BC + GTb) of the case study

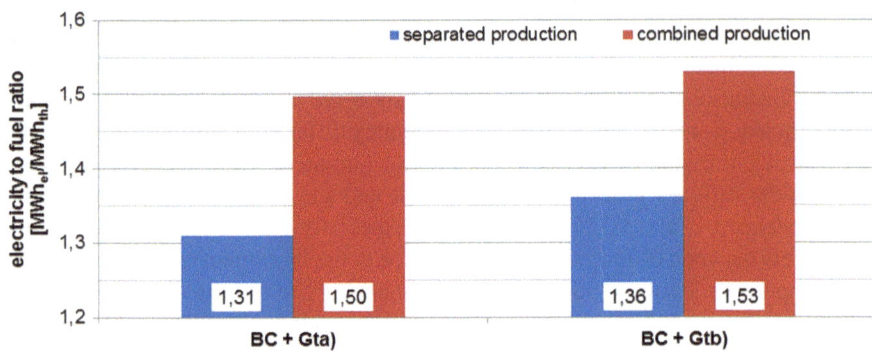

Fig. 36.10 Fuel to efficiency ratio for the two BC + GT configurations for separated and combined production

of GTb) at the design-point into account, the absolute advantage is lessened through the integrated configuration. This effect is mainly caused by the fact, that the gas turbine GTa) with a lower electrical efficiency offers at design point a 19 % higher waste heat potential than the GTb). Following this, the amount of bled steam for the configuration BC + GTa) could be lowered, which leads to the higher steam turbine gross electricity of round about 1.2 GWh_{el}/a.

As mentioned in the introduction a certain amount of peaker gas turbines is installed in the UAE. For the evaluation of the advantage of a combined usage in comparison to a separated configuration the electricity to fuel ratio is used. For the separated production the gas turbine's waste heat is not integrated into the base case power plant. Due to missing information on the operation scheme of the peaker gas turbines in UAE, this evaluation was simplified in that way, that the operating hours of the peaker gas turbine are similar with the base case power plant. The following values are also based on the yearly values.

As Fig. 36.10 shows, within a combined production the electricity to fuel ratio could be elevated depending on the configuration round about 0.17–0.19 $MWh_{el}/$

MWh_{th}. Based on the above mentioned assumptions for the two BC + GT configurations 4.4–5 Mt_{CO2}/a could be saved towards a separated production.

Conclusions

In the present publication the hybridization of parabolic trough power plants using state of the art thermal oil as HTF with natural gas has been evaluated based on the power plant configuration of Shams One. In a first step for this base case model a detailed operational logic has been developed in order to fulfill all technical constraints. In a second step a further hybridization through integrating gas turbines in the performance class of 10–20 MW_{el} has been done, whereby two different gas turbines has been used in order to cover the range. For this purpose of evaluating the fuel to electricity conversion efficiency in addition a solar only reference model has been prepared. For all these configurations annual yield simulations have been performed.

It is shown, by example of the base case model, that through the developed operational logic the identified main technical boundary conditions are satisfied.

Through the hybridization done in the base case model, based on Shams One, the net efficiency could be raised by 2.76 %-points for design-point and 1.92 %-points for annual view in comparison to the solar only reference model. The annual fossil fuel to electricity conversion efficiency amounts to 40.74 %. By integrating additionally a gas turbine this efficiency value could elevated by another 2.2 %-points up to 42.94 %. Supplementary to this absolute efficiency improvement, the combined production has versus to the separate production a better electricity to fuel ratio and thereby saves CO_2. Here it should be noted, that therefore the assumptions for gas turbine operation are deciding. Furthermore, the operation flexibility of the peaker gas turbine is in the combined approach not affected.

All in all the hybridization of parabolic trough power plants using a HTF limited to 400 °C with natural gas offers the opportunity to improve the power plant in terms of live steam parameters, efficiency and flexibility.

As brief outlook on the economy it has to be noted, that a gas turbine integration in the evaluated arrangement offer a reduction potential for the levelized costs of electricity, which amounts with the used assumptions to round about 5–6 %.

Acknowledgements The authors would like to thank the German Federal State of North Rhine-Westphalia and the European Regional Development Fund for the financial support of the project TURIKON in the frame of the program progress. NRW and the goal 2-program 2007–2013, Phase VI (Grant No. 64.65.69-EN-2019).

Nomenclature

ACC	Air-cooled condenser
BC	Base case
CSP	Concentrating solar power
DNI	Direct normal irradiance, W/m^2
ELEP	Expansion line end point
GT	Gas turbine
HTF	Heat transfer fluid
HP	High pressure
FH	Feed heater
IP	Intermediate pressure
ISCC	Integrated solar combined cycle
LP	Low pressure
p	Pressure, bar or kPa
P	Power output, W
Q	Thermal output, W
SSG	Solar steam generator
ST	Steam turbine
T	Temperature, °C
UAE	United Arab Emirates

Greek Letters

Δ	Difference
η_{el}	Efficiency, %

Subscripts

out	Outlet
in	Inlet
hybrid	Hybrid configuration
net	Net
solar	Solar configuration
el	Electrical
fuel	Fuel
mech	Mechanical
s	Isentropic

References

Al Mirfa Power Company. http://www.ampc.ae/en/production/. Retrieved November 27, 2014.

Balan, R. (2014). *Simulation des transienten Verhaltens von Parabolrinnenkraftwerken*. Master thesis. Essen: University of Duisburg-Essen.

BP. (2014). *BP energy outlook 2035*. bp.com/energyoutlook. Retrieved March 16, 2015.

Gas Turbine World. (2011). *Gas turbine world 2011-Performance Specs*. Pequot Publishing. ISSN 0746–4134, (27th ed., Vol. 41 No.1).

Goebel, O. (2010). Shams one 100 MW CSP plant in Abu Dhabi—Update on Project Status. *Proceedings of the SolarPACES 2010 Conference*. Perpignan, France.

Goebel, O., & Luque, F. (2012). Shams One 100 MW CSP Plant in Abu Dhabi. Update on Proceedings of the SolarPACES 2012 Conference. Marrakesch: Marocco.

Lechner, C., & Seume, J. (2010). *Stationäre Gasturbinen*. e-ISBN 978-3-540-92788-4. (2nd ed.). Heidelberg: Springer-Verlag Berlin.

MAN. http://www.corporate.man.eu/man/media/content_medien/images/global_corporate_website/presse_und_medien/mdt/2013_2/T_Turbinenlaeufer.jpg. Retrieved December 29, 2014.

Pawellek, R., Löw, T., & Hirsch, T. (2009). *EbsSolar—a solar library for Ebsilon®Professional*. Proceedings of the SolarPACES 2009 Conference. Berlin, Germany.

Reuß, N. (2012). *Modern gas and steam turbines in solarthermic applications*. 14th International Symposium on Transport Phenomena and Dynamics of Rotating Machinery (ISROMAC). Honolulu, HI, USA.

Schröder, K. (1966). *Große—Dampfkraftwerke—Planung Ausführung und Bau, Dritter Teil A*. Berlin: Springer.

Servert, J. F., Cerrajero, E., Lopez, D., Yagüe, S., Gutierrez, F., Lasheras, M., & San Miguel, G. (2014). *Base case analysis of a HYSOL power plant*. Proceedings of the SolarPACES 2014 Conference. Bejing, China.

SOLUTIA. *Product data sheet—therminol VP-1*. http://twt.mpei.ac.ru/TTHB/HEDH/HTF-VP1.PDF. Retrieved May 6, 2013.

TRANSCO. (2014). *Seven year electricity planning statement (2014–2020)—report*. http://www.transco.ae/media/docs.htm. Retrieved November 27, 2014.

Turchi, C. S., Ma, Z., Erbes, M. (2011). *Gas turbine/solar parabolic trough hybrid designs*. ASME Turbo Expo 2011. Vancouver, Canada.

United Arab Emirates. (2012). *Annual statistical report—electricity and water 2012*. Ministry of Energy. Abu Dhabi, Dubai.

Vogel, T., Oeljeklaus, G., Görner, K. Dersch, J., & Polklas, T. (2013). *Hybridization of parabolic trough power plants with natural gas*. Energy Procedia 49 (2014)—Proceedings of the SolarPACES 2013 Conference. (pp. 1238–1247). Las Vegas, USA.

Part X
Energy Technologies and Their Effect on Global Warming

Thermodynamic System Analysis and Optimization

Specific cases of system analysis and optimization

Chapter 37
Performance Analyses of CO_2-N_2O Cascade System for Cooling

Fatih Yılmaz, Reşat Selbaş, Arif Emre Özgür, and M. Tolga Balta

Introduction

Because of environmental issues associated with the global warming and ozone layer depletion attributed to the application of synthetic refrigerants chlorofluorocarbons (CFC's), hydrochlorofluorocarbons (HCFC's) and hydrofluorocarbons (HFC's), the return to the usage of natural materials for cooling system appears to be an appropriate alternative. Accordingly, the natural refrigerants such as ammonia, carbon dioxide and hydrocarbons, have recently received increasing attentions (Dopazo et al. 2009). Amongst the natural refrigerants, the suggested of the use of CO_2 seems to be the most promising especially as the natural refrigerant (Lorentzen and Petterson 1993). The CO_2 including that non-toxic, easily available, excellent thermophysical properties and not explosive. N_2O refrigerant, which is alternative of CO_2 in terms of thermophysical properties. N_2O refrigerants' disadvantage is, its global warming potential (GWP) is higher than CO_2.

The cooling systems involving a high temperature difference between the condenser and the evaporator, employing a single stage cooling system is not

F. Yılmaz (✉)
Department of Energy, Vocational Schools of Technical Sciences, Aksaray University, 68100 Aksaray, Turkey
e-mail: fatiyilmaz7@gmail.com

R. Selbaş • A.E. Özgür
Department of Energy Systems Engineering, Faculty of Technology, Suleyman Demirel University, 32100 Isparta, Turkey
e-mail: resatselbas@sdu.edu.tr; emreozgur@sdu.edu.tr

M.T. Balta
Department of Mechanical Engineering, Faculty of Engineering, Aksaray University, 68100 Aksaray, Turkey
e-mail: mustafatolgabalta@hotmail.com

P. Grammelis (ed.), *Energy, Transportation and Global Warming*, Green Energy and Technology, DOI 10.1007/978-3-319-30127-3_37

economical because of the corresponding high pressure ratio leads to a low volumetric efficiency of the compressors and consequently low coefficient of performance of the system. In conditions cascade cooling cycle is more appropriate. Cascade cooling cycles are generally in low evaporating temperature applications, which is evaporating temperature between −40 °C and − 130 °C. A commonly used refrigerants pair in the past has been R12, R502 in high temperature cycle and R13 in low temperature cycle of cascade refrigeration system. These refrigerants have been phased out since 1996 in the developed countries, and should be totally phased out by 2010 in developing countries as per Montreal Protocol and its amendments from the United Nations Environment Programme (UNEP 1987, 2007). Currently, many manufacturers tend to use natural refrigerants due to environmental problems and harmful effects synthetic refrigerants. CO_2/N_2O cascade system uses N_2O in the high temperature cycle and CO_2 in the low temperature cycle. N_2O is natural refrigerant.

Ratts and Brown (2000) shows that the entropy generation method to analyze the cascade cycle. In their work, relationships were developed for the specific heat and temperature ratio terms and the results were investigated for a cascade system in two low temperatures of 0.684 and 0.681, with the refrigerant R-134a and finally, the optimum temperature distribution was found. The study the optimum temperature in a cascade condenser, maximized the coefficient of performance and minimized the exergy destruction of the system, CO_2 is using the evaporation temperature and NH_3 is condensation temperature carried out by (Lee et al. 2006).

Energy and irreversibility analysis of a cascade refrigeration system employing various refrigerant couples using a computer code carried out by (Kilicarslan and Hosoz 2010). Bansal and Jain (2007) shown that evaluated the optimum cascade condensing temperature of CO_2 when different refrigerants such as NH_3, propane, propylene and ethanol are used in the high temperature circuits of a subcritical cascade system. Bhattacharyya et al. (2005) shows that an optimization research of the CO_2/C_3H_8 cascade system for cooling and heating. Analyzed a vapor compression heat pump working with R11 and a vapor compression refrigeration system operating with R12 cascaded to produce heating and cooling effects simultaneously carried out by (Murthy and Murthy 1985). Bhattacharyya et al. (2009) give in that thermodynamic analysis and optimization of a novel N_2O–CO_2 cascade system for refrigeration and heating. They study, a natural refrigerant based cascaded system, with nitrous oxide as the low temperature fluid and carbon dioxide as the high temperature fluid, is analyzed for simultaneous cooling and heating applications.

In literature generally CO_2 and NH_3 refrigerants in cascade cooling system were compared. In this study, a cascade system for cooling with N_2O as the high temperature fluid and CO_2 as the low temperature fluid is theoretical analyzed. Thermodynamic analysis of cooling system as evaporator temperature variations and pumps inlet pressure variations are investigated. Also, the effects of operating parameters on system performance are studied and illustrated in tables. The effect on the system performance of the heat exchangers are also investigated and given in figures.

Performance Analyses of CO_2-N_2O Cascade

A schematic diagram of a cascaded CO_2 and N_2O cooling system is given in Fig. 37.1 and the corresponding P–h diagram is shown Fig. 37.2. This cooling system includes two separate circuits, one high temperature circuit (HTC) with N_2O as the refrigerant and the other one, low temperature circuit (LTC) with CO_2 as the refrigerant. The analysis was carried out as subcritical cycle. HTC of condenser as water cooler condenser is discussed. These two circuits are thermally coupled using a heat exchanger which acts as an evaporator for the HTC and a condenser for the LTC.

The natural fluid N_2O has a triple point −90.82 °C. Furthermore similarity between critical temperature N_2O and CO_2. Table 37.1 given in thermodynamic properties of CO_2 and N_2O.

Each refrigeration system consists of a compressor, a condenser, an expansion valve, and an evaporator. In this study N_2O is the refrigerant in HTC, whereas CO_2 is the refrigerant in LTC. The circuits are thermally connected to each other through a heat exchanger, which acts as an evaporator for the HTC and a condenser for the LTC. The evaporator of LTC absorbs the cooling load from the cooling space. The condenser in HTC rejects heat flow at condensing temperature.

The cascade cooling system components shown in Fig. 37.1 can be treated as a control volume. The thermodynamic analysis of the cascade cooling system was performed based on the following general assumptions;

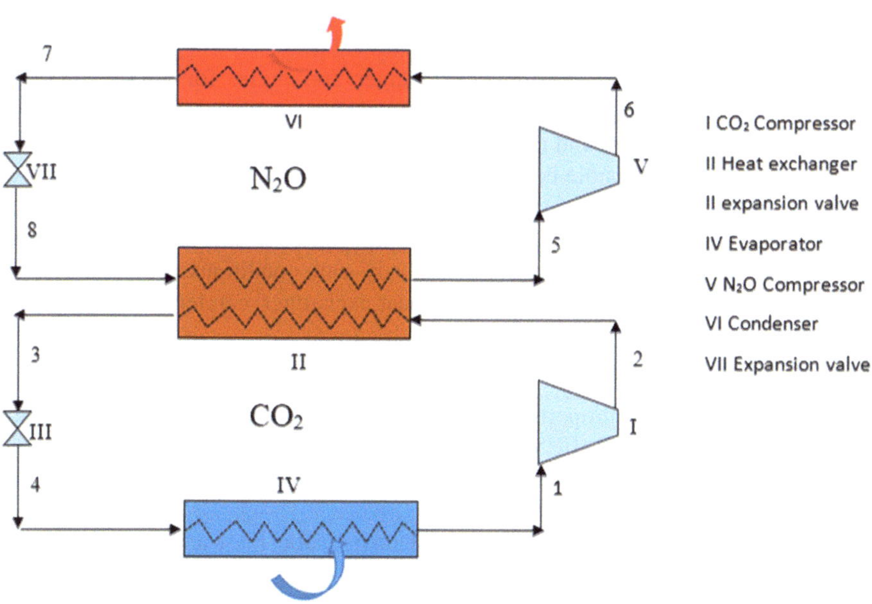

Fig. 37.1 Schematic diagram of the CO_2-N_2O cascade refrigeration system

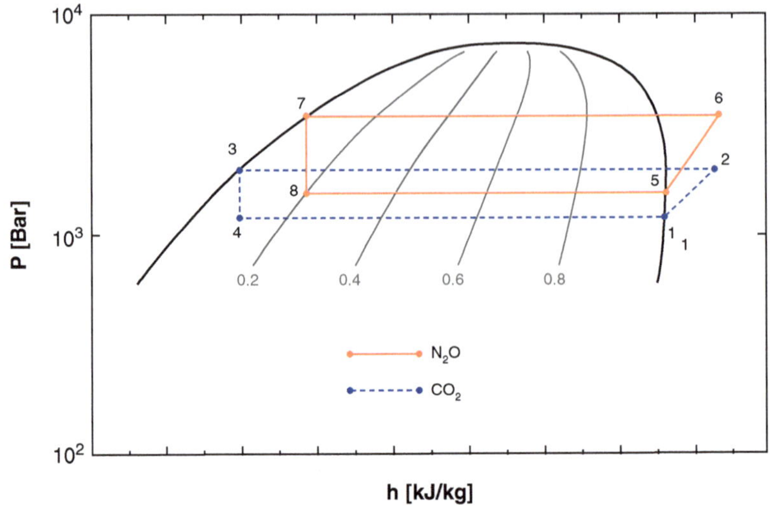

Fig. 37.2 A schematic P-h diagram of the CO_2-N_2O cascade refrigeration system

Table 37.1 Thermodynamic properties of CO_2 and N_2O

Property	CO_2	N_2O
Critical pressure (bar)	73.77	72.45
Critical temperature (°C)	30.98	36.37
Triple point temperature (°C)	−56.56	−90.82
Molecular weight (kg/kmol)	44.01	44.01

- Refrigerants at the cascade heat exchanger and outlet for HTC cycle and evaporator for LTC cycle outlet are saturated.
- The changes in kinetic and potential energy are negligible.
- Adiabatic and irreversible compression in HTC with an isentropic efficiency of 0.8.
- Negligible pressure and heat losses in the pipe networks or system components.
- All system components operate under steady-state conditions.
- NH_3Heat transfer processes in cascade heat exchanger, condenser and evaporator are isobaric.
- Ambient (dead state) temperature (T_0) 21 °C and ambient pressure (P_0) 1 bar.

The thermodynamic properties of CO_2 and NH_3 were determined using Engineering Equations Solver (EES). EES can generate publication-quality plots, do optimization, provide linear and non-linear regression, solve equations and simplify uncertainty analyses.

Taking into account the assumptions previously made, mass and energy balances are given by Eqs. (37.1) and (37.2), respectively.

For the steady-state process, the mass, energy and exergy balance for the each control volume can be expressed as follows;

$$\sum \dot{m}_{in} = \sum \dot{m}_{out} \tag{37.1}$$

where $\dot{m}$ is the mass flow rate, and the subscript in stands for inlet and out for outlet. The mass balance equation can be expressed in the rate form as with all energy terms as follows (Cengel and Boles 2008)

$$\dot{Q} - \dot{W} + \sum \dot{m}_{in} h - \sum \dot{m}_{out} h = 0 \tag{37.2}$$

The general exergy balance can be expressed in the rate form as;

$$\dot{Ex}_{in} - \dot{Ex}_{out} = \dot{Ex}_{dest} \tag{37.3}$$

$$\dot{Ex}_{heat} - \dot{Ex}_{work} + \dot{Ex}_{mass,\ in} - \dot{Ex}_{mass,out} = \dot{Ex}_{dest} \tag{37.4}$$

Using Eq. (37.4), the rate form of the general exergy balance can also be written as

$$\sum \left(1 - \frac{T_0}{T_k}\right) \dot{Q}_k - \dot{W} + \sum \dot{m}_{in} \psi_{in} - \sum \dot{m}_{out} \psi_{out} = \dot{Ex}_{dest} \tag{37.5}$$

where $\dot{Q}_k$ is the heat transfer rate through the boundary at temperature, T_k at location k, $\dot{W}$ is the work rate, I ? is the flow (specific) exergy, h is enthalpy, s is entropy, and the subscript zero indicates properties at the restricted dead state of P_0 and T_0.

The specific exergy (flow exergy) of refrigerant (or water) is calculated by;

$$\psi = h - h_0 - T_0(s - s_0) \tag{37.6}$$

The exergy rate is calculated by;

$$\dot{Ex} = \dot{m}\ \psi \tag{37.7}$$

Energy (or first law) efficiency of the cascade cooling system and exergy efficiency are determined as follows, respectively,

$$COP = \frac{\dot{Q}_{ev}}{\dot{W}_{N2O} + \dot{W}_{CO2}} \tag{37.8}$$

$$\varepsilon = \frac{\dot{Ex}_{out}}{\dot{Ex}_{in}} \tag{37.9}$$

where out refers to "net output" or "product" or "desired value", and in refers to "given" or "used".

For LTC compressor (I);

$$\dot{m}_{CO_2} = \dot{m}_1 = \dot{m}_2 \tag{37.10}$$

$$P_{rCO_2} = \frac{P_2}{P_1} \tag{37.11}$$

$$\eta_{CO_2\ isen} = 0,815 + 0,022 * P_{rCO_2} - 0,0041 * P_{rCO_2}^2 + 0,0001 * P_{rCO_2}^3 \tag{37.12}$$

$$\eta_{CO_2\ isen} = \frac{h_{2s} - h_1}{h_2 - h_1} \tag{37.13}$$

where the heat transfer versus the environment was neglected. Equation (37.12) depending on the LTC of compressor isentropic efficiency the compressor outlet enthalpy is calculated (Robinson and Groll 1998).

$$\dot{W}_{CO_2} = \dot{m}_{CO_2}(h_2 - h_1) \tag{37.14}$$

$$\dot{Ex}_{dest,\ comp} = \dot{m}_{CO_2}(\psi_1 - \psi_2) + \dot{W}_{CO_2} \tag{37.15}$$

For LTC condenser–HTC evaporator (heat exchanger) (II);

$$\dot{m}_{CO2} = \dot{m}_2 = \dot{m}_3 \tag{37.16}$$

$$\dot{m}_{N2O} = \dot{m}_8 = \dot{m}_5 \tag{37.17}$$

$$\dot{Q}_{hexc} = \dot{m}_{N2O}(h_5 - h_8) = \dot{m}_{CO_2}(h_2 - h_3) \tag{37.18}$$

$$\dot{Ex}_{dest,\ hexc} = \dot{m}_{CO_2}(\psi_3 - \psi_2) + \dot{m}_{N2O}(\psi_8 - \psi_5) \tag{37.19}$$

The heat rate exchanged and exergy destruction from LTC to HTC in the heat exchanger is called $\dot{Q}_{hexc}$ and can be calculated by Eqs. (37.18) and (37.19).

For expansion valve (III)

$$\dot{m}_{CO2} = \dot{m}_3 = \dot{m}_4 \tag{37.20}$$

$$(h_3 = h_4) \tag{37.21}$$

$$\dot{Ex}_{exp} = \dot{m}_{CO_2}(\psi_3 - \psi_4) \tag{37.22}$$

The expansion valve exergy destruction is calculated with Eq. (37.22). Where expansion valve of heat transfer is neglected.

For LTC evaporator (IV);

$$\dot{m}_{CO2} = \dot{m}_4 = \dot{m}_1 \tag{37.23}$$

$$\dot{Q}_{ev} = \dot{m}_{CO_2}(h_1 - h_4) \tag{37.24}$$

$$\dot{Ex}_{dest,\ evap} = \dot{m}_{CO_2}(\psi_4 - \psi_1) + \dot{Q}_{ev}\left(1 - \frac{T_o}{T_{EV}}\right) \tag{37.25}$$

The evaporator heat rate and exergy destruction is calculated with Eqs. (37.23) and (37.24). Where is the T_0 is ambient temperature (dead state), T_E is evaporator temperature.

For HTC compressor (v);

$$\dot{m}_{N2O} = \dot{m}_5 = \dot{m}_6 \tag{37.26}$$

$$\dot{W}_{N2O} = \dot{m}_{N2O}(h_6 - h_5) \tag{37.27}$$

$$\dot{Ex}_{dest,\ comp} = \dot{m}_{N2O}(\psi_6 - \psi_5) + \dot{W}_{N2O} \tag{37.28}$$

The HTC compressor exergy destruction is calculated with Eq. (37.28).

For HTC condenser (VI);

$$\dot{m}_{N2O} = \dot{m}_6 = \dot{m}_7 \tag{37.29}$$

$$\dot{Q}_{con} = \dot{m}_{N2O}(h_6 - h_7) \tag{37.30}$$

$$\dot{Ex}_{con} = \dot{m}_{N2O}(\psi_4 - \psi_1) - \dot{Q}_{con}\left(1 - \frac{T_o}{T_{CON}}\right) \tag{37.31}$$

The condenser heat rate and exergy destruction is calculated with Eqs. (37.30) and (37.31). Where is the T_0 is ambient temperature (dead state), T_c is condenser temperature.

For expansion valve (VII)

$$\dot{m}_{N2O} = \dot{m}_7 = \dot{m}_8 \tag{37.32}$$

$$(h_7 = h_8) \tag{37.33}$$

$$\dot{Ex}_{exp} = \dot{m}_{N2O}(\psi_7 - \psi_8) \tag{37.34}$$

The expansion valve exergy destruction is calculated with Eq. (37.34). Where expansion valve of heat transfer is neglected.

Results and Discussion

A performance analysis of the cascade cooling with CO_2-N_2O system is investigated using energy and exergy analysis. Under the assumptions, the calculated properties for cascade cooling system are given in Table 37.2. The cooling coefficient of performance and exergy efficiency of the cascade system (COP) is calculated from Eqs. (37.8) and (37.9) and is found to be 2.061 % and 35 %, respectively.

Figure 37.3 displays the effect of evaporator temperature on COP and exergy efficiency cycle. As can be seen in Fig. 37.3. The cycle of evaporator temperature 25 °C, by increasing evaporator temperature the system of COP and exergy efficient has increased. While evaporator temperature −50 °C, COP and exergy efficiency

Table 37.2 Exergy analysis results of the cascade cooling system studied for one representative unit

State no	Fluid	Phase	Temperature (°C)	Pressure (bar)	Specific enthalpy (kJ/kg)	Specific entropy (kJ/kg K)	Exergy rate (kW)
1	CO_2	saturated vapor	−23	14.28	−62.17	−0.7095	11.81
2	CO_2	Super-heated vapor	66.58	45.02	0.2552	−0.6806	14.28
3	CO_2	Comp. liquid	15.72	45.02	−264.6	−1.596	20.42
4	CO_2	Mixture	−30	14.28	−264.6	−1.542	19.28
5	N_2O	saturated vapor	10	33.7	408.9	1.594	13.45
6	N_2O	Super-heated vapor	52.14	56.51	435.4	1.611	14.88
7	N_2O	Comp. liquid	20	56.51	212.5	0.87	15.58
8	N_2O	Mixture	3	33.7	212.5	0.8839	15.29

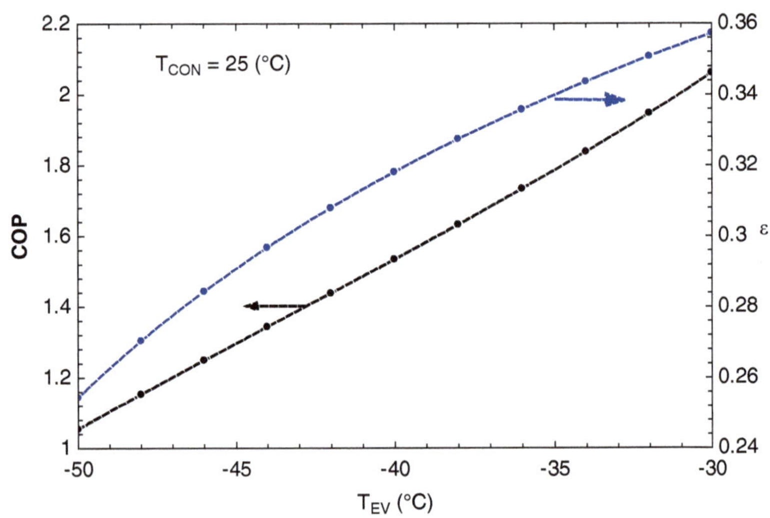

Fig. 37.3 Effect of evaporator temperature on COP and exergy efficiency

are 1.055 % and 25 %, respectively. In Fig. 37.3, the systems are represented COP and the exergy efficiency trends with increases in evaporator temperature.

The systems of the COP and exergy efficiency effect on condenser temperature can be observed in Fig. 37.3.

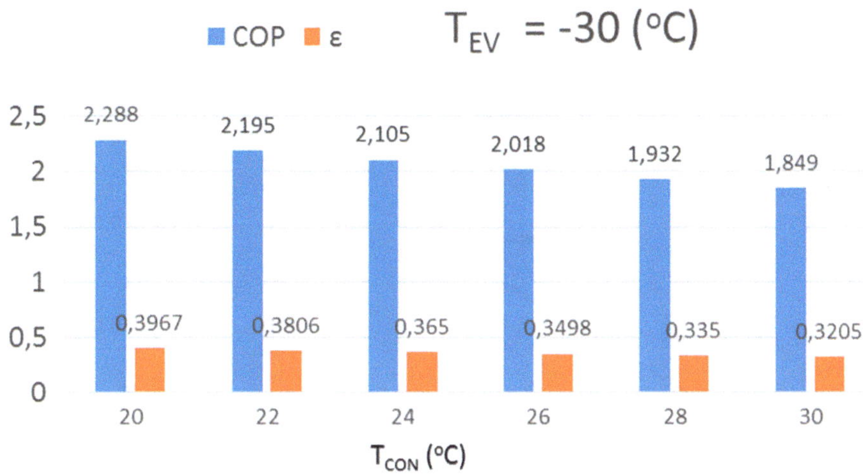

Fig. 37.4 Effect of condenser temperature on COP and exergy efficiency

In this system with increases condenser temperature, COP and exergy efficiency is decreases shown in Fig. 37.3. While evaporator temperature −30 (°C) and the condenser temperature is 20 (°C), the cycle of COP and exergy efficiency are shown 2.28 % and 39 %, respectively.

Figure 37.4 given in the effect of N_2O compressor isentropic efficiency on the cycle COP and exergy efficiency. While condenser and evaporator temperature are fixed respectively 25 (°C), −30 (°C), with HTC compressor isentropic efficiency increases, COP and exergy efficiency are increases can be observed in Fig. 37.5. As the compressor isentropic efficiency up to 90 % on, the cycle of COP value rise up to 2.14. The figurate result in understand that if the compressor isentropic efficiency is increases, the system of increases COP and exergy efficiency.

The system effect of heat exchanger efficiency on COP and exergy efficiency illustrated in Fig. 37.6. When the refrigeration cycle of heat exchanger efficiency is rise up from 60 % to 95 %, the system of increases COP and exergy efficiency. Figure 37.5 shows that evaporator temperature and condenser temperature are fixed "−30 (°C), 25 (°C)", while heat exchanger efficient is 75 %, the systems of COP and exergy efficiency are calculated 1.986 34 %, respectively.

Figure 37.6 shows that heat exchanger efficiency and evaporator temperature changed on the system COP and exergy efficiency. The system of condenser temperature 25 (°C) is fixed, as the heat exchanger efficiency is increases, which is fixed evaporator temperature, the system increases of COP. When the system of heat exchanger efficiency and evaporator temperature are increased, the system is increases of COP value observed in Fig. 37.6.

Figure 37.7 demonstrates the variation in the exergy destruction of different components of the cascade refrigeration system. According to this figure, the

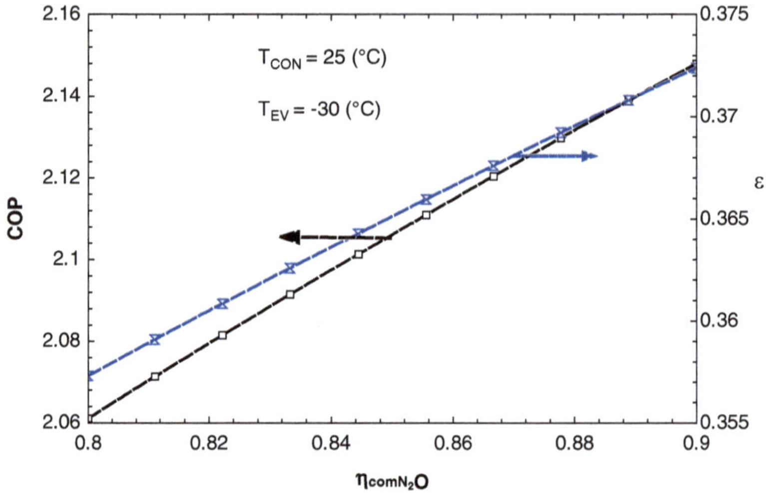

Fig. 37.5 Effect of N_2O compressor isentropic efficiency on COP and exergy efficiency

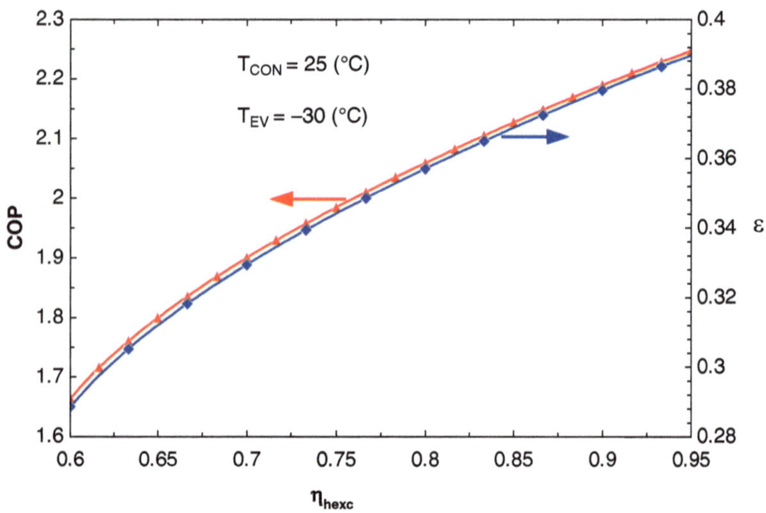

Fig. 37.6 Effect of heat exchanger efficiency on COP and exergy efficiency

highest amount of exergy destruction rate takes place in the evaporator (18.08 kW). The lower amount of destruction rate take place in HTC and LTC expansion valves is are calculated 0.029 and 0.06 respectively.

The effect on evaporator temperature on the exergy destruction at the system components and COP the whole system at different evaporator temperature values are demonstrated at Fig. 37.9. The exergy destruction occurs given in figurate HTC

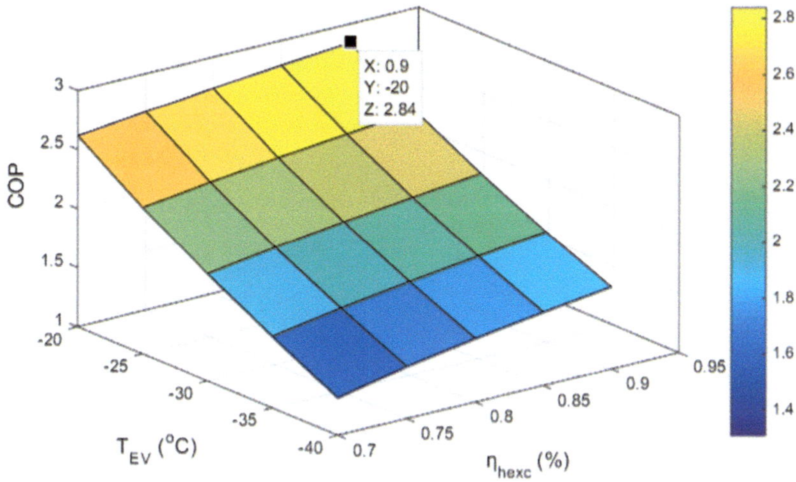

Fig. 37.7 Effect of heat exchanger efficiency and evaporator temperature on COP and exergy efficiency

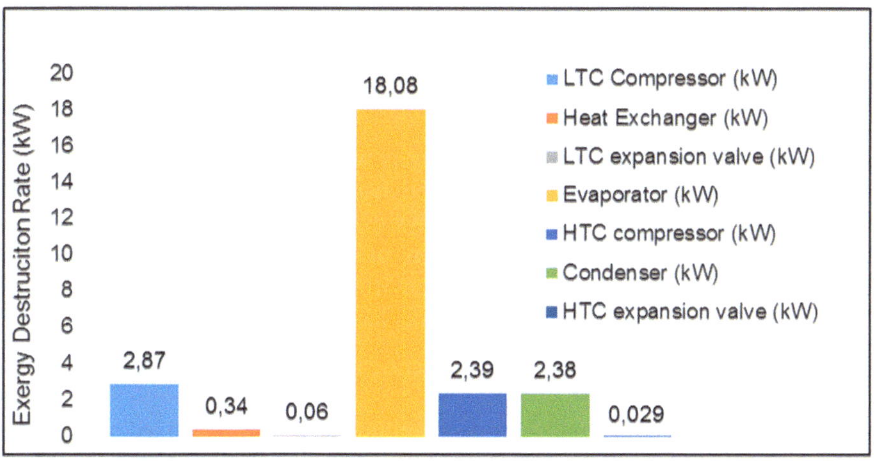

Fig. 37.8 The exergy destruction values in different components of cascade refrigeration system

compressor, LTC compressor and heat exchanger. The system COP increases from 1.33 to 2.88 with evaporator temperature increase from −40 °C to −10 °C. But the system of HTC and LTC compressor exergy destruction rate are decreases with increase of evaporator temperature shown that Fig. 37.9.

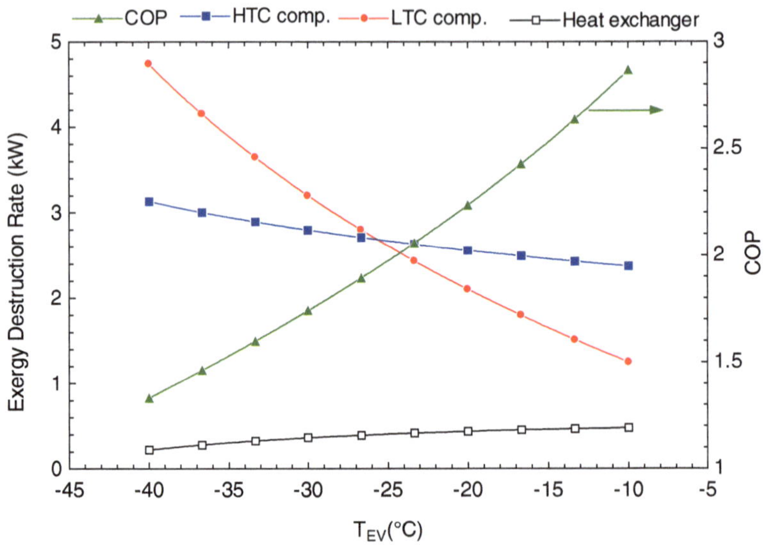

Fig. 37.9 Exergy destruction at the system components and COP the whole system at different evaporator temperature values

Conclusion

In this paper, a cascade system for cooling is with N_2O-CO_2 fluids as theoretical performance analyzed according to thermodynamic assessment in order to investigate cascade system COP and exergy analysis. According to condenser and evaporator temperature exchanges the system of COP and exergy efficiency are investigated. This paper as a refrigerants environmentally friendly fluids CO_2 and N_2O are used. The main concluding remarks have been given as follows;

- This cascade refrigeration cycle is subcritical and therefore water-cooled condenser was adopted.
- While the system of condenser temperature at 25 (°C) is fixed, evaporator temperature −30 (°C), the system of COP and exergy efficiency are respectively 2.1 % and 36 %.
- As the system of evaporator temperature is increased, COP and exergy efficiency are increases.
- Evaporator temperature in −30 (°C), if the condenser temperature is increases the system of COP and exergy efficiency are decreased.
- Exergy efficiency have been calculated to be between 28 % and 39 %.
- COP and exergy efficiency are increased as the heat exchanger efficiency is increased. Therefore, the more highly efficient heat exchanger are used, the more exergy efficiency highly.

Nomenclature

COP	Coefficient of performance
Ex	Exergy flow rate (kW)
Exdes	Exergy destruction
h	Specific enthalpy (kJ/kg)
$\dot{m}$	Mass flow rate (kg/s)
$\dot{Q}$	Heat flow rate (kW)
s	Specific entropy (kJ/kg K)
T	Temperature (K or °C)
$\dot{W}$	Work rate or power (kW)
ψ	Specific exergy (kJ/kg)
η	Efficiency (dimensionless)
ε	Exergy (second law) efficiency (dimensionless)
in	Inlet
out	Outlet
des	Destruction
ev	Evaporator
con	Condenser
exp	Expansion valve
comp	Compressor
hexc	Heat exchanger
isen	Isentropic
LTC	Low temperature cycle
HTC	High temperature cycle

References

Bansal, P. K., & Jain, S. (2007). Cascade systems: Past, present, and future. *ASHRAE Transactions, 113*(1), 245–252.

Bhattacharyya, S., Garai, A., & Sarkar, J. (2009). Thermodynamic analysis and optimization of a novel N_2O–CO_2 cascade system for refrigeration and heating. *International Journal of Refrigeration, 32*, 1077–1084.

Bhattacharyya, S., Mukhopadhyay, S., Kumar, A., Khurana, R., & Sarkar, J. (2005). Optimization of a CO_2–C_3H_8 cascade system for refrigeration and heating. *International Journal of Refrigeration, 28*, 1284–1292.

Cengel, Y. A., & Boles, M. A. (2008). *Thermodynamics: An engineering approach* (6th ed.). New York: McGraw-Hill.

Dopazo, J. A., Fernández-Seara, J., Sieres, J., & Uhía, F. J. (2009). Theoretical analysis of CO_2-NH_3 cascade refrigeration system for cooling applications at low temperatures. *Applied Thermal Engineering, 29*, 1577–1583.

Kilicarslan, A., & Hosoz, M. (2010). Energy and irreversibility analysis of a cascade refrigeration system for various refrigerant couples. *Energy Conversion and Management, 51*(12), 2947–2954.

Lee, T. S., Liu, C. H., & Chen, T. W. (2006). Thermodynamic analysis of optimal condensing temperature of cascade-condenser in CO_2/NH_3 cascade refrigeration system. *Refrigeration, 29*, 1100–1108.
Lorentzen, G., & Petterson, J. (1993). A new efficient and environmentally benign system for car air-conditioning. *International Journal of Refrigeration, 16*(1), 4–12.
Murthy, S. S., & Murthy, M. V. K. (1985). Experiments on a cascaded R11–R12 vapour compression system for cogeneration of heat and cold. *Journal of Heat Recovery Systems, 5* (6), 519–526.
Ratts, E. B., & Brown, J. S. (2000). A generalized analysis for cascading single fluid vapor compression refrigeration cycles using an entropy generation minimization method. *Refrigeration, 23*, 353–365.
Robinson, D. M., & Groll, E. A. (1998). Efficiencies of transcritical CO_2 cycles with and without an expansion turbine. *International Journal of Refrigeration, 21*(7), 577–589.
UNEP. (1987). *Montreal protocol on substances that deplete the ozone layer*. United Nations Environment programme.
UNEP. (2007). *Assessment report of the technology and economic assessment panel*. Nairobi, Kenya: UNEP Ozone Secretariat.

Chapter 38
Comparison of Thermal Repowering Alternatives for Thermal Power Plants

M. Zeki Yilmazoglu

Introduction

Electricity consumption increases rapidly due to the increasing population and industrialization. Greenhouse gas emissions, mainly CO_2, increase in order to offset the increasing energy demand. Approximately half of the electricity demand is obtained from fossil fuels combustion. Efficient use of energy, renewable energy, clean energy generation systems, CO_2 capture and storage, energy harvesting systems etc. current research topics in power generation. Many regulations have been implemented by the governments to decrease the greenhouse gases. On the one hand, the energy demand increases rapidly, and on the other hand GHGs have to be decreased. In Turkey, low rank lignite is generally used for power generation in thermal power plants which most of them were installed in 80s. These old thermal power plants are still in operation with lower net electrical efficiencies. The net electrical efficiency of these power plants decrease with time due to the aging and operational problems. Therefore, repowering of these power plants can increase their energy efficiency and reduce their contribution to the global warming potential.

Repowering can be defined as increasing the installed capacity, the net electrical efficiency, and decreasing the emissions per installed capacity of an existing thermal power plant (Yilmazoglu et al. 2012, 2013a, b). Generally, a gas turbine is added to the cycle in thermal repowering applications. Feedwater heating, hot windbox and parallel repowering are three of the most commonly implemented repowering options (Escosa and Romeo 2009; Yilmazoglu and Durmaz 2013a, b).

M.Z. Yilmazoglu (✉)
Department of Mechanical Engineering, Faculty of Engineering, Gazi University, Maltepe, Ankara, Turkey
e-mail: zekiyilmazoglu@gazi.edu.tr

P. Grammelis (ed.), *Energy, Transportation and Global Warming*, Green Energy and Technology, DOI 10.1007/978-3-319-30127-3_38

In feedwater heating, steam turbine extraction points are repealed and heating of feedwater is supplied from a heat recovery steam generator or a solar field (Popov 2011; Yilmazoglu et al. 2012). In thermal power plants, repowering reduces CO_2 emissions per installed capacity (Schenk and Ehren 2003; Walters 2008). The most important parameter in repowering applications is the expected life time of the equipment. Therefore, a detailed life expectancy analysis has to be carried out before repowering. In addition, gas turbine and heat recovery steam generator selection are crucial for the operation of thermal power plant after repowering (D'Yakov et al. 1998; Mathieu 1998).

Gas turbine leverage and repowering efficiency are the decisive parameters in the analysis of a repowering application. Repowering efficiency can be defined as the rate of increment in electricity generation to increment in heat added to the cycle which is given in Eq. (38.1). The subscripts, ar and br symbolize after repowering and before repowering, respectively. After repowering the electricity generation is increased due to the additional power from gas turbine and steam turbine. Moreover, the natural gas consumption and consequently, the heat added to the cycle are increased. Gas turbine leverage can be defined as the rate of increment in electricity generation to the gas turbine installed capacity which is given in Eq. (38.2). In a typical combine cycle power plant with an installed capacity of 400 MW_e, approximately 67 % of electricity generation capacity is supplied from the gas turbine. However, in the case of repowering it is approximately 10–25 % as a result of smaller gas turbine selection.

$$\eta_{rep} = \frac{\Delta P_{el}}{\Delta Q_g} = \frac{P_{ar} - P_{br}}{Q_{in,ar} - Q_{in,br}} \tag{38.1}$$

$$\lambda_{GT} = \frac{\Delta P_{el}}{P_{el,GT}} = \frac{P_{ar} - P_{br}}{P_{el,GT}} \tag{38.2}$$

Lignite is the main fuel source in electricity generation in Turkey and low rank lignite is used generally for power generation. Most of the thermal power plants are firstly operated in mid 80s. Therefore, the net electric efficiency and availabilities of these power plants are decreased with time due to the ageing of components.

In this study, Soma A thermal power plant was examined and the application of different repowering alternatives, feedwater heating, hot windbox, and parallel repowering, were compared by a commercial software, Thermoflex. Soma A thermal power plant was decommissioned in 2010. Currently, the installed capacity of the power plant is 44 MW_{el} with two units. One unit of the power plant was simulated for different repowering alternatives. This study sums up and summarizes the results of previous studies and compares different alternatives in terms of the installed power increment, CO_2 reduction potential per installed capacity and unit electricity generation price.

Repowering Alternatives

Feedwater Heating

In Fig. 38.1, the connection schema for feedwater heating repowering application is shown. A gas turbine and two feed water heaters are added to the thermal power plant. Dashed lines indicate components added to the thermal power plant after repowering. Bled steam is only taken for deaerator and others are repealed. The waste heat of gas turbine is utilized by two economizers. Natural gas or syngas can be used as the fuel source in gas turbine. Also, a gasifier island can be added to the cycle to obtain syngas. In this study natural gas is accepted as the fuel source in gas turbine.

Hot Windbox

Hot windbox repowering can be implemented in three different methods. In the first method, the exhaust gas of gas turbine is fed to the burners. In this method, feedwater heating is supplied by two economizers installed after the boiler. In direct hot windbox application, burners and other related components have to be modified due to the high temperatures of exhaust gases of gas turbine. As a result, first investment cost of repowering increases. In addition, due to the low O_2 content which is approximately 13–14 %, some combustion problems i.e. flame stabilization, can occur in the steam boiler. Therefore, fresh air dilution is a necessity to lower the investment cost and

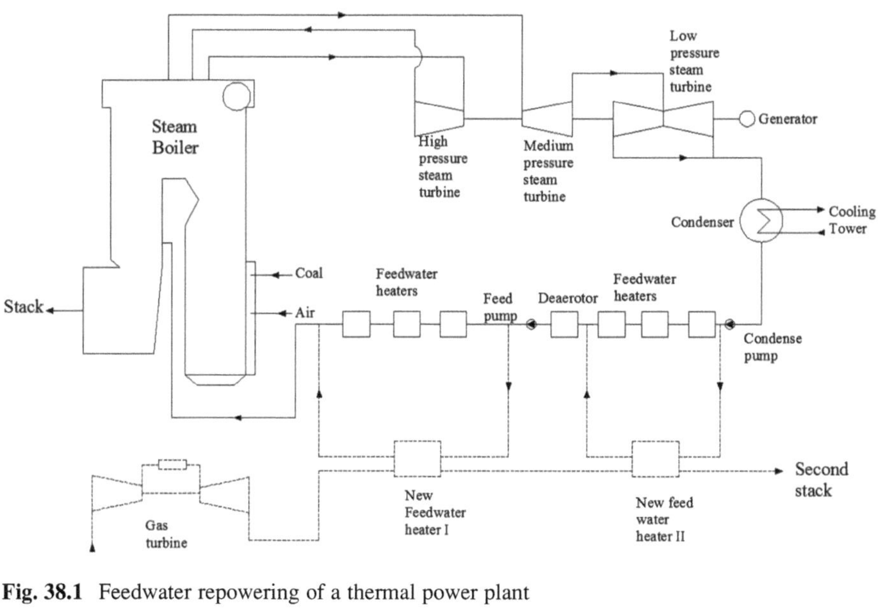

Fig. 38.1 Feedwater repowering of a thermal power plant

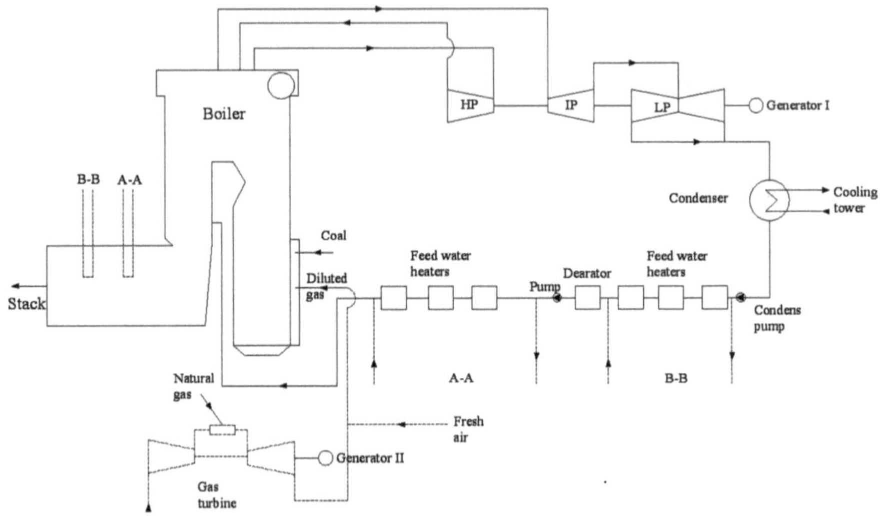

Fig. 38.2 Hot windbox repowering with fresh air dilution

prevent combustion problems. The fresh air dilution application is shown in Fig. 38.2. In this method fresh air is mixed with flue gases of GT to increase the O_2 content and to decrease the burner inlet temperature. In the third method, a pre-cooling system is installed after the gas turbine. This precooling system acts as an evaporator and produces extra steam for the steam turbine. It is directly connected to the drum. The temperature of the flue gas of GT is reduced to an acceptable range. However, the first investment cost of pre-cooling system is higher than fresh air dilution application. Therefore, fresh air dilution is used in the simulations.

Parallel Repowering

In parallel repowering, a gas turbine and a HRSG are installed and additional steam is fed to the steam turbine. Parallel repowering application with additional gas turbine stack brings the power plant operation flexibility (Yilmazoglu and Durmaz 2011). In all applications installed capacity of the power plant is increased with a properly selected gas turbine and heat recovery steam generator (HRSG) combination (Carapellucci 2009; Elmasri 2008).

Design Parameters of Soma A Thermal Power Plant

Soma A thermal power plant was designed according to the data, given in Table 38.1, in 1957. The power plant was operated at constant maximum load, at

Table 38.1 Design data of Soma A thermal power plant

Operation conditions[a]	1	2	3	4	5
Turbine power [MW]	7	12	17	22	NA
Water/steam pressures [bar]					
Inlet of economizer	62	62.6	65.7	70	72
Outlet of dome	59.6	61.5	64	68	69.8
Outlet of superheater	59.4	60.7	62.4	65	66.2
Temperatures [°C]					
Steam temperature at outlet of superheater	489.7	487.7	487	486.5	486.4
Water temperature at inlet of economizer	139	165	180	192	196
Water temperature at outlet of economizer	197	216	230	242	245
Gas temperature at inlet of superheater	826	872	922	980	1000
Gas temperature at outlet of economizer	239	261	278	296	302
Air temperature at outlet of air pre-heater	206.5	213	222.5	226.5	228
Stack temperature	129	142	152	160	162.5
Mass flow rates [t/h]					
Fuel	6.96	11.4	15.6	20.3	22
Combustion gas	60.6	93.3	121.4	150.8	161.5
Steam mass flow rate	30	51	72	96	105

[a]Operation conditions—1: Technical minimum, 2: Constant minimum load, 3: Normal load, 4: Constant maximum load, 5: Transient maximum load

Table 38.2 Ultimate analysis of Soma coal

Ultimate analysis [wt.%]	As received	Dry
C	39.48	52.8
H	2.95	3.94
O + N	13.42	18
S	0.53	0.71
W	25.22	0
A	18.4	24.55

operation #4. The ultimate analysis of coal, used in the power plant, is given in Table 38.2. The lower heating value and ash melting temperature of the coal are 3550 kcal/kg and 1400 °C, respectively. Three closed type feed water heaters and a deaerator is designed to preheat the condensed working fluid. Condenser pressure is 0.063 bar and a cross flow type cooling tower with fans is used to transfer the condenser heat to the environment. An air preheater unit is also designed to send the air into the combustion chamber at 220 °C. Flue gases are emitted to the atmosphere by a 55 m stack at 160 °C. The thermal efficiency of the boiler is 87.2 % and stack losses are 8.4 % of the total energy input. Bled steam, for preheating of the water in feed water heaters, are taken from 4th, 8th, 11th and 13th stages, of a 15 stage turbine. First bled steam is taken from 10–13 bar at 350 °C. The others are taken 4.5–4.8 bar at 250 °C for 8th stage, 1.7–1.8 bar at 175 °C for 11th stage and 0.3–0.35 bar at 110 °C for 13th stage (Fig. 38.3).

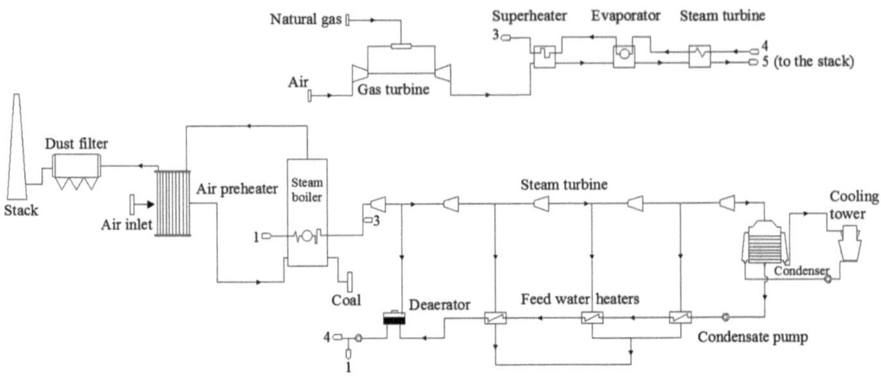

Fig. 38.3 Parallel repowering application

Results

Aero-derivative type gas turbines, with a capacity of 12–35 % of the existing power plant, give the optimum results for feedwater heating repowering applications. The limit is determined by the condenser mass flow rate increment percentage which is taken as 20 % greater than the initial total mass flow rate for all repowering applications. In Figs. 38.4, 38.5, and 38.6 the effects of the feedwater repowering application was presented. The ratio of gas turbine power to initial power indicates the installed power or capacity of the selected gas turbine for repowering. For instance, 20 % ratio means that the GT installed power is 4.4 MW_{el}. It is clear from Fig. 38.4 the net power of the TPP was increased substantially after repowering. Also, the condenser mass flow rate was increased due to the repealing of extraction points of the steam turbine which also increased the mass flow rate of steam entering the steam turbine and consequently, a 10 % increment in net power generation was obtained by steam turbine. For 30 % GT ratio the total plant capacity was found to be 30.2 MW_{el} and the net electrical efficiency was increased from 31.7 % to 33.87 %. Around 5 % GT ratio ST net power and net electrical efficiency was decreased as shown in Fig. 38.4. The reason of this decrement can be explained by the lower capacity GT selection. Repowering efficiency and GT leverage for different ratios were given in Fig. 38.5. It is shown that the minimum gas turbine capacity has to be greater than 10 % of the initial power. Figure 38.6 shows the CO_2 emissions of the TPP after repowering. In the modeling of the current state the CO_2 emission was found to be 7.019 kg/s. After repowering with a 30 % GT ratio, the CO_2 emission was found to be 8.146 kg/s. It is an expected result that the installed capacity of the TPP was increased. However, the CO_2 emission per installed capacity shows an interesting result. In the current state the CO_2 emission per installed capacity was found to be 0.321 and after repowering it was found to be 0.27 which indicates that repowering of old TPP can be used to decrease the CO_2 emissions in short term.

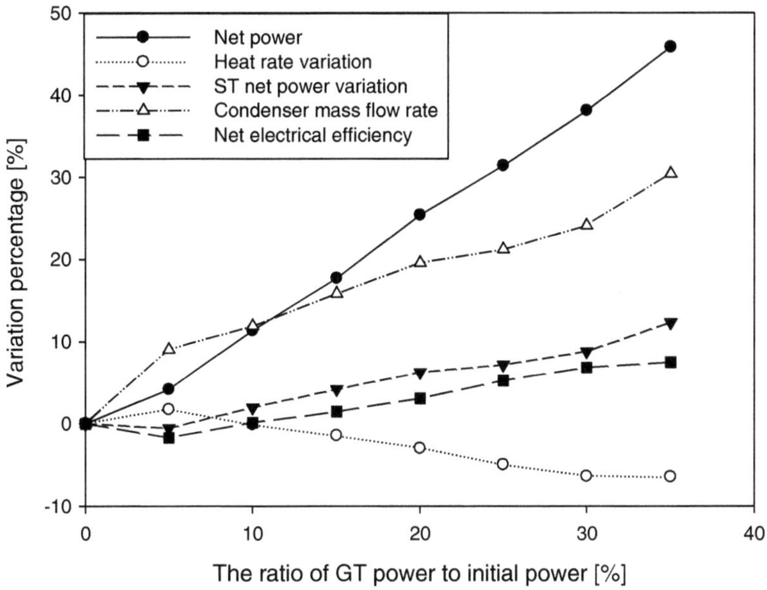

Fig. 38.4 Effects of the feedwater heating repowering application in TPP

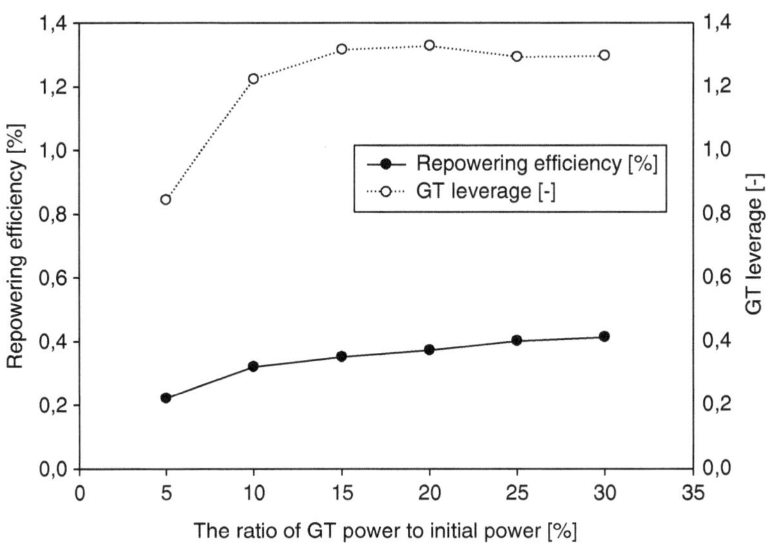

Fig. 38.5 Variation of repowering efficiency and gas turbine leverage

For hot windbox repowering application a GT was installed. Flue gas was diluted by fresh air to support combustion or flame stability. The results were given in Figs. 38.7, 38.8, and 38.9. In Fig. 38.7 the effects of hot windbox repowering application on TPP performance was shown. The net electrical efficiency was slightly changed. However, when Figs. 38.4 and 38.7 were compared

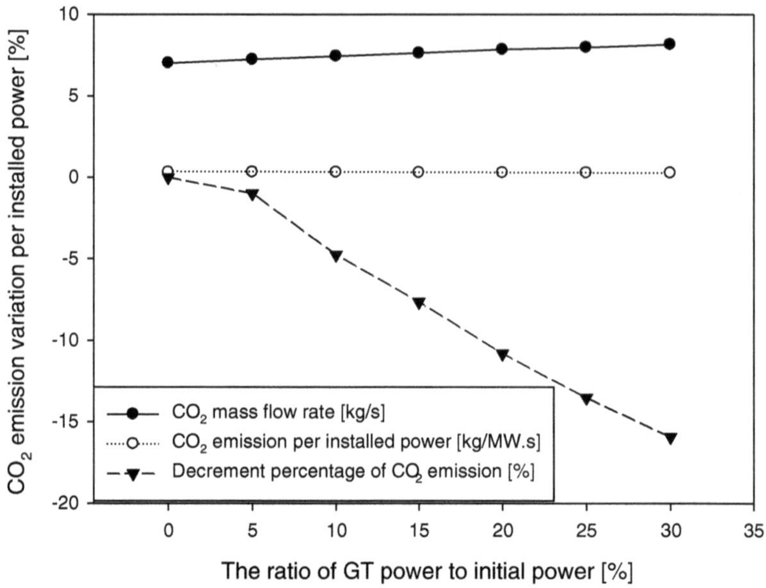

Fig. 38.6 Variation of CO_2 emissions and fuel consumption

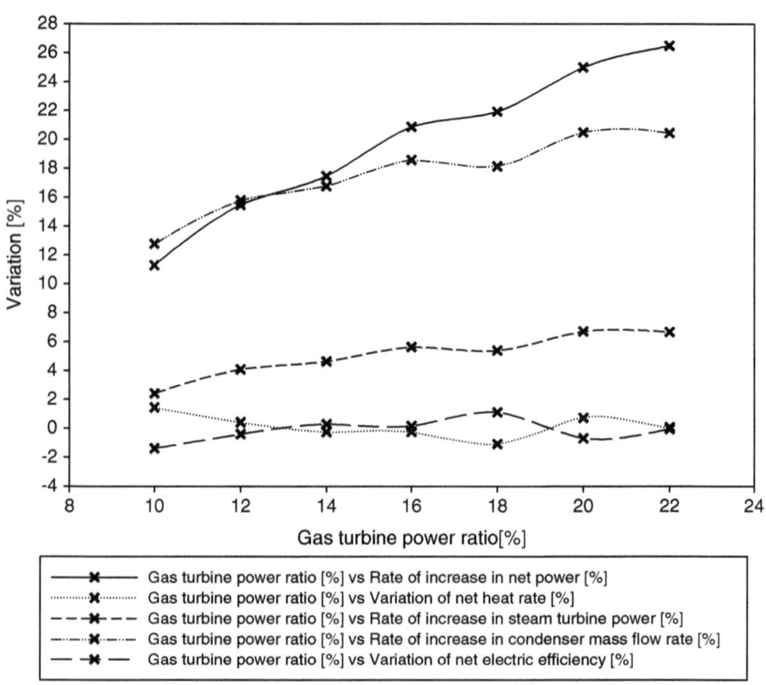

Fig. 38.7 Effects of hot windbox repowering application in TPP

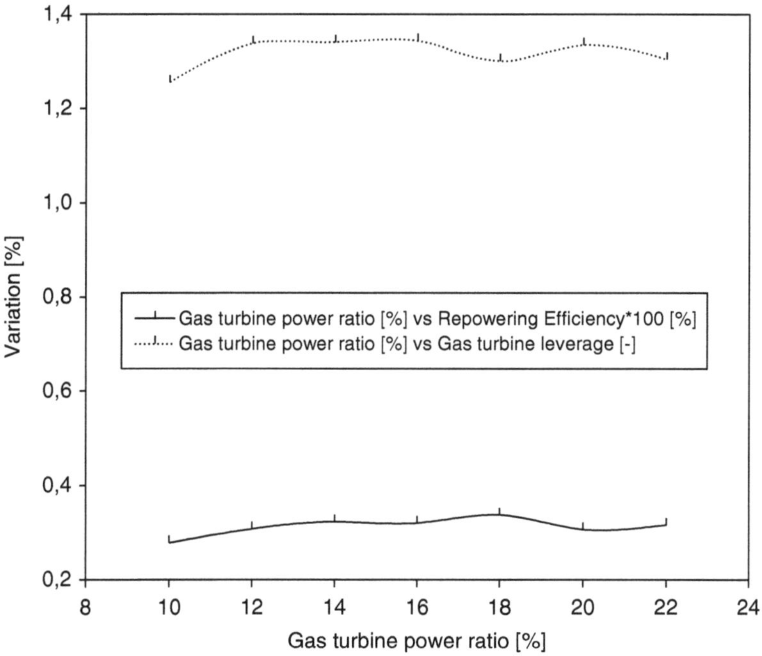

Fig. 38.8 Variation of repowering efficiency and gas turbine leverage

the increment in net electrical efficiency is distinctive in feed water heating application. The auxiliary power consumption of the hot windbox application is greater than feed water heating application due to the increased forced draught fan power. GT leverage and repowering efficiency was found to be 1.3 and 0.32, respectively for 22 % GT ratio. In the case of CO_2 emissions the coal consumption was increased due to the increased mass flow rate of air. Approximately half of the total mass flow rate of inlet air was selected as fresh air to decrease the temperature of the flue gas and increase the O_2 content of the inlet air. It is also obvious from Fig. 38.9 that CO_2 emission per installed power was decreased.

Parallel repowering of TPP was also investigated and the effects of repowering of TPP performance were presented in Fig. 38.10. In parallel repowering it is clear that the selection range of GT power was significantly increased. Because of the structure of the system steam was generated separately via a heat recovery steam generator (HRSG), located after the GT. Therefore, steam mass flow rate increment is the limiting factor in this kind of repowering application. Steam can be produced with different pressure levels in a multi-pressure HRSG. However, in this case initial and operational costs will be significantly increased. Instead of multi-pressure HRSG, single pressure HRSG was located and same pressure with live steam was obtained in order to mix these streams before steam turbine inlet. According to Fig. 38.10 the net power of the TPP was doubled and the net electrical efficiency was increased from 31.7 % to 38.26 %. In this case repowering efficiency and GT leverage were found to be 47.3 % and 1.174, respectively. CO_2 mass flow rate was increased. However, CO_2 emission per installed power was significantly decreased (Fig. 38.11).

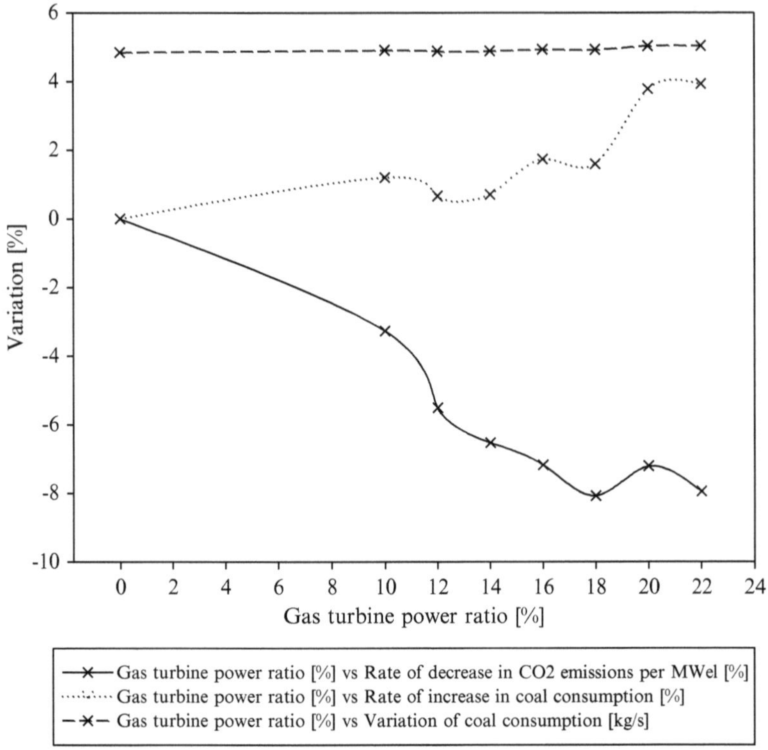

Fig. 38.9 The variation of CO_2 emissions and fuel consumption

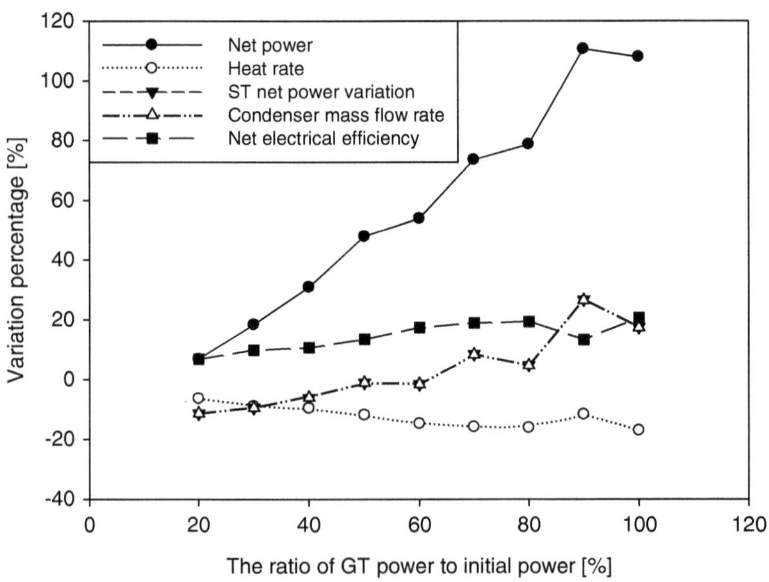

Fig. 38.10 The effects of parallel repowering application in TPP

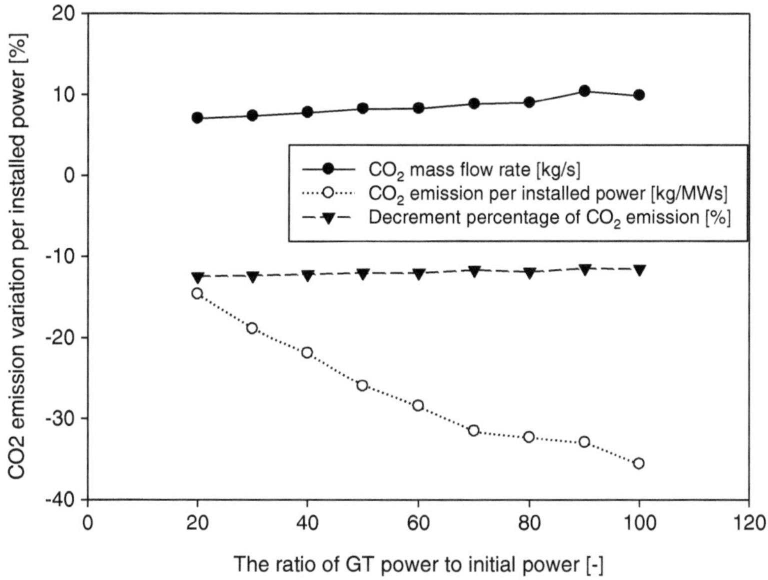

Fig. 38.11 The variation of CO_2 emissions and fuel consumption

Table 38.3 Economic comparison of repowering alternatives

Technoeconomic parameters	Current case	FWH	HWB	PR
Net power [MW]	21.877	30.207	27.67	45.5
Net electrical efficiency [%]	31.70	33.87	31.68	38.26
Heat rate [kJ/kWh]	11,355	10,609	11,363	9410
CO_2 mass flow rate [kg/s]	7.02	8.15	8.17	9.18
CO_2 mass flow rate per inst. power [kg/s/MW]	0.321	0.270	0.295	0.202
Coal consumption [kg/s]	4.843	4.821	5.033	3.813
First investment cost (FIC) [1000 × USD]	10,940	5,700	4,210	15,940
Specific repowering cost [USD/kWel]	500	685	726.4	674.6
Unit electricity generation cost [cent/kWh]	4.24	5.38	5.36	6.57
Unit electricity gen. cost without FIC [cent/kWh]	0.00	4.9	4.83	6.24

According to the results given above, repowering of the old TPPs can significantly increase the net electrical efficiency, net energy generation and decrease the CO_2 emissions per installed power. Repowering offers a short term solution for CO_2 decrement while new power plants have been installed.

An economical comparison was carried out to determine the unit electricity cost. Table 38.3 summarizes the situation after repowering for different cases. The cost of repowering, specific cost of repowering, unit electricity generation cost and with/without first investment costs were given. According to the limiting factors best GT ratios were compared. For feed water heating, hot windbox and parallel repowering applications GT ratios were selected as 30 %, 22 % and 100 %, respectively.

It is clear from the table that CO_2 mass flow rate increases. However, CO_2 mass flow rate per installed power decreases. In parallel repowering application coal consumption was found to be the lowest. However, it must be noted that natural gas consumption was found to be four times more than FWH and HWB repowering. In the economic analyses first investment cost of the TPP was calculated according to the design data with specific investment cost value. After, additional costs of each repowering application were calculated. It is obvious that the calculated highest repowering investment cost is parallel repowering application due to GT and HRSG costs. Unit electricity generation costs were also calculated with and without FIC. According to the results, HWB repowering with fresh air dilution slightly increases the unit electricity cost when compared to current case costs.

Conclusions

In this study comparison of repowering applications were performed. Net power, net electrical efficiency, CO_2 emissions and unit electricity generation costs were compared. Feedwater heating, hot windbox and parallel repowering applications were simulated in order to compare the results. Soma A thermal power plant with a capacity of $2 \times 22\ MW_{el}$ was selected as the case study. According to the results, net power and net electrical efficiency were increased. Total CO_2 mass flow rate was increased. However, CO_2 mass flow rate per installed capacity was sharply decreased. Therefore, repowering of old thermal power plants can be a short term solution in decreasing the CO_2 emissions while the electricity demand increasing continuously.

Nomenclature

P	Electric power, kW
Q	Thermal power, kW
FWH	Feedwater heating
HWB	Hot windbox
PR	Parallel repowering

Greek Letters

η	Efficiency, %
λ	Leverage, –

Subscripts

El	Electricity
ar	After repowering

br Before repowering
rep Repowering
GT Gas turbine
g Inlet

References

Carapellucci, R. (2009). A unified approach to assess performance of different techniques for recovering exhaust heat from gas turbines. *Energy Conversion and Management, 50*, 1218–1226.

D'Yakov, A. F., Nechaev, V., Olkhovsky, R., & Gurgen, G. (1998). Repowering existing thermal power stations. *Proceedings of the American Power Conference, 2*, 1033–1037.

Elmasri, M. A. (2008). *Design of gas turbine combined cycle and cogeneration systems*. Milan, Italy: Thermoflow Inc. Seminar.

Escosa, M. J., & Romeo, M. L. (2009). Optimizing CO_2 avoided cost by means of repowering. *Applied Energy, 86*, 2351–2358.

Mathieu. P. F. (1998). Repowering options for existing power plants. In *Proceedings of the NATO advances study institute on thermodynamics and optimization of complex energy systems* (pp. 251–260).

Popov, D. (2011). An option for solar repowering of fossil fuel fired thermal power plants. *Solar Energy, 85*, 344–349.

Schenk, H., & Ehren, G. (2003). Gas turbine based power plants repowering reduces emissions and increase efficiency of existing plants while re-utilizing available assets. In *Proceedings of the international gas turbine congress*.

Walters, A. B. (2008). Power plant topping cycle repowering. *Energy Engineering, 92*, 49–71.

Yilmazoglu, M. Z., & Durmaz, A. (2011). Parallel repowering of Soma A thermal power plant, 18. In *National conference on thermal science and technology*, Zonguldak, Turkey.

Yilmazoglu, M. Z., & Durmaz, A. (2013a). Hot windbox repowering of coal-fired thermal power plants. *Turkish Journal of Engineering and Environmental Sciences, 37*, 33–41.

Yilmazoglu, M. Z., & Durmaz, A. (2013b). Technoeconomic analysis of feed water heating repowering of Soma A thermal power plant. *Journal of Science and Technology of Dumlupınar University, 32*, 79–90.

Yilmazoglu, M. Z., Durmaz, A., & Baker, D. (2012). Solar repowering of Soma A thermal power plant. *Energy Conversion and Management, 64*, 232–237.

Chapter 39
Experimental Study of Heat Transfer for a Non-Newtonian Fluid in a Heated Cylindrical Pipe

Mounir Mellal, Hacina Abchiche, and Sabrina Ait Ouazzou

Introduction

During these last decades, the authors were interested in analyzing the problem of transfer of heat during the flows of the non-Newtonian fluids, in cylindrical pipes [Benslimane, 2012]. Thanks to its implication in numerous natural phenomena and industrial process, the researchers are more and more interested in this problem and more specifically in the transfers of heat which develop in the cylindrical pipes, in forced convection. Numerous researchers dealt with the understanding of the phenomenon of transfer of heat in laminar flow which the object of our work.

Materials and Methods

The setup of eviction which was of use to our experiment was specially designed within our laboratory, for the rheologic, thermo-rheologic study of the laminar flows of the various fluids. It consists of:

A storage tank made of stainless steel with a capacity of 15 L in double wall (isotherm) equipped with a sensor of pressure (manometer) with a membrane, an electric resistance, two gates of subjection, followed by a 3 m long cylindrical conduct, a heating conduct in the form of a tube in stainless steel altered by five prickings of temperature and two prickings of pressure, an electric resistance is wrapped around the tube to maintain a constant flow at the end of this installation.

M. Mellal (✉) • H. Abchiche • S. Ait Ouazzou
Laboratoire des Phénomènes de Transfert, Faculté de Génie Mécanique et Génie des Procédés, Université des Sciences et de la Technologie Houari Boumediene (USTHB), BP 32 El Alia, Bab Ezzouar, 16111 Algiers, Algeria
e-mail: m_rinmou@yahoo.fr

P. Grammelis (ed.), *Energy, Transportation and Global Warming*, Green Energy and Technology, DOI 10.1007/978-3-319-30127-3_39

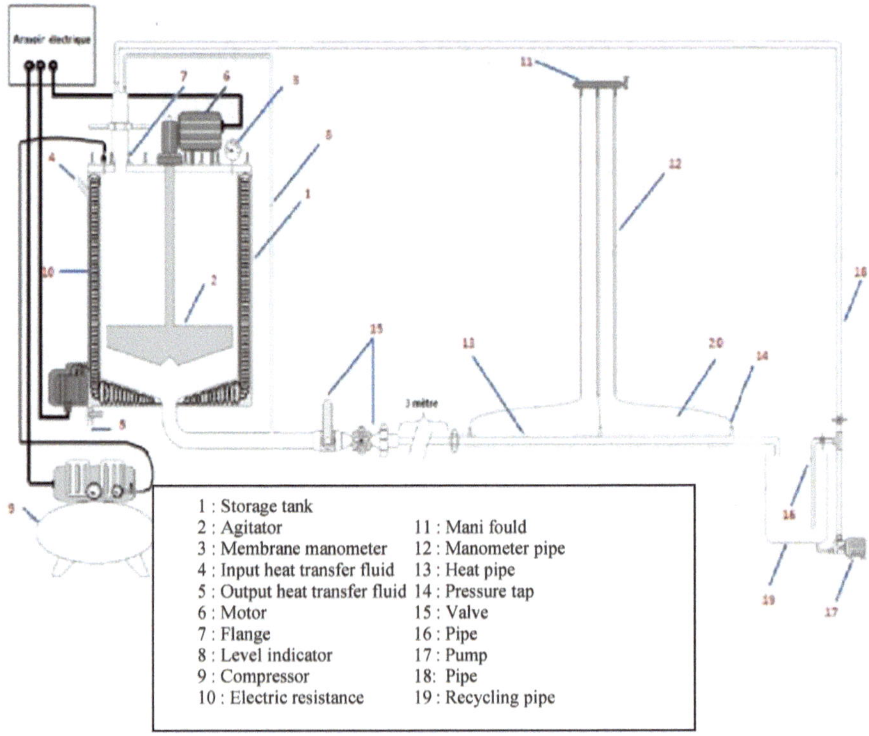

Fig. 39.1 The installation of eviction

A recycling tank for the product is placed out of the section of measure provided with a centrifugal pump allowing to feed again the main tank through a feeding pipe and a conduct of expulsion allowing the mixture of the product before its use, also, a thermo regulator is there in order to regulate and to fix the temperature of the external wall of the heating conduct (Fig. 39.1).

Conception of the Heating Conduct

We designed a heating conduct with resistance to warm the model fluid we had used for the thermo-rheologic study also to favor the flow by reducing the effects of sliding into the conduct; it is also convenient with the study of the transfer of heat to make a heating conduct which answers our expectations, we used a number of equipments, which the most important are:

A 2 m tube in stainless steel with prickings of pressure and temperature, around the tube we, a heat insulator was placed which is the Teflon in the form of an adhesive tape, afterward it was wrapped into an electric resistance of first class type Nickel/Chrome, then aluminum in the form of an adhesive tape, we provided the tube with electric cables for the supply, plastic and metallic rings for the tightening,

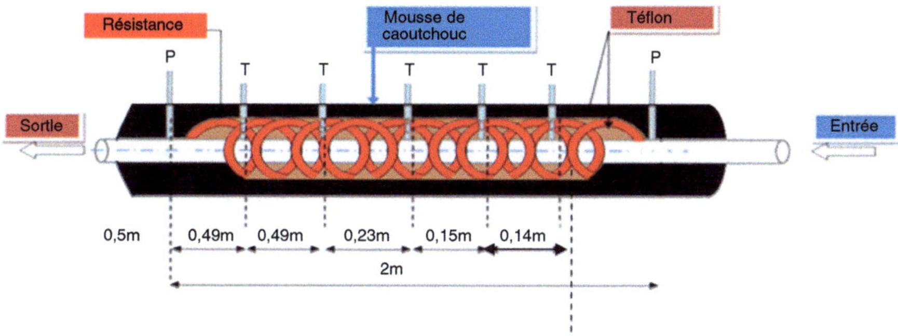

Fig. 39.2 Generalized plan by the heating conduct

then we wrapped up the whole system in an insulating material which is the glass wool and for the new conducts we put the foam rubber, then sheets of aluminum, also, a sensor for thermocouple and finally of two transformers and a thermoregulator.

Our study was made on three heating conducts of different diameters, one of the conceptions was later changed; the main insulating material, the glass wool was replaced by the foam rubber, a comparative study was also made. The conception of two new heating conducts was made according to one method as follows:

We began with the making of the prickings, then we set up sensors of thermocouple, followed by a primary insulation then the implantation of the resistance, then we placed the secondary insulation, then we prepared the grips of supply we finalized the whole by a tertiary insulation.

Figure 39.2 represents a generalized plan by the heating conduct which was of use to our thermo-rheologic study, it to study the thermal transfer which occurs there.

Used Product

The product used for the thermo-rheologic study which was of use to the study of the thermal transfer is the Carboxymethyl Cellulose. The study was made on concentrations of 0.5 %, 0.8 %, 1 % and 1.5 %.

Results and Discussion

Rheological Characterization of the Flowing Suspension

Having drawn the graphs of the logarithm of the difference of pressure according to the flow, we deducted that the fluid presents a rheo-fluidifying behavior of Oswald's model. The results which follow were obtained from the relations of Rabinowitch

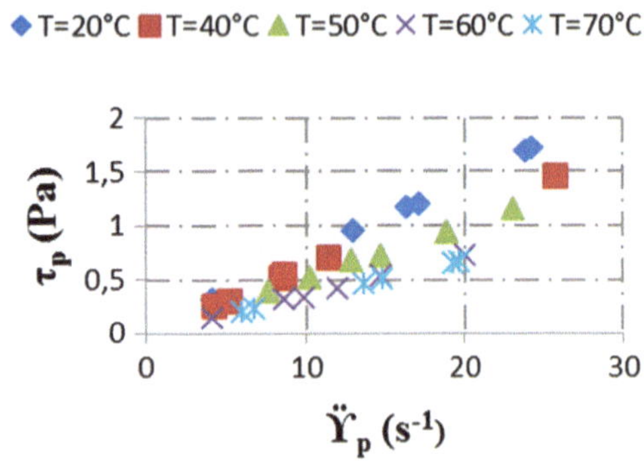

Fig. 39.3 Evolution of the constraint of cutting according to the speed of parietal cutting for a 0.5 % concentration, a diameter 10 mm

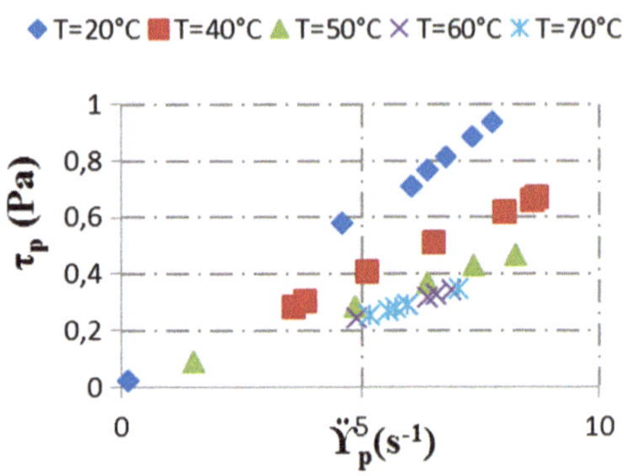

Fig. 39.4 Evolution of the constraint of cutting according to the speed of parietal cutting for a 0.8 % concentration, a diameter 10 mm

Mooney, from there we succeeded in drawing rheogrammes giving the evolution of the constraint of parietal cutting according to the speed of parietal cutting for various concentrations to several temperatures and various diameters.

Figures 39.3, 39.4, 39.5, and 39.6 represents the evolution of the constraint of parietal cutting according to the speed of parietal cutting for a diameter of 10 mm "we chose to present the results of a single diameter of 10 mm".

From the results shown on Figs. 39.4, 39.5, and 39.6 we notice ascending curves passing by the origin towards an asymptotic value, these curves take a look in the form of curve of power which answers the mathematical model which follows

$$\tau_p = K\dot{\gamma}^{n'} \tag{39.1}$$

From this equation we can deduct the type of the rheologic model which follows the used fluid, this model is the model of OSTWALD.

The treatment of the experimental results by the relation of Rabinowitch Mooney shows that the suspension in the range of the rate of studied cutting can have a behavior rheo-fluidizing of Oswald's model.

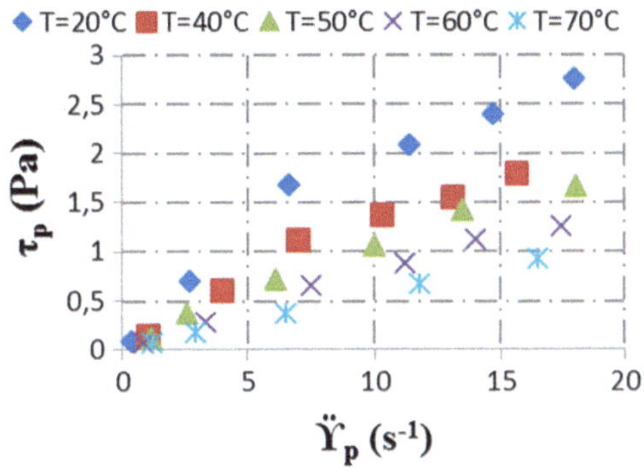

Fig. 39.5 Evolution of the constraint of cutting according to the speed of parietal cutting for a 1 % concentration, a diameter 10 mm

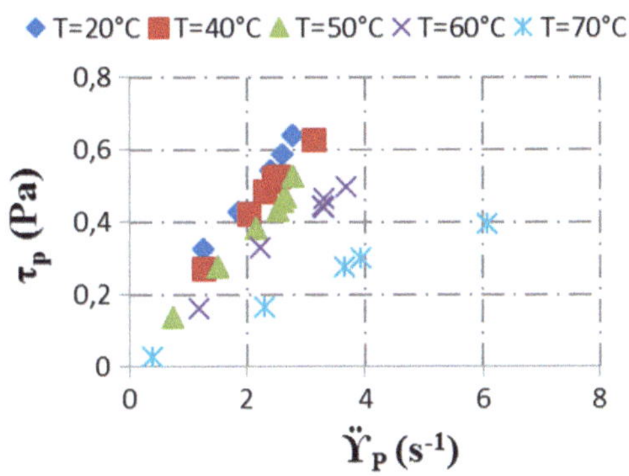

Fig. 39.6 Evolution of the constraint of cutting according to the speed of parietal cutting for a 1.5 % concentration, a diameter 10 mm

This behavior can be modeled using the following relation:

$$\tau = k\ddot{\gamma}^n \tag{39.2}$$

The pile of curves shows that the more we increase in the temperature of the regulator; the more there is decrease of the speed of cutting, thus the viscosity decreases.

Heat Transfer of in the Flow of Non Newtonien Fluids

Because of the number of Richardson, we were able to demonstrate that it is the forced convection which takes the flow on the natural convection.

The correlation $Nu = \left[\frac{11}{48} - \frac{1}{2}\sum_{m=1}^{\infty} \frac{\exp\left(-\gamma_m^2\, x^+\right)}{A_m\, \gamma_m^4}\right]^{-1}$ allowed us to confirm the constant parietal flow, we found a value of $Nu = 4{,}36$ with one m varying from 1 to 1000 where m represents a constant of precision, the higher m is, the better the precision will be, thus according to the literature this value of Nusselt found using FORTRAN (software), is represented for a flow of telling heat.

The thermal study was made on typical correlations for a forced convection on Oswald's fluid to constant parietal flow and it to study the evolution of the number of Nusselt according to the length, to Graetz and to the indication of the flow.

Nusselt Number Effect

Having calculated the number of Nusselt and the number of Graetz, we were able to draw graphs Figs. 39.7 to 39.8, which represents the evolution of the number of Nusselt according to the length for a diameter of 10 mm.

The number of Nusselt represents the total thermal transfer with regard to the thermal transfer by conduction throughout the conduct, thus according to the graphs of Figs. 39.8, 39.9, and 39.10, we notice the decrease of the number of Nusselt with the increase of the length, also we distinguish two zones. The first zone into which the number of Nusselt drops until a value of L = 0,7 m, this is because of the consequent exchange which is made in the entry of the zone with the aim of the

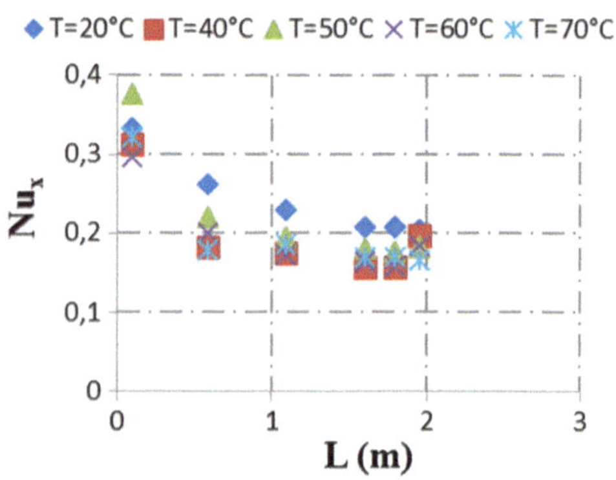

Fig. 39.7 Evolution of the number of Nusselt according to the distance for a 0.5 % concentration and one diameter of 10 mm

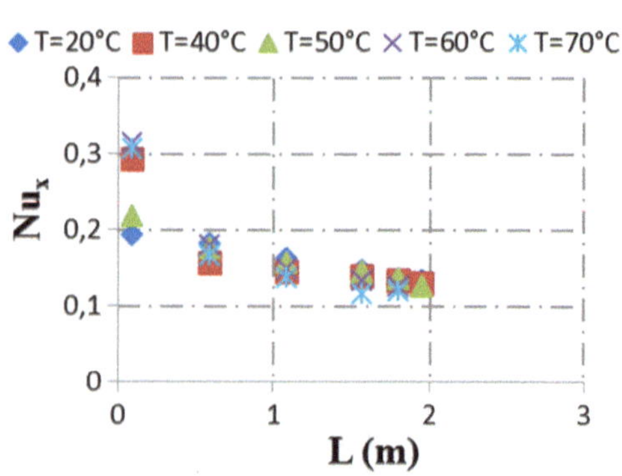

Fig. 39.8 Evolution of the number of Nusselt according to the distance for a 0.8 % concentration and one diameter of 10 mm

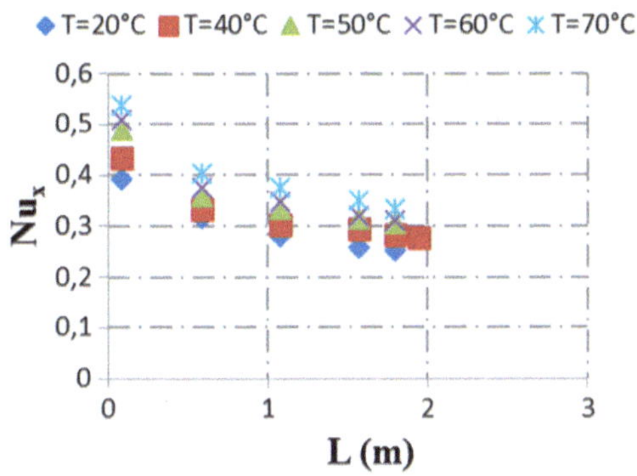

Fig. 39.9 Evolution of the number of Nusselt according to the distance for a 1 % concentration and one diameter of 10 mm

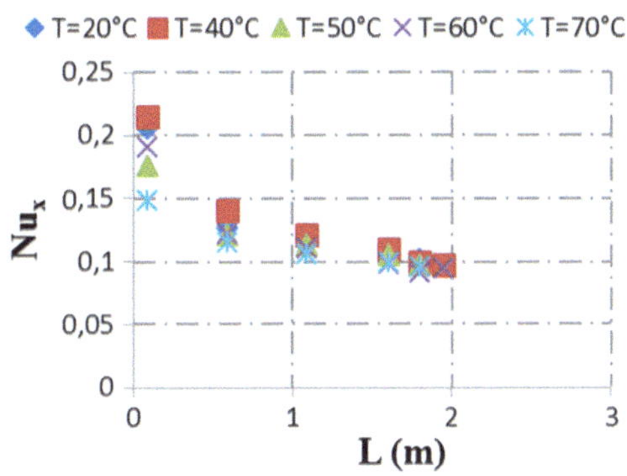

Fig. 39.10 Evolution of the number of Nusselt according to the distance for a 1.5 % concentration and one diameter of 10 mm

importance of the gradient of temperature, between the fluid and the wall. The second zone where the number of Nusselt stabilizes until an asymptotic value, we also notice that according to these graphs the number of Nusselt is important with the increase of the temperature. Thus the warmer it is, the bigger the exchange is.

Profile of the Number of Nusselt According to the Number of Graetz

According to Figs. 39.11, 39.12, 39.13, and 39.14 the evolution of the number of Nusselt according to the number of Graetz, we notice an increase of the number of Nusselt with the increase of the number of Graetz, also a superimposing of curves with various temperatures, concentrations and various diameters. We deduct that the length has no effect, seen that Graetz depends of the length. It confirms the works done before which also concluded that the length has no influence on the thermal transfer.

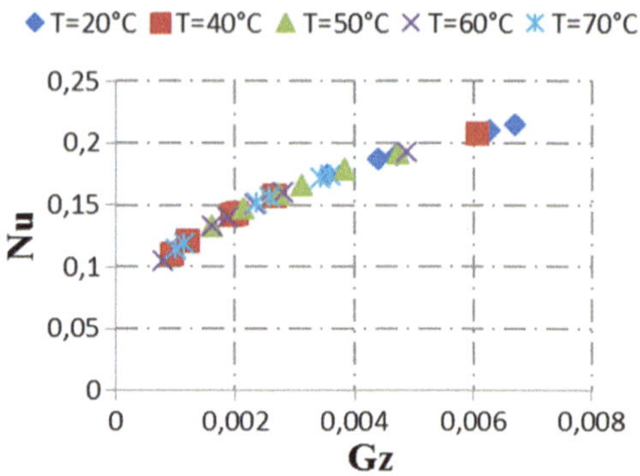

Fig. 39.11 Evolution of the number of Nusselt according to the number of Graetz for a 0.5 % concentration and one diameter of 10 mm in various temperatures

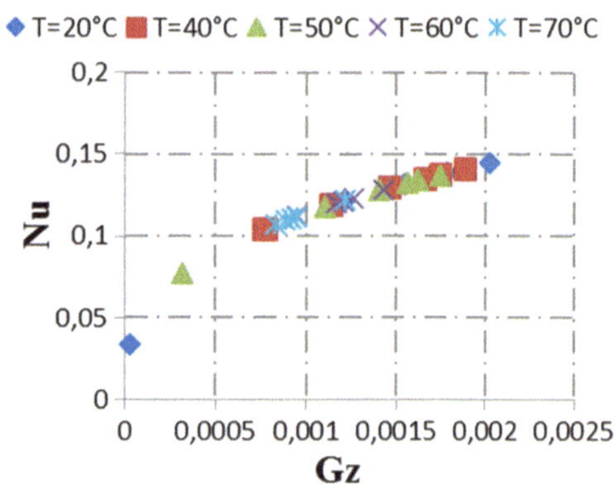

Fig. 39.12 Evolution of the number of Nusselt according to the number of Graetz for a 0.8 % concentration and one diameter of 10 mm in various temperatures

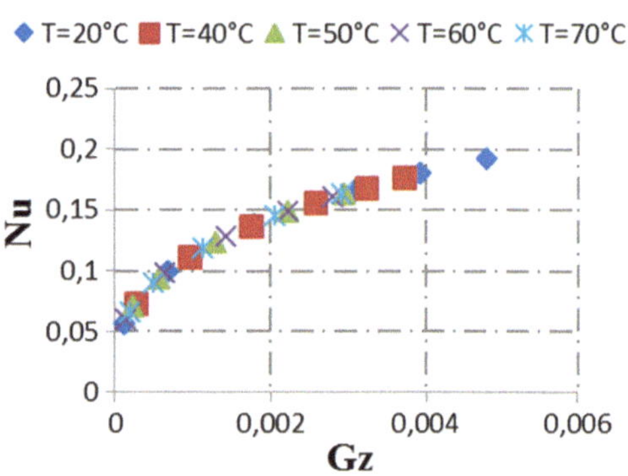

Fig. 39.13 Evolution of the number of Nusselt according to the number of Graetz for a 1 % concentration and one diameter of 10 mm in various temperatures

Determination of the Number of Asymptotic Nusselt

We studied the number of asymptotic Nusselt according to the index of the flow for all the concentrations, the temperatures and the diameter, the following results (profits) concern all the concentrations for a diameter of 10 mm.

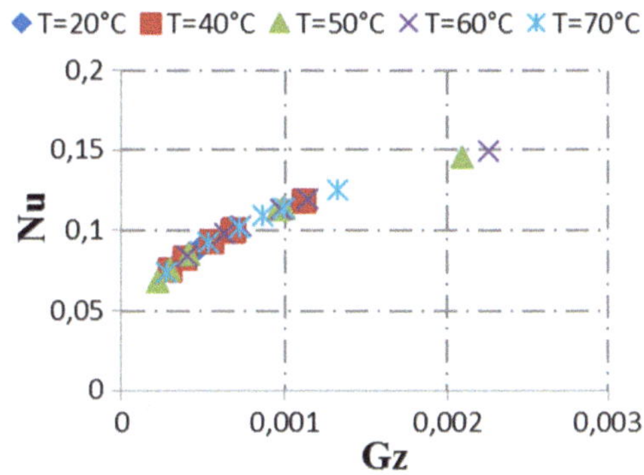

Fig. 39.14 Evolution of the number of Nusselt according to the number of Graetz for a 1.5 % concentration and one diameter of 10 mm in various temperatures

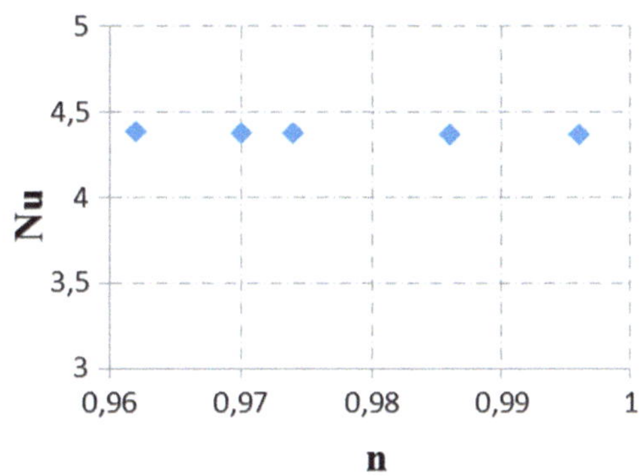

Fig. 39.15 Evolution of the number of Nusselt according to the index of flow for a 0.5 % concentration and one diameter of 10 mm

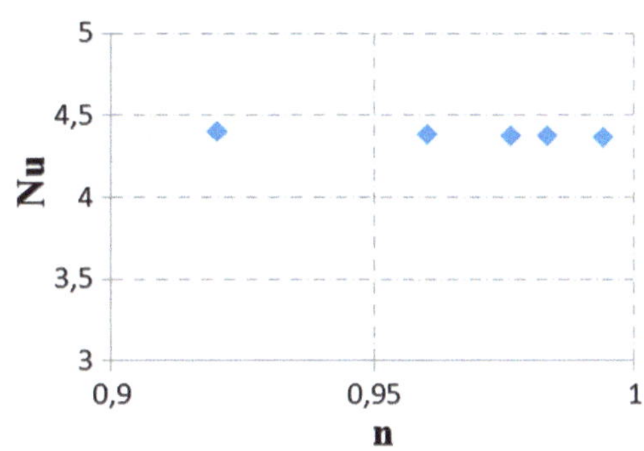

Fig. 39.16 Evolution of the number of Nusselt according to the index of flow for a 0.8 % concentration and one diameter of 10 mm

Figures 39.15, 39.16, 39.17, and 39.18 express the variation of the number of asymptotic Nusselt for the case "isoflux" according to the index of flow n, for the concentrations of 0.5, 0.8, 1 and 1.5 %, diameter 10 mm and for various temperatures of the thermoregulator, we notice that n is between $0 < n < 1$, which corresponds to the pseudoplastic fluids, we also notice that Nu_{∞} decreases with the increase of the index of flow (n) until stabilization. We can confirm that the rate of transfer of heat is more important within the pseudoplastic fluids $n < 1$.

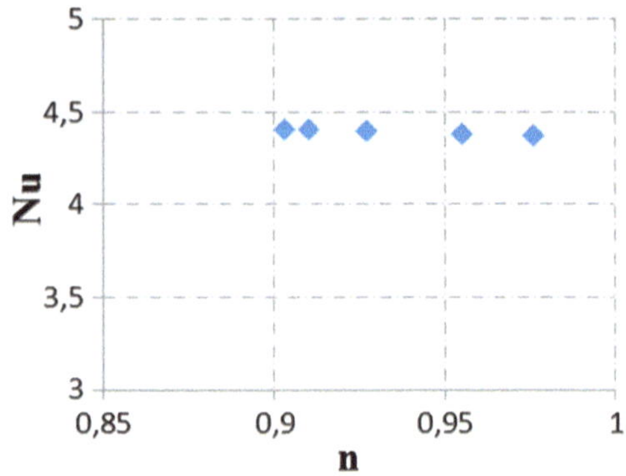

Fig. 39.17 Evolution of the number of Nusselt according to the index of flow for a 1 % concentration and one diameter of 10 mm

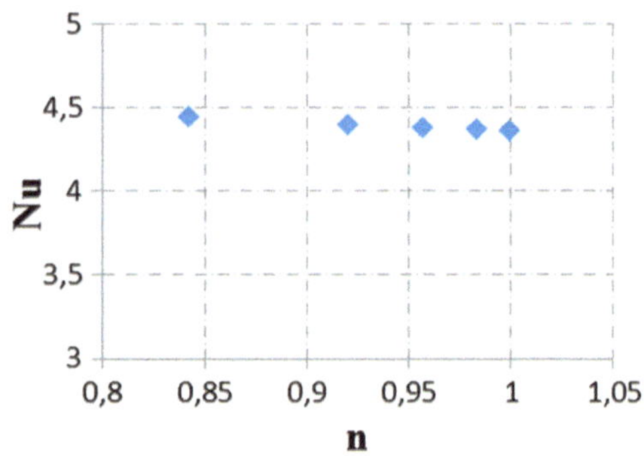

Fig. 39.18 Evolution of the number of Nusselt according to the index of flow for a 1.5% concentration and one diameter of 10 mm

Table 39.1 Flux lost through insulation

T = 70 °C				
C %	D mm			
	8	10 « rubber foam »	10 « glass wool »	12
0.5	3.8535	11.0690883		11.6255
0.8	3.3732	8.8967		10.8866
1	3.3843	11.9229125		9.2730
1.5	3.1618	10.8272789	3.239845	10.4824

Study of the Conduction

For this study, we proceeded by the method of harbinger's Law, for reasons of non-availability of all the parameters to study the conduction of the source towards the ambient air, we made an reverse study, we studied at first the flow absorbed by the fluid, by using harbinger's law of the source towards the fluid, afterward knowing the loose flow we proceeded with a subtraction to find the lost flow through insulations. The results are grouped in Table 39.1.

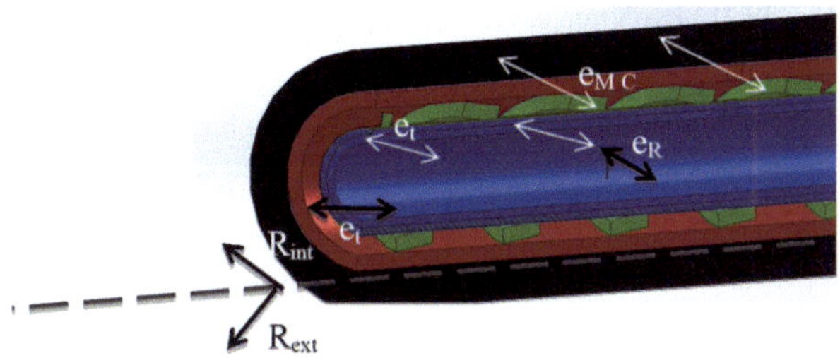

Fig. 39.19 Plan representing a cylindrical conduct

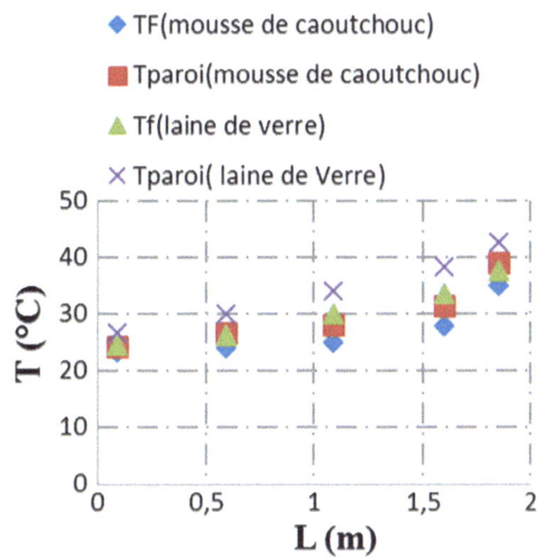

Fig. 39.20 Evolution of the fluid temperature and the wall for the glass wool and foam rubber according to the distance x, to a concentration of 1.5 %, a diameter of 10 mm to $T_{TR} = 50$ °C

Figure 39.19 represents a multi-cylindrical conduct sleep, where, R_{int} internal radius of the conduct and Rext the external radius, the distance between them representing the thickness of the teflon, e_R the thickness of the resistance and finally e_{MC} the thickness of the foam rubber. The following table represents the lost flow through insulations in different diameters and concentrations as well as the comparison of both insulations made on a diameter of 10 mm and for a concentration of 1.5 % in 70 °C.

Another study was made to confirm that the insulating material of glass wool is more successful than the foam rubber. We compared the temperatures of the fluid and the wall of two various insulations according to the distance, for a temperature of the thermoregulator of 50 and 60 °C [Gacem, 2010].

Figures 39.20 and 39.21 represent the evolution of the temperature of the fluid and the wall for both cylindrical conducts, we notice an increase of the temperature with the increase of the length, on the other hand we notice that for the temperature of the fluid and the wall for the glass wool, insulation is higher than foam rubber, thus the fluid stores more heat in glass wool insulation than foam rubber this is the same for the temperature of the wall, what leads us to say that the glass wool is more hermetic.

Fig. 39.21 Evolution of the fluid temperature and the wall for the glass wool and foam rubber according to the distance x, to a concentration of 1.5 %, a diameter of 10 mm to $T_{TR} = 60\ °C$

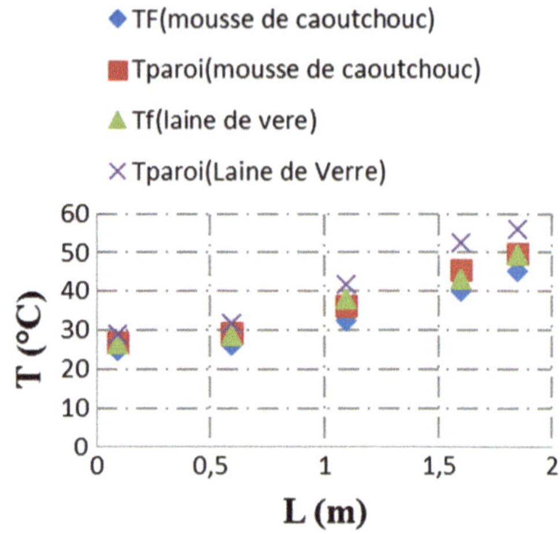

Fig. 39.22 Evolution of the wall temperature averages according to the distance x, for both insulations “glass wool” and “foam rubber”, for a 1,5 % concentration, one diameter of 10 mm and $T_{TR} = 50\ °C$

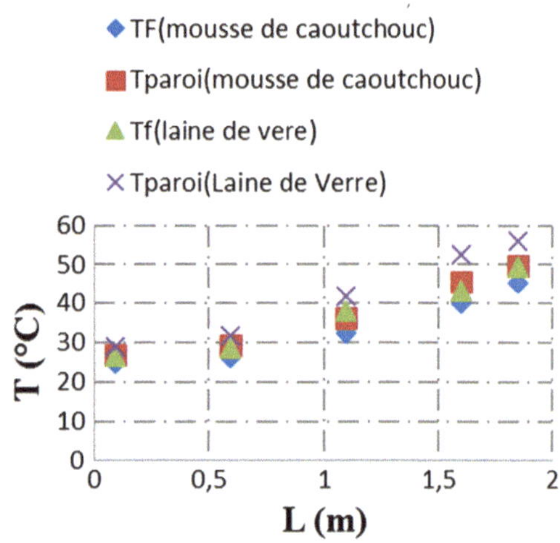

Figures 39.22 and 39.23 represent the evolution of the average temperature of wall according to the length for two various insulations “Foam rubber” and “Glass wool”. We notice an increase of the temperature with the increase of the length by both insulations, we also notice that the curve which represents the glass wool insulating material is higher than the one which represents the foam rubber insulating material to the temperature of 50 °C and 60 °C of the thermoregulator, that leads us to confirm that the glass wool insulation is more hermetic.

Figures 39.24 and 39.25 present the evolution of the average temperature of the wall according to the flow, for both insulations “glass wool” and “foam rubber”, for a 1.5 % concentration, one diameter of 10 mm and the temperature of the thermoregulator of 50 and 60 °C, these curves present a decrease of the temperature of the

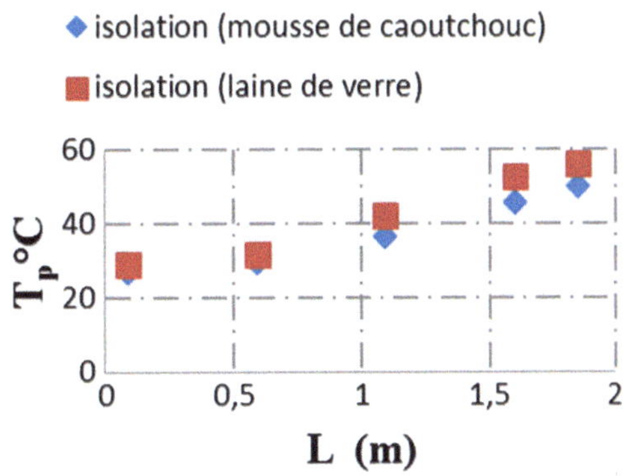

Fig. 39.23 Evolution of the wall temperature averages according to the distance x, for both insulations “glass wool” and “rubber foam”, for a 1,5 % concentration, one diameter of 10 mm and $T_{TR} = 60\ °C$

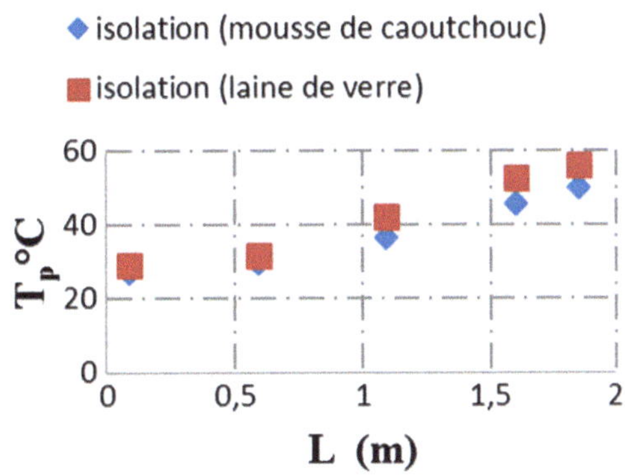

Fig. 39.24 Evolution of the temperature of wall averages according to the outflow, for both insulations “glass wool” and “foam rubber”, for a 1,5 % concentration, one diameter of 10 mm and $T_{TR} = 50\ °C$

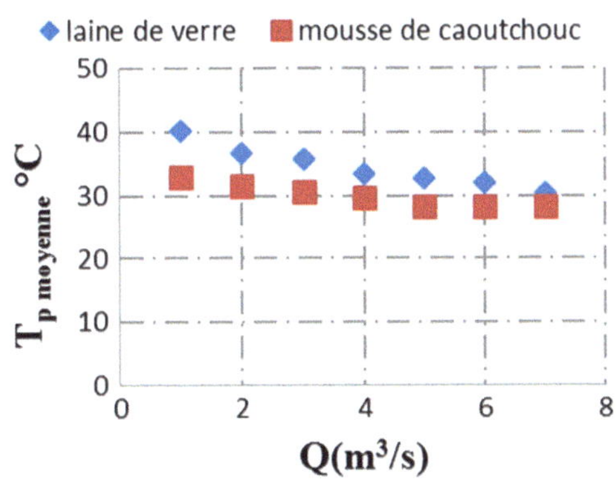

Fig. 39.25 Evolution of the temperature of wall averages according to the outflow, for both insulations “glass wool” and “foam rubber”, for a 1.5 % concentration, one diameter of 10 mm and $T_{TR} = 60\ °C$

wall average with the increase of the flow, we also notice that the temperature of wall average for the glass wool insulation is bigger than that of the foam rubber insulation, it can be explained by the fact that the glass wool isolates better that the foam rubber, thus heat losses are lesser.

Conclusion

The thermo-rheological study allowed us to show that the non-Newtonian fluid used, can be defined by establishing Oswald Devaele pseudoplastic type.

The testing made in the heated cylindrical pipe (HCP) has allowed us to promote transfer mode through the relation of Richardson, which happens to be less than 1, so it is the forced convection that outweighs the natural convection.

Therefore, this study was conducted on a forced convection using Oswald fluid for a laminar flow regime in a cylindrical pipe with constant heat and parietal flux at different concentrations 0.5 %, 0.8 %, 1 % and 1.5 % and different diameters 8, 10 and 12 mm. From these results we were able to assess the Nusselt number with different correlations, from one correlation we confirmed that we are in constant flux compared with the literature, we obtained a value. We have also plot graphs of the Nusselt number depending upon the length, and the Graetz number of flow.

To study the evolution of the Nusselt number in reference to the length, we found that the Nusselt number drops with increasing length, and then stabilizes at an asymptotic value. This result in the exchange therefore occurs at the entrance to the heating pipe and caused by the elevation of the temperature gradient in the etching area between fluid and wall.

The study of the evolution of the Nusselt number according to Graetz, allowed us to observe an increase in the Nusselt number with the increase in the number of Graetz, as we noticed a stack of curves at different temperatures the temperature controller we found that the length of the pipe has no effect on the thermo-rheological behavior.

Two of our pipes were insulated with two different insulation at the main insulation, one insulated glass wool, the other rubber foam, for this we have made a comparative study between the two insulators, by studying the fluid temperature, the wall and the flow lost through these insulators, the results allowed us to infer that the glass wool insulation is more airtight than the foam rubber.

Acknowledgement Thermal transfer, thermo rheology, carboxymethylcellulose, cylindrical pipe.

References

Benslimane, A. (2012). « Rhéologie et Ecoulement de Fluides Chargés: Application aux Réseaux d'Assainissement Urbains. Etude Expérimentale et Modélisation ». Mémoire de Doctorat, décembre 2012.

Gacem, M. (2010/2011). « comparaison entre l'isolation thermique Extérieure et Intérieure d'une pièce d'un Habitat situé dans le Site de Ghardaïa », mémoire de magister, 2010–2011 (convection forcée).

Lachemet, A. (1997). « Rheologie et Transfert de Chaleur en Milieu non-Newtonien », mémoire Magister, avril 1997.

Sacadura, J. F. (1995). « Initiation aux Transferts Thermiques ». Centre d'Actualisation Scientifique et Technique INSA de LYON, janvier 1995.
Taine, J., Iaconna, E., Petit, J. P. (2009). « Transfert Thermique », 4eme Edition.

Chapter 40
The SOC Estimation of LCO Battery Based on BP Neural Network

Sy-Ruen Huang, Yen-Huai Ma, Jheng-Shyun Li, and Jun-Han Chan

Introduction

Improvements in battery technology have led to the development of batteries with greater energy yield, efficacy, and capacity. Currently, the weight and energy density of batteries, in addition to battery capacity and weight, have facilitated in providing sufficient energy for machines requiring high-power output (e.g., electric vehicles, electric boats, and power motors). Previous batteries for use in high energy-consumption machines were generally too heavy or large for most practical applications. However, the energy density of conventional lithium batteries can exceed 120 W·h/kg. Consequently, this type of battery has been used as an energy source in various types of machine.

Current battery technologies have led to the development of lithium ion batteries that provide up to 190 W·h/kg in energy density and 1000 charge cycles (80 % depth of discharge). The lifetime of a lithium ion battery varies according to how it is used. State of charge (SOC) is typically used to represent battery capacity and is equivalent to the percentage of energy stored in a battery. Generally, maintaining the SOC of a lithium ion battery at 20–80 % can prolong its lifetime. Therefore, accurately determining battery SOC is a crucial goal in battery research. Previous studies on batteries have indicated that accurately estimating a battery's SOC is a crucial topic in the fields of battery balancing and battery management systems (Burkett et al. 1970; Linden and Reddy 2007).

S.-R. Huang • J.-S. Li • J.-H. Chan
Department of Electrical Engineering, Feng Chia University, No. 100, Wenhwa Road, Seatwen, Taichung 40724, Taiwan, ROC

Y.-H. Ma, Ph.D. (✉)
Ph. D. Program of Electrical and Communications Engineering, Feng Chia University, No. 100, Taichung 40724, Taiwan ROC
e-mail: ahma5653@gmail.com

P. Grammelis (ed.), *Energy, Transportation and Global Warming*, Green Energy and Technology, DOI 10.1007/978-3-319-30127-3_40

Artificial neural networks (ANNs) are a family of equation models and algorithms proposed by McCulloch and Pitts in 1943. These mathematic models imitate the neuron structures and behaviors of organisms, and perform calculations according to a network comprising numerous artificial neurons. ANNs are a type of parallel calculation system that utilizes equations to emulate the processes of a biological neural system and various types of numerical estimation, including classification, identification, and prediction. Previous studies have used ANNs in numerical analysis in various fields such as engineering, science, finance, and literature. An ANN is composed of an input layer, one or more hidden layers, and an output layer. Each hidden layer consists of a parallel array of neurons, in which an input is multiplied by the weights assigned to the layer, added to the bias values of the layer, and then transferred to the next layer through a transfer function. The hidden layers of an ANN continually transfer input values to the lower layers until the value reaches the output layer, at which point the calculated result is obtained (Hagan).

Types of Lithium Ion Batteries

The prevalent use of lithium ion batteries have led to improvements to the characteristics of such cathode materials as lithium cobalt oxide (LCO), lithium manganese oxide, lithium iron phosphate, and lithium nickel manganese cobalt oxide. Table 40.1 illustrates a comparison of various cathode materials. The safety of lithium ion batteries has been improved, and the charge cycle of such batteries can exceed 1000 charges. Consequently, lithium ion batteries have been applied in electric cars and hybrid electric cars. Studies on lithium ion batteries have been conducted since 1980. The oxide structure of such batteries is classified as an α-$NaFeO_2$ layered structure, which is relatively stable compared with other battery types. Characteristics such as high battery density, low discharge rate, absence of memory effect, long cycle life, antioxidation, and acid-corrosion resistance enable various types of electrolyte to be used in lithium ion batteries. Additionally, such batteries have a long storage life when not in use. Lithium ion batteries have become a mature and primary cathode material product that features higher capacity and more stable performance compared with other cathode material products.

However, lithium ion batteries also possess several disadvantages. Overcharging such a battery can permanently fixate lithium ions in its crystal lattices, thereby

Table 40.1 Comparisons of cathode materials

	$LiCoO_2$	$LiMn_2O_4$	$LiMnO_2$	$LiFeO_4$
Density of energy	190–240	100–120	200	130–140
Voltage	3.6	3.8–3.9	3.4–4.3	3.2–3.7
Cut-off voltage	4.2	4.2	4.2	3.6
Life cycle	>500	>500	Bad	>1000

reducing the lifetime of the battery. By contrast, overdischarging such batteries can release excess lithium ions and collapse the crystal lattices, thereby decreasing the battery lifetime. Therefore, when charging and discharging a lithium ion battery, its SOC must be accurately measured for the protection mechanism to prevent overcharging and overdischarging.

Methods for Measuring SOC

Current methods for measuring SOC include the discharge method, open circuit voltage (OCV) method, coulomb counting method (CC method), direct current (DC) impedance method, and ANNs. The discharge method is advantageous because of its accuracy; under a constant current, a battery is discharged and its current and voltage are then multiplied by time duration to measure the amount of energy expended by the battery for a specific period (W·h). However, the discharge method is an offline approach, and thus cannot be applied to measure the SOC while a battery is in use. The OCV method involves measuring the relationship between a battery's capacity and voltage. Under an open-circuit condition, battery voltage represents the battery's remaining capacity. To measure the OCV of a battery, the target battery must stand until the reaction of the battery plate is complete, and then the voltage becomes stable and is measured as the OVC of the battery. This method is time-consuming and cannot be applied to measure the SOC of a battery in real time. The discharge method and OCV method are commonly used to estimate the full capacity of a battery (Lee et al. 2007).

The CC method is a simple and prevalent method that involves multiplying battery current by time duration to measure the cumulative capacity (A·h), which represents the amount of electric charge in a battery for 1 A of current to flow for 1 h discharge. However, this method can build up cumulative errors that reduce its accuracy. Specifically, when the CC method is applied to continually estimate the SOC of a battery, the method can yield values exhibiting small but consistent errors; eventually, the cumulative error reduces the accuracy of the method (Leksono et al. 2013). The DC impedance method involves measuring how the functioning region of a battery equivalent circuit is reduced when the electrolyte level decreases. This method is commonly used to determine a battery's state of health (SOH).

ANNs involves various methods. Generally, a battery's voltage, current, and capacity can be used as training data for an ANN. Battery capacity can serve as a target matrix for training the weight and bias value of each neuron, and an ANN can then be constructed to estimate the SOC of the battery (Liu et al. 2011; Eddahech et al. 2011; Sarvi and Adeli 2010).

Table 40.2 Battery parameters of NCR18650B

Rated capacity	3200 mAh
Capacity	Min: 3250 mAh Std: 3350 mAh
Rated voltage	3.6 V
Charger condition	CC-CV, 1625 mA, 4.20 V, 4.0 h
Weight	48.5 g
Temperature	Charging: 0 to + 45 °C Discharging: −20 to + 60 °C Steady: 20 to + 50 °C
Density of energy	Capacity: 676 Wh/l Weight: 243 Wh/kg

Battery Charge/Discharge Test and Experimental Data Storage

NCR18650 LCO batteries were used in the present study. The output voltage of LCO batteries is stable and relatively high compared with that of other lithium batteries. The rated capacity and maximum voltage of an NCR18650 are 3200 mAh and 4.2 V, respectively (Table 40.2).

This study employed the constant current–constant voltage (CC–CV) method to charge the target battery. During the initial stage of charging, a constant current was applied to the battery, and the voltage of the battery increased. Once the battery voltage value reached the preset voltage value, the CC–CV charging method was initiated, during which the current decreased until reaching the preset minimum value, at which point the charging process was completed.

To accurately measure the capacity of the target battery, the OCV method and CC method were applied to acquire its basic data. Specifically, we plotted the current–voltage curve of the battery. The OCV values of the battery estimate the corresponding capacity values. The cumulative capacity values of the battery were determined using the CC method to measure the current flowing through the battery. LabVIEW was used to collect the data during the process of charging. Figure 40.1 depicts the configuration of the system architecture and the human machine interface. During the charging process, 0.34 A of current was applied to charge the battery from 2.8 V (the minimum voltage of the battery) to 4.2 V. The voltage of a battery can increase or decrease during the process of charging and discharging. Consequently, the current was set at a low value (0.34 A) to ensure that the differences in voltage were sufficiently small to approximate an OCV–capacity curve. Figure 40.1 illustrates the current–voltage curve of the battery.

The arithmetic mean of the charging curve and discharging curve were applied to plot the average voltage–capacity curve, in which the voltage values can be perceived as the OCV values of the target battery (Fig. 40.2). The capacity values (A·h) in Fig. 40.2 were calculated using the CC method to determine the battery's cumulative current across a time duration. During the charging process, the initial capacity value of the battery was incremented. By contrast, the initial capacity

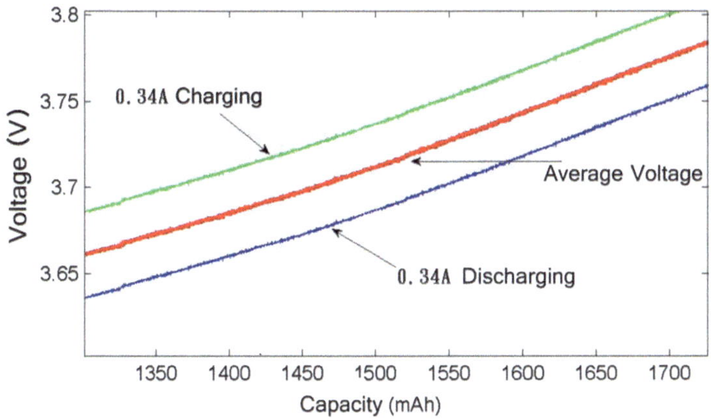

Fig. 40.1 Battery charging, discharging, and average voltage curves at 0.34 A

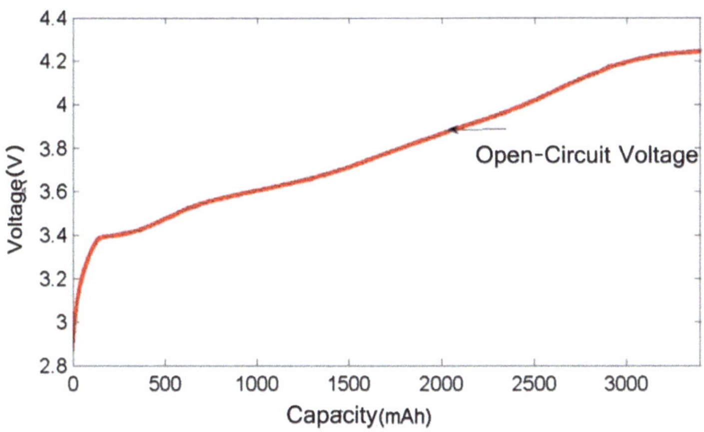

Fig. 40.2 OCV–capacity curve

value was decreased during the discharging process. Equation (40.1) illustrates how the capacity values of the battery were determined through the CC method.

$$Ah(t_n) = Ah\ t_0 \pm \int itdt \tag{40.1}$$

Design and Training of Back Propagation Neural Network

In a back-propagation neural network (NN), an input unit is multiplied by the weights of the layer and the error of the target value. Next, a stochastic gradient descent method is applied to minimize the input-value error, and then the input

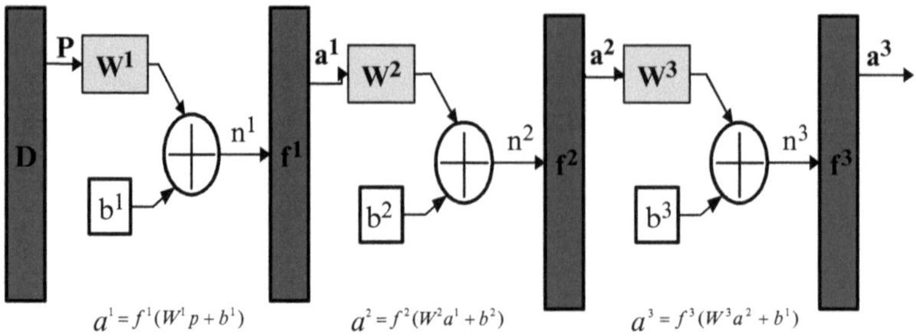

Fig. 40.3 The architecture of a multilayer ANN

value is multiplied by the learning rate. Subsequently, the modified input value is transferred to the previous layer to correct the weights and bias values of that layer. The input value propagates backward in the network until it reaches the input layer, thereby modifying the overall network and ensuring that the errors of the network are within a given tolerance range. The initial weights of the neurons can be modified through the backward propagation of the input value; hence, such a network learning process is called a back-propagation NN.

A back-propagation NN can be regarded as a multilayer network (Fig. 40.3). Such NNs can be used to accurately train any continuous function, of which the accuracy is determined according to the number of neurons present in the hidden layers of the network. Generally, a network with 10 nodes of neurons can result in undesirable training errors. Increasing the number of neurons can improve the accuracy of the output functions. Specifically, the presence of 20–50 neurons in a network can effectively increase the training speed and ensure that the accuracy of output functions is within a given tolerance range. However, a high amount of memory is required to process a network containing many neurons. Therefore, a network comprising 20 neurons was used in the present study.

This study employed a single-layer NN. The data were inputted through the input layer, and then transferred through the neurons of the hidden layers and modified according to the neuron weights. Through a hyperbolic tangent transfer function, the data were successively transferred to the subsequent layers until reaching the output layer. Finally, the data were modified according to the output layer weights, yielding the output function $\mathbf{a}^3$, which is expressed as (40.2).

$$\mathbf{a}^3 = \mathbf{f}^3\left(\mathbf{W}^3\mathbf{f}^2\left(\mathbf{W}^2\mathbf{f}^1\left(\mathbf{W}^1\mathbf{P} + \mathbf{b}^1\right) + \mathbf{b}^2\right) + \mathbf{b}^3\right) \tag{40.2}$$

Expanding the hidden layers of the network yields (40.3):

$$\begin{aligned} &\mathbf{a}^{\mathrm{m}+1} = \mathbf{f}^{\mathrm{m}+1}\left(\mathbf{w}^{\mathrm{m}+1}\mathbf{a}^{\mathrm{m}} + \mathbf{b}^{\mathrm{m}+1}\right) \\ &m = 0, 1, \ldots, M - 1 \end{aligned} \tag{40.3}$$

where M is the number of layers in the NN.

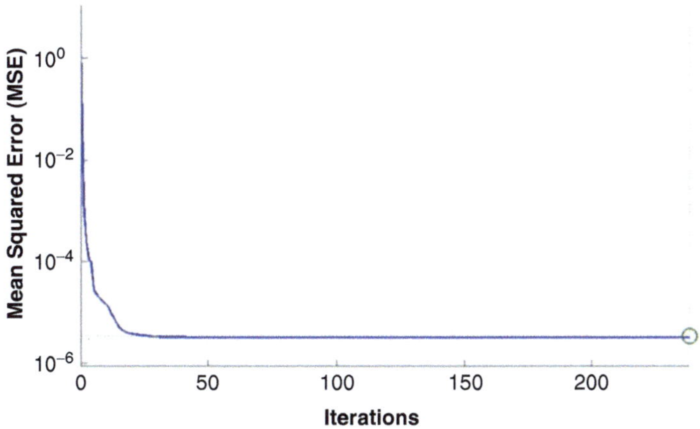

Fig. 40.4 The MSE curve of the iteration number

Through training, a back-propagation NN can generate accurate functions. The NN training parameters are inputted into the network and the mean squared error (MSE) of the training samples can then be obtained using (40.4). The MSEs are used to modify the weights and bias values of the neurons.

$$F(x) = E\left[\mathbf{e}^{T}\mathbf{e}\right] = E\left[(\mathbf{t}-\mathbf{a})^{T}(\mathbf{t}-\mathbf{a})\right] \tag{40.4}$$

Through network back-propagation, the updated weights and bias values are obtained using (40.5) and (40.6).

$$\mathbf{s}^{M} = -2\dot{\mathbf{F}}^{M}\left(\mathbf{n}^{M}\right)(\mathbf{t}-\mathbf{a}) \tag{40.5}$$

$$\begin{aligned}\mathbf{s}^{m} &= \dot{\mathbf{F}}^{m+1}\left(\mathbf{n}^{m}\right)\left(\mathbf{W}^{m+1}\right)^{T}\mathbf{s}^{m+1}\\ m &= M-1,\ \ldots,\ 2,\ 1\end{aligned} \tag{40.6}$$

The least mean squares (LMS) algorithm and an estimated gradient value was used to approximate the LMS value. As shown in (40.7) and (40.8), this method is similar to the least gradient descent method and is used to update the weights and bias values of the layers (Fig. 40.4).

$$\mathbf{W}^{m}(k+1) = \mathbf{W}^{m}(k) - \alpha\mathbf{s}^{m}\left(\mathbf{a}^{m-1}\right)^{T} \tag{40.7}$$

$$\mathbf{b}^{m}(k+1) = \mathbf{b}^{m}(k) - \propto \mathbf{s}^{m} \tag{40.8}$$

where α is the learning rate.

Figure 40.5 depicts the MSE of the training process. When the number of iterations reached 240, the LMS was reduced to $9.41 \times e^{-0.9}$. Network training took 1 min and 26 s to complete.

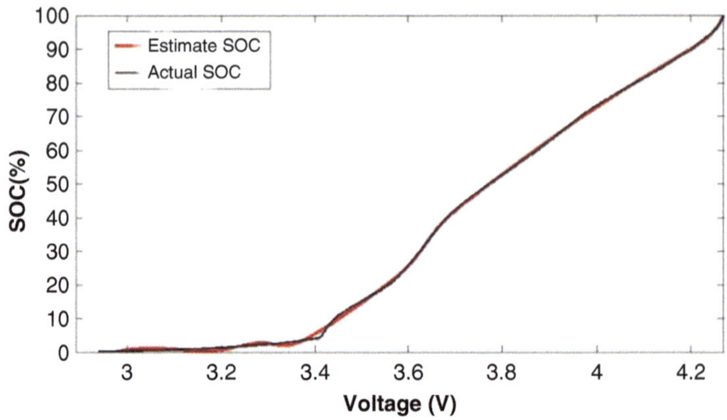

Fig. 40.5 Estimated capacity versus actual capacity of the target battery

Estimation of Battery SOC Through ANN

The weights and bias matrices obtained through the back-propagation training method were inputted into the NN, and a mathematical model for estimating the SOC of the target battery was then obtained. The current and voltage of the battery were measured and then inputted into the NN to estimate its real-time capacity. Subsequently, the real-time capacity was divided by the peak capacity to obtain the SOC of the target battery, as expressed in (40.9).

$$SOC = \frac{\mathbf{a}(V, I)}{Q_0} \tag{40.9}$$

$\mathbf{a}(V, I)$: Real-time capacity

Q_0: Maximum capacity

To evaluate the applicability of the proposed estimation method, we used a 1-A constant current to charge the NCR18650 battery. Subsequently, the NN estimation and CC methods were applied to obtain the SOC of the target battery. Figure 40.5 shows the voltage–SOC curves, in which the red curve represents the SOC values (estimated through the NN), whereas the black curve indicates the actual SOC values (measured through the CC method). The estimated and actual curves exhibited similar trends, indicating that the proposed NN method can accurately estimate battery SOC. Table 40.3 lists the numerical values for plotting the voltage–SOC curves. The errors of the estimated SOC values and the actual SOC values were negligible (Table 40.3).

Table 40.3 Estimated capacity versus actual capacity of the target battery

Voltage (V)	Current (A)	Estimate SOC	Actual SOC	Error
3.34	0.9887	2.10 %	2.20 %	0.10 %
3.57	0.9885	21.70 %	21.40 %	0.30 %
3.68	0.9824	40.50 %	40.30 %	0.20 %
3.86	0.9881	59.70 %	59.20 %	0.50 %
4.08	0.9879	80.00 %	79.90 %	0.10 %

Conclusion

The present study was conducted to develop a back-propagation NN for estimating battery SOC. The proposed model can be employed to estimate the SOC of a battery in a closed-loop system. Specifically, the measured voltage and current of a target battery can be inputted into the model to calculate the real-time capacity of the battery, and the SOC of the battery can then be estimated. Compared with the CC method, the NN method is more complex; however, the NN method can be applied to continuously measure battery SOC without generating cumulative errors. The results of the present study show that the errors of the estimated SOC and actual SOC are within the tolerance range. Consequently, increasing the number of neurons in the NN is unnecessary unless the number of input items is increased.

Through measuring various types of battery parameters for modifying the weights and bias values of a trained NN, a mathematical model can be obtained for estimating battery capacity. Such an estimation method can be applied to measure the capacity of batteries of various characteristics. Future studies should consider applying temperature parameters to estimate a change in battery capacity relative to a change in temperature. In addition, the number of parameters can be increased to improve the accuracy of battery capacity estimation.

Acknowledgement This study received financial support from the Ministry of Science and Technology under the grants NSC101-2632-E-035-001-MY3.

References

Burkett, W. B., Palisades, P., & Jackson, R. V. (1970, June). Rapid charging of batteries. U.S. Patent 3,517,219.

Eddahech, A., Briat, O., & Vinassa, J.-M. (2011). Neural networks based model and voltage control for lithium polymer batteries. In *IEEE international symposium*, 5–8 Sept 2011.

Hagan, M. T., Demuth, H. B., & Beale, M. H. Neural network design. Princeton University Press.

Lee, S. J., Kim, J. H., Lee, J. M., & Cho, B. H. (2007). The state and parameter estimation of an li-ion battery using a new OCV-SOC concept. In *IEEE power electronics specialists conference*, 17–21 June, 2007.

Leksono, E., Haq, I. N., Iqbal, M., Nugroho Soelami, F. X., & Merthayasa, I. G. N. (2013). State of charge (SoC) estimation on $LiFePO_4$ battery module using Coulomb counting methods with modified Peukert. In *Rural information & communication technology*, 26–28 Nov 2013.
Linden, D., & Reddy, T. B. (2007). Battary Handbook
Liu R.-h., Sun Y.-k., & Ji X.-f. (2011). Battery state of charge estimation for electric vehicle based on neural network. In *IEEE 3rd international conference* (p. 493–496).
Sarvi, M., & Adeli, S. (2010). A neural network method for estimation of battery available capacity. In *UPEC2010*, 31st Aug–3rd Sept 2010.

Chapter 41
Investigating the Effect of Different Refrigerants on the Performance of a Supercritical Organic Rankine Cycle

Duygu Melek Çakıcı and Can Özgür Çolpan

Introduction

Consumption of fossil fuels leads to some environmental problems such as pollutions, depletion of the ozone layer and global warming. There has been a high interest in the renewable energy resources after the oil crisis in the 1970s. Geothermal, biomass, wind, solar, and ocean energy resources instead of fossil fuel such as coal, natural gas, and oil are preferred as they are domestic, clean, economic, and sustainable resources. One of these renewable energy sources is geothermal energy. Geothermal energy is thermal energy under the earth. The thermal energy (heat) from the Earth's core continuously flows outward. Temperature close to the center of the Earth is around 5500 °C. Scientists estimate that 42 million megawatts (MW) of power flow through the Earth's interior (http://www.geothermal.org/what.html). The potential of geothermal energy in Turkey is 31,500 MW (http://www.eie.gov.tr/yenilenebilir/turkiyede_jeo.aspx). Turkey lies in a highly active tectonic region. It is located on the Alpine-Himalayan orogenic belt, which has high geothermal potential. In Turkey, there are high enthalpy grabens and high reservoir temperature fields. Some of these fields are Büyük Menderes Graben, Gediz Graben and Simav Graben (DiPippo 2012).

Geothermal energy has been used for different purposes since ancient times such as bathing, cooking, balneological and health spas, district heating, cooling, greenhouses heating, heat pump and power generation. Geothermal energy can be classified according to the temperature of the resources. Geothermal resources can be divided into three groups according to the temperature: Low temperature fields (<70 °C), moderate temperature fields (70–150 °C), and high temperature

D.M. Çakıcı (✉) • C.Ö. Çolpan
Department of Mechanical Engineering, Dokuz Eylul University,
Tınaztepe Yerleskesi, Buca, Izmir 35397, Turkey
e-mail: duygu.melek.cakici@gmail.com

P. Grammelis (ed.), *Energy, Transportation and Global Warming*, Green Energy and Technology, DOI 10.1007/978-3-319-30127-3_41

fields (more than 170 °C) (Acar 2010). High temperature fields are suitable for electricity generation applications. However, it is possible to generate electricity at lower temperatures with the new technologies. Most of the world's geothermal power plants were built in the 1970s and 1980s after the oil crisis. The first geothermal researches in Turkey were started by Directorate of the Mineral Resource and Exploration of Turkey (called MTA in Turkey) in 1962 (Acar 2010). In Turkey, the first geothermal power plant was built in Kızıldere, in Denizli, in 1984. According to 2013 data, total installed generating capacity in the world is 11,772 MWe (http://www.geothermal-energy.org/electricity_generation.html). The potential for electricity generation from these fields are 1500 MWe in Turkey (http://www.eie.gov.tr/yenilenebilir/turkiyede_jeo.aspx). The short-term forecast indicates an installed capacity of 18,500 MWe by the year 2015 (Zarrouka and Moon 2014). The major types of power plants are as follows: Dry-steam power plants, flash-steam power plants and binary power plants. Binary power plants are appropriate for electricity production from low and medium temperature geothermal sources. Binary power plant consists of two different cycles. In these plants, it is aimed to increase the efficiency by increasing the temperature of the geothermal fluid. Organic Rankine cycle and Kalina cycle can be given as examples to this cycle. Organic Rankine cycle (ORC) is a Clausius Rankine cycle in which an organic working fluid is used instead of water-steam. In the last years it has become quite popular due to its suitability for electricity generation in the low-medium temperatures fields. The processes in ORC are similar to the processes in the Rankine cycle with steam as the working fluid. In the cycle, thermal energy of geothermal brine is transferred to the second working fluid.

Some parameters must be considered for optimization of organic Rankine cycle such as energy and exergy efficiencies of the cycle, the type of the working fluid, mass flow rate of a working fluid, temperature, pressure and chemical composition of the geothermal fluid, the reinjection temperature, the ambient temperature, heat supplied at evaporator or heat exchanger, heat transfer coefficients, pressures drops, net work output of turbine, cooling load of the condenser, subcritical and supercritical parameters. The reason for the consideration of these parameters is to find the optimum operating condition. Franco and Villani (2008) studied medium and low-temperature water-dominated system. The authors introduced the geothermal field analyzed, and proposed a methodology for optimizing the geothermal binary plants. They considered the fundamental variables as follows: Selection of the working fluid; temperature, pressure and chemical composition of the geothermal fluid, the reinjection temperature, the ambient temperature and the maximum rate of energy destruction. Borsukiewicz-Gozdur and Nowak (2007) studied several ways of increasing the power output of a geothermal power plant based on organic working fluid. Several types of organic working fluid were used in the analysis. Hettiarachchia et al. (2007) presented optimum design criterion for organic Rankine cycle using the low temperature geothermal sources. During the optimization process, objective function is minimized by varying evaporation and condensation temperature, geothermal water velocity in the evaporator and the cooling water velocity in the condenser. After their analyses, heat rejection at the condenser, the

Rankine cycle efficiency, working fluid mass flow rate, heat supplied at evaporator, exergy efficiency for binary power plant, heat transfer coefficients, pressures drops, total heat exchanger area, geothermal and cooling water pumping power and net-power for the cycle were found.

Fluid selection is important for design of the organic Rankine cycle. Basaran and Ozgener (2013) investigated different refrigerants on the performances of the cycle. Energy and exergy efficiency were calculated for 12 refrigerants on binary cycle R236ea, R600, R600a and R227ea (dry type fluid) showed higher energy and exergy efficiencies. R143a, R415A, R290, and R413a that are wet fluids indicated lower energy and exergy efficiencies. However, the investigation of supercritical fluid parameters is high importance. Schuster et al. (2010) compared subcritical and supercritical fluid parameters. They chose more than one refrigerants for geothermal power plant using organic Rankine cycle such as R134a, R227ea, R152a, isobutane, isopentane. Calculations showed supercritical cycle yields better than efficiency subcritical cycle. Vetter et al. (2012) worked with different refrigerants for comparing subcritical and supercritical parameters. In another study, Borsukiewicz-Gozdur and Nowak (2010) presented theoretical calculations of an Organic Rankine cycle power plant with supercritical parameters. The temperature of geothermal water ranged from 95 °C to 120 °C and mass flow rate was taken as 15 kg/s. Four organic fluids were selected as the working fluids in ORC (R115, R125, R143a, and Propylene). As a result of the calculations, propylene was found to give the best performance.

In this study, the effect of different refrigerants on performance of supercritical organic Rankine was investigated. Power output and electrical efficiency of the supercritical cycle were found for different refrigerants (R134a, R143a, R142b, R124, R227ea). The equations based on the energy balances of the components of the system were solved using Engineering Equations Solver (EES). The refrigerant giving the best performance for the organic Rankine cycle was found.

System Description

Gümüşköy geothermal power plant is located in Aydın, Turkey. The plant has been operated and managed by BM Holding since 2013. The geothermal fluid is produced at a temperature of 178 °C and a pressure of 7.1 bar and the net power output is 5.5 MW. Power plant is based on a supercritical organic Rankine cycle in which R134a is the working fluid. Operating data is given Table 41.1.

Working principle is as follows: Geothermal wells produce a mixture of steam-liquid. The temperature of brine is 178 °C and 7.1 bar, respectively. High pressure geothermal fluid from the well is separated as saturated liquid and superheated vapor in a separator. Superheated vapor enters the heat exchanger. The geothermal brine transfers its heat to R134a stream in the heat exchanger. The temperature of R134a reaches the supercritical state before entering the turbine. Then R134a enters the turbine at a temperature of 151.8 °C and a pressure of 43.7 bar, and power is

Table 41.1 Operating data of Gümüşköy power plant

Parameter	Value
Temperature of production well	178 °C
Temperature of geothermal fluid entering the heat exchanger-1	164 °C
Temperature of R134a entering the turbine	151.8 °C
Temperature of geothermal fluid entering the heat exchanger-2	160.6 °C
Temperature of air entering the air cooled condenser	18 °C
Temperature of air outlet the air cooled condenser	30 °C
Pressure of geothermal fluid entering the heat exchanger-1	6.82 bar
Pressure of geothermal fluid entering the heat exchanger-2	7.8 bar
Pressure of R134a entering the turbine	43.7 bar
Pressure of R134a outlet the turbine	9.47 bar
Mass flow of R134a	213.33 kg/s
Mass flow of air	3240 kg/s
Isentropic efficiency of the turbine	0.8
Isentropic efficiency of the pump	0.8

produced. R134a leaves the turbine and enters the desuperheater (regenerator). The aim of desuperheater is to increase the efficiency of the ORC by increasing the temperature of the working fluid entering the heat exchanger. Working fluid then enters the air cooled condenser and leaves as saturated liquid. Air enters the condenser at 18 °C and air leaves at 30 °C. The R134a enters the pump with the 9.47 bar pressure. R134a is pumped into the desuperheater. Then refrigerant enters the heat exchanger group. Finally, geothermal liquid is sent back to the reinjection wells. Reinjection wells provide the longevity of geothermal resources. Process flow diagram is shown in Fig. 41.1.

Modeling

Mathematical modeling of the system is presented in this section. The equations developed were programmed using Engineering Equations Solver (EES). The input data in the code used the actual operating data taken from the power plant. Modeling of the heat exchanger is presented first. Then the energy analysis of the overall system is presented.

Three shell and tube heat exchangers are used in the system. ε-NTU (effectiveness-number transfer units) method is used in the modeling of the heat exchanger. To define effectiveness of heat exchanger, we must determine the maximum possible heat transfer rate, $\dot{q}_{max}$ for the heat exchanger. In the ε-NTU method (Bergman et al. 2006; Ramesh and Dušan 2003), the heat transfer rate from the hot fluid to the cold fluid in the exchanger is defined as:

$$\dot{q}_{max} = C_{min} \cdot (T_{h,i} - T_{c,i}) = C_{min} \cdot \Delta T_{max} \qquad (41.1)$$

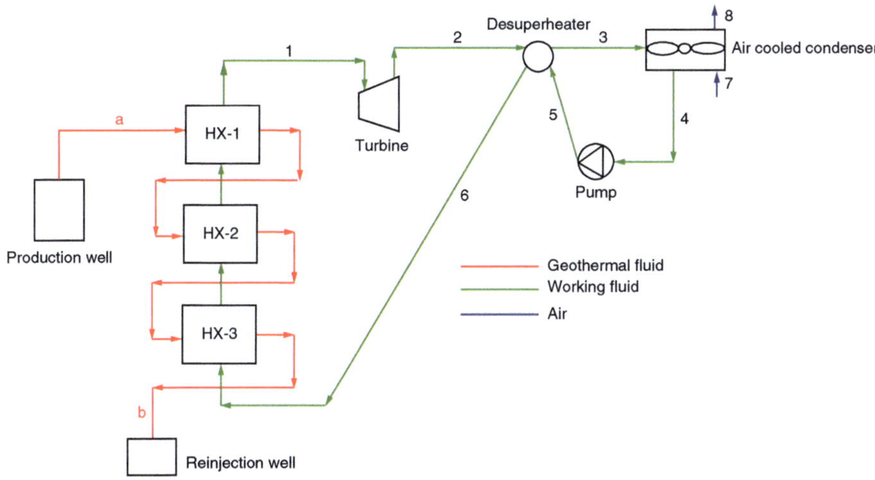

Fig. 41.1 Schematic diagram of Gümüşköy geothermal power plant

$T_{h,i}$ is inlet temperature of hot fluid, $T_{c,i}$ is inlet temperature of cold fluid. C_{min} is equal to C_c or C_h, whichever is smaller.

$$C_c = \dot{m}_{wf} \cdot c_{p,wf} \tag{41.2}$$

$$C_h = \dot{m}_{geofluid} \cdot c_{p,geofluid} \tag{41.3}$$

C_c is heat capacity of cold fluid, and C_h is heat capacity of hot fluid. Cold fluid is R134a and hot fluid is the geothermal brine. Effectiveness ε is a measure of thermal performance of a heat exchanger. ε is define as a ratio of the actual heat transfer rate from the hot fluid to the cold fluid to the maximum possible heat transfer rate:

$$\varepsilon = \frac{q}{q_{max}} \tag{41.4}$$

Then, for any heat exchanger it can be shown that:

$$\varepsilon = f\left(NTU, \frac{C_{min}}{C_{max}}\right) \tag{41.5}$$

$\frac{C_{min}}{C_{max}}$ is equal to $\frac{C_c}{C_h}$ or $\frac{C_h}{C_c}$. NTU is dimensionless parameter that is widely used for heat exchanger analysis and is defined as:

$$NTU = \frac{A \cdot U}{C_{min}} \tag{41.6}$$

A is area; U is overall heat transfer coefficient.

$$\frac{1}{U} = \frac{1}{h_o} + \left(\frac{d_o}{d_i} \cdot \frac{1}{h_i}\right) + \left(\ln\frac{d_o}{d_i}\right) \cdot \left(\frac{d_o}{2k_{tube}}\right) \quad (41.7)$$

d_o is outer diameter of the tube, d_i inner diameter of the tube, k_{tube} is conductivity of the tube, h_i is convection heat transfer coefficient inside the tube, h_o is convection heat transfer coefficient outside the tube.

The main assumptions made in the energy analysis are as follows:

- The system runs at steady state conditions.
- Pressure drop and heat loss in pipeline are all neglected.
- Kinetic and potential energies are ignored.

The supercritical organic Rankine cycle is modeled based on the mass and energy conservation laws. Energy balance of the control volume can be given as:

$$\dot{Q} - \dot{W} = \sum_e \dot{m} \cdot h - \sum_i \dot{m} \cdot h \quad (41.8)$$

The power output of the turbine is given by:

$$\dot{W}_T = \dot{m}_{wf} \cdot (h_1 - h_2) \quad (41.9)$$

In the air cooled condenser, $\dot{W}_{fan}$ is calculated as follows:

$$\dot{W}_{fan} = \dot{m}_{wf} \cdot (h_4 - h_3) + \dot{m}_{air} \cdot (h_8 - h_7) \quad (41.10)$$

The power output of the pump is given by:

$$\dot{W}_P = \dot{m}_{wf} \cdot (h_5 - h_4) \quad (41.11)$$

Net work output the system:

$$\dot{W}_{net} = \dot{W}_T - \dot{W}_P - \dot{W}_{fan} \quad (41.12)$$

The total heat input of the system:

$$\dot{Q}_{in} = \dot{m}_{geo} \cdot (h_a - h_b) \quad (41.13)$$

The net electric power generated is defined as:

$$n_{el} = \frac{\dot{W}_{net}}{\dot{Q}_{in}} \quad (41.14)$$

Results and Discussion

The effect of different fluids on the performance of the system is studied. R134a, R143a, R124, R227ea, and R142b are chosen as the working fluids. Selection of working fluid has been done considering the turbine inlet temperature as super-critical. In addition, reinjection well temperature should be at least 60 °C as the minerals found in the brine precipitate below this temperature (Yıldırım and Şimşek 2003). T-s diagrams for these working are shown fluids in Figs. 41.2, 41.3, 41.4, 41.5, and 41.6.

Thermal analysis was conducted separately for each fluid type. Result of calculations, reinjection well temperature (T_b), entering temperature of the turbine (T_1), net work output of the system, power input of the pump, power output of the turbine, electrical efficiency of the system were found. These results are shown in Table 41.2 for all working fluid.

A parametric study was done with EES. In Fig. 41.7, the effect of condenser pressure on the work output of the system is shown. W_{net} values were calculated for condenser pressure between to 9 and 15 bars. It is found that as condenser pressure increases, power output is decreased for all chosen working fluid. The effect of condenser pressure on the electrical efficiency of the system is shown in Fig. 41.8. The results show that R143a gives the best performance in Figs. 41.7 and 41.8. R227ea, R124 and R142b are similar values in Fig. 41.8. R227ea gives the worst performance. Condenser pressure increases and net fan work increases for the cool. So that net work output and electrical efficiency decreased.

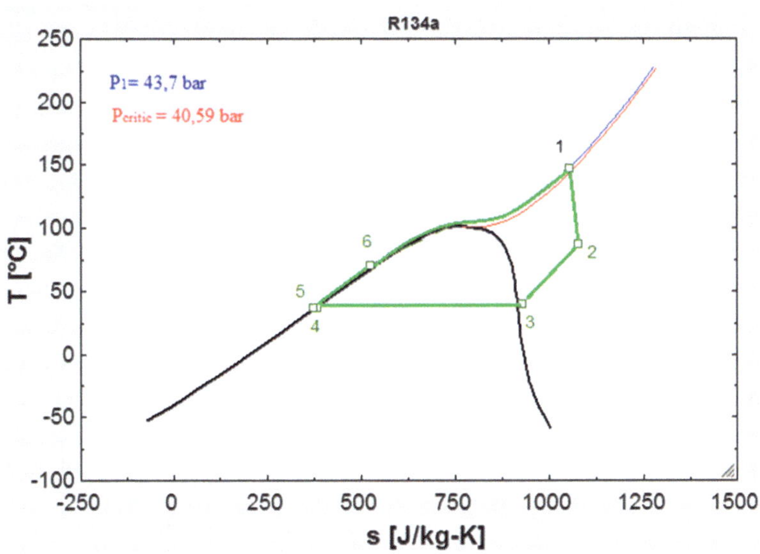

Fig. 41.2 T-s diagram of the ORC that uses R134a as the working fluid

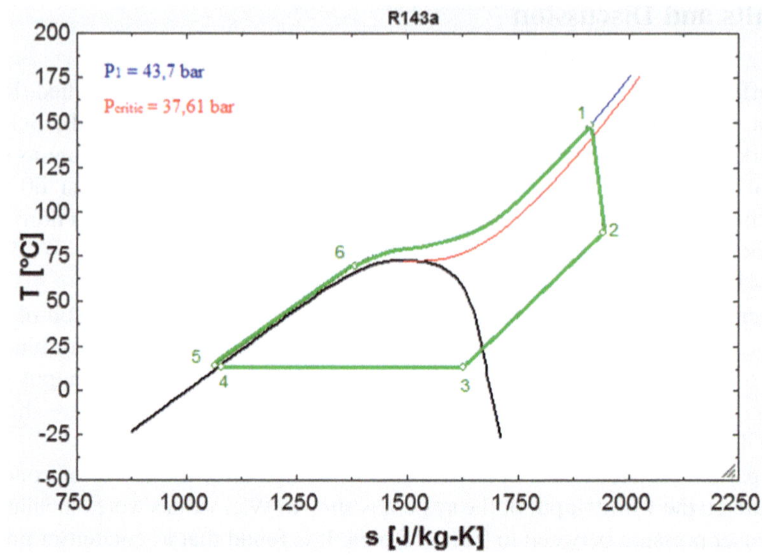

Fig. 41.3 T-s diagram of the ORC that uses R143a as the working fluid

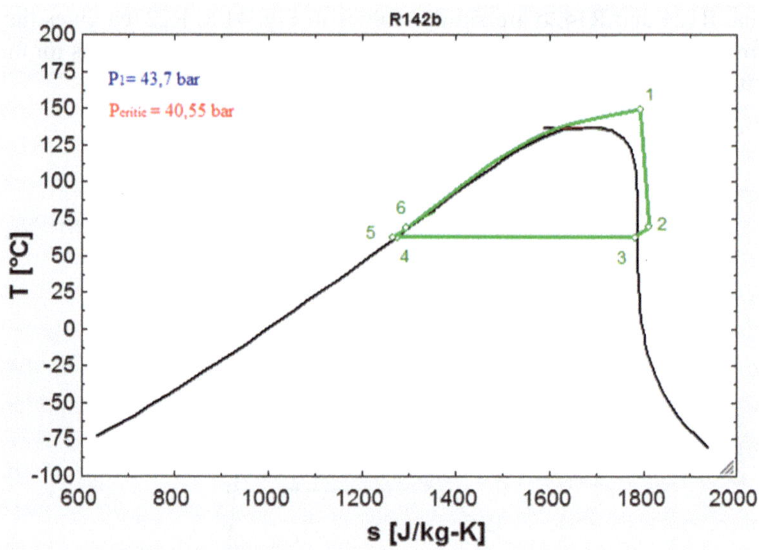

Fig. 41.4 T-s diagram of the ORC that uses R142b as the working fluid

The effect of production well temperature on the net work output of the system was studied. W_{net} values were calculated for temperature of production well between 100 and 200 °C in Fig. 41.9. The results show that as the production well temperature increases; power output also increases for all working fluid. The effect of production well temperature on the electrical efficiency of the system is

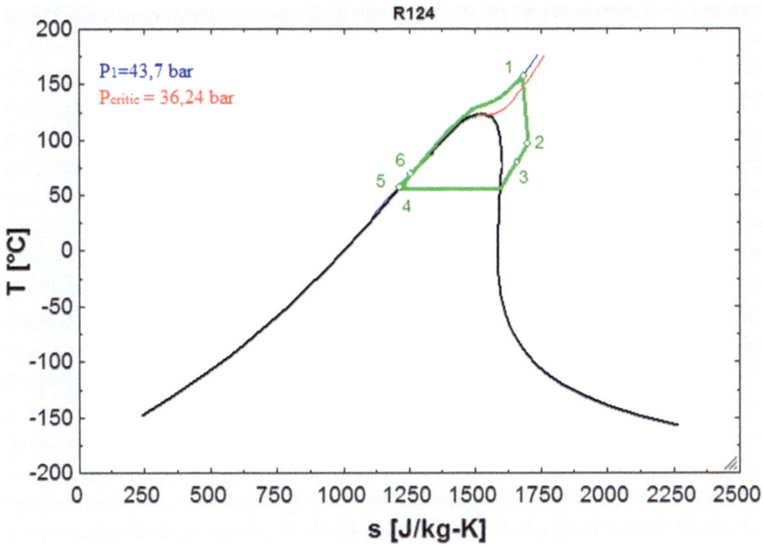

Fig. 41.5 T-s diagram of the ORC that uses R124 as the working fluid

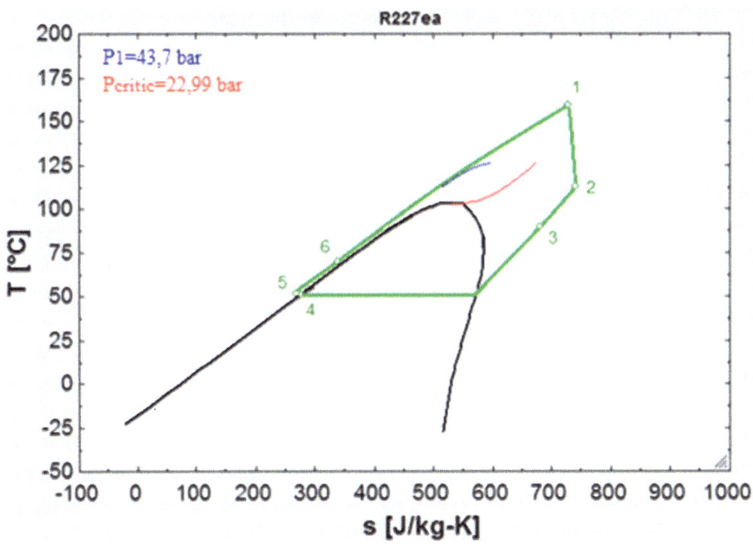

Fig. 41.6 T-s diagram of the ORC that uses R227ea as the working fluid

Table 41.2 The results of the thermal analysis

Working fluids	T_b [°C]	T_1 [°C]	T_{critic} [°C]	P_1 [bar]	P_{critic}	W_{net} [Mw]	η_{el}
R134a	77.57	147.7	101	43.7	40.59	6.49	0.1522
R143a	78.55	149.3	72.7	43.7	37.61	8.29	0.1969
R142b	79.06	150	137.1	43.7	40.55	5.30	0.1266
R124	90.77	158.9	122.3	43.7	36.24	4.62	0.1286
R227ea	97.78	159.8	102.8	43.7	29.99	3.94	0.1218

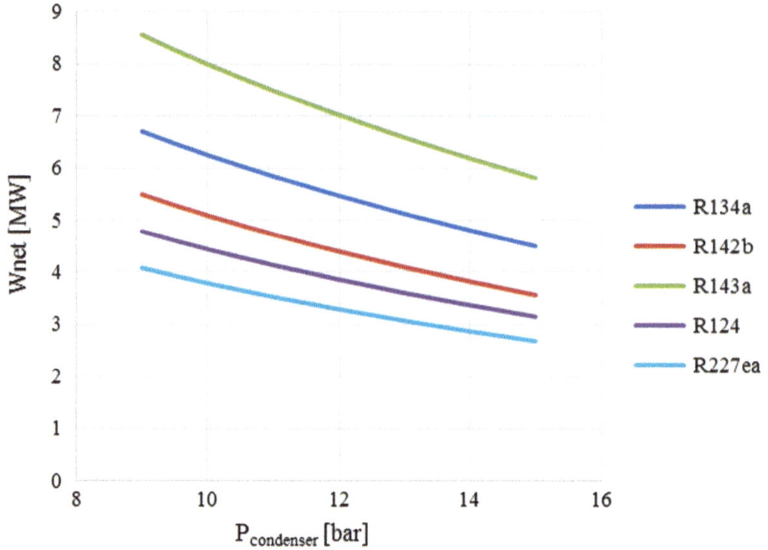

Fig. 41.7 The effect of condenser pressure on the net work output

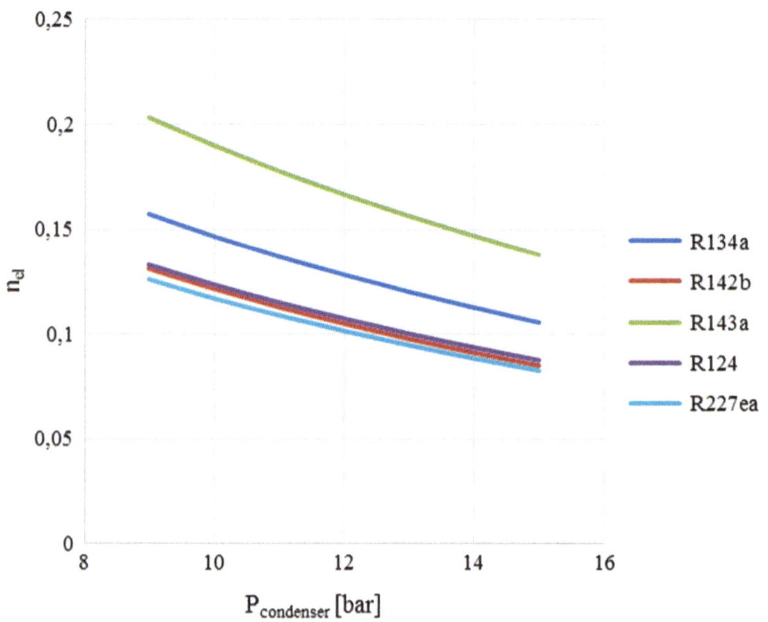

Fig. 41.8 The effect of condenser pressure on the electrical efficiency

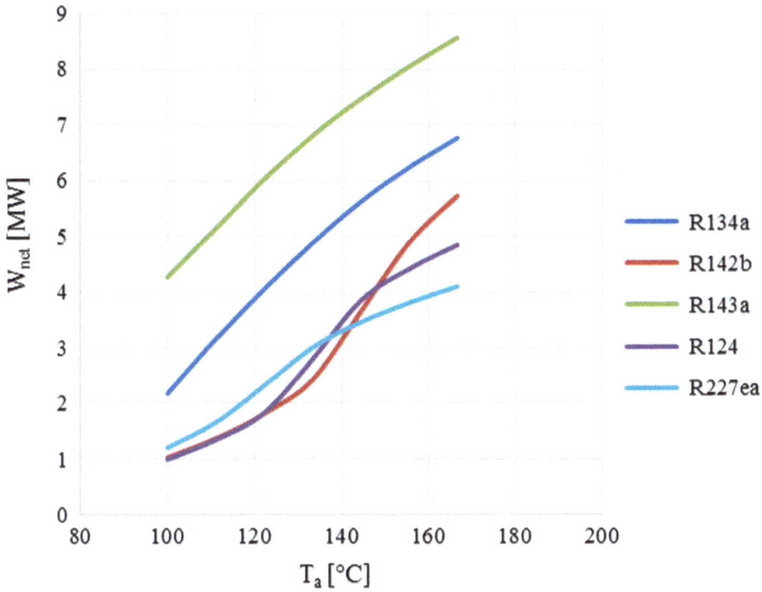

Fig. 41.9 The effect of production well temperature on the net work output

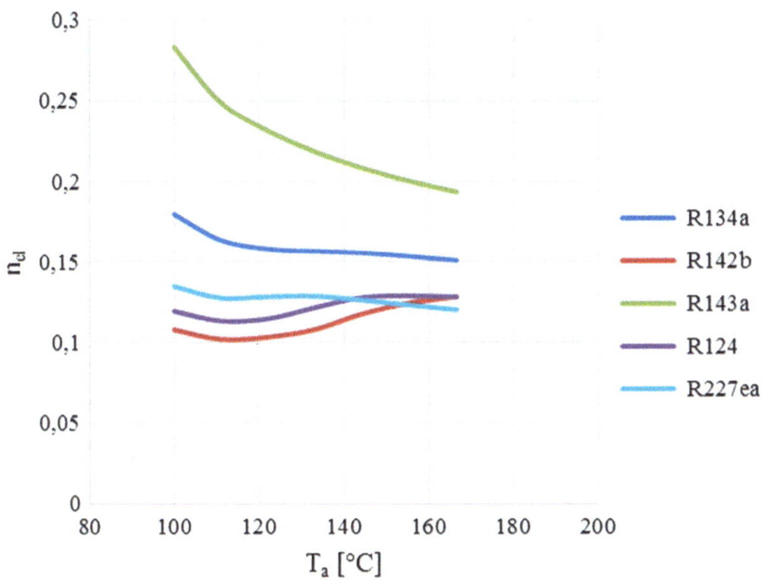

Fig. 41.10 The effect of production well temperature on the electrical efficiency

given in Fig. 41.10. The results show that R143a gives the best performance. The electrical efficiency of system was calculated for different temperatures of the production well. This efficiency decreases because of the increase in the reinjection well temperature.

Conclusions

In this study, the effect of different working fluids on the performance of an existing geothermal power plant is investigated. Energy balances developed were programmed using EES. Using this code, the power output of the turbines, the power demand of the pumps, the heat input to the heat exchanger, and the net power output and the electrical efficiency of the cycle are determined. The main conclusions derived from the results of the analyses conducted are listed below:

- The results for net work of the system and electrical efficiency, entering temperature of the turbine and reinjection well temperature are given in Table 41.2. It is found that R143a gives the best performance.
- Electrical efficiency from highest to lowest, respectively, is as follows: R143a, R134a, R124, R142b, R227ea.
- As production well temperature increases, electrical efficiency decreases for all working fluids. As condenser pressure increases electrical efficiency decreases for all working fluid.
- Although R143a gives the best performance, we suggest that R134a that gives the second highest performance should be selected as the working fluid in the Gümüşköy power plant. This is due to the fact that the working fluid R134a is non-flammable, non-toxic and yields better performance compared to other refrigerants (R124, R142b, R227ea).

Acknowledgments The author would like to thank Anıl Erdoğan, Gökhan Fidan, Ömer Katırcı and the engineers and managers in the Gümüşköy plant for their technical contributions.

Nomenclature

A	Heat exchanger area, m^2
c	Specific heat capacity, kJ/kg K
d	Diameter of heat exchanger tube, mm
ε	Effectiveness
h_i	Convection heat transfer coefficient inside the tube, W/m^2 K
h_o	Convection heat transfer coefficient outside the tube, W/m^2 K
h_{air}	Specific enthalpy of air, J/kg
h_{geo}	Specific enthalpy of the geothermal fluid, J/kg
k	Conductivity of exchanger tube
$\dot{m}$	Mass flow rate, kg/s
NTU	Number transfer units
U	Overall heat transfer coefficient, W/m^2 K
$\dot{W}_{net}$	Net work output of the system, W
$\dot{W}_T$	Power output of the turbine, W
$\dot{W}_P$	Power input of the pump, W

$\dot{W}_{fan}$ Power input of the fans in the air cooled condenser, W
$\dot{q}$ Heat transfer rate per tube area, W/m^2
$\dot{Q}$ Net heat gain rate, W
Q_{in} Total heat input rate, W
n_{el} Electrical efficiency
w_f Working fluid

References

Acar, H. I. (2010). A review of geothermal energy in Turkey. *Energy Sources, 25*, 111083–111088.

Basaran, A., & Ozgener, L. (2013). Investigation of the effect of different refrigerants on performances of binary geothermal power plants. *Energy Conversion and Management, 76*, 483–498.

Bergman, T. L., Lavine, A. S., Incropera, F. P., & DeWitt, D. P. (2006). *Fundamentals of heat and mass transfer*. New York: Wiley.

Borsukiewicz-Gozdur, A., & Nowak, W. (2007). Maximising the working fluid flow as a way of increasing power output of geothermal power plant. *Applied Thermal Engineering, 27*, 2074–2078.

Borsukiewicz-Gozdur, A., & Nowak, W. (2010). Geothermal power station with supercritical organic cycle. In *Proceedings world geothermal congress*, Bali, Indonesia.

DiPippo, R. (2012). *Geothermal power plants: Principles, applications, case studies and environmental impact*. Amsterdam: Elsevier.

Franco, A., & Villani, M. (2008). Optimal design of binary cycle power plants for water-dominated, medium-temperature geothermal fields. *Geothermics, 38*, 379–391.

Hettiarachchia, H. D. M., Golubovica, M., Woreka, M. W., & Ikegamib, Y. (2007). Optimum design criteria for an organic Rankine cycle using low-temperature geothermal heat sources. *Energy, 32*, 1698–1706.

Ramesh, K. S., & Dušan, P. S. (2003). *Fundamentals of heat exchanger design*. New York: Wiley.

Retrieved December 5, 2014, from http://www.geothermal.org/what.html.

Retrieved December 5, 2014, from http://www.eie.gov.tr/yenilenebilir/turkiyede_jeo.aspx.

Retrieved November 30, 2014, from http://www.geothermal-energy.org/electricity_generation.html.

Schuster, A., Karellas, S., & Aumann, R. (2010). Efficiency optimization potential in supercritical Organic Rankine Cycles. *Energy, 35*, 1033–1039.

Vetter, C., Wiemer, H., & Kuhn, D. (2012). Comparison of sub- and supercritical Organic Rankine Cycles for power generation from low-temperature/low-enthalpy geothermal wells, considering specific net power output and efficiency. *Applied Thermal Engineering, 51*, 871–879.

Yıldırım, N., & Şimşek, Ş. (2003). Determination of appropriate injection conditions for Kizildere geothermal waste fluid to avoid scale formation and cooling. In *European geothermal conference*.

Zarrouka, S. J., & Moon, H. (2014). Efficiency of geothermal power plants: A worldwide review. *Geothermics, 51*, 142–153.

Part XI
Environmental Technologies Related to Global Warming

Carbon Capture and Storage (CCS) and Geotechnology Issues

Chapter 42
CaO-Based Sorbents for Post Combustion CO_2 Capture via Carbonate Looping

Zinovia Skoufa, Andy Antzara, Ioannis Milios, Eleni Heracleous, and Angeliki A. Lemonidou

Introduction

In 2007, the Intergovernmental Panel on Climate Change (IPCC) concluded that worldwide temperature rise is primarily caused by the increase in anthropogenic greenhouse gases. CO_2 is by far the most important of these gases, mainly due to its abundance (Blockstein and Shockley 2006, Fan 2010). Among several industrial processes, electricity production sector holds the lion share in CO_2 emissions; in Greece its contribution in CO_2 emissions climbs to ~50 % (United Nations 2012). In that context, research aiming at mitigating CO_2 emissions particularly from power plants has received a lot of attention, so that a pure stream of CO_2 is collected from flue gases and subsequently sequestrated in appropriate geological formations. Among several available processes for carbon dioxide capture, CO_2 capture with amines is a well-established technology that however suffers from serious problems

Z. Skoufa (✉) • A. Antzara • I. Milios
Department of Chemical Engineering, Aristotle University of Thessaloniki, Thessaloniki 54124, Greece
e-mail: zskoufa@auth.gr; zskoufa@gmail.com; aantzara@auth.gr; johnnymil62@gmail.com

E. Heracleous
Chemical Process and Energy Resources Institute (CPERI), Centre for Research and Technology Hellas (CERTH), 6th km Charilaou-Thermi Road, P.O. Box 361, Thessaloniki 57001, Greece
e-mail: eheracle@cperi.certh.gr

A.A. Lemonidou
Department of Chemical Engineering, Aristotle University of Thessaloniki, Thessaloniki 54124, Greece

Chemical Process and Energy Resources Institute (CPERI), Centre for Research and Technology Hellas (CERTH), 6th km Charilaou-Thermi Road, P.O. Box 361, Thessaloniki 57001, Greece
e-mail: alemonidou@cheng.auth.gr

P. Grammelis (ed.), *Energy, Transportation and Global Warming*, Green Energy and Technology, DOI 10.1007/978-3-319-30127-3_42

of energetic/economic and environmental importance. First of all, high energy demands, leading to high operational costs, are associated with the regeneration of the amine solution. Moreover, there are problems associated with the degradation of sorbents at higher temperatures, the reactions of amines with compounds present in the flue gases such as oxygen and SO_2 and finally the corrosive nature of amines and their degradation products (Mandal and Bandyopadhyay 2006, Blamey et al. 2010, Wang et al. 2014). Therefore, alternative processes for CO_2 separation are extensively investigated, including reactive CO_2 separation using dry, solid sorbents such as limestone, alkali zirconates/silicates and sodium carbonates (White et al. 2003, Park et al. 2010). Unlike liquid adsorbents, solid adsorbents can be used over a wider temperature range from ambient temperature to 700 °C. Moreover, they yield less waste during cycling, while the spent solid adsorbents can be disposed more easily and in an environmentally friendly way (Harrison 2005). The choice of suitable materials depends mainly on the type of the application that CO_2 capture is intended for. Considering CO_2 removal from flue gases, sorbent properties such as heat tolerance, sorption capacity and stability, fast kinetics are very important for the economic viability of the process. Cost is also a major factor for such large scale applications. In this context, calcium oxide is a very attractive material for carbon dioxide capture, mainly due to high theoretical CO_2 capture capacity of 0.786 g CO_2/g sorbent (17.8 mmol CO_2/g sorbent) and fast kinetics of the CO_2 capture and release reaction (Broda et al. 2014). Carbonate Looping (CaL) is an ex-situ, post-combustion CO_2 capture technology in which CO_2 from flue gases is captured by a CaO-based sorbent in the carbonator. The carbonate cycle for post combustion CO_2 capture has been chosen by The Technology Task Force of the European Technology Platform for Zero-Emission Power Plants as one of the two highest priorities for future research and development (EU 2014). Moreover, preliminary techno-economic calculations show that calcium looping can separate CO_2 from flue gases at a cost of 25 USD/ton CO_2 compared to 50 USD/CO_2 that removal with amine scrubbing currently costs (Epple 2010).

CO_2 capture with calcium looping (CaL) is based on the reversible reaction of calcium oxide with carbon dioxide to produce calcium carbonate, according to the following reaction:

$$CaO + CO_2 \rightleftharpoons CaCO_3$$

In a calcium looping cycle (CaL), the CaO-based sorbent is continuously transferred between two reactors: the carbonator, where the flue gases are stripped from CO_2 by the forward reaction of CO_2 uptake, and the calciner, where the formed $CaCO_3$ is decomposed to regenerated CaO producing at the same time a concentrated CO_2 stream ready to be sequestrated or used as raw material in downstream chemical processing units. The carbonation reaction is exothermic and the heat released can be integrated for the realization of the endothermic reverse reaction (calcination) for sorbent regeneration. The optimal operating temperatures of both steps are thermodynamically determined from CO_2 partial pressure. Achieving fast CO_2 sorption and desorption rates is also a matter of concern. The conditions in the carbonator must strike a balance between the increased equilibrium conversion obtained at lower temperatures and the faster reaction rate at higher temperatures. The conditions in the calciner must compromise between the increased rate of the decomposition reaction at higher temperatures and the reduced rate of degradation of sorbent at lower temperatures (Blamey et al. 2010). Based on the above, the typical temperature window for CO_2 capture from a gas-fired flue gas stream ranges between 600 and 650 °C in the carbonator and above 850 °C in the calciner (Park et al. 2010; Broda et al. 2014).

Key for the large scale deployment of this process is sorbent development. In view of the carbonation reaction, CO_2 sorption process can be divided into two distinct steps: (1) an initial fast surface reaction controlled by the reaction kinetics and (2) a slower reaction restricted by the diffusion of CO_2 in the $CaCO_3$ product (Mandal and Bandyopadhyay 2006). An important aspect of the carbonation reaction is the difference in the molar volume of CaO and the product $CaCO_3$. Therefore, the formation of $CaCO_3$ results in the formation of a product layer of high molar volume which in turn leads to pore blockage (Broda et al. 2014). Taking the above into consideration, it is obvious that the sorption capacity of a CaO-based material greatly depends on its morphological and structural characteristics, mainly its surface area and porosity. Moreover, taking into account that in practice such a process comprises several consecutive carbonation/calcination cycles, sorbent cyclic stability in multi-cycle operation is prerequisite for the successful industrial implementation of the calcium looping process. Good mechanical properties are also required for an efficient sorbent material, due to a number of important issues of solids handling, associated with cohesive particles, attrition and agglomeration, electrostatic charge effects, etc. (White et al. 2003). CaO derived from the decomposition of naturally occurring limestone, although cheap, has the major drawback of rapidly decreasing CO_2 capture capacity with a number of repeated carbonation and calcination cycles (Grasa and Abanades 2006). Several efforts to enhance physicochemical and sorption characteristics of calcium oxide have been employed: preparation of synthetic CaO, improvements in the synthesis methods and synthesis of CaO-based mixed oxides. These efforts, recently reviewed thoroughly (Liu et al. 2012), include preparation of synthetic pure CaO sorbents via different Ca-precursors, different preparation methods as well as doping with a second metal and/or incorporation of inert materials. These efforts have indeed proved to be efficient, especially in terms of cyclic stability enhancement (Martavaltzi and Lemonidou 2008; Angeli et al. 2014).

The majority of published literature on the field reports multi-cyclic evaluation of novel synthetic and/or natural sorbents in a thermogravimetric apparatus (TGA), while a few recent papers concern pilot-scale operation of a carbonator/calciner system (Diego et al. 2014; Dieter et al. 2014; Ströhle et al. 2014). It is generally accepted that operating conditions, including for example reactor system employed, carbonation/calcination temperature and feed composition, affect to a significant extent the evaluation results on sorption capacity and stability (Alonso et al. 2014). Steam, which is anyway present in flue gases, is reported to enhance sorption capacity (Manovic and Anthony 2010; Kavosh et al. 2014), whereas no positive effect was reported in other studies (Sun et al. 2008). Steam has been also used for regeneration attempts of the sorbents, usually via the incorporation of a hydration step in the calcium looping cycle (Sun et al. 2008; Li et al. 2014; Coppola et al. 2014).

In this work, we report the development, preliminary evaluation and bench-scale testing of synthetic and natural mixed CaO-based sorbents for post-combustion CO_2 capture from industrial plants flue gases. The synthetic materials consist of CaO-based mixed oxides with Al, Zr, La and Mg, prepared via sol-gel auto-combustion synthesis using citric acid as a combustion agent. Concerning the natural CO_2 sorbents, we investigated the effect of different pre-treatments on the sorption capacity and stability of industrial hydrated lime. Mixing with inert minerals, such as kaolin, bauxite and magnesia, was attempted in an effort to reduce sintering and degradation under consecutive sorption/desorption cycles. The most promising materials were further studied at bench-scale in a fixed bed reactor, under a flow of CO_2, O_2, H_2O and N_2 simulating typical flue gases composition from a power plant. The possibility of hydration-induced regeneration and the effect of steam in flue gases are also investigated for selected materials.

Experimental

Sorbent Preparation

Four CaO-based powders promoted with Al, Zr, La and Mg were prepared by sol-gel auto-combustion synthesis using citric acid as combustion agent. The starting material was $Ca(NO_3)_2{\cdot}4H_2O$ (J.T. Baker) and the corresponding precursor for each promoter ($Al(NO_3)_3{\cdot}9H_2O$ (Carlo Erba), $La(NO_3)_3{\cdot}6H_2O$ (Merck), $Mg(NO_3)_2{\cdot}6H_2O$ (J.T. Baker) and $ZrO(NO_3)_2{\cdot}xH_2O$ (Acros)). All sorbents were prepared with a constant 66 wt% concentration of free CaO. The required amounts of calcium nitrate and the precursor of each promoter were dissolved in distilled water under continuous heating and stirring. Subsequently, the required amount of citric acid was added to the aqueous solution in order to maintain a constant nitrate salts/combustion agent molar ratio of 3. The new solutions were heated on a heating plate to promote water evaporation. After gelation occurred, the formed gel was transferred to a preheated furnace at 300 °C, where after a few minutes the gel was combusted in a self-propagating combustion manner. The as synthesized powders were then calcined at 900 °C for 1.5 h under air flow. The sorbents are denoted as Ca-X, where X refers to the promoter (Zr, La, Mg or Al).

Hydrated lime, $Ca(OH)_2$, provided by *CaO Hellas* was investigated as CaO source for natural CO_2 sorbents. Three different pre-treatment protocols were tested for the activation of raw $Ca(OH)_2$: *Protocol 1*: the raw material was calcined at 900 °C for 1.5 h under air flow ($Ca(OH)_2$-C900), *Protocol 2*: the calcined sample was then hydrated with water at 70 °C for 4 h to increase the surface area and as a

Table 42.1 Chemical composition of MgO, Kaoline and Bauxite binders

MgO-Grecian Magnesite	wt%	Kaoline-Prolat	wt%	Bauxite-S&B	wt%
MgO	97.30	SiO_2	46.55	SiO_2	0.87
SiO_2	0.50	Al_2O_3	39.50	Fe_2O_3	15.27
CaO	1.90	H_2O	13.95	TiO_2	3.12
Fe_2O_3	0.05			CaO	0.09
Al_2O_3	0.25			L.O.I.	14.46
				Al_2O_3	65.77
				S	0.03
				Zn (ppm)	26

result the sorption capacity, and was finally recalcined at 600 °C for 3 h in order to decompose the formed $Ca(OH)_2$ during hydration ($Ca(OH)_2$-W-C600) and *Protocol 3*: same as protocol 2, excluding final calcination step which was conducted at 900 °C ($Ca(OH)_2$-W-C900). Three natural heat resistant binders were used as means to increase cyclic stability, namely kaolin, bauxite and magnesia obtained from *Prolat, S&B* and *Grecian Magnesite* respectively. Three methods were investigated for the incorporation of the binders: *Protocol 4*: $Ca(OH)_2$-W-C600 sample obtained from *protocol 2* was mixed with Kaolin ($Ca(OH)_2$-KAOL), Bauxite ($Ca(OH)_2$-Baux) or MgO ($Ca(OH)_2$-MgO) in distilled water at 70 °C under continuous stirring for 4 h. The materials were filtered, dried overnight at 120 °C and calcined at 900 °C for 1.5 h under air flow; *Protocol 5:* raw $Ca(OH)_2$ instead of $Ca(OH)_2$-W-C600 was used as the CaO-precursor, and MgO ($Ca(OH)_2$-MgO-RW) or Bauxite ($Ca(OH)_2$-Baux-RW) was added following the same procedure as in *protocol 4*; *Protocol 6*: raw $Ca(OH)_2$ and MgO were grinded in a mortar for 20 min and the resulting mechanical mixture was calcined at 900 °C for 1.5 h under air flow.

Mixed natural sorbents were all prepared with a constant 66 wt% concentration of free CaO. Kaolin is a mineral composed mainly of Al_2O_3 and SiO_2 with a general formula of $Al_2O_3{\cdot}2SiO_2{\cdot}2H_2O$, bauxite is composed mainly of Al_2O_3 and Fe_2O_3 while the magnesia was mainly MgO with less than 3 % impurities, as tabulated in Table 42.1.

Physicochemical Characterization

Surface areas of the samples were determined by N_2 adsorption at 77 K, using the multipoint BET analysis method, with an Autosorb-1 Quantachrome flow apparatus. Prior to the measurements, the samples were dehydrated in vacuum at 250 °C overnight.

X-Ray diffraction (XRD) patterns were obtained using a Siemens D500 diffractometer, with Cu-Kα radiation. The morphology of the synthesized materials was examined by scanning electron microscopy (SEM) on a JEOL 6300 microscope, coupled with energy-dispersive X-ray analysis (EDX; Oxford Link ISIS-2000) for local elemental composition determination.

Preliminary Evaluation in TGA

An SDT Q600 (TA Instrument) thermal gravimetric analysis (TGA) instrument was used for the preliminary evaluation of all sorbent in 100 consecutive carbonation/calcination experiments. SDT Q600 works in conjunction with a controller and associated software to make up a thermal analysis system. A small quantity of the material (5–10 mg), placed in an aluminum sample cup, was pretreated at 850 °C, in the presence of 100 cm^3/min pure N_2, for 10 min to remove possible humidity and CO_2 absorbed. The CO_2 capacity of the sorbents was tested under 15 % CO_2 flow in N_2 for 30 min at 650 °C while desorption took place under 100 % N_2 flow for 5 min at 850 °C. Results are presented in terms of sorption capacity of the materials (mol CO_2/kg of sorbent) and CaO conversion. CaO conversion can be expressed as a function of the weight increase of the sorbent due to CO_2 absorption:

$$X(t),\% = \frac{MW_{CaO}}{MW_{CO_2}} * \%\text{weight increase}(t) * \frac{1}{w_{CaO}} \tag{42.1}$$

where MW_{CaO} and MW_{CO2} are the molecular weight of CaO and CO_2 respectively, and w_{CaO} is the weight fraction of free CaO present in the sorbent.

Bench-Scale Testing

The most promising synthetic and natural sorbents according to preliminary evaluation in TGA were tested in a bench scale flow unit for 20 consecutive CO_2 sorption/desorption cycles. A fixed bed quartz reactor with 18 mm external diameter of the reactor tube in the reaction zone and equipped with coaxial thermocouple for temperature monitoring was used for testing. The reactor was heated electrically by a tubular furnace, with three independently controlled temperature zones. Carbonation was conducted at 650 °C under a 10 % CO_2/20 % H_2O/3.2 % O_2/N_2 feed. Calcination was conducted at 800 °C in a 30%H_2O/N_2 flow. The CO_2 concentration in the reactor exit was monitored online by a CO_2 analyzer (Horiba, VIA 510). Sorption capacity of the sorbents, expressed as mol CO_2/kg of sorbent, % CaO conversion and sorption rate (mmol CO_2/kg/s) were calculated as follows:

$$\text{Sorption capacity, } mol_{CO_2}/kg_{sorbent} = \frac{\int_0^t M_{CO_2} dt}{W_{init}} \tag{42.2}$$

$$X(t),\% = \frac{\int_0^t M_{CO_2} dt}{n_{CaO}} * 100\% \tag{42.3}$$

$$\text{Sorption rate, mmol/kg}_{\text{sorbent}}/\text{sec} = \frac{M_{CO_2} * 10^3}{w_{init}} \tag{42.4}$$

where M_{CO2} are the moles of CO_2 captured (mol CO_2/s), w_{init} is the initial weight of the sorbent material (kg) and n_{CaO} are the initial moles of CaO in the sorbent sample.

Results and Discussion

Physicochemical Characterization

According to XRD analysis (not shown here for brevity), the main crystal phase formed in all synthetic CaO-based sorbents is CaO, accompanied by minor peaks of low intensity corresponding to $Ca(OH)_2$ in Ca-Mg, Ca-Al and Ca-Zr samples. Formation of mixed $Ca_3Al_2O_6$ and $CaZrO_3$ was observed in Ca-Al and Ca-Zr sorbents, respectively. On the contrary no mixed phase was formed between CaO and Mg or La, where the precursor nitrate salts decomposed to MgO and La_2O_3 respectively (Antzara et al. 2015). Concerning the textural properties of the synthetic sorbents, summarized in Table 42.2, it can be observed that the nature of the dopant affects the surface area of the final material. Surface areas ranging from 28 m^2/g for Mg-doped CaO to 9 m^2/g for La-doped CaO were recorded.

The as delivered industrial material ($Ca(OH)_2$) is composed of almost pure calcium hydroxide, presenting only a small XRD reflection corresponding to calcium carbonate. After thermal pre-treatment at 900 °C ($Ca(OH)_2$-C900), Ca $(OH)_2$ was almost completely decomposed to CaO with minor $Ca(OH)_2$ reflections. When the calcined sample was subjected to water treatment and calcined, only

Table 42.2 BET surface area and pore volume of the synthetic and natural CaO-based sorbents

Sorbent	Surface area (m^2/g)	Pore volume (cm^3/g)
Ca-Mg	28.37	0.112
Ca-La	8.78	0.138
Ca-Zr	15.28	0.051
Ca-Al	11.93	0.062
$Ca(OH)_2$-C900	5.04	0.041
$Ca(OH)_2$-W-C600	17.24	0.181
$Ca(OH)_2$-W-C900	16.82	0.149
$Ca(OH)_2$-KAOL	3.22	0.023
$Ca(OH)_2$-Baux	5.22	0.0198
$Ca(OH)_2$-MgO	23.25	0.427
$Ca(OH)_2$-Baux-RW	5.29	0.029
$Ca(OH)_2$-MgO-RW	25.49	0.199
$Ca(OH)_2$-MgO-GM	18.7	0.13

reflections at the angular positions of $Ca(OH)_2$ appear, despite the high temperature used in calcination either at 600 °C or at 900 °C, that was expected to favor hydroxide decomposition to CaO. The addition of kaolin leads to the formation of a mixed phase $Ca_3Al_2Si_2$, together with pure Al_2O_3 and SiO_2. Regardless of the preparation method, in the case of the use of bauxite or MgO as binders, the XRD patterns revealed the existence of CaO and MgO or Al_2O_3 crystalline phases, together with low intensity peaks corresponding to $Ca(OH)_2$ in the case of $Ca(OH)_2$-Baux and $Ca(OH)_2$-Baux-RW samples, without any formation of mixed components. The surface area and pore volume of all natural CaO-based sorbents is presented in Table 42.2. Hydration of the calcined sample leads to an increase in surface area and porosity compared to Ca $(OH)_2$-C900 sample (Li et al. 2014). Concerning the mixed materials, the nature of the binder seems to affect the textural properties of the final sorbent. The introduction of kaolin or bauxite in industrial hydrated lime decreases the surface area. On the contrary, the use of MgO greatly increases the surface area to ~25 m^2/g which can be partly attributed to the high surface area of pure MgO (70 m^2/g). Overall, it can be concluded that basic physicochemical characterization of the natural sorbents revealed no important effect of the preparation method, whereas doping with MgO seems to greatly affect both surface area and porosity.

Preliminary Evaluation in TGA

The performance of the four promoted synthetic CaO sorbents is presented in Fig. 42.1. Although a wide range of surface areas was observed, all the promoted sorbents (with the exception of Ca-La) achieved very high CaO conversions

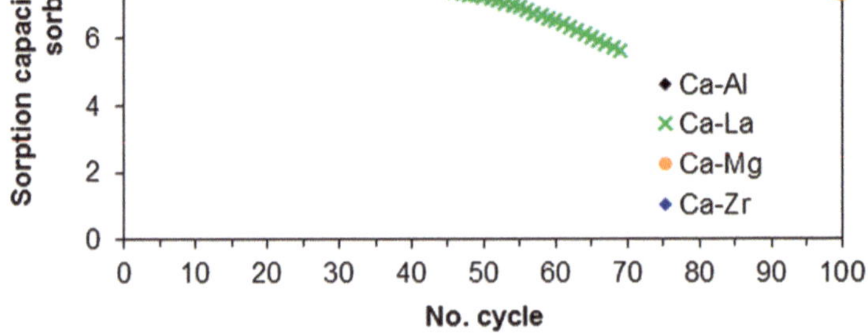

Fig. 42.1 CO_2 sorption capacity of the synthetic CaO-based sorbents as a function of the number of cycle in TGA

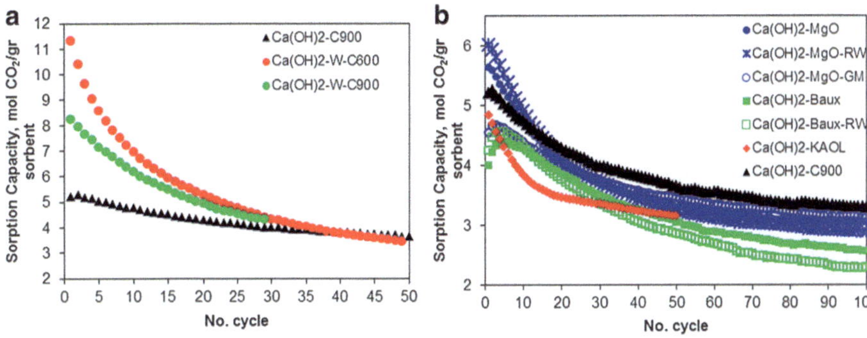

Fig. 42.2 CO_2 sorption capacity of the natural CaO-based sorbents as a function of the number of cycle in TGA

(>80 %) that could be attributed mainly to the nature of the dopants and the formed structure of the final materials during combustion synthesis (Antzara et al. 2015). All of the doped sorbents, except Ca-La, exhibit a very stable performance during the first 70 cycles, unlike the pure CaO prepared with citric acid which deactivates very fast in the first 20 cycles. Among the different dopants, doping with Zr leads to the development of the most stable sorbent, followed closely by Al. Moreover, Al exhibits the highest initial sorption capacity.

Natural sorbents consisting of pure CaO were tested in TGA for 50 carbonation/calcination cycles. It can be observed (Fig. 42.2a) that water treatment of the industrial hydrated lime greatly increases the initial CO_2 sorption capacity compared to the sample directly calcined at 900 °C, but unfortunately in expense of stability. Sample $Ca(OH)_2$-C900 which presents the highest sorption capacity after 50 cycles was further tested for 100 cycles and the results are compared to those obtained with doped natural sorbents in Fig. 42.2b. The use of kaolin, bauxite and MgO binders leads to a decrease initial CO_2 sorption capacity due to the lower content of CaO in the materials, with the exception of $Ca(OH)_2$-MgO-RW and Ca $(OH)_2$-MgO-GM samples. Nevertheless, CaO conversion increases for all doped sorbents, indicating the beneficial effect of binders that seem to increase the accessibility of active calcium oxide surface. Overall, the most promising results were recorded for samples $Ca(OH)_2$-C900 and $Ca(OH)_2$-MgO-GM, that were prepared via a solvent-free method. These two sorbents, together with Ca-Al and Ca-Zr synthetic materials were chosen for further testing in a fixed bed reactor. Table 42.3 summarizes the preliminary evaluation results for the above sorbents. It becomes clear that the synthetic CaO-based sorbents exhibited almost twice the initial sorption capacity of the natural sorbent and a very stable performance with much lower deactivation after the same number of sorption/desorption cycles. However, the use of natural sorbents presents important economic and environmental advantages, since they are based on naturally occurring minerals. A feasibility study of the use of either natural or synthetic sorbents exceeds the scope of the present paper, and should in any case be based on sorbent properties derived after tens of thousands of cycles (Abanades et al. 2004).

Table 42.3 Preliminary evaluation results obtained from TGA the most promising synthetic and natural sorbents

	Ca-Al	Ca-Zr	$Ca(OH)_2$-C900	$Ca(OH)_2$-MgO-GM
Initial capacity (mol CO_2/kg sorbent)	11.64	10.63	5.22	4.54
Initial CaO conversion, %	99	90.3	29.6	38.6
Deactivation after 100 cycles (%)	20.7	13.7	36.5	30.9

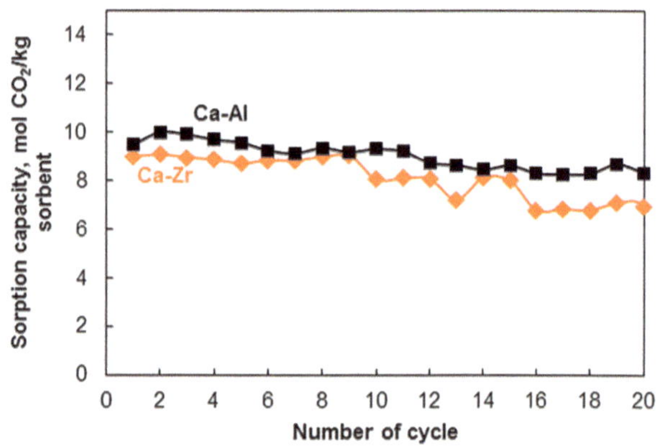

Fig. 42.3 Sorption capacity versus number of cycle for Ca-Zr (GHSV = 318 h^{-1}) and Ca-Al (GHSV = 311 h^{-1}) synthetic sorbents in fixed bed experiments

Bench-Scale Testing in a Fixed Bed Reactor

After initial sorbent screening in TGA, the most promising synthetic and natural sorbents were tested in a fixed bed reactor, under realistic feed conditions. Figure 42.3 presents the results for Ca-Al and Ca-Zr testing in 20 consecutive carbonation/calcination experiments. It should be noted that the sorption capacity refers to the amount of CO_2 captured in the pre-breakthrough period, i.e. in the carbonation stage controlled by surface reaction as will be discussed later on. Both synthetic sorbents present high sorption capacity and satisfactory stability. Given the differences between experimental set-ups, the sorption capacity recorded in these experiments is similar to the results obtained in TGA, which refer to the overall sorption capacity (not just in the pre-breakthrough period. Al-doped CaO based sorbent presents only 12.5 % deactivation after 20 cycles, compared to 23 % deactivation for Ca-Zr.

Carbonation profiles for the 1st and 20th cycle for Ca-Zr sorbent are presented in Fig. 42.4, where CO_2 concentration in reactor outlet and sorption rate, defined as mmol CO_2/kg/s, are plotted versus time. According to the carbonation profiles, CO_2 capture proceeds via two distinctive stages. In the first stage, also known as the pre-breakthrough period, the rate determining step of carbonation is the surface reaction between CO_2 and CaO. During this period, the CO_2 concentration is maintained at very low levels (<2%vol.), indicating a constantly high CO_2 capture rate. In the experimental conditions used, both synthetic sorbents approach thermodynamic equilibrium CO_2 conversion per pass. In the first cycle the duration of

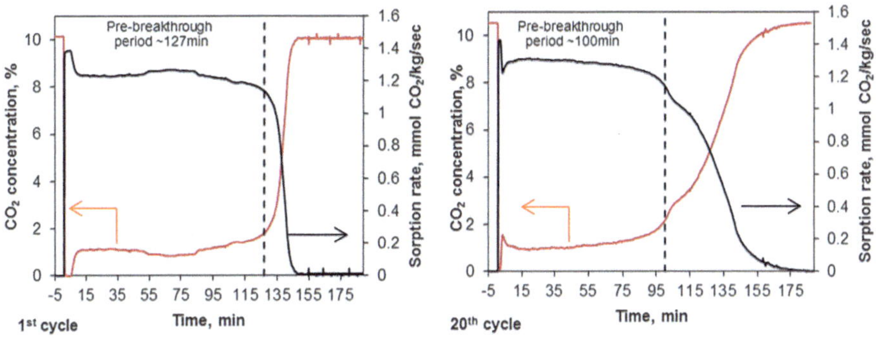

Fig. 42.4 Carbonation profiles for the 1st (*left*) and 20th (*right*) cycle for Ca-Zr synthetic sorbent

pre-breakthrough period for Ca-Zr sample is ~127 min (Fig. 42.4-left). At that time ~80 % CaO conversion is achieved, corresponding to a sorption capacity of 9.02 mol CO_2/kg sorbent. As the formation of $CaCO_3$ proceeds, the overall CO_2 capture process is governed by diffusion of CO_2 through the $CaCO_3$ layer. This step is generally slower, and Fig. 42.4, it is characterized by fast decrease in the reaction rate, accompanied by a fast increase of CO_2 concentration in reactor outlet. As already discussed, in a possible industrial implementation of the CaL concept, the sorbent would exit the carbonator before gradual diffusion becomes rate-controlling.

As shown in Fig. 42.4-right for Zr-doped CaO-based sorbent, after 19 cycles the duration of the kinetically controlled period has decreased to 100 min in the 20th cycle, leading to sorbent deactivation by 23 %. Characterization of the spent sorbent evidenced a 50 % reduction in surface area (surface area of used sample: 11.1 m^2/g). Characterization of the fresh and used samples with SEM revealed significant agglomeration of the material after 20 cycles. As shown in Fig. 42.5a,b, the fresh material is composed of small particles in the range of 10 μm, whereas in the used sample agglomeration led to the formation of larger particles, some of which are as large as 500 μm (Fig. 42.5e). Nevertheless, it seems that the porous structure of the sorbent is retained to a significant extent, in line with the moderate degree of deactivation that was recorded for Ca-Zr sorbent. Formation of $CaZrO_3$ mixed phase seems to be the key for the avoidance of extended sintering phenomena, as previously reported (Reddy et al. 2014), due to its high Tamman temperature (Ruff et al. 1929). Characterization results of the used and fresh samples may constitute the basis for enlightening the deactivation mechanism of the sorbent. Due to surface area reduction, the carbonation reaction more quickly falls into diffusion-controlled regime. Nevertheless, total sorption capacity, calculated as the total amount of CO_2 that was captured throughout carbonation and not just in the pre-breakthrough period, is the same as in the first cycle. This result indicated that sorbent did not suffer any loss of intrinsic activity. The stable porous structure enables CO_2 diffusion to the bulk of Ca-Zr particles, even if this step is completed in a longer time after 20 cycles, as evidence by comparing the diffusion-controlled period in the 1st and 20th cycle (Fig. 42.4).

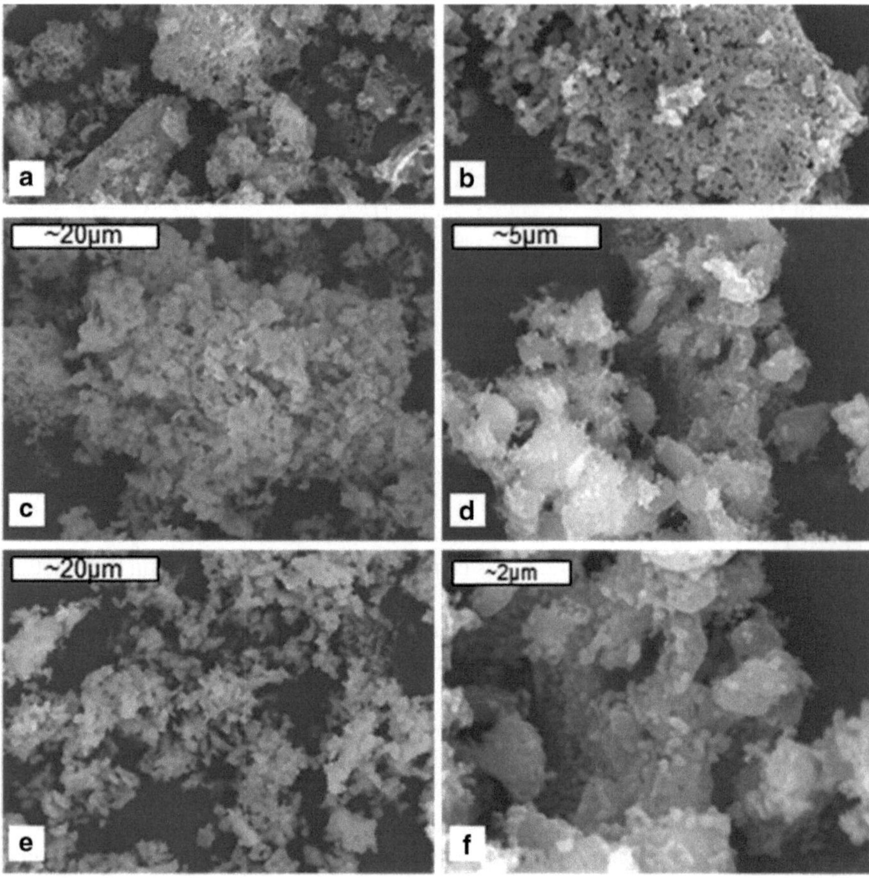

Fig. 42.5 SEM micrographs of fresh (**a**, **b**) and used (**c**–**f**) Ca-Zr sorbent

Ca-Al synthetic sorbent presented high sorption capacity, equal to 9.49 mol $CO_2/kg_{sorbent}$ in the pre-breakthrough period which lasted for 110 min, as shown in Fig. 42.6-left. Compared to Ca-Zr sample, Ca-Al presents higher sorption rate, leading to higher sorption capacity despite the smaller pre-breakthrough period. In the 20th cycle, the surface reaction controlled step is reduced to 90 min. As discussed for the Zr doped sample, the deactivation of Ca-Al sorbent is attributed to a decrease in surface activity. The total active sites of the sorbent material do not decrease judging from the fact that the same total amount of CO_2 was captured in cycles 1 and 20. Characterization of the used sample showed a marginal decrease in surface area (from 12.5 m^2/g to 9.3 m^2/g) in line with a small deactivation rate of 12.5 % after 20 carbonation/calcination cycles. The porous structure of the material seems to remain almost intact, despite the formation of small agglomerates, as shown in SEM micrographs (Fig. 42.7).

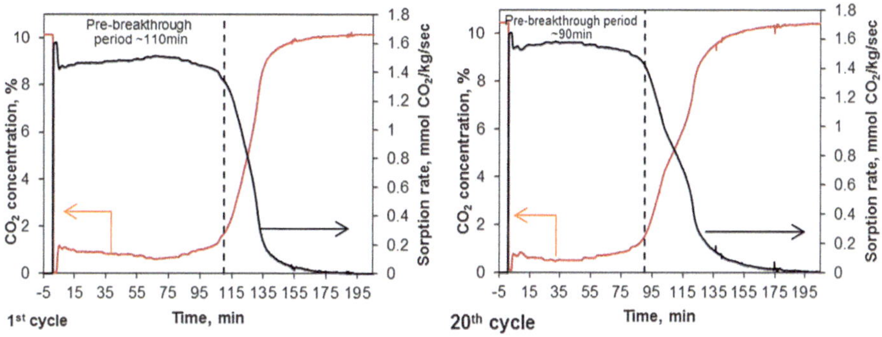

Fig. 42.6 Carbonation profiles for the 1st (*left*) and 20th (*right*) cycle for Ca-Al synthetic sorbent

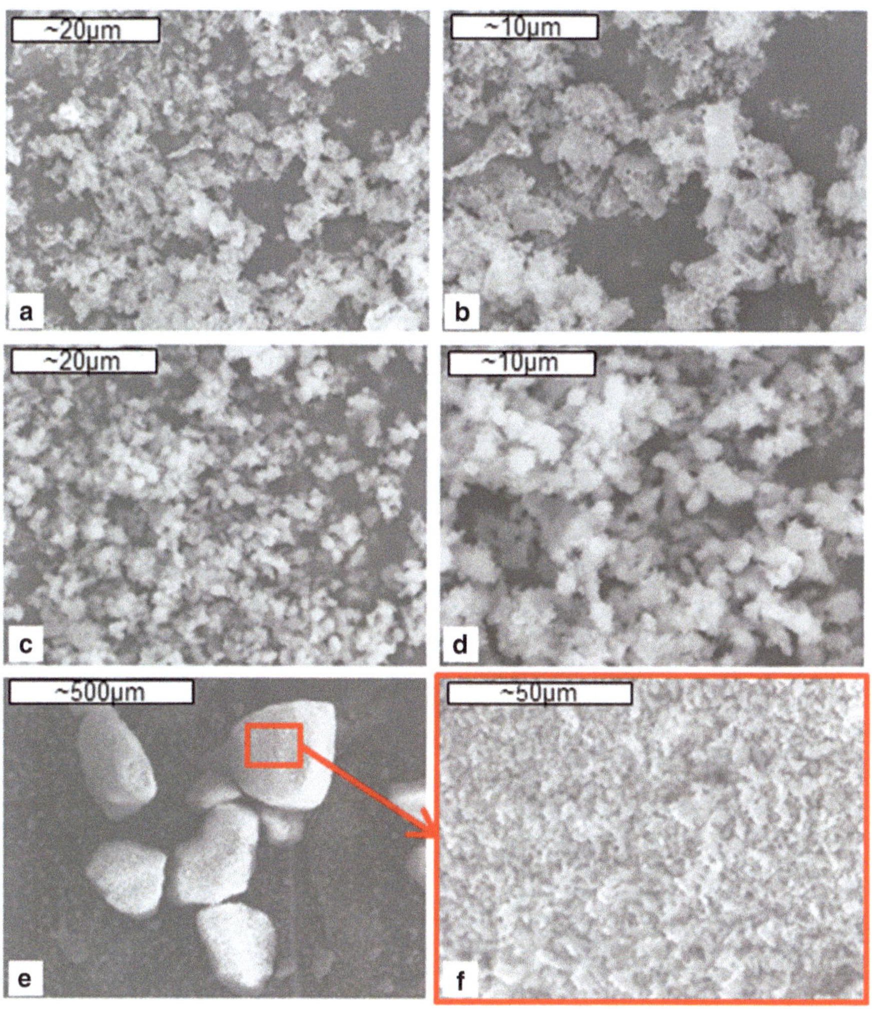

Fig. 42.7 SEM micrographs of fresh (**a**, **b**) and used (**c–f**) Ca-Al sorbent

As expected, $Ca(OH)_2$-C900 and $Ca(OH)_2$-MgO-GM natural sorbents presented significantly lower sorption capacity compared to synthetic sorbents, and intense deactivation in the first five cycles, as shown in Figs. 42.8 and 42.9. The carbonation profiles for both natural sorbents in the 1st cycle are shown in Fig. 42.10. Unlike

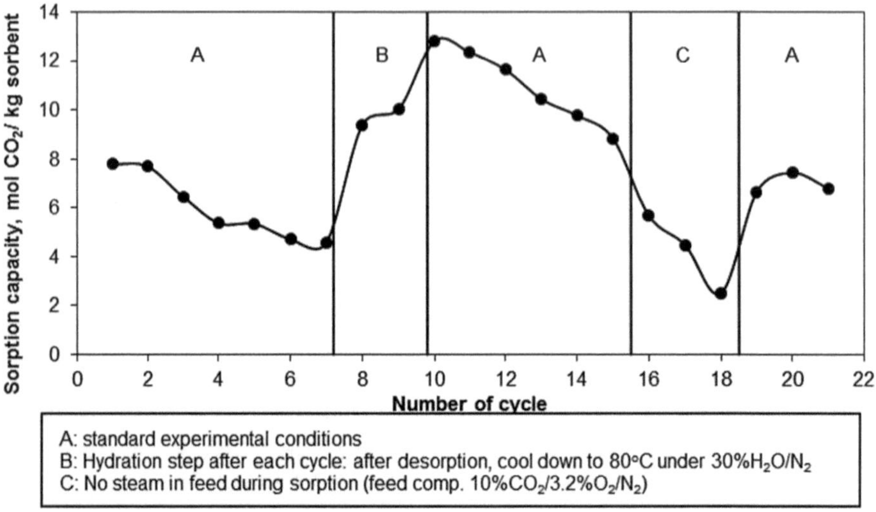

Fig. 42.8 Sorption capacity versus number of cycle for $Ca(OH)_2$-C900 (GHSV = 286.5 h^{-1}) natural sorbent in fixed bed experiments

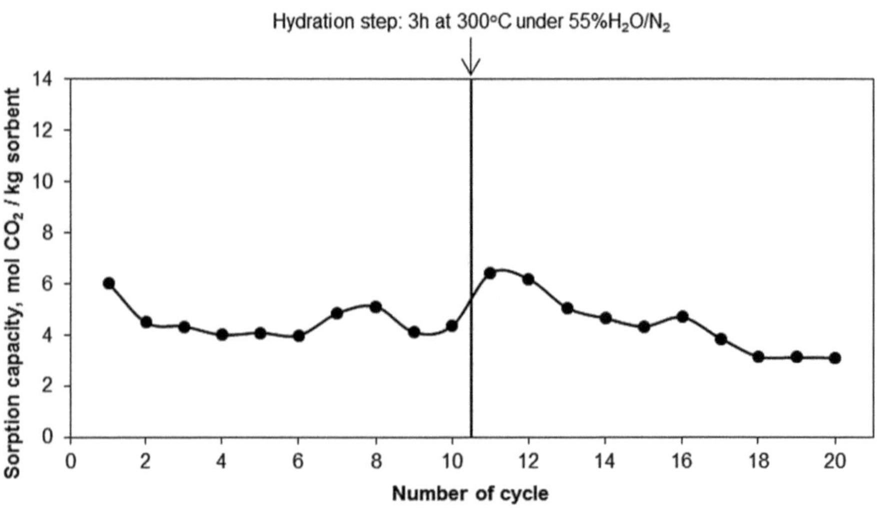

Fig. 42.9 Sorption capacity versus number of cycle for $Ca(OH)_2$-MgO (GHSV = 427 h^{-1}) natural sorbent in fixed bed experiments

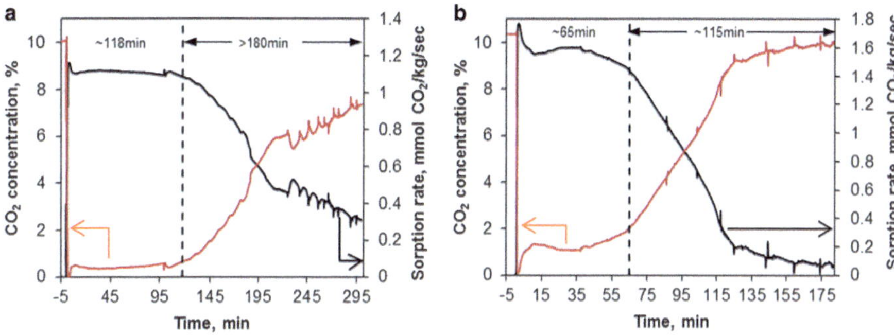

Fig. 42.10 Carbonation profiles for the 1st cycle for (**a**) $Ca(OH)_2$-C900 (GHSV = 286.5 h^{-1}) and (**b**) $Ca(OH)_2$-MgO-GM (B) (GHSV = 427 h^{-1}) natural sorbents

synthetic sorbents, it can be observed that for the natural sorbents, the breakthrough period lasts longer than the pre-breakthrough period. In other words, the amount of CO_2 that is captured in the diffusion controlled regime is greater than in the kinetics-controlled regime, indicating different morphology between natural and synthetic sorbents. Indeed, pore size distribution evidenced the occurrence of smaller pores (mesopores) for natural sorbents compared to synthetic samples that mainly present macroporosity. Consequently, in natural sorbents the easily accessible surface sites are presumably less compared to synthetic sorbents. Moreover, unlike synthetic sorbents, for which the same activity results were obtained both in TGA and bench scale testing, both natural sorbents presented higher activity in the fixed bed reactor experiments compared to the preliminary results in TGA, in terms of total sorption capacity. This increase of total sorption capacity results from both surface reaction- and diffusion controlled steps; however the pre-breakthrough period contributes the most to the observed total capture activity of the solids. In the TGA setup, the 15 % CO_2/N_2 feed is inserted in the oven in a "flow over" mode as opposed to "flow-through" conditions in the fixed bed reactor. Moreover, in bench scale testing the presence of steam might affect the carbonation reaction, as previously suggested (Manovic and Anthony 2010; Kavosh et al. 2014). According the results reported in the present study, it can be deduced that the presence of steam enhances both steps of the carbonation reaction. Overall, based on the above, enhanced CO_2 diffusion to less easily accessible surface sites is probably the reason for the higher CO_2 capture efficiency recorded in the bench-scale experiments for natural sorbents.

Since the natural sorbents were significantly deactivated during the first five cycles, two different approaches were studied in an attempt to investigate the possibility of sorbent regeneration. Concerning $Ca(OH)_2$-C900 sorbent, after the 7th, 8th and 9th cycle, the sample was cooled down from 800 °C (desorption temperature) to 80 °C, under a 30 % H_2O/N_2 flow. According to the thermodynamic equilibrium curve for CaO hydration, calcium oxide is hydrated to calcium hydroxide at temperatures lower than 400 °C. The reaction is accompanied by an increase in specific volume, from 16.9 cm^3/mol for CaO 33.1 cm^3/mol for $Ca(OH)_2$, ultimately resulting to an increase in surface area and porosity (Li et al. 2014). As shown in Fig. 42.8, the hydration procedure led not only to containment of sorbent deactivation, but also to a significant increase (~180 %) in sorption capacity from cycle 7th to 10th. It should be noted here, that in case of process scale-up the addition of such a regeneration procedure would not be economically efficient, in terms of energy needs for cooling down and re-heating the regenerated sorbent. Cycles 10–15 were conducted without a hydration step, and as shown in Fig. 42.8, the sample was again deactivated following roughly the same trend as in the first six cycles. In order to investigate the effect of steam presence in the flue gases, cycles 16, 17 and 18 were performed in the absence of steam in the feed, leading to a sharp decrease of the sorption capacity from 8.85 mol CO_2/kg in cycle 15–2.51 mol CO_2/kg in cycle 18. Overall, the above results indicate that the presence of steam in the feed increases CO_2 capture efficiency and promotes sorbent stability, most probably by facilitating CO_2 diffusion.

As shown in Fig. 42.9, $Ca(OH)_2$-MgO-GM sorbent presents an initial sorption capacity of 6.03 mol CO_2/kg, and is deactivated by ~33 % within the first five cycles. The second approach to regenerate the sorbents was then investigated for $Ca(OH)_2$-MgO-GM sorbent. After calcination at 800 °C, the sample was cooled down to 300 °C and left to react towards $Ca(OH)_2$ under a 55 % H_2O/N_2 flow for 3 h. This hydration process resulted in an increase in sorption capacity, as shown in Fig. 42.9. For the next cycles up until cycle 20 the carbonation/calcination procedure was repeated without any intermediate regeneration step and the sorbent continued to deactivate; an overall ~49 % deactivation was finally recorded.

Table 42.4 summarizes the bench-scale results for CO_2 capture. It is obvious that the synthetic sorbents prepared by sol-gel auto-combustion synthesis presented superior capture activity and stability compared to the natural sorbents derived from industrial hydrated lime. In view of the important conclusions derived from the comparison of TGA and fixed bed reactor testing and characterization of spent solids, it can be deduced that the superiority of synthetic sorbents lies in their porous structure. On one hand, this porous structure enhances CO_2 capture by maximizing the available CaO surface. On the other hand, the addition of Zr and Al leads to the formation of thermally stable $CaZrO_2$ and $Ca_3Al_2O_6$ mixed phases that provide stability. The final choice on whether synthetic or natural sorbents should be used in a potential scale up will be determined by techno economic factors, taking into account the cost of sorbent development versus the need for sorbent make-up due to deactivation.

Table 42.4 Summary of CO_2 capture results (pre-breakthrough time) obtained in the fixed bed reactor for the four selected sorbents

	$Ca(OH)_2$-C900	$Ca(OH)_2$-MgO-GM	Ca-Zr	Ca-Al
Initial sorption capacity, mol $CO_2/kg_{sorbent}$	7.80	6.03	9.02	9.49
Initial CaO conversion, %	43.7	51.2	76.6	80.6
Deactivation after 20 cycles, %	13[a]	48.7[a]	12.5	23

[a]After (several) hydration steps, see text for details

Conclusions

CO_2 capture from flue gases via carbonation calcination cycles over CaO-based sorbents is a promising technology. In the present study, several synthetic and natural sorbents were developed and preliminary evaluated in TGA. Synthetic sorbents prepared by sol-gel auto-combustion technique exhibited high sorption capacity and stability. Natural sorbents derived from industrial hydrated lime presented promising results; the preparation procedure seems not to significantly affect activity and stability. The most promising synthetic, Ca-Zr and Ca-Al, and natural, pure CaO derived from $Ca(OH)_2$ direct calcination and MgO-doped Ca $(OH)_2$, were tested in a fixed bed reactor under realistic flue gas feed composition. Zr and Al-doped CaO based sorbents exhibited very high sorption capacity and stability over 20 consecutive carbonation/calcination cycles, owing to the porous structure obtained by the auto-combustion technique and retained due to the presence of thermally stable Ca-Zr and Ca-Al mixed phases. The natural sorbents presented inferior results, however their regeneration via hydration seems possible, although most likely not economically viable. The presence of steam in the flue gases seems to enhance sorption capacity and stability. Ca-Al synthetic sorbent exhibited the most promising performance and shows great potential for process scale up.

Acknowledgement We acknowledge the financial support from European Union-European Regional Development Fund and Ministry of Development, ESPA 2007–2013/EPAN II programme, Action Synergasia11'.

Nomenclature

f	Total flow rate, cm^3/s
M_{CO2}	Moles of CO_2 captured, mol CO_2/s
MW_{CaO}	Molecular weight of CaO
MW_{CO2}	Molecular weight of CO_2
n_{CaO}	Initial moles of CaO in the sorbent sample
w_{CaO}	Weight fraction of free CaO present in the sorbent
w_{init}	Initial weight of the sorbent, kg

References

Abanades, J. C., Rubin, E. S., & Anthony, E. J. (2004). Sorbent cost and performance in CO_2 capture systems. *Industrial & Engineering Chemistry Research, 43*, 3461–3466.

Alonso, M., Criado, Y. A., Abanades, J. C., & Grasa, G. (2014). Undesired effects in the determination of CO_2 carrying capacities of CaO during TG testing. *Fuel, 127*, 52–61.

Angeli, S. D., Martavaltzi, C. S., & Lemonidou, A. A. (2014). Development of a novel-synthesized Ca-based CO_2 sorbent for multicycle operation: Parametric study of sorption. *Fuel, 127*, 62–69.

Antzara, A., Heracleous, E., Lemonidou, A. A., et al. (2015). Improving the stability of synthetic C_aO-based CO_2 sorbents by structural promoters. Applied Energy, 156(2015), 331–343.

Blamey, J., Anthony, E. J., Wang, J., & Fennell, P. S. (2010). The calcium looping cycle for large-scale CO_2 capture. *Progress in Energy and Combustion Science, 36*, 260–279.

Blockstein, D. E. & Shockley, M. A. (Eds.) (2006). National Council for Science and the Environment. 2006. Energy for a Sustainable and Secure Future: A Report of the Sixth National Conference on Science, Policy and the Environment. Washington, D.C.

Broda, M., Pacciani, R., & Muller, C. R. (2014). CO_2 capture via cyclic calcination and carbonation reactions. In A.-H. Lu & S. Dai (Eds.), *Porous materials for carbon dioxide capture*. Heidelberg, Berlin: Springer.

Coppola, A., Salatino, P., Montagnaro, F., & Scala, F. (2014). Hydration-induced reactivation of spent sorbents for fluidized bed calcium looping (double looping). *Fuel Processing Technology, 120*, 71–78.

Diego, M. E., Arias, B., Grasa, G., Abanades, J. C., Díaz, L., Lorenzo, M., et al. (2014). Calcium looping with enhanced sorbent performance: Experimental testing in A large pilot plant. *Energy Procedia, 63*, 2060–2069.

Dieter, H., Bidwe, A. R., Varela-Duelli, G., Charitos, A., Hawthorne, C., & Scheffknecht, G. (2014). Development of the calcium looping CO2 capture technology from lab to pilot scale at IFK, University of Stuttgart. *Fuel, 127*, 23–37.

Epple B. (2010). *2nd IEA Meeting on High Temperature Solid Looping Cycles*, Netherlands 2010.

EU Technology Platform for Zero Emission Fossil Fuel Power Plants. (2014). The EU technology platform for zero emission fossil fuel power plants, recommendations for RTD, support actions and international collaboration activities within FP7 energy workprogrammes in support of deployment of CCS in Europe.

Fan, L.-S. (2010). *Chemical looping systems for fossil energy conversions*. Hoboken, NJ, USA: Wiley.

Grasa, G. S., & Abanades, J. C. (2006). CO_2 capture capacity of CaO in long series of carbonation/calcination cycles. *Industrial and Engineering Chemistry Research, 45*, 8846–8851.

Harrison, D. P. (2005). The role of solids in CO_2 capture: A mini review. In E. S. Wilson, D. W. Rubin, C. F. Keith, M. Gilboy, T. Thambimuthu, J. Morris, & K. Gale (Eds.), *Greenhouse gas control technologies*. Oxford: Elsevier Science Ltd.

Kavosh, M., Patchigolla, K., Anthony, E. J., & Oakey, J. E. (2014). Carbonation performance of lime for cyclic CO2 capture following limestone calcination in steam/CO_2 atmosphere. *Applied Energy, 131*, 499–507.

Li, C.-C., Cheng, J.-Y., Liu, W.-H., Huang, C.-M., Hsu, H.-W., & Lin, H.-P. (2014). Enhancement in cyclic stability of the CO_2 adsorption capacity of CaO-based sorbents by hydration for the calcium looping cycle. *Journal of the Taiwan Institute of Chemical Engineers, 45*, 227–232.

Liu, W., An, H., Qin, C., Yin, J., Wang, G., Feng, B., et al. (2012). Performance enhancement of calcium oxide sorbents for cyclic CO_2 capture-a review. *Energy and Fuels, 26*, 2751–2767.

Mandal, B. P., & Bandyopadhyay, S. S. (2006). Absorption of carbon dioxide into aqueous blends of 2-amino-2-methyl-1-propanol and monoethanolamine. *Chemical Engineering Science, 61*, 5440–5447.

Manovic, V., & Anthony, E. J. (2010). Carbonation of CaO-based sorbents enhanced by steam addition. *Industrial & Engineering Chemistry Research, 49*, 9105–9110.

Martavaltzi, C. S., & Lemonidou, A. A. (2008). Development of new CaO based sorbent materials for CO_2 removal at high temperature. *Microporous and Mesoporous Materials, 110*, 119–127.

Park, A.-H. A., Gupta, P., Li, F., Sridhar, D., & Fan, L.-S. (2010). Novel applications of chemical looping technologies. In L.-S. Fan (Ed.), *Chemical looping systems for fossil energy conversionsm*. Hoboken, NJ, USA: John Wiley & Sons, Inc.

Reddy, G. K., Quillin, S., & Smirniotis, P. (2014). Influence of the synthesis method on the structure and CO_2 adsorption properties of Ca/Zr sorbents. *Energy and Fuels, 28*, 3292–3299.
Ruff, O., Ebert, F., & Stephan, E. (1929). An article on highly fire resistant ceramic substances II The ZrO_2-CaO system. *Zeitschrift fur anorganische und allgemeine Chemie, 180*, 215–224.
Ströhle, J., Junk, M., Kremer, J., Galloy, A., & Epple, B. (2014). Carbonate looping experiments in a 1MWth pilot plant and model validation. *Fuel, 127*, 13–22.
Sun, P., Grace, J. R., Lim, C. J., & Anthony, E. J. (2008). Investigation of attempts to improve cyclic CO_2 capture by sorbent hydration and modification. *Industrial & Engineering Chemistry Research, 47*, 2024–2032.
United Nations, C. C. S. (2012). Summary of GHG emissions for Greece. url:unfccc.int/files/ghg_emissions_data/application/pdf/grc_ghg_profile.pdf
Wang, J., Huang, L., Yang, R., Zhang, Z., Wu, J., Gao, Y., et al. (2014). Recent advances in solid sorbents for CO_2 capture and new development trends. *Energy and Environmental Science, 7*, 3478–3518.
White, C. M., Strazisar, B. R., Granite, E. J., Hoffman, J. S., & Pennline, H. W. (2003). Separation and capture of CO2 from large stationary sources and sequestration in geological formations. *EM: Air and Waste Management Association's Magazine for Environmental Managers* 29–34.

Chapter 43
Kinetics of CO_2 Capture by Carbon Dioxide Binding Organic Liquids

Ozge Yuksel Orhan, Hakan Kayi, and Erdogan Alper

Introduction

CO_2 emissions from human activities are consistently rising with the increase of population growth and industrialization (Couchaux et al. 2013). Industries such as chemical and petrochemical manufacturing, cement and steel production and natural gas processing are responsible for the largest sources of CO_2 emissions (Liu et al. 2014). Since CO_2 is regarded as a major greenhouse gas that causes the global warming and climate change problems, effective and economic CO_2 capture technologies have gained importance. The most common and proven approach to capture CO_2 from gas streams is chemical absorption (Kumar et al. 2014). Therefore, aqueous solutions of alkanolamines, such as monoethanolamine (MEA), diethanolamine (DEA) and methyldiethanolamine (MDEA) are widely used solvents for the removal of acid gas impurities such as CO_2 and H_2S (Alexia Finotello et al. 2008; Kierzkowska-Pawlak et al. 2013; Garđarsdóttir et al. 2015). Solvents for such absorption processes absorb CO_2 from gas streams efficiently at low temperatures by an exothermic reaction resulting either carbamate or bicarbonate salts. Regeneration of traditional amine solutions can be achieved by desorption at near atmospheric pressures but high temperatures such as 120–130 °C (Finotello et al. 2008) are required. Therefore, the energy-intensive stripping requirement (especially, the reboiler duty) is one of the disadvantages for using aqueous alkanolamines arising from high specific and latent heats of water. In addition,

O. Yuksel Orhan (✉) • E. Alper
Department of Chemical Engineering, Hacettepe University, Beytepe, 06800 Ankara, Turkey
e-mail: oyuksel@hacettepe.edu.tr; ealper@hacettepe.edu.tr

H. Kayi
Chemical Engineering and Applied Chemistry Department, Atilim University, İncek, 06836 Ankara, Turkey
e-mail: hakan.kayi@atilim.edu.tr

P. Grammelis (ed.), *Energy, Transportation and Global Warming*, Green Energy and Technology, DOI 10.1007/978-3-319-30127-3_43

thermal stripping of CO_2 from aqueous alkanolamines at temperatures higher than boiling point of the water leads to large evaporative losses of solvent. Further, amines (especially MEA) at high concentrations have been reported to be responsible for stress corrosion cracking. Accordingly, solvent loss, corrosive nature and high energy intensive properties are major drawbacks of amine based technologies (Guo et al. 2013).

Consequently, there are considerable on-going research and development to design new solvents to reduce (or eliminate) the high latent heat requirement of the aqueous systems. The key factors for solvent selection are high absorption capacity, fast absorption rate, cost effectiveness, low corrosiveness, high solvent stability and low energy requirement for regeneration (Liu et al. 2014). A new approach to handle these conflicts is to employ a liquid absorbent medium of an organic base in a non-aqueous, polar solvent. For instance, Jessop et al. (2005) developed CO_2-binding organic liquids (CO_2BOLs).

CO_2BOLs are smart solvents comprised of amidine/guanidine bases and organic alcohols that capture CO_2 and form amidinium or guanidinium alkyl carbonate salts. It is presumed that the base of CO_2BOL does not react directly with CO_2, as amines react to form carbamates. Conversely, the alcohol group in the CO_2BOL reacts firstly with CO_2 to form an alkyl carbonic acid, and then protonates the base to form a liquid alkylcarbonate (Mathias et al. 2013). Alkyl carbonate salts formed are non-volatile ionic compounds and they do not contain as much hydrogen bond as carbamate and bicarbonate salts do. Therefore, binding enthalpy of CO_2 decreases and high stripping temperatures that amine systems need are no longer required (Gordesli et al. 2013).

CO_2BOLs can reversibly switch from non-ionic form to ionic form when exposed to carbon dioxide and causes a notable increase in polarity (Jessop et al. 2012). Then, CO_2 is removed from CO_2BOLs below the boiling point of the mixture by simple heating or sweeping with an inert gas such as nitrogen. They have tunable physicochemical properties; the solvent reverts back to its less-polar non-ionic form and is ready for future CO_2 uptake (Tao Yu et al. 2008). The main advantages of these systems are high CO_2 binding capacities, low heat capacities and lower energy requirement for regeneration as compared to aqueous alkanolamine solutions (Mathias et al. 2013). Wang et al. developed several strategies to reduce volatility of superbase-derived absorbents by making them such as alcohol-functionalized ionic liquids (Wang et al. 2010). Currently, several CO_2BOL systems are being investigated by us and one of these studies focuses on reaction kinetics of CO_2BOLs, composed of a mixture of 1,8-diazabicyclo [5.4.0] undec-7-ene (DBU) amidine base in 1-hexanol or 1-propanol (Ozturk et al. 2012). The others are focused on reaction performance and kinetics of CO_2BOLs, composed of a mixture of 1,1,3,3-Tetramethylguanidine (TMG) guanidine base in 1-hexanol and 1-propanol, with carbon dioxide (Ozturk et al. 2014; Yüksel Orhan et al. 2015). However, the observed rate constants reported in these works are relatively low which can be enhanced by addition of small amounts of activators such as piperazine and its derivatives (Moioli and Pellegrini 2015).

Analysis

Generally, the mechanism of the reaction between CO_2 and amine systems can be explained by two established mechanisms. These are the zwitterions mechanism, originally proposed by Caplow in 1968, and then reintroduced by Danckwerts in 1979 and termolecular reaction mechanism proposed by Crooks and Donnellan (1989) and modified by Ozturk et al. (2014).

The zwitterion mechanism, also known as a two-step mechanism, involves two sequential reactions. In the first step, CO_2 reacts with the amine (denoted here as RNH_2) and forms zwitterion intermediate product as in Eq. (43.1). In the next step, this zwitterion reacts with a base (or bases) and a proton from zwitterion is transferred to the base (B), resulting in a protonated base and carbamate ion. As shown in Eq. (43.2), a water molecule, an additional amine, or any other basic species can act also as the base.

$$CO_2 + RNH_2 \underset{k_{-1}}{\overset{k_2}{\rightleftharpoons}} RN^+H_2COO^- \tag{43.1}$$

$$RN^+H_2COO^- + B \xrightarrow{k_B} RNHCOO^- + BH^+ \tag{43.2}$$

When the base is amine and the carbamate ion is stable, the resulting net reaction is given in Eq. (43.3).

$$CO_2 + 2\,RNH_2 \rightleftharpoons RNHCOO^- + RNH_3^+ \tag{43.3}$$

The termolecular reaction mechanism assumes that an amine reacts with both one molecule of carbon dioxide and one molecule of a base in a single step as illustrated in Fig. 43.1. It is also assumed that the reaction takes place via the weakly-bound intermediate product as shown in Eq. (43.4).

$$CO_2 + RNH_2 \cdots B \rightleftharpoons RNHCOO^- \cdots BH^+ \tag{43.4}$$

The modified termolecular reaction mechanism for CO_2BOLs, containing amidine/guanidine base and a linear alcohol, is shown in Eq. (43.5). While a

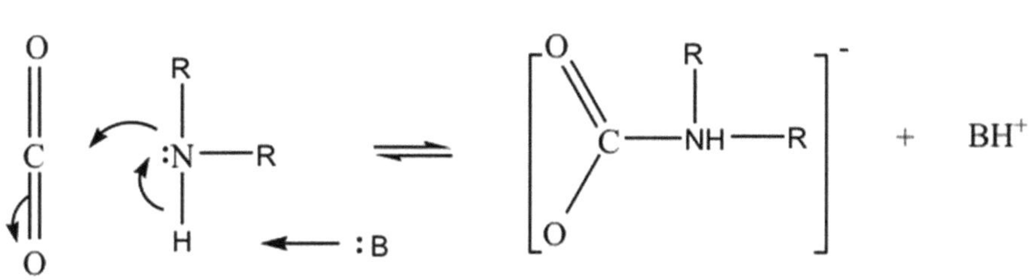

Fig. 43.1 Schematic drawing of a termolecular reaction mechanism (Crooks and Donnellan 1989)

fraction of this complex reacts with a second molecule of the amine or alcohol molecule to form alkyl carbamate salt, a fraction of resulting intermediate breaks up to form reactant molecules (CO_2 and amine).

$$CO_2(g) + B_{(l)} + ROH_{(l)} \rightleftharpoons [BH^+][ROCOO^-]_{(l)} \tag{43.5}$$

Under pseudo-first order conditions where the solvent would be in excess, the observed forward reaction rate can be expressed as in Eq. (43.6).

$$r_{obs} = k_o[CO_2] \tag{43.6}$$

The observed reaction rate constant (k_o) can be expressed by Eq. (43.7) in terms of CO_2BOL/alcohol system.

$$k_o = \{ k_B[B] + k_{ROH}[ROH] \} [B] \tag{43.7}$$

Since, alcohol concentration is assumed to be in excess for the pseudo-first order conditions, ROH can be considered constant and a new rate constant, k, can be defined by Eq. (43.8).

$$k = k_{ROH}[ROH] \tag{43.8}$$

$$k_o = \{ k_B[B] + k\} [B] \tag{43.9}$$

As seen in Eq. (43.9), if the amine and alcohol are comparably dominant bases, degree of the reaction can change between 1 and 2. Also if the alcohol is the dominant base, system exhibits a first-order reaction and Eq. (43.9) reduces to Eq. (43.10).

$$k_o = k[B] \tag{43.10}$$

In summary, the rate constants of CO_2BOLs were obtained by using Eqs. (43.9) and (43.10).

Material and Method

Reagents

Reagent grade 1,5,7-Triazabicyclo[4.4.0]dec-5-ene (TBD) with purity of 99 % (CAS no. 5807-14-7) and 1,5-Diazabicyclo[4.3.0]non-5-ene with purity of 98 % (CAS no. 3001-72-7) were supplied by Sigma-Aldrich (St. Louis, MO, USA). 1-Hexanol with 98 % purity (CAS no. 111-27-3) was also obtained by Sigma-Aldrich. Carbon dioxide with a purity of 99.99 % was obtained from Linde (Munich, Germany). Reagent grade chemicals were used without further purification.

Kinetic Measurement

The observed reaction rate constants of the amines with CO_2 in 1-hexanol were measured over a range of temperatures (278, 298 and 308 K) using a customised stopped-flow instrument (Hi-Tech Scientific, UK, Model SF-61SX2). The conductivity-detection system of the apparatus calculates directly the intrinsic rate of a rapid homogeneous reaction by measuring the conductivity change during the reaction. All parts of the flow circuit, with the exception of the stop/waste syringe, are thermostated and the temperature control is better than ± 0.1 K. During an experimental run, amine (TBD or DBN)/1-hexanol solution and carbon dioxide dissolved 1-hexanol solution were placed in the sealed drive syringes in the sample unit. For each experimental run, equal volumes of the solutions were mixed instantaneously in a stainless-steel cell for the reaction to occur and the flow was stopped. The ion formation initiates a voltage change which is monitored as function of time. Then, the equipment software calculates the pseudo-first order reaction rate constants of the rapid homogeneous reaction based on least squares regression. A typical experimental output from the standard stopped flow system is shown in Fig. 43.2. To obtain consistent pseudo-first-order rate constants (k_o), each experimental set was repeated for an average ten times. The amine and alcohol concentrations were always much in excess to that of CO_2 (usually the molar ratio was about 20:1) to satisfy the pseudo-first-order conditions. The main advantages of the stopped-flow technique are easy operation, quick experiment run (~0.05 s), small amount of solvent consumption for each experimental run (~ 0.1 mL) and no effect of gas phase resistance (Liu et al. 2014). Further details of the equipment and experimental procedure can be found elsewhere (Ume et al. 2012; Yüksel Orhan et al. 2015).

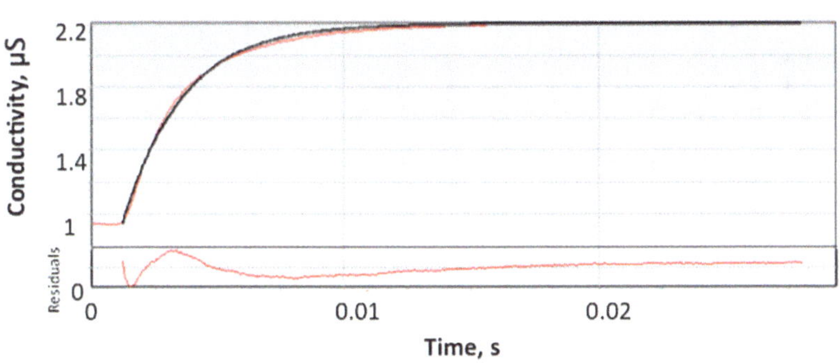

Fig. 43.2 Combined avg. graphs of 5 wt% DBN/1-hexanol system at 298 K

Results and Discussion

In order to evaluate the potential integration of these switchable solvents to industrial carbon dioxide capture applications, the reaction kinetics between CO_2BOLs and carbon dioxide were examined. In this work, as an amidine; DBN (1,5-Diazabicyclo[4.3.0]non-5-ene) and as a guanidine; TBD (1,5,7-Triazabicyclo [4.4.0]dec-5-ene) were used in 1-hexanol medium. Intrinsic reaction rates were measured in "stopped flow" equipment for a temperature range of 298–308 K. Experiments were carried out by varying organic base (amidine or guanidine) percentage in a 1-hexanol medium from 5.0 wt% to 20.0 wt%.

Table 43.1 shows a summary of observed pseudo-first order reaction rate constants for the CO_2/TBD/1-hexanol system versus the weight-percent concentration of TBD at temperatures ranging from 288 K to 308 K. Table 43.1 shows clearly that k_o increases by increasing concentration or temperature.

In order to determine the reaction order of the CO_2/TBD/1-hexanol system, the natural logarithms of observed reaction rate constants versus TBD concentrations were plotted as shown in Fig. 43.3. Empirical power law kinetics was fitted to lines

Table 43.1 Observed pseudo-first order rate constants for the CO_2/TBD/1-hexanol system at different temperatures

	k_o [s^{-1}]			
TBD [wt%]	5	10	15	20
288 K	253.1	420.6	526.4	690.7
298 K	463.5	645.4	912.8	1190.1
308 K	621.7	969.6	1190.1	1490.2

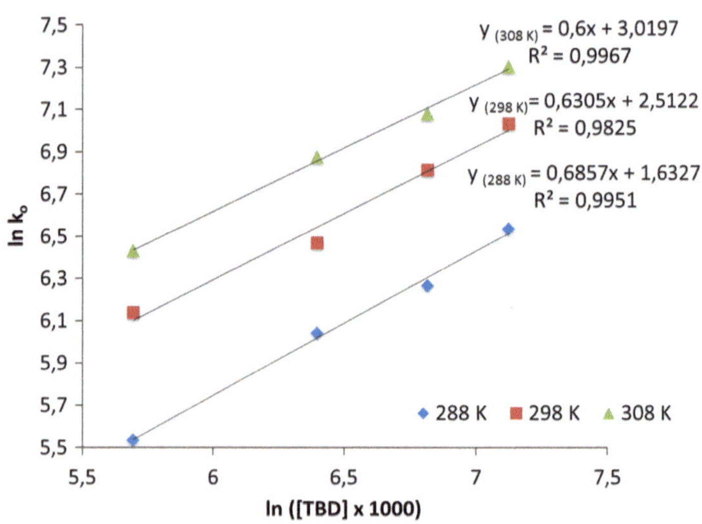

Fig. 43.3 Determination of the apparent reaction order for the CO_2/TBD/1-hexanol system at various temperatures

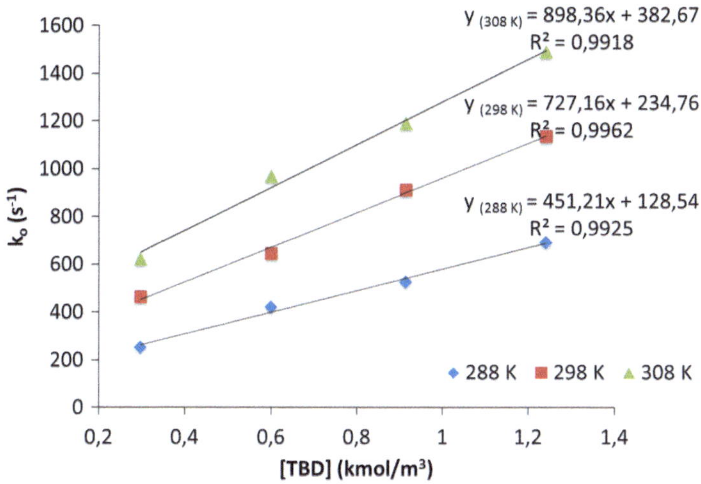

Fig. 43.4 Pseudo-first order rate constant as a function of TBD concentration at various temperatures

Table 43.2 Observed pseudo-first order rate constants for the CO_2/DBN/1-hexanol system at various temperatures

	k_o [s^{-1}]			
DBN [wt.%]	5	10	15	20
288 K	79.8	211.5	271.9	360.5
298 K	145.4	409.3	521.9	699.4
308 K	216.1	548.5	689.1	891.2

in Fig. 43.3 by using the least square method. Their slopes correspond to the reaction orders of the CO_2/TBD/1-hexanol system which are determined to be approximately 0.6 with regression values of $R^2 = 0.99$ for the 5.0–20.0 weight percentages at a temperature range of 288–308 K. The experimentally observed k_o values were correlated using the termolecular mechanism to determine the forward reaction rate constant k [m^3 $kmol^{-1}$ s^{-1}]. The reaction rate constants vs. TBD concentration were plotted according to Eq. (43.6) in a very satisfactory pseudo-first-order plot as seen in Fig. 43.4.

From the slopes of the fitted lines in Fig. 43.4, the first-order forward reaction rate constants for CO_2/TBD/1-hexanol systems were determined to be 451.21 m^3 $kmol^{-1}$ s^{-1} at 288 K, 727.16 m^3 $kmol^{-1}$ s^{-1} at 298 K and 898.36 m^3 $kmol^{-1}$ s^{-1} at 308 K.

Similarly, observed pseudo-first order rate constants for the CO_2/DBN/1-hexanol system at 288, 298 and 308 K were determined. The summary of obtained observed reaction rate constants versus the weight-percent concentration of DBN at temperatures ranging from 288 K to 308 K are given in Table 43.2.

In Fig. 43.5, the natural logarithms of the observed rate constants (k_o) are shown as a function of versus DBN concentrations for the 5.0–20.0 weight percentages at a temperature range of 288–308 K. By fitting empirical power law kinetics to lines in

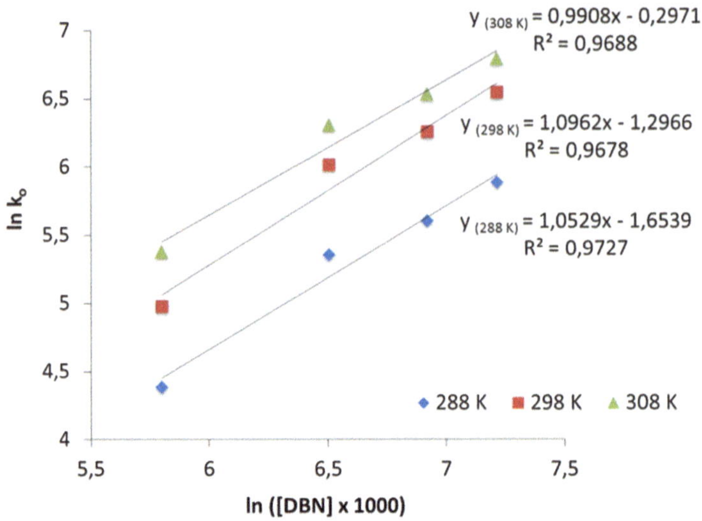

Fig. 43.5 Determination of the apparent reaction order for the CO_2/DBN/1-hexanol system at various temperatures

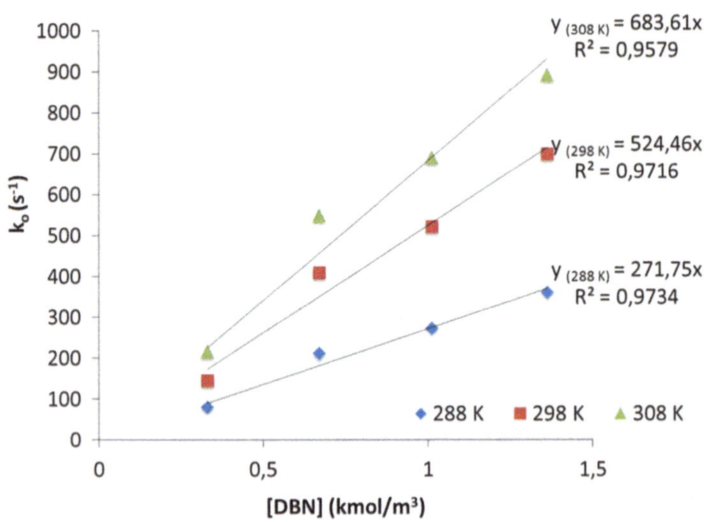

Fig. 43.6 Pseudo-first order rate constant as a function of DBN concentration at various temperatures

Fig. 43.5, the reaction orders of CO_2/DBN/1-hexanol system were found about 1.00 with high R-squared value. This result is in agreement with the pseudo-first-order reaction suggestion of a single-step termolecular reaction mechanism for CO_2/DBN/1-hexanol system. In order to determine the forward reaction rate constants k [$m^3\ kmol^{-1}\ s^{-1}$], the experimentally observed k_o values were correlated as seen in Fig. 43.6.

From the slopes of the fitted lines in Fig. 43.6, the first-order forward reaction rate constants for CO_2/DBN/1-hexanol systems were determined to be 271.75 m^3 $kmol^{-1}$ s^{-1} at 288 K, 524.46 m^3 $kmol^{-1}$ s^{-1} at 298 K and 683.61 m^3 $kmol^{-1}$ s^{-1} at 308 K.

Activation Energies

In order to determine the activation energies, the Arrhenius diagrams were plotted as shown in Figs. 43.7 and 43.8. Activation energies for both CO_2/TBD/1-hexanol system and CO_2/DBN/1-hexanol were calculated by evaluating the Arrhenius equation as in Eq. (43.11).

$$k = A \exp\left(-\frac{E_a}{RT}\right) \tag{43.11}$$

Figure 43.7 shows the Arrhenius diagram for CO_2/TBD/1-hexanol system at 5.0, 10.0, 15.0 and 20 wt%, respectively. Using the slopes of these plots, activation energies for the CO_2/TBD/1-hexanol system were calculated as 33.25 kJ/mol at 5.0 wt%, 30.79 kJ/mol at 10.0 wt%, 30.46 kJ/mol at 15.0 wt% and 28.44 kJ/mol at 20.0 wt%. Figure 43.8 shows the Arrhenius diagram for CO_2/DBN/1-hexanol system at 5.0, 10.0, 15.0 and 20 wt%, respectively. Using the slopes of these plots, activation energies for the CO_2/DBN/1-hexanol system were calculated as 37.18 kJ/mol at 5.0 wt%, 35.37 kJ/mol at 10.0 wt%, 34.43 kJ/mol at 15.0 wt% and 33.54 kJ/mol at 20.0 wt%.

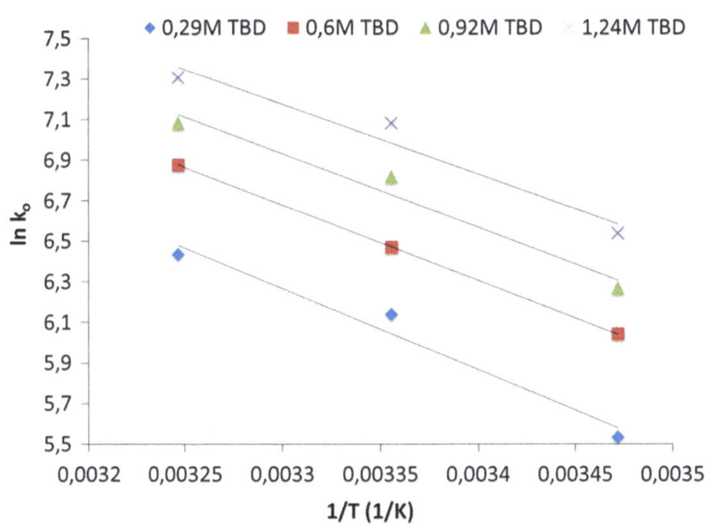

Fig. 43.7 Arrhenius diagram for the CO_2/TBD/1-hexanol system

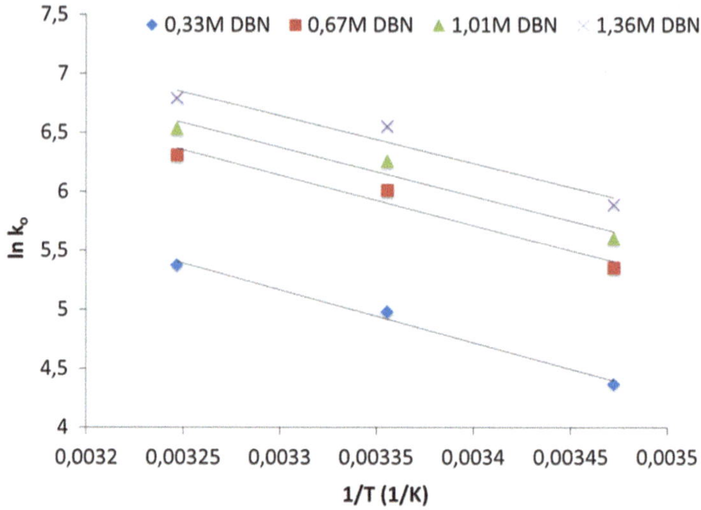

Fig. 43.8 Arrhenius diagram for the CO_2/DBN/1-hexanol system

Fig. 43.9 B3LYP/6-31G (d) optimized lowest energy conformers of 1,5,7-Triazabicyclo[4.4.0]dec-5-ene (TBD) (atom types: *white* = H, *gray* = C, *blue* = N) (Color figure online)

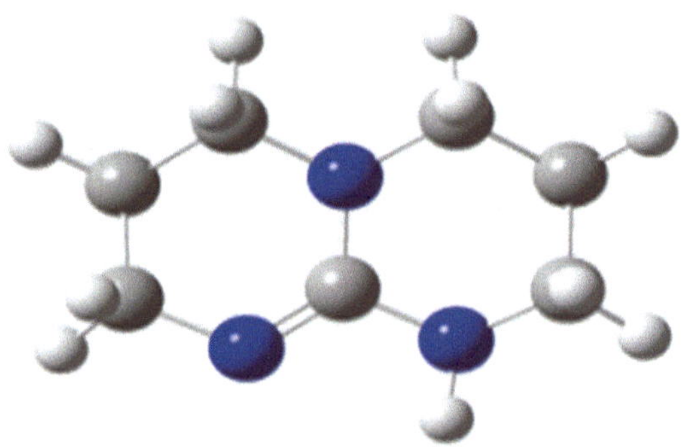

The stable lowest energy conformers of organic bases (TBD and DBN) obtained from the B3LYP/6-31G(d) calculations are given in Figs. 43.9 and 43.10. As expected, the activation energy of TBD is lower than the activation energy of DBN since the amino group of TBD is not hindered (Figs. 43.9 and 43.10).

Finally, the obtained results of this work were compared with the published papers of other CO_2BOLs in 1-hexanol medium at 298 K are shown in Table 43.3.

However, the k_o values are generally low in comparison with MEA or PZ systems (Ume et al. 2012; Gordesli et al. 2013; Yuksel Orhan and Alper 2015), they are comparable to DEA systems (Marta Siemieniec 2011).

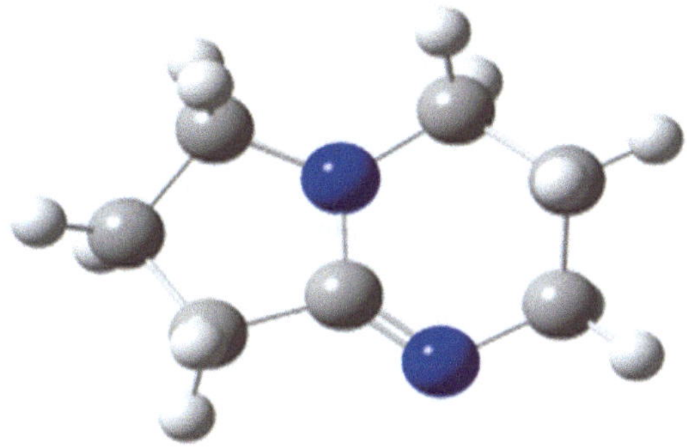

Fig. 43.10 B3LYP/6-31G (d) optimized lowest energy conformers of 1,5-Diazabicyclo[4.3.0] non-5ene (DBN) (atom types: *white* = H, *gray* = C, *blue* = N) (Color figure online)

Table 43.3 Comparison of kinetic properties of various CO_2BOLs at 298 K

Amines	TBD/ 1-hexanol	DBN/ 1-hexanol	TMG/ 1-hexanol	DBU/ 1-hexanol
Reference	This work	This work	Ozturk et al. (2014)	Ozturk et al. (2012)
Reaction order	0.63	1.09	0.98	1.21
k ($m^3\ kmol^{-1}\ s^{-1}$)	727.16	524.46	64.10	627.0
Ea ($kJ\ mol^{-1}$)	30.74	35.13	9.76	13.7

Conclusion

The reaction between organic bases (TBD or DBN) and CO_2 was investigated and modeled using a single termolecular reaction mechanism. Although, the observed rate constants obtained in this work are lower than the other commercial carbon dioxide capture agents, this can be enhanced by addition of small proportion of activators such as piperazine and its derivatives as shown by Yuksel Orhan and Alper (2015). CO_2BOLs are potential candidates for industrial carbon dioxide capture with beneficial properties, such as requiring a smaller reboiler duty in the desorber. This work provided kinetic data for two of these novel CO_2BOL systems which are needed for rate-based modeling.

Acknowledgement This work was supported by a Turkish Scientific and Technological Research Council (TUBITAK) through a research project (Project No.: 213M390). Authors gratefully acknowledge this financial support.

Nomenclature

CO_2BOLs	Carbon dioxide binding organic liquids
B	Base (i.e., amine, water, or hydroxyl ion)
CO_2	Carbon dioxide
DBN	1,5-Diazabicyclo[4.3.0]non-5-ene

DBU	1,8-Diazabicyclo[5.4.0]undec-7-ene
DEA	Diethanolamine
GHG	Greenhouse gas
Ea	Activation energy
k_{OH}	Rate constant for alcohol, m^3/kmol s
k_B	Rate constant for base according to Eq. (43.7), m^3/kmol s
k_o	Observed pseudo–first-order-rate constant, s^{-1}
MDEA	Methyldiethanolamine
MEA	Monoethanolamine
SCC	Stress corrosion cracking
TBD	1,5,7-Triazabicyclo[4.4.0]dec-5-ene
TMG	1,1,3,3-Tetramethylguanidine

References

Alexia Finotello, J. E. B., Camper, D., & Noble, R. D. (2008). Room-temperature ionic liquids: Temperature dependence of gas solubility selectivity. *Industrial & Engineering Chemistry Research, 47*, 3453–3459.

Caplow, M. (1968). Kinetics of carbamate formation and breakdown. *Journal of the American Chemical Society, 90*, 6795–6803.

Couchaux, G., Barth, D., Jacquin, M., Faraj, A., & Grandjean, J. (2013). Kinetics of carbon dioxide with amines. I. Stopped-flow studies in aqueous solutions. A review. *Oil & Gas Science and Technology—Revue d'IFP Energies nouvelles, 69*, 865–884.

Crooks, J. E., & Donnellan, J. P. (1989). Kinetics and mechanism of the reaction between carbon-dioxide and amines in aqueous-solution. *Journal of the Chemical Society-Perkin Transactions, 2*, 331–333.

Danckwerts, P. V. (1979). Reaction of CO_2 with ethanolamines. *Chemical Engineering Science, 34*, 443–446.

Garđarsdóttir, S. Ó., Normann, F., Andersson, K., & Johnsson, F. (2015). Postcombustion CO_2 capture using monoethanolamine and ammonia solvents: The influence of CO_2 concentration on technical performance. *Industrial & Engineering Chemistry Research, 54*, 681–690.

Gordesli, F. P., Ume, C. S., & Alper, E. (2013). Mechanism and kinetics of carbon dioxide capture using activated 2-amino-2-methyl-1,3-propanediol. *International Journal of Chemical Kinetics, 45*, 566–573.

Guo, D., Thee, H., Tan, C. Y., Chen, J., Fei, W., Kentish, S., et al. (2013). Amino acids as carbon capture solvents: Chemical kinetics and mechanism of the glycine + CO_2 reaction. *Energy & Fuels, 27*, 3898–3904.

Jessop, P. G., Mercer, S. M., & Heldebrant, D. J. (2012). CO2-triggered switchable solvents, surfactants, and other materials. *Energy & Environmental Science, 5*, 7240.

Kierzkowska-Pawlak, H., Siemieniec, M., & Chacuk, A. (2013). Investigation of CO_2 and ethylethanolamine reaction kinetics in aqueous solutions using the stopped-flow technique. *Chemical Papers, 67*, 1123–1129.

Kumar, S., Cho, J. H., & Moon, I. (2014). Ionic liquid-amine blends and CO_2BOLs: Prospective solvents for natural gas sweetening and CO_2 capture technology—A review. *International Journal of Greenhouse Gas Control, 20*, 87–116.

Liu, H., Sema, T., Liang, Z., Fu, K., Idem, R., Na, Y., et al. (2014). CO_2 absorption kinetics of 4-diethylamine-2-butanol solvent using stopped-flow technique. *Separation and Purification Technology, 136*, 81–87.

Siemieniec, M., Kierzkowska-Pawlak, H., & Chacuk, A. (2011). Reaction kinetics of carbon dioxide in aqueous diethanolamine solutions in the temperature range of 293–313 K using the topped-flow technique. *Ecological Chemistry and Engineering, 19*(1), 55–56.

Mathias, P. M., Afshar, K., Zheng, F., Bearden, M. D., Freeman, C. J., Andrea, T., et al. (2013). Improving the regeneration of CO_2-binding organic liquids with a polarity change. *Energy & Environmental Science, 6*, 2233.

Moioli, S., & Pellegrini, L. A. (2015). Improved rate-based modeling of the process of CO_2 capture with PZ solution. *Chemical Engineering Research and Design, 93*, 611–620.

Ozturk, M. C., Yuksel Orhan, O., & Alper, E. (2014). Kinetics of carbon dioxide binding by 1,1,3,3-tetramethylguanidine in 1-hexanol. *International Journal of Greenhouse Gas Control, 26*, 76–82.

Ozturk, M. C., Ume, C. S., & Alper, E. (2012). Reaction mechanism and kinetics of 1,8-diazabicyclo[5.4.0]undec-7-ene and carbon dioxide in alkanol solutions. *Chemical Engineering & Technology, 35*, 2093–2098.

Jessop, P. G., Heldebrant, D. J., Xiaowang Li, C. A. E., & Liotta, C. L. (2005). Green chemistry: Reversible nonpolar-to-polar solvent. *Nature, 436*, 1102.

Tao Yu, T. Y., Gaviola, G. C., & Weiss, R. G. (2008). Carbon dioxide and molecular nitrogen as switches between ionic and uncharged room-temperature liquids comprised of amidines and chiral amino alcohols. *Chemistry of Materials, 20*, 5337–5344.

Ume, C. S., Ozturk, M. C., & Alper, E. (2012). Kinetics of CO_2 absorption by a blended aqueous amine solution. *Chemical Engineering & Technology, 35*, 464–468.

Wang, C., Mahurin, S. M., Luo, H., Baker, G. A., Li, H., & Dai, S. (2010). Reversible and robust CO_2 capture by equimolar task-specific ionic liquid–superbase mixtures. *Green Chemistry, 12*, 870.

Yuksel Orhan, O., & Alper, E. (2015). Kinetics of carbon dioxide binding by promoted organic liquids. *Chemical Engineering and Technology, 38*(8), 1485–1489.

Yüksel Orhan, Ö., Öztürk, M. Ç., Şeker, A., & Alper, E. (2015). Kinetics and performance studies of a switchable solvent TMG (1,1,3,3-tetramethylguanidine)/1-propanol/carbon dioxide system. *Turkish Journal of Chemistry, 39*, 13–24.

Chapter 44
Flue Gas CO_2 Sequestration by Turkish Coal Fly Ashes and Anatolian Geothermal Hot Waters

Yıldırım İsmail Tosun

Introduction

The increasing CO_2 concentration in the Earth's atmosphere, mainly caused by fossil fuel combustion, has led to concerns about global warming. A technology that could possibly contribute to reducing carbon dioxide emissions is the in-situ mineral sequestration (long term geological storage) or the ex-situ mineral sequestration (controlled industrial reactors) of CO_2.

Carbonation of CO_2 gas has many various advantages. Most distinct fact is that carbonates have a lower energy state than CO_2. Mineral carbonation is thermodynamically occurs in nature (i.e., the weathering of rock over geologic time periods). Further, the raw materials such as magnesium based minerals are abundant on earth. The produced carbonates are basely stable and thus re-evolve of CO_2 into the atmosphere is not an issue. However, conventional carbonation reactions are slow under ambient temperatures and pressures.

Forty percent of global electricity is generated in fossil fuel power plants per annum, with emissions of about 33 % (9.5 billion metric tons) of global energy related CO_2 emissions of approximately 28.8 Gt in 2010 (Minchener and McMullan 2007; IEA 2012). Over half of the electricity demand of the Turkey is supplied by coal-fired power plants, with emissions of about 16 million metric tons of CO_2 per annum, representing about 33 % of Turkey energy-related CO_2 emissions (TKI 2009; TTK 2009).

Therefore, developing effective CO_2 sequestration is one of the critical components in addressing global climate change. Note that improving the efficiency of energy production and utilization, and developing renewable energy sources will certainly play a very important role in reducing CO_2 emissions (5), however these

Y.İ. Tosun (✉)
Mining Engineering Department, Şırnak University, Şırnak, Turkey
e-mail: yildirimtosun@sirnak.edu.tr; yildirimismailtosun@gmail.com

P. Grammelis (ed.), *Energy, Transportation and Global Warming*, Green Energy and Technology, DOI 10.1007/978-3-319-30127-3_44

measures alone cannot address the greenhouse emissions issue mainly because world energy consumption will increase significantly as the living standard improves in many parts of the world.

Three boreholes, one injection well, and two observation wells will be drilled, about 50–100 m (160–330 ft) apart, into the flank of an anticlinal structure. A total of up to 1 t/day of CO_2, in gaseous state at the well head, will be injected at about 700 m (2300 ft) depth into a saline mudstone aquifer. Injection is planned to start at the end of 2013 and last for approximately 2 years, during which the distribution and fate of the injected gas will be monitored by surface geophysical surveys and by downhole instrumentation. Various processes and subprojects and work packages address the science and operational aspects of CO_2 (IPCC 2005; Stangeland 2007; Solomon 2006; O'conner et al. 2000; Siegenthaler and Oeschger 1987; Keeling et al. 1995; Plasynski et al. 1999). The proposed part of the project is restricted to the baseline survey and the injection and basic monitoring of CO_2. The supply of CO_2 will be provided by near Coal Power Station in Şırnak. Other choice will be the purchase of a pure CO_2 originating from flue gas of hydrocarbon production at an oil refinery. The CO_2 will be delivered by truck in liquid phase, temporarily stored at the well site, and conditioned before injection. The project proposed plans to inject the CO_2 at the well head in gaseous phase at a slightly supercritical pressure and slightly supercritical temperature. The anticline at Şirnak-Silopi was used for gas storage in the past in a shallower depth interval in Şırnak-Silopi pit mine field, which also is an mining industry partner in CO_2 sequestration and will be responsible for the CO_2-injection operations.

This paper discussed progress on reactor achieved by tests and search for fast reaction methods using exhaust gas containing waste sulfur and carbon gases at the stack of Power Stations. This experimental study demonstrates that 1 ton of fly-ash could sequester up to near 20 kg of CO_2, i.e. 50 ton of fly ash per ton of CO_2 sequestered. This confirms the possibility to use this alkaline residue for CO_2 mitigation. Other alkaline sources containing calcium, magnesium and magnesium salts, supercritical CO_2, water slurry, and additives were searched for optimum sequestration methods and also in order to enhance mineral reactivity; and in analyzing the structural changes to identify reaction paths and potential barriers. Carbonation liquid and gaseous products may change to near 20 %–45 % yield performances by time increase from 1 to 6 h.

CO_2 Sequestration and Carbonation

Options for storing CO_2 in deep underground geological formations need adequate porosity and thickness for storage capacity and permeability for gas injection are critical. The storage formation should be capped by extensive confining units such as shale, salt caves or anhydrite beds to ensure that CO_2 does not escape into overlying, shallower rock units and ultimately to the surface. Geological storage of CO_2 requires compression of CO_2 to allow injection. This is done by compressing the CO_2 to a dense fluid state known as 'supercritical'. This supercritical state is achieved by exposing the CO_2 to temperatures higher than 31.1 °C and pressure greater than 73.9 bars. The density of CO_2 will increase with depth, until about 800 m or greater, where the injected CO_2 will be in a dense supercritical state (DOE 1998; Simbeck 2004; Burruss 2004; Oldenburg et al., 2001).

The mineral carbonation, a process of converting CO_2 into stable minerals—mineralization has been studied extensively to capture and store CO_2 (Stevens et al. 2001). However, most of the mineral carbonation studies have been largely investigated at lab scale. Preliminary and pilot scale studies for accelerated mineral carbonation (AMC) were conducted at one of the largest coal-fired power plants (2120 MW) in the USA by reacting flue gas with fly ash particles in a fluidized bed reactor. In the preliminary experiments, flue gas CO_2 and SO_2 concentrations decreased from 13.0 % to 9.6 % and from 107.8 to 15.1 ppm, respectively, during the first 2 min of reaction (Stevens et al. 1999). The flue gas treated by fly ash particles, even mineralization hold high mercury (Hg) concentration of 0.22 mg/kg in flue gas (Myer 2003; Zweigel et al. 2001).

Geothermal brine waters utilization and using CO_2 as feed material in mineral carbonation produce various environmentally benign products. However, many challenges to any solution include technical feasibility, economic viability, environmental soundness and long term sustainability for mineralization of CO_2 (Pacala 2003; Hepple and Benson 2003).

Method of Mineral Sequestration

Mineral sequestration involves the reaction of CO_2 with minerals to form geologically stable carbonates. This mineralization of CO_2 was searched by different studies using various materials (Lindeberg 2003; Rubin and Rao 2003). General and specific global mineral carbonation reaction pathway is shown in Fig. 44.1. The goals of the current work were to design an ash–water suspension carbonation process in a continuous mode laboratory-scale plant and to search for potential means of intensifying the water neutralization process (Davis 2004; Seifritz 1990; Kojima et al. 1997; Gunter et al. 1993; Lackner et al. 1995). The carbonation process was optimized by cascading columns in which the pH progressed from

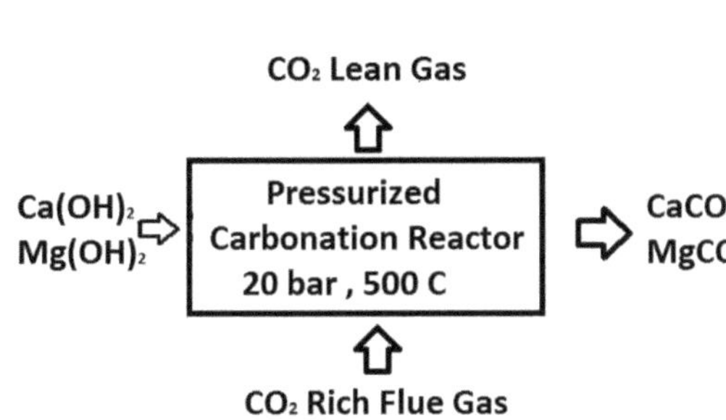

Fig. 44.1 General chemical mineralization path

alkaline to almost neutral. The amount of CO_2 captured from flue gases can reach 0,01–0,12 kg/l at the 500 °C temperature and 10 bar pressure level (O'Connor 1998). Laboratory-scale neutralization experiments were carried out to compare the reactor designs. Sedimentation of carbonate particles was observed and their main characteristics were determined.

Mineral carbonation reactions are known to geologists and occur spontaneously on geological time scales. For example, the reaction of CO_2 with common mineral silicates to form carbonates like calcite and magnesite or calcite is exothermic and thermodynamically favored (O'Connor 1998; Butt et al. 1997; Drägulescu et al. 1972).

The family of reactions represented by Reaction 44.1 has the potential to convert naturally occurring silicate minerals to geologically stable carbonate minerals and silica (Goldberg et al. 2000). This process follows natural chemical transformations such as weathering of rocks to form carbonates over geologic time periods. Reaction 44.2 illustrates the transformation of the common silicate mineral serpentine (Goff et al. 1997), $Mg_3Si_2O_5(OH)_4$, and CO_2 into magnesite, $MgCO_3$, silica and water. Using this ideal case, 1 ton of serpentine can dispose of approximately one-half ton of CO_2. Reaction 44.3 illustrates the transformation of forsterite, which is the end member of the common silicate mineral olivine. One ton of olivine can dispose of approximately two-thirds of a ton of CO_2. Again, the reaction is exothermic and releases 90 kJ/mole of CO_2.

$$\begin{aligned}&(\mathrm{Mg,Ca})_x\mathrm{Si}_y\mathrm{O}_x + 2y + z\mathrm{H}_{2z} + x\mathrm{CO}_2\\&\quad\rightarrow x(\mathrm{Mg,Ca})\mathrm{CO}_3 + y\mathrm{SiO}_2 + z\mathrm{H}_2\mathrm{O}\end{aligned} \tag{44.1}$$

$$\begin{aligned}&1/3\,\mathrm{Mg_3Si_2O_5(OH)_4} + \mathrm{CO_2} \rightarrow \mathrm{MgCO_3} + 2/3\,\mathrm{SiO_2}\\&\quad + 2/3\,\mathrm{H_2O} \quad + 64\ \mathrm{kJ/mole.}\end{aligned} \tag{44.2}$$

$$1/2\,\mathrm{Mg_2SiO_4} + \mathrm{CO_2} \rightarrow \mathrm{MgCO_3} + 1/2\,\mathrm{SiO_2} + 90\ \mathrm{kJ/mole} \tag{44.3}$$

A conceptual illustration of the projected Fly ash mineralization process is presented in Fig. 44.2. As illustrated, CO_2 from one or more power plants is transported to a carbonation reactor, combined with fly ash slurry tank and held at the appropriate reaction conditions until the desired degree of carbonation is

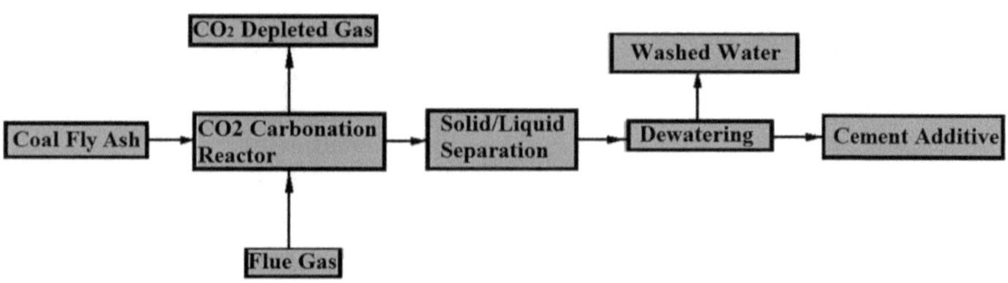

Fig. 44.2 Chemical reaction process of fly ashes for sequestration in the diagram

reached. Careful control of solution chemistry yielded olivine conversions of 90 % in 24 h and 83 % within 6 h. The most recent results show further modifications of the same basic reaction can achieve 65 % conversion in 1 h and 83 % conversion in 3 h. A recent literature review indicated that weak carbonic acid treatments had also been suggested for Mg extraction in the prior literature (O'Connor 1998). Carbonation tests performed at ARC employing heat pretreated serpentine have resulted in up to 83 % conversion in 30 min lower than 115 bars (O'Connor 1998). Indeed, by increasing sodium bicarbonate concentration the carbonation reaction of serpentine can reach 62 % completion under 50 bars. Waters contains various natural elements. Substantial fractions of these elements in springs could be lost during clay rock pores leak. 50–60 % of arsenic, lead, manganese, mercury and selenium could be removed by salt transforming in cooling.

Then products of the reaction, which might be slurry of carbonated minerals and residues and lean gas CO_2 in aqueous, are separated. The residual CO_2 is recycled. Useful materials are collected for construction works or the carbonated materials and residue are returned to the mine site. Almost calcium and magnesium oxide (MgO) content in the magnesium silicate ore mineral of 40 % and 60–70 % chemical efficiency of the carbonation reaction; mass balance of equation 44.1. A 100 MW power plant in Sırnak-Silopi, generating approximately 200 tons/day of CO_2, would require just over 140 tons/day of calcium and magnesium containing fly ash. Several fly ash types in Turkey containing sufficient calcium and magnesium oxide quantity in silicate mineral to provide raw materials for the mineral carbonation.

Experimental Work

In this research, representative specimens the different types of Turkish geothermal hot water sources Sirnak, were classified to calcium content and bicarbonate by chemical analysis. Gas samples of 10–20 kg tubes from each different coal power stations were used. Screen analysis of Turkish lignite samples were made by standard Tyler Screens and particle size distributions and normal distributions of Turkish fly ash samples are respectively illustrated from Fig. 44.3.

Number citations consecutively in alkali waters were reacted by waste gas in the process reactor (Fig. 44.2) and solution distributions were investigated at each stage of sequestration. Specific surface area of fly ash minerals was about 4.76–6.2 m^2/g determined by BET surface analyzer and highly sufficient in order to react with gaseous CO_2. 80 % of weights of fly ash were reacted in 10 bar alkali solution. Main reactive salt structure is widely distributed and pores structures are associated with clay minerals. Coarse alkali oxides are also seen (Table 44.1).

From these results, we designed and developed laboratory scale process unit consisting a three consecutive autoclave reactors (1 m $\Phi \times 0.7$ m) to capture and carbonize flue gas CO_2. Flue gas was filtered and withdrawn from the gas tube and was fed to the autoclave with fly ash slurry at about 7 l/s. The cyclone and decanter

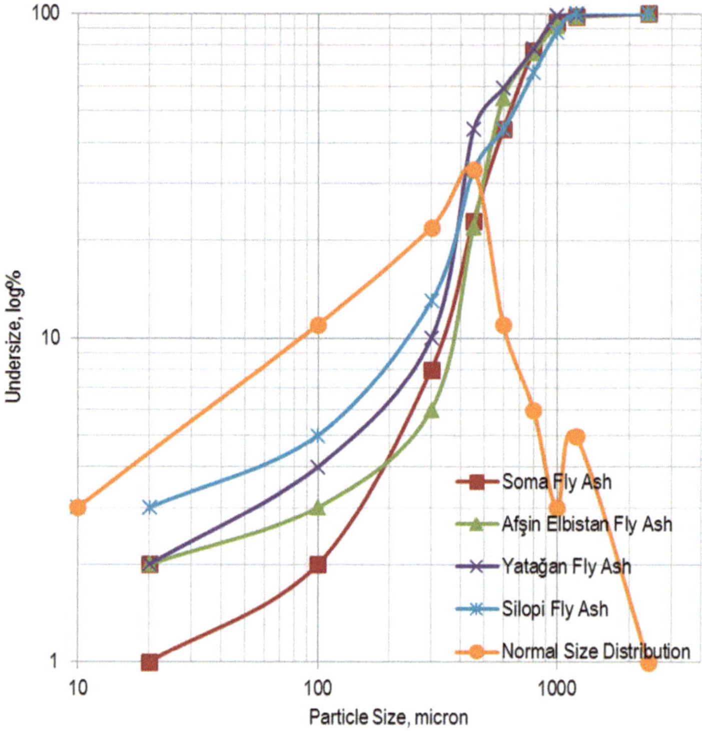

Fig. 44.3 Particle size distribution and normal size distribution of the fly ash used in carbonation process

Table 44.1 Chemical analysis of Turkish fly ash used in the experiment

Chemical composition	Soma fly ash F	Yatagan fly ash F	Silopi fly ash F	Afşin Elbistan fly ash F
SiO_2 (S)	42.14	52.05	40.4	20.4
Al_2O_3 (A)	17.12	23.69	4.1	7.1
Fe_2O_3(F)	4.34	5.94	15.96	4.96
CS + CAF	63.26	81.68	81	88
CaO	29.6	19.60	24.7	64.7
MgO	1.53	2.38	2.1	3.1
K_2O	1.15	2.51	1.55	0.55
Na_2O	1.29	2.56	1.25	0.250
Loss of ignition	0.34	0.90	5.22	4.22
SO_3	3.43	1.45	2.2	2.4
Density (g/cm^3)	2.22	2.12	1.9	2.0
Loose unit weight (g/cm^3)	0.72	0.88	0.71	0.69

unit separates ash and aqueous water. The salt particles were fluidized by flow of slurry through a slurry tank and pump. The laboratory scale studies were conducted at a controlled pressure (115.1 kPa) by controlling the flue gas content. The flue gas was continuously monitored to measure flue gas CO_2, SO_2 and NO_x concentrations by an industrial grade gas analyzer, while the fresh and spent ash were analyzed for calcium carbonate ($CaCO_3$), sulfur (S), and mercury (Hg) content.

Results and Discussion

The major technical challenge now hindering the use of minerals to sequester CO_2 is theirs low reaction rate. Weathering of rock is extremely slow. The highest priority is given to identifying faster reaction pathways. The optimized process has to be economical. Although many carbonation reactions are exothermic, it is generally very difficult to recover the low-grade heat while the long reaction time and demanding reaction conditions.

The environmental impact from mining, mineralization and carbonation processes must be considered in sequestration. We succeeded in achieving shortened carbonation reaction times employing fly ash containing lime and calcium magnesium silicates such as gehlenite, mehlenite. Reaction took 4 h to reach 40–50 % completion of carbonation. The reaction required temperatures of 450–550 °C, pressures of 10–20 bar, and mineral particles in the −100 μm size range. Because the high pressure requirement of the carbonation reaction will certainly lead to high process costs, the team is modifying solution chemistry to allow reaction to proceed at a lower pressure and temperature. The research showed that the concentration of HCO_3^- in the solution is critical to the reaction rate. The high CO_2 pressure will lead increased CO_2 absorption in the solution and thus enhance the HCO_3^- concentration. Adding bicarbonate such as sodium bicarbonate in the solution will significantly increase the HCO_3^- concentration even at a relatively lower CO_2 pressure.

In experiments with 3 h reaction period at reactor temperature of 500 °C and fly ash slurries were kept under changing CO_2 pressure ranged 5 bar to 40 bar. Products were subjected to analysis for carbonate yield determination. Test results of yields for different Turkish fly ash sources were seen in Fig. 44.4.

That quantity in the carbonation amounts were determined for different source evaluation and reduce the effect of gas content of waste gas in order to optimize carbonation rates. As given in Fig. 44.4 gas conversion yield for steam and ash samples were greatly differed. Yatağan and Elbistan Fly ash showed high carbonation yields reaching 32 % and 30 % respectively. Because of active lime amounts were over 20 % and even alkali Na and K contents were over totally over 5 %, the higher carbonation yields were sufficiently provided by three consecutive carbonation columns at 20 bar.

In the carbonation experiments with addition alkali, reactor temperature changed between 200 and 650 °C and ash slurries were mixed with active lime additionally by 10 % rate. Products received from carbonation of ash specimens were subjected

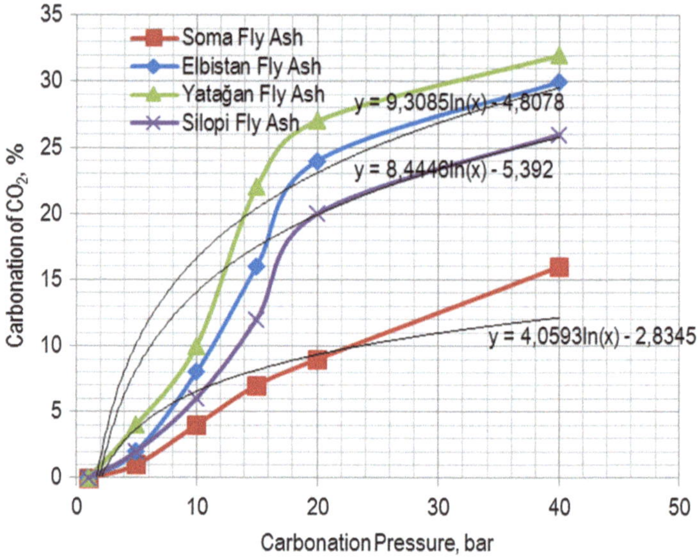

Fig. 44.4 Effect of carbonation time over conversion yield rates in hot 500 °C steam used

to analysis for gas hold-up determination. Test results of carbonation by fly ash and also alkali lime contained over active lime at 10 % weight rate were investigated.

In experiments with 3 h reaction period of ash specimens, at reactor temperature changed between 200 and 600 °C and all ash samples contained over active lime at 10 % weight rate. Products were subjected to analysis for carbonate yield determination. Test results of carbonation yields of Turkish fly ash sources and additional 10 % lime were seen in Fig. 44.5.

As given in Fig. 44.5 carbonation yield for steam and ash samples were linearly changed between 200 and 650 °C. Yatağan and Elbistan Fly ash showed high carbonation yields reaching 44 % and 24 % at 650°C, respectively. Because of low reaction kinetics below 500 °C and even the effect 10 % over dosage of active lime content in the reactors were ranged over 3 % carbonate yield per hour, the higher carbonation yields were sufficiently provided over 500 °C at 20 bar.

In the carbonation experiments, the experimental conditions were controlled on the basis of the gas composition in the ambient state. So neither the contained water vapor nor the others, slurry was taken into account. In order to determine the effect of reaction period on carbonation yield for Turkish fly ash samples, the tests were carried out in three consecutive carbonation columns at 500 °C and 20 bar between1min to 6 h. Yatağan and Elbistan Fly ash showed high carbonation yields reaching 33 % and 26 % at 500°C at 3 h reaction period, respectively. Because of low reaction kinetics below 30 min reaction period in the reactors, yield rates were ranged over 1 % carbonate yield per hour, the higher carbonation yields were sufficiently provided over 2 h reaction period at 20 bar. Carbonation of slurries and yield products changed to 37 % and 29 % for Yatağan and Elbistan Fly ash, respectively, by time increase from 1 to 6 h (Fig. 44.6).

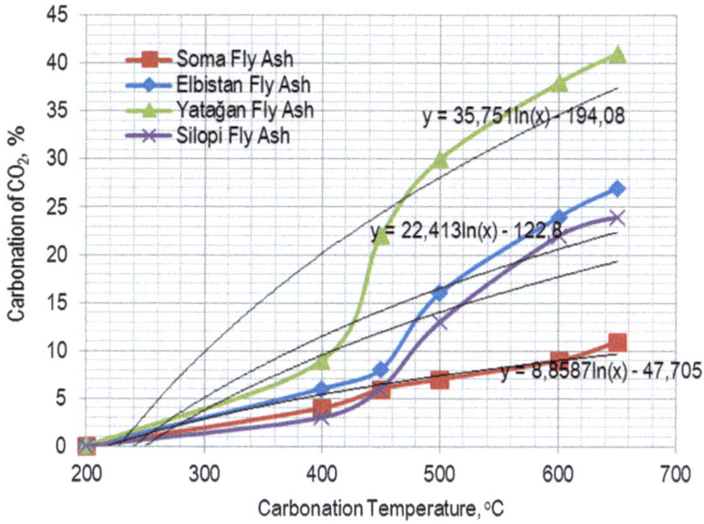

Fig. 44.5 Effect of time of carbonation over conversion yield rates in steam with 10 % additional lime used at 20 bar

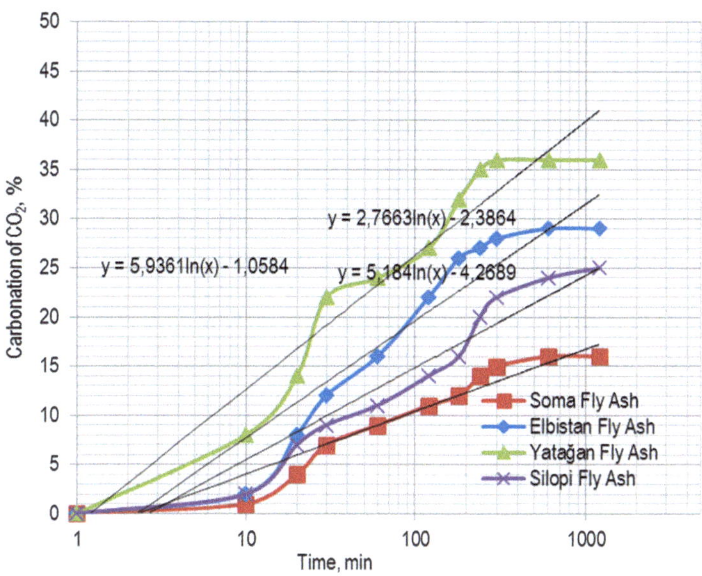

Fig. 44.6 Effect of time of carbonation over conversion yield rates in hot waters used at 500 °C and 20 bar

Conclusion

The qualitative comparison of the carbonation for three different Turkish fly ash reactants and products revealed a complete MgO, CaO–MgO to $MgCO_3$-$CaCO_3$ conversion. The carbonation efficiency of CaO was dependent on the initial pressure of CO_2 of 20 bar. This was significantly affected by reaction temperature 500 °C and by the fly ash dose 100 g/l. The kinetic data demonstrated that the initial rate of CO_2 transfer was enhanced by carbonation process for our experiments. The precipitate calcium carbonate was characterized by isolated micron sized particles and micron agglomerates of calcite.

This study reveals suitable large scale operating units in order to achieve the carbonation method as a viable sequestration tool at industrially relevant scales by using Turkish fly ashes in 500–600 °C waters. Carbonation liquid and gaseous products may change to near 20–40 % yield performances by time increase from 1 to 6 h.

While there is a potential to utilize other types of flue ashes in mineralization, lime or similar alkali can be evaluated to sequester CO_2 allowing clearly significant amounts. Even there are researches succeeded usage of serpentine and olivine. Consequently, the flue gas should be continuously monitored to measure flue gas flow at depleted gas outlet in order to reprocess it.

Carbonation of the ash slurries and carbonation yield products changed to 37 %, 29 %, 25 % and 19 % for Yatağan, Elbistan, Silopi and Soma Fly ash, respectively, by time increase from 1 to 6 h.

Other harmful emissions caused by flue gas containing high sulfur (S), and mercury (Hg) content can be eliminated by this method. In that study, results suggested that an appreciable amount of flue gas CO_2 and significant amounts of SO_2 and Hg can be directly captured and mineralized by the fly ash particles.

Even with progress made so far, to develop an economical method to sequester CO_2 with minerals is still a challenging task, because the process is still relatively slow, and most reactions require high pressure and moderately elevated temperature.

Acknowledgment The author would like to thank the Mining Engineering Department of Süleyman Demirel University, Isparta.

References

Burruss, R. (2004). *Geologic sequestration of carbon dioxide in the next 10 to 50 years: An energy resource Perspective*. Prepared for The 10-50 solution: Technologies and policies for a low-carbon future, Pew Center on Global Climate Change and the National Commission on Energy Policy, March 25–26, Washington, DC.

Butt, D. P., Lackner, K. S., Wendt, C. H., Park, Y. S., Benjamin, A., Harradine, D. M., et al. (1997). A method for permanent disposal of CO_2 in solid form. *World Resource Review, 9*(3), 324–336.

Davis. (2004). *Gasification and carbon capture and storage: The path forward.* Prepared for the 10-50 solution: Technologies and policies for a low-carbon future, Pew Center on Global Climate Change and the National Commission on Energy Policy, March 25–26, Washington, DC.

Drägulescu, C., Tribunescu, P., & Gogu, O. (1972). Lösungsgleichgewicht von MgO ausSerpentinen durch Einwirkung von CO_2 und Wasser. *Revue Roumaine de Chimie, 17*(9), 1518–1524.

DOE. (1998, November). Vision 21, Clean energy for the 21st century. U.S. Department of Energy, Office of Fossil Energy. DOE/FE-0381. www.fetc.doe.gov/publications/brochures/.

Goff, F., Guthrie, G., Counce, D., Kluk, E., Bergfeld, D., & Snow, M. (1997). *Preliminary investigations on the carbon dioxide sequestering potential of ultramafic rocks.* Los Alamos, NM: Los Alamos National Laboratory. LA-13328-MS.

Goldberg, P. M., Chen, Z.-Y., O'Connor, W., Walters, R., Lackner, K., & Ziock, H. (2000). *CO_2 mineral sequestration studies.* Paper presented at GlobeEx 2000, August, Las Vegas, NV.

Gunter, W. D., Perkins, E. H., & McCann, T. J. (1993). Aquifer disposal of CO_2 rich gases reaction design for added capacity. *Energy Conversion and Management, 34*, 941–948.

Hepple, R., & Benson, S. (2003). Implications of surface seepage on the effectiveness of geologic storage of carbon dioxide as a climate change mitigation strategy. In J. Gale & Y. Kaya (Eds.), *Proceedings of greenhouse gas control technologies 6th international conference (GHGT-6)* (Vol. 1, pp. 261–266). Amsterdam: Elsevier.

IEA. (2012). *World energy outlook.*

IPCC. (2005). *IPCC special report on carbon dioxide capture and storage.* Prepared by working group III of the intergovernmental panel on climate change [B. Metz, O. Davidson, H. C. de Coninck, M. Loos, & L. A. Meyer (Eds.)]. Cambridge/New York: Cambridge University Press, 442p.

Keeling, C. D., Whorf, T. P., Wahlen, M., & van der Plicht, J. (1995). Interannual extremes in the rate of rise of atmospheric carbon dioxide since 1980. *Nature, 375*, 666–670.

Kojima, T., Nagamine, A., Ueno, N., & Uemiya, S. (1997). Absorption and fixation of carbon dioxide by rock weathering. In *Proceedings of the third international conference on carbon dioxide removal*, Cambridge, Massachusetts, September 9–11, 1996. *Energy and Conservation Management, 38*(Suppl.), S461–S466.

Lackner, K. S., Wendt, C. H., Butt, D. P., Sharp, D. H., & Joyce, E. L. (1995). Carbon dioxide disposal in carbonate minerals. *Energy (Oxford), 20*(11), 1153–1170.

Lindeberg, E. (2003). The quality of a CO_2 repository: What is the sufficient retention time of CO_2 stored underground. In J. Gale & Y. Kaya (Eds.), *Proceedings of greenhouse gas control technologies 6th international conference (GHGT-6)* (Vol. 1, pp. 255–260). Amsterdam: Elsevier.

Minchener, A. J., & McMullan, J. T. (2007). IEA Coal Research Ltd, Clean Coal Technology, London.

Myer, L. (2003). Sensitivity and cost of monitoring geologic sequestration using geophysics. In J. Gale & Y. Kaya (Eds.), *Proceedings of greenhouse gas control technologies 6th international conference (GHGT-6)* (Vol. 1, pp. 377–382). Amsterdam: Elsevier.

O'Connor, W. K. (1998). Investigations into carbon dioxide sequestration by direct mineral carbonation. Presentation at second meeting of mineral sequestration working group, November 3. Albany, Oregon: Albany Research Center.

O'Connor, W. K., Dahlin, D. C., Nilsen, D. N., Walters, R. P., & Turner, P. C. (2000). Carbon dioxide sequestration by direct mineral carbonation with carbonic acid. Presentation at 25th international technical conference on coal utilization & fuel systems, March 7, Clearwater, FL.

Oldenburg, C., et al. (2001). Process modeling of CO_2 injection into natural gas reservoirs for carbon sequestration and enhanced gas recovery. *Energy & Fuels, 15*(2), 293–298.

Pacala, S. (2003). Global constraints on reservoir leakage. In J. Gale & Y. Kaya (Eds.), *Proceedings of greenhouse gas control technologies 6th international conference (GHGT-6)* (Vol. 1, pp. 267–272). Amsterdam: Elsevier.

Plasynski, S. I., Bose, C. B., Bergman, P. D., Dorchak, T. P., Hyman, D. M., Loh, H. P., et al. (1999). Carbon mitigation: A holistic approach to the issue. Paper presented at the 24th intl. tech. conf. on coal utilization and fuel systems, March 8–11, Clearwater, FL.
Rubin, E., & Rao, A. (2003). Uncertainties in CO_2 capture and sequestration costs. In J. Gale & Y. Kaya (Eds.), *Proceedings of greenhouse gas control technologies 6th international conference (GHGT-6)* (Vol. 1, pp. 1119–1124). Amsterdam: Elsevier.
Seifritz, W. (1990). CO_2 disposal by means of silicates. *Nature, 345*, 486.
Siegenthaler, U., & Oeschger, H. (1987). Biospheric CO_2 emissions during the past 200 years reconstructed by deconvolution of ice core data. *Tellus, 39B*, 140–154.
Simbeck, D. (2004). CO_2 capture economics. Prepared for the 10-50 solution: Technologies and policies for a low-carbon future, Pew Center on Global Climate Change and the National Commission on Energy Policy, March 25–26, Washington, DC.
Solomon, S. (2006). *Carbon dioxide storage: Geological security and environmental issues—Case study on the Sleipner gas field in Norway*. Oslo, Norway: The Bellona Foundation. http://www.bellona.no/artikler/notat_solomon.
Stangeland, A. (2007). A model for the CO_2 capture potential. *International Journal of Greenhouse Gas Control, 1*, 418. http://www.bellona.no/artikler/notater_stangeland_solomon.
Stevens, S., Fox, C. E., & Melzer, L. S. (2001). Sequestration of CO in depleted oil and gas fields: Global 2 capacity, costs, and barriers. In D. J. Williams, R. A. Durie, P. McMullan, C. A. J. Paulson, & A. Y. Smith (Eds.), *Proceedings of greenhouse gas control technologies 5th international conference (GHGT-5)* (pp. 278–283).
Stevens, S., Rigg, A. J., Pearce J. M. (1999). CO_2 sequestration in deep coal seams: Pilot results and worldwide potential. In B. Eliasson, P. Riemer, & A. Wokaun (Eds.), *Proceedings of greenhouse gas control technologies 4th international conference (GHGT-4)* (pp. 175–180).
TKI. (2009). The Turkish Ministry of Energy, Energy, Dept., Lignite Coal Report.
TTK. (2009). The Turkish Ministry of Energy, Energy, Dept., Hard Coal Report.
Zweigel, P., Hamborg, M., Arts, R., Loethe, A., Sylta, F., & Tomeras, A. (2000). Prediction of migration of CO2 injected into and underground depository: Reservoir geology and reservoir modeling in the Sleipner Case (North Sea). In D. J. Williams, R. A. Durie, P. McMullan, C. A. J. Paulson & A. Y. Smith (Eds.), *Proceedings of greenhouse gas control technologies 5th international conference (GHGT-5)* (pp. 360–365), Cairn, August 13–16, 2000.

Chapter 45
Geo.: Gas Production in Offshore Reservoirs in Brazil's Pre-salt Region

Isabella Vaz Leal da Costa, Pedro Rochedo, Mariana Império, Alexandre Salem Szklo, and Roberto Schaeffer

Introduction

The expectations regarding the oil sector in Brazil are very promising, both by the extent of the sedimentary basins of interest for hydrocarbon exploration untapped (about 7 million square kilometers), but also because of the recent discoveries in the pre-salt region (BNDES 2010; Wood Mackenzie 2013). Petrobras estimates that there are between four billion and eight billion recoverable barrels in Lula field and four billion barrels in Iara field. Besides these, there are the Jubarte field (Campos Basin) and (the ring-fence) Golfinho in the Espírito Santo Basin. The estimates of the total reserves in the pre-salt region (between 70 billion and 100 billion barrels), if confirmed, make Brazil a oil exporting country in the next few years (BNDES 2008; ANP 2014; Wood Mackenzie 2013).

However, we must also consider the difficulties linked to the pre-salt, mainly because of its geographical location. The pre-salt reserves are located at a depth of over 6000 m, with about 2000 m corresponding to a layer of salt. So there are a lot of difficulties to extract the hydrocarbons deposited in the pre-salt layer, requiring special materials and equipment, with technologies that aren't fully developed yet (Petrobras 2014). Some of these difficulties are showed below:

- Going through a thick layer of salt in deep waters is not a trivial task;
- Plasticity and the solubility of the pre-salt layer will require new technologies to maintain the well stability;

I.V.L. da Costa (✉) • P. Rochedo • M. Império • A.S. Szklo • R. Schaeffer
Energy Planning Program, Graduate School of Engineering, Centro de Tecnologia, Federal University of Rio de Janeiro, Bloco C, sala 211, Cidade Universitária, Ilha do Fundão, Rio de Janeiro 21941-972, Brazil
e-mail: isabella_costa@msn.com; rochedopedro@gmail.com; marianaimperio@poli.ufrj.br; szklo@ppe.ufrj.br; roberto@ppe.ufrj.br

P. Grammelis (ed.), *Energy, Transportation and Global Warming*, Green Energy and Technology, DOI 10.1007/978-3-319-30127-3_45

- Hydrocarbons are located in carbonate rocks, environment that is unknown by Petrobras activities because their experience is concentrated in other types of rocks such as sandstones;

And finally, the production occurs in a high temperature and pressure, at the presence of large CO_2 content between 10 % and 45 mol% (Formigli 2008; Dino 2014; Credit Suisse 2014), which may be stored in the same underground field.

Petrobras believes that these difficulties can be overcome in the following years (Petrobras 2011). In addition to the technological difficulties, there is the challenge of producing hydrocarbons at an average distance of 170 km from the Brazilian coast, distributing over a range of about 800 km of coastline extension of the states of Espírito Santo and Santa Catarina (BNDES 2010).

According to estimates made by Petrobras (2014), Brazil will become self-sufficient in oil products in the year 2020. In addition, the domestic supply of natural gas will increase from 41 million cubic meters per day in 2013 to 86 million m^3/day, in 2020. Imports of Bolivia will remain at 30 million cubic meters/day (contract expiring in 2019), plus the import via liquefied natural gas (LNG). By 2013, two LNG regasification plants were already in operation phase (27 million m^3 /day) and the additional third plant started operation in 2014 (14 million cubic meters/day) (Petrobras 2014).

The pre-salt oil fields contain large amounts of carbon dioxide (CO_2), which prevents the transportation of natural gas to the coast. CO_2, when in contact with water, forms acids such as carbonic acid, making it very corrosive that can cause damages to the pipelines and equipment (Nascimento 2012). The presence of CO_2 decreases the calorific value of natural gas, and helps the formation of hydrates at low temperatures and high pressures, which are precisely the flow conditions of the natural gas present in the pre-salt area (Nascimento 2012).

Thus, the need for separation of CO_2 from natural gas produced in the pre-salt becomes necessary to their production. The process of carbon capture and storage (carbon capture and storage—CCS) has been used worldwide by large oil and natural gas companies. In the oil extracting sector, this process is aimed precisely to separate CO_2 from the natural gas extracted. However, a high CO_2 content in the natural gas, reduces the amount of natural gas that can be produced. In the case of the pre-salt region, the Brazilian natural gas production is related to the amount of carbon dioxide present in the pre-salt fields.

So, this paper presents an estimate of the natural gas production in the pre-salt fields, drawn from a model developed in HYSYS software, considering the amount of CO_2 estimated for this region.

Carbon Capture and Geological Storage Review

Carbon Capture and Geological Storage (CCGS)

CO_2 Capture and Geological Storage systems (CCGS) are recognized worldwide as alternatives to reduce carbon dioxide emissions from stationary sources (IEA 2008; IPCC 2007; Rochedo 2011; Nogueira et al. 2014; Costa 2009, 2014). These practices are used in the global oil industry to reduce emissions, for example, in the offshore fields of exploitation in Norway (Sleipner Field) and onshore fields in Algeria (Costa 2009, 2014; Roddy 2011; Statoil 2013a).

The first capture demonstration plant that is operating on an industrial scale is the plant installed at Sleipner, Norway, located in the North Sea and operated by Statoil since 1996 (Statoil 2014a). This plant is also the main CCGS offshore project in the world. The Sleipner T platform was built especially for separating CO_2 from natural gas and has being able to separate one million tons of CO_2 per year. These platforms use chemical absorption technology with amine based solvents for the separation of gases, used in post-combustion (Rochedo 2011). It is worth mentioning that for the pre-salt fields, FPSOs (Floating Production, Storage, and Offloading) are being used. In the pre-salt area the conditions are more severe due to the distance from the coast and the depth, so it would be best not to use chemical absorption technology that has large equipment that can generate instability on the floating platforms.

The In Salah project in Algeria is another CO_2 capture project, conducted by Statoil, which uses the same separation technology that was already applied in the Sleipner field, based on the process of amines (Statoil 2014b). The project became operational in 2004, and contains a storage industrial scale. Several fields with a carbon dioxide content between 1 and 10 mol%, in order to adapt the natural gas to the specifications of 0.3 mol% of CO_2 present in the composition of the natural gas, are injecting CO_2 into deep saline formations, between 1850 and 1950 feet deep (Mathieson et al. 2010). By 2010 were injected over three million tons of carbon dioxide (Mathieson et al. 2010). However, the CO_2 injection was discontinued in 2011 as a safety measure, mainly related to reservoir capacity. New data and seismic results are being studied to create a more appropriate strategy for this reservoir injection (Statoil 2014b).

CO_2 Capture in the Pre-salt Fields

The CO_2 capture can be performed before (pre-combustion), or after the use of natural gas (post-combustion). This study is considering that the capture of CO_2 is directly made of associated gas extracted in the pre-salt fields (pre-combustion). It is important to mention that the capture of CO_2 from gas flaring and from gas combustion for power generation did not occur. It is, also, worth mentioning that is not possible to capture the total amount of CO_2 emitted by oil and gas production platforms. Here comes to light the concept of "capturable" CO_2. The "capturable" CO_2 is the portion of CO_2 emitted mainly in the oil and natural gas extraction due to generation of heat and/or burning fossil fuels, due to the industrial sector and due to the use of fossil fuels in plants for generation of electricity. In these situations it is feasible to separate CO_2 using at least one of the technological routes available. Technological routes includes pre-combustion, post-combustion and oxy-combustion. In addition, it is only captured about 85 % mol of "capturable" CO_2 due to technical limitations of existing methods of gas separation (Costa 2014; Roddy 2011; Rochedo 2011; Kuramochi et al. 2012).

The capture method most suitable for the new pre-salt platforms (FPSOs) would be the membranes (Dino 2014). Membranes are structures that act as selective barriers for molecules that are able to cross from one side to another. In the gas separation, membrane technology is based on chemical or physical interaction of the gas with the membrane which allows certain gases to go through the membrane at higher speeds than others.

Membrane performance was determined by the permeability and selectivity of the materials, which the membranes are made of. The permeability is defined by the ability to be penetrated with a certain driving force at once. Permeability is generally indicated by units such as barrer or flux and, selectivity by the ratio of the permeability of the gas components. The membrane separation takes place through different mechanisms, which have important influence on the selectivity and permeability.

Membranes are industrially applied in the form of flat sheets or hollow fibers. In the case of hollow fibers, the fibers are joined into a tubular beam. Membranes in flat form are usually produced as a sheet of spiral modules, where two membrane sheets are placed back-to-back, in the middle of a material which, ensures the minimum space required to keep the gas flow.

These models have extremely high surface-volume ratios and offer advantages related to costs and footprint (Schendel 1984).

Therefore, in the case of the platforms to be used in the exploitation of the pre-salt fields, the use of membranes is justified mainly due to its footprint, which is significantly smaller than the footprint of conventional absorption technology. A membrane module has a smaller footprint than a capture plant using, for example, the chemical absorption method. The space required to accommodate all the equipment used to capture by chemical absorption is not usually available on a FPSO. The layouts of the FPSO are well defined and restrict. Furthermore, this kind of platform is not able to deal with the structure instability by installing equipment for chemical absorption (Rochedo 2011; Dino 2014; Castelo Branco 2008).

After the capture method selection (membranes), the UOP Separex™ system, was chosen as a standard module for the separation of CO_2 in the pre-salt FPSOs (UOP 2013; Petrobras 2014; Echt and Meister 2009). According to UOP (2013), these modules were bought by PETROBRAS to separate CO_2 from seven million m^3 of natural gas per day in the pre-salt fields. In addition, eight modules will be installed on production platforms (FPSO) in the Santos Basin. It is estimated that a UOP Separex™ system (membrane + skid) module occupies an area of approximately 347 m^2 and weigh about 582 tons (Harness and Sharma 2012). This is a compact module suitable for oil and gas production platforms because it can reduce the occupied area for gas separation equipment in a platform (footprint) by 70 % (Harness and Sharma 2012).

According to Petrobras (2014) and UOP (2013), will be used the same type of membrane module in all FPSOs exploiting the pre-salt region. So, according to this information, in this study, we made the assumption that all membrane modules would be the same. Moreover, it is worth mentioning that the number of modules should vary depending on the daily production capacity of each FPSO.

Methodology

Oil and natural gas exploration and production in Brazil occurs mostly offshore and in deep waters. For the accomplishment of upstream sector analysis in Brazil, in this study, information about typical marine installations of oil and natural gas production located in Campos basin were used. These installations have electric power system, using their own produced natural gas as an input. Using typical installations was an assumption taken in this study due to the difficulty in the acquisition of process data about all marine installations in all Brazilians basins. Campos Basin represents more than 75 % of Brazilian production.

The methodology used to estimate pre-salt natural gas production was divided in three phases. Initially, it was adopted that pre-salt natural gas derives only from the associated gas. Thereby, natural gas production is linked to pre-salt oil production. Hence, the first phase was estimating oil production and number of platforms required in pre-salt. For this, the oil curve developed by Saraiva (2013), where a Multi-Hubbert modeling was made, was adopted. This modeling considers the total volume of ultimately recoverable resources (URR), according to different probabilities of adding reserves, to accomplish the oil production projection. Base year of 2010 has been updated with real oil production data (ANP et al. 2013a; Credit Suisse 2014) and then an interpolation was made to reach oil production projection made by Saraiva (2013) for 2015.

This Saraiva's (2013) result, with a production of 3.8 million bpd, is clearly above numbers that will be achieved in 2015, which probably will be around 2.8 million bpd (Petrobras 2014). However, the aim of this study is building a long-term scenario, where the current adversities undergone by Brazilian oil industry should, by hypothesis, be overcome. This evaluation is even reported by Goldemberg et al. (2014) projection, which clearly distinguishes the role of production arising out the oil under the onerous assignment tax regime, for which Petrobras has expended considerable resources.

In this sense, it's worth noting that Petrobras has been focusing its investments for pre-salt region (Petrobras 2013). Therefore, even with large production increases in pre-salt region, total production in 2012 and 2013 showed a decline, returning to 2010 levels (BDEP-ANP 2014; Reuters Brasil 2014). In 2012, for example, Roncador field showed a 27 % decrease in its production. In 2014, Brazilian production increased again (BDEP-ANP 2014), but it was still far below the post-salt region forecast developed by Saraiva (2013), which was used in this work.

As a result of oil production, an estimate for pre-salt platforms was also made, considering that new offshore facilities will be FPSO due to exploitation in deep and ultra-deep waters. All platforms implemented in the same year will have the same average capacity. This average capacity increases over time, due to increased production resulting from pre-salt, which requires larger platforms to process the estimated oil volumes for this region. Therefore, three different capacities platforms had been used, with a daily production of 100, 150 and 200 thousand barrels of oil.

Based on pre-salt oil production, the second phase was accomplishing the estimated natural gas gross production. According to IEA (2014), the extracted gas amount from the processing of a pre-salt barrel of oil is around 40 m^3. Total gas extracted comes from the multiplication of 40 m^3 of natural gas by the estimated number of platforms, considering the average capacity of each platform. Bypass is subtracted from extracted gas, resulting in natural gas gross production. Therefore, each platform will have its respective gross gas production average capacity. Since there is still much uncertainty about the precise amount of CO_2 present in pre-salt natural gas, a wide range was used, between 10 % mol of CO_2 and 45 % mol of CO_2, as proposed by the available scientific literature (Echt and Meister 2009; IEA 2014; Credit Suisse 2014). Thus, three classifications have been proposed for estimated natural gas quantities associated with pre-salt resources: (1) total amount of extracted gas; (2) natural gas gross production (extracted gas minus bypass); and (3) natural gas net production (gross production minus platform auto-consumption and energy penalty from membrane system used in CO_2 capture). Equations (45.1) and (45.2) refer to the amounts of natural gas represented by aforementioned classifications.

$$\text{NG GROSS PROD} = \text{EXTRACTED GAS} - \text{BYPASS} \tag{45.1}$$

$$\text{NG NET PROD} = \text{GROSS NG PROD} - (\text{AC} + \text{EP}) \tag{45.2}$$

Note: NG Gross Prod = Natural gas gross production (m^3); NG Net Prod = Natural gas net production (m^3);

$$\text{AC} = \text{Auto} - \text{consumption on platforms}\ \left(m^3\right); \quad \text{EP} = \text{energy penalty}\ \left(m^3\right).$$

Regarding the gross natural gas produced and bypass, they were calculated from a modeling made in HYSYS software, which took into account the membranes' properties used in UOP Separex™ membrane module and CO_2 content range considered in this study. Reinjection volumes and gas production for a single module were also obtained through HYSYS model, as well as the composition of each of these gases. These data are different for different CO_2 contents considered.

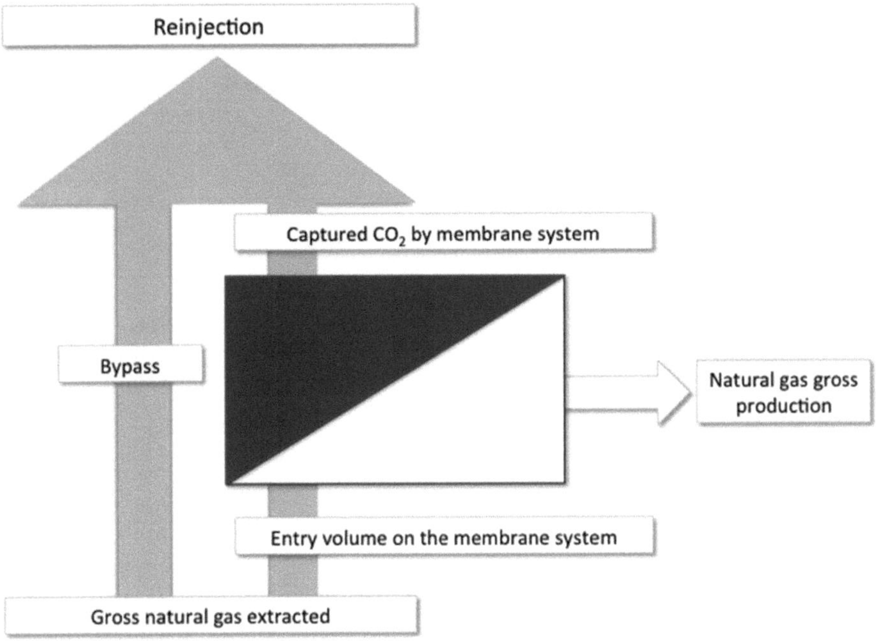

Fig. 45.1 Membrane functioning scheme

According to platform capacity, a gross extracted gas volume from the field may exceed membrane capacity. This volume does not pass by the membrane (bypass), and it is directly injected into the reservoir along with the gas rich in CO_2 captured in the membrane, as shown schematically in Fig. 45.1.

The third phase consists in estimating natural gas net production, which is the amount that actually can be transported to the coast and be available to natural gas processing units (NGPU). Energy penalty considered by using membranes was the electrical energy used by compressors for CO_2 reinjection in reservoirs. This energy was estimated based on compressor power, dimensioned by HYSYS® model, to process gross natural gas volumes on membranes for each platform capacity. Compressors powers are shown in Table 45.1.

Then, considering compressors operating 24 h a day throughout the year, energy that would be used by compressors is given in GWh and m^3 of natural gas. These values are calculated according to platforms estimates, their capacities and compressors powers. They are presented in Table 45.2.

Platform auto-consumption, refers to energy consumption in platform, which includes activities such as gas reinjection for oil recovery, transportation to pipelines, power generation for operational activities and accommodations in offshore platforms, heat and steam production, among others. Auto-consumption used in this study was 3.2 m^3/bbl, derived from oil and gas ratio and auto-consumption percentage (Maia 2007; ANP 2013b).

Table 45.1 Estimated compressor's powers on the membrane capture

Compressor power (MW)					
CO_2 content: 10 %			CO_2 content: 45 %		
100 kbpd	150 kbpd	200 kbpd	100 kbpd	150 kbpd	200 kbpd
4.62	9.24	9.24	11.14	22.28	22.28

Table 45.2 Consumption/energy penalty of membranes (compressors)

	CO_2 content: 10 %		CO_2 content: 45 %	
Year	Electric consumption (GWh)	Gas consumption (Mm^3)	Electric consumption (GWh)	Gas consumption (Mm^3)
2010	40.47	12.38	97.59	29.85
2015	323.77	99.02	780.69	238.77
2020	526.13	160.91	1268.62	388.00
2025	849.90	259.93	2049.31	626.76
2030	1254.61	383.71	3025.18	925.22
2035	1659.32	507.49	4001.04	1223.68
2040	1861.68	569.37	4488.97	1372.91
2045	1821.20	557.00	4391.39	1343.06
2050	1618.85	495.11	3903.46	1193.83

Then, with natural gas gross production and number of platform for each average capacity in operation, oil and gas ratio was calculated for each year, weighted by number of platforms in operation. Therefore, annual natural gas production was reached. To arrive at platform net production, energy auto-consumption and energy penalty related to membrane system were discounted from gross production. Auto-consumption was considered constant over time and equal for all average platform capacities.

Thus, for each CO_2 content and for each platform capacity considered, the required modules number, volumes captured in membrane, reinjected volumes (i.e., CO_2 captured and bypass) and pre-salt natural gas volumes produced, were possible to be calculated.

Results

Pre-salt Natural Gas Production

Natural gas gross production projections by typology of platform are presented in Tables 45.2 and 45.3, separated by respective CO_2 content in gross natural gas. CO_2 Content: 10 % mol of CO_2 (Table 45.4).

Table 45.3 Production and reinjection volumes of natural gas with content of 10 % mol of CO_2

Average capacity platform (k bpd)	Extracted gas (M m^3/day)	Membrane modules	Bypass (M m^3/day)	Natural gas gross production (M m^3/day)	
100	4.0	1	1	2.73	
150	6.0	2	0	5.46	
200	8.0	2	2	5.46	
Average capacity platform (k bpd)	**Reinjection (M m^3/day)**	**CO_2 content in reinjected gas**	**CO_2 capture on membrane (kt/day)**	**Capture tax**	
				%	**(tCO_2/bbl)**
100	1.37	26 %	0.66	89 %	0.0066
150	0.74	70 %	0.95	86 %	0.0064
200	2.74	26 %	1.31	89 %	0.0066

CO_2 content: 45 % mol of CO_2

Table 45.4 Production and reinjection volumes of natural gas with content of 45 % mol of CO_2

Average capacity platform (k bpd)	Extracted gas (M m^3/day)	Membrane modules	Bypass (M m^3/day)	Natural gas gross production (M m^3/day)	
100	4.0	1	1	1.82	
150	6.0	2	0	3.64	
200	8.0	2	2	3.64	
Average capacity platform (k bpd)	**Reinjection (M m^3/day)**	**CO_2 content on reinjected gas**	**CO_2 Capture on membrane (kt/day)**	**Capture tax**	
				%	**(tCO_2/bbl)**
100	2.28	75 %	3.15	96 %	0.0315
150	2.56	95 %	4.48	91 %	0.0299
200	4.56	75 %	6.30	96 %	0.0315

Thus, auto-consumption and energy penalty were deducted from natural gas gross production, resulting in a new projection curve of natural gas net production. A CO_2 content range mentioned above—10 % mol of CO_2 and 45 % mol of CO_2—for pre-salt was considered, so, a natural gas net production range was estimated.

Membrane area calculated by HYSYS model was 32,000 m^2. Below, Table 45.5 presents extracted gas estimates, gross and net natural gas production for pre-salt contents of 10 and 45 % mol of CO_2. Figure 45.2 shows the same results.

Table 45.5 Pre-salt natural gas production (10 and 45 % mol of CO_2)

Year	10 % mol CO_2			45 % mol CO_2		
	Extracted gas (M m^3)	Gross production (M m^3)	Net production (M m^3)	Extracted gas (M m^3)	Gross production (M m^3)	Net production (M m^3)
2010	1460	428	283	1460	286	123
2015	11,680	6994	4724	11,680	4663	2252
2020	18,250	11,569	7914	18,250	7713	3830
2025	27,010	19,117	13,481	27,010	12,744	6,742
2030	37,960	28,790	20,692	37,960	19,194	10,554
2035	48,910	38,534	27,992	48,910	25,689	14,431
2040	56,210	43,654	31,585	56,210	29,103	16,230
2045	54,750	44,409	32,441	54,750	29,606	16,852
2050	48,180	38,980	28,672	48,180	25,987	14,980

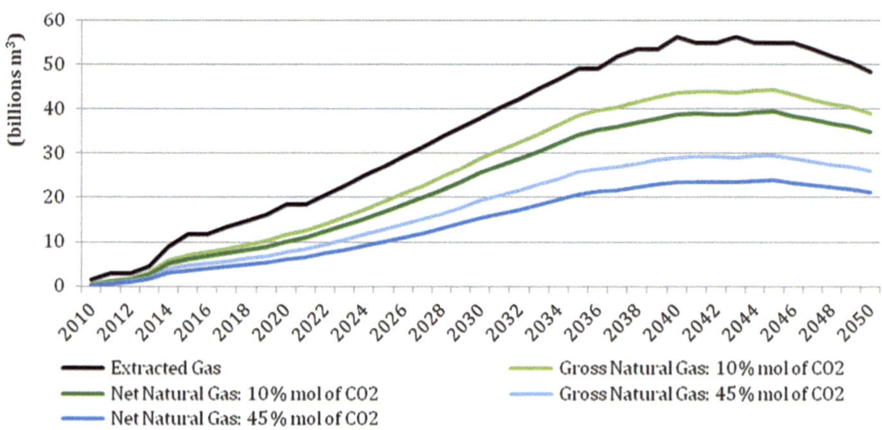

Fig. 45.2 Pre-salt natural gas production (10 and 45 % mol of CO_2)

Concluding Remarks

CO_2 content in reinjected gas is higher for 150 thousand bpd platforms. Its production volume coincides with capacity from two membrane modules, so it became possible to treat all extracted gas. In other words, there is no bypass, which would dilute CO_2 in reinjected gas and reduce hydrocarbon production. For 100 and 200 thousand bpd platforms there is a gas quantity, which is rich in natural gas that cannot be treated by membranes. Therefore, it is reinjected back into the reservoir. This wasted natural gas decreases the reinjected gas CO_2 content. For 45 % mol of CO_2 gas, 95 % CO_2 content in reinjected gas has been reached, which is configured as a very pure gas, justifying the membrane area calculated.

Regarding the capture taxes, when they are compared, it can be seen that 150 thousand bpd platform has a lower tax than the 100 and 200 thousand bpd.

This difference can be explained due to the passage through the membrane of all extracted gas from the well. Once in gas separation process there is a small CO_2 loss, capture tax of 150 thousand bpd platform is penalized. At the same time, because the other platforms have a volume of bypass, they directly reinjected the CO_2 volume from bypass, without any loss, which increases the capture tax.

As regards the capture, it occurs more efficiently in 45 % mol of CO_2 gross natural gas, naturally because of the increased presence of CO_2 in extracted gas. Therefore, there is both a greater CO_2 capture in membrane system, as a larger amount of carbon dioxide in gross natural gas that does not pass through the membrane, but is directly reinjected. So, for an extracted gas from pre-salt with higher CO_2 content, natural gas production that would be available for the market decreases, which can be seen from natural gas production curves estimates. Natural gas gross production is 50 % higher for gas with 10 % mol of CO_2 content, while net production (discounting energy penalty and auto-consumption) is approximately twice as large.

CCS implementation is a way to enable pre-salt exploration, as the associated natural gas contains large CO_2 amounts, estimated here between 10 and 45 % mol of CO_2, but can reach higher values. When in contact with water, CO_2 can produce acids that may cause wear on pipes and equipment. It can also reduce natural gas calorific value and it can favor the formation of hydrates at low temperatures and high pressures. Thus, if there is no separation of CO_2, it impedes natural gas transportation, and therefore, their commercialization. From CCS implementation and from separation of the largest portion of CO_2 contained in natural gas, natural gas production becomes technically feasible.

Analysis presented in this study is unprecedented and, therefore, in the absence of public data, lack of relevant historical pre-salt oil and natural gas production and complexity of the problem portrayed, many assumptions and hypotheses were adopted to make the presented simulations. The high degree of uncertainty that was considered here is clear and, thus, these simulations and their results could be improved in the future, with new available data.

References

ANP. (2013b). Anuário estatístico brasileiro do petróleo, gás natural e biocombustíveis. http://www.anp.gov.br/?pg=66833.

ANP, Agência Nacional do Petróleo, Gás Natural e Biocombustíveis. (2013a). *Dados Estatísticos Mensais*. http://www.anp.gov.br/?pg=56046&m=&t1=&t2=&t3=&t4=&ar=&ps=&cachebust=1311692440338.

BDEP-ANP. (2014). Base de dados. http://www.bdep.gov.br/?lng=br.

BNDES. (2008). Mercado de embarcações de apoio marítimo às plataformas de petróleo: oportunidades e desafios.

BNDES. (2010). Perspectivas e desafios no setor de petróleo e gás.

Castelo Branco, D. (2008). *Análise Técnica e Econômica da Aplicação da Tecnologia GTL de pequena escala para monetização do gás natural associado remoto offshore no Brasil*. Master Dissertation, Energy Planning Program—COPPE/UFRJ, Rio de Janeiro, RJ, Brasil.

Costa, I. V. L. (2009). *Análise do Potencial Técnico do Sequestro Geológico de CO_2 no Setor Petróleo no Brasil*. Master Dissertation, Energy Planning Program—COPPE/UFRJ, Rio de Janeiro, RJ, Brazil.

Costa, I. V. L. (2014). *Proposta de Estrutura Regulatória para Sequestro Geológico de CO_2 no Brasil e uma Aplicação para o Estado do Rio de Janeiro*. PhD Thesis, Energy Planning Program—COPPE/UFRJ, Rio de Janeiro, RJ, Brazil.

Credit Suisse. (2014). *Brazil Upstream Tracker*. In: Connections Series.

Dino, R., (2014). Personal communication. CENPES, PETROBRAS.

Echt, W., & Meister, P. (2009). Design, fabrication and startup of an offshore membrane CO_2 removal system. 88th Annual Convention, Gas Processors Association, P-28, Texas, EUA, Retrieved March 8–11, 2009.

Formigli, J. (2008). O pólo Pré-sal da Bacia de Santos—Desafios Tecnológicos para a área de Engenharia. Presented at Rio Oil and Gas, November/2008.

Goldemberg, J., Schaeffer, R., Szklo, A., & Lucchesi, R. (2014). Oil and natural gas prospects in South America: Can the petroleum industry pave the way for renewables in Brazil? *Energy Policy, 64*, 58–70.

Harness, J., & Sharma, B. (2012). UOP Separex TM membrane systems. *2012 AIChE Spring Meeting*, Houston, TX.

IEA. (2008). *CO_2 capture and storage—A key carbon abatement option*. International Energy Agency.

IEA. (2014). *Brazil Energy Outlook, in: World Energy Outlook 2013*. International Energy Agency.

IPCC. (2007). *Special Report on Carbon Dioxide Capture and Storage*. Cambridge: Cambridge University Press, 2005. Preparado pelo Grupo de Trabalho III do IPCC.

Kuramochi, T., Ramirez, A., Tukenburg, W., & Faaji, A. (2012). Comparative assessment of CO_2 capture technologies for carbon-intensive industrial processes". *Progress in Energy and Combustion Science, 38*, 87–112.

Maia, J. L. P. (2007). *Separação e captura do dióxido de carbono em instalações marítimas de produção de petróleo*. PhD Thesis. School of Engineering of the University of São Paulo, São Paulo, SP, Brazil.

Mathieson, A., Midgely, J., Wright, I., Saoula, N., & Ringrose, P. (2010). In Salah CO_2 Storage JIP: CO_2 sequestration monitoring and verification technologies applied at Krechba, Algeria. *Energy Procedia, 4*, 1063–1070.

Nascimento, J. T. (2012). Avaliação técnico-econômica de ultracentrífugas a gás para remoção de CO_2 em poços do pré-sal. Master Dissertation. Rio de Janeiro: School of Chemistry, Federal University of Rio de Janeiro.

Nogueira, L., Lucena, A., Rathmann, R., Rochedo, P., Szklo, A., & Schaeffer, R. (2014). Will thermal power plants with CCS play a role in Brazil's future electric power generation? *International Journal of Greenhouse Gas Control., 24*, 115–123.

Petrobras. (2011). *Energia e Tecnologia*. http://www.petrobras.com.br/pt/energia-e-tecnologia/.

Petrobras. (2013). Relatório de Sustentabilidade 2013. Disponível em http://www.investidorpetrobras.com.br/pt/plano-de-negocios-e-gestao/apresentacao-plano-estrategico-2030-e-png-2014-2018.htm. Acessado em 22/07/2014.

Petrobras. (2014). Plano Estratégico Petrobras 2030. http://www.investidorpetrobras.com.br/pt/plano-de-negocios-e-gestao/apresentacao-plano-estrategico-2030-e-png-2014-2018.htm.

Reuters Brasil. (2014). Petrobras perde participação na produção brasileira de petróleo e gás. http://br.reuters.com/article/topNews/idBRKBN0EE22V20140603.

Rochedo, P. R. R. (2011). *Análise Econômica sob Incerteza da Captura de Carbono em Termoelé tricas a Carvão: Retrofitting e Capture-Ready*. Dissertação de Mestrado, Programa de Planejamento Energético—COPPE/UFRJ, Rio de Janeiro, RJ, Brasil.

Roddy, D. (2011). Development of a CO_2 network for industrial emissions. *Applied Energy, 91*(1), 459–465.

Saraiva, T. A. (2013). *Uma Aplicação de um Modelo Multi-Hubbert Modificado para a Elaboração de Cenários de Produção de Petróleo no Brasil*, Dissertação de Mestrado, Programa de Planejamento Energético—COPPE/UFRJ, Rio de Janeiro, RJ, Brasil.

Schendel, R., (1984). Gas separation membranes and the gas industry. Apresentação em PCGA Transmission Conference. San Jose, CA.

Statoil. (2013a). http://engeprojnews.blogspot.com.br/2013/10/armazenamento-geologico-de-co2-eor-no.html.

Statoil. (2014a). Sleipner Vest. http://www.statoil.com/en/EnvironmentSociety/Sustainability/2007/Environment/Climate/CarbonCapture/Capture/Pages/SleipnerVest.aspx.

Statoil. (2014b). In Salah. http://www.statoil.com/en/technologyinnovation/newenergy/co2capturestorage/pages/insalah.aspx.

UOP. (2013). http://www.uop.com/?press_release=honeywells-uop-technology-selected-by-petrobras-for-natural-gas-processing-in-brazil.

Wood Mackenzie. (2013). *Brazil Country Overview*. Report on oil and gas production areas.

Part XII
Environmental Technologies Related to Global Warming

Water Resources and Management Issues

Water management issues reflected on specific test cases

Chapter 46
Spatial and Temporal Patterns of the Water Quality in the Hammam Boughrara Reservoir in Algeria

Belkheir DJELITA, Souaad Bouzid-Lagha, and Kheira Camellia NEHAR

Introduction

Algeria is one of the countries where water availability is limited while the needs are growing strongly. This growth is due to the population explosion, the industrial and agricultural development as well as drought. To address this shortage, dam construction seems to be the most suitable solution to meet different needs. If these structures constitute a necessity to ensure, in any season, the water supply essential for our country, it should control and protect the quality of water dams against various sources of pollution, particularly those due to human activities, namely agricultural practices, industrial and urban waste. Note that these discharges loaded pollutants often lead to the natural environment without any treatment. Added to this, the soil erosion natural processes are likely to bring different elements that may be the cause of the deterioration of the water quality. One consequence of this is the enrichment of water by nutrients, especially phosphorus and nitrogen, indicators of an advanced eutrophication of water status (El-Ghachtoul et al. 2005).

Eutrophication is an increase in the fertilization of the lake water by a contribution of nutrients favoring their proliferation of phytoplankton and aquatic plants. Gradually, this process accelerates sedimentation: the lake is shrinking, is closing and finally disappears (Ramade 1982). This phenomenon was classified in 1997 by

B. DJELITA (✉) • S. Bouzid-Lagha
Laboratory of Environment, Water, of Geomechanics and Structures (LEEGO), Faculty of Civil Engeniering (FCE), University of Science and Technology Houari Boumediene (USTHB), BP 32 El-Allia, Bab-Ezzouar, Algiers 16111, Algeria
e-mail: belkheirdjelita@yahoo.fr; bouzidsouad@yahoo.fr

K.C. NEHAR
University of Zian Achour, Djelfa BP 3117, Algeria
e-mail: camellia90@hotmail.fr

P. Grammelis (ed.), *Energy, Transportation and Global Warming*, Green Energy and Technology, DOI 10.1007/978-3-319-30127-3_46

the International Lake Environment Committee (ILEC), the fourth of the six major risks that may affect the lakes and reservoirs of the planet.

The present study focuses on the dam Hammam Boughrara and aims to establish a diagnosis of the current trophic state of the lake to contribute to the knowledge of the evolution of this ecosystem. The first aim of this study is to assess the trophic status of the dam, including eutrophication seems to be an evolutionary process, natural, nutrient salts and therefore more and more rich in living organisms and organic matter. Indeed, the lakes are pitfalls to nutritive substances they are simultaneously victims. The Second is to determine the influence of water quality, by its physico-chemical properties, chemical and biological aquatic populations. As The third, it aims both to establish a better understanding of the quantitative relationship between the intake of nutrient materials and trophic response of our dam, the other to determine the impact of morphology, hydrology and physico-chemical characteristics on the latter.

Materials and Methods

Presentation of the Study Site

Hammam Boughrara Dam, located in the extreme northwest of the Wilaya of Tlemcen (West Algeria), belongs to the watershed area of river Mouillah, whose area (largely shared with Morocco) is 2000 km^2. This basin is bounded by a perimeter of 241 km and composed in its majority by Angad Plains (located in Oujda Moroccan territory) and that of Zrigua (located in Maghnia Algerian territory). The length of its main trough is 104.4 km (Fig. 46.1).

The study area is characterized by a semi-arid climate with mild winter, where the average annual rainfall is 325 mm, the average annual temperature is 18.5 °C and potential evapotranspiration of 1167 mm. Monthly liquid intake is highly variable with maximum averages in November (6.29 hm^3) and lowest in June (0.88 hm^3). Solid contributions are also variable, the highest values are observed in March with 0.15 MT and the lowest in July with 0.003 MT. From a geomorphological point of view, the catchment area of the river Mouillah is characterized by an elongated shape ($k = 1.5$), an index of overall slope $Ig = 7.5$ % and a concentration time $Tc = 12$ h, which means speeds low, favoring the settling of solids in streams. Therefore, the runoff reaches the watershed outlet, low in suspended solids.

On the geological basin, a terrain succession from Paleozoic to Quaternary has been identified. The ancient land, dominated by carbonate formations (limestones and dolomites very cracked and karstified) and sandstone are the reliefs. The depressions are covered with soft terrain represented by clays, gypsum marl, alluvium and silt.

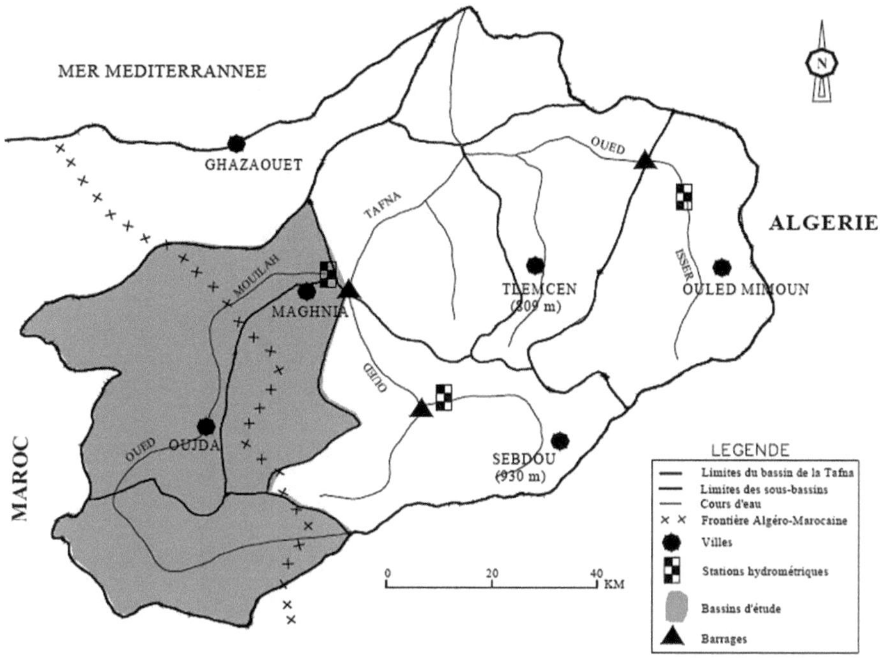

Fig. 46.1 Situation of the watershed of Oued Mouilah

Built at the confluence of oueds Tafna and Mouillah the dam Hammam Boughrara is intended mainly to meet the drinking water supply needs (AEP) of cities of Oran (33 hm^3) and Maghnia (17 hm^3) ; 09 hm^3 are provided for irrigation. It is characterized by a total capacity of 177 hm^3, volume 59 hm^3 regularized and a dead volume of 23.30 hm^3. The area of the lake is 984 ha. Liquid intake of Wadi Mouillah to the barrier Hammam Boughrara, as measured by the National Agency for Water Resources (ANRH) during the 1999–2008 period an average of 79 % of the total intake. The rest is provided by its tributaries. As for the operation of the dam, monitoring was conducted during the 1999–2008 period. The main results are shown in the following table (Table 46.1).

They testify to a renewal rates (or dilution) Water ($D = Q/V$) relatively small, rarely exceeding unity. Using the formula ($D = 1/Ts$) largely confirmed these results by order of values $0.39 < D < 1.14\ y^{-1}$. With Q: flow in m^3/y; V: lake volume in m^3. Ts values: mean residence time of the water in the dam (yr), are shown in Table 46.2.

Table 46.1 Situation of Dam Hammam Boughrara in operation (1999–2008)

Years	Pluviometry (mm)	Tributary (hm^3)	Consommation-AEP (hm^3)		Irrigation (hm^3)	Lacher V.FOND	Leak (hm^3)	Evaporation (hm^3)	Deffluent (hm^3)
			Oran	Maghnia					
			Oran	Maghnia					
1999–2000	215.0	11.593	–	–	8.958	–	0.010	3.663	12.612
2000–2001	238.6	36.651	8.888	–	9.626	–	0.014	4.681	23.208
2001–2002	308.6	43.917	39.75	–	9.273	0.324	0.030	4.569	53.946
2002–2003	293.6	38.15	31.98	–	1.916	0.027	0.036	4.173	38.132
2003–2004	346.6	48.371	39.378	–	4.409	0.360	0.036	3.863	48.011
2004–2005	229.3	15.202	6.044	0.342	–	1.023	0.036	3.441	12.196
2005–2006	189.6	38.449	6.909	7.008	–	0.540	0.036	5.093	19.643
2006–2007	253.0	32.325	1.992	6.490	–	0.343	0.045	6.252	15.122
2007–2008	203.0	30.12	4.465	3.210	–	0.087	0.026	3.834	16.422

Source: National Agency for Dams and Transfers (ANBT)

Table 46.2 Results of the application of Eqs. (46.1) and (46.2) of dam Hammam Boughrara

N°	(1)	(2)	(3)	(4)	(5)	(6)
Years	Z (m)	A_0 (m^2)	T_S (yr)	Lt(P) (t yr^{-1})	Ld(P) (t yr^{-1})	L. entré (t yr^{-1})
1999	10.6	2,538,983	2.32	0.37	0.49	6.34
2000	14.0	3,803,621	1.45	0.75	1.11	33.05
2001	15.3	4,097,382	1.52	0.82	1.23	14.21
2002	13.5	3,645,454	2.48	0.56	0.76	6.85
2003	12.6	3,397,891	1.18	0.70	1.06	21.73
2004	12.0	3,137,695	0.87	0.75	1.18	32.62
2005	11.0	2,813,500	1.19	0.54	0.80	35.89
2006	14.5	3,893,022	1.52	0.76	1.13	43.94
2007	18.8	4,790,995	2.57	0.83	1.18	31.42

Source of parameters Z, A0, Ts: ANRH & ANBT

Sources of Pollution

Domestic pollution

Generally, this pollution is from discharges domestic waste water of all cities and towns, located in the algerian part of the basin, namely the communes of Maghnia, Bougherara Hammam, Sidi Medjahed and Bouhlou. A wastewater treatment plant, with a total capacity of 150,000 population equivalents, was performed to treat wastewater from the city of Maghnia. after several tests conducted from June 1999 to January 2000, the station has gradually achieved acceptable results.

Industrial pollution

Industrial effluents comes from four companies:

The ENOF: Company nonferrous minerals and useful substances. It produces bleaching earth and bentonite foundry and used as raw materials: clay, and calcium carbonate ($CaCO_3$), and as secondary material: 98 % sulfuric acid (H_2SO_4) and sodium carbonate (Na_2CO_3). It rejects 500 m^3/day of water containing suspended particles of clay, metal and heavy metals. All of these releases are in the wadi Ouerdifou.

The ENCG: fatty substances Complex. It produces oil, soap and glycerin. It uses as raw material: crude oil, animal and vegetable fats, and as secondary materials: sodium hydroxide (NaOH), phosphoric acid (H_3PO_4), sodium chloride (NaCl) and ferric chloride ($FeCl_3$). The ENCG rejects 528 m^3/day of water containing crude oils, tallow and glycerin. Releases are in the wadi Abbes (tributary of the wadi Ouerdifou).

The ERIAD: Company food and derivatives industries. The unit produces products derived from corn. It uses as raw materials corn, and secondary materials such as: sulfur water at 0.2 % (SO_2), hydrochloric acid (HCl), sodium carbonate and sodium bisulfite. It rejects 600 m^3/day of water containing gluten in Wadi Abbes.

The ECVO: ceramic tableware company. It uses as raw materials: clay, kaolin, sand, water, calcite, dolomite, gypsum, mineral dyes and feldspars. It rejects 130 m^3/day of highly contaminated water with minerals in the wadi Abbes.

Agricultural pollution

Agricultural activities in the studied watershed are located mainly in the irrigated area of Maghnia and especially in the plain of Angad.

The major sources of pollution from these activities are: fertilizer application, soil treatment and plant well as livestock, especially poultry. Added to this, a potential source of pollution that can cause serious damage to health and the environment, in this case, storage of obsolete pesticides and maintenance of agricultural machinery (drain) well as fuel and lubricants

Pollution from the Moroccan side

This pollution is caused by urban and industrial wastewater from the Moroccan town of Oujda, discharged directly into the Oued Bounaim, the main tributary of the Oued Mouillah, without any treatment.

Data and Methods

Faced with the problem of eutrophication, assessment of water quality requires knowledge, as complete as possible, physicochemical and biological water. To this end, nineteen (19) parameters (temperature, pH, conductivity, Ca, Mg, Na, K, Cl, SO_4, HCO_3, NO_2, NO_3, PO_4, P_{tot}, O_2dissous, NH_4, TSS, BOD_5, COD) and biotic parameters (chlorophyll a, phytoplankton) were selected.

Monthly samples were taken at the surface, at a station located in the tower measures water intake dam over a period from 1999 to 2008. The samples used for assays of chemical parameters and chlorophyll (a) have been achieved through a bottle of overturning 2 L capacity. Various methods of analysis were used (Rodier et al. 2005). Thus, the temperature, the salinity and the pH were measured by a standard field probe WTW Multiline P3 PH/LF-SET. Dissolved oxygen was measured by the chemical method of Winkler. Chemical parameters, Ca, Mg, Na, K, Cl, HCO_3, NO_3, NO_2, NH_4, PO_4, and P_{tot} were analyzed by colorimetric methods using a spectrophotometer. The gravimetric method was used for the determination of SO_4. Chlorophyll (a) was measured by the fluorimetric method described by Neveux (1976). The analyzes were performed in the laboratories of the ANRH. The results were the fundamental database for this study.

Results and Interpretations

Study of Physicochemical and Biotic Parameters

The graphical representation of different results and reading of changes in physical and chemical parameters (temperature, conductivity, Ph, Ca, Mg, Na, K, Cl, SO_4, HCO_3, NO_2, NO_3, NH_4, PO_4, P_{tot}, O_2dissous, TSS, BOD_5, COD) imposes the following comments:

- the monthly temperatures range between 10 and 12.5 °C in winter and between 25 and 31 °C in summer. The annual average is around 18.5 °C.
- the electrical conductivity fluctuates between 977 and 2750 µs/cm and indicates a highly mineralized water. This high mineralization would be due, firstly, to leaching of the crossed fields and, secondly, to the pollution of waters from industrial, domestic and agricultural.
- the pH measured varying between 7.5 and 9.1, indicating low to moderately alkaline water. the value of 9.1 recorded in September 2004 coincided with high photosynthetic activity.

Graphical representations and interpretations of concentrations of dissolved oxygen (O_2), suspended solids (TSS) and biochemical and chemical oxygen demand (BOD_5, COD), are as follows (Fig. 46.2):

- The Summer dissolved oxygen levels, ranging from 1.3 to 23 mg/L and is abnormally proportional to temperature. This could be due, firstly, to the intense photosynthetic activity, on the other hand, to the exchange of oxygen between air and water, as our sampling station is located on the surface.
- The values of TSS, below 75 mg/L, show that the dam water is of acceptable quality, with the exception of those of June 2002 and April 2004 (200 and 175 mg/L) due to major floods.
- The BOD_5 is generally greater than 15 mg/L. These high values correspond to high organic load mainly from the food industry. As to the DCO concentrations vary between 19 and 110 mg/L. High values are associated with an increase in

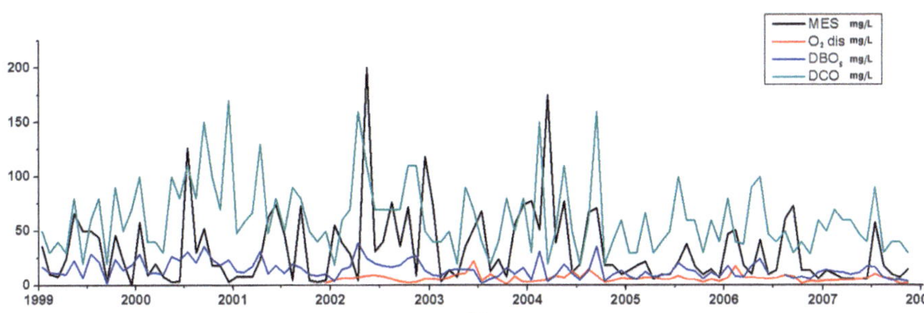

Fig. 46.2 Monthly evolution of dissolved oxygen, TSS, BOD_5 and COD (1999–2008)

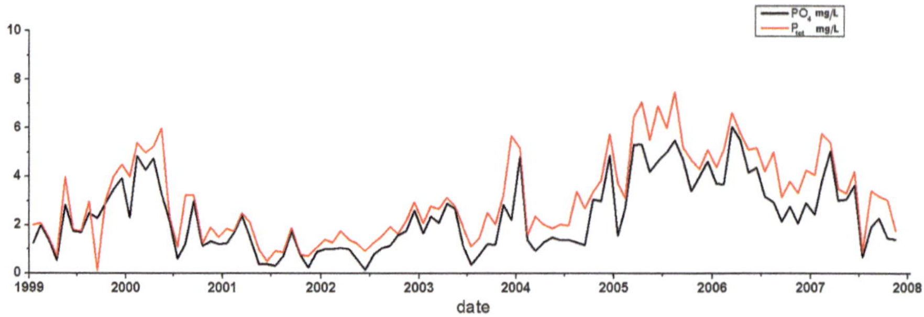

Fig. 46.3 Monthly variation in concentrations of total phosphorus and phosphates (1999–2008)

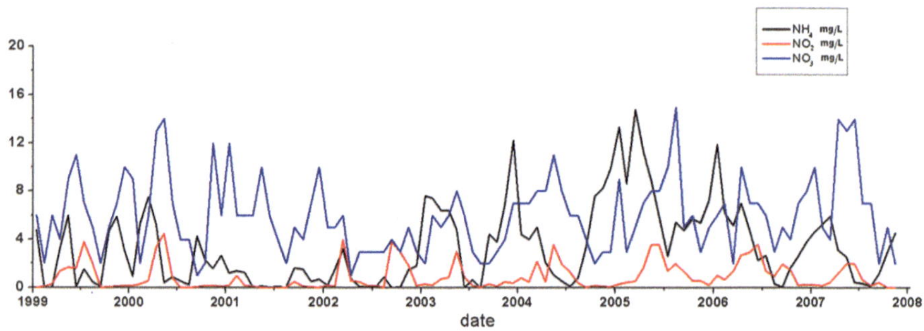

Fig. 46.4 Monthly variation in concentrations of ammonia Nitrogen, Nitrites and Nitrates (1999–2008)

the amount of organic material and the oxidizable substances. The BOD_5/COD ratio of between 0.2 and 0.6 indicates the possibility of a biological treatment.

- Total phosphorus contents (P_{tot}), much higher than 0.5 mg/L (Standard ANRH) indicate mediocre water quality at extremely polluted. They are due on the one hand, for industrial discharges including those containing detergents, secondly, to the phosphate fertilizers used in agriculture. It should be noted that the values of phosphate (PO_4) are consistent with those of total phosphorus and are always less, with an average value of 2.37 mg/L. This confirms the excessive pollution of water studied (Fig. 46.3).
- The concentrations of ammoniacal nitrogen (NH_4) are the highest recorded in January 2004, February 2005, April 2005, May 2005 and February 2006. They are 12.2, respectively; 13.3; 14.8; 11.35 and 11.9 mg/L.

Concentrations of nitrite (NO_2) vary between 0.01 and 4.5 mg/L.

These high values are due to industrial waste (ERIAD, corn mill and ENCG). The values of nitrate (NO_3) fluctuate between 1 and 15 mg/L (Fig. 46.4).

As to major cations and anions, we can see that:

- Values of chloride (Cl) showed a significant difference between the month and vary between 213 and 510 mg/L. which ranks our water in Class 6 characterized

by specific water more or less polluted. Their presence is exclusively related to discharges of ENCG (soap Workshop) dominated by sodium chloride (NaCl), used to activate saponification.
- The average concentrations of sulphates (SO_4) recorded is 155 mg/L and proves that these waters are heavily polluted or very selenitic. This is due to the increase in sulfur waste urban waste and especially industrial (ENOF unit with sulfuric acid rejection of 98 %).
- Bicarbonates range from 127 to 487 mg/L. They are due to the predominance of limestone in the watershed.
- The calcium contents (Ca) sometimes exceed 100 mg/L, which may be related to discharges of ENOF and the predominance of limestone in our site.
- Sodium values (Na) fluctuate between 113 and 416 mg/L. High levels recorded in the summer, would be mainly due to the scarcity of precipitation combined with increased evaporation.
- The potassium concentration (K) generally vary between 10 and 21 mg/L with peaks of up to 32 mg/L. They result from the contributions of groundwater and runoff. The concentration of magnesium (Mg) vary between 35 and 109 mg/L.
- The Values of chlorophyll "a" vary between 0.5 and 23.8 mg/L. The lowest recorded concentrations may be due to the decrease of the water temperature and the reduction of the light intensity. Those recorded in the summer would likely due to high levels of suspended solids that increase the turbidity of the medium and reduce, thus the photosynthetic activity. It should be noted, however, that these values are unusually low for the season. The peak values recorded are due to the elevation of the algal activity stimulated by environmental enrichment in nutrients including phosphorus and nitrogen.

Modelisation of Water Plane

In studies of eutrophication, a multitude of models have been developed and applied to various problems. We will concentrate on models of reservoir lakes. This is justified by the fact that these models are used for two distinct purposes, but closely related. They can, on one hand, help in the understanding of the processes that further determine the overall behavior of the lake and on the other hand, used to predict the response of the lake at a given change, such as a change in the burden of nutrients or liquids contributions. Thus, they are a very useful tool for the evaluation of eutrophication and planning and water resources management.

We considered the two main approaches to modeling (Riley 1946):

- The empirical approach: In empirical models (development of an equation from statistical correlations of data), system variables are related to the input variables without the detailed mechanisms that cause their relationships are known. The relationship between trophic status, average depth of the lake and phosphorus load, developed by Vollenweider (1968), is an example of this type of model.

- Conceptual approach: conceptual models (development of an expression on theoretical grounds and its subsequent combination with constants derived data) attempt to explain the interactions between different components of the system. As an example of simple models in this category include nutrient balance models, such as phosphorus, also developed by Vollenweider (1975, 1976a, b). The model gives a differential equation for the average concentrations of nutrients, based on a number of input parameters and the lake. In practice, the two approaches, empirical and conceptual, are usually combined.

Based on these considerations and in order to proceed to the modeling of the water dam Hammam Boughrara, we followed the following steps:

Estimate of Tolerable and Dangerous Loads of Phosphorus Concentration in the Dam Hammam Boughrara According Vollenweider

The equation of Vollenweider (1976b) permits to determine if phosphorus loading constitutes a risk of eutrophication, since the charges "tolerable" and "dangerous" a water body are set at 10 and 20 μg P/L, these loads are represented by the following equations:

$$Lt\ (P)\ = A_0 \left(100 + \frac{10Z}{TS}\right) \tag{46.1}$$

$$Ld\ (P)\ = A_0 \left(100 + \frac{20Z}{TS}\right) \tag{46.2}$$

Where: Lt(P) = tolerable loads; Ld(P) = dangerous loads; A_0: the area of the tank (m^2); TS: the mean residence time of the water in the tank (year); The entered. : Estimated annual phosphorus load into our dam (t/year); Z: the average depth of the container (m).

It should be noted that:

A0: was determined from the curve "height- capacity -surface" of the dam, established by ANBT collaboration with the research department FUGRO GEOID SAS-LEM (Laboratory for Maritime Studies), Algiers.

TS = V/Q = lake volume (m^3)/flow at the outlet (m^3 an^{-1})

The results of the application of these equations waters Boughrara dam are shown in the following table (Table 46.2).

The values of tolerable and dangerous loads calculated using the equation Vollenweider (1976b) (Table 2) show that the quantity of phosphorus in the dam is 9–45 times greater than the load "dangerous" set by this equation when these values were recorded in 2002 and 2005 respectively. The dam would therefore have the characteristics of a hyper-eutrophic lake.

It is emphasized that the application of these equations is not very reliable, knowing they have been established according to water bodies located in temperate zones at different trophic states; and that the dam studied (Hammam Boughrara) is located in a Mediterranean climate zone and receives too loaded water nutrient urban and industrial original features.

Simple Model of the Phosphorus Cycle (Vollenweider Model)

Vollenweider (1975, 1976a) has developed its models on the conceptual basis that the concentration of phosphorus in lakes, although largely determined by the rate of contributions is modified by losses to the sediments and in the flow of lake. Hence the new formula Vollenweider (1976a):

$$P = \frac{\left(\frac{L_p}{qs}\right)}{(1 + \sqrt{T_s})} \tag{46.3}$$

Where Ts = l/Γw, is the average residence time of water (yr).
Γw : annual frequency of renewal (yr^{-1}).

The OCDE Model, 1982

This model represents the equation (3) after being perfected on the base of the results of the international OCDE study (1982) and is expressed as follows:

$$[P]_\lambda = 1.55\left[\frac{[P]j}{(1 + \sqrt{T_s})}\right]^{0.82} \tag{46.4}$$

Where: $[P]_\lambda$: average annual concentration of total phosphorus in the lake (μg/L);
$[P]j = \frac{L_p}{qs}$: average concentration of phosphorus in annual water intake (μg/L).

Application of Models in the Dam of Hammam Boughrara

The data are based on the results of analyzes made during the period from 1999 to 2007 by the services of the ANRH (Table 46.3).

The results of the application of Eqs. (46.3) and (46.4) obtained from the data of Table 46.3 are summarized in Table 46.4.

The application of models, Vollenweider and OCDE, for the various annual loads gave values close and near each other, but below the actual measured concentrations.

Table 46.3 Recapitulation of data (1999–2007)

	1999	2000	2001	2002	2003	2004	2005	2006	2007
L: total annual intake of phosphorus (tonnes)	6.34	33.05	14.21	6.85	21.73	32.62	35.89	43.94	31.42
Average volume of the reserve (hm^3)	23.33	40.55	45.32	38.16	35.12	31.49	27.24	42.01	23.33
Surface water ($10^6 m^2$)	2.54	3.80	4.10	3.65	3.40	3.14	2.81	3.89	2.54
Liquid annual intake (hm^3)	11.59	36.65	41.25	19.85	36.14	43.28	25.97	37.08	35.11
Z: average depth of the dam (m)	10.60	14.00	15.30	13.50	12.60	12.00	11.00	14.50	18.80
Ts: Residence time	2.32	1.45	1.52	2.48	1.18	0.87	1.19	1.52	2.57
qs: Annual water load per unit area	4.57	9.64	10.07	5.44	10.63	13.79	9.23	9.53	7.33
Lp: annual charge of P per unit area	2495.7	8688.3	3469.2	1879.3	6396.4	103,96.9	127,55.4	112,85.7	6557.1
Coast in m NGA (**N**ational **G**eospatial-**I**ntelligence **A**gency)	277.75	283.26	284.47	282.62	281.76	280.64	279.21	283.64	287.95

Source: ANRH & ANBT

Table 46.4 Results of models application of Vollenweider (1976a) and of OCDE (1982)

Year	$Y = [P]_y$ (μg/L)	$P = \frac{\left(\frac{Lp}{qs}\right)}{\left(1 + \sqrt{T\ s}\right)}$ (μg/L)	$[P]_\lambda = 1,55\left[\frac{[P]_j}{\left(1+\sqrt{T\ s}\right)}\right]^{0,82}$ (μg/L)
1999	2186.3	216.59	127.51
2000	3606.6	408.85	214.69
2001	1378.3	154.33	96.57
2002	1380.8	134.08	86.05
2003	2405.8	287.99	161.07
2004	3015.0	389.97	206.53
2005	5528.3	660.70	318.22
2006	4739.1	530.40	265.77
2007	3579.1	343.93	186.31

The difference between the values calculated by these models and the measured values could be several explanations for namely:

- the hypotheses of the models suggest that the lake is a homogeneous reactor, there are no differences in concentrations in the water column. However, differences exist of concentration in the same column of water especially at the bottom. This is due to release of phosphorus at the sediment-water interface. These hypotheses therefore become difficult to apply to the dam studied.
- One of the conditions for the application of the two models assume constant inputs of phosphorus during the year. The observed irregular contributions do not verify this hypothesis.
- The difference in high phosphorus concentration would be anthropogenic. Urban and industrial waste undergoes primary treatment that degrades the organic materials and leave the water containing phosphorus.

Adaptation of the OCDE Model for Estimating Phosphorus Concentrations Dam Hammam Boughrara

The differences in phosphorus concentrations observed between the calculated and measured values can't be just a consequence of regional climate character (Mediterranean), it is appropriate to add the effect of human activity. Thus has established itself the adaptation of the model in question the reality of the waters of the dam studied. For this purpose we used a non-linear estimation method Rosenbrock and Quasi Newton, resolvable by "STATISTICA 6.1" software.

Remember that the OCDE equation is of the form:

$$Y = a^*X^b \tag{46.5}$$

knowing that: $Y = [P]_y$; $X = \left[\frac{[P]_J}{\left(1+\sqrt{TS}\right)}\right]$

Table 46.5 Results of the application of Eq. (46.6) Dam Hammam Boughrara

N°	(1)	(2)	(3)
Years	$Y = [P]_y$ (μg/L)	$X = \left[\frac{[P]_l}{(1+\sqrt{TS})}\right]$	$[P]_\lambda = 1,48*X^{1,28}$
1999	2186.36	216.59	1489.75
2000	3606.67	408.85	3371.86
2001	1378.33	154.33	963.58
2002	1380.83	134.08	804.14
2003	2405.83	287.99	2148.84
2004	3015.00	389.97	3172.96
2005	5528.33	660.70	6249.58
2006	4739.17	530.40	4711.90
2007	3579.17	343.93	2699.69

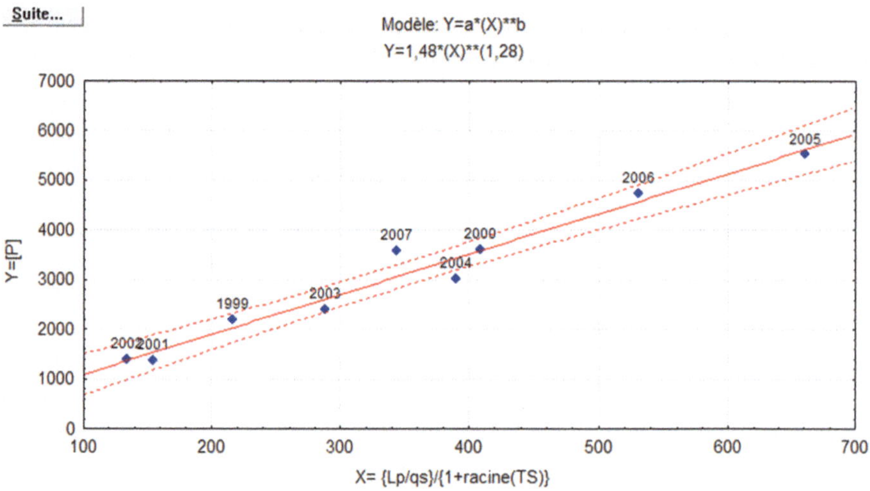

Fig. 46.5 Relationship between the intake and the Phosphorus [from Eq. (46.6)] at the dam Hammam Boughrara

We will try to find again the coefficients a and b, based on the measured values. The derived equation is of the form:

$$[P]_\lambda = 1,48*X^{1,28} \tag{46.6}$$

This equation permits to determine the values of concentration, average annual total phosphorus in the dam, from the average total phosphorus loads.

The results obtained are shown in column (3) of the following table (Table 46.5).

The majority of the points are within the confidence interval of 95 % with a correlation coefficient $R^2 = 0.87$, which shows that the model fits correctly (Fig. 46.5).

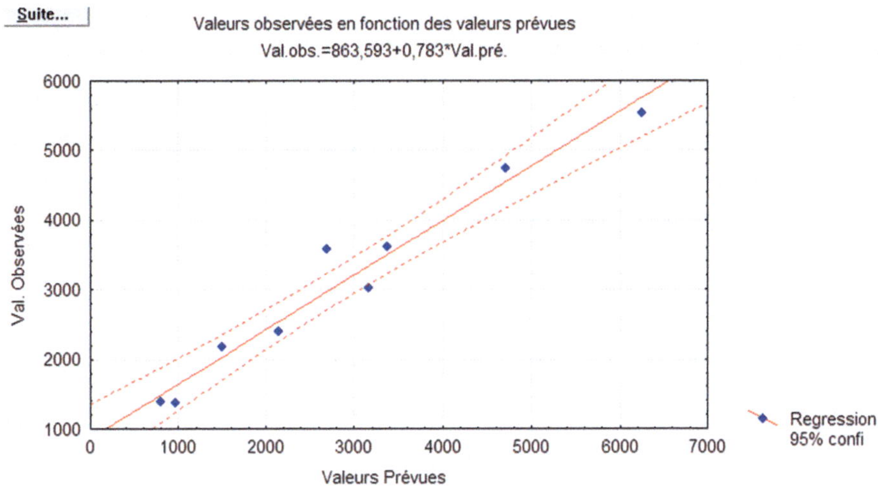

Fig. 46.6 The regression line Eq. (46.6), indicating the relationship between the observed and predicted values of the mean concentration phosphorus water dam Hammam Boughrara

The regression line, indicating the relationship of observed values based on theoretical values calculated by the equation (7) shows good linear fit as most points fall within the confidence interval of 95 % (Fig. 46.6).

Qualitative Predictions

The management of dam requires the mastery of water quality forecasting system. The open simple classification system is based on the probabilistic method, it allows to define five categories namely: ultra-oligotrophic, oligotrophic, mesotrophic, eutrophic and hyper-eutrophic.

This classification was developed from a large database of more than a hundred of water qualities and different trophic states (OCDE 1982). This system allows to state a prediction closer to the reality that facilitates the evaluation and management of eutrophication.

Figures 46.7, 46.8 and 46.9 illustrate the classification system and the recovery of values to different thresholds obtained by projection, concentration values of the different curves of trophic states then the probabilistic condition, in order to have the percentage of each trophic state.

The parameters presented in the following table (Table 46.6) characterize the probabilistic trophic status of the dam Hammam Boughrara. They are based on the figures of distribution of trophic states.

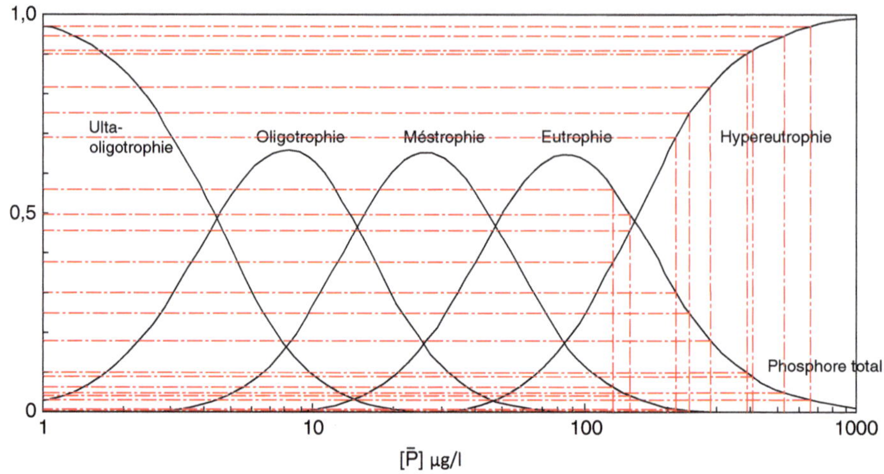

Fig. 46.7 Trophic state distribution function of the total average concentration of phosphorus dam Hammam Boughrara (according Vollenweider 1976a)

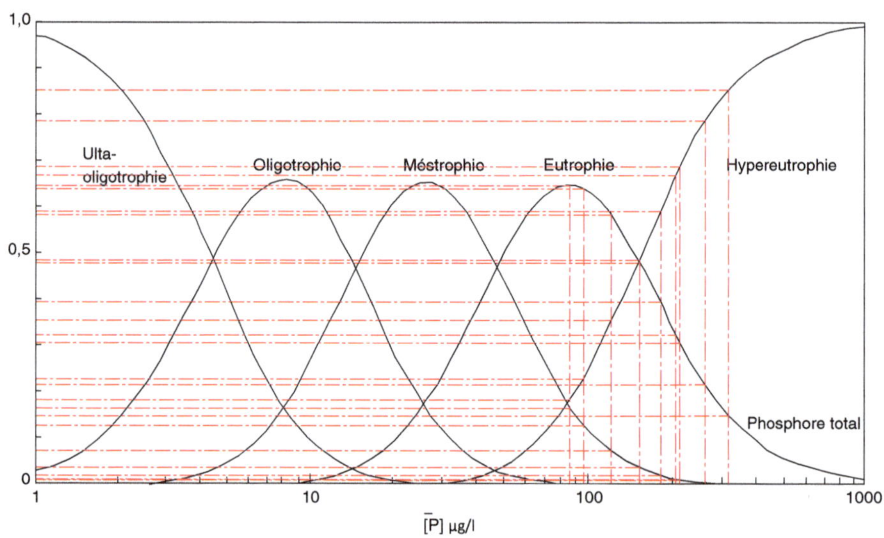

Fig. 46.8 Distribution of trophic states based on the total average concentration of phosphorus dam Hammam Boughrara (based on OCDE 1982)

From the open classification system based on the charts of the OECD (1982) and according to the results of the models Vollenweider and OECD applied to our lake, the probabilistic trophic status usually varies from eutrophic to hyper eutrophic. Otherwise, the results of appropriate equations show a state 100 % hyper eutrophic.

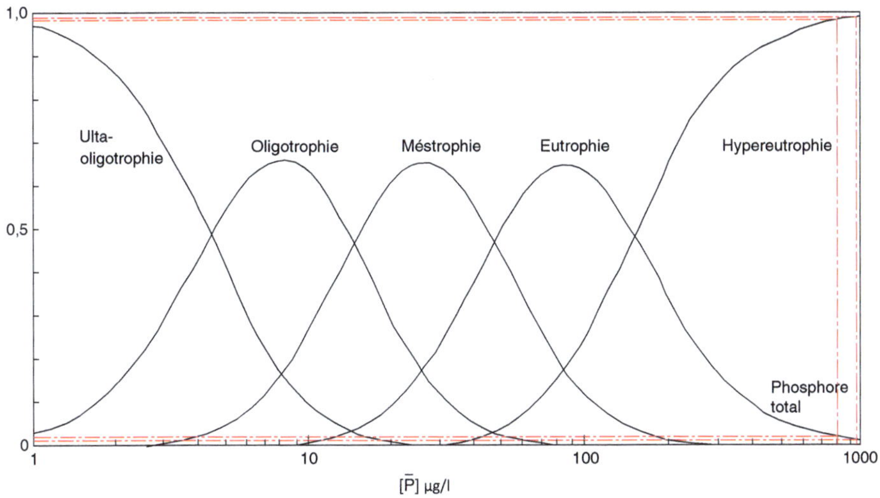

Fig. 46.9 Distribution of trophic states based on the total average concentration of phosphorus dam Hammam Boughrara (from Eq. 46.6)

Conclusion

The study of physico-chemical and biological characteristics of water collected in the dam Hammam Boughrara reveals high concentrations of certain parameters such as Cl, SO_4, NH_4, NO_2, Na, P, BOD_5 and COD, which characterize water highly mineralized and selenitic with conductivities ranging from 977 to 2750 µs/cm. These waters are rich in nutrients, such as phosphorus, and with a slightly alkaline pH (between 7.5 and 9.1) favor the proliferation of phytoplankton. This proliferation causes a decrease in transparency and dissolved oxygen are also signs of eutrophication. However, the BOD_5/COD ratio, ranging between 0.2 and 0.6, indicating the possibility of a biological treatment. The observed concentrations of K, Mg, HCO_3 result of a contribution by groundwater and leaching of carbonate terrains. The knowledge of the trophic status of a dam is not enough. To protect it, we must be able to predict the one hand, changing its state department overlooked time changes in phosphorus concentrations and their impacts on the ecosystem, on the other hand, better manage discards based on water use.

Thus, the application of models Vollenweider and OCDE annual loads to different phosphorus, gave close but lower than actual concentrations estimates. These models cannot be applied as such to our site, especially as climate and anthropogenic conditions on which these models are not verified in our case. Therefore, we adapted the parameters a and b of equation : $Y = a{*}X^b$ (OCDE 1982) and in our case, by using the non-linear estimation method Rosenbrock and Quasi Newton. The regression line obtained using the above mentioned equation, equipped with the new parameters a and b, shows good fit all the points with

Table 46.6 Probabilistic trophic states dam Hammam Boughrara within the parameters of simple classification system

Years		1999	2000	2001	2002	2003	2004	2005	2006	2007
Pt (μg/L) (Vollenweider 1976a) Eq. (46.3)	%	0 oligo. 1 méso. 30 eu 69 hyper	0 oligo 0 méso 09 eu 91 hyper	0 oligo 4 méso 50 eu 46 hyper	0 oligo 6 méso 56 eu 38 hyper	0 oligo 0 méso 18eu 82 hyper	0 oligo 0 méso 10 eu 90 hyper	0 oligo 0 méso 3 eu 97 hyper	0 oligo 0 méso 5 eu 95 hyper	0 oligo 0 méso 25 eu 75 hyper
Pt (μg/L) (OCDE 1982) Eq. (46.4)	%	0 oligo 7 méso 58 eu 35 hyper	0 oligo 1 méso 31 eu 68 hyper	0 oligo 13 méso 64 eu 23 hyper	0 oligo 16 méso 65 eu 18 hyper	0 oligo 4 méso 48 eu 48 hyper	0 oligo 1 méso 32 eu 67 hyper	0 oligo 0 méso 15 eu 85 hyper	0 oligo 0 méso 22 eu 78 hyper	0 oligo 2 méso 39 eu 59 hyper
Pt (μg/L) Eq. (46.6)	%	0 oligo 0 méso 0 eu 100 hyper	0 oligo 0 méso 0 eu 100 hyper	0 oligo 0 méso 1 eu 99 hyper	0 oligo 0 méso 2 eu 98 hyper	0 oligo 0 méso 0 eu 100 hyper	0 oligo 0 méso 0 eu 100 hyper	0 oligo 0 méso 0 eu 100 hyper	0 oligo 0 méso 0 eu 100 hyper	0 oligo 0 méso 0 eu 100 hyper

Oligo oligotrophic, *méso* mesotrophic, *eu* eutrophic, *hyper* hyper-eutrophic correlation coefficient R^2 equal to 0.87. The model thus obtained gives concentrations near the measured values. However, it would be interesting to test it on a longer observation period. The trophic status of the studied waters varies from eutrophic to hyper-eutrophic according to the models of Vollenweider and OECD. While the classification according to the adapted models is 100 % hyper eutrophic for phosphorus.

The probabilistic model obtained will contribute to better management of water quality in terms of trophic response. It allows to set specific goals for reducing phosphorus inputs depending on the intended use of the waters. Thus, it can be used in the context of sustainable development and decision support.

References

Agence Nationale des Ressources Hydriques (ANRH). (2008). *Bulletins mensuels de la qualité des eaux superficielles*. Algérie: Oran.

El Ghachtoul, Y., Alaoui Mhamidi, M., & Gabi, H. (2005) Eutrophisation des eaux des retenues des barrages Smir et Sehla (Maroc): Causes, conséquences et consignes de gestion. *Revue des sciences de l'eau/Journal of Water Science, 18*, 75–89.

ILEC. (1997). Water issue and Global Warming, International Lake Environment Committee Foundation. http://www.ilec.or.jp/database/Gl.html.

Neveux, J. (1976). Dosage de la chlorophylle a et de la phéophytine a par fluorimétrie. In Annales Institut Océanographique *52*, pp. 165–174.

OCDE (Organization for Economic Cooperation and Development). (1982). Eutrophication of Waters. Monitoring assessment and Control, Final Report. OCDE, Cooperative Programme on Monitoring of Inland Waters (Eutrophication control), *Environment Directorate, OCDE, Paris,* France.

Ramade, F. (1982). *Elément d'écologie: Ecologie Appliquée. Action de l'homme sur la biosphère*. Paris, France: Macgraw-Hill.

Riley, G. A. (1946). Factor controlling phytoplankton populations on Georges Bank. *Journal of Marine Research, 6*, 54–73.

Rodier, J., et al. (2005). *L'analyse de l'eau: eaux naturelles, eaux résiduaires et eau de mer* (8th ed.). Paris, France: Dunod.

Vollenweider, R. A. (1968). Scientific fundamentals of the eutrophication of lakes and flowing waters with particular reference to nitrogen and phosphorus as factors in eutrophication. *OECD Technical Report DAS, CSI, 68*, 27–54.

Vollenweider, R. A. (1975). Input-output models with special reference to the phosphorus loading concept in limnology. *Schweiz Journal of Hydrology, 3*, 53–84.

Vollenweider, R. A. (1976a). Advances in defining critical loading levels for phosphorus in eutrophication. *Memorie della Societa Entomologica Italiana, 33*, 53–83.

Vollenweider, R. A. (1976b). Rotsee, a source, not a sink for phosphorus? A comment to and a plea for nutrient balance studies. *Journal of Hydrology, 38*, 29–34.

Chapter 47
Natural Tracers for Identifying Causes of the Quality Reduction in Groundwater Emerging Along the Aegean Volcanic Arc (Greece)

E. Dotsika and P. Chantzi

Introduction

The Santorini and Nisyros volcano in Greece located along the Aegean volcanic arc, which developed during the Quaternary time as a result of a lithosphere subduction process (Le Pichon and Angelier 1979). The Aegean volcanic arc comprises, from west to east, Loutraki, Egina, Milos, Santorini, Kos and Nisyros. In the Aegean Sea, this arc (Fig. 47.1) represents the youngest example of volcanism and the only one that can be related to the zone of subduction. The morphological situation of such a geodynamic setting is found in the south (200 km) of the arc where the African and Eurasian plate's collision is in progress (Mckenzie 1970; Lort et al. 1974). The collision takes place from the Pliocene along the south Aegean island arc and is responsible of volcanism and seism city (Makris 1978). All the volcanic centers of the south Aegean islands are situated across the Benioff zone at depth from 130 to 150 km, suggesting that subduction controls magma generation. Several studies (Papazachos et al. 1995; Jolivet et al. 2013; Chatzipetros et al. 2013) have shown that tectonic activity is very vigorous in the Aegean region and all these areas and islands were affected by numerous tectonic events that are still active.

E. Dotsika (✉)
Institute of Geosciences and Earth Resources, Via G. Moruzzi 1, 56124 Pisa, Italy

Laboratories of Stable Isotopes and Radiocarbon, Institute of Nanoscience and Nanotechnology, National Centre for Scientific Research "Demokritos", 15310 Agia Paraskevi, Attica, Greece
e-mail: e.dotsika@inn.demokritos.gr

P. Chantzi
Laboratories of Stable Isotopes and Radiocarbon, Institute of Nanoscience and Nanotechnology, National Centre for Scientific Research "Demokritos", 15310 Agia Paraskevi, Attica, Greece
e-mail: p.chantzi@inn.demokritos.gr

P. Grammelis (ed.), *Energy, Transportation and Global Warming*, Green Energy and Technology, DOI 10.1007/978-3-319-30127-3_47

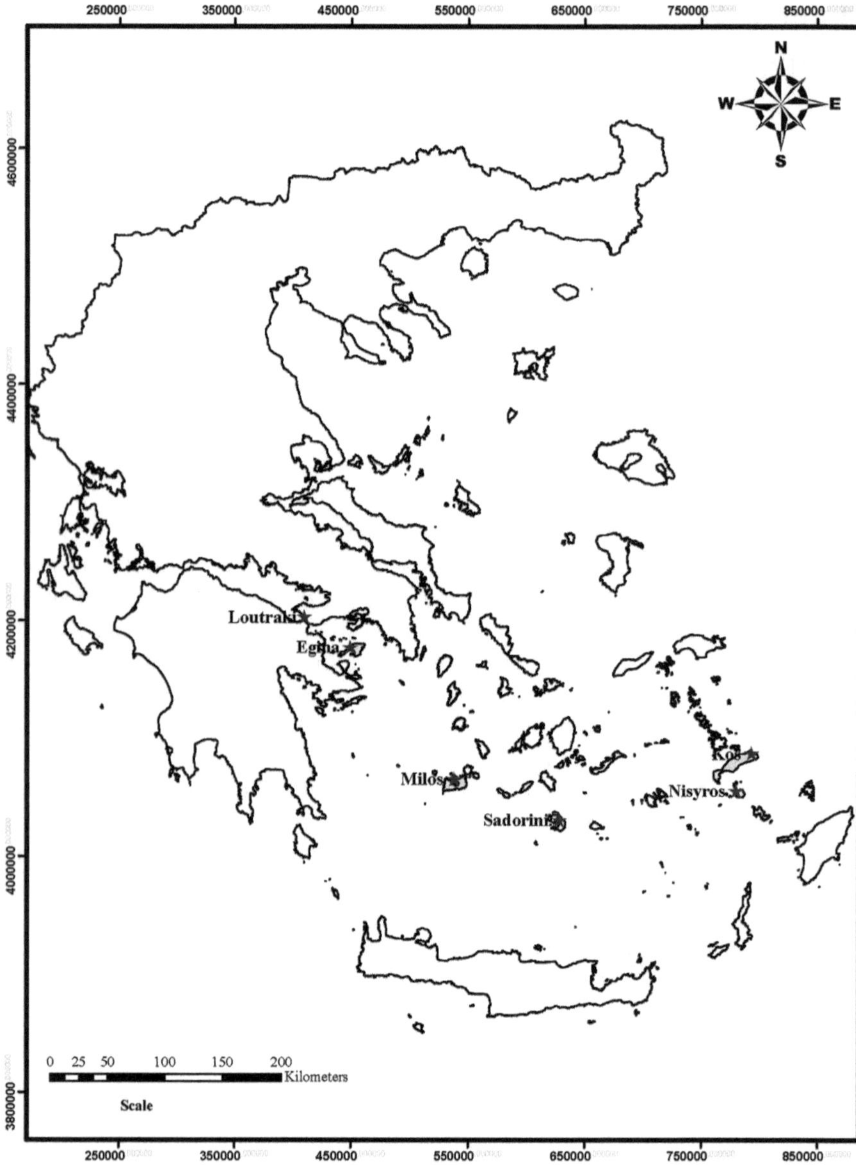

Fig. 47.1 Study areas of Egina, Milos, Santorini, Kos and Nisyros of the Aegean Volcanic arc

A significant number of thermal water springs emerge in the areas and islands of the Aegean arc. The temperature of these waters springs ranges from 30 to 53 °C. Also an important number of fresh water springs, with degrading quality, emerge in the vicinity of hot waters of these islands. These fresh water resources frequently suffer from severe salinisation problems that put in danger present and future utilization as well economic and social development of the concerned islands. The origin of this salinisation is very often attributed to the sea water intrusion. In the islands, marine water might enter the aquifer systems through different pathways e.g., atmospheric transport of marine aerosols, direct seawater ingression, particularly near the coastal zones, by seawater and modified seawater. However the vicinity of cold springs with the thermal fields suggests that part of the salinisation is due to the deep thermal water, coming from geothermal aquifers that intrude in the fresh aquifers.

The mobile constituent, chloride, is one of the most important geochemical tools used in the hydro geochemical investigations. The high chloride concentration, in most groundwater aquifers, in potable and irrigation waters is attributed to evaporative concentration, intrusion of seawater, dissolution of evaporates and human pollution. Although chloride can be added to deep aquifers by different sources the mixing processes could be identifiable. Additionally the correlation $\delta^{18}O$-$\delta^{2}H$ and mobile specie are particularly useful to investigate the mixing, because they provide information about the water sources and the different mixing processes that can operate between the sources. Especially, the $\delta^{18}O$-$\delta^{2}H$ isotope composition of groundwater is mainly affected by recharge altitude, water-rock interaction and mixing of different fluids. Additionally, the use of isotopes of sulphate $\delta^{34}S$, $\delta^{18}O$, consents to identify the origin of the dissolved sulphates and the mixing between sulphates of different origin. The ^{13}C values of bicarbonates ions in thermal and fresh waters are used to identify modification processes of equilibration between water and CO_2. The combined use of $\delta^{34}S$, $\delta^{18}O_{(SO4)}$, $\delta^{18}O$, $\delta^{2}H$, ^{13}C and Cl^- serves as tracers for identifying mixing processes in groundwater system and is used as an important tool for recognizing the causes of the quality reduction of coastal water systems.

In this study we used Cl^-, HCO_3^- and SO_4^{2-} with isotopes of water, carbonate and sulphate to identify the spatial distribution of pollution and its possible origin.

Sampling and Analysis

Seventy-six samples were collected for chemical (Cl^-, HCO_3^-, SO_4^{2-}) and isotopic analyses of water (^2H, ^{18}O), carbonate (^{13}C) and sulfate (^{34}S, ^{18}O). The isotopic composition of water was determined in all collected waters samples while the isotopic composition of carbonate and sulfate was determined in selected waters samples. $\delta^{13}C$ were also determined in two samples of carbonate rocks and ^{34}S were also measured in three samples of native sulfur from Santorini. Seventy one samples of water were collected for this study: 3 from Loutraki, 7 from Egina, 21 from Santorini, 19 from Kos, 15 from Milos, 7 from Nisyros and one from sea water. Temperature and HCO_3^- were measured directly in situ. Untreated samples were collected for analyses of Cl^- and SO_4^{2-}. The temperatures of water samples varied between 16 and 53 °C. Emissions of gases from these waters are abundant in Santorini, Nisyros, Kos and Milos. The major chemical constituents were analyzed with standard methods described in Apha (1989). The anions of the water samples were analyzed with ion chromatography. The $\delta^{18}O$ compositions of water in the samples were determined from CO_2 equilibrated with the water (Epstein and Mayeda 1953). The $\delta^{2}H$ compositions of water were determined from the H_2 generated by the Zn-reduction method (Coleman et al. 1982). The dissolved SO_4^{2-} in the water samples was precipitated as $BaSO_4$ for the analysis of S isotopic compositions (Rafter 1957). The dissolved HCO_3^{-} in the water, for the analysis of ^{13}C, is collected as $BaCO_3$ precipitates. Determination of the different isotope ratios carried out in Laboratories of Stable Isotopes and Radiocarbon of Nanoscience and Nanotechnology Institute in N.C.S.R. in Demokritos with the following precision: $\delta^{2}H \pm 1‰$; $\delta^{18}O \pm 0.2‰$; $\delta^{34}S \pm 1‰$; $\delta^{18}O_{(SO4^{2-})} \pm 0.3‰$; $\delta^{13}C \pm 0.5‰$.

Results and Discussion

The Origin of the Waters and the Sources of the Quality Reduction in Ground Water

The stable isotope contents show quite large variations of $\delta^{18}O$ ranging from –7.3‰ to 1.4‰ and δD from –47.4‰ to 6.5‰. The limits of maximum and minimum values were drawn by Nisyros and Loutraki respectively. In general, the water samples from thermal springs show an isotopic enrichment in relation to the fresh waters.

The stable isotopic composition ($\delta^{2}H$ *versus* $\delta^{18}O$) of all sampled waters from areas and islands of the Aegean Volcanic arc is shown in Fig. 47.2. At the same diagram, the global meteoric water line (GMWL) (Craig 1961) is also marked. Generally, an isotope relationship between $\delta^{2}H$ and $\delta^{18}O$ with a slope of about 8 is normal for fresh precipitation of all types (Craig 1961), as well as for surface waters not subjected to excessive evaporation relative to input. In Greece the correlation function for rainwater (IAEA 1981) is: $\delta D = 7.2 * \delta^{18}O + 9.3$. A regression equation similar to the above and slightly greater (7 and 7.5 respectively) is reported for the northern part of Eastern Macedonia and for the Eastern Peloponnesus (Leontiadis et al. 1984). A slope inferior to 7 is reported for Central Macedonia (Christodoulou et al. 1993), the southern part of Eastern Macedonia and Thrace (Leontiadis et al. 1996), as well as the northern part of Epirus (Leontiadis and Nikolaou 1999). The local meteoric line that correspond to Greece (LMWL) has a correlation equation $\delta^{2}H = 8.7\delta^{18}O + 19.5$ (Dotsika et al. 2009).

In Fig. 47.2, two linear correlations appear, apart from the water of mainly meteoric origin (Loutraki, Egina, Poros), between $\delta^{2}H$ and $\delta^{18}O$ suggesting a different origin for both elements. Line A, which expresses all the water samples coming from the islands, is an ideal line mixing between seawater and fresh water, suggesting that the isotopic data of waters are the result of a mixing at different degree of meteoric and seawater. This is the case mainly of Santorini Island. The

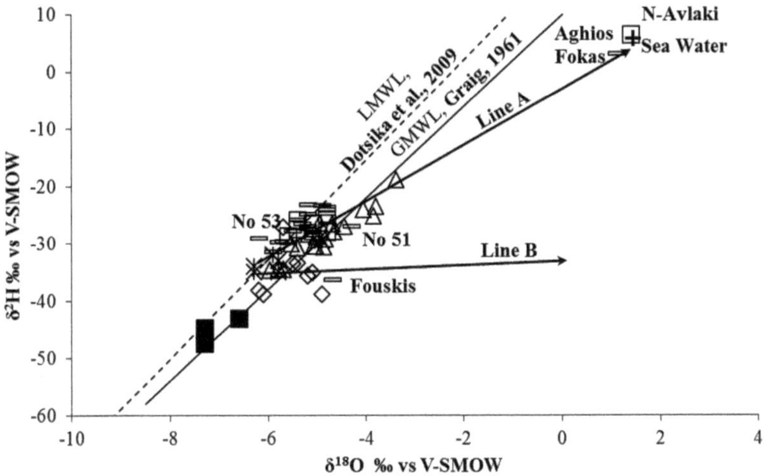

Fig. 47.2 The stable isotopic composition (δ^2H ‰ vs V-SMOW versus $\delta^{18}O$ ‰ vs V-SMOW) of all sampled waters from Egina, Milos, Santorini, Kos and Nisyros of the Aegean Volcanic arc. *Closed square*: Loutraki;✱: Egina; *open triangle*: Santorini; *open parallelogram*: Kos; *open diamond*: Milos; *open square*: Nisiros; *cross*: sea water

second line (B) groundwater, from Milos Island, presents an enrichment of $\delta^{18}O$ in relation to the $\delta^{18}O$ of meteoric waters of the area, while the δ^2H remains about stable. This change could suggest the intervention of modification processes. So, for better identification of the causes of the quality reduction in ground water we study separately the areas of Santorini, Kos and Nisyros islands, Loutraki-Egina and Milos Island.

The Case of Santorini, Kos and Nisyros Islands

Figure 47.3 shows the relationship δ^2H and $\delta^{18}O$ of all the sampled waters from Santorini, Kos and Nisyros islands. The waters from Santorini are distributed along a line joining local seawater with local fresh water, suggesting that the hot and cold waters result from a mixing at different degree of meteoric and seawater. These isotopic data confirm the sea intrusion in fresh aquifers of Santorini Island. In fact in Santorini a significant number of water boreholes was found, which presented high Cl^- and SO_4^{2-} values. Nevertheless, such brackish waters, with relatively high temperature, Cl^- and SO_4^{2-} contents, are used, in summer for irrigation and drinking purposes, and could therefore have an antagonistic effect on crop yield and health.

In the $\delta^{18}O$ versus Cl^- content (Fig. 47.3) diagram the samples from Santorini located along a straight line, which intercepts the x-axis at the $\delta^{18}O$ content of −5.2 ‰. This value corresponds to the mean content of the local meteoric waters as the

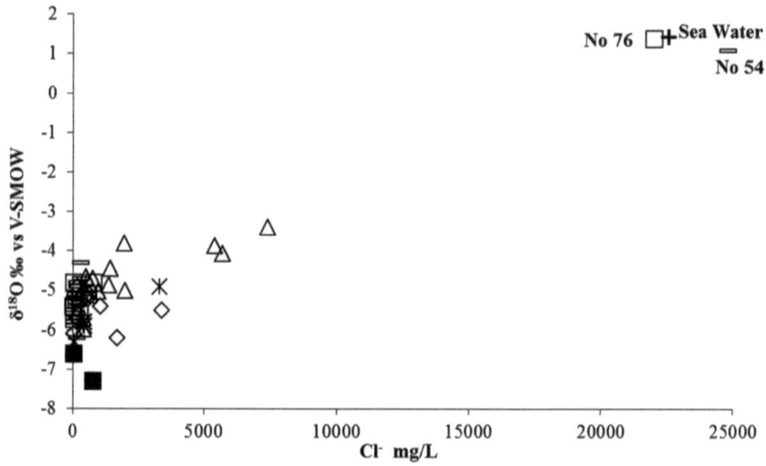

Fig. 47.3 The relationship between $\delta^{18}O$ ‰ vs V-SMOW and Cl^- (mg/L) content of Egina, Milos, Santorini, Kos and Nisyros of the Aegean Volcanic arc. Symbols as Fig. 47.2

rainwater samples collected from the area have been found to have similar value for the $\delta^{18}O$.

The position of sample N-Avlaki from Nisyros in the diagram δ^2H versus $\delta^{18}O$ (Fig. 47.2) is very close to that of mean seawater, confirming its purely marine origin. Other samples from Nisyros show very strong meteoric contributions. Similar positions, in the diagram δ^2H versus $\delta^{18}O$, to the one of the N-Avlaki sample show also the water from Kos (Aghios Fokas) suggesting its marine origin as well. In reverse the $\delta^{18}O$ and δ^2H contents of the other samples from Kos Island show a meteoric origin for these waters. However the dispersion of some data suggests intervention of modification processes. Figures 47.4 and 47.5 show the relationship between $\delta^{18}O$ versus SiO_2 (Fig. 47.4) and δ^2H versus SO_4^{2-} (Fig. 47.5). Despite the small number of samples the positive correlation between δ ^{18}O and SiO_2 (Fig. 47.4) is of major importance as concerns mainly the samples which present the major dispersion of $\delta^{18}O$ and δ^2H. This correlation, $\delta^{18}O$-SiO_2, could be explained from the equilibration between water-rock or/and water-CO_2 of deep origin. The augmentation of SiO_2 (and probably of HCO_3^-) could show the "aggressiveness" of the water correlated to the partial pressure of CO_2. In the diagram δ^2H versus SO_4^{2-} (Fig. 47.5) the enrichment of δ^2H content is accompanied by an increase of SO_4^{2-} content. This correlation could be attributed to the equilibration between water and H_2S of volcanic origin. Figure 47.2, which summarizes these modification processes, indicates the isotopic change of the fresh waters due to the modification processes of the equilibration between water and gases—CO_2 and H_2S—of deep origin. The isotopic values of the Kos meteoric water are −6.5‰ and −37‰ for the ^{18}O and ^2H respectively. So these two hypotheses, equilibration between H_2O-CO_2 and H_2O-H_2S could explain the $\delta^{18}O$ and δ^2H of all waters sampled in Kos Island (Fig. 47.2). In Fig. 47.2 the lines A and

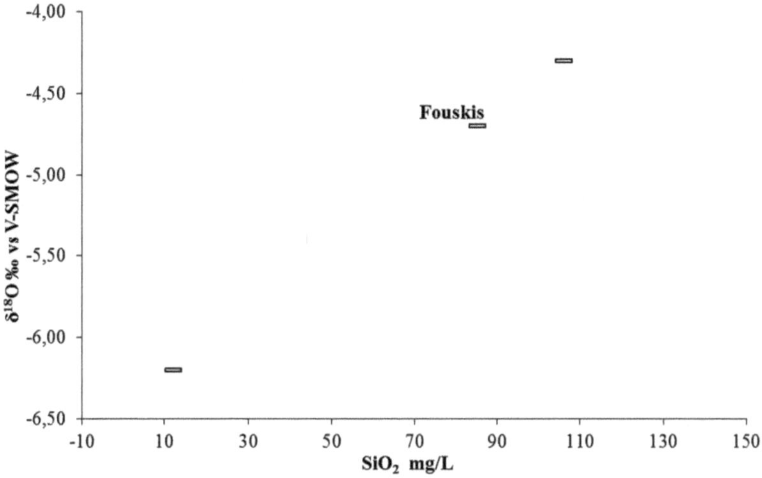

Fig. 47.4 Relationship between $\delta^{18}O$ ‰ vs V-SMOW versus SiO_2 (mg/L) of Santorini samples. Symbols as Fig. 47.2

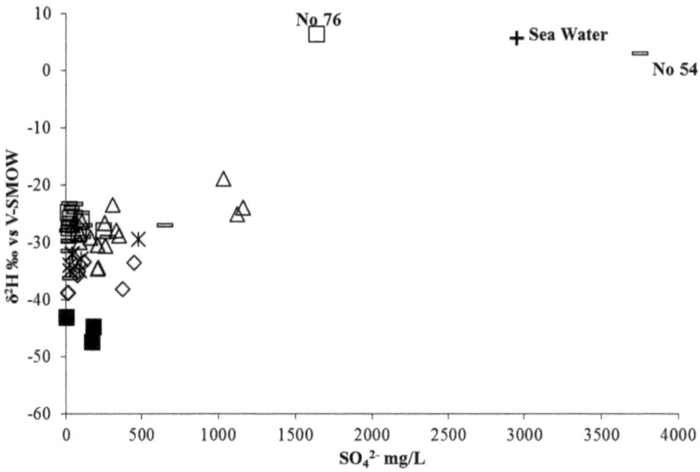

Fig. 47.5 Relationship between $\delta^{18}O$ ‰ vs V-SMOW versus SO_4^{2-} (mg/L) of Egina, Milos, Santorini, Kos and Nisyros of the Aegean Volcanic arc. Symbols as Fig. 47.2

B show the exchange with H_2S and CO_2. The intermediate area illustrates the exchange with both gases, H_2S and CO_2, of deep origin. The two-$\delta^{13}C$ values from the bicarbonate ions in the waters from Kos are −8.4‰ (Aghios Fokas) and 11.9‰ (Fouskis). The very positive $\delta^{13}C$ values of bicarbonates ions from Kos (Fouskis) confirm the equilibration between water and CO_2 of deep origin.

The negative $\delta^{13}C$ data, similarly as data from the Santorini samples (range from −13.7‰ to −8.6‰) probably suggest that the main part of dissolved CO_2 is of biogenic origin; even a small contribution of deep CO_2 cannot be excluded.

In Fig. 47.6 the ^{34}S and ^{18}O $_{(SO4)}$ contents of the samples, from Santorini, Nisyros and Kos, are compared with those of marine sulphate, which are very constant over the world [$\delta^{34}S = 20‰$ CD, $\delta^{18}O = 9.5$ ‰ SMOW (Longinelli 1989; Loyd 1968; Rafter and Mizutani 1967, Mizutani and Rafter 1969)]. In the same diagram, it is also reported the S-bearing sublimate of the Santorini Volcano (Nea Kameni). In total, ten water samples were analyzed from Santorini and three samples of native sulfur ($\delta^{34}S$ of native sulfur 7.9‰, 5.7‰ and 8.8‰) of the Santorini Volcano, three water samples from Kos, one from Nisyros and one seawater sample. It appears, in general, that the isotope contents of the samples are different from those of the marine sulphates. Only one sample from Kos and one from Nisyros do not differ significantly from those of seawater, suggesting that the dissolved sulphates of these waters are of purely marine origin. All the Santorini samples contain sulphate that is depleted in heavy isotopes relative to marine sulphate. The isotopic data suggest that these sulphates could come from the mixing between sulphates of marine origin and sulphates, probably, contained in the volcanic rocks of the island, which have been formed by the oxidation of the primary sulfur of deep origin. Michaud et al. 2000 has reported the presence of sulfur in various volcanic products of Santorini. In fact this $\delta^{34}S$ depletion was attributed to a mixing of marine sulphate with ‘light’ sulphate (Michelot et al. 1993), deriving from the oxidation of sulphur minerals. This contribution of sulphur oxidation is confirmed by the decrease of ^{34}S concentration and the increase of the sulphate contents in the water (Fig. 47.6). The very low $\delta^{34}S$ content of the Kos waters is also attributed to oxidation of primary sulfur of deep origin and defends the condition of existence of strong pressure of H_2S of volcanic origin.

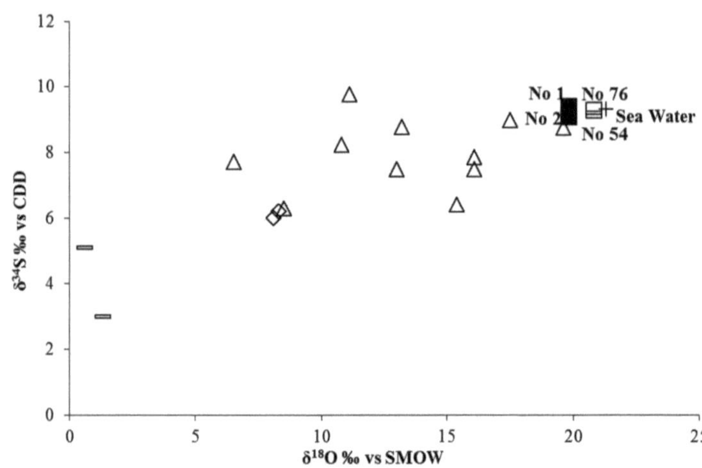

Fig. 47.6 Relationship between $\delta^{34}S$ ‰ vs CDD versus $\delta^{18}O$ (SO_4) vs SMOW of Egina, Milos, Santorini, Kos and Nisyros of the Aegean Volcanic arc. Symbols as Fig. 47.2

The Case of Loutraki and Egina Waters

Figure 47.2 shows the relationship between δ^2H and $\delta^{18}O$ contents of the water of Loutraki and Egina and the range of local meteoric water. The δ^2H and $\delta^{18}O$ values of Loutraki and Egina waters indicate a negligible contribution of seawater. The position of these samples in the diagram δ^2H versus $\delta^{18}O$ confirms their mainly meteoric origin. However in Fig. 47.6 the ^{34}S and $^{18}O_{(SO4)}$ contents of the Loutraki samples are compared to those of marine sulphate. It appears that the isotope contents of the samples do not differ significantly from those of seawater, suggesting that the dissolved sulphates of these waters are of purely marine origin. This marine participation is the cause for the augmentation of salinization.

The Case of Milos Waters

The δ^2H and $\delta^{18}O$ values of Milos's waters (Fig. 47.2) are shifted to the right of the local meteoric water line.

The isotopic data of Milos's water show an ^{18}O enrichment with respect to those observed for the local fresh water. In opposite strong meteoric contributions appear in the ^2H contents. These changes could suggest that the stable isotope contents of Milos's water are controlled by geothermal exchange. However an exchange between water-rock increases the $\delta^{18}O$, conserving "firm" the Cl^- contents. In Fig. 47.3 it is observed a modest enrichment in chloride, probably by mixing with a solution rich in heavy isotope. So the position of the samples in Fig. 47.3 could be attributed to the mixing between fresh water and deep thermal waters modified jointly by water-rock interaction and evaporation. Regardless of the origin of deep water the modification processes results the quality degradation of waters. The temperature increase and the salinisation of waters contribute to the unacceptable values of these waters according to the European standards.

In Fig. 47.6 the ^{34}S and $^{18}O_{(SO4)}$ contents of Milos's samples are compared to those of marine sulphate. It appears that the isotope contents of the samples, especially the Milos samples, differ significantly to those of seawater, suggesting that the dissolved sulphates of these waters come from oxidation of sulphur minerals. In fact the contribution of sulphur oxidation is confirmed by the decrease of ^{34}S concentration and the increase of the sulphate contents in the water.

Conclusions

Interpretation of groundwater geochemical data, including isotopic data of water, carbonate and sulphates, in terms of mixing and origin of solutes, and therefore origin of pollution, is often complicated by the non-conservative nature of dissolved components resulting from the mixing with gases, different type of fluid and from interactions with the aquifer matrix. These fluids are of different origin and, in particular, in groundwater emerging along the Aegean volcanic arc; result from actual seawater and water that vary by interactions with the aquifer matrix, by the exchange with the gases and the geothermal exchange. In the water emerging along the Aegean volcanic arc, salinisation enrichment is a problem to the quality of the intensely exploited groundwater, especially during the summer on account of tourist increase. In general, this salinisation enrichment in coastal areas conjoins with seawater intrusion. In fact the results of this study indicate that the waters examined from Loutraki, Egina are essentially of meteoric origin, but some of them are affected by a slight mixing with seawater. The seawater intrusion is the main cause for the quality reduction of Nisyros and Santorini Island.

However, the enrichment in HCO_3^- and SO_4^{2-} concentrations in groundwater with the parallel increase of water isotopes indicates that the salinisation enrichment in Kos Island is due to the intrusion of gases in the Island aquifers. Therefore the salinisation enrichment in coastal areas in Kos Island is due to the modifications processes of equilibration water- gases, CO_2 and H_2S, of deep origin.

The salinisation enrichment in groundwater in Milos, does not derive from a contamination source, but is due to mixing between local fresh water and deep water. In Milos Island the salinisation is controlled probably by mixing of fresh water with a solution rich in heavy isotope and geothermal exchange. In both cases the deuterium isotopes are applied in investigations on the origin of the salinisation in groundwater and the use of δ^2H is proved to be an effective approach for understanding the origin of quality reduction in geothermal areas.

In summary, for the water emerging along the Aegean volcanic arc, the salinisation enrichment is due to sea intrusion, exchange with gas, geothermal exchange and mixing of fresh water with a deep solution rich in heavy isotope.

References

Apha. (1989). *Standard methods for the examination of water and wastewater* (19th ed.). Washington, DC: American Public Health Association.

Christodoulou, T., Leontiadis, I. L., Morfis, A., Payne, B. R., & Tzimourtas, S. (1993). Isotope hydrology study of Axios River plain in northern Greece. *Journal of Hydrology, 146*, 391–404.

Chatzipetros, A., Kiratzi, A., Sboras, S., Zouros, N., & Pavlides, S. (2013). Active faulting in the north-eastern Aegean Sea Islands. *Tectonophysics, 597–598*(19), 106–122.

Coleman, M. L., Shepard, T. J., Durham, J. J., Rouse, J. E., & Moore, G. R. (1982). Reduction of water with zinc for hydrogen analysis. *Analytical Chemistry, 54*, 993–995.

Craig, H. (1961). Isotopic variations in meteoric waters. *Science, 133*, 1702–1703.

Dotsika, E., Lykoudis, S., & Poutoukis, D. (2009). Spatial distribution of the isotopic composition of precipitation and spring water in Greece. *Global and Planetary Change, 71*, 141–149.

Epstein, S., & Mayeda, T. (1953). Variation of ^{18}O content of waters from natural sources. *Geochimica et Cosmochimica Acta, 4*(5), 213–224.

International Atomic Energy Agency (IAEA). (1981). Statistical treatment of environmental isotope data in precipitation. IAEA Tech. Rep. Ser. No. 206. Vienna: IAEA.

Jolivet, L., Faccenna, C., Huet, B., Labrousse, L., Pourhiet, L., Lacombe, O., et al. (2013). Aegean tectonics: Strain localisation, slab tearing and trench retreat. *Tectonophysics, 597–598*(19), 1–33.

Leontiadis, I., Payne, B. R., Letsios, A., Papagianni, N., Kakarelis, D., & Chadjiagonakis, D. (1984). Isotope hydrology study of Kato Nevrokopi of drama. In *Proc. of the symposium on isotope hydrology 1983* (pp. 193–206). Vienna: International Atomic Energy Agency.

Leontiadis, I. L., Vergis, S., & Christodoulou, T. (1996). Isotope hydrology study of areas in Eastern Macedonia and Trace, Northern Greece. *Journal of Hydrology, 182*, 1–17.

Leontiadis, I. L., & Nikolaou, E. (1999). Environmental isotopes in determining groundwater flow systems, northern part of Epirus, Greece. *Hydrogeology Journal, 7*, 219–226.

Le Pichon, X., & Angelier, J. (1979). The hellenic arc and trench systems a key to the neotectonic evolution of the eastern Mediterranean area. *Tectonophysics, 60*, 1–42.

Longinelli, A. (1989). Oxygen-18 and Sulphur-34 in dissolved oceanic sulphate and phosphate. In P. Fritz & J. C. Fontes (Eds.), *Handbook of environmental isotope geochemistry* (Vol. 3, pp. 219–255). Amsterdam: Elsevier. Chapter 7.

Loyd, R. M. (1968). Oxygen isotope behaviour in the sulfate-water system. *Journal of Geophysical Research, 73*, 6099–6110.

Lort, J. M., Limond, W. Q., & Gray, F. (1974). Preliminary seismic studies in the eastern Mediterranean. *Earth and Planetary Science Letters, 21*(4), 355–366.

Makris, J. (1978). The crust and upper mantle of the Aegean region from deep seismic soundings. *Tectonophysics, 46*, 269–284.

McKenzie, D. (1970). Plate tectonics of the Mediterranean region. *Nature, 226*, 269–284.

Michaud, V., Clocchiatti, R., & Sbrana, S. (2000). The Minoan and post-Minoan eruptions, Santorini (Greece), in the light of melt inclusions: Chlorine and sulphur behaviour. *Journal of Volcanology and Geothermal Research, 99*(1–4), 195–214.

Michelot, J. L., Dotsika, E., & Fytikas, M. (1993). A hydro chemical and isotopic study of thermal waters on Lesbos Island (Greece). *Geothermics, 22*(2), 91–99.

Mizutani, H., & Rafter, T. A. (1969). Oxygen isotopic composition of sulfates: 3. Oxygen isotopic fractionation in the disulfate-water system. *New Zealand Journal of Science, 12*, 54–59.

Papazachos, C., Hatzidimitriou, P., Panagiotopoulos, D., & Tsokas, G. (1995). Tomography of the crust and upper mantle in southeast Europe. *Journal of Geophysical Research, 100*, 12405–12422.

Rafter, T. A. (1957). Sulphur isotopic variations in nature: Part 1. The preparation of sulphur dioxide for mass spectrometer examination. *The New Zealand Journal of Science and Technology, B38*, 849.

Rafter, T. A., & Mizutani, H. (1967). Oxygen isotopic composition of sulfates: 2. Preliminary results on oxygen isotopic variation in sulphates and relationship of the environment and to their $\delta^{34}S$ values. *New Zealand Journal of Science, 10*, 816.

Chapter 48
Experimental Study of Longitudinal Dispersion on Trapezoidal Open Channel

Ali Mansour Lagoun and Salim Benziada

Introduction

Water demand does not cease to increase, especially with demography growth, industry and agricultural development. Global warming and climate changing has affected directly water resources, rains becomes rare and several areas becomes arid.

Water, this vital source, that it becomes today very precious, is affected by pollution, and providing water with quantity and quality is becoming very difficult and expensive. Governments have to make all procedures to preserve this source of life.

Pollution in shallow water resources and river is often punctual (Gerhard et al. 2005), when pollutants discharged in streams, is subjected to several processes of mixing and transportation, a lot of scientists are interesting today to grow knowledge of its mechanisms and to develop models for studying phenomenon of pollutants transport and predicting its evolution in space and time, to develop helpful decision tools.

A.M. Lagoun (✉)
Research Laboratory of Water Sciences, Polytechnic National School of Algiers, Algiers, Algeria

Laboratory of Environment, Water, Geomechanics and Structures, Faculty of Civil Engineering (FGC), University of Sciences & Technology Houari Boumediene (USTHB), Bab Ezzouar, Algiers, Algeria

Scientific and Technical Research Center on Physical and Chemical Analyses, Bou-Ismail, Tipaza, Algeria
e-mail: lagounali@yahoo.fr

S. Benziada
Research Laboratory of Water Sciences, Polytechnic National School of Algiers, Algiers, Algeria

P. Grammelis (ed.), *Energy, Transportation and Global Warming*, Green Energy and Technology, DOI 10.1007/978-3-319-30127-3_48

Researches at this field are in situ (at rivers and streams), at reduced physical models or mathematical and numerical developments, at this manuscript, we have conceived an experimental device with trapezoidal open channel, that is the most responded, usual and similar at rivers, a lot of experimentation was applied by changing hydrodynamical parameters of flow to determine its effects and understanding the phenomenon of contaminations dispersion.

Phenomenon of Transport in Rivers

When any substance is injected in flow, a cloud with high concentration forms, this cloud occupies a wide area at moving and concentrations become lower. Moving of substance injected in flow and decreasing in concentration are effected by several processes such as: diffusion, advection, dispersion (Fischer 1966; Shen et al. 1995; Gharbi 1998; Hibbs et al. 1999; Czernuszenko et al. 2005; Steve et al. 2005).

Experimental Setup

Experimental setup consists essentially of feed tank, a trapezoidal channel, an injection device and sampling and collecting systems.

Installation was conceived, and experiments were performed at Research Laboratory of water Sciences (LRS eau) in Polytechnic National School of Algiers with the aim to measure concentration profiles of contaminant (phenol) at downstream stations resulting due different injections scenario of phenol in the upstream of channel.

Figures 48.1 and 48.2 shows equipment of experimental setup, characteristics of flume are presented in Table 48.1.

Measurement Tools

Discharge Measurement: Discharge measurement was made with V notch weir, after weir calibration, Discharge is given as function of head flow with:

$$Q = 1,35.{h_d}^{5/2} \tag{48.1}$$

Concentration Measurement: Phenol concentration was measured by Spectrophotometer, after calibration of phenol, concentration is given as function of optical densities by:

$$C = 80\ D_0 \tag{48.2}$$

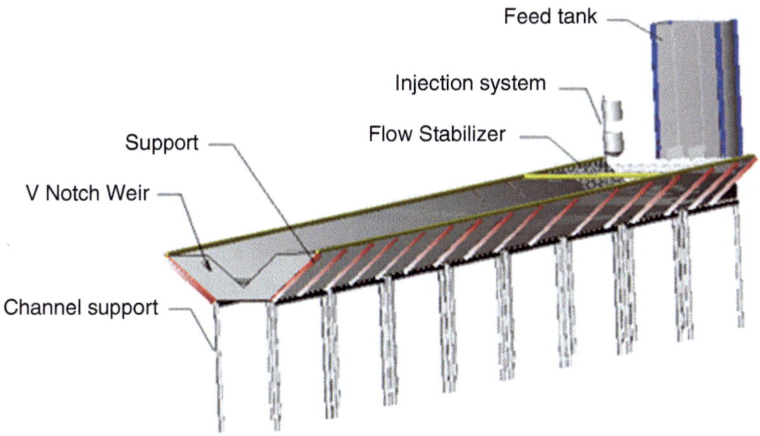

Fig. 48.1 3D schema of experimental setup

Fig. 48.2 Photography of experimental setup (LRS. Eau, ENP)

Table 48.1 Summary of flume characteristics

Designation	Parameters	Values
Channel	Form	Trapezoidal
	Bed slope [%]	5
	Lench [m]	10,4
	Bottom Width [m]	0,4
	Side Slope [°]	60
	Slope head [cm]	50
Weir	Form	Triangular (V notch)
	Notch angle [°]	90
	Crest elevation [cm]	10

Experimental Procedure

Discharge flow is regulated at upstream by valve and controlled and measured at downstream by triangular weir.

The water was supplied to the flume through overhead tank with constant water level in order to have a constant discharge of flow; Flume was market at every 1 m length.

Once flow is stabilized by passing of water through a flow stabilizer at upstream, injections of phenol for 30 s with three values of concentration ($C_0 = 500$, $C_0 = 1000$ and $C_0 = 1500$ mg/l) were applied at upstream.

Experimental measurements of water head was made, and samples from every 1 m length of flume was taken at surface of water for different scenario of injection and a several flow values (Q = 0.52; Q = 0.78, Q = 1.6, Q = 1.75 and Q = 1.95 l/s). Samples were immediately analyzed at laboratory and concentrations were determined.

Hydraulic Study of Fume Flow

Reynolds number: is given as: (Lencastre 2005).

$$Re = \frac{UD_h}{\nu} \tag{48.3}$$

Flow regimes are classified according Reynolds number as follows: $Re < 2000$: laminar flow, $2000 < Re < 4000$: transient flow, $Re > 4000$: turbulent flow (Lencastre 2005).

Froude Number: is given by:

$$Fr = \frac{U}{\sqrt{g \times h}} \tag{48.4}$$

Flow regimes are classified according Froude number as follows: $Fr > 1$: super-critical flow, $Fr < 1$: subcritical flow, $Fr = 1$: critical flow (Lencastre 2005) (Table 48.2).

Table 48.2 Flow regimes

Q [l/s]	U [m/s]	Re	Fr	Flow regimes
0.52	0.008	2848.03	0.006	Subcritical and transient
0.79	0.011	4220.03	0.009	Subcritical and turbulent
1.60	0.019	8144.88	0.015	Subcritical and turbulent
1.75	0.021	8830.58	0.016	Subcritical and turbulent
1.95	0.023	9754.51	0.017	Subcritical and turbulent

Results and Discussion

Concentration Profiles of Longitudinal Dispersion

Concentration profiles of phenol for every 1 m of flume length at the surface of water are shown at Figs. 48.3 and 48.4 (for Q = 0.79 l/s and Q = 1.75 l/s).

(**Instantaneous injection of phenol for 30 s**):

Discussion

- Concentration peaks obtained (Figs. 48.3 and 48.4) are very smaller than initial concentrations injected, this considerable decrease is mainly due to ***dilution phenomenon*** that occurs just after discharging pollutant at water flow.

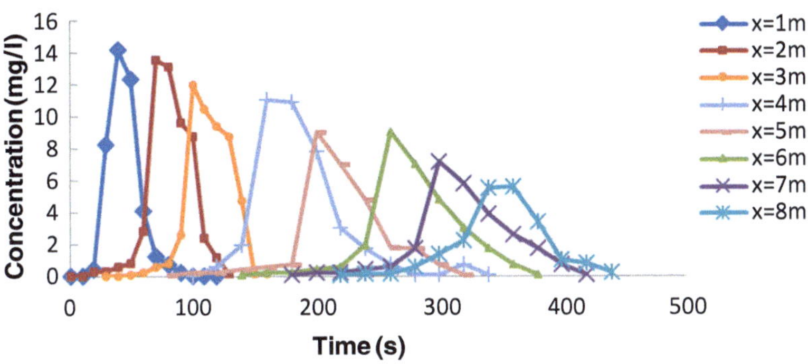

Fig. 48.3 Concentration profiles of phenol for Q = 0.79 l/s and C_0 = 500 mg/l

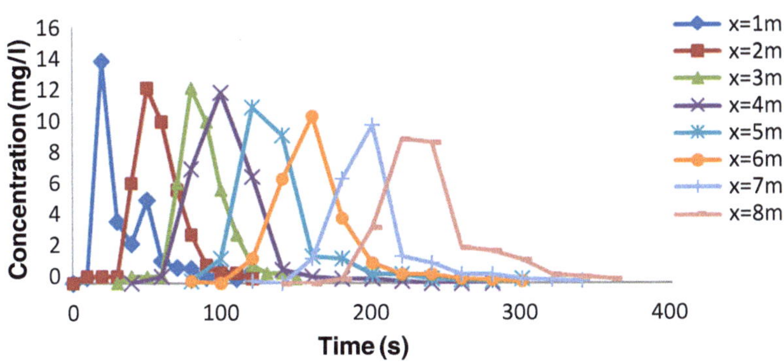

Fig. 48.4 Concentration profiles of phenol for Q = 1.75 l/s et C_0 = 1000 mg/l

- In the Vicinity of injection area, concentrations profiles are pointed and sharp. At the downstream direction, profiles bearing are becoming much more important and profiles takes the Gaussian form with decrease in concentrations.

This is explained by the predominance of *advection* in the vicinity of the injection area, away from the pollutant discharge point, the transportation becomes ***diffusive***.

Flow Discharge Influence

Figures 48.5 and 48.6 show a comparison of phenol concentration profiles for different discharge flow. We note that for all initial phenol concentration injected (C_0):

- Concentrations obtained are much more important for low flow rates.
- Stays time of pollutant is more important for low flow rates.

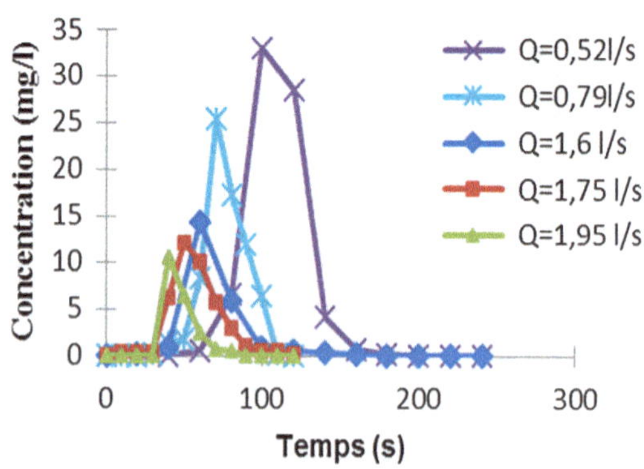

Fig. 48.5 Concentration profiles of phenol for different flow discharges at $\boldsymbol{x = 2}$ **m** $C_0 = 1000$ mg/l

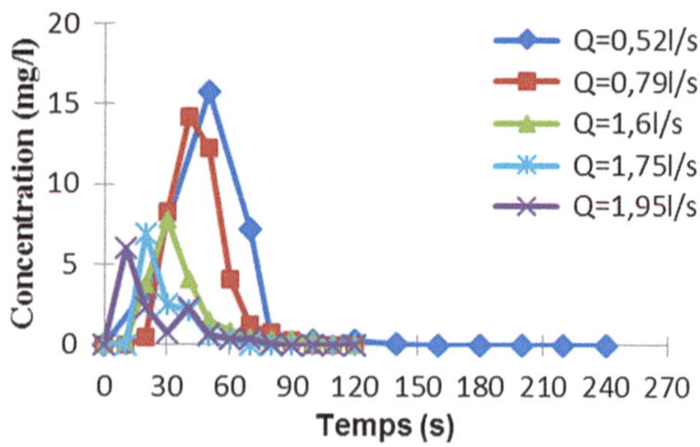

Fig. 48.6 Concentration profiles of phenol for different flow discharges at $\boldsymbol{x = 1}$ **m**, $C_0 = 500$ mg/l

For determination of flow rates and stays times of pollutants on dispersion longitudinal phenomenon, we analyze concentration peaks according discharge flows and distances length at downstream.

Concentration Peaks According Discharge Flow

Concentration peaks as a function of discharge flow are presented in Figs. 48.7 and 48.8.

Peaks obtained are lower for important flow; this is due to the *dilution rate* that increases with flow discharge.

Longitudinal evolution of concentration peaks for different flow:

Discussion

By analyzing Figs. 48.9 and 48.10, we observe that decreasing of concentration peaks is much more important for flow rate Q = 0.52 l/s, this is effect of **stays time** that it is

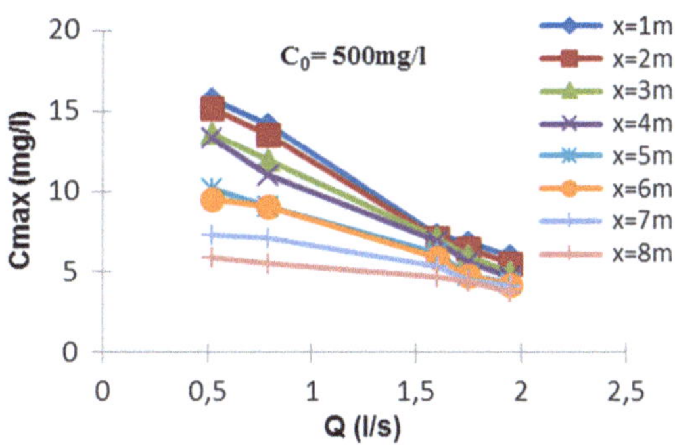

Fig. 48.7 Concentration peaks as function of discharge flow for $C_0 = 500$ mg/l for different distances x

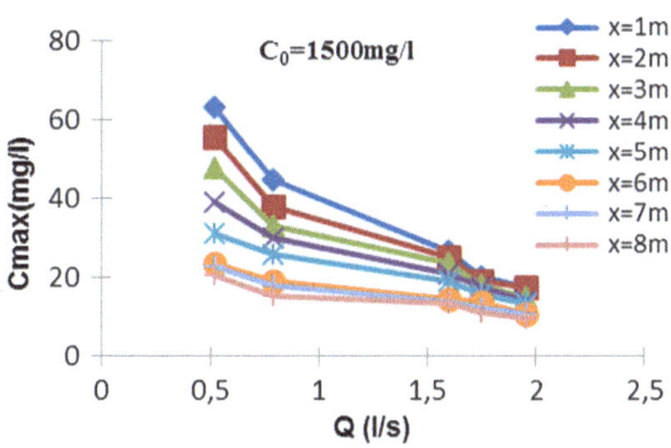

Fig. 48.8 Concentration peaks as function of discharge flow for $C_0 = 500$ mg/l for different distances x

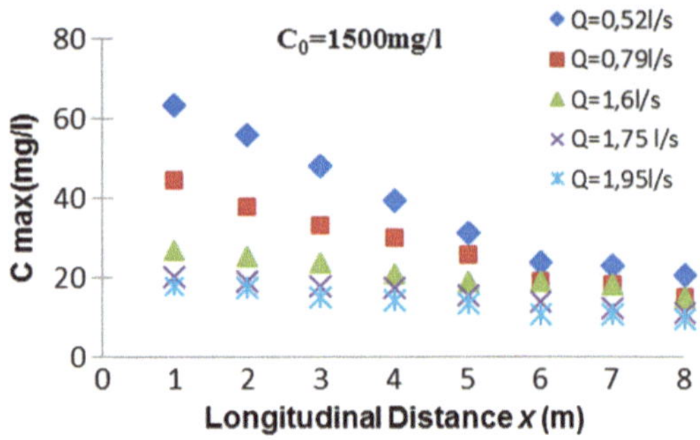

Fig. 48.9 Evolution of concentration peaks at flume for different flows. $C_0 = 1500$ mg/l

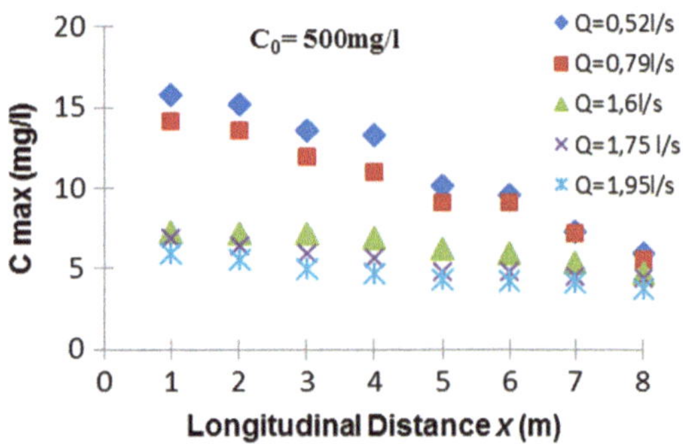

Fig. 48.10 Evolution of concentration peaks at flume for different flows. $C_0 = 500$ mg/l

more important for this flow discharge, dispersion have had more time to manifest, contrary to the case of flow with Q = 1.95 l/s when predominance was for advection.

Conclusion

Qualitative study of concentration profiles obtained experimentally, allowed to the good understanding of dispersion longitudinal phenomenon, and to identify mechanisms that govern it.

For instantaneous injection of pollutant case, the analysis of the spatiotemporal evolution of concentration of phenol at flume has identified two areas:

- An *advection area*, near the injection where profiles obtained are pointed; the pollutant transport is convective.
- A *diffusive area*, in the far field, where the concentration profiles bearing become important and the pollutant transport begin to be increasingly diffusive.

The quantitative study and the comparison between the results for different injection scenarios shows that the transport phenomenon is related to the flow rate, since its influence on the properties of the flow that transports the pollutant:

- The average velocity of flow, therefore the advection processes. Increased discharge results in higher velocity and pollutant transport becomes more convective.
- The residence (stays) time is more important for low flows (low velocity), so the decreasing rate of longitudinal dispersion concentration will be higher.
- Dilution rate, which it is function as flow rate.

Nomenclature

C	Pollutant Concentration, mg/l
D_h	Hydraulic diameter, m
D_o	Optical density
g	Gravity, m/s^2
h	Hydraulic depth, m
h_d	Head, m
Q	Flow discharge (m^3/s)
U	Flow velocity (m/s)

Dimensionless Numbers

Fr	Froude number
Re	Reynolds number

Greek Letters

ν	Cinematic viscosity of water, m^2/s

References

Czernuszenko, W., & Alexey, R. (2005). Three-dimensional model of flow and mixing processes in open channels . In *Water quality hazards and dispersion of pollutants* (pp. 35–54). Library of Congress Cataloging-in-Publication Data. New York: Springer.

Fischer, H. B. (1966) . Longitudinal dispersion in laboratory and natural streams. Report No. KH-R-12. *Journal of Water Resources Division*, 250p.

Gerhard, H. J., & Volker, W. (2005). Mixing models for water quality management in rivers. In *Water quality hazards and dispersion of pollutants* (pp 1–34). Library of Congress Cataloging-in-Publication Data. New York: Springer.

Gharbi, S. (1999). Évaluation des coefficients de mélange longitudinal et transversal des polluants dans les cours d'eau: proposition de nouvelles formules. PhD thesis, University Laval Québec, 197p.

Hibbs, D., Gulliver, J., Voller, V., & Chen, Y. F. (1999). An aqueous concentration model for riverine spills. *Journal of Hazardous Materials, A64*, 37–53.

Lencastre, A. (2005). Hydraulique Général . Eyrolles Edition. 633.

Shen, H. T., Yapa, P. D., & Zhang, B. Z. (1995). A simulation model for chemical spills in the upper St Lawrence River. *Journal of Great Lakes Research, 21*(10), 652–664.

Steve, W., & Russell, M. (2005). On the theoretical prediction of longitudinal dispersion coefficients in a compound channel. In *Water quality hazards and dispersion of pollutants* (pp. 69–84). Library of Congress Cataloging-in-Publication Data. New York: Springer.

Chapter 49
Mygdonia Basin (N. Greece) in the View of Isotope Geochemistry

P. Chantzi and E. Dotsika

Introduction

Mygdonia basin located 10 km N-E to Thessaloniki in northern Greece and it is consisted by two sub-basins: Koroneia and Volvi lakes. Both lakes constitute an important wetland which is being protected by International Ramsar Convention. Mygdonia basin constitutes an elongated EW tectonic depression which combined with adjusted minor basins (Zagliveri, Marathousa, Doubia) are the remains of an initial basin called Promygdonia (Koufos et al. 1993). The initial basin was formed by tectonic activity probably in the late Paleogene to early Neogene. In late Neogene and early Pleistocene a set of river and lake surface sediments deposited in this basin. A new tectonic activity at the end of the period "early-Pleistocene" resulted to many faults of Promygdonia basin were several smaller basins formed (Mygdonia, Zagliveri, Marathousa, Doubia). Among them the largest was Mygdonia basin forming the Mygdonia Lake which gradually drained during the medium-Pleistocene. Sedimentation continued in this basin mainly by lacustrine deposits (Psilovikos 1977; Sotiriadis et al. 1983). Koronia and Volvi lakes are the remnants of the initial Mygdonia Lake.

P. Chantzi (✉)
Laboratories of Stable Isotopes and Radiocarbon, Institute of Nanoscience and Nanotechnology, National Centre for Scientific Research "Demokritos", 15310 Agia Paraskevi, Attica, Greece
e-mail: p.chantzi@inn.demokritos.gr

E. Dotsika
Laboratories of Stable Isotopes and Radiocarbon, Institute of Nanoscience and Nanotechnology, National Centre for Scientific Research "Demokritos", 15310 Agia Paraskevi, Attica, Greece

Institute of Geosciences and Earth Resources, Via G. Moruzzi 1, 56124 Pisa, Italy
e-mail: e.dotsika@inn.demokritos.gr

P. Grammelis (ed.), *Energy, Transportation and Global Warming*, Green Energy and Technology, DOI 10.1007/978-3-319-30127-3_49

Mygdonia basin is bended in metamorphic layers from Serbomacedonian massif (paleozoic gneisses, amphibolites, schists, marbles and granitic intrusions) with groundwater circulating in metamorphic rocks and discharging in loose formations in basin. Moreover it covers a basin about 2026 km^2 where the terrestrial part covers about 656 km^2 and particularly extends west of Koronia Lake. The basin is defined by the mountains of Volvi (627 m) and the lower vertices of Vertiskos mountain (1103 m) in the north, in the east by the mountains of Psili Rachi (341 m) and Soulgianodiou (746 m) which are interposed between the mountains Kerdylion and Stratoniko and through the Strait of Rendina communicates with Strymoniko Bay and in the south side by Chortiatis mountain (1206 m) the lower vertices of Cholomontas (1165 m) and Stratoniko (918 m) with S-E direction. West boundaries of basin are not well defined as there are a set of low hills and depressions except Camila Mountain (569 m). The annual precipitation in the basin ranges 283–721 mm in the last 30 years. The climate varies from Mediterranean and continental. For the entire basin the average annual temperature is estimated about 13.55 °C and an average annual precipitation about 584.9 mm (Vatseris 1992).

Both Koroneia and Volvi subcatchments have been undergone sever environmental and human impacts resulting their ecological death (Nimfopulos, M. K. et al., 2002). An attempt was made in order to fingerprint the isotopic signature of groundwater in order to strengthen the knowledge on the hydrology regime of Mygdonia basin. Cold water samples from boreholes were collected from Mygdonia basin for stable isotope analyses of $\delta^{18}O$, $\delta^{2}H$, $\delta^{13}C$ which was performed in Laboratories of Stable Isotopes and Radiocarbon at NCSR "Demokritos". The network of boreholes that were sampled corresponded to shallow (30–60 m) and moderate deep water circulation (60–100 m). The sampling took place in June 2014 where the boreholes were operated for at least one month resulting in the water sample to correspond to the actual groundwater body. Moreover two water samples from lakes Koroneia and Volvi were taken however that wasn't possible for fresh spring water as no such springs were detected (Fig. 49.1).

Results and Discussion

Oxygen $\delta^{18}O$ and $\delta^{2}H$ Isotopes

The values of stable isotopes ranging from −10.3‰ to −7.3‰ and −70.4‰ to −45.7‰ for $\delta^{18}O$ and δD. In more detail on the sub-basin of Lake Koronia the isotope values of water molecules vary from −9.0‰ to −7.5‰ (range 1.5‰) and from −59.2‰ to −45.7‰ (range 13.5‰) for $\delta^{18}O$ and δD respectively, while on the sub-basin of Lake Volvi values vary from −10.3‰ to −7.3‰ (range 3‰) and from −70.4‰ to −47.2‰ (range 23.2‰) for $\delta^{18}O$ and δD respectively. These

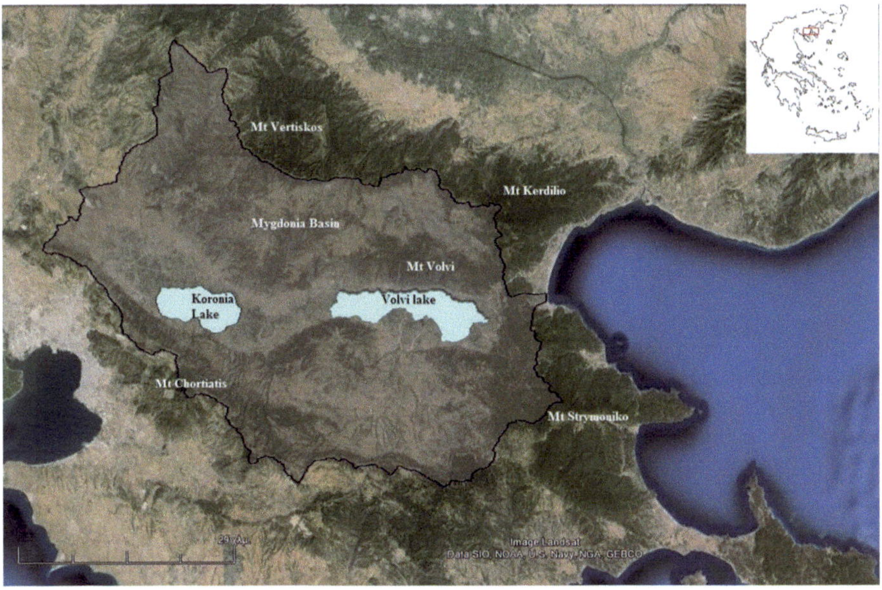

Fig. 49.1 Study area of Mygdonia basin with Koronia and Volvi sub-basins (Google earth modified picture)

values are shown in diagram δ^2H‰ V-SMOW versus δ^{18}O‰ V-SMOW (Fig. 49.2) where global meteoric water line (GMWL, Craig 1961) and local meteoric water line (LMWL) for Greece as proposed by Dotsika et al. (2010) are pictured as well.

The correlation equation of groundwater for Mygdonia basin is:

$$\delta\mathrm{D} = 7.8 * \delta^{18}O + 10.30 \tag{49.1}$$

Groundwater in Mygdonia basin presents an excellent correlation. Moreover Koronia is grouped with groundwater in the south part of Volvi sub-basin, sites where the network of boreholes is strong, supporting the hydraulic communication (Fig. 49.3).

The correlation equation of groundwater for Koronia and Volvi sub-basins are:

$$\delta\mathrm{D} = 7.1 * \delta^{18}O + 4.11 \tag{49.2}$$

$$\delta\mathrm{D} = 7.8 * \delta^{18}O + 10.95 \tag{49.3}$$

The correlation equations of Mygdonia, Koronia and Volvi basins present slopes «7.8», «7.8» and «7.1» respectively. For Central Macedonia a slope inferior to 7 is reported (Christodoulou et al., 1993) while for East Macedonia is reported slightly greater «7.5» (Leontiadis et al., 1984). Generally fresh precipitation of all types and surface water that have not undergone evaporation exhibit slopes between 7–8 (Craig 1961), while waters that have undergone secondary evaporation exhibit

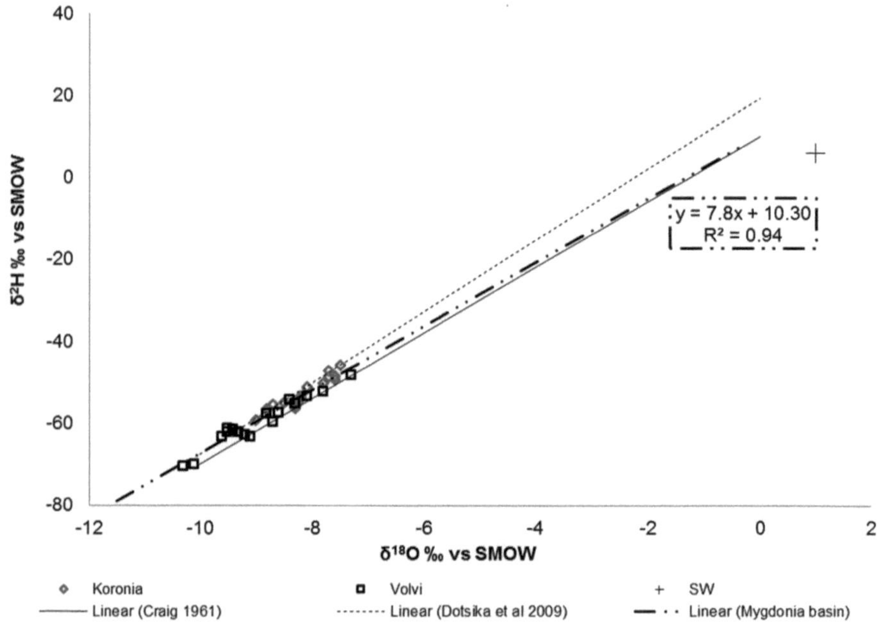

Fig. 49.2 δ^2H‰ V-SMOW versus δ^{18}O‰ V-SMOW of cold groundwater in Mygdonia basin

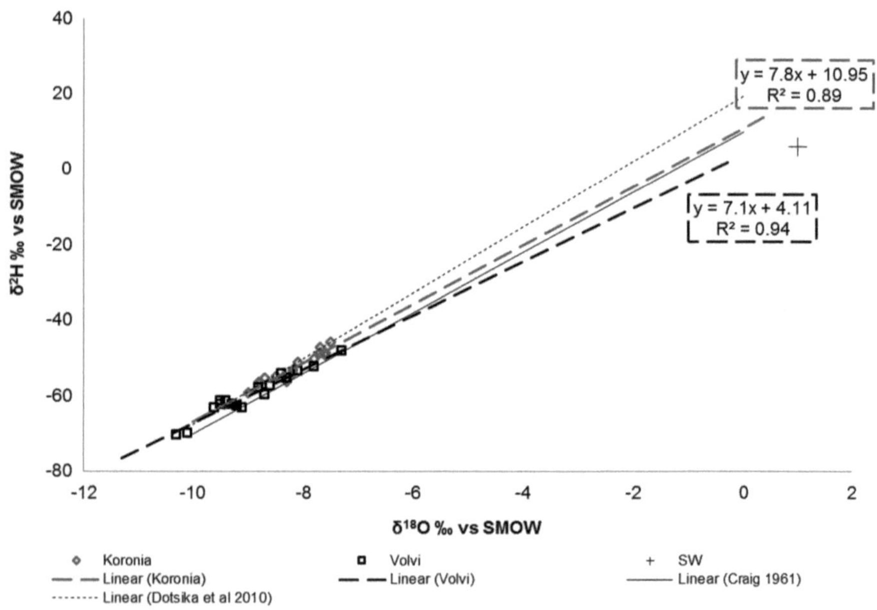

Fig. 49.3 δ^2H‰ V-SMOW versus δ^{18}O‰ V-SMOW of cold groundwater in Koronia and Volvi sub-basins

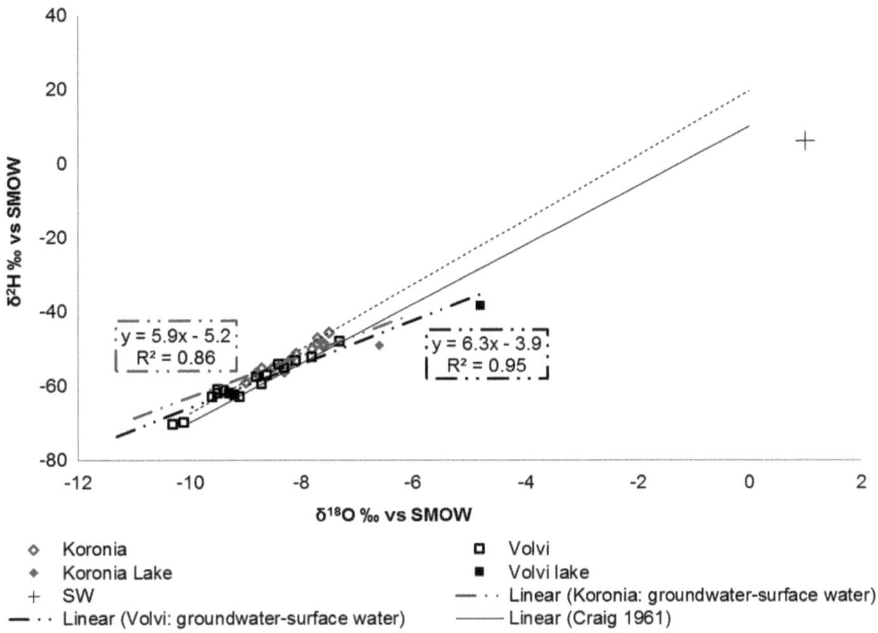

Fig. 49.4 δ^2H‰ V-SMOW versus δ^{18}O‰ V-SMOW of cold groundwater and surface water in Koronia and Volvi sub-basins

lower slopes about «6» (Gat 1980). Therefore it is concluded that groundwater of Mygdonia basin doesn't present any evaporation or mixing episodes with different isotopic water.

Both oxygen and deuterium isotopic values are characterized by homogeneity without significant fluctuations implying that no isotopic exchange with the geological environment takes place. This conclusion is in agreement with geological regime of the basin where groundwaters circulate in metamorphic rocks and discharge in loose formations in basin without presenting long retention time in groundwater aquifers.

The correlation between sub-basins and surface water of the corresponding lake is pictured in Fig. 49.4 with the following equations:

$$\delta \mathrm{D} = 5.9 * \delta^{18}O - 5.2 \quad (49.4)$$

$$\delta \mathrm{D} = 6.3 * \delta^{18}O - 3.9 \quad (49.5)$$

The slopes of lines (49.4) and (49.5) with values «5.9» and «6.3» for the Koronia and Volvi respectively reflect evaporation processes occurring in surface reservoirs with Koronia exhibiting a higher rate. In both sub-basins groundwater and surface lake water correlates excellent however the linking trend is evident in Volvi catchment implying their direct hydraulic communication but in Koronia catchment surface lake water placed slightly away from the corresponding trend-line. Isotopic values are strongly influenced by temperature and the evaporation processes but also by the humidity (Fritz and Fontes 1980). Therefore this slightly deviation could be attributed to the humidity above lake water layers. These observations possible imply the communication of surface water with groundwater in Koronia basin through lateral shallow formations as the lacustrine bottom is covered by a clay layer that makes it impermeable

Carbon Isotopes $\delta^{13}C$

The values of carbon isotopes $\delta^{13}C$ can detect the different origins of organic or inorganic carbon. The value of carbon in marine carbonate rocks is around 0‰, the soil CO_2 around −25‰ (similar to that of plants) and about −7‰ for atmospheric CO_2. However, carbonates evaporation may exhibits higher values about +10‰. The carbon isotopic ratio of fresh water is strongly affected by carbonate mineral dissolution in the case of carbonate bicarbonate aquifers: $CaCO_3 + H_2CO_3 \rightarrow Ca^{2+} + 2HCO_3^-$ (Dotsika 2015). In temperate climate the contribution of dissolved carbonates ($\delta^{13}C = -2$‰ to +1‰, Clark and Fritz, 1997) resulting groundwater $\delta^{13}C$ values about −11‰ (Jin et al. 2009) however in shallow aquifers soil CO_2 and atmospheric CO_2 (water-rock interaction product) contribute to the values formation resulting aquifers with $\delta^{13}C$ values between −11‰ and −22‰ (Jin et al. 2009). Carbon isotope values $\delta^{13}C$ range between −13.6‰ and −10.2‰ for Mygdonia basin. In greater detail range from −13.6‰ to −10.2‰ and from −13.2‰ to −10.4‰ for Koronia and Volvi groundwater respectively, reflecting both the dissolution of carbonate rocks and the participation of organic carbon from soils.

Conclusions

Stable isotope values of cold water samples ranged from −10.3‰ to −7.3‰, from −70.4‰ to −45.7‰ and from −13.6‰ to −10.2‰ for $\delta^{18}O$, δD and $\delta^{13}C$ respectively reflecting the isotopic signature of local precipitation water. In the entirely basin subsurface water circulation exhibits short retention time influenced by the strong network of irrigation wells without taking place any isotopic exchange with the geological environment. Mechanism of neither evaporation nor mixing with waters of different isotopic origin was detected. Dissolution of carbonate rocks and the participation of CO_2 from soils taking place through the circulation of groundwater to the shallow and middle shallow aquifers.

References

Christodoulou, Th., Leontiadis, I.L., Morfis, A., Payne, B.R., Tzimourtas, S., 1993: Isotope hydrology study of Axios River plain in northern Greece. J. Hydrol. 146, 391–404.

Clark, I. D., & Fritz, P. (1997). *Environmental isotopes in hydrogeology*. Boca Raton, FL: Lewis Publishers.

Craig, H. (1961). Isotopic variations in meteoric waters. *Science, 133*, 1702–1703.

Dotsika, E., Lykoudis, S., & Poutoukis, D. (2010). Spatial distribution of the isotopic composition of precipitation and spring water in Greece. *Global and Planetary Change, 71*, 141–149.

Dotsika, E. (2015, March). H–O–C–S isotope and geochemical assessment of the geothermal area of Central Greece. *Journal of Geochemical Exploration, 150*, 1–15.

Fritz, P., & Fontes, J. (1980). *Handbook of environmentai lsotope geochemistry*. New York: Elsevier.

Gat, J. R. (1980). Isotope hydrology of very saline lakes. In A. Nissenbaum (Ed.), *Developments in sedimentology* (Vol. 28, pp. 1–7). New York: Elsevier. Chapter 1.

Jin, L., Ogrinc, N., Hamilton, S., Szramek, K., Kanduc, T., & Walter, L. (2009, June). Inorganic carbon isotope systematics in soil profiles undergoing silicate and carbonate weathering (Southern Michigan, USA). *Chemical Geology, 264*(1–4), 139–153. ISSN: 0009-2541.

Koufos, G., Syridis, G., Kostopoulos, D., & Koliadimou, K. (1993). Preliminary results about the stratigraphy and the palaeoenvironment of Mygdonia Basin, Macedonia, Greece. In *Proceedings of the 1st congress of paleontological association*, Geobios.

Leontiadis, I., Payne, B.R., Letsios, A., Papagianni, N., Kakarelis, D., Chadjiagonakis, D., 1984: Isotope Hydrology study of Kato Nevrokopi of Drama. Proc. Of the Symposium on Isotope Hydrology 1983. International Atomic Energy Agency, Vienna, pp. 193–206.

Nimfopulos, M. K., Mylopoulos, N., & Katirtzoglou, K. G. (2002). A qualitative-quantitative study of water and environmental pollution at the broader area of the Mygdonia Basin, Thessaloniki, Northern Greece. In *Proceedings of the 6th Pan-Hellenical geographical conference of the Hellenic geographical society*, Thessaloniki, 3–6 October 2002 (Vol. II, pp. 436–444).

Psilovikos, A. (1977). Paleogeographic development of the basin and lake of Mygdonia (Lagada—Volvi area, Greece). Ph.D. thesis, Department of Geology, Aristotle University of Thessaloniki (in Greek).

Sotiriadis, L., Psilovikos, A., Vavliakis, E., & Syrides, G. (1983). Some tertiary and quaternary basins of Macedonia/Greece. Formation and evolution, *Clausthaler Geologische Abhandlungen* (p. 21).

Vatseris, C. (1992). Hydrogeologie des Mygdonias-Beckens (Nord-Griechenland) under Berucksichtigung der hydrochemischen und isotopen physikalischen komposition der thermalwasse. Ph.D. thesis, University of Munster, 171p.

Chapter 50
Sustainable Management of Sewage Sludge Conditioning and Valorization

S. Igoud, F. Souahi, and C.-E. Chitour

Introduction

Between 1999 and 2010, the Algerian wastewater treatment data showed that the wastewater volume has doubled. It increased from 600 million m^3 to 1.2 billion m^3 corresponding to a ratio of 32.34 m^3/inhabitant. This has required the building of 105 new wastewater treatment plant (WWTP) and the increase of the wastewater collection which passed from 79 % to 86 % (Hammouche 2011). But, in the same manner, this performance has increased the electricity consumption, the public spending and the greenhouse gas emission.

The increase of the electricity consumption has been also recorded because the wastewater treatment uses mainly the activated sewage sludge and the aerated lagoon processes applied at 67.7 % even if they allowed significant remediation efficiency (Igoud et al. 2014).

This situation has been maintained particularly between 2010 and 2013 with increases estimated at 35.2 % of wastewater volume, 45.8 % of electricity consumption and 58 % of electricity budget. During 2013, the volume of wastewater treated by the most important treatment operator: the National Office of Sanitation (ONA) was estimated at 161 million m^3. Its remediation has consumed 21.68 GWh of electricity for pumping and 60.28 GWh for treatment. Financially, the electricity cost has been estimated at 2.6 million €. Also, the electricity consumption has emitted 185.61 tons of CO_2-eq (Igoud et al. 2014). At 2015, ONA estimated that the

S. Igoud (✉)
Unité de Développement des Equipements Solaires, UDES/Centre de Développement des Energies Renouvelables, CDER, 42415 Bou Ismail, W. Tipasa, Algerie
e-mail: igoud.sadek@udes.dz

F. Souahi • C.-E. Chitour
Ecole Nationale Polytechniques (ENP), Hacène Badi, 16200 El-Harrach, W. Alger, Algerie
e-mail: fsouahi@yahoo.fr; lavalef@yahoo.fr

P. Grammelis (ed.), *Energy, Transportation and Global Warming*, Green Energy and Technology, DOI 10.1007/978-3-319-30127-3_50

total volume of treated wastewater reached 197 millions m^3 and its treatment consumed 97.73 GWh of electricity (ONA 2015).

The development of the wastewater treatment sector which occurred this last decade will continuous to increase. That because it is planned the systematic treatment of all collected wastewater and its reuse to irrigate 1 million hectares. This trend is also projected at the international level. The International Energy Agency expects that between 2006 and 2030 the energy needs for the wastewater treatment will increase by 44 % (International Energy Agency 2006). This rate will reach 60 % in developing countries (UNESCO-WWAP 2012).

Contributing to reduce the electricity consumption, the aim of the study is to suggest a sustainable wastewater treatment approach as it is planned by ONA. This wastewater management was recommended since 2002 in the Johannesburg world summit (United Nations 2002) to anticipate the significant energy consumption and GHG emission.

Currently, in order to reduce the electricity consumption, two trends are prospected: the generation of renewable energy inside the wastewater treatment plants (WWTP). This requires the use of renewable resources available at the plants level especially sewage sludge and solar energy. Also, the use of sustainable treatment processes.

In the study, three proposal of a sustainable sewage sludge treatment are investigated: the sludge methanization for the cogeneration of electricity and thermal energy, the solar sludge dryness and the agricultural valorization by land application in forest areas.

Also, in order to evaluate the electricity and cost saving the sequestration of the greenhouse gas (GHG) emission, the yearly electricity consumption has been conducted in an activated sludge wastewater treatment plant.

Methodology

Sewage Sludge Methanization

The energetic bioconversion of sewage sludge requires two steps: the sludge methanization applied for the biogas production. Then the biogas cogeneration used for both of electricity and thermal energy production.

The biogas production has been estimated using theoretical and experimental approaches. The theoretical step consisted on use of Eq. (50.1) as follows:

$$BV = SSQ.DM.RVM.CC \tag{50.1}$$

where:

BV: Biomethane volume
SSQ: Sewage Sludge Quantity = 3040 tons
DM: Dry Matter = 65.1 kg/m^3
RVM: Removed Volatile Matter = 41.23 %

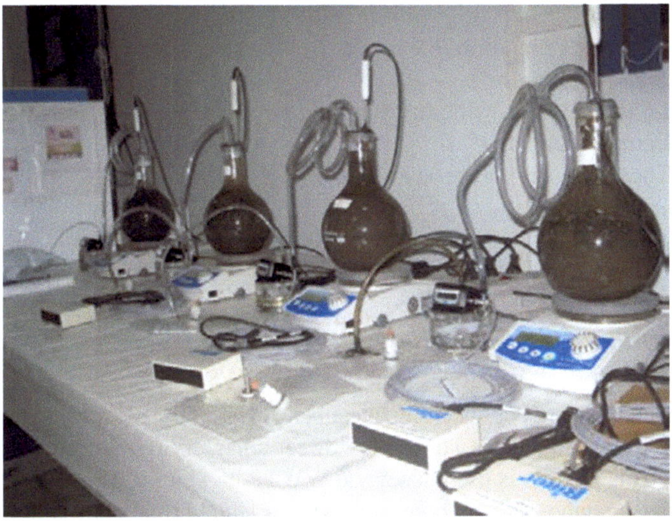

Fig. 50.1 Test bench of sewage sludge methanization (Igoud 2015)

Removal biodegradable pollution rate = 30 %

CC: Conversion Coefficient = ranging from 0.5 to 0.7 m^3 of biomethane/tons of removed dry volatile matter (Record 2009; Lagrange 1979, 1995).

The experimental study of the biogas production has also been conducted. It has been experimented using a test bench (Fig. 50.1) composed of four 2 L flat-bottom flasks.

During the experiments, each flask has been filled by 1 kg sewage sludge recovered from Tipasa WWTP. The hermetic closure of the flasks allowed the maintaining of the anaerobic conditions. To ensure the optimum methanization of sludge, this last has been diluted at 100 % of its weight. Then the flasks were tightly closed and putted on controlled magnetic stirrer hotplates which allowed 37 °C sludge heat and 800 rpm sludge agitation.

The produced biogas has been quantitatively evaluated using gas meters then stored in plastic bags for its analyze.

The second step evaluated the use of biogas for the cogeneration of electricity and thermal energy. This estimation used technical characteristics of a 65-kW biogas microturbine. Its electric efficiency (EE) is fixed at 29 % and its thermal efficiency (TE) at 50 % (Product Catalogue 2010). The calculations were performed as follows:

$$EG = BV.LCV.EE \quad (50.2)$$

$$TEG = BV.LCV.TE \quad (50.3)$$

where:
EG: Electricity generation (kWh)
BV: Biogas volume (m^3)
LCV: Lower calorific value (kWh/m^3)
EE: Electric efficiency (%)
TEG: Thermal energy generation (kWh)
TE: Thermal efficiency (%)

Solar Drying of Sewage Sludge

The experiments were conducted in a test bench (Fig. 50.2) according to a protocol that allowed a comparative study of a direct solar drying using a "drying bed" prototype. This process is used in several WWTP in Algeria. The second prototype used a solar greenhouse prototype; this type of sludge treatment is not applied in Algeria.

The efficiency evaluation of these two drying processes was undertaken using two types of sludge recovered from the studied WWTP: (1) biological sludge recovered from the clarifier and containing 99.5 % of humidity, (2) thickened sludge containing 95.5 % of humidity which was recovered from the thickener.

During the experiments, a sludge thickness of 2 cm has been maintained on the drying bed and a volume of 15 L of sludge has been dried. The dryness kinetic has been estimated through several daily samples. These last were weighed and then dried at 105 °C for 24 h to be weighed again; this in order to determine their moisture content.

In the studied WWTP, the sludge conditioning consists firstly, on the gravity thickening step secondary on the mechanical dehydration step using a belt press to reach finally 24 % of moisture.

Fig. 50.2 Test bench of the sewage sludge solar drying (Igoud 2015)

Sewage Sludge Agricultural Valorization

The essay (Ouanouki and Igoud 1993) has been conducted by the planting of forest young seedlings: *Pinus maritima* and *Acacia cyanophyla* in an area of 3672 m^2. This last was split into four plots. The first consisted on the witness. And the remaining plots were planting after the digested sludge spreading as follows: 2 tons of digested sludge (21.76 tons/ha) in the first plot, 8 tons (87.14 tons/ha) in the second and 15 tons (163.40 tons/ha) in the third.

For the experimentation, 576 young seedlings were planted with an equidistance of 3 m × 3 m in a staggered rows distribution.

Results and Discussion

During 2010, the yearly electricity consumed in the activated sludge WWTP has been estimated at 1.25 GWh. The electricity has been consumed at 89.63 % by the wastewater treatment process, 5.77 % by the outdoor lighting and 4.6 % by the management department and laboratory (Fig. 50.3).

The electricity has cost 24,766 € and emitted 757.28 tons of CO_2-eq with an average of 607.18 g of CO_2-eq/kWh.

Renewable Electricity Generated from Sewage Sludge

The theoretical approach of the sewage sludge methanization estimates that 338,103 m^3 of biogas could be produced. And its cogeneration could produce 392.78 MWh of electricity and 667.22 MWht of thermal energy recoverable for the digesters heating.

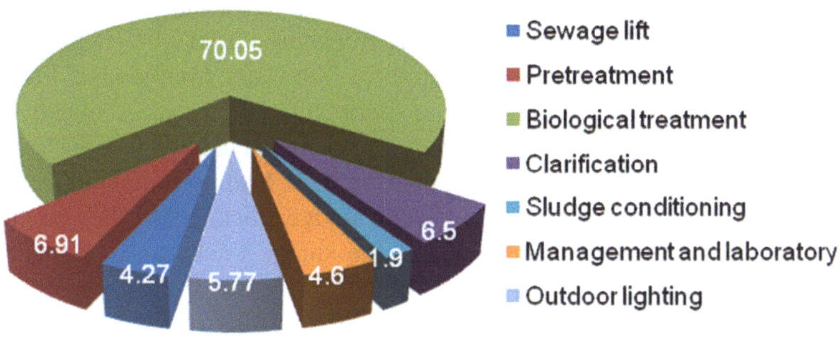

Fig. 50.3 Yearly electricity consumption in Tipasa WWTP (Igoud et al. 2014)

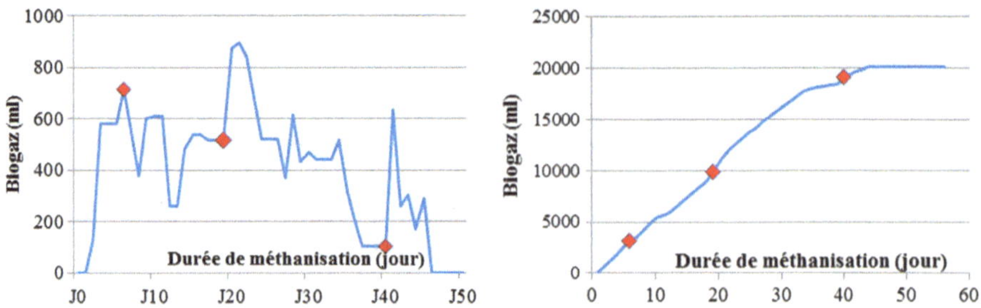

Fig. 50.4 Total biogas production (Igoud 2015)

This amount of electricity would have to substitute 31.5 % of conventional electricity consumed by Tipasa WWTP. The experimental results have recorded a lower biogas production (Fig. 50.4). The sludge methanization lasted 45 days and cumulated a biogas volume of 20.12 L/kg of sludge. The biogas has been characterized by an average methane content of 48 %. This totalized a yearly production of 61.16 103 m^3 of biogas whose combustion could cogenerate 106.42 MWh of electricity and 183.5 MWht of thermal energy.

The conventional electricity saving has been evaluated at 8.5 % of the total WWTP electricity consumption. This result is lower comparatively to the theoretical. This could be induced by the experimental conditions of the sludge methanization which require to be optimized. It could be also the result of the use of mature sludge recovered from the storage area rather than fresh one when available excess.

Sewage Sludge Dryness Induced by Solar Energy

During the trials, the dryness of the biological sludge characterized by 0.5 % of moisture reached 93 %. The drying evolution has showed relatively similar kinetics drying in the bed drying and under the solar greenhouse until 73 h (Fig. 50.5). This duration is characterized by the evaporation of free water from sludge. From 73 h which corresponded to the 3rd day of the experimentation, the speed of dehydration has increased, especially in the drying bed. This is explained by the reduction of the thickness of the sludge blade which increased their heating temperature. Indeed, simulations studies performed for different sludge thicknesses showed that the energy saving, about 20 %, is obtained if the sludge thickness is reduced from 20 cm to 10 cm (Slim 2007).

Compared to the drying beds, the drying rate of the solar greenhouse was slower. This is due to the low moisture wicking saturated air. The openings in the upper part of the solar greenhouse were not sufficient for the natural wick moisture. For this,

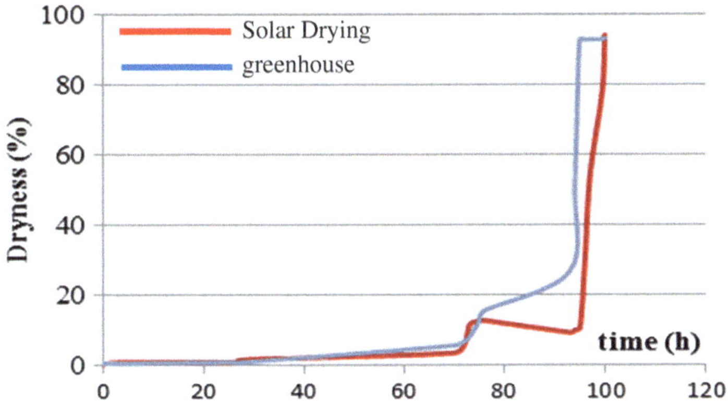

Fig. 50.5 Dryness kinetic of biologic sludge (Igoud 2015)

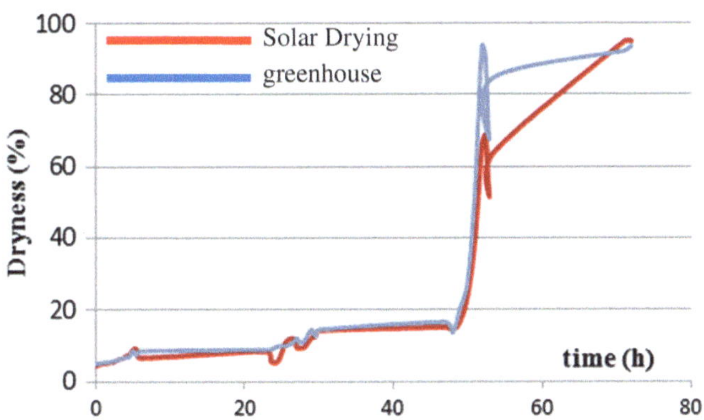

Fig. 50.6 Dryness kinetic of thickened sludge (Igoud 2015)

the optimization and control of the drying parameters would require the use of a forced air circulation. Indeed, in industrial plants, the evacuation is carried out through the air extractors to the ambient environment.

The solar drying of the thickened sludge (Fig. 50.6) containing 4.5 % of moisture also reached 93.6 % of dryness. This result approaches that obtained after the biological sludge treatment but this last was obtained during a longer drying time (100 h against 72 h).

The solar drying of sludge could allow the substitution of the conventional electricity estimated at 23,630 kWh (Hammouche 2011) required to achieve dryness from 0.5 % to 24 %. This result corresponds to a saving of 2.1 % of the electricity consumption in the WWTP, 1.90 % of the energy budget and 2.1 % of greenhouse gas emissions.

Biomass Production Induced by Sewage Sludge Spreading

The sludge spreading in the plantation showed a positive correlation between the height plants growth and the digested sludge spreading.

The best results have been obtained in the plot were the high quantity of sludge was used. In this plot, the increase of the growth rate has been estimated at 71.87 % for pine, 73 % for acacia and 68.38 % for the mixed plantation using both pine and acacia. The influence of acacia on pine growth has not shown a positive result. This has been induced by the short duration of the experimentation which would not allow the soil enrichment in nitrogen through the acacia nodules (i.e. root outgrowth involved in the fixation of atmospheric nitrogen).

Conclusion

The obtained results showed the existence real opportunities for the sustainability management of sewage sludge which benefit to the fossil energy saving and the environment preservation.

The theoretical bioconversion of sewage sludge evaluated that 392.78 MWh of electricity and 667.22 MWht of thermal energy could be generated. This amount of renewable electricity would have to substitute 31.5 % of conventional electricity consumed in the WWTP. However, the experimental assessment allowed a lower electricity generation which estimated a saving of 8.5 %. This result has been induced by the uncontrolled parameters of the experimentation which need to be optimized.

Concerning the sewage sludge conditioning in the WWTP it is restricted at 24 % of dryness. This result has been improved by the use of solar drying process. The obtained results showed the possibility to reach d up 93 % dryness. A part this improvement, the obtained results could allow saving a saving 2.1 % of the electricity consumption in the WWTP, 1.90 % of the energy budget and 2.1 % of greenhouse gas emissions. By the improving the dryness from 24 % to 93 %, the savings will be most significant. Also, this sustainable treatment reduces the cost of sludge storage and transport, increases their stabilization, sanitation and deodorization sludge and facilitates energy recovery by burning.

The spreading of digested sewage sludge also has been beneficial for forest plants growth. It has allowed an increase of the height plants growth which ranged from 68.33 % to 73 %. A part this advantage, this kind of valorization should particularly successes the reforestation undertaken in Algeria since the seventies' to fight against desertification which threatening all the African countries. This reforestation policy should be encouraged for the increase of the afforestation rate which constitutes carbon sinks in fight against climate change.

Acknowledgement The authors would like to thank colleagues and all people who contributed to this studies especially Mrs. Ahmed and Toufik Medjiah from Tipasa's wastewater treatment plant, Miss Amina Chalal from the Development Unit of Solar Equipment at Bou-Ismaiïl.

References

Hammouche, H. (2011, March). *Algeria report country. Experts consultation on wastewater management in the Arab World*. Dubai: The Arab Water Council.

Igoud, S. (2015). Integration of renewable energy and sustainable processes for the urban wastewater treatment. Ph.D. thesis. National Polytechnic School ENP, Algeria.

Igoud, S., Souahi, F., Chitour, C. E., Adjrad, A., Habchi, M., & Chouikh, A. (2014). Diagnostic of electricity consumption, its cost and greenhouse gas emission in the wastewater treatment sector of Algeria. *Desalination and Water Treatment, 55*(7), 2015 1725-1735.55, 1725–1734. doi:10.1080/19443994.2014.928236.

International Energy Agency. (2006). *World energy outlook 2009*. Paris: IEA.

Lafarge, B. (1995). Procédés de fermentation méthanique. Ed. Masson.

Lagrange, B. (1979). Biométhane 2. Principes-Techniques-Utilisations. Ed. Edisud.

ONA (2015). Tableau de bord mensuel, Direction de l'exploitation et de la maintenance. Office National de l'Assainissement, Alger. ww.ona-dz.org.

Ouanouki, B., & Igoud, S. (1993). Contribution à l'étude des boues résiduaires issues de stations d'épuration urbaine dans les plantations forestières. *Revue Forestière Française, 45*(2), 153–158.

Product Catalogue. (2010). Capstone Turbine Corporation. Documentation technique microturbine à biogaz. Capstone.

Record. (2009). Record N°07-0226/1A. Techniques de production d'électricité à partir de biogaz et de gaz de synthèse.

Report of the world summit on sustainable development, Johannesburg, South Africa, 26 August–4 September 2002. Ed. United Nations, New York, NY, 2002. Available from: www.unmillenniumproject.org/projects/131302_wssd_report_reissued.pdf.

Slim, R. (2007). Etude et conception d'un procédé de séchage combiné de boues de stations d'épuration par énergie solaire et pompe à chaleur. Thèse de Doctorat, Ecole des Mines de Paris, France.

UNESCO-WWAP. (2012). Managing water under uncertainty and risk. Facts and figures from the United Nations world water development report 4 (WWDR4). United Nations World Water Assessment Programme.

Chapter 51
Photocatalytic Degradation of Tylosin and Spiramycin in Water by Using TiO_2 and ZnO Catalysts Under UV Radiation

D. Tassalit, N. Chekir, O. Benhabiles, F. Bentahar, and N.A. Laoufi

Introduction

Some organic substances are either extremely stable or they have à very slow rate of natural degradation. Pharmaceuticals for example are widely used by humans and for food production and may react differently and persist in the environment (Sung-Suh et al. 2004). During their life cycle, they create a new and emerging problem and significant effects on the human health and ecological risks. For this reason, it's important to develop a new efficient processes which can deteriorate these recalcitrant pollutants (Neelavannan and Ahmed Basha 2010; Assabane et al. 2000). The use of UV or visible light as a source of energy, especially sunlight, to promote photocatalytic reactions is of great importance (Campion et al. 1999; Turchi and Ollis 1989). The intensity of solar radiation reaching earth can be an alternative source used especially in developing countries. Indeed, this renewable energy is free and inexhaustible; it is the most abundant energy on Earth, especially in Algeria (Herrmann et al. 2002; Simon and Dauby 2008).

D. Tassalit (✉) • N. Chekir
Unité de Développement des Equipements Solaires, UDES/Centre de Développement des Energies Renouvelables, CDER, 42415 Bou Ismail, W. Tipaza, Algeria

Laboratoire des Phénomènes de Transfert, Faculté de Génie Mécanique et de Génie des Procédes, Université des Sciences et de la Technologie Houari Boumediene (USTHB), Bab-Ezzouar, 16111 Algiers, Algeria
e-mail: tassalit2003@gmail.com

O. Benhabiles
Unité de Développement des Equipements Solaires, UDES/Centre de Développement des Energies Renouvelables, CDER, 42415 Bou Ismail, W. Tipaza, Algeria

F. Bentahar • N.A. Laoufi
Laboratoire des Phénomènes de Transfert, Faculté de Génie Mécanique et de Génie des Procédes, Université des Sciences et de la Technologie Houari Boumediene (USTHB), Bab-Ezzouar, 16111 Algiers, Algeria

P. Grammelis (ed.), *Energy, Transportation and Global Warming*, Green Energy and Technology, DOI 10.1007/978-3-319-30127-3_51

In the environment, pharmaceuticals have been detected in hospital wastewaters (Jose-Gomez et al. 2006), in surface water such as rivers and lakes (Ternes and Joss 2006), in marine waters as well as in soil matrices (Benner et al. 2009; Santos et al. 2009). Recent investigations have shown that a number of pharmaceutical compounds, such as Naproxen, Ibuprofen, Tylosin, spiramycin, Diclofenac, Naproxen, Amoxicillin and Ibuprofen (Yu et al. 2006; Kemper et al. 2008), are not completely eliminated in wastewater treatment, they are often toxic and non-biodegradable, becoming extremely dangerous for the ecosystems (Vıctor et al. 2011; Westergaardn et al. 2001).

New waste water purification technology leading to the complete mineralization of organic pollutants is now considered as being the most suitable solution concerning the decrease of the contamination level in water and wastewater (Lanao et al. 2012; Magalhaes and Lago 2009). Semiconductor-assisted photocatalysis has attracted considerable attention because it uses advanced oxidation process as a tool for implementing large-scale purification of waste waters at low cost. Most studies have focused on large bandgap semiconductors oxides, such as TiO_2 and ZnO, whose photo-excitation by UV light provides electron-hole pairs capable of initiating the production of hydroxyl radicals in water (Radjenovic et al. 2009; Gumy et al. 2006; Oyama et al. 2004; Mathews et al. 2009). Although TiO_2 is universally recognized as being the most photoactive catalyst (Farooq et al. 2009; Shifu and Gengyu 2005), many reports have highlighted the effectiveness of ZnO in removing organic compounds in water matrices (Bahnemann 2004). ZnO photocatalysts in most studies are generally applied in the form of aqueous suspensions (Carlos et al. 2000).

The aim of this work was to investigate the kinetic and efficiency parameters of the photo-oxidation of mixture of two antibiotics models (tylosine and spiramycin) in an aqueous solution under UV illumination. The bibliographic research indicated that assisted photocatalysis degradation of the tylosin and the spiramycin in the same solution has never been reported. On the other hand, biological methods were used for the elimination of this pollutant but the elimination ratio is not important. Biodegradation and biosorption of tetracycline and tylosin were conducted by Amrane. Tetracycline presented a good adsorbability while tylosin remained mostly present in the soluble phase. The Langmuir maximum adsorption capacity was found to be 72 and 7.7 mg g^{-1} for tetracycline and tylosin respectively (Amrane et al. 2009). Ching-Hua et al. estimated the tylosin concentrations in animal's liquid waste to be 496 mg/day-animal. The detection limit of tylosin in waste water is 0.01 mg L^{-1} (Ching-Hua et al. 2001).

Materials and Methods

All the experiments were carried out under an artificial UV light radiation using Pyrex glass helical reactor with an illuminated volume of 800 mL. A peristaltic pump permitted the circulation of water from a glass reservoir to the reactor with a flow rate of 3.787 mL s^{-1}. The volume of the solution in the system (1000 mL) is divided into two parts: the 800 mL irradiated volume and the dead volume present in connecting tubing and in the reservoir. The reactor was illuminated from the axe with a UV lamp. This lamp has a spectral light that distributes the emitted photons at wavelengths less than 365 nm.

Chemical and Reagents

Tylosin and spiramycin were obtained from Sinochem Corporation. The semiconductors employed as photocatalysts are TiO_2 Dugussa P25 obtained from Merck, with a BET surface area of 50 m^2 g^{-1} and an average particle size equal to 2 μm and Zinc oxide was purchased from Aldrich Chemical Company of 99 % purity, with a BET surface area of 9 m^2 g^{-1} and an average particle size equal to 1.5 μm. This kind of particles is small enough to be transported by water suspension. Solutions were prepared by dissolving requisite quantity of Tylosin and/or Spiramycin in distilled water before each experiment. The pH of the solutions was not adjusted (free pH of the solution) and the catalyst was introduced in solution of Tylosin or spiramycin.

The solution was stirred for at least 20 min in the dark to allow the equilibration of the system, so that the problem of the loss of compound caused by the adsorption can be solved. The zero time reading was obtained from a blank solution kept in the dark but otherwise treated similarly to the irradiated solution. Irradiations were obtained out using a Phillips ATLD 24 W lamp ($\lambda max = 365$ nm). 2 mL of samples were collected before and after at regular intervals during irradiation from the reservoir by a plastic syringe then filtered in a Millipore disk. The tylosin concentration was determined by measuring the absorption of pollutants, on a UV–vis spectroscopy (UV-Visible Lambda 25, commercialized by Perkin Elmer Company). A correlation curve between tylosin, spiramycin concentrations and the absorption was set-out.

Results and Discussion

The organic pollutants are never found alone in the environment; wastewater is a complex mixture of several chemical compounds. To get closer to reality, we conducted a study of a mixture of two pollutants (spiramycin and tylosin) to improve the influence of the coexistence of pollutants on the photocatalytic kinetics degradation.

The presence of a second pollutant may inhibit the conversion of the first one, which can be explained by a competition and a greater affinity of the second pollutant for the catalyst than the first one (Jose-Gomez et al. 2006). The study of the photodegradation of tylosin and spiramycin mixture was carried out with

Table 51.1 Different solutions considered during the codegradation

C_{0Spi} (mg L^{-1})	C_{0tyl} (mg L^{-1})	C_{Spir}/C_{tyl}
2.5	10	0.25
5	10	0.50
10	10	1.00
15	10	1.50
20	10	2.00

reports of mass concentrations from 0.25 to 2. To achieve this study, the solutions were prepared with different weight ratios of tylosin/spiramycin (Table 51.1) keeping the operating conditions of TiO_2 and ZnO concentrations of 0.05 mg L^{-1}, flow rate Q equal to 3.787 mL s^{-1} and a free pH of the solution.

Effect of TiO_2 on the Co-degradation

The photocatalytic degradation of an initial concentration of 10 mg L^{-1} of tylosin with different amounts of spiramycin was performed using the titanium dioxide suspension as catalyst. The study showed that the adsorption of these two pollutants on the surface of catalyst was negligible. The kinetics of the two degradation curves of pollutants during photocatalytic treatment are shown in Figs. 51.1 and 51.2 for tylosin and spiramycin respectively. The concentration of pollutants decreases exponentially involving first-order kinetics. Mineralization of spiramycin and tylosin is complete after 100 and 120 min of the treatment respectively.

In the presence of spiramycin, tylosin degradation is slowed. The idea of a competition for adsorption sites between tylosin and spiramycin is required. The results of the degradation kinetics of tylosin are shown in Table 51.2. We observed that the apparent rate constant of degradation increases with increasing concentration of spiramycin to a ratio equal to 1, then it remains practically constant for ratios greater than 1. This can be explained by the competitiveness of the pollutants adsorption which minimizes the amount of tylosin adsorbed on the catalyst surface.

The results revealed in Fig. 51.2 and in Table 51.2 shows that the presence of an amount of tylosin affects the degradation kinetics of the spiramycin. The degradation of tylosin and spiramycin mixture was compared with the photodegradation of the spiramycin alone in the solution; the results show that the total mineralization time of the spiramycin alone (40 min) is shorter than in the tylosin/spiramycin mixture which presents 120 min of radiation. The degradation of spiramycin is slowed with increasing the concentration of this pollutant but the difference is insignificant for the various studied reports except for the 0.25, where the reaction kinetic follow a pseudo Zero order. The degradation kinetic of spiramycin for other reports follows a pseudo first order (Table 51.3).

To compare between the different calculated rate constants, we plotted in Fig. 51.3, the k/k_0 report based on the concentrations of spiramycin and tylosin (C_{Spi}/C_{Tyl}) with k_0, the apparent speed constants of tylosin and spiramycin alone.

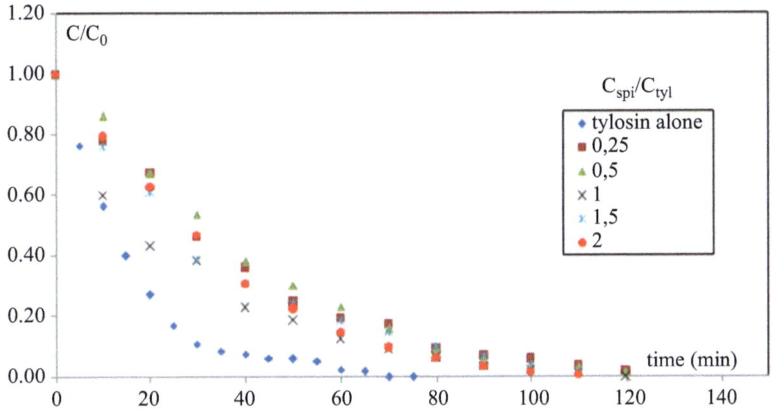

Fig. 51.1 Photodegradation of tylosin in the presence of spiramycin with different ratio C_{Spi}/C_{tyl},, $C_{TiO2} = 0.05$ g L^{-1}, $Q = 3.787$ mL s^{-1}, free pH

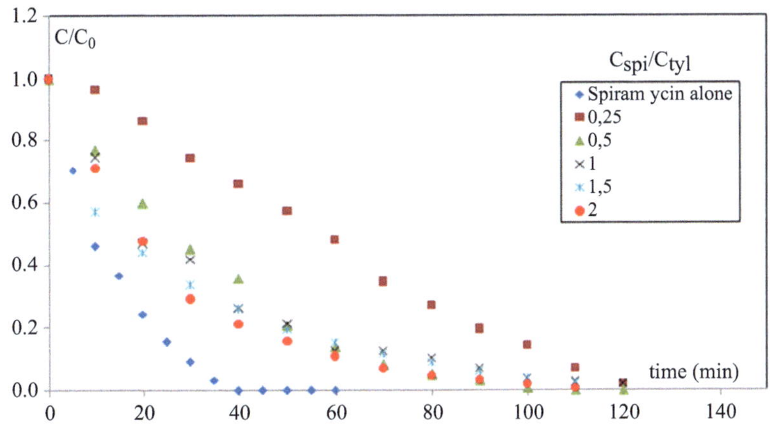

Fig. 51.2 Photodegradation of the spiramycin for different ratio C_{Spi}/C_{tyl}, $C_{TiO2} = 0.05$ g L^{-1}, $Q = 3.787$ mL s^{-1}, free pH

Table 51.2 Rate constant, initial rate and the half time of the degradation the tylosin on the presence of TiO_2

C_{spi}/C_{tyl} (mg/mg)	k_{app} (min^{-1})	$r_0.10^3$ (mmol L^{-1} min^{-1})	$t_{1/2}$ (min)
0	0.068 ± 0.002	0.739 ± 0.004	10
0.25	0.021 ± 0.001	0.229 ± 0.005	33
0.5	0.024 ± 0.003	0.262 ± 0.002	29
1	0.035 ± 0.002	0.382 ± 0.008	20
1.5	0.031 ± 0.003	0.338 ± 0.011	22
2	0.032 ± 0.001	0.349 ± 0.010	22

Table 51.3 Rate constant, initial rate and the half time of the degradation of the spiramycin on the presence of TiO_2

C_{Spi}/C_{tyl}	k_{app} (min^{-1})	$r_0.10^3$ (mmol L^{-1} min^{-1})	$t_{1/2}$ (min)
Spi. alone	0.078 0.003 min^{-1}	0.908 ± 0.005	9
0.25	$0.254.10^{-4} \pm 0.001$ mol L^{-1} min^{-1}	0.025 ± 0.002	57
0.5	0.030 ± 0.003 min^{-1}	0.175 ± 0.001	23
1	0.031 ± 0.001 min^{-1}	0.361 ± 0.001	22
1.5	0.032 ± 0.002 min^{-1}	0.559 ± 0.003	22
2	0.038 ± 0.005 min^{-1}	0.885 ± 0.007	18

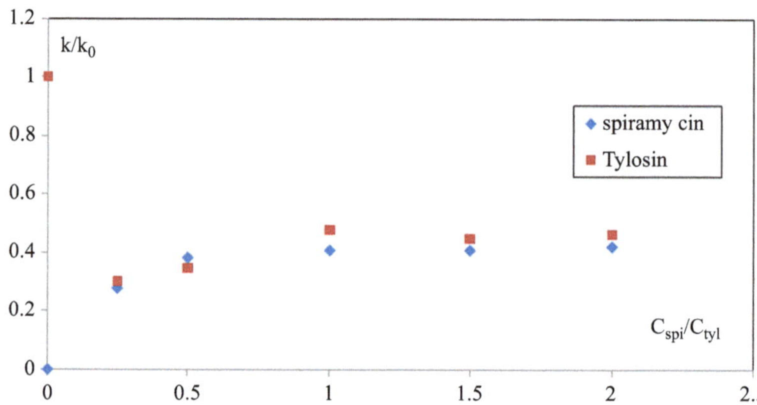

Fig. 51.3 Evolution of the constant rate for different ratios C_{Spi}/C_{tyl}, Q = 3.787mL s^{-1}, $C_{TiO2} = 0.05$ g L^{-1}

When using the 0.25 report, we found that there is degeneration of the order of the reaction, the rate constant relative to the apparent reaction order equal to the unit was estimated by dividing it by the initial concentration of the pollutant. The plot k/k_0 based C_{Spi}/C_{Tyl} shows that from a concentration ratio equal to 1, the two curves are superimposed indicating that the degradation kinetics of the two pollutants is practically the same.

Effect of ZnO on Co-degradation

It is interesting to study the behavior of a solution containing a mixture of tylosin and spiramycin in competition in the reaction medium in the presence of zinc oxide (ZnO). Thus, we studied the mineralization of a solution composed of a mixture of spiramycin and tylosin in optimal conditions previously defined ($C_{TiO2} = 0.05$ g L^{-1}, Q = 3.787 mL s^{-1}, pH and free $C_{polluant} = 10$ mg L^{-1}). Concentrations measured during the photodegradation are plotted in Figs. 51.4 and 51.5.

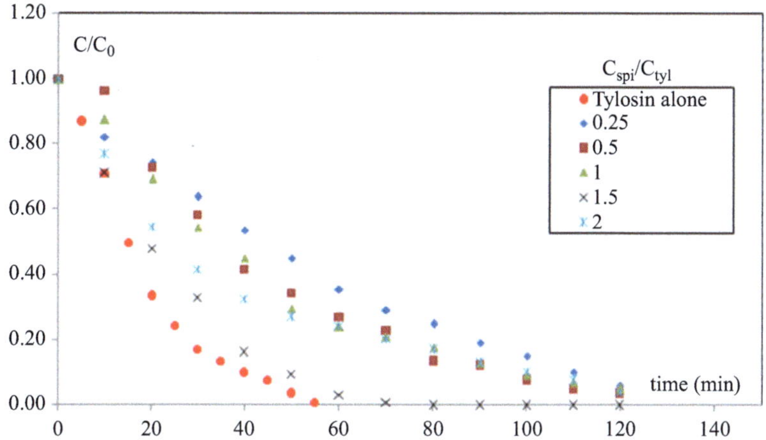

Fig. 51.4 Degradation of tylosin for different ratios C_{tyl}/C_{Spi}. $C_{ZnO} = 0.05$ g L^{-1}, $Q = 3.787$ mL s^{-1}

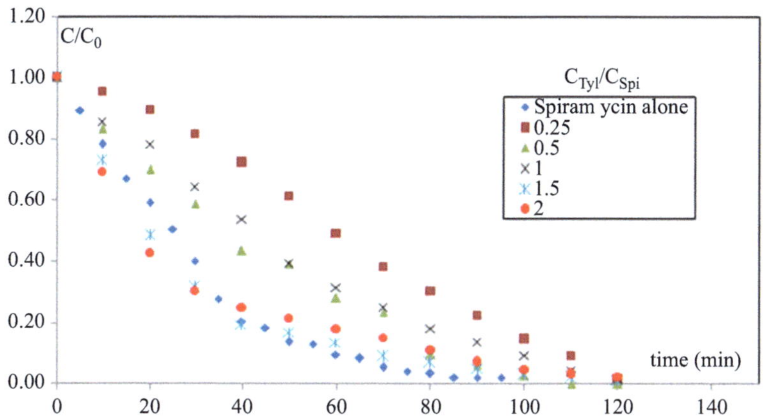

Fig. 51.5 Degradation of the spiramycin for different ratios C_{tyl}/C_{Spi}, $C_{ZnO} = 0.05$ g L^{-1}, $Q = 3.787$ mL s^{-1}

The photodegradation of tylosin is delayed by the coexistence of spiramycin. Mineralization of tylosin is complete after 120 min for the different ratios studied except for the 1.5 and tylosin alone. It's concluded that the photo-oxidation constant rate of tylosin alone is higher than k_{app} obtained in the case of tylosine mixed with the spiramycin. However, the decomposition of spiramycin is strong when it is alone or mixed with equal ratios at 1.5 and 2 where the kinetic curves codégradation appear very close together and the pseudo-first order. The apparent rate constants for degradation of tylosin increase with increase in the amount of 1.5 relative to the spiramycin, and then decreases to a ratio equal to 2 (Table 51.4).

Table 51.4 Values of the rate constants, initial speed and half-degradation reaction time of tylosin by using ZnO as catalyst

C_{tyl}/C_{Spi}	k_{app} (min^{-1})	$r_0.10^3$ ($mmol\ L^{-1}\ min^{-1}$)	$t_{1/2}$ (min)
Tyl. alone	0.056 ± 0.001	0.611 ± 0.002	12
0.25	0.017 ± 0.001	0.185 ± 0.002	41
0.5	0.022 ± 0.001	0.240 ± 0.003	32
1	0.023 ± 0.001	0.251 ± 0.001	30
1.5	0.044 ± 0.002	0.480 ± 0.004	16
2	0.026 ± 0.001	0.393 ± 0.001	19

Table 51.5 Values of the rate constant, initial rate and the half time of the degradation of the spiramycin on the presence of ZnO

C_{tyl}/C_{Spi}	k_{app} (min^{-1})	$r_0 10^{-3}$ ($mmol\ L^{-1}\ min^{-1}$)	$t_{1/2}$ (min)
Spi alone	0.028 ± 0.004	0.356 ± 0.003	25
0.25	$0.025 \pm 0.003.10^{-3}$ $mmol\ L^{-1}\ min^{-1}$	0.025 ± 0.002	57
0.5	0.038 ± 0.003 min^{-1}	0.064 ± 0.008	45
1	0.049 ± 0.004 min^{-1}	0.128 ± 0.011	45
1.5	0.069 ± 0.002 min^{-1}	1.204 ± 0.031	10
2	0.074 ± 0.005 min^{-1}	1.373 ± 0.025	12

The curves in Fig. 51.5 correspond to the temporal evolution of the reduced spiramycin concentration for different ratios of tylosin/spiramycin. From these curves, the time of degradation of the pollutant is the lowest for the mixtures of spiramycin and tylosin ratios equal to 1.5 and 2 in the case of the spiramycin. The curves are almost super-imposable for these other values, while the highest degradation time recorded is when using the tylosin/spiramycin ratio equal to 0.25. The co-degradation of the spiramycin follows a pseudo zero order for a ratio of 0.25 and becomes a pseudo-first order kinetics for higher values of C_{Tyl}/C_{Spi} ratios (Table 51.5).

Figure 51.6 present the values of k/k_0 versus the ratio of concentrations of spiramycin and tylosin (C_{Spi}/C_{Tyl}) in the case of two pollutants photodegradation in the presence of zinc oxide as catalyst, k0 is the apparent rate constant for tylosin and spiramycin alone in the solution. When 0.25 ratio is used, it's found that there is a degeneration of the reaction order. It's concluded that the reduced constant k/k_0 speed increases with increasing the ratio of these two pollutants in the case of spiramycin but decreases in the case of tylosin. Furthermore, it's noted that the values of rate constants for the degradation of tylosin are located below that of spiramycin.

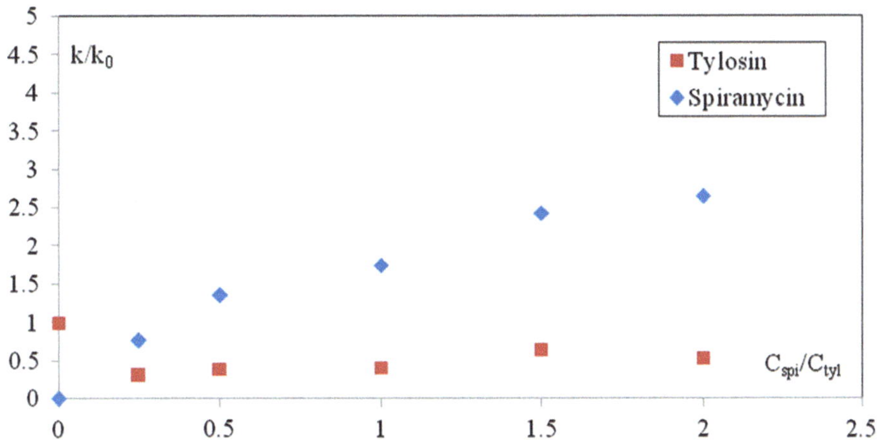

Fig. 51.6 Evolution of the constant rate of the spiramycin and tylosin for different ratios C_{Spi}/C_{tyl}, $Q = 3.787$ mL s^{-1}, $C_{ZnO} = 0.05$ g L^{-1}

Comparison Between ZnO and TiO_2 Efficiency on the Photodegradation Process

The tylosin and Spiramycin degradation have been studied by heterogeneous photocatalysis process with Titanium dioxide (TiO_2), Degussa P-25, and zinc oxide (ZnO). The *comparison Between TiO_2* and ZnO Photocatalysis rate is reported in Figs. 51.7 and 51.8. In the case of the tylosin, the results show that the degradation with TiO_2 presents a better photocatalytic behavior than the ZnO for ratios lower than 0.25 and 1.2, for highest ratios the photodegradation rate is more important when using Zinc oxide.

The titanium dioxide and Zinc oxide are also reported as good photocatalysts for the degradation of the Spiramycin. The degradation rates obtained by using TiO_2 are also important than ZnO values for ratios less than 1.25 and decrease for the highest ratios. In other hand, two picks are obtained for rates equal to 0.48 and 0.38 10^{-3} mmol L^{-1} min^{-1} when using TiO_2 and ZnO respectively.

Conclusion

In this study, ZnO and TiO_2 are used as photocatalysts to degrade two pharmaceutical pollutants. It has been shown in the literature review that many of contaminants are present in the water to be treated. It would then be appropriate to assess the effectiveness of the catalysts in the case of mixtures of target molecules. To investigate this situation, the photodegradation of a mixture of tylosin and spiramycin was conducted with different mass ratios. The obtained results show a

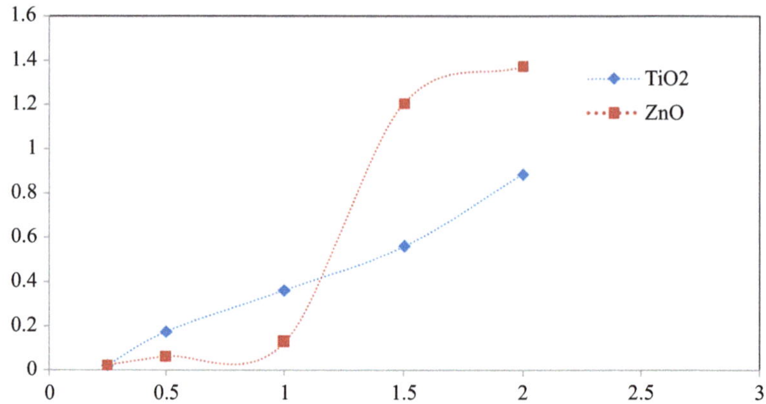

Fig. 51.7 Tylosin photodegradation rate in the presence of TiO_2 and ZnO catalysts

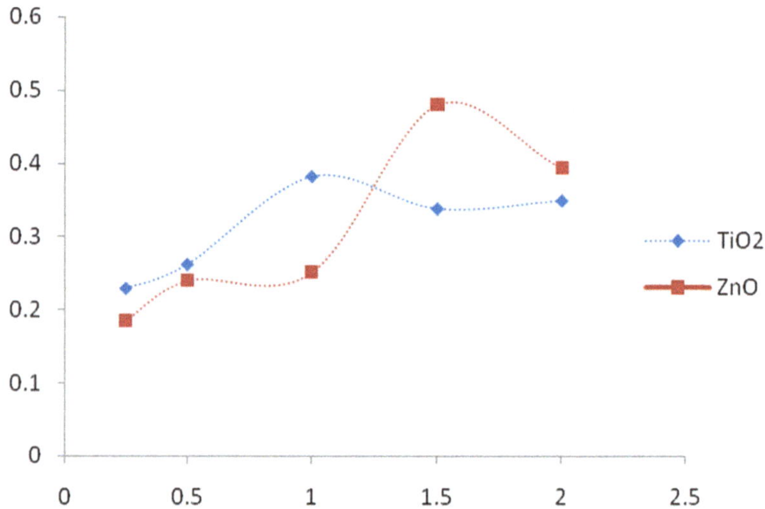

Fig. 51.8 Spiramycin photodegradation rate in the presence of TiO_2 and ZnO catalysts

decrease of the removal rate when other pollutant is present in the same solution. The presence of TiO_2 photocatalysis is effective for the complete degradation of the tylosin and the spiramycin pollutants. Photocatalytic degradation by a suspension of ZnO also showed a good performance for the removal of these pollutants. Degradation ratio of both pollutants in the presence of ZnO is substantially important than TiO_2 for ratios less than 1.25, this ratio increases for the highest ratios of pollutants. The effectiveness of the reactor is considerable giving the advantage of being used in real cases upstream or downstream of the biological treatment to a complete water treatment of urban and industrial waste by using TiO_2 and ZnO as catalysts.

Nomenclature

C_0	Initial concentration of the pollutant, mg/L
C	Concentration of the pollutant, mg/L
k_{app}	Kinetic rate constant, min^{-1}
r_0	Rate of the degradation of pollutant, mmol/L min
$T_{1/2}$	Half time of the reaction, min
Q	Flowrate of the water solution, mL s^{-1}
UV	Ultraviolet light
Tyl.	Tylosin
Spi.	Spiramycin

References

Amrane, A., Prado, N., & Ochoa, J. (2009). Biodegradation and biosorption of tetracycline and tylosin antibiotics in activated sludge system. *Process Biochemistry, 44*, 1302–1306.

Assabane, A., Ichou, Y., Tahiri, H., Guillard, C., & Hermann, J. M. (2000). Photocatalytic degradation of polycarboxylic benzoic acids in UV-irradiated aqueous suspensions of titania. Identification of intermediates and reaction pathway of the photo-mineralization of trimellitic acid (1,2,4-benzene tricarboxylic acid). *Applied Catalysis B: Environmental, 24*(2), 71–87.

Bahnemann, D. (2004). Photocatalytic water treatment: Solar energy applications. *Solar Energy, 77*, 445–459.

Benner, J., Ternes, T. A., & Scholz, J. (2009). Ozonation of beta-blockers: kinetic studies, identification of oxidation products and pathways. *Environmental Science & Technology, 42*, 10–18.

Campion, L., Giannotti, C., & Ouazzani, J. (1999). Photocatalytic degradation of 5-nitro-1,2,4-triazol-3-one (NTO) in aqueous suspension of TiO_2 comparison with Fenton oxidation. *Chemosphere, 38*(7), 1561–1570.

Carlos, S., Gouvêa, A. K., Wypych, F., Moraes, S. G., Durán, N., Nagata, N., et al. (2000). Semiconductor-assisted photocatalytic degradation of reactive dyes in aqueous solution. *Chemosphere, 40*, 433–440.

Ching-Hua, H., Karen, E. R. J. P., & David, L. S. (2001). Occurrence and fate of antibiotic compounds in municipal wastewater and animal's waste. *Proceedings of the Water Environment Federation, 60*, 686–697.

Farooq, M., Raja, I. A., & Pervez, A. (2009). Photocatalytic degradation of TCE in water using TiO_2 catalyst. *Solar Energy, 83*, 1527–1533.

Gumy, D., Rincon, A. G., Hajdu, R., & Pulgarin, C. (2006). Solar photocatalysis for detoxification and disinfection of water: Different types of suspended and fixed TiO_2 catalysts study. *Solar Energy, 80*, 1376–1381.

Herrmann, J. M., Guillard, C., Disdier, J., Lehaut, C., Malato, S., & Blanco, J. (2002). New industrial titania photocatalysts for the solar detoxification of water containing various pollutants. *Applied Catalysis B: Environmental, 35*(4), 281–294.

Jose-Gomez, M., Petrovic, M., Fernandez-Alba, A. R., & Barcelo, D. (2006). Determination of pharmaceuticals of various therapeutic classes by solid-phase extraction and liquid chromatography tandem mass spectrometry analysis in hospital effluent wastewaters. *Journal of Chromatography A, 1114*, 224–233.

Kemper, N., Färber, H., Skutlarek, D., & Krieter, J. (2008). Analysis of antibiotic residues in liquid manure and leachate of dairy farms in Northern Germany. *Agricultural Water Management, 95*, 1288–1292.
Lanao, M., Ormad, M. P., Mosteo, R., & Ovelleiro, J. L. (2012). Inactivation of Enterococcus sp. By photolysis and TiO_2 photocatalysis with H_2O_2 in natural water. *Solar Energy, 86*, 619–625.
Magalhaes, F., & Lago, M. R. (2009). Floating photocatalysts based on TiO_2 grafted on expanded polystyrene beads for the solar degradation of dyes. *Solar Energy, 83*, 1521–1526.
Mathews, N. R., Morales, E. R., Cortes-Jacome, M. A., & Toledo Antonio, J. A. (2009). TiO_2 thin films-Influence of annealing temperature on, optical and photocatalytic properties. *Solar Energy, 83*, 1499–1508.
Neelavannan, M. G., & Ahmed Basha, C. (2010). Ag–TiO_2 doped photo catalytic degradation of Procion blue H-B dye in textile washwater. *Toxicological & Environmental Chemistry, 92*(8), 1423–1434.
Oyama, T., Aoshima, A., Horikoshi, S., Hidaka, H., & Zhao, J. (2004). Solar photocatalysis, photodegradation of a commercial detergent in aqueous TiO_2 dispersions under sunlight irradiation. *Solar Energy, 77*, 525–532.
Radjenovic, J., Petrovic, M., & Barcel, D. J. (2009). Fate and distribution of pharmaceuticals in wastewater and sewage sludge of the conventional activated sludge (CAS) and advanced membrane bioreactor (MBR) treatment. *Water Research, 43*, 8–31.
Santos, J. L., Aparicio, I., Callejon, M., & Alonso, E. (2009). Occurrence of pharmaceutically active compounds during 1-year period in wastewaters from four wastewater treatment plants in Seville (Spain). *Journal of Hazardous Materials, 164*, 1509–1519.
Shifu, C., & Gengyu, C. (2005). Photocatalytic degradation of organophosphorus pesticides using floating photocatalyst $TiO_2.SiO_2$/beads by sunlight. *Solar Energy, 79*, 1–9.
Simon, J. C., & Dauby, I. R. (2008). Evaluation de l'efficacité de l'oxydation avenacée par photocatalyse hétérogène. U.V./TiO_2 sur un effluent industriel contamine par des composes organiques non biodégradables (colorants). *Revue scientifique des ISILF, 22*, 18–20.
Sung-Suh, H. M., Choi, J. R., Hah, H. J., Koo, S. M., & Bae, Y. C. (2004). Comparison of Ag deposition effects on the photocatalytic activity of nanoparticulate TiO_2 under visible and UV light irradiation. *Journal of Photochemistry and Photobiology A: Chemistry, 163*(1–2), 37–44.
Ternes, A., & Joss, A. (2006). *Human pharmaceuticals, hormones and fragrances: The challenge of micropollutants in urban water management* (1st ed., p. 453). London: IWA Publishing.
Turchi, C. S., & Ollis, D. F. (1989). Mixed reactant photocatalysis: Intermediates and mutual rate inhibition. *Journal of Catalysis, 119*(2), 483–496.
Vıctor, J. P., Lıvia-Pinho, X., & Ariana, M. A. (2011). Treatment of textile wastewaters by solar-driven advanced oxidation processes. *Solar Energy, 85*, 1927–1934.
Westergaardn, K., Muller, A. K., Christensen, S., Bloem, J., & Sorensen, S. J. (2001). Effects of tylosin as a disturbance on the soil microbial community. *Soil Biochemistry, 33*, 2061–2071.
Yu, J. T., Bouwer, E. J., & Coelhan, M. (2006). Occurrence and biodegradability studies of selected pharmaceuticals and personal care products in sewage effluent. *Agricultural Water Management, 16*, 72–80.

Chapter 52
Comparison Between the Photocatalytic Degradation of a Textile Dye Under Sun Light and Artificial Irradiation

N. Sahraoui, N. Chekir, and D. Tassalit

Introduction

Dyes occupy an important place in synthetic organic compounds. They are used extensively in industries including textiles, ink, plastics, cosmetics, tannery, and are therefore, automatically found in water as an industrial pollutants (Schwarzenbach et al. 2006). Their rejection into aquatic systems is causing damage to the environment due to their toxicity, which requires treatment (Liu et al. 2009).

Some of these pollutants are very stable and therefore difficult to be degraded naturally or by using classic processes. Others are able to be degraded partially but producing highly stable intermediate compounds which can have more toxicity than the first pollutant.

The complexity of these pollutants and their color greatly affect the efficiency of the classical treatments application, hence the need to develop more efficient processes than conventional methods which can degrade recalcitrant pollutants.

Among the most recent progress, based on advanced oxidation techniques, heterogeneous photocatalysis is nowadays, an emerging solution to the problems of pollution by organic pollutants (Gaya and Abdullah 2008).

The Advanced Oxidation Processes are defined by (McGuigan et al. 1999) as the very interesting alternative for the degradation of non-biodegradable organic pollutants by biological treatment processes. These POAs are based on the generation and use of the hydroxyl radical (OH°).

N. Sahraoui (✉)
Département de Génie des Procédés, Laboratoire des phénomènes de transfert, USTHB, BP32, El Alia, Bab Ezzouar, 16111 Alger, Algeria
e-mail: sahraouinaima@yahoo.fr

N. Chekir • D. Tassalit
Unité de Développement des Equipements Solaires (UDES)/Centre de Développement des Energies Renouvelables CDER, Bou-Ismail 42415, W. Tipaza, Algeria

P. Grammelis (ed.), *Energy, Transportation and Global Warming*, Green Energy and Technology, DOI 10.1007/978-3-319-30127-3_52

Any process requires, in addition to the products for the treatment, a source of energy. Solar energy available throughout the year with varying intensities can be a useful alternative source, especially in developing countries. Indeed, this renewable energy source is free and inexhaustible; it is the most abundant energy on Earth.

Photocatalytic treatment is thus presented as a technology of choice for the pollution control and the integration of wastewater reuse, because it is an efficient, simple and economical method. These attractive factors generate a great interest from researchers for understanding, optimizing the industrial application of this process (Helali 2012).

The objective of our study is to demonstrate the effectiveness of the photocatalytic process for treating contaminated water with a Sandocryl Red (BR46) textile dye. For this purpose, optimization of the operating conditions was carried out. Photocatalytic degradation experiments were carried out at room temperature by irradiating Zinc oxide ZnO semiconductor using a batch photoreactor under ultraviolet (UV) radiation via sunlight or artificially by using a UV lamp. The temporal evolution of the initial concentration of the contaminant was studied with different concentrations of catalyst. The light intensity is measured with a digital UV pyranometre.

Materials and Measurement Method

Chemical Reagents

The used pollutant in this study is the Sandocryl Red (BR46), a pollutant that belongs to the family of azo dyes. It is a synthetic red dye powder which is very soluble in water (Table 52.1).

The used photocatalyst in this study is the commercial dioxide Zinc (ZnO), it is a white powder supplied by Sigma Aldrich. It has a 100 % zincite structure with a specific surface area of 9 $m^2\ g^{-1}$ and a particle diameter of 1.5 μm.

Experimental Facility

The photocatalysis treatment of polluted water is conducted by using used batch photoreactor of 1 L volume capacity. The installation consists essentially of a batch reactor exposed to either direct sunlight or the Ultraviolet radiation. This last is UVC radiation of a lamp with a power of 15 W. The experimental setup is shown in Fig. 52.1.

Table 52.1 Physical and chemical properties of the BR46

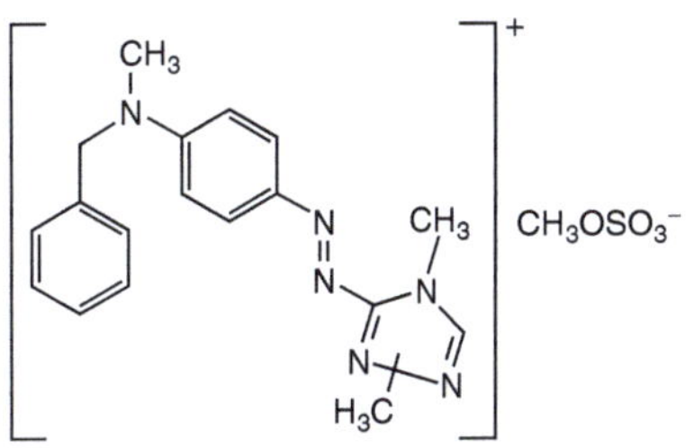 **Chemical structure of the BR46 dye**	**Physical properties**	
	Physical State	Solid
	Appearance	Powder
	Chemical properties	
	Molar mass	432 g mol^{-1}

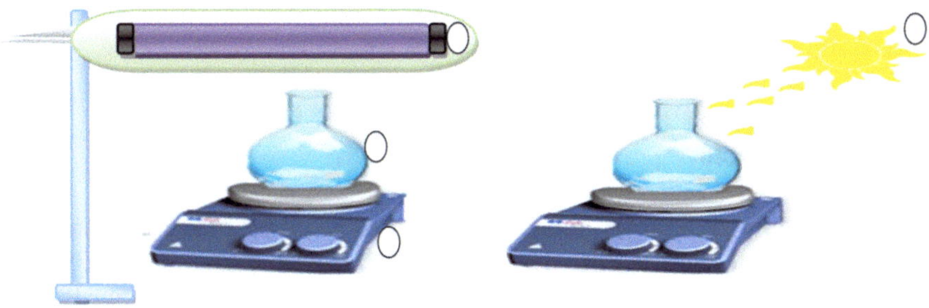

Fig. 52.1 Experimental setup

Measurement of the Pollutant Concentration

The temporal evolution of the pollutant concentration of Sandocryl Red is given by the measurement of absorbance by a UV-visible spectrometer with dual beam of type Shimadzu UV1800m, coupled to acquisition results software (UV PROB). The contaminant has a maximum absorbance at a wavelength $\lambda = 530$ nm. The solar flux is measured at 5-min intervals of time between 9 h am and 15 h, using a pyranometer coupled to software (Kipp & ZONZN, CMP11) with inclination of 36°.

The aim of this work is to perform the parametric study on the photocatalytic degradation of BR46 in the presence of zinc oxide. The reaction mixture was stirred to homogenize the solution. Three milliliters of samples are taken every 15 min of reaction. Before analysis by UV spectrophotometer, the mixture is filtered using a cellulose acetate filter to remove any traces of ZnO before analysis. At the end of each experiment, the system is immediately rinsed with distilled water.

Results and Discussion

Determination of the Wavelength of the BR46

The absorbance of BR46 is committed to their maximum adsorption. The absorption spectrum was obtained by scanning the solution of BR46 with concentration of 10 mg L^{-1} between 300 and 600 nm. Figure 52.2 shows the absorption spectrum of the pollutant for different times of the photodegradation reaction. It is clearly observed that the BR46 maximum absorption coincide with a wavelength $\lambda = 530$ nm.

Direct Photolysis and Photocatalysis

Photolysis of BR46

We conducted a preliminary study to see the degradation of the BR46 under UV radiation in the absence of ZnO. To study the degradation of the pollutant by direct photolysis, a solution of this pollutant with a concentration of 10 mg L^{-1} was exposed to UV irradiation in the photoreactor. The obtained results show that the dye concentration decreases insignificantly, demonstrating that the irradiation with

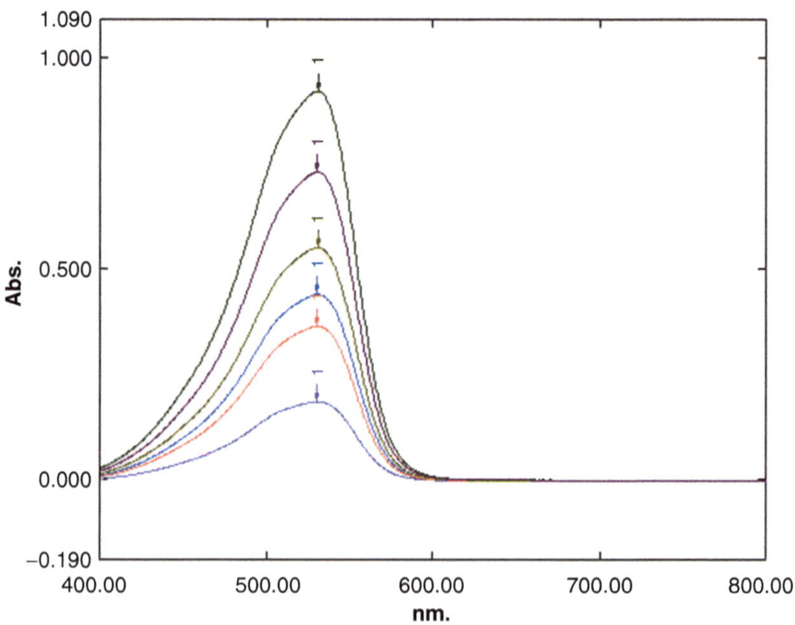

Fig. 52.2 Absorption spectra for different concentrations of BR46

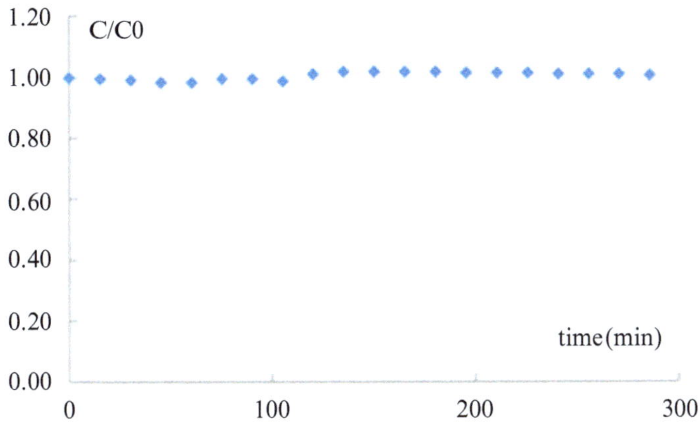

Fig. 52.3 Photolysis of BR46, $C_{BR46} = 10$ mg L^{-1}, free pH

UV in the absence of catalyst have no effect on the degradation of the dye. The ratio of degradation is about 2 % after 300 min of irradiation as shown in Fig. 52.3.

Photocatalytic Degradation of BR46 by Sun Radiation and UV

The photodegradation study of BR46 was performed using zinc oxide as catalyst under solar and UV radiation. In order to optimize the operating conditions of the photodegradation of BR46, several experiments were conducted by varying the following parameters:

- Initial concentration of the pollutant;
- Catalyst concentration.

For the better understanding of the phenomenon of photocatalytic degradation of BR46, knowledge of the degradation reaction order of is essential. For this reason, we took inspiration from the results found in the literature explaining that the kinetics of the phenomenon of photodegradation of many organic pollutants follows a first order reaction whose speed is described by the following equation:

$$r = \frac{dC}{dt} = k_{app} C \quad (52.1)$$

with:

r: Photocatalytic degradation rate (mg L^{-1} min^{-1})
k_{app}: Apparent degradation constant (min^{-1})
C: Dye concentration (mg L^{-1})
t: Irradiation time (min)

The solution concentration of the dye when $t = 0$, generates another equation which gives the linearization:

$$-\ln\frac{C}{C0} = k_{app}\,t \qquad (52.2)$$

Photocatalytic tests were then carried out in the presence of ZnO assuming that the reactions are pseudo-first order.

Influence of the Initial Concentration of the Pollutant Under Sunlight

The concentrations of Sandocryl Red (BR46) was varied between 5 and 40 ppm at a free pH of the solution (between 6, 5 and 7) while maintaining the amount of used catalyst constant and equal to 1 g L^{-1}. Figure 52.4 shows the temporal evolution of the reduced concentrations of Sandocryl Red (BR46). The results confirm that the degradation of BR46 is important when the initial concentration of this pollutant decreases.

The plot of $-\ln C/C_0$ versus time in Fig. 52.5 and the values of K_{app} reported in Table 52.2 shows that the rate constant (Kapp) decreases as the substrate concentration increases. This is explained by a lower participation of photons when the color of the solution becomes more intense and therefore a low activation of ZnO (low production of OH° radicals).

The obtained results show that the rate constant for the degradation (Table 52.2) decreases with increasing of the concentration of BR46, the rate constants of the values are between 0.143 min^{-1} for the lower concentration of pollutant (5 ppm) and 0.048 min^{-1} for the highest concentration (40 ppm).

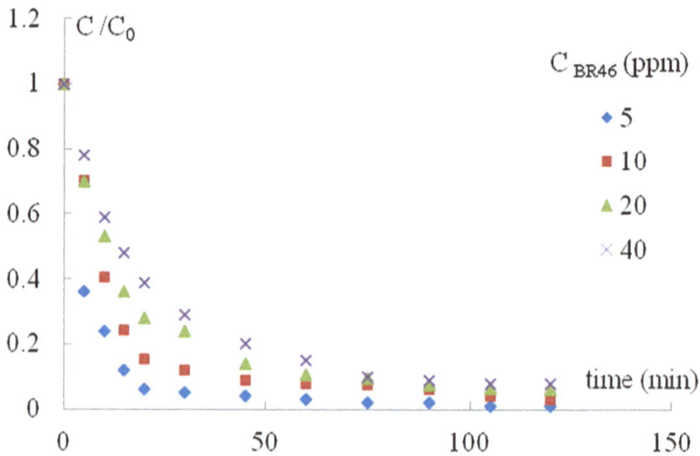

Fig. 52.4 Evolution of the temporal different reduced pollutant concentrations with free pH and $C_{ZnO} = 1$ g L^{-1}

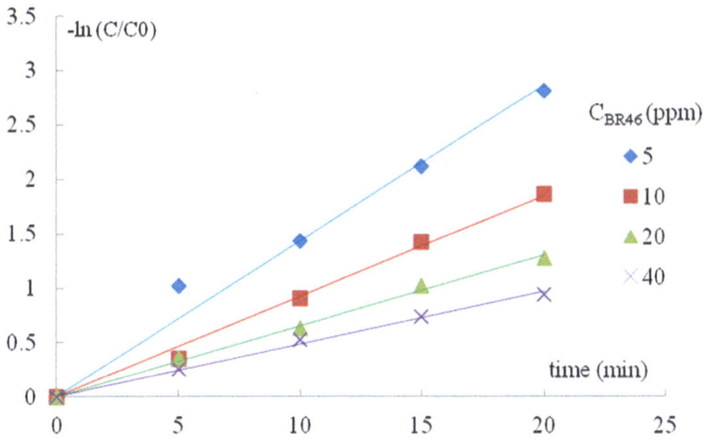

Fig. 52.5 Temporal evolution of ln (C/C0) for different pollutant concentrations with free pH and $C_{ZnO} = 1\ g\ L^{-1}$

Table 52.2 Values of the rate constants for different values of the pollutant concentrations

BR 46 (ppm)	K_{app} (min^{-1})	r_0 ($mg\ L^{-1}\ min^{-1}$)	$t_{1/2}$ (min)	R^2
5	0.143	0.715	5	0.99
10	0.092	0.920	7.5	0.99
20	0.065	1.300	10	0.99
40	0.048	1.921	14	0.99

Effect of the Initial Concentration of the Pollutant Under UV Radiation

The same experiments with sun light were repeated in the presence of UV radiation in order to compare the two modes of radiation maintaining the same operating conditions:

- Free pH of the solution (between 6.5 and 7);
- Initial concentrations of dye range are from 5 to 40 ppm;
- Amount of catalyst used constant are equal to $1\ g\ L^{-1}$.

Figure 52.6 shows the evolution of reduced concentrations of Sandocryl Red (BR46) versus time.

From the curves of Fig. 52.6, the photodegradation is strongly influenced by the initial concentration of the substrate. The degradation of the BR46 is even more important when the initial concentration of this pollutant decreases. By using a concentration of 5 ppm, the degradation is very fast and it reaches an elimination ratio of 97 %.

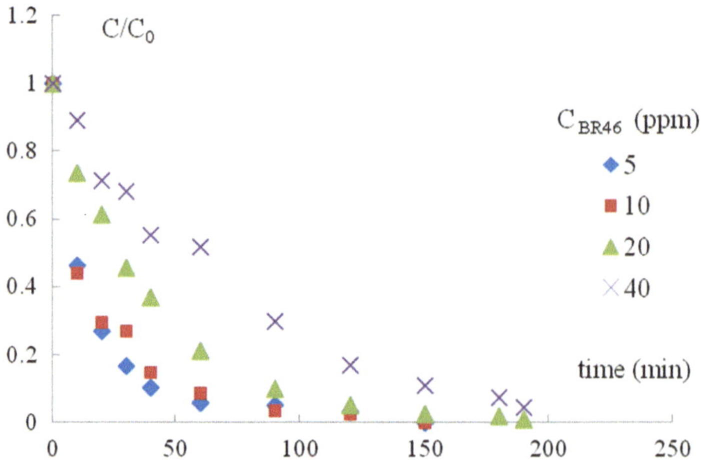

Fig. 52.6 Temporal evolution of different reduced pollutant concentrations with free pH and mass of catalyst = 1 g L^{-1}

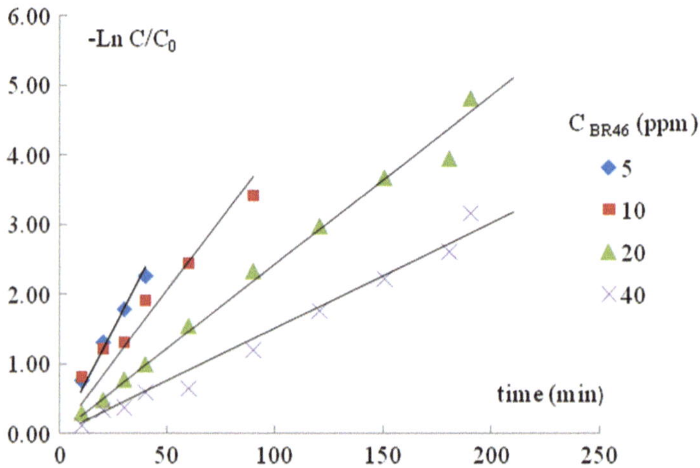

Fig. 52.7 Temporal evolution of $-\ln(C/C_0)$ for different pollutant concentrations with free pH and catalyst mass = 1 g L^{-1}

The plot of $-\ln(C/C_0)$ versus time for various initial concentrations of Sandocryl Red (BR46) (Fig. 52.7) show that the kinetics is pseudo first-order, the concentration satisfactory follows the expression of the first order equation with a slope equal to the apparent speed constant kapp. The values are shown in Table 52.3.

The obtained results (Fig. 52.7 and Table 52.3) shows that the rate constant of degradation decreases with increasing the concentration of BR46, the rate constants are situated between 0.059 min^{-1} for the lowest concentration of pollutant (5 ppm)

Table 52.3 Values of rate constants for the different values of pollutant concentrations

C_{BR46} (ppm)	K_{app} (min^{-1})	r_0 ($mg\ L^{-1}\ min^{-1}$)	$t_{½}$ (min)	R^2
5	0.059	0.295	11.748	0.95
10	0.041	0.410	16.906	0.90
20	0.24	0.480	28.881	0.98
40	0.015	0.600	46.210	0.98

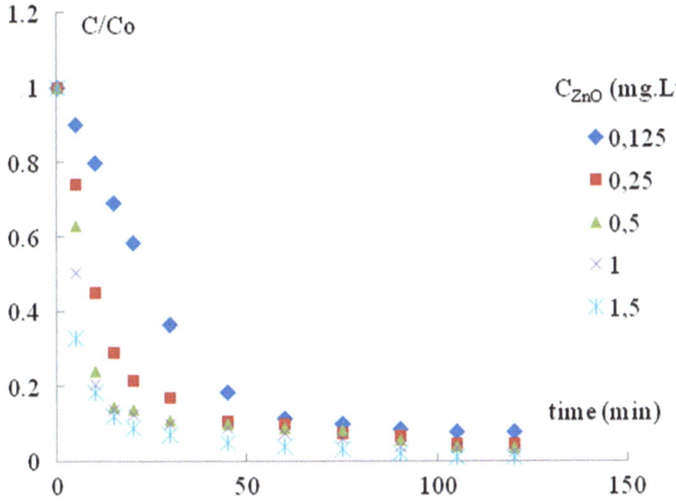

Fig. 52.8 Evolution of the concentration of the pollutant over time with free pH, $C_{pollutant} = 10$ ppm

and 0.015 min^{-1} for the highest concentration (40 ppm). It can be observed that the rate constant (k_{app}) decreases as the substrate concentration increases. This is explained by a lower participation of photons because of the screen effect.

Influence of the Catalyst Concentration Under Sun Radiation

The effect of catalyst concentration (ZnO) on the photo-oxidation of the BR46 was studied using amounts ranging from 0.125 g L^{-1} to 1.5 g L^{-1} under the sun and UV radiation. The experiments were performed with a solution of 10 g L^{-1} of BR46 and free pH (pH of the solution without adjustment). The obtained results are illustrated in Fig. 52.8.

The results presented in Fig. 52.9 show that the rate of deterioration of BR46 ZnO for different concentrations are very high and can reach a value of 100 % of removal, like in the case of high concentrations of catalyst (1,5 g L^{-1}). The rate of degradation and the rate constants increase with increasing the amount of ZnO as

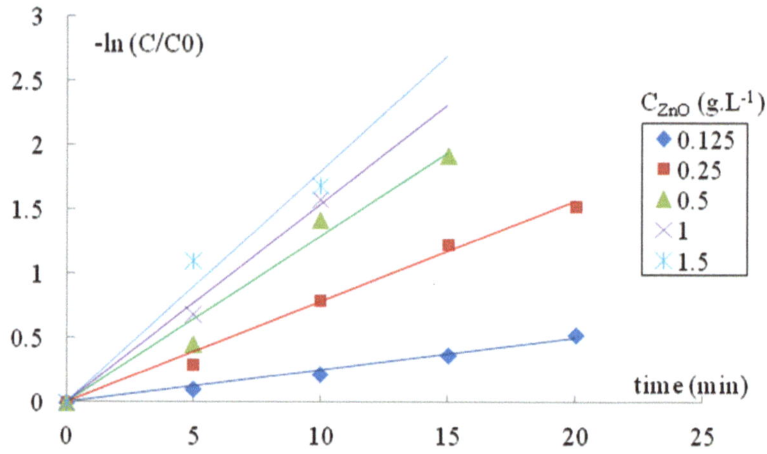

Fig. 52.9 Temporal evolution of ln(C/C0) for different concentrations of ZnO, C0BR46 = 10 mg L^{-1}, free pH

Table 52.4 Values of rate constants, initial speed and time of half-reaction of BR46 degradation for different concentration ZnO

C_{ZnO} (g L^{-1})	Kapp (min^{-1})	r_0 (mg L^{-} min^{-1})	$t_{1/2}$ (min)	R^2
0.125	0.029	0.29	23	0.99
0.250	0.078	0.78	9	0.99
0.500	0.129	1.29	5	0.98
1.000	0.139	1.39	5	0.99
1.500	0.179	1.79	4	0.96

shown by the results in Table 52.4 and Fig. 52.10 which present the temporal evolution of $-\ln C/C_0$ at different concentrations in the catalysts.

Influence of the Catalyst Concentration Under UV

It is important to know the optimal catalyst load for the efficient removal of a pollutant. The effect of catalyst concentration (ZnO) on the photo-oxidation of BR46 under UV light was also conducted under the same operating conditions as for the study of solar radiation using catalyst amounts ranging from 0.125 g L^{-1} to 1.5 g L^{-1}. The experiments were performed with a solution of 10 g L^{-1} of BR46 and free pH (pH of the solution without adjustment). The results obtained are illustrated in Fig. 52.11.

The results shown in Fig. 52.11 show that BR46 deterioration rate by ZnO with different concentrations are very high and can reach a value of 100 % removal, in case of concentrations of catalyst (1.5 g L^{-1}). It is also found that the rate of

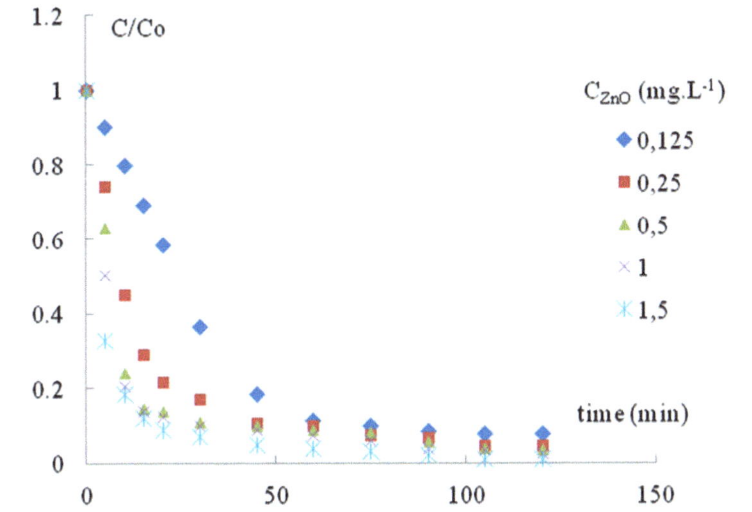

Fig. 52.10 Evolution of different reduced concentrations of the catalyst over time with free pH and $C_{pollutant} = 10$ ppm

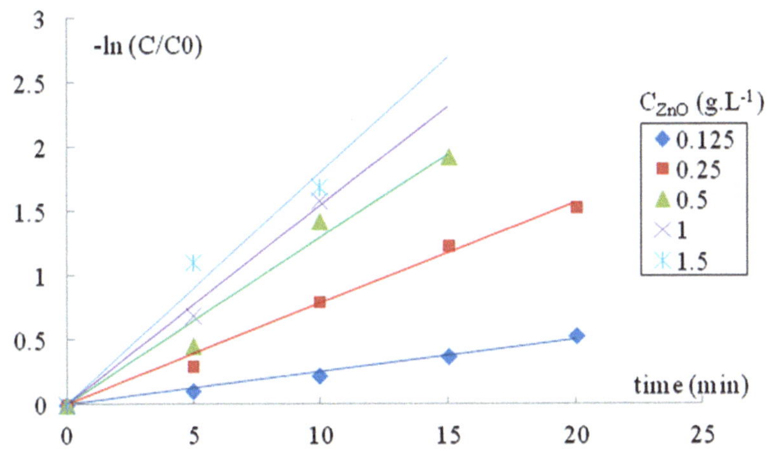

Fig. 52.11 Temporal evolution of ln(C/C0) for different concentrations of ZnO, $C_{OBR46} = 10$ mg L^{-1}, free pH

degradation and the rate constants increase with increasing the amount of ZnO as shown by the results in Fig. 52.11 and the temporal evolution of −ln C/C0 for different catalysts concentrations and Table 52.5.

The increase in catalyst concentration generally favors the absorption of the contaminant to the surface, which increases the degradation rate. However too much affects the light scattering in the solution to treat involving a reduction of the degradation kinetics

Table 52.5 Values of rate constants, initial speed and time BR46 degradation half-reaction for different concentration ZnO

C_{ZnO} (g L^{-1})	K_{app} (min^{-1})	r_0 (mg L^{-1} min^{-1})	$t_{1/2}$	R^2
0.125	0.014	0.14	49.51	0.99
0.25	0.024	0.24	28.88	0.98
0.5	0.026	0.26	26.65	0.95
1	0.027	0.27	25.672	0.94
1.5	0.052	0.52	13.330	0.81

Conclusion

The carried studies in this work show that the advanced oxidation processes is effective for the treatment of water polluted by textile dyes. Degradation of the Sandocryl Red: BR46 is performed by the oxidizing action of the hydroxyl radicals generated in the medium to be treated. These radicals are capable of degrading almost all organic pollutants because of their high oxidizing power and their reactivity and non selectivity against the organic substances. The study of the influence of different parameters has shown that photodegradation of this pollutant is favored for low pollutant concentrations and with increasing the concentration of catalyst.

However, the results of treatment using sun radiation gave better results compared to the treatment by using UV light radiation.

References

Gaya, U. I., & Abdullah, A. H. (2008). Heterogeneous photocatalytic degradation of organic contaminants over titanium dioxide: A review of fundamentals progress and problems. *Journal of Photochemistry and Photobiology C: Photochemistry Reviews, 9*(1), 1–12.

Helali, S. (2012). *Application de la photocatalyse pour la dégradation des polluants chimiques et bacteriologiques dans l'eau en utilisant des catalyseurs irradies par des photons de lumiere naturelle ou arti_cielle (UV-A/UV-B)*. France: Université Claude Bernard—Lyon I.

Liu, Z., Kanjo, Y., & Mitzutani, S. (2009). Removal mechanisms for endocrine disrupting compounds (EDCs) in wastewater treatment—physical means, biodegradation, and chemical advanced oxidation: A review. *Science of the Total Environment, 407*, 731–748.

McGuigan, K. G., Joyce, T. M., & Conroy, R. M. (1999). Solar disinfection: Use of sunlight to decontaminate drinking water in developing countries. *Journal of Medical Microbiology, 48*, 785–787.

Schwarzenbach, R. P., Escher, B. I., Fenner, K., Hofstetter, T. B., Johnson, C. A., Von Gunten, U., et al. (2006). The challenge of micropollutants in aquatic systems. *Science, 313*(5790), 1072–1077.

Chapter 53
Vulnerability and Impact of Climate Change Processes on Water Resource in Semi-Arid Areas: In Essaouira Basin (Morocco)

Salah Ouhamdouch, Mohammed Bahir, Abdellatif Souhel, and Carreira Paula

Introduction

Since the 1950s, the earth's climate has undergone significant changes represented mainly by the warming due to greenhouse gas emissions, in close relation with human activity (IPCC, 2013). These changes have a negative influence on the environment and man. In most semi arid areas, seasonal and long-term climate variability changes the dynamics of water cycle. Groundwater levels are also falling at alarming rates. Nowadays, groundwater withdrawal and environmental trends may bring about serious problems in some parts of the world. The availability of sufficient fresh waters has become a limiting factor for development. Among the consequences of these changes, include: (1) the increase of the temperature, (2) decrease in precipitation as well as the mass of the ice sheets, (3) rise in sea level and (4) the scarcity of water resources. In addition, the countries of the Maghreb (Algeria, Morocco and Tunisia) have not been spared the impact of these changes that negatively affect the most of their natural resources, including water resources. Locally, the bay of Essaouira also has its share of these effects,

S. Ouhamdouch, Ph.D. (✉)
Department of Geology, Laboratory 3 GEOLAB, Faculty of Science Semlalia, Marrakech, Morocco
e-mail: salah.ouhamdouch@edu.uca.ma

M. Bahir • A. Souhel
Department of Geology, E.N.S. Marrakech, Marrakech, Morocco
e-mail: bahir@uca.ma; souhel.a@uca.ma

C. Paula
Centro de Ciências e Tecnologias Nucleares, CTN/IST, Universidade Técnica of Lisboa, Lisboa, Portugal
e-mail: carreira@itn.pt

P. Grammelis (ed.), *Energy, Transportation and Global Warming*, Green Energy and Technology, DOI 10.1007/978-3-319-30127-3_53

which are represented mainly in the scarcity of water resources see also salinity following the invasion of sea water and the increase in the salinity of the soil, degradation of groundwater quality and the extermination of a number of water sources.

Climate Change on a Global Scale

According to its report of September 2013, the Intergovernmental panel on Climate change (IPCC) states that warming of the climate system of the earth is unequivocal, adding that the human influence is the primary cause of this warming. Among the results of this global warming, we note that the three examples below:

- The warming of the atmosphere: The last three decades are successively warmer at the earth's surface than all preceding decades since 1850. The average temperature of the globe shows a warming of the order of 0.85 °C during the period 1880-2012 (Fig. 53.1). For precipitation, they have experienced an

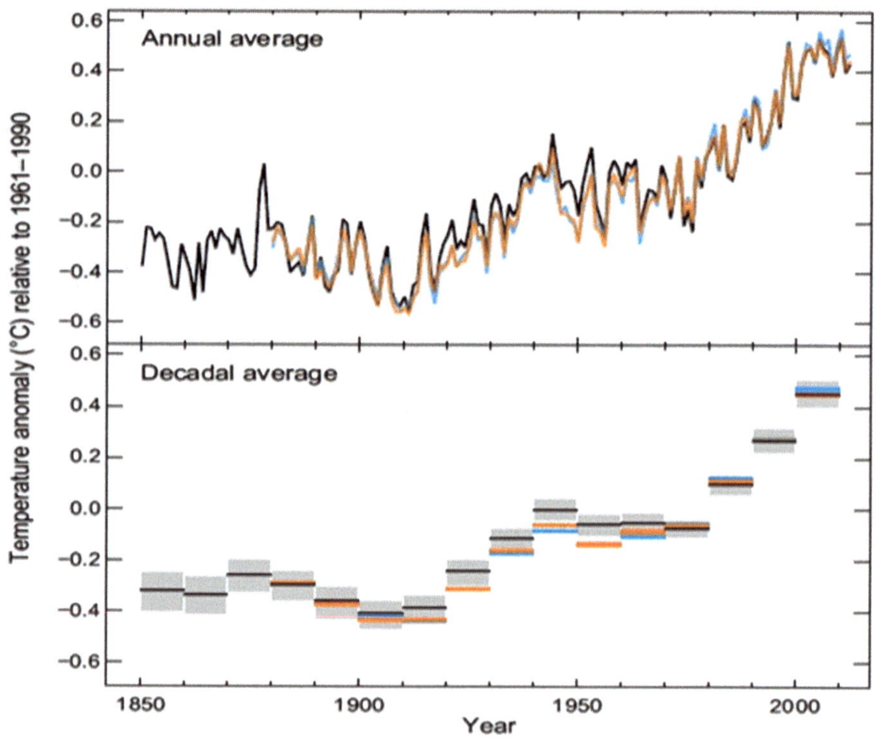

Fig. 53.1 Observed Anomalies of average temperatures on the surface of the globe (IPCC, 2013)

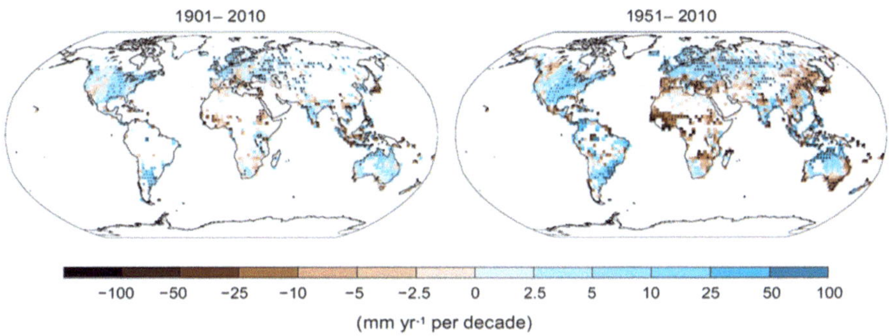

Fig. 53.2 Maps of the observed changes in precipitation between 1901 and 2010, and between 1951 and 2010 (IPCC, 2013)

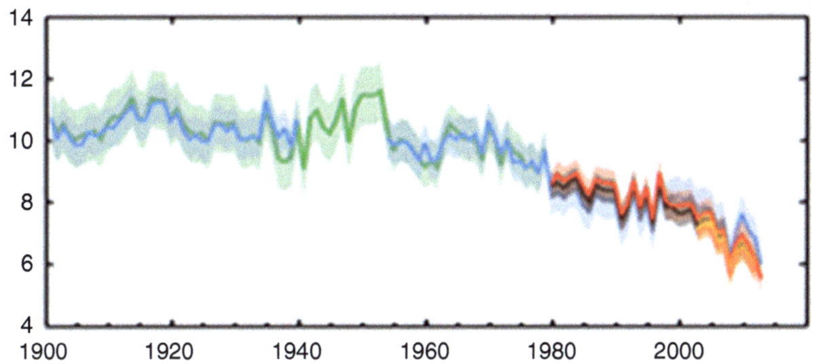

Fig. 53.3 Arctic summer sea ice extent (in million km^2) (IPCC, 2013)

increase since 1901 on the continental regions of the middle latitudes of the northern hemisphere, as well as eastern South America, and a decrease in the Sahel; in the Mediterranean; in southern Africa and in parts of south Asia (Fig. 53.2) (IPCC, 2013).

- The reduction of the surface area of the ice caps and glaciers: everywhere in the world, there is a reduction of the surface area of the ice caps, and ices. Over the past three decades, the extent of annual average arctic sea ice has been decreasing at a rate of 3.5–4.1 % per decade (Fig. 53.3).
- Sea level: since the early 1970s, the decrease of the glacier masses and the thermal expansion of the oceans contribute to them only about 75 % of the rise global mean sea level. Between 1901 and 2010, the average level of the sea shows an increase of 0.19 cm (Fig. 53.4).

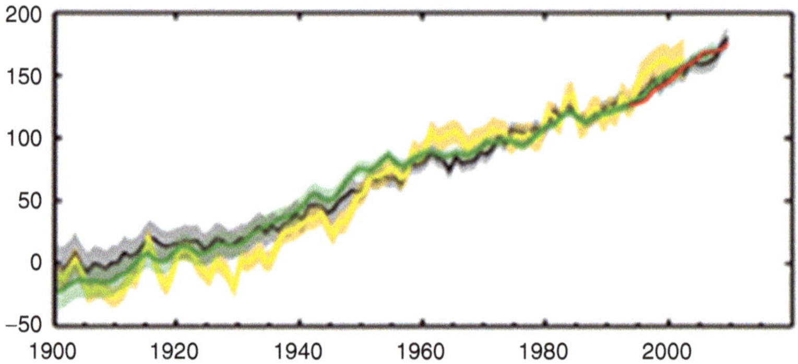

Fig. 53.4 Global average sea level evolution (in mm) (IPCC, 2013)

Climate Change in North African

As in most world countries, the three countries of the Maghreb (Algeria, Morocco and Tunisia) have not been spared the impact of climate change, which has affected and still affects negatively the most of their natural resources. Among these resources, the resource water. For example, the north-west Algeria has experienced a reduction of the order of 40 % of annual precipitation from the first half of the 1970s (Nouaceur et al. 2013; UNESCO, 2010). This reduction is carried out in a temperature increase, which corresponds to the signal of global climate change. This change has affected the river flows of this region. According to the fourth report of the IPCC (Bates et al., 2008), the blades of waters passed the annual average of the period 1976–2002 are 28–36 % lower than those of the period 1949–1976, and this reduction is due to climate change. According to the national institute of meteorology Tunisia, the rainfall pattern in the Tunisia during the twentieth century has been marked by relative stability. In contrast to the rest of the century was marked by the alternation of dry periods (1940–1950) and wet periods (1950–1960). However, the temperature tends to increase (1.1 °C) since the sixties. In addition to this increase, Tunisia has suffer amplification erosion processes dry up the soil and consequently augment and increase the salinity of the water tanks. In addition, the elevation of accelerated sea level due to global warming will have serious impacts on the shoreline such that the contamination of the aquifers in the coastal waters of freshwater by the marine waters of salt. In Morocco, the water resources are limited and have a potential water that can be mobilised estimated at 20 billion m^3, giving an average of 700 m^3/capita/year, which is recognized as a relatively high water stress. The number of years of deficit rainfall everywhere is more important than that of surplus years (Fig. 53.5). This decline coincides with an increase in the frequency of droughts also visible from the 1980s (Driouech et al., 2013; Sebbar et al., 2011; Driouech, 2010; Sinan et al., 2009). As found by various previous studies (including that of the minister of public works, 2007), we distinguish two dry periods: 1980–1985 and 1990–1995.

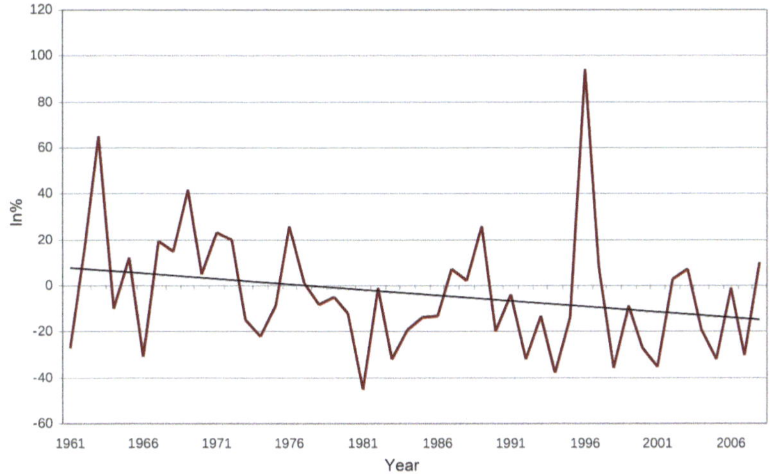

Fig. 53.5 National averages (in %), of relative anomalies of annual rainfall, calculated over the 14 Moroccan meteorological stations. 1961–2008 period (Driouech, 2010)

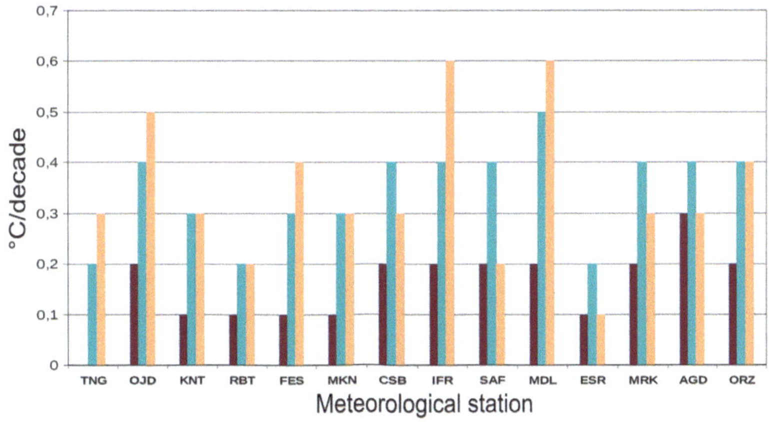

Fig. 53.6 Trends in seasonal mean temperature calculated for the 1961–2008 period at the 14 meterological stations (in °C / decade). (Driouech, 2010) (Color figure online). Note bene: winter: *red bars*, spring: *green bars*, autumn *yellow bars*

By contrast, the evolution of the temperature is constant, and tends to increase (Fig. 53.6), which corresponds to the signal of global climate change, and makes Morocco among the countries affected by climate change. Among the results of this change, we quote: (1) Erosion of the soil under the effect of desertification, which resulted in the silting up of dams. (2) Accentuation of soil salinization and water resources. (3) Lowering of the piezometric levels of the groundwater on one side and the increase in the level of the sea on the other side, which creates in coastal areas, a reversal of current, resulting in the invasion of seawater and the destruction of the hydraulic potential.

Climate Change and Essaouira Basin

In Morocco, most studies of the impact of climate change on water resources focused on coastal areas (Bouchaou et al., 2011; Bahir et al., 2003; Hsissou et al., 2001). These regions that are densely populated and subjected tourism activities are characterized by a growing demand for water.

Geographic Location

The Essaouira basin is located on the Atlantic coast of Morocco, specifically at the western end of the high atlas chain, where it occupies an area of 1827 km^2. This basin includes several aquifer systems whose structures and resources are often poorly known. The most important of these systems belong to two synclinal units: the unit of Bouabout located in the eastern part of the basin and the unit of Essaouira synclinal who represents the Western part of the basin. These two units are separated by the Tidzi diaper (Fig. 53.7).

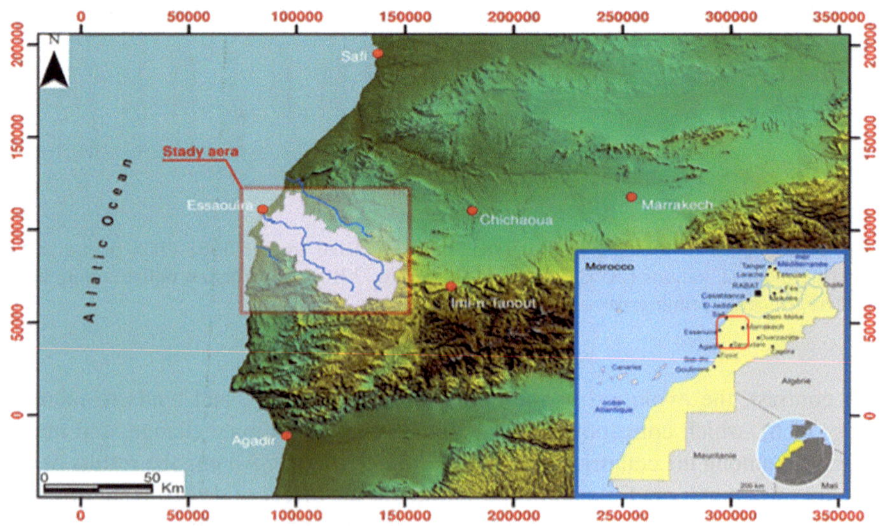

Fig. 53.7 Location of the study area

Geological Cadre

At the basin level of Essaouira, the Triassic and the Jurassic have outcrops very reduced and localized to the anticlines core (jbel Hadid NW, jbel Amstétene to the SW and Tidzi diaper). While the tertiary formations and the quaternary are found in the basins syncline. The Triassic is made up of red clay salt, basalt diabase and pelites sandstone. The Jurassic is represented by an alternation of carbonate deposits (limestone's and dolomites) and marls rich in evaporite (gypsum and anhydrites). While the cretaceous formations and quaternary rocks are grouped in the stratigraphic log synthetic (Duffaud et al., 1966) (Fig. 53.8).

The Essaouira basin is a vast area synclinal open on the Atlantic Ocean, affected by several folds and accident, which allow for the individualization of many

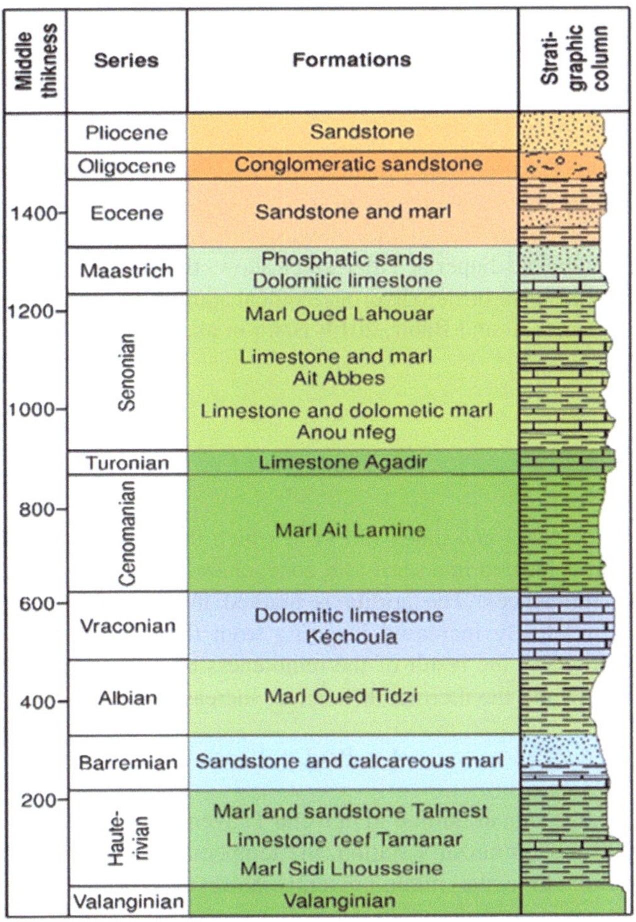

Fig. 53.8 Log synthetic stratigraphic Cretaceous and Tertiary of Essaouira basin

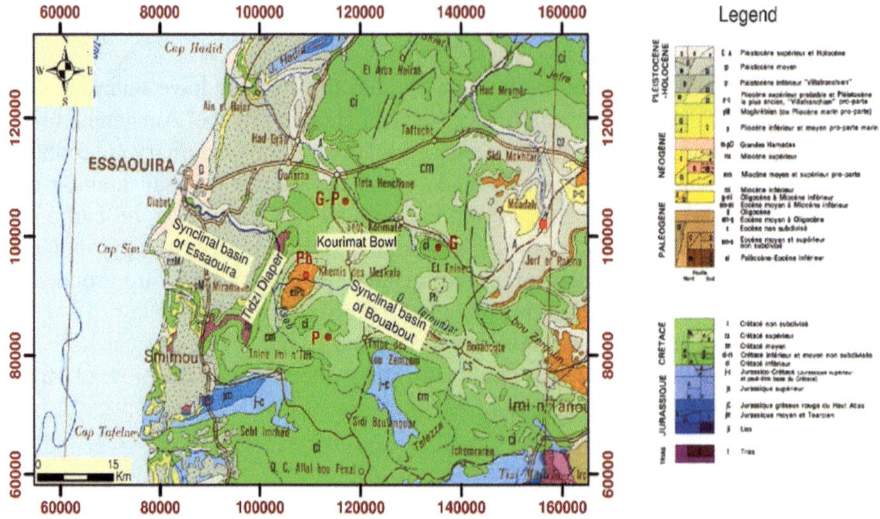

Fig. 53.9 Geological map (1:1,000,000) of study area

synclinal basins: (1) the synclinal basin of Bouabout occupying the upstream part of the basin, crossed by the river Igrounzar. (2) The bowl of Kourimat and (3) the synclinal basin of Essaouira synclinal (downstream part of the basin), separate from that of Bouabout by the diaper of Tidzi which shows the outcrop of the subterranean triassic. These cups give birth to hydrogeological aquifer systems (Fig. 53.9) (Bahir et al., 2013; Chamchati and Bahir, 2013; Bahir et al., 2008; Jalal et al., 2001).

Climate Cadre

According to the Martonne index (53.1) in which the equation stated below, the Essaouira basin is located in a semi-arid zone, characterized by ocean, continental and mountain influences. The aridity is marked in the basin, especially in the summer; it is relatively increased in going from the Atlantic to the continent. This aridity growth is the result of the remoteness of the ocean influence, where the rain decreases and the thermal differences increase.

$$I = P/(T + 10) \tag{53.1}$$

With P: average annual rainfall in mm and T: average annual temperature in °C. At the level of the Igrounzar's station, the temperatures show a very significant seasonal variation. In winter, minimum temperatures can reach up to −11 °C, while summer highs are in the order of 40–45 °C (Chamchati and Bahir, 2011). From the Ombrothermic diagram of Igrounzar's station, the hot season runs from March until

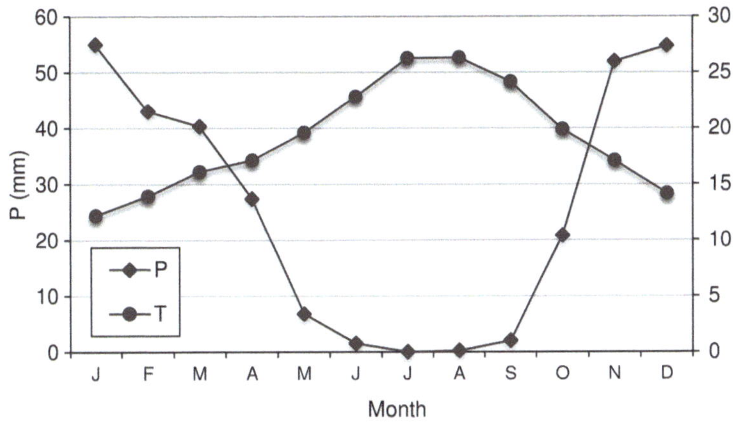

Fig. 53.10 Ombrothermic diagram of Igrounzar station

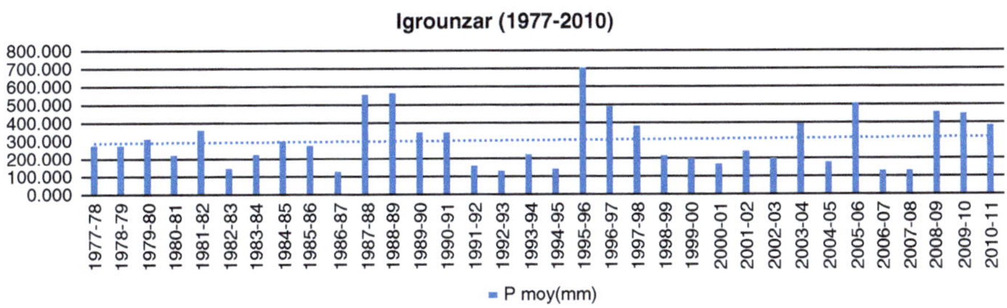

Fig. 53.11 Evolution of the annual rains in Igrounzar station from 1977 to 2010

October or even November. While the wet period occurs from November to march (Fig. 53.10). The rainfall varies from year to year around an average of 306 mm (Fig. 53.11). For the series of years stripped (1977–1978 to 2010–2011), are found in wet years, the height of which exceeds the average (1987–1989, 1995–1997, 2005–2006, 2008–2010), therefore, the rush and show an irregular interannual striking.

Piezometry

To have an idea about the evolution of the piezometric level in the Essaouira basin, we will treat in a first time the water cenomanian-turonian of Bouabout (upstream) and in the second time the water plio-quaternary (downstream):

– Cenomanian-turonian water (Bouabout): The companions of measuring the level piezometric carried out from 2006 to 2010 were used to establish the evolution

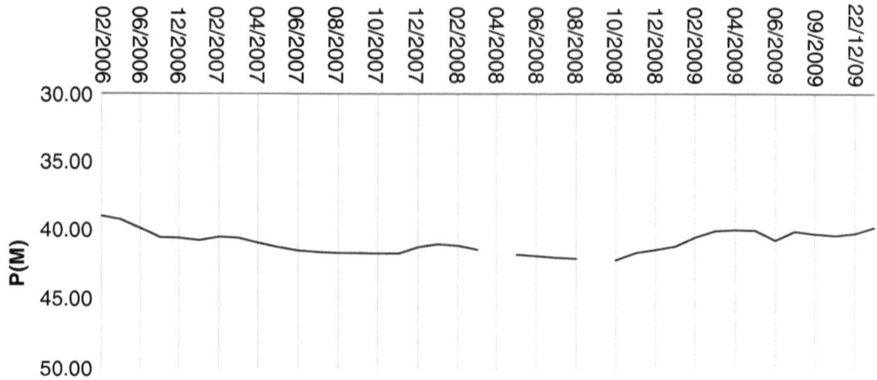

1726/52

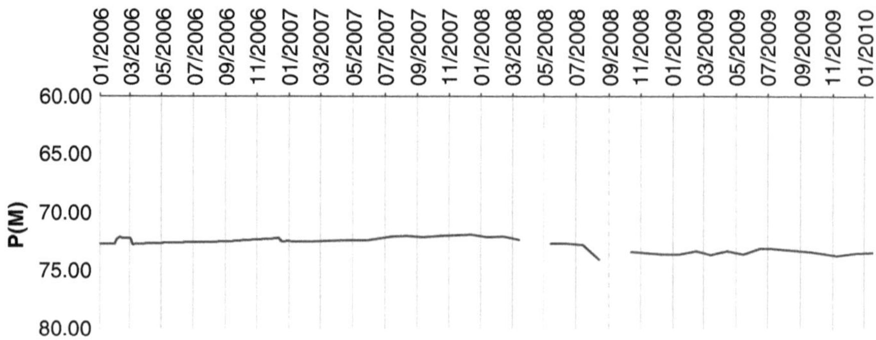

1126/52

Fig. 53.12 Evolution of the piezometric level in the period 2006–2010 at the level of the water points 1126/52 and 1726/52 (2006–2010)

curves of the water level of a water points number such as 1126/52 and 1726/52 (Fig. 53.12). La représentation graphique de l'évolution du niveau piézométrique dans la période 2006–2010 à ces points, montre une diminution du niveau de l'eau de Juin 2006 à Octobre 2008. Cette baisse se explique par la sécheresse qui affecte la région au cours des deux années 2007 et 2008 et aussi par la surexploitation de l'aquifère pour l'irrigation afin de compenser l'impact de la sécheresse. Après Octobre 2008, le niveau de l'eau commence à monter pendant une période de huit mois, puis recommence à baisser.

– Plio-quaternary water: the companions of measurement of the piezometric level of the Plio-quaternary aquifer, during 1990,1995, 2000, 2004 and 2009 resulted in the piezometric maps (Fig. 53.13). They show a sense of overall flow of the South-East to North-West, conditioned by the recovery of its substratum to the Is a result of the lifting of the Tidzi diaper. A comparison of these piezometric maps reveals a decrease of the piezometric levels in 1995 (the driest year of the century in Morocco) within a certain wells have experienced measures of level

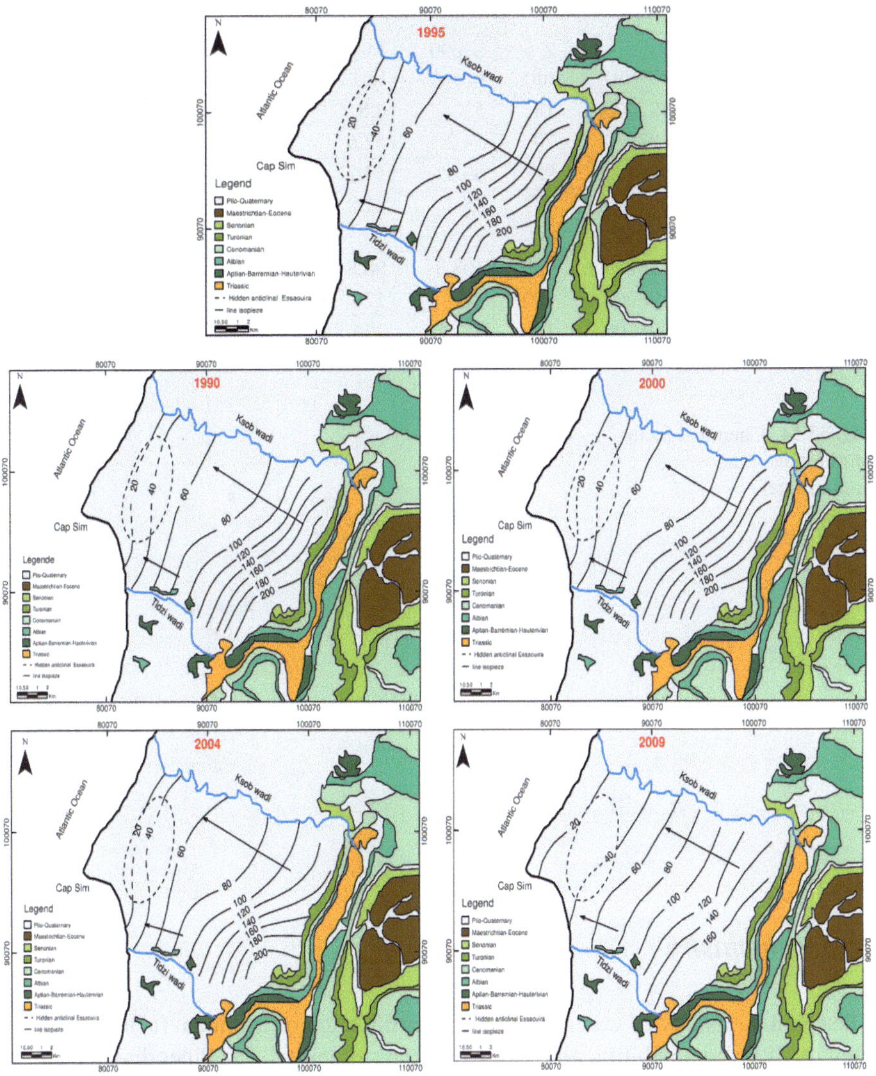

Fig. 53.13 Maps piezometric groundwater Plio-Quaternary of the synclinal area of Essaouira in 1990, 1995, 2000, 2004 and 2009

in the five periods (Table 53.1). After the dry year of 1995, a number of wet years, the occurrence of 1996–1997, 2002, 2003, 2004, 2006 and 2009 can restore and go back in the piezometric level in the aquifer. This variation of the piezometric level can only be explained by the vagaries of the weather and the over-exploitation for irrigation in order to disguise the impact of the drought.

Table 53.1 Variation in piezometric within the water Plio-Quaternary

N° IRE	X (m)	Y (m)	Z (m)	Sep90 (m)	May95 (m)	Jul00 (m)	March04 (m)	Oct09 (m)
3/51	81400	93400	18	5.4	4.8	5.2	5.5	4
15/51	86000	97970	70	54.2	52.6	54	65	63
93/51	92370	101900	98	66.9	64.7	66.5	67.4	54
138	88275	92825	109	85	81.8	84.1	85.3	104
149/51	85100	105800	40	2	1	1.5	4	2
327/51	88800	88800	130	108.3	98.8	107.2	108.2	106
116/51	100650	96000	200	179	176	178.2	sec	sec
117/51	100450	98500	180	137.4	134.5	136.7	sec	sec

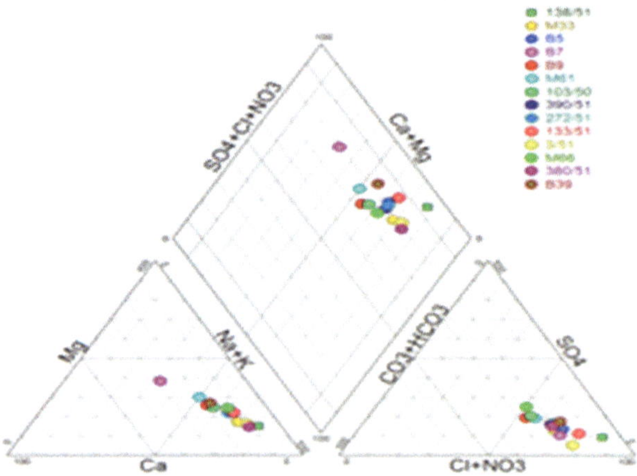

Fig. 53.14 Chemical facies of plio-quaternary groundwater

Hydrochemistry

The chemistry study of water allows the identification of chemical facies of waters, their quality and potability, and their irrigation suitability. The Piper diagram shows the water of plio-quaternary aquifer has a chloride-sodium facies (Fig. 53.14). Based on electrical conductivity and chlorides of plio-quaternary groundwater, we can have an idea on water quality and its evolution. In the temporary scale, the groundwater quality is becoming more and more good by going to the driest year that knows Morocco (1995) in 2009, through the years modestly dry and rainy. Spatially, its takes a positive direction from the North to South of the groundwater table (Fig. 53.15 and Table 53.2). The temporal and spatial evolution of this quality have closely linked to climatic change.

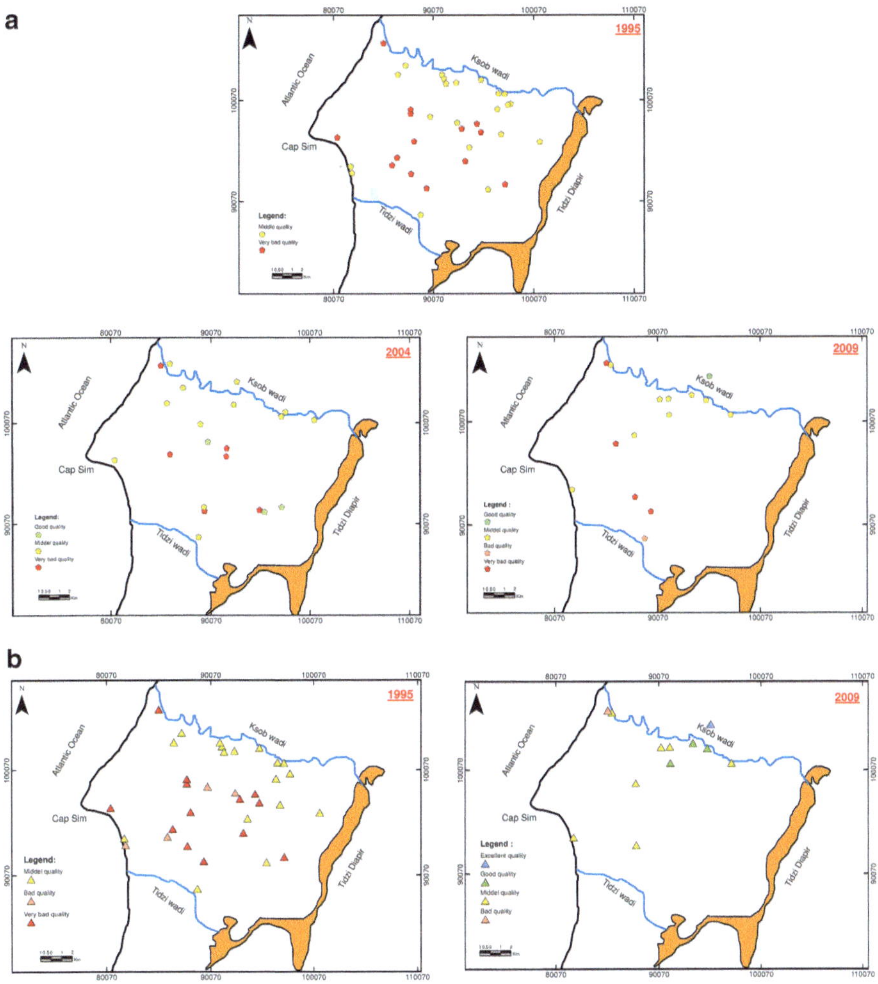

Fig. 53.15 (**a**) Temporal and spatial distribution of the Electrical conductivity of the groundwater in the aquifer plio-quaternary (1995, 2004 and 2009). (**b**) Chlorides (1995 and 2009)

Table 53.2 Simplified grid for groundwater

Quality parameter	C.E. (µs/cm)	Cl^- (mg/l)	NO_3^- (mg/l)	NH_4^+ ($mgNH_4^+/l$)	MO (mgO_2/l)
Excellent	<400	<200	<5	≤0,1	<3
Good	400–1300	200–300	5–25	0,1–0,5	3–5
average	1300–2700	300–750	25–50	0,5–2	5–8
Bad	2700–3000	750–1000	50–100	2–8	>8
Very bad	>3000	>1000	>100	>8	

Stable Isotopes

In the Essaouira basin, the hydrodynamic is influenced by the structure (folds and faults) which determines the flow (Bahir, 2007). In the study area, the isotopes of the water molecule, the oxygen-18 (^{18}O), deuterium (^{2}H), tritium (^{3}H) and carbon-14 (^{14}C), may provide part of the answer in the determination of the areas and conditions aquifer recharge, as well as their relationship and their age. The isotopic analyses were carried out in the institute of technology and nuclear Lisbon. The isotopic content of the water plio-quaternary are between −3.72 and −4.56 δ‰ v-smow. Those of groundwater turonienne are between −4.17 and −4.55 δ‰ v-smow. The results of the analyses of ^{2}H to allow to set on a diagram of correlation ^{18}O-^{2}H the local meteoric right (LMW) equation $^{2}H = 7.72 \times {}^{18}O + 10.53$ ($r^2 = 0.82$), close to the Global meteoric water (GMW) slope 8 with excess ^{2}H close 10 (Craig, 1961). It characterizes the precipitation of oceanic origin. The equation of this line was calculated without taking into account the three points 390/51, 27251, and Ksob wadi, identified as evaporated, because they fall below GMW. The other points are aligned on the LMR, which indicates that the supply of water aquifers is completed quickly without significant evaporation. The ^{3}H is a radioactive element in the cycle of water through rainfall, its presence with concentrations above 2 units tritium (TU) means a charging current (Olive et al., 1996). By contrast, the water having concentrations of less than 2 UT are confederated without active recharge (Olive, 1999). The results are compiled in Table 53.3. The levels of ^{3}H varied between a minimum value of 0.2 and a maximum of 4.2 UT. The water points 272, 327, 65, 386, and 203/51 have a content ^{3}H greater than 2 UT, which means a charging current. While the other points are levels very low. This uses the technique of dating by ^{14}C to confirm or affirm the hypothesis. The ^{14}C is the radioelement the most used to date ancient waters at very low levels of ^{3}H.the results obtained are grouped in Table 53.3. The analysis makes it possible to identify that: the waters of the two points 65 and 386/51 have significant assays in ^{3}H and percentages in ^{14}C higher than 85 %, which makes them as current. The drilling M98, which does not, ^{3}H detectable but has a ^{14}C content of 80 % shows a power supply prior to the nuclear tests from 1962 to 1963, and its radiocarbon age does not exceed a 2000 years. The radiocarbon age of the drilling 390/51 is of the order of 6500 years ago, and one of 380/51 is greater than 2000 year. These two boreholes capturing the turonian aquifer, feed, respectively, the Essaouira city and the village of Si Ahmed Ou Hmad (5000 inhabitants). This reflects that we are in the presence of an ancient water and that it is in the process of exhausting a strategic resource, which could be used in extreme cases (Fig. 53.16).

Conclusion

In addition to the demographic pressure, the cultural practices and pastoral, climate change put the countries of the Maghreb (Algeria, Morocco and Tunisia) in alarming water situation manifested by: (1) the drought, which accentuates the

Table 53.3 Physical parameters and analysis of radioactive isotopes of groundwater in the coastal area of Essaouira

N°IRE	Nature	Origin	Altitude	Prof (m)	T (°C)	PH	Alc Méq/l	^{3}H UT	^{14}C pcm	^{13}C °/°°
272/51	Well	Plio-Q	105.5	38.4	20.6	7.65	3.30	3.9		
21/51	Well	Plio-Q	135	29	21.7	7.13	2.80	2.0		
327/51	Well	Plio-Q	130		22.2	7.23	4.03	3.2		
65/51	Well	Turon	15	20.3	22.5	7.48	4.64	2.8	88.9 ± 0.5	−9.8
390/51	borehole	Turon	95	200	26.7	7.35	4.94	2	32.5 ± 0.4	−9.4
386/51	borehole	Turon	105	100	23.2	7.56	4.26	4.2	84.8 ± 0.6	−10.3
380/51	borehole	Turon	135	194	26.1	7.54	4.67	1	3.0 ± 0.5	−9.0
M98	borehole	Bar-Apt	90	100	22.0	7.59	3.55	1.1	79.6 ± 0.6	−9.3
A. Aghbalou	borehole	Bar-Apt	80		23.1	7.50	3.58	1	72.0 ± 0.4	−10.0
216/51	Source	Por-Ber	160		23.1	7.28	4.53	1		
218/51	Source	Por-Ber	308		22.5	7.29	4.24	1	68.5 ± 0.6	−9.9
361/51	borehole	Lias	382	90	23.5	7.22	4.45	2		
203/51	borehole	Callow	14	50	22.7	7.11	4.26	3.8		
346/51	borehole	Turon	105					1.5		
363/51	borehole	Turon	150					1.2		
149/51	borehole	Turon	40					0.2		

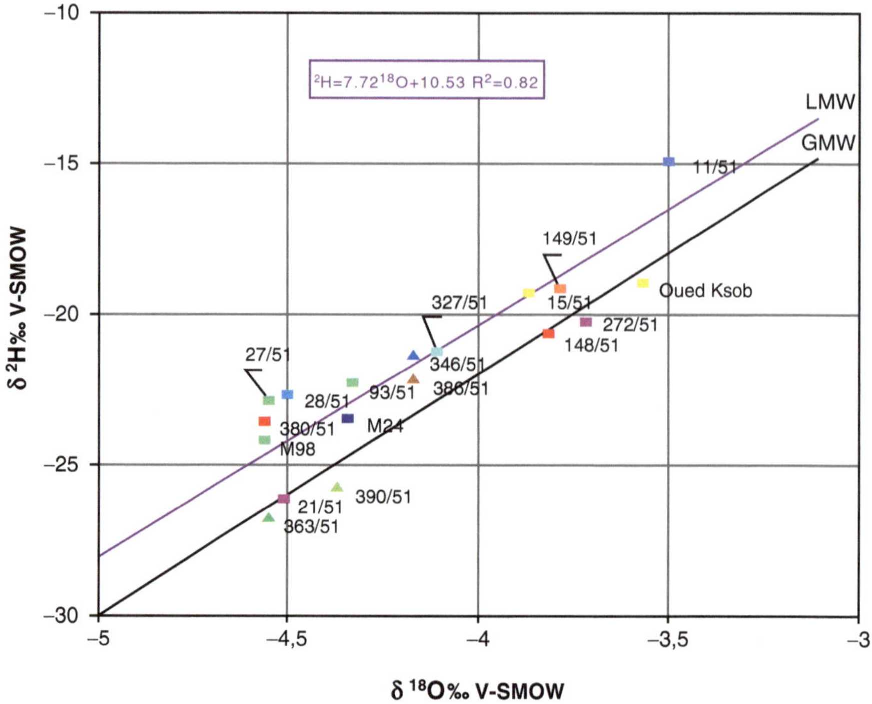

Fig. 53.16 Corrélation ^{18}O-^{2}H dans les eaux souterraines de la zone synclinale d'Essaouira (2006)

phenomenon of desertification, (2) increased salinity of the soil and the lowering of the piezometric levels of the groundwater. It encourages the countries of the Maghreb to develop the strategy of the increased mobilization of non-conventional, such as the desalination of seawater and brackish water. The complementarity of the approaches hydrochemical, piezometric and isotopic may lead to the diagnosis of the state of vulnerability of the aquifers in the basin of Essaouira in the face of climate change. However, the development of a strategy for the rational exploitation could thus enable the value of its waters, while preserving its potential in the long term. As well as the use of non-conventional resources such as desalinated seawater for the supply of drinking water or treated wastewater for agriculture must be seen as a priority in order to avoid trigger situations of water scarcity.

Acknowledgment This research was carried out within the Laboratory 3GEOLAB science faculty Semlalia of Marrakech Cadi Ayyad University, in collaboration with the ENS of Marrakech and the Institute of Technology Nuclear Lisbon, Portugal. Our acknowledgments will go to the agents of the agency of the Basin Hydraulic Tensift that contributed to the realization of this work. We also thank the research group of Professor Paulà Carreira at the institute of technological and nuclear Lisbon, who has done isotope analyses.

References

Bahir, M. (2007). Isotopes et ressources en eaux en zones-arides et semi-arides: exemple du bassin d'Es saouira (p. 175).

Bahir, M., Mennani, M., Olivreira Da Silva, M., & Blavoux, B. (2003). *Impact de la sècheresse sur la ressource en eau en zone semi-aride : cas de l'aquifère barremien-aptien de la zone côtière d'Essaouira (Maroc)*. Países Mediterráneos: Tecnología De La Intrusión De Agua De Mar En Acuíferos Costeros.

Bahir, M., Carreira, P., Oliveira da Silva, M., & Fernandes, P. (2008). Caractérisation hydrodynamique, hydrochimique et isotopique du système aquifère de Kourimat (Bassin d'Essaouira, Maroc). *Estudios Geológicos, 64*(1) enero-junio 2008, 61–73. ISSN: 0367-0449.

Bahir, M., & Chamchati, H. (2011). Caractérisation du système aquifère plio-quaternaire et turonien (Bassin d'Essaouira, Maroc). *geohydro* (pp. 1–6).

Bates, B. C., Kundzewicz, Z. W., Wu, S., & Palutikof, J. P. (2008). *Le changement climatique et l'eau. Document technique VI* (p. 236). Genève: du GIEC. ISBN 978-92-9169-223-1.

Bouchaou, L., Tagma, T., Boutaleb, S., Hssaisoune, M., & ElMorjani, Z. A. (2011). *Climate change and its impacts on groundwater resources in Morocco: The case of the Souss-Massa basin* (pp. 129–144). Boca Raton, FL: CRC Press. H. Treidel, J. L. Martin-Bordes, & J. J. Gurdak (Eds.).

Chamchati, H. & Bahir, M. (2011). Contribution of climate change on water resources in semi-aride areas: Example of the essaouita basin. *Geographia Technica, 1*, 1–8.

Chamchati, H. & Bahir, M. (2013). Potential hydrogeological, environment and vulnerability to pollution of the plio-quaternary aquifers of the coastal basin of Essaouira (Morocco). *Journal of Environment and Earth Science, 3*(10), 170–185.

Craig, H. (1961). Standarts for reporting concentration of deuterium and oxygen 18 in natural waters. *Science, 133*, 1833–1834.

Driouech, F. (2010). *Distribution des précipitations hivernales sur le Maroc dans le cadre d'un changement climatique: descente d'échelle et incertitudes*. Thèse de doctorat de l'université de Toulouse, p. 163.

Driouech, F., Ben Rached, S., & ElHairech, T. (2013). *Climate change and food security in West Asia and North Africa* (pp. 161–171). Dordrecht: Springer. doi:10.1007/978-94-007-6751-5_9.

Hsissou, Y. Bouchaou, L., Krimissa, M., Mudry, J. (2001). *Caractérisation de l'origine de la salinité des eaux de la nappe côtière d'Agadir (Maroc)*. First International Conference on Saltwater Intrusion and Coastal Aquifers—Monitoring, Modeling, and Management. Essaouira, Morocco, 23–25 April, 2001.

IPCC. (2013). *Climate Change 2013: The physical science basis* (Contribution of Working Group I to the Fifth Assessment Report of the Intergovernmental Panel on Climate Change, p. 1535). Cambridge, UK and New York, NY: Cambridge University Press. T. F. Stocker, D. Qin, G.-K. Plattner, M. Tignor, S.K. Allen, J. Boschung, A. Nauels, Y. Xia, V. Bex & P.M. Midgley (Eds.).

Jalal, M., Bahir, M., & Mennani, A. (2001). Pollution nitratée des eaux souterraines du bassin synclinal d'Essaouira (Maroc) (Nitrate in groundwater of the Essaouira Synclinal Basin, Morocco). *Journal of Environmental Hydrology, 9*, 1–10.

Nouaceur, Z., Laignel, B., & Turki, I. (2013). Changements climatiques au Maghreb: vers des conditions plus humides et plus chaudes sur le littoral algérien. *Physio-Geo, 7*, 307–323.

Olive, P., Hubert, P., & Ravailleau, S. (1996). Estimation pratique de l'age des eaux souterraines en Europe par le tritium. *Revue des Sciences de l'eau, 4*, 523–533.

Olive, P. (1999). Datation des eaux souterraines à long de temps de résidences par le radioacarbone. Mode d'emploi. Hydrogéologie. N°1, 1999.

Sebbar, A., Badri, W., Fougrach, H., Hsaine, M., & Saloui, A. (2011). Etude de la variabilité du régime pluviométrique au Maroc sep-tentrional (1935–2004). *Sécheresse, 22*(3), 139–148. doi:10.1684/sec.2011.0313.

Sinan, M. Boussetta, M., El Rherari, A. (2009). Changements climatiques: causes et consequences sur le climat et les ressources en eau. *Revue HTE N°142*. Mars–Juin 2009, PP. 21–30

UNESCO. (2010). *Changement climatique enjeux et perspectives au Maghreb*. GEB-Environment, pp. 258.

Part XIII
Environmental Technologies Related to Global Warming

Waste Management

Chapter 54
Experimental Study of Transverse Mixing of Pollutants in Trapezoidal Open Channel

Ali Mansour Lagoun and Salim Benziada

Introduction

Natural streams receive usually domestic, industrial and agricultural wastes. In some cases, variously incidental discharges find way to rivers, which are generally punctual and cause different problems for water quality. It is important to know the capacity of rivers to mix those pollutants, and to determine the of mixing and dispersing rates of rivers.

When miscible product is injected instantly into a river, it propagates and mixes in three directions under the effect of several processes such as: diffusion, advection and dispersion (Fischer 1966; Shen and Yapa 1995; Gharbi 1999; Hibbs et al. 1999; Czernuszenko and Alexey 2005; Rutherford 1994; Steve and Russell 2005)

A.M. Lagoun (✉)
Research Laboratory of Water Sciences, Polytechnic National School of Algiers, Algiers, Algeria

Laboratory of Environment, Water, Geomechanics and Structures, Faculty of Civil Engineering (FGC), University of Sciences & Technology HouariBoumediene (USTHB), Bab Ezzouar, Algiers, Algeria

Scientific and Technical Research Center on Physical and Chemical Analyses (CRAPC), Bou-Ismail, Tipaza, Algeria
e-mail: lagounali@yahoo.fr

S. Benziada
Research Laboratory of Water Sciences, Polytechnic National School of Algiers, Algiers, Algeria

P. Grammelis (ed.), *Energy, Transportation and Global Warming*, Green Energy and Technology, DOI 10.1007/978-3-319-30127-3_54

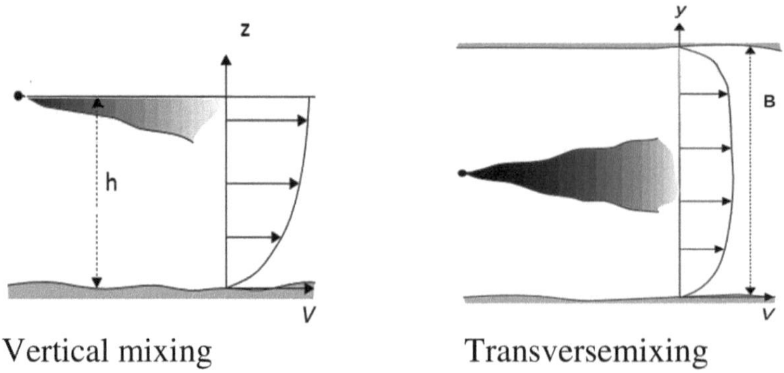

Fig. 54.1 Mixing processues in rivers (Chanson 2004)

Transverse Mixing

The *mixing* term is reserved to express the global result of a miscible fluid transportation processes without accounting mechanism sources (Gharbi 1999).

Generally, the term of the transverse mixing refers to the lateral spread of contaminants (horizontal) caused by turbulence (Fischer and John 1979; Rutherford 1994; Van Prooijen and Uijttewaal 2005) (Fig. 54.1).

Notion of Mixing Zones

Generally, Pollutants transport processes is consisted of three distinct areas (Fischer 1967; Yotsukura et al. 1976; Jobson 1997; Jabour 2006) which are:

Initial Mixing Zone: where the process is effected in the three dimensions of space, this area is achieved relatively very quickly since the water depth is often less than the width of the watercourse (Demetracopoulos and Stelan 1983a, b; Rutherford 1994; Sanchez-Cabeza and Pujol 1999).

Complete Mixing Zone: The complete mixing zone begins at the limit of the initial mixing zone. It ends when the rejected product becomes homogeneous throughout the section of the river, in the vertical and lateral directions (Jabour 2006; Rutherford 1994).

The Far-Field Zone: It extends from the full mixing zone to the infinity. Dispersion is one-dimensional in flow direction (Jabour 2006).

Experimental Setup and Procedures

Tests were performed in the Research Laboratory of Water Sciences (LRS water) at the Polytechnic National School in Algiers, in a trapezoidal channel that was designed specifically to study the problem of transport of pollutants in rivers, especially transverse mixing, it consists essentially of:

Feeding System: allows adjusting the flow through block valve, and constant level tank to maintain the flow invariable.

Trapezoidal Channel: 10 m of length, provided with:

- *A triangular weir*: to measure and control the flow. After calibration of the weir across flow measurement by the volumetric method according to water head at weir, discharge is given by:

$$Q = 1,35.{h_d}^{5/2} \quad (54.1)$$

- A flow stabilizer.
- An injection system of pollutant (phenol): consists of a double tanks with a constant rate for 30 s. The injection is done at the channel center.
- A sampling system for each 1 m longitudinally and for each 10 cm laterally (see Figs. 54.2 and 54.3) photography of the experimental setup is shown at Fig. 54.4.

Samples for different injection scenarios have been analyzed and the concentrations were determined by spectrophotometer (Table 54.1).

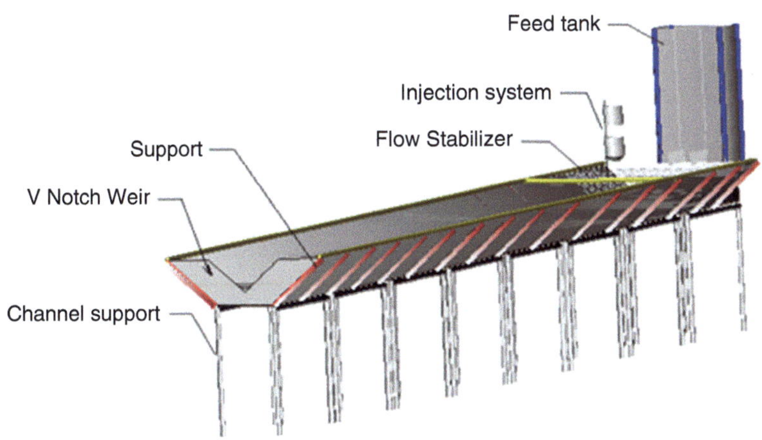

Fig. 54.2 3D schema of experimental setup

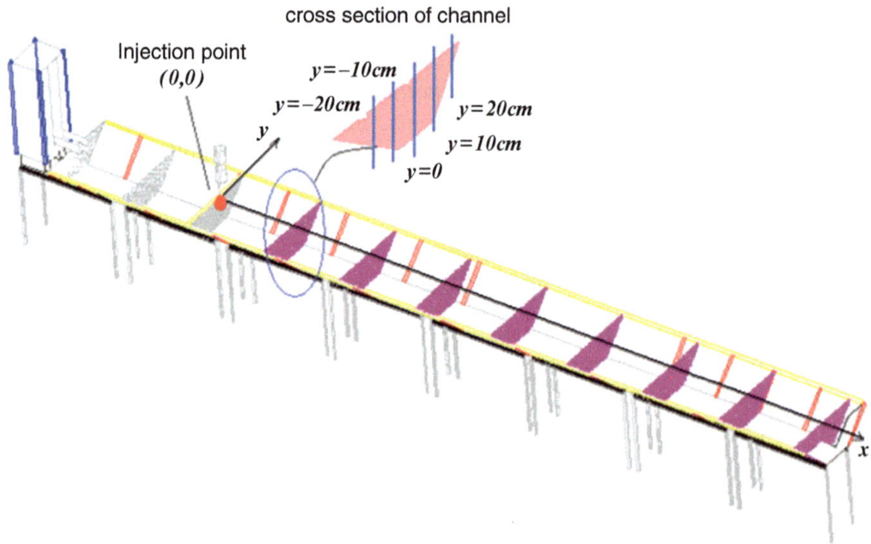

Fig. 54.3 Injection system and sampling points

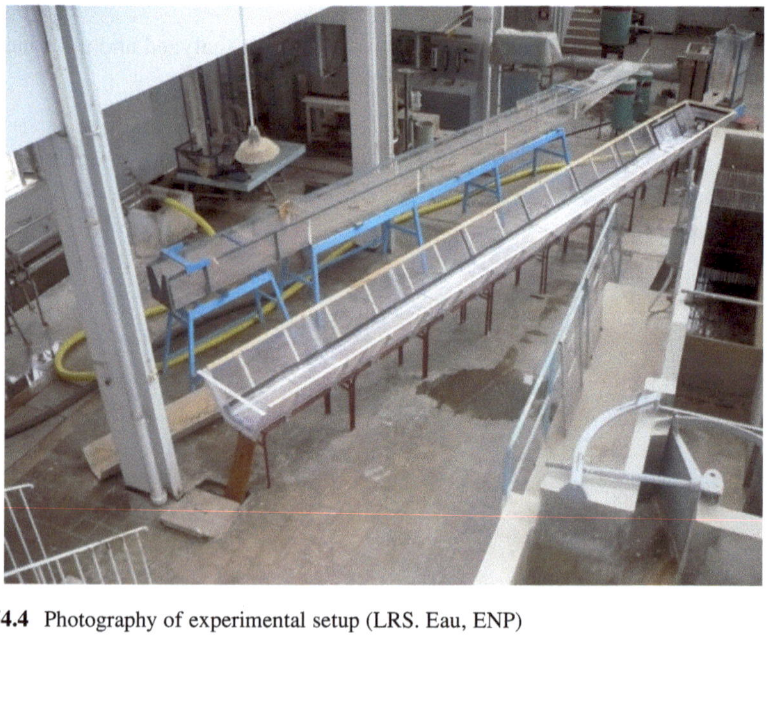

Fig. 54.4 Photography of experimental setup (LRS. Eau, ENP)

Table 54.1 Summary of flume characteristics

Designation	Parameters	Values
Channel	Form	Trapezoidal
	Bed slope [%]	5
	Length [m]	10,4
	Bottom width [m]	0,4
	Side slope [°]	60
	Slope head [cm]	50
Weir	Form	Triangular (V notch)
	Notch angle [°]	90
	Crest elevation [cm]	10

Variables Studied

(a) **Discharge flow**:

The most hydrodynamic variables cited in the literature are the flow discharge and mean flow velocity (Gharbi 1999; Chaudhry 2008). We selected the first variable to study it; it is a general variable that covers several hydrodynamic parameters (velocity and Reynolds number Re), and geometric parameters (water depth, surface width ...). This variable has a practical advantage, since its value is relatively easy to measure, with V-notch weir at our case.

The selected flow rates studied are: 0.52, 0.79, 1.6, 1.75 and 1.95 l/s.

(b) **Initial concentration of pollutant**:

Three initial concentrations C_0were studied: 500, 1000 and 1500 mg/l.

Results and Discussion

Flow Regimes

The different flow regimes are determined by the Reynolds number and Froude number as follows:

Reynolds Number: Is defined as the ratio of inertial forces to viscous forces (Lencastre 2005; Gualtieri and Dragutin 2008):

$$\mathrm{Re} = \frac{\mathrm{UD_h}}{\nu} \tag{54.2}$$

Flow regimes are classified according Reynolds number as follows: $\mathrm{Re} < 2000$: laminar flow, $2000 < \mathrm{Re} < 4000$: transient flow, $\mathrm{Re} > 4000$: turbulent flow (Ernest and Horace 1996).

Table 54.2 Flow regimes according Reynolds number Re and Froude number Fr

Q [l/s]	U [m/s]	Re	Fr	Flow regimes
0.52	0.008	2848.03	0.006	Subcritical and transient
0.79	0.011	4220.03	0.009	Subcritical and turbulent
1.60	0.019	8144.88	0.015	Subcritical and turbulent
1.75	0.021	8830.58	0.016	Subcritical and turbulent
1.95	0.023	9754.51	0.017	Subcritical and turbulent

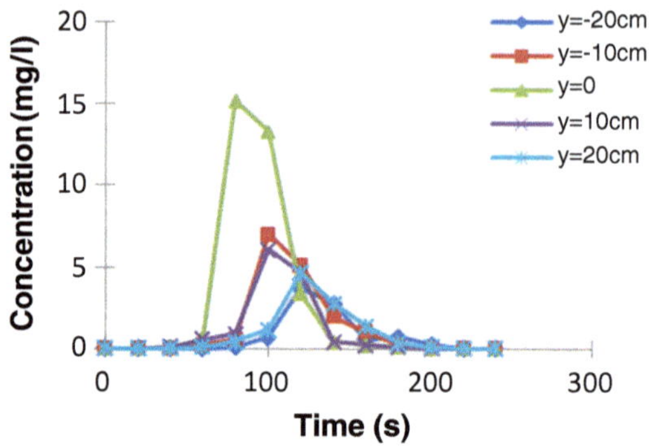

Fig. 54.5 Evolution of concentration profiles of phenol in time for different transverse distances at $x = 2$ m—Q = 0,52 l/s, $C_0 = 500$ mg/l—Instantaneous injection

Froude Number: Is the ratio between the forces of inertia and forces of gravities:

$$Fr = \frac{U}{\sqrt{g \times h}} \tag{54.3}$$

Flow regimes are classified according Froude number as follows: Fr > 1: supercritical flow, Fr < 1: subcritical flow, Fr = 1: critical flow (Dingman 2009).

Flow regimes for the different flow rates examined are shown in Table 54.2.

Transverse Concentration Profiles

Lateral (horizontal) concentration evolution of phenol according time, at 2 m of injection point of phenol with $C_0 = 500$ mg/l, for different flows, are given at Figs. 54.5 and 54.6.

Remarks and Interpretation

Transverse concentration profiles are identical for the two symmetrical point from center of flow direction (y = 20 cm and y = −20 cm/ y = 10 cm and y = −10 cm), it shows that the pollutant injected at the center of channel is mixed to both lateral sides of the channel (symmetric concentration profile); it is the effect of the

Fig. 54.6 Evolution of concentration profiles of phenol in time for different transverse distances at $x = 2$ m—Q = 1,95 l/s, $C_0 = 500$ mg/l—Instantaneous injection

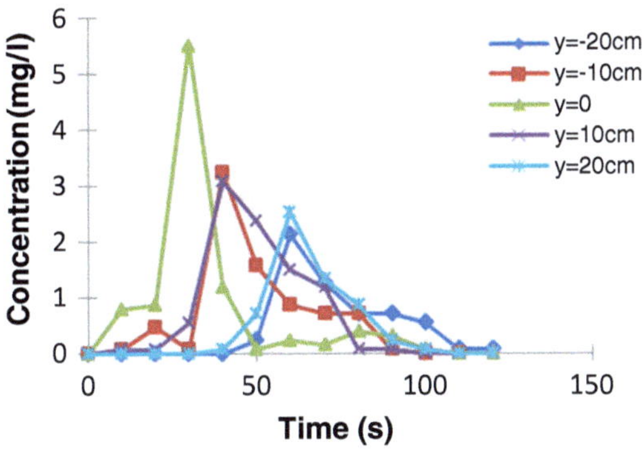

Fig. 54.7 Evolution of concentration profiles of phenol in time for different transverse distances at x = 1 m—Q = 0,52 l/s, C0 = 500 mg/l—Instantaneous injection

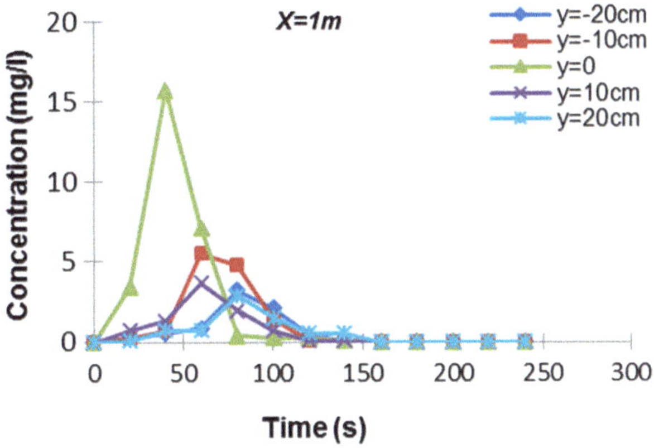

horizontal velocity profile, which is not yet well defined (Chaudhry 2008; Gharbi 1999; Chanson 2004), but it is maximal at center (y = 0) and minimum at both banks of the canal.

Transverse Concentration Profiles Evolution at Downstream in Flow Direction

The concentration profiles obtained experimentally with deferent transverse distances (y = −20 cm, y = −10 cm, y = 0, y = 10 cm and y = 20 cm), at different distances from the injection point in downstream (x-direction) for a flow rate Q = 0.52 l/s, with an injection of phenol with initial concentration $C_0 = 500$ mg/l, are shown in Figs. 54.7, 54.8, 54.9, and 54.10.

Fig. 54.8 Evolution of concentration profiles of phenol in time for different transverse distances at x = 2 m—Q = 0,52 l/s, C0 = 500 mg/l—Instantaneous injection

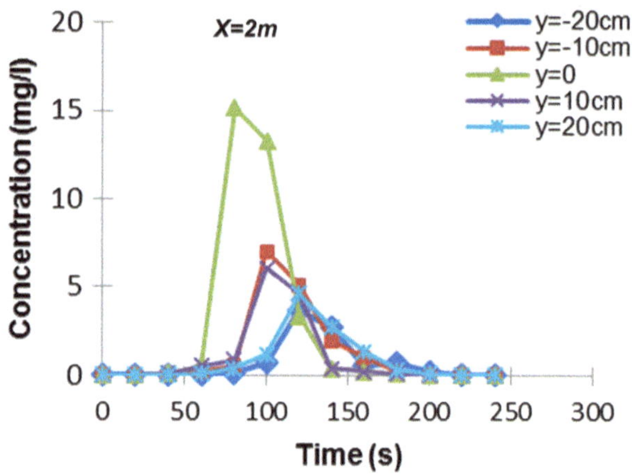

Fig. 54.9 Evolution of concentration profiles of phenol in time for different transverse distances at $x = 5$ m—Q = 0,52 l/s, $C_0 = 500$ mg/l—Instantaneous injection

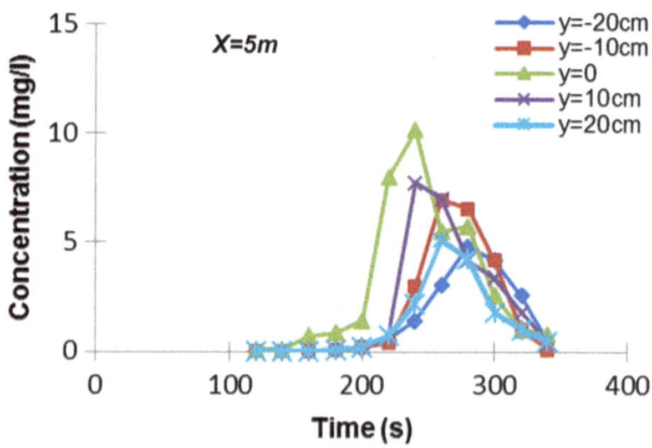

Fig. 54.10 Evolution of concentration profiles of phenol in time for different transverse distances at x = 6 m—Q = 0,52 l/s, C0 = 500 mg/l—Instantaneous injection

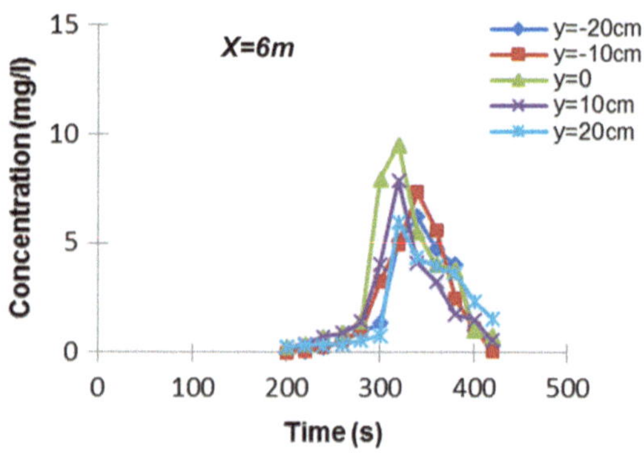

Remarks and Interpretation

From Figs. 54.7–54.10, we note that:

(a) There has a cloud movement of the pollutant (phenol) in the flow direction (x direction) by the flow velocity effect; it is the process of *advection.*
(b) There is a significant gradient of concentration profiles for different horizontal points, concentrations are maximal at the center of the channel and becomes less important for both boundaries of the channel (Figs. 54.7 and 54.8). Away from the injection site at downstream, concentration profiles gradient becomes less important (Figs. 54.9 and 54.10), and concentration started to be uniform throughout the section of the channel at transverse direction (Fig. 54.10), this is explained by the manifestation the transverse mixing of phenol.

Effect of Flow Discharge and Turbulence Phenomena:

Remarks and Interpretation

From Figs. 54.11–54.13 below, at different point in the downstream of injection point, we note that concentration profiles for different horizontal distances (y) are more homogeneous for flow with discharge Q = 1.95 l/s than for Q = 0.52 l/s. this means that the contaminant has been completely mixed at transverse direction, concentration distribution becomes homogeneous and uniform in the cross section of flow. This is can be explained by rate of turbulence that increase with flow discharge, turbulent diffusion will be important that increase transverse mixing rate.

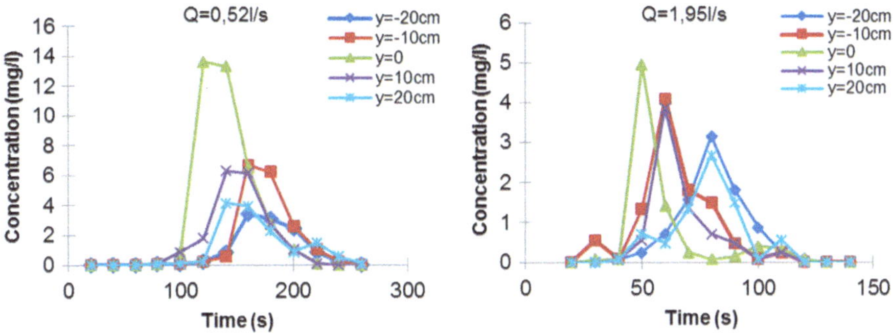

Fig. 54.11 Concentration Profiles for different transverse distances (y) at $x = 3$ m for Q = 0,52 l/s and Q = 1,95 l/s, $C_0 = 500$ mg/l

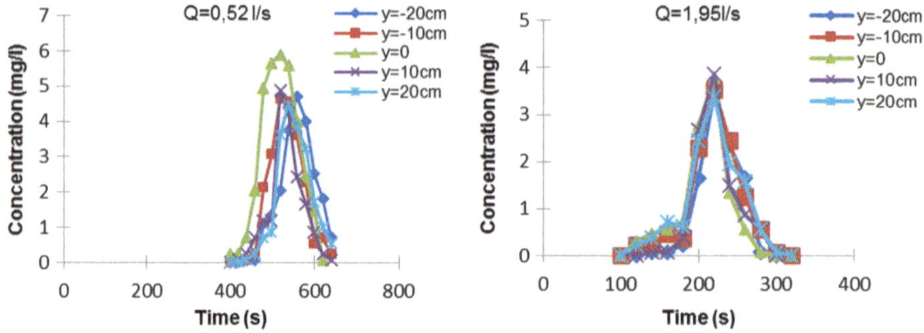

Fig. 54.12 Concentration Profiles for different transverse distances (y) at x = 8 m for Q = 0,52 l/s and Q = 1,95 l/s, C0 = 500 mg/l

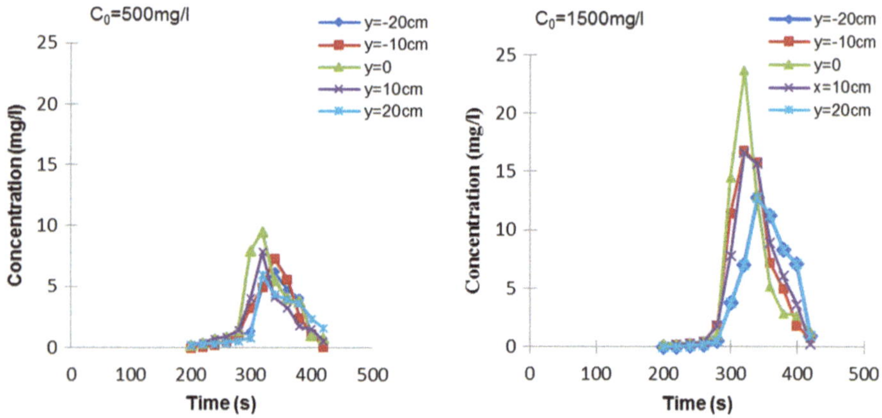

Fig. 54.13 Concentrations profiles observed for different transverse distances (y) at x = 6 m for flow discharge Q = 0.52 l/s, for C0 = 500 mg/l and C0 = 1500 mg/l

Effect of Pollutant Quantity (Initial Phenol Concentration)

To determine the influence of the initial concentration on the transverse mixing, making a comparative analysis between the concentration profiles obtained in (x = 6 m, x = 7 m and x = 8 m) and at several transverse distances (y), for flow Discharge Q = 0.52 l/s for two initial concentrations $C_0 = 500$ mg/l and $C_0 = 1500$ mg/l:

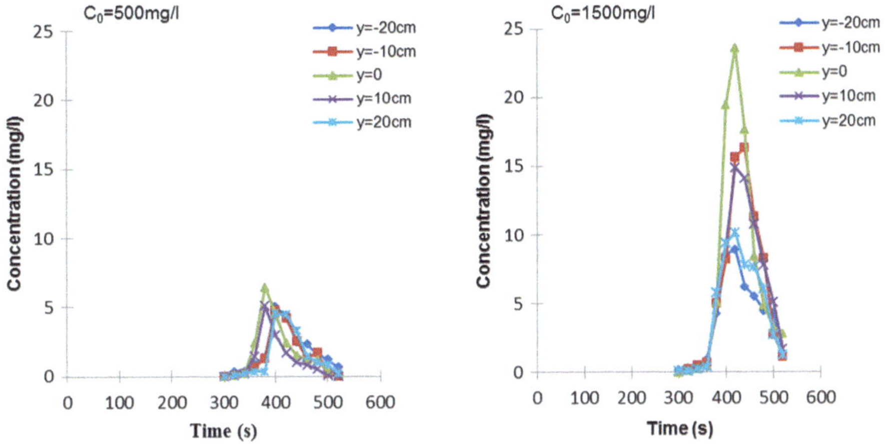

Fig. 54.14 Concentrations Profiles observed for different transverse distances (y) at x = 7 m for flow discharge Q = 0.52 l/s, for C0 = 500 mg/l and C0 = 1500 mg/l

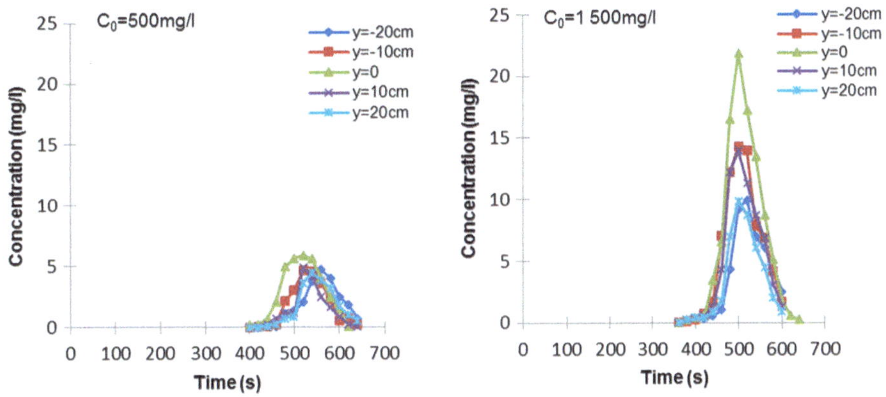

Fig. 54.15 Concentrations Profiles observed for different transverse distances (y) at x = 8m for flow discharge Q = 0.52 l/s, for C0 = 500 mg/l and C0 = 1500 mg/l

Remarks and Interpretation

From Figs. 54.14 and 54.15, we remark that the concentration profiles gradient for different transverse distances is more important for an injection of phenol with C0 = 1500 mg/l that for C0 = 500 mg/l, where the quantity of pollution has been well mixed, and the concentration distribution is more homogeneous and uniform laterally for phenol injection $C_0 = 500$ mg/l then $C_0 = 1500$ mg/l. Mixing favors low amount.

Conclusion

Experimental model of flume designed at Research Laboratory of Water Sciences at National Polytechnical School, and analyses of concentration profiles obtained from several tests, have allowed to well study transverse mixing and to define the mechanisms that govern it (advection with flow velocity and propagation with velocity profile, turbulent diffusion).

Concentration profiles examination for different transverses sections at several point in downstream allowed to understand the transverse mixing processes, to distinguish the mixing zones and shows the effect of flow discharge in the advection rate.

Flow discharge has a great effect on transverse mixing, turbulent diffusion that becomes predominate away from injection area is accentuated with increasing flow discharge, with growth the transverse mixing rate.

Pollutants and contamination rejected at flow will be easily mixed when their amount and concentrations were low.

Nomenclature

C	Pollutant Concentration, mg/l
D_h	Hydraulic diameter, m
D_o	Optical density
g	Gravity, m/s^2
h	Hydraulic depth, m
h_d	Head, m
Q	Flow discharge (m^3/s)
U	Flow velocity (m/s)

Dimensionless Numbers

Fr	Froude number
Re	Reynolds number

Greek Letters

ν	Cinematic viscosity of water, m^2/s

References

Chanson, H. (2004). *Environmental hydraulics of open channel flow*. Library of Congress Cataloguing in Publication Data, 423p.

Chaudhry, M. H. (2008). *Open channel flow* (2nd ed., 523p). Library of Congress, Control Number: 2007936602. New York: Springer.

Czernuszenko, W., & Alexey, R. (2005). Three-dimensional model of flow and mixing processes in open channels. In *Water quality hazards and dispersion of pollutants* (pp. 35–54). Library of Congress Cataloging-in-Publication Data. New York: Springer.

Demetracopoulos, A. C., & Stelan, H. G. (1983a). Transverse mixing in wide and shallow rivers: Case study. *Journal of Environment Engineering, 109*(3), 685–699.

Demetracopoulos, A. C., & Stelan, H. G. (1983b). Model of Mississipi river pool: Mass transport. *Journal of Environment Engineering, 109*(5), 1006–1019.

Dingman, S. L. (2009). *Fluvial hydraulics*. Oxford: Oxford University Press. 559p.

Ernest, F. B., & Horace, W. K. (1996). *Handbook of hydraulics* (7th ed.). New York: McGraw-Hill. 611p.

Fischer, H. B. (1966). Longitudinal dispersion in laboratory and natural streams. Report N°. KH-R-12 . *Journal of Water Resources Division*, 250p.

Fischer, H. B. (1967). The mechanisms of dispersion in natural streams. *Journal of Hydraulic Division. ASCE, 93*(HY6), 187–215.

Fischer, H. B. & John, E. L. (1979). *Mixing in inland and coastal waters*. Academic Press. Library of Congress Cataloging in Publication Data, 458p.

Gharbi, S. (1999). Évaluation des coefficients de mélange longitudinal et transversal des polluants dans les cours d'eau: proposition de nouvelles formules. Thèse Doctorat à *l'université Laval*, Québec, 197p.

Gualtieri, C., & Dragutin, T. M. (2008). *Fluid mechanics in environmental interfaces*, Taylor & Francis e-Library, 332p.

Hibbs, D., Gulliver, J., Voller, V., & Chen, Y. F. (1999). An aqueous concentration model for riverine spills. *Journal of Hazardous Materials, A64*, 37–53.

Jabour, D. (2006). Etude expérimentale et modélisation de la dispersion en champ lointain suite à un rejet accidentel d'un polluant miscible dans un cours d'eau. Application à la gestion de crise. Thèse Doctorat à *l'Université de Provence*, 246p.

Jobson, H. E. (1997). Predicting travel time and dispersion in rivers and streams. *Journal of Hydraulics Engineering. ASCE, 123*(11), 971–978.

Lencastre, A. (2005). *Hydraulique Général*. Edition Eyrolles, 633p.

Rutherford, J. C. (1994). *River mixing*. New York: Willey. 347p.

Sanchez-Cabeza, J. A., & Pujol, L. (1999). Study on the hydrodynamic of the Ebro River lower course using tritium as radiotracer. *Water Research, 33*(10), 2345–2356.

Shen, H. T., & Yapa, P. D. (1995). A simulation model for chemical spills in the upper St Lawarence River. *Journal of Great Lakes Research, 21*(10), 652–664.

Steve, W., & Russell, M. (2005). On the theoretical prediction of longitudinal dispersion coefficients in a compound channel. In *Water quality hazards and dispersion of pollutants* (pp. 69–84). Library of Congress Cataloging-in-Publication Data. New York: Springer.

Van Prooijen, B. G., & Uijttewaal, W. S. J. (2005). Horizontal mixing in shallow flows; Physical aspects and numerical modelling. In *Water quality hazards and dispersion of pollutants* (pp 55–68). Library of Congress Cataloging-in-Publication Data. New York: Springer.

Chapter 55
Elimination of Micropollutent Lysine Acetylsalicylate by Adsorption on Natural and Synthetical Supports

Amel Djouadi and Fatiha Bentahar

Introduction

The existence of several micropollutents in wastewater has been long identified in broad categories in persistent organic pollutants that remain active after conventional wastewater treatment. This study conducts a biotechnological strategy, the adsorption phenomena, using natural supports as algae, black tea, or synthetical supports as the powered activated carbon or the titan oxide. Its main focus is to produce water that may be reutilized after treatment (Gupta and Rastogi 2008). Adsorption phenomena have been identified for its ability to degrade micropollutents. The pollutant, a common painkiller, acetylsalicylic acid also goes by the name aspirin. Lysine acetylsalicylate is a slightly different molecule in structure to the original aspirin molecule, but scientists still use the name aspirin to refer to it (Aronson 2006).

Aspirin Properties

Lysine acetylsalicylate is an analgesic substance. Its different properties are the following:

- Formule: $C_9H_8O_4$
- Poids moléculaire: 180.1574
- Structure Chimique

A. Djouadi (✉) • F. Bentahar
Laboratoire de phénomènes de transfert, Faculté de Génie des Procèdes et de Génie Mécanique, USTHB, Bab Ezzouar, BP 32 El Alia, 16111 Alger, Algeria
e-mail: adjouadi@usthb.dz; djouad2@voila.fr; fbentahar@usthb.dz

P. Grammelis (ed.), *Energy, Transportation and Global Warming*, Green Energy and Technology, DOI 10.1007/978-3-319-30127-3_55

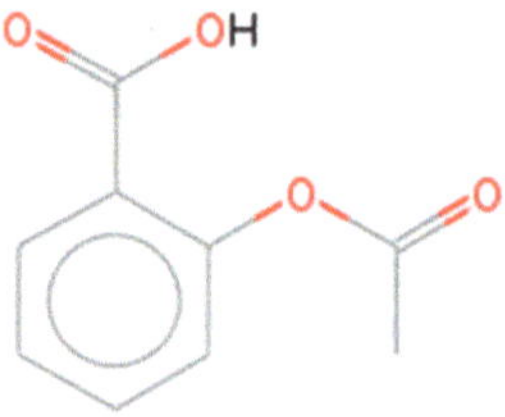

Method and Procedure

A UV-Vis Absorbance Spectroscopy is used in this experiment in order to determine the absorbance of the sample and, in turn, determine purity. This is done by weighing out a carefully measured amount of product and fully dissolving it in a certain amount of water. If some of this resulting solution is placed in a cuvette and is analyzed by the UV-Vis machine, it is then possible to determine the concentration of the dissolved substance from the given absorbance values using Beer's law. The IR analysis is established to determine the characterization of each utilized adsorbent giving their own absorbance bands and to identify the experimental adsorption protocol (Sarat Chandra et al. 2015; Salima et al. 2013). The parametric study is based on varying the initial concentration of the adsorbate, Lysine acetylsalicylate, and the temperature of the aqueous solution, pH and the mass of adsorbent. The obtained results are summarized giving the adsorption efficiencies of these adsorbents for the degradation of this adsorbate: The activated carbon is considered as the best one followed by titan oxide, algae and black tea (Fig. 55.1).

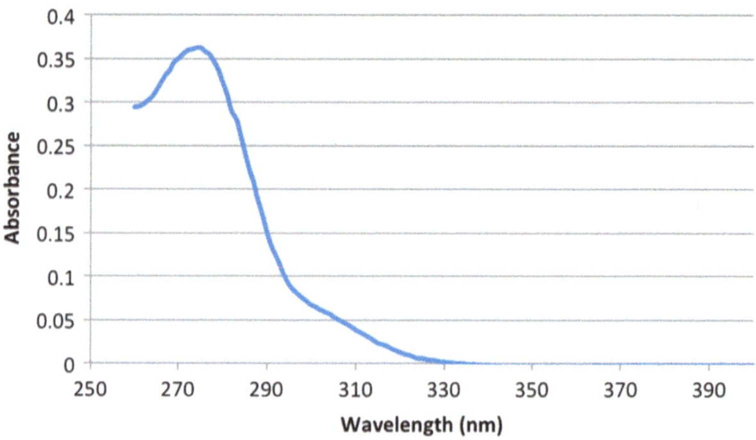

Fig. 55.1 Absorbance versus wavelength

Table 55.1 Absorbance values of aspirin

Equivalent conc. (mg/ml)	Absorbance
0.0125	0.232
0.0250	0.466
0.0501	0.942

Results and Discussion

The obtained results are summarized giving the adsorption efficiencies of these adsorbents for the degradation of this adsorbate: The activated carbon is considered as the best one followed by titan oxide, algae and black tea. The adsorption isotherms of each couple (adsorbate-adsorbent) resulting from experimental data show that they describe well the model of Langmuir with the coefficient of correlation R2 equal to 0.996. The kinetic model pseudo-second order represents faithfully the experimental results, coefficient of correlation R2 = 0.996. The thermodynamic study has revealed that the adsorption of Lysine acetylsalicylate on the natural supports is an exothermic reaction characterized by an increase of molecular disorder at interface solid/ adsorbate. Adsorption reaction is endergonic on natural supports, algae and black tea but spontaneous reaction using synthetical supports, activated carbon or titan oxide. At the end of this work, we can conclude that the waste plant studied have an important adsorbing power of this pollutant, Aspirin.

The absorbance values of the three working standard solutions were measured, the measured values are shown in Table 55.1.

Conclusions

The batch studies conducted in the present study provides significant information regarding biosorption of the pollutent on green algae Spirogyra species in terms of optimum pH and biomass dose from the aqueous solution. The studies indicate that Spirogyra species is an effective biosorbent for Lysine acetylsalicylate removal. The maximum biosorption capacity has been found of dry weight of biomass at an algal dose of 0.5 g/L in 150 min of contact time with initial Lysine acetylsalicylate concentration of 200 mg/L and optimum pH of 5.0. The Langmuir and Freundlich adsorption model were used for the mathematical description of the biosorption of Lysine acetylsalicylate onto algal biomass and it was found that the adsorption equilibrium data fitted well to the Langmuir model. The biosorption of Lysine acetylsalicylate on the algal biomass follows second-order biosorption kinetics. With the advantage of high biosorption capacity, the biomass of Spirogyra has the potential to be used as an efficient and economic biosorbent microorganism for the removal of Lysine acetylsalicylate from aqueous solutions.

References

Aronson, J. K. (2006). Lysine acetylsalicylate. In J. K. Aronson (Ed.), *Meyler's side effects of drugs: The international encyclopedia of adverse drug reactions and interactions* (15 ed., p. 2179). Amsterdam: Elsevier.

Gupta, V. K., & Rastogi, A. (2008). Biosorption of lead from aqueous solutions by green algae Spirogyra species: Kinetics and equilibrium studies. *Journal of Hazardous Materials, 152*(1), 407–414.

Salima, A., et al. (2013). Application of Ulva lactuca and Systoceira stricta algae-based activated carbons to hazardous cationic dyes removal from industrial effluents. *Water Research, 47*(10), 3375–3388.

Sarat Chandra, T., et al. (2015). Defatted algal biomass as a non-conventional low-cost adsorbent: Surface characterization and methylene blue adsorption characteristics. *Bioresource Technology, 184*, 395–404.

Chapter 56
Integrated System for Optimized Data Collection and processing of End of Life Tires: Case of Greece

Panagiotis Vounatsos, John Vournas, George Mavrias, and Panagiotis Grammelis

Introduction

The tires used from all types of vehicles, like private vehicles, trucks, motorbikes, bikes, construction and agricultural machinery, turn into waste at the end of their life cycle. According to European Tyre and Rubber Manufacturers Association (ETRMA 2011), about 3.3 million tonnes of end of life tyres were gathered in Europe, of which 2.5 million tonnes were recycled or recovered. In the international scale, each year about 1 billion tyres reach the end of their life cycle. The treatment schemes followed in order to utilize the amounts of collected End of Life Tyres (ELT) vary (ETRA 2013). As presented in Fig. 56.1, the situation in EU is led to a continuous increase of methods that are more environmental friendly than landfilling which, as shown, represents less than 10 % of the total treatment methods in 2012. On the contrary, methods such as energy utilization and tyre recycling grew the past decade to such a degree, that more than 70 % of the collected tyres are treated by one of these methods.

The material recovery applications include among others the recycling of ELTs as rubber granulated and powder (83 %), utilization in civil engineering works (~12 %), and other applications which cover the rest of the percentage. The energy recovery options include the utilization in cement industry at a devastating percentage of 91 % and the majority of the rest of the quantities are used for electricity

P. Vounatsos (✉) • P. Grammelis
Chemical Process and Energy Resources Institute (ISFTA Sector), Centre for Research and Technology-Hellas, 6th Km, Charilaou-Thermi Road, 57001 Thermi-Thessaloniki, Greece
e-mail: vounatsos@certh.gr

J. Vournas
Motion Hellas Ltd., 8, Philippou Litsa Str., Halandri, 15234 Athens, Greece

G. Mavrias
Ecoelastika SA, 14 Sorou Str., Marousi, 15125 Athens, Greece

P. Grammelis (ed.), *Energy, Transportation and Global Warming*, Green Energy and Technology, DOI 10.1007/978-3-319-30127-3_56

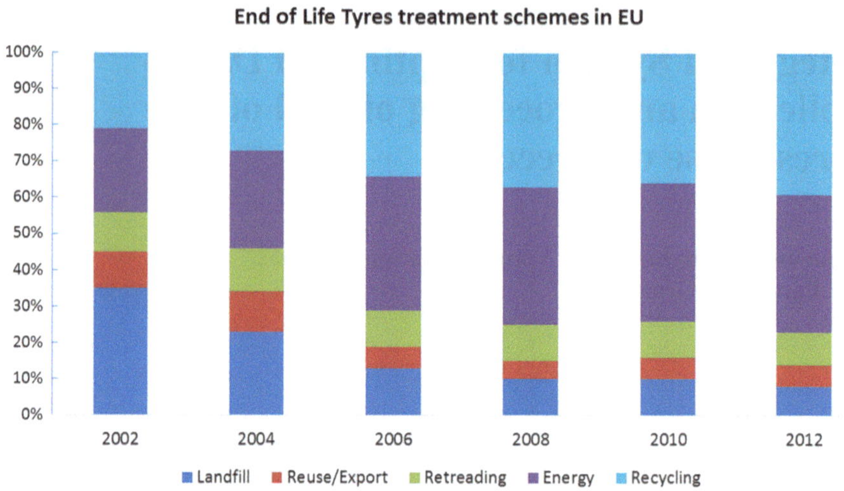

Fig. 56.1 End of Life Tyres treatment schemes in EU

generation (ETRMA 2014). Likewise, as presented from the Rubber Manufacturers Association (Rubber Manufacturers Association 2013) the percentage of end of life tires that is led to landfill is reduced to around 12,6 %, while the percentage of tyres utilized in energy or recycling sector (as Tyre-Derived Fuel, Ground Rubber and Civil Engineering) covers more than 70 %. The available data are presented in Table 56.1.

As far as the best treatment methods for ELT is concerned, it has been a case of study from many different teams around the world in terms of financial and environmental viability and sustainability. Xingfu et al. (2010) compared four tire treatment methods in China, were it was found that the most eco-effective ELT treatment technology is pyrolysis (with product recovery) followed by dynamic devulcanization and ambient grinding. It is noted that in future, ambient grinding for crumb rubber may be more advantageous with less energy consumption and less emissions of pollutants. Moreover, regarding the pyrolysis process, the viability depends on the variation of the prices that the products will be sold in the future. The work elaborated from Corti and Lombardi (2004) supports the aforementioned utilizing life cycle assessment (LCA) tools, they found that the use of ELT as fuel substitute in cement kilns and in Waste to Energy (WTE) processes allow better environmental behavior than other alternatives. More specifically cryogenic and mechanical pulverization of ELT for reuse as filling materials presented the worst results of environmental impact comparing to all other treatment methods. Pyrolysis was studied and is proposed as an alternative method for waste tyres treatment (Williams 2013), proving that high added value products can derive from this treatment method. Sometimes the produced products may require upgrading in order to fit in specific standards for commercial use.

Table 56.1 Market trends in U.S.

Product (thousand tons)	2005	2007	2009
Tire-derived fuel	2144.64	2484.36	2084.75
Ground rubber	552.51	789.09	1354.17
Land disposed	590.81	593.98	653.38
Used tires	–	–	371.25
Civil engineering	639.99	561.56	284.92
Reclamation projects	–	132.58	130
Exported	111.99	102.08	102.1
Baled tires/market	–	–	27.79
Electric arc furnace	18.88	27.14	27.1
Baled/no market	42.22	9.31	15.57
Agricultural	47.59	7.13	7.1
Punched/stamped	10.51	1.85	1.9
Generated	4410.73	4595.72	5170.5
Land disposed/generated	13.39 %	12.92 %	12.64 %

Samolada and Zabaniotou (2012) studied the application of pyrolysis in Greece and found that it is environmentally and financial a more sound application from combustion-incineration (which is a destructive one). Specifically, pyrolysis can be considerably attractive if all of the final products are effectively used. It was also studied from Zabaniotou et al. (Zabaniotou et al. 2014), the barriers and drivers for ELT pyrolysis in industrial application, and it was noted that the current EU legislation prevents the implementation of large scale pyrolysis plants. As another means for waste ELT utilization, the production of acoustic material from tire fluff was studied (Jimenez-Espadafor et al. 2011) and it was found that it is a technically possible and sound approach.

Apart from the studying of the after use treatment methods, there have been studies for the replacement of metal parts (alumimium) of tires with other rubber or plastic parts, in order to improve the carbon footprint depending on the final treatment method that will be used. Specifically, it was found that aluminum wore is better for landfilling while plastic and rubber parts are far better for thermal utilization (Simoes et al. 2013). However Arroyo et al. (2011) found that steel corrosion is the main source of heat production in Tires that have been landfilled or in tire derived aggregate that is actually ELT that have been reduced in size and used as lightweight fill alternative solution. This leads to possibilities of occurring uncontrolled fires in landfills. Shakya et al. (2008) studied the emissions of uncontrolled open-air burning of waste vehicle tyres, as happens in landfills. It was found that CO and SO_2 emissions were significantly higher than controlled thermal utilization of tyres. Moreover the emission of other pollutants such as NO_2, CO_2, polyaromatic hydrocarbons and smoke were also higher than controlled utilization in specially designed facilities.

As presented from all of the aforementioned studies, all of the technologies and treatment schemes present advantages and disadvantages. It was clearly observed that despite the wide technological study for environmental and financial viability, there is not a widely accepted scheme or combination of technologies in order to achieve the optimum treatment of ELTs. There are some studies (Uruburu et al. 2013; Ene and Ozturk 2015; Subulan et al. 2015) were the product recovery operations network as well as the final treatment for environmental optimization has been investigated. As presented in these studies the first step is data collection, in order to acquire certain parameters for processing.

Scope of the present study is to present a model for optimized data collection and processing of end of life tires produced in Greece. This will lead to an integrated system for controlling the fate of ELT, leading the treatment to most environmental friendly solutions and therefore reducing the emissions and pollution created from ELT utilization.

Methodology

In Greece the products from ELT treatment can be utilized in new products and replace in some cases other raw material with equal or better properties. In the following table (Table 56.2) the tyres categories that are collected in Greece are presented.

The monitoring of data from the collection, transportation and final utilization of end of life tyres is organized from ECOELASTIKA S.A. which is the Greek body responsible for tyres recovery. Afterwards, the data are registered in a designed database. Each party that collaborates with Ecoelastika, fills in periodically (every week) an electronic file in which the following parameters are detailed:

- Time of collection
- The collection point
- Tyre category and quantities of used tyres per collection point
- The tyre quantities carried forward to the final users

Table 56.2 Tyres Categories that are collected in Greece

	Description	Mean weight of each tyre
Category A′	– Passenger car tires (passenger tires and tires 4 × 4) – Commercial vehicle tires – Conventional (diagonal) agricultural tractor tyres for front wheels – Tyres for industrial lift trucks	8 kg
Category B′	– Commercial vehicle tires with nominal (inner) rim diameter greater than or equal to a certain diameter – Agricultural tractor tyres for rear drive wheels – Tires for industrial lift trucks with nominal (inner) rim diameter greater than a certain diameter – Tires for earthmoving vehicles	50 kg
Category C′	– Motorcycle tires	2,5 kg

The database can afterwards provide information regarding the following points:

- The collected amounts in pieces or mass per region, for specific periods
- The frequency of transportation from certain collection points
- The collected quantities per collection point for specific periods
- The number of serviced collection points
- The amounts of tyres stored in temporary storages
- The amounts of utilized (recycled, thermal utilization) tyres in final users

The active collection points have been reduced in the latter years. From 3093 active points in 2010 only 2603 remain in operation in 2014.

The final users vary depending on the treatment method and include:

1. Mechanical grinding
2. Recycling
3. Utilization in construction projects, and
4. Thermal utilization

From the above end users 51.55 % end in material recovery, 28.21 % end in Energy Recovery, while the rest 20.24 % end as Tyre Derived Fuels (TDF). The main products from the recovered tyres include textiles, wire, shredded and crumb rubber. The percentages that derive from the mechanical grinding treatment is presented in Fig. 56.2.

However, the business as usual may present deviations between the real data and the ones collected. This produces two weaknesses. The first is that the calculations derived from the database may not fully depict the reality and the second is that it is not real time and automated. These reasons are the source for delays in processing of the data and decision making for optimum ELT treatment options.

Therefore the data collection system determines the user requirements, which include the following:

- Procedures for production and quality
- Integration of the Green WTMS system with other systems

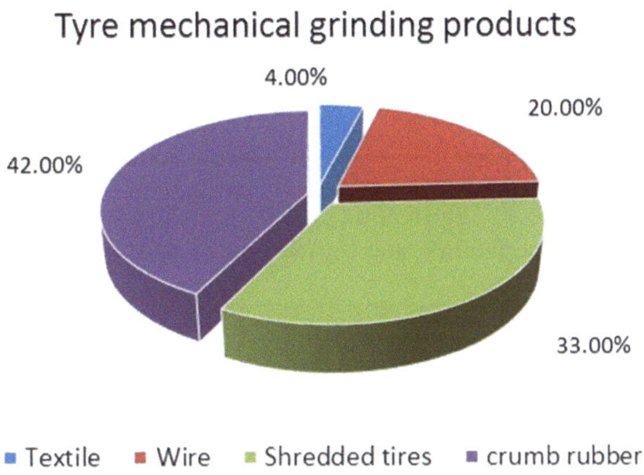

Fig. 56.2 Products from mechanical grinding process

- List the number and types of relevant equipment
- System size including spare capacity e.g., number of workstations
- Requirements for data presentation, records and reports
- Any networking requirements to local/wide area networks LANs/WANs
- Communication links

An automated online system will be installed in the final users, with an automated weighing system which will continuously communicate with the central database in order to inform real time about the tyres received and about to be treated. The online system will be composed of a network of stations installed in the weight bridges of end users which will directly communicate with the server located in Ecoelastika S.A. providing detailed information on:

- Mixed, net weight and tare of each ELT load that is input or output of the final user plants for end of life tyres utilization
- Information on the trucks transporting the ELTs entering/leaving end use plant areas (e.g. registration number, net truck volume, photos)
- Time and day of each transaction

Moreover photos of the front and the rear of each truck will be taken during the first and second weighting (mixed weight and tare) of each truck that enters or leaves the final use plant areas.

Utilizing this type of systems, the capacity is provided in order to:

- Online check the quantities of whole tyres that enter or leave the end use facilities
- Check the quantities of final or intermediate products that leave from the end use facilities in order to be sold
- Crosschecking the data that are gathered through the manual system of weighting the loads and sending the data via a form to the system manager (Ecoelastika S.A.)

The loads will be registered as following:

1. Input load of whole end of life tyres
2. Output load
   (a) Shredded tyres
   (b) Tyre crumb
   (c) Wire
   (d) Textile

Moreover information will be included for the user that will receive the load exiting the facility, and for the type of final treatment.

Results and Discussion

Taking into consideration the requirements of the data collection system designed, a demonstration was implemented in an already existing tyre recycling plant located in North Greece. The incoming material (end of life tires) are brought in the plant through closed trucks. The trucks leaving the plant and carrying the products are mostly three-axle or four-axle vehicles with open or closed containers and lifting hook type vehicles with trailers (hook lifts). The plant is divided in six different modules, which are the following:

1. *Shredding module*: It is consisted of a tyre shredder and a separator for the shredding product
2. *Grinding module*: It is consisted of two grinders for the reduction of the mean size of the tyres leaving the shredder
3. *Metal separation*: It is consisted of two stage magnetic separation
4. *Module for tire crumb milling*: It is consisted of two mills
5. *Textile separation module*: it is consisted of a series of aspirators and air filters
6. *Sieving and packaging module*: It is consisted of machinery were the separation in different categories of final products takes place, according to particle size of the product

The designed system was studied in order to include 32 different requirements as they were set from the users. The more important of them were the following:

1. Potential for a detailed recording of all sensors, metering and measurement systems in the database
2. Potential for continuous and automatic recording of the measurements for the existing analogue and digital signals
3. Capability to connect with RFID readers
4. Ability to connect to message boards
5. Ability to connect to cameras for photographing vehicles
6. Capacity to connect with photocells for vehicle presence control
7. Ability to connect to traffic lights for the management of incoming and exiting vehicles
8. It must be based on internet/intranet architecture, operate in a Windows environment and be based on a GUI
9. The database in which the data will be saved and inserted will be SQL
10. It must provide the opportunity for continuous monitoring and presentation of the measurements in a PC, and give the results for all the previous periods
11. Capability to record the following data on input and output vehicles
    (a) Vehicle, type, owner
    (b) Date, time and system operator
    (c) Mix weight, net and tare
    (d) Collector

Fig. 56.3 Weight bridge installed in plant in order to collect and transmit online data

In the entrance of the plant, the control room as well as the weight bridge for weighting the vehicles is located. The weight bridge is electronic industrial type with reinforced structure, according to European Regulations. It is connected through an electronic alphanumeric weighting scale which is installed in the control room. In Fig. 56.3, the weight bridge installed is presented.

The weight bridge was made to communicate with the system that was described before and gather, store and transmit the gathered information. In Fig. 56.4 the data acquisition system with the online statistic analysis is presented.

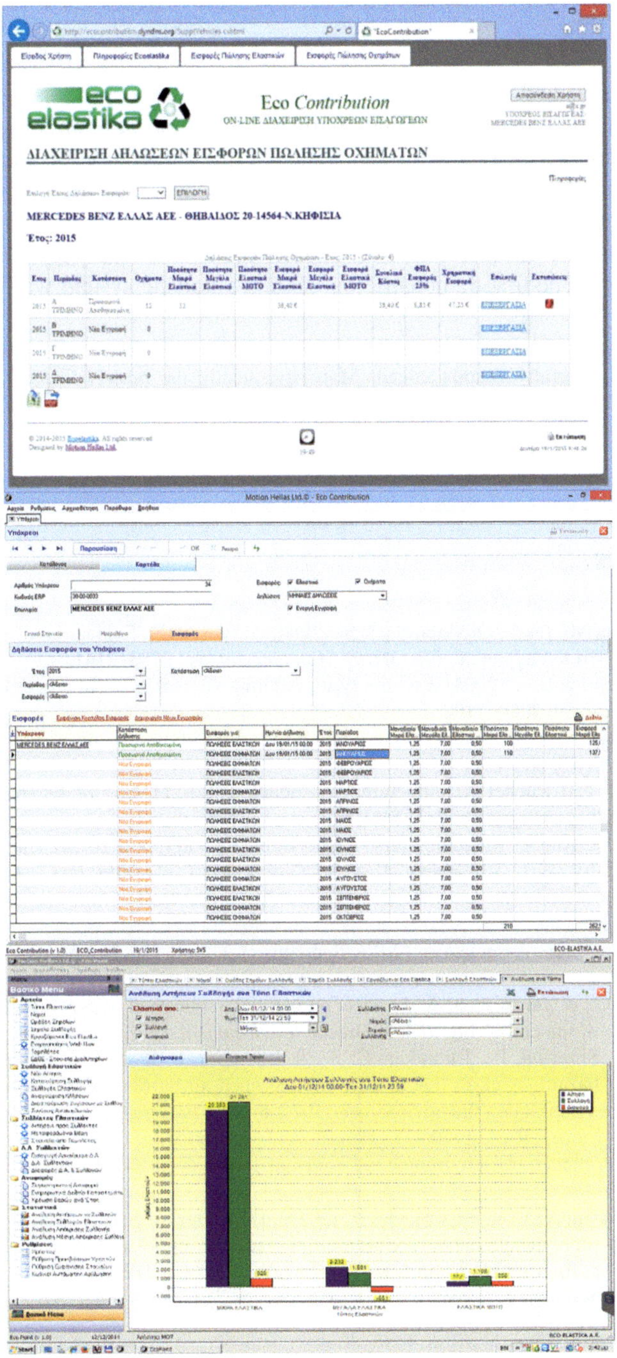

Fig. 56.4 Weight bridge installed in plant in order to collect and transmit online data

Conclusions

The installation and operation of a data acquisition and processing system was found to be critical for the implementation of the first stage of an integrated control of End of Life Tyres treatment. As presented, the integration of a data recording system was implemented in a weight bridge located in a ELT treatment plant on the Northern Greece. Its purpose was to record and transmit data relevant with the type and quantities of the input stream for treatment, as well as the output stream, led to the final users. The first demonstration of such a holistic system was successful.

Acknowledgement The support of the Greek National Strategic Reference Framework 2007–2013 is gratefully acknowledged. More specifically the project GreenWTMS (contract number: 27-BET-2013) in the framework of the Program for Development of Industrial Research and Technology 2013.

References

Arroyo, M., San Martin, I., Olivella, S., & Saaltink, M. W. (2011). Evaluation of self-combustion risk in tire derived aggregate fills. *Waste Management, 31*, 2133–2141.

Corti, A., & Lombardi, L. (2004). End life tyres: Alternative final disposal processes compared by LCA. *Energy, 29*, 2089–2108.

Ene, S., & Ozturk, N. (2015). Network modeling for reverse flows of end-of-life vehicles. *Waste Management, 38*, 284–296.

European Tyre and Rubber Manufacturers Association. (2011). *End of life tyres: A valuable resource with growing potential* (2011 ed.).

European Tyre & Rubber Manufacturers' Association. Annual report 2013/2014.

Jimenez-Espadafor, F. J., Villanueva, J. A. B., Garcia, M. T., Trujillo, E. C., & Blanco, A. M. (2011). Optimal design of acoustic material from tire fluff. *Materials and Design, 32*, 3608–3616.

Rubber Manufacturers Association. (2013, September). U.S. Scrap Tire Management Summary 2005–2009.

Samolada, M. C., & Zabaniotou, A. A. (2012). Potential application of pyrolysis for the effective valorisation of the end of life tires in Greece. *Environmental Development, 4*, 73–87.

Shakya, P. R., Shrestha, P., Tamrakar, C. S., & Bhattarai, P. K. (2008). Studies on potential emission of hazardous gases due to uncontrolled open-air burning of waste vehicle tyres and their possible impacts on the environment. *Atmospheric Environment, 42*, 6555–6559.

Simoes, C. L., Simoes, R., Carvalho, J., Pontes, A. J., & Bernardo, C. A. (2013). The quest for a sustainable product: An environmental study of tyre recyclates. *Materials and Design, 52*, 196–206.

Subulan, K., Tasan, A. S., & Baykasoglu, A. (2015). Designing an environmentally conscious tire closed-loop supply chain network with multiple recovery options using interactive fuzzy goal programming. *Applied Mathematical Modeling, 39*(9), 2661–2702.

European Tyre Recycling Association: Introduction to Tyre Recycling: 2013. (2013). *Twenty years of tyre recycling in the EU*.

Uruburu, A., Ponce-Cueto, E., Cobo-Benita, J. R., & Ordieres-Mere, J. (2013). The new challenges of end-of-life tyres management systems: A Spanish case study. *Waste Management, 33*, 679–688.
Williams, P. T. (2013). Pyrolysis of waste tyres: A review. *Waste Management, 33*, 1714–1728.
Xingfu, L., Gao, Y., & Yijun, T. (2010). Comparison of end-of-life tire treatment technologies: A Chinese case study. *Waste Management, 30*, 2235–2246.
Zabaniotou, A., Antoniou, N., & Bruton, G. (2014). Analysis of good practices, barriers and drivers for ELTs pyrolysis industrial application. *Waste Management, 34*, 2335–2346.

Chapter 57
Smart Recovery of Materials and Upgrade of Organic Compost and RDF in Existing Mechanical Biological Treatment Plants by Using NIR Technology

Dimitrios-Sotirios Kourkoumpas, Georgios Kontopoulos, Ioannis Vournas, Dimitrios Koulocheris, Panagiotis Grammelis, and Emmanouel Kakaras

Introduction

The first near infra-red (NIR) sorting systems were used to process recyclable materials in the early 1990s. Until then, hand-sorting for the separation of the recyclables materials from the MSW had been used. Nowadays modern sorting systems, where e.g. paper, plastics or other recyclable materials can be sorted, cannot be run economically without near infra-red technology.

The NIR spectroscopy is applied as a quick and reliable method for the analysis of sample mixtures during the process analysis. Regarding the working method of the technology, the infrared radiation generated by halogen lamps is directed on the sample surface. Specific wavelength ranges are particularly absorbed depending on the sample material, since this is directly linked with the resonance frequency of the excited molecules. In what concerns the radiation reflected by the sample, this is detected as a function of the wavelength. A measurement of the sample in the relevant wavelength band with a high spectral resolution is required, in order to get a detailed analysis about the sample composition (Habich 2007).

Sorting systems using NIR technology have applied in the waste processing sector in the last 15 years. Especially, such of type of automatic sorting systems are

D.-S. Kourkoumpas (✉) • P. Grammelis • E. Kakaras
Centre for Research & Technology Hellas/Chemical Process and Energy Resources Institute, 52, Egialias Str., Maroussi, 15125 Athens, Greece
e-mail: kourkoumpas@certh.gr; grammelis@certh.gr; kakaras@certh.gr

G. Kontopoulos • I. Vournas
HELECTOR S.A., 25, Ermou Str., N. Kifissia, 14564 Attica, Greece
e-mail: g.kontopoulos@helector.gr

D. Koulocheris
National Technical University of Athens, 9, Heroon Polytechniou, 15780 Athens, Greece
e-mail: dbkoulva@mail.ntua.gr

P. Grammelis (ed.), *Energy, Transportation and Global Warming*, Green Energy and Technology, DOI 10.1007/978-3-319-30127-3_57

applied for the separation of the different plastic fractions, in order to get high purity materials, which means higher selling price. In the MBT plant, mechanical separation is achieved by using several components as optical (NIR) separators, ballistic separators and magnets. In addition a hand picking step for the removal of heavy and large objects is found in the most of the existing MBT plants in order to remove the heavy and large materials. For the largest part these plants operate automatically. In order to facilitate the separation, waste is usually shredded to the average size of a few hundred mm. The shredder is usually followed by a drum screen, creating two waste streams of different sizes, (Böhm et al. 2010).

The aim of this paper is to present the scope of the project "Smart Recovery of materials and upgrade of organic compost and RDF in existing mechanical biological treatment plants by using NIR technology" funded by the GSRT.

Methodology

Scope of the Project

The scope of the project is to design, develop and integrate a Smart System for Recovery of materials and upgrade of the organic Compost and RDF quality in existing Mechanical Biological Treatment Plants by using NIR technology, state of the art technology systems for automation and measurement and algorithms for waste process and recycling optimization. The main goals for the SmartWasteTech project are:

- the development and integration of an online monitoring software for the MBT plant operation based on the continuous information obtained from the modern online sensor technology
- to increase the valuable plastic fraction (PET, PE/PP, LDPE film) sorted out from MSW, towards increasing recycling efficiency and overall environmental performance of the plant
- the process optimization based on the Greek MSW quality
- to increase the compost quality through application of innovative technology for sorting out inert materials (glass, inert materials) and other unwanted materials such as heavy metals and the production of a high purity compost
- to perform continuous quality monitoring of the produced Refused Derived Fuel to investigate the economic, feasibility and environmental benefits of the proposed technology
- to disseminate the project results to all relevant stakeholders including among others manufacturers and operators of waste treatment facilities, local authorities, engineering and planning companies in the waste treatment sector, end users of recycled plastics and compost

Innovation Activities

Through the latest development of online sensor technology (Near Infrared, laser, X-Ray etc.) different machines have been developed and are currently applied as complete solutions for sorting of mixed Municipal Solid Waste. Based on this development, modern waste treatment plants have the appropriate process technology installed to efficiently recover valuable materials from mixed Municipal Solid Waste (metals, specific plastic fractions like PET), while the remaining high calorific fraction of MSW is usually led to other energy recovery processes by producing high quality Solid Recovered Fuels (SRF) or Refused Derived Fuels (RDF) (Nasrullah et al. 2014; Cimpan et al. 2015). Nevertheless, recently developed waste sorting techniques cannot be applied in the existing Mechanical Biological Treatment Plants (MBTs), without additional research work and carefully planned modifications in the current waste treatment processes. Due to the difficulty of this task, no integrated concept solution for the enhancement of environmental performance in existing MBT plants has been developed and demonstrated up to now. This is exactly the first innovative aspect in the present project.

This conventional type of waste treatment plants usually includes a separation step for the fine organic fraction, which is usually followed by a biological drying and a composting step towards the production of a material with characteristics argued to be similar with these of organic "compost". However, after a thorough evaluation of the compost composition it can be noticed that the compost material derived from mixed MSW has usually a high amount of inert materials and a high concentration in heavy metals. Hence the utilization of this type of material is rather limited, thus it cannot be promoted as fertilizer. Therefore, the goal of the project is the upgrade of the compost quality up to a level that will be comparable with the organic fertilizer. This will be achieved by the development and optimization of a process for sorting out inert, heavy metals and other unwanted materials, which will be the second innovative point of the project. The particular process has been applied for other types of waste in the past (commercial waste) but it has never been tested for mixed MSW. Therefore, a further development of the process based on the particular characteristics of compost derived from mixed MSW is required and will be carried out. The remaining stream with the bigger size particles, including packaging materials is then utilized for the production of a refused derived fuel (RDF). Nevertheless, the particular type of RDFs, produced at MBT plants operating with an aged technology, can hardly meet the high quality criteria required by today's incineration or co-incineration plants (cement, brick kilns) in terms of calorific value, chlorine content and heavy metals concentration. For this reason, their marketability is limited and only a view of them are standardized according to the recently published European standards (EN 15357, EN 15358, EN 15359). Furthermore, the recovery of paper and plastic fractions is a target of the project. In specific, this type of recyclables is usually not recovered in mixed waste treatment plants. An ambitious goal of recovering about 85 % of the recyclables still present in the packaging material stream is set. This will be achieved through the customization of new sorting equipment and optimization based on the specific parameters of mixed MSW, which is the third innovative point of the project.

The concept under development was implemented in the Athens Mechanical and Biological Waste Treatment plant located in the main Athens land fill site of Ano Liosia and operated by Helector S.A. and to the MBT plant in Larnaca, Cyprus which was designed, constructed and operated by Helector S.A. since 2010.

The daily input stream of the Athens MBT plant is about 1200 tn of mixed MSW. The production of RDF and compost rise to 450 t/d and 180 t/d respectively. The recyclable fraction of the input mixed MSW is about 1.5 % PET, 1.5 % PE/PP, 8 % recyclable paper, 9 % glass and 5 % metals. The proposed work intends to provide a complete solution for the quality improvement of MBT plant's output streams and the increase of its recycling efficiency by the application of modern, innovative sensor and separation technology.

The technologies that were incorporated in the Integrated System SmartWasteTech are related with industrial research (IR) and aim at obtaining expertise by design and install:

- special electromechanical structures
- specific type and measurement sensors
- automation and control systems
- optimization algorithms for measurement and process
- quantitative and qualitative management system and evaluation assessment models (LCA, feasibility study for environmental, social and economic benefits) of the collected data.

The Technology

The Titech equipment uses an NIR sensor to detect the characteristic infrared spectrum of light reflected by an illuminated object. The NIR spectrum of each material is unique and can be used to identify specific materials and then separate them (Fig. 57.1). The system has

- Modular design
- Modular build

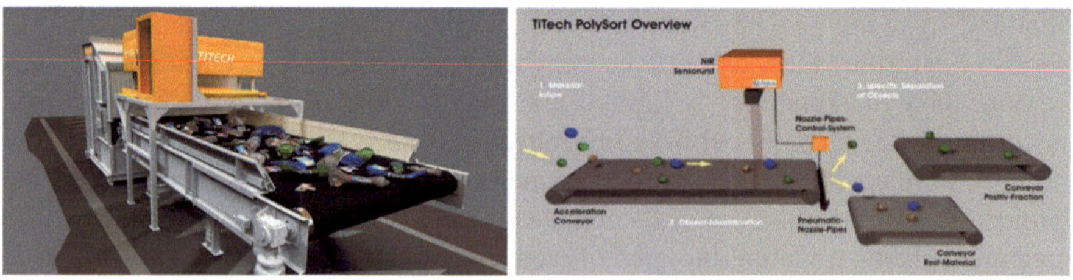

Fig. 57.1 NIR technology of TITECH (TOMRA)

- Maximize cost benefit process plant to product recovery

The basic characteristics of the system are:

- High tech sensors are utilized to identify objects on a conveyor belt
- High speed processing of information: material, shape, size, color, defect, damage and location objects
- Precise sorting by air jets

The system works by scanning the material as it travels along a conveyor belt. The PolySort UHR uses a fast near infrared analyzer along with a computer-controlled air jet ejection system. Material is identified across the entire width of the conveyor belt. The type of material, position and projected area of every single object on the belt is determined by the machine. A computer then rapidly processes the information and controls a series of air jets situated at the end of the conveyor belt. The air jets are activated accordingly to remove the identified material from the main product stream. The projected area determines which air jets need to be activated and only those which the object will pass over are used. The duration for which the air jet is active corresponds to the overall projected length of the object.

Results

NIR's municipal waste sorting systems identify and sort grain sizes from 10 to 500 mm, separated according to preferred size ranges. These systems provide inert material elimination of up to 95 % in municipal waste, producing material suitable for biomethanisation. Inert material, such as glass and stones, can cause significant problems during the treatment process and therefore has to be reduced to a minimum, while as little organic content as possible is removed. In addition, materials such as PET, PE/PP, LDPE film etc. can be recovered successfully.

The NIR system is operated entirely by a computer with an LCD screen display from which the system can be controlled. Most settings and functions can be changed via the screen such as the belt speed, material selected for ejection. The SmartWasteTech system communicates directly to the NIR in order to monitor via a SCADA and MIS system the operating parameters. The integration of the SmartWasteTech system in the MBT plant leads to the following results:

- Process analysis
- Real time process and material information
- Batch process and reporting
- Real time analysis and process tuning
- Input composition over a period
- Material timetable which shows how much material was detected in a period
- Distribution of the material which shows how much material was detected on different parts of the conveyor
- Temperature and pressure

- Management
- Database capture of process and material
- Overall equipment effectiveness

The integration of the NIR technology in the MBT plant in Larnaca, operated by Helector S.A. have increased the recovery of the material and the efficiency of the technology performance. These results are presented in the following diagrams.

In specific, the purity of the materials are shown in Figs. 57.2 and 57.3. As it is shown, the purity of the recovered materials are high. This means that the environmental benefit is high, as well as the selling price of the materials to the recycling industries. In Fig. 57.3, the purity of the materials in Larnaca is compared with the green Dot system in Germany. The purity of materials is higher than the respective purity of the German Dot system, thanks to the NIR technology applied in the plant.

The NIR online analysis has been applied successfully for the identification of the input waste. The results are shown in Fig. 57.4. The results from NIR online analysis are very closed to the results from the experimental analysis (mass balancing). In summer months, there is a significant deviation of the results, since the waste composition is quite different due to the change of the consumer's behaviour e.g. increase of tourism.

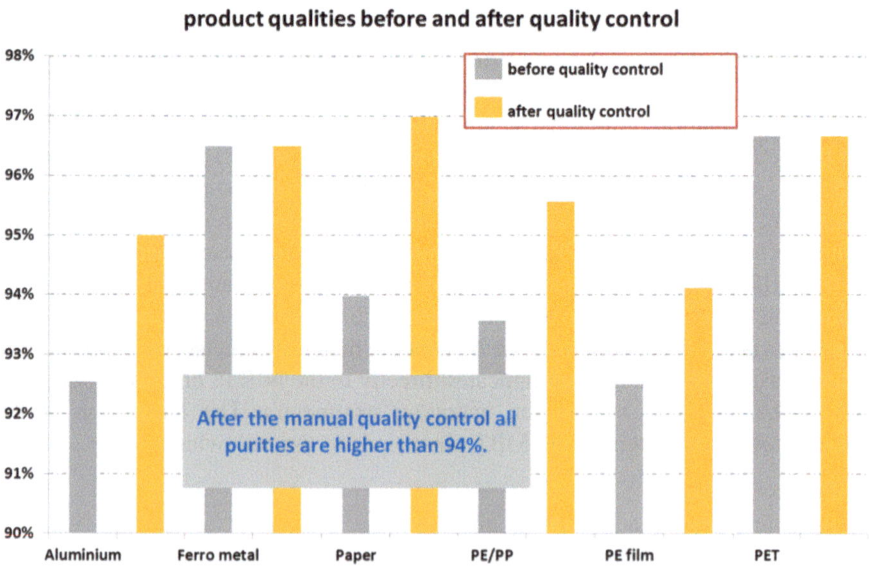

Fig. 57.2 Indicative purity results in MBT plant at Larnaca

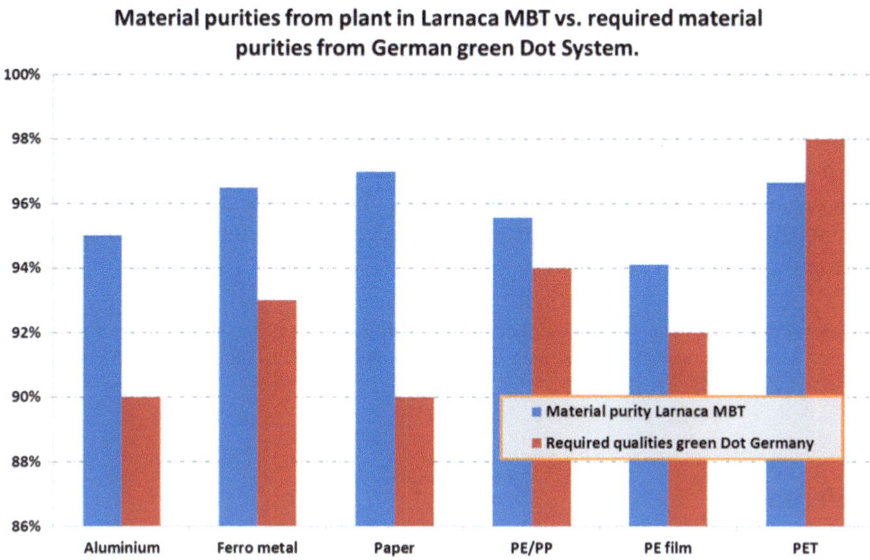

Fig. 57.3 Comparison between Larnaca's MBT plant with a typical plant in Germany

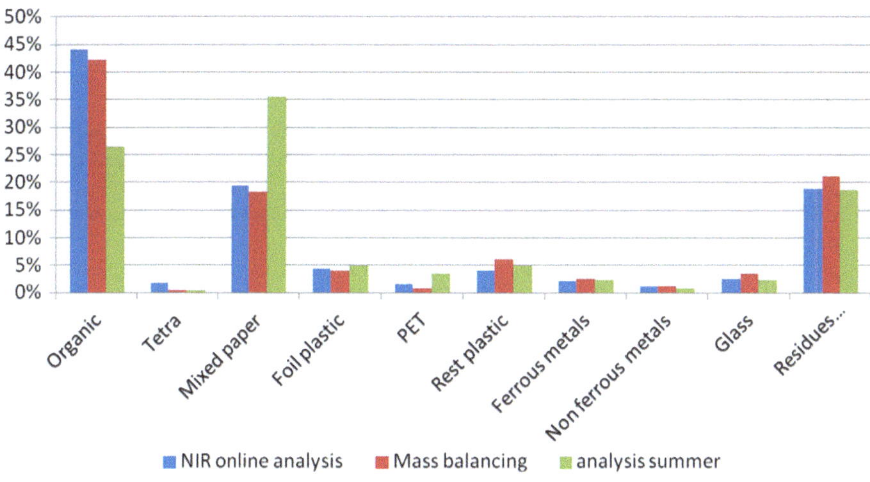

Fig. 57.4 Composition of input waste based on NIR online analysis and experimental analysis

Conclusions

The SmartWasteTech proposal, using NIR technology for recycling, improves the operation of the existing MBT plants and the SCADA system communicates directly to the NIR in order to monitor the process and the operating parameters. The SmartWasteTech system can be applied in existing MBT plant, giving a solution to the improvement of the final products (compost, RDF, PET, PP, etc.). In addition, despite the fact that the input waste corresponds to mixed municipal waste, the proposed system can recover materials in high purity. The overall environmental footprint of the plant is improved, as well as its economic feasibility. The identification of the input waste over a time period is also another outcome of the proposed system. Based on this identification, appropriate configurations of the NIR separators can be carried out, in order to maximize the recovery efficiency of the materials, as well as to improve the final quality of the produced materials (compost, RDF, PET, PP, etc.).

Acknowledgement The study described in this publication was financially supported by the Greek General Secretariat for Research and Technology (GSRT). The enumerated code of the Project is 1115-BET-2013.

References

Böhm, K., Smidt, E., Binner, E., Schwanninger, M., Tintner, J., & Lechner, P. (2010). Determination of Mbt-waste reactivity – An infrared spectroscopic and multivariate statistical approach to identify and avoid failures of biological tests. *Waste Management, 30*, 583–90

Cimpan, C., Maul, A., Jansen, M., Pretz, T., & Wenzel, H. (2015). 'Central sorting and recovery of Msw recyclable materials: A review of technological state-of-the-art, cases, practice and implications for materials recycling'. *Journal of Environmental Management, 156*, 181–99

Habich, U. (2007). Sensor-based sorting systems in waste. In *Processing international symposium MBT*.

Nasrullah, M., Vainikka, P., Hannula, J., Hurme, M., & Kärki, J. (2014). Mass, energy and material balances of Srf production process. Part 1: SRF produced from commercial and industrial waste. *Waste Management, 34*, 1398–407

TOMRA. *Brochures and Guidelines for Titech equipment*. http://www.tomra.com/en/solutions-and-products/sorting-solutions/recycling.

Part XIV
Securing Sustainable Mobility to Mitigate Climate Change

Transport Operation and Resilience Issues

Chapter 58
Road Transport Induced GHG Emissions Calculation for Urban Transportation Networks: The Case of Athens and Thessaloniki in Greece

Christos Samaras, Iraklis Stamos, Leonidas Ntziachristos, Evangelos Mitsakis, Zissis Samaras, and Georgia Ayfantopoulou

Introduction

The threat of climate change is not only well recognized, but is an important item on the political agenda of all countries due to the severity and intensity with which the impacts are already occurring (Graus and Blomen 2008). The world is experiencing extreme weather conditions, such as lengthy droughts, heat waves, changing rainfall patterns, changing seasonal patterns, severe hurricanes, etc. (IPCC 2015). Climate change is mainly induced by human activity and is primarily caused by the combustion of fossil fuels (coal, oil and gas). The average global temperature has already risen by 0.8 °C since the beginning of the Industrial Revolution (NOAA 2005). According to researchers (IPCC 2015; Hare 2003), an average global warming of 2 °C or above compared to the pre-Industrial Revolution level would result in dangerous and probably irreversible impacts in human ecosystems.

It is evident that GHG emissions are strongly related with climate change. According to the annual European Union (EU) greenhouse gas inventory, total GHG emissions, without Land Use, Land-Use Change and Forestry (LULUCF), in the EU-15 decreased by 15.1 % between 1990 and 2012. However, at the same time frame, GHG emissions produced by road transport increased by 72 million tons of CO_2 equivalent. Furthermore, in 2012, the contribution of road transport alone in

C. Samaras (✉) • L. Ntziachristos • Z. Samaras
Laboratory of Applied Thermodynamics, Aristotle University of Thessaloniki, 54124 Thessaloniki, Greece
e-mail: cesamara@auth.gr; leon@auth.gr; zisis@auth.gr

I. Stamos • E. Mitsakis • G. Ayfantopoulou
Center for Research and Technology Hellas—Hellenic Institute of Transport, 6th Km Charilaou Thermis Road, 57001 Thessaloniki, Greece
e-mail: stamos@certh.gr; emit@certh.gr; gea@certh.gr

P. Grammelis (ed.), *Energy, Transportation and Global Warming*, Green Energy and Technology, DOI 10.1007/978-3-319-30127-3_58

the EU-15 total CO_2 emissions—the main greenhouse gas—was about 24 % (EEA 2014).

In respect of GHG emissions, the situation in Greece is worse than the average EU-15; total GHG emissions—without LULUCF—were increased by 5.8 % between 1990 and 2012 (EEA 2014). The peak years were 2005 and 2007, but, after 2007 there was a substantial decrease, mainly due to the economic crisis (MEECC 2014). In the annual EU emissions inventory for 2012, GHG emissions in Greece were calculated equal to 111 million tons of CO_2 equivalent (EEA 2014).

The Greek Ministry of Environment, Energy and Climate Change (MEECC) estimates that in 2010 the contribution of transport in the total national GHG emissions was about 24.7 %, making transport the second most important source of GHG emissions (MEECC 2012). In the same report, MEECC states that approximately 84 % of total GHG emissions of the transport sector were derived from road transport (83.8 % for CO_2, 94.5 % for CH_4 and 57 % for N_2O emissions).

From the aforementioned statistical data it is evident that road transport is a major source of GHG emissions in Greece. In an effort to gain a clear and more detailed understanding of the amount of GHG emitted in Greece, this paper maps GHG pollutants in the two largest cities of the country, namely Athens and Thessaloniki. According to the latest national census that was conducted in 2011, the examined cities accommodate almost half of the country's population (Hellenic Statistical Authority 2014); therefore, the calculation conducted herein will largely reflect the national status in respect to GHG emissions.

Compared to national and European inventories, as well as to other studies (Progiou and Ziomas 2012), this paper focuses solely on road transport. Apart from daily GHG emissions values, the paper tries to identify the road sections that produce the majority of emissions (hot-spots). Moreover, emphasis is given to the hourly variation of GHG emissions, as well as to the contribution of different vehicle categories on total emissions.

Materials and Methods

In the next section, the methodology for both traffic and emissions modelling is analyzed. It should be highlighted that due to the lack of available data the study focuses only on hot emissions.

Traffic Modelling

The transportation model for Athens has been developed with the PTV VISUM software (Gentile and Noekel 2009), a traffic assignment tool for urban and regional operational planning analysis that has been used in several studies (Stamos et al. 2011; Ayfadopoulou et al. 2012; Mitsakis et al. 2013). The network used in this study consists of a detailed representation of the urban and regional road network of the metropolitan region of Athens, based on open-source GIS, fused with traffic related parameters.

The network consists of 81.880 directed links and 36.725 nodes. The links contain information about the number of lanes, the road type and its hierarchy in the network, width, length, free flow speed, design and effective capacity, direction and allowed transport systems. The link delays are calculated with the use of volume-delay functions, the parameters of which rely on previous studies and have been updated through travel time measurements for the purposes of the work presented herein. The nodes contain detailed information about the junction's geometry, allowed movements and control type of the node.

The network consists of 359 traffic analysis zones (TAZ) connected to physical nodes of the road network via 3.468 connectors, according to their accessibility index (Friedrich and Galster 2009), avoiding connections with nodes belonging to high hierarchy links. The demand side is comprised by 24 hourly Origin-Destination (OD) matrices and the travel demand for a typical weekday is within the range of 3.8 million vehicle trips. The obtained OD matrices are corrected using the hourly volume data measured by inductive loop detectors installed at 557 locations across the city. The OD matrix correction is performed with a fuzzy-set based matrix correction procedure (Rosinowski 1994).

Since traffic measurements are available only at a number of locations which is smaller than the number of trips, then the problem of determining the OD matrix which reproduces trips that result to traffic volumes equal to the ones measured is underdetermined. The matrix correction procedure described herein is of bi-level nature, where at the upper level user equilibrium traffic flows are computed, subject to the corrected OD matrices at the lower level. The upper level user equilibrium traffic flow estimation, known as the Traffic Assignment Problem, based on Wardrop's user equilibrium principle (Wardrop 1952), is solved with an implementation of the Linear User Cost Equilibrium algorithm (Gentile and Noekel 2009), with an average goodness-of-fit of 0.94.

A similar approach has been followed for modelling traffic in the network of Thessaloniki. The final output concerns a large-scale model with 137.938 links, 47.838 nodes and 339 TAZ accounting for a daily total of about 1.3 million vehicle trips. The OD matrix correction is conducted via real-time measurements of cameras, radars and inductive loop detectors located in 37 signalized junctions of the road network, measuring volumes at a frequency of 1.5 min. Figure 58.1 depicts instances of the road transportation network in Athens and Thessaloniki respectively.

Emissions Modelling

The emission calculations were performed with COPERT Micro (Samaras et al. 2014), a specially developed version of COPERT 4 (Ntziachristos et al. 2009) for urban areas. COPERT Micro is an average speed emissions model

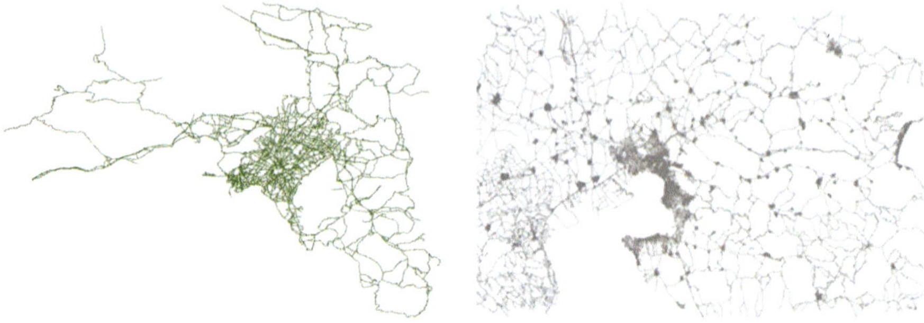

Fig. 58.1 Graphical representation of the road network in Athens (*left*) and Thessaloniki (*right*)

that is able to calculate fuel consumption, as well as several important pollutants (CO, NOx, VOC, PM exhaust, GHG emissions and others). It is a bottom-up model, hence, it can calculate the emissions from a single traffic link up to an entire city. COPERT Micro was developed focusing primarily on hot exhaust emissions. It includes 230 different vehicles categories and incorporates several emission factors that were obtained experimentally as a function of the average vehicle speed.

The equations below summarize the main calculations conducted by COPERT Micro during hot exhaust emissions calculation. First of all, based on the average speed of each traffic link the corresponding emission factors are calculated for each vehicle category and for every pollutant:

$$EF_{i,k}(V) = f_{EF,i,k}(V_j) \tag{58.1}$$

where:

$EF_i(V)$: hot exhaust emission factor per vehicle category k, for average speed V and for pollutant i [g/km]
i: pollutant of interest (CO_2, N_2O and CH_4)
$f_{EF,\ i,\ k}$: polynomial function derived from measured data (as trendline); unique for every vehicle category k and for each pollutant i
V_j: average speed of the vehicles circulating on the traffic link j [km/h]

After calculating the emission factors for all pollutants, the hot exhaust emissions of the traffic link j are calculated by the formula:

$$Ehot_{i,j} = L_j \times N_j \times \sum \left(P_{j,k} \times EF_{i,k}(V)\right) \tag{58.2}$$

where:

$Ehot_{i,\ j}$: hot emissions of pollutant i produced by N_j vehicles that circulate on the link j [g]
L_j: length of the traffic link j [km]
N_j: number of vehicles circulating on link j

$P_{j,\,k}$: percentage of vehicles of the specific category k on the overall vehicle fleet that circulate on link j, e.g. passenger cars 0.8–1.4 l gasoline Euro 3 = 5 % etc. This is not constant, but it changes form link to link.

Finally, the total hot exhaust emissions of the pollutant i for the entire area ($Ehot_{i,\ area}$) are calculated by summing the emissions of individual traffic links j:

$$Ehot_{i,area} = \sum Ehot_{i,j} \tag{58.3}$$

Since the hourly activity data are available from the traffic model, the calculation is repeated for each hour of the day and the hourly and daily emissions are calculated for all pollutants and for all traffic links.

Results and Discussion

The results are divided into two sections. In the first one, the emissions at micro/link level are presented upon the GIS grid, whereas in the second one, the hourly and daily emissions for both cities are given.

Emissions on Micro Level

In general, the visual representation of emissions on GIS grid is extremely helpful for identifying the localized hot-spots throughout a road network. This is the reason why in Fig. 58.2 the daily CO_2 hot emissions of both Athens and Thessaloniki are

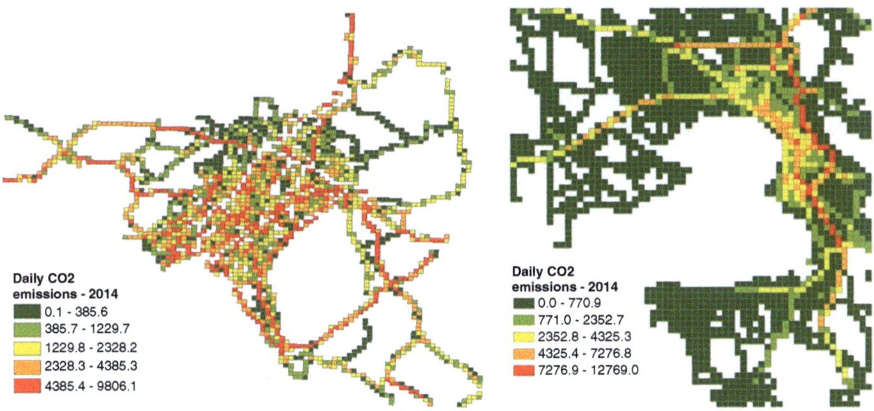

Fig. 58.2 Daily CO_2 hot emissions (kg) on a 100 × 100 (500 × 500 m^2) grid for a typical weekday of October 2010 in Greater Athens Area (*left*) and for a typical weekday of November 2014 (*right*) in metropolitan area of Thessaloniki

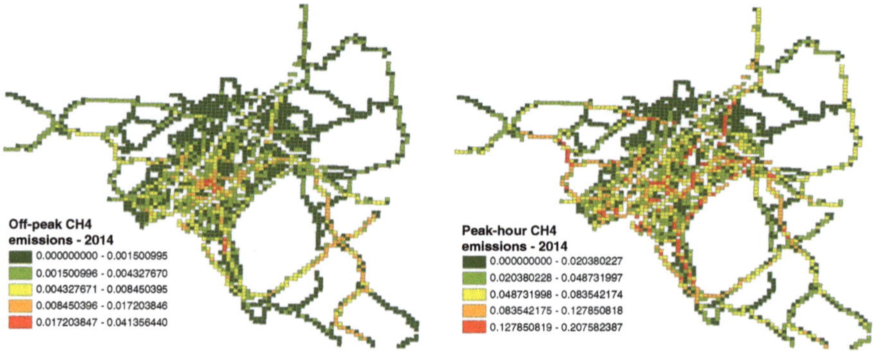

Fig. 58.3 Off-peak—3–4 A.M.—(*left*) and peak—8–9 A.M.—(*right*) CH_4 hot emissions (kg) on a 100×100 (500×500 m^2) grid for a typical weekday of October 2010 in the Greater Athens Area

presented. It is evident that in both cities there are specific road sections—marked with red color—that strongly affect CO_2 emissions. In the case of Athens, sections of large boulevards, such as Attica Road, Kifisou Avenue, Mesogion Avenue, Athinon Avenue, along with Kifisias Avenue, Vouliagmenis Avenue and Poseidonos Avenue, are major CO_2 emitters. In Thessaloniki the situation is much clearer; Inner Ring Road constitutes the major source of CO_2 emissions in the city, while large sections of Egnatia Avenue, Tsimiski Avenue and Karamanli Avenue follow. In both cases the results are not surprising, since the number of vehicles circulating in these boulevards is quite high, so, as a consequence, the emissions will be high.

Since transport activity is not constant throughout the day, the local hot-spots will vary from hour to hour. This is demonstrated in Fig. 58.3, where the off-peak (3–4 A.M.) and peak (8–9 A.M.) CH_4 hot emissions for Athens are shown upon the GIS grid. During the off-peak hour the highest values of CH_4 hot emissions are found close to city center, mainly in Stadiou, Panepistimiou, Agiou Konstantinou and P. Tsaldari streets, as well as in a section of Vouliagmenis Avenue near Argyroupoli. During the rush hour in the morning, there are multiple hot-spots, including also the off-peak hour hot-spots. Note, that although the colors in both pictures are the same, the legends have different values.

The variation of N_2O hot emissions in Thessaloniki during peak and off-peak hours is shown in Fig. 58.4. In both cases the Inner Ring Road, again, plays the most important role. However, late at night, the city center seems to have a significant contribution too. More precisely, the N_2O hot emissions caused by transport activity in Egnatia and Tsimiski Avenue are quite high during the off-peak hour, while, contrary to Athens, these hot-spots disappear in the morning (peak hour). It should be highlighted that the Inner Ring Road is probably one of the country's busiest highway sections. It is a dual carriageway ring road encircling most of the urban area of Thessaloniki. It consists of two sections; the Western Ring and the main Inner Ring. It has three lanes per direction and carries over 100.000 vehicles per day (Ewen et al. 2009).

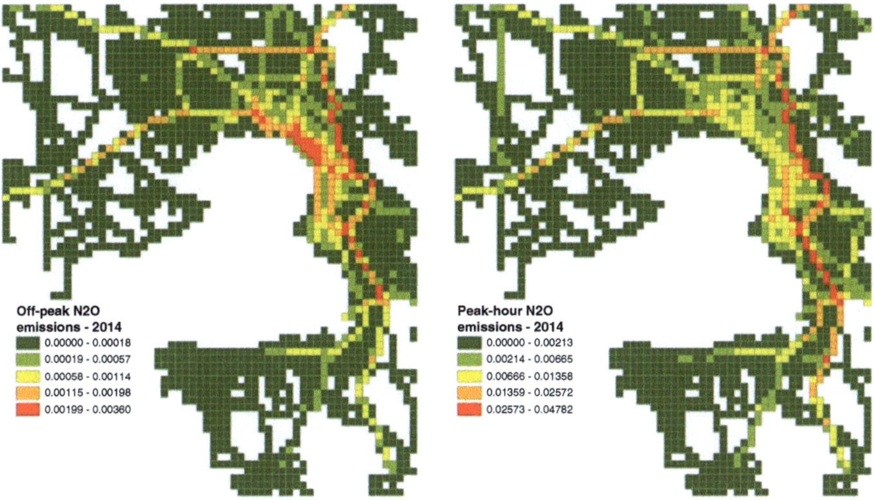

Fig. 58.4 Off-peak—3–4 A.M.—(*left*) and peak—8–9 A.M.—(*right*) N_2O hot emissions (kg) on a 100 × 100 (500 × 500 m^2) grid for a typical weekday of November 2014 in metropolitan area of Thessaloniki

Emissions on Macro Level

Figure 58.5 shows the hourly GHG emissions for both Athens and Thessaloniki. It is evident that GHG emissions are not constant during the day, but follow proportionally the transport activity. In both cities, GHG emissions attain their highest values during the rush hour in the morning (8–9 A.M.), when most citizens start their typical activities (work, school etc.). Emissions decreasing until about noon (12–1 P.M.), and, then, start rising again during afternoon (2–6 P.M.), when the majority of people finish their activities and return back to their origin. From the evening until late at night (6 P.M.–4 A.M.) the traffic activity drops, so do GHG emissions. Finally, after the off-peak hour (3–4 A.M.), the emissions start increasing again.

The emission patterns for both cities have many similarities. This is expected, since most residents have similar daily schedules; they usually drive to work between 7–9 A.M., and, after about 8 h, they drive back to their houses in the afternoon (3 P.M.–5 P.M.). It's worth mentioning that due to the increased leisure activities after the midnight in Thessaloniki, there is a small increase of GHG emissions between 1 A.M. and 3 A.M., when people usually return to their homes. Furthermore, between 9 A.M. and 1 P.M., GHG emission values in Thessaloniki vary significantly, indicating that transport activity in the city is more variable compared to Athens.

In the next three figures, the daily hot GHG emissions are presented both for Athens and Thessaloniki, as well as the share of those emissions in the main vehicle categories. Starting with CO_2 in Fig. 58.6, it is evident that passenger cars are the

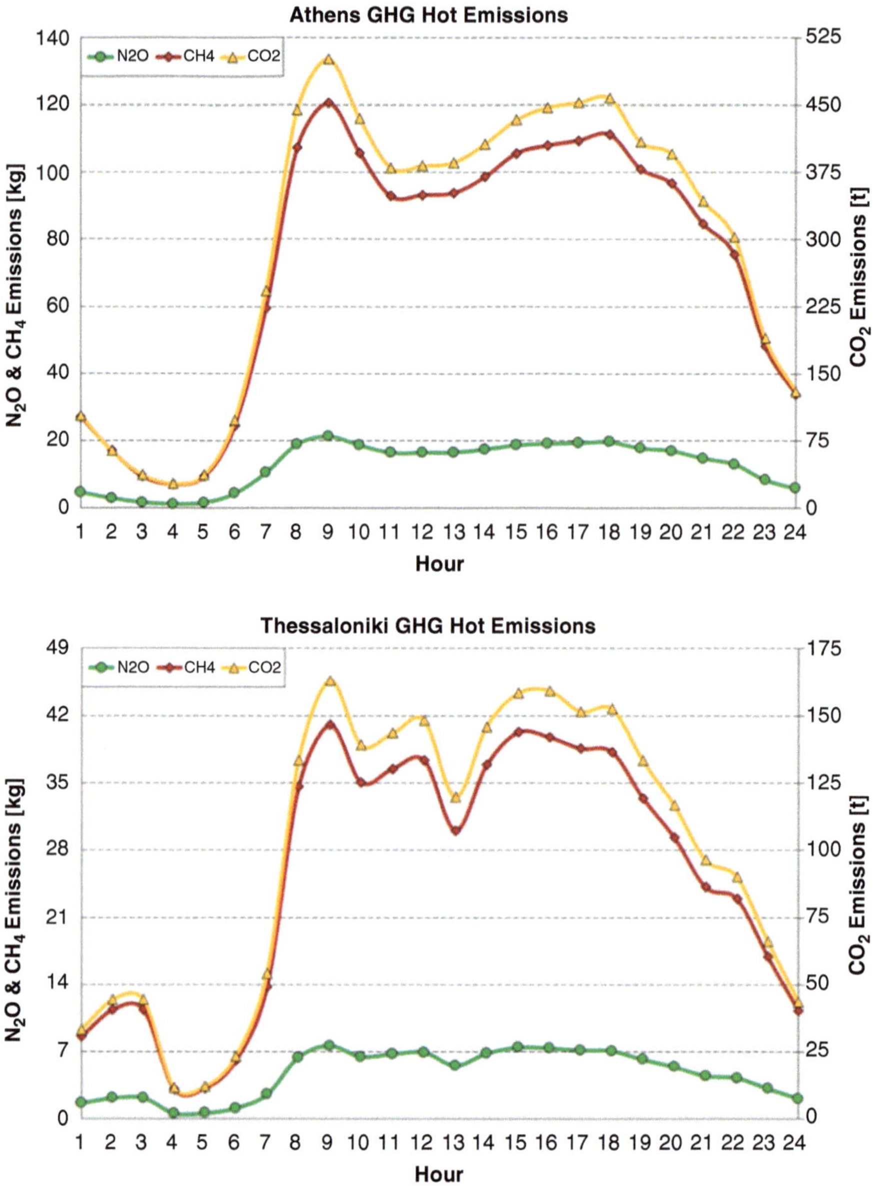

Fig. 58.5 Hourly CO_2, N_2O and CH_4 hot emissions for a typical weekday of October 2010 in Greater Athens Area (*up*) and for a typical weekday of November 2014 (*down*) in metropolitan area of Thessaloniki

main source of CO_2 emissions in both cities. The contribution of light and heavy duty vehicles in the case of Athens is similar—close to 15 %, whereas in Thessaloniki, light duty vehicles have more impact on CO_2 emissions than heavy

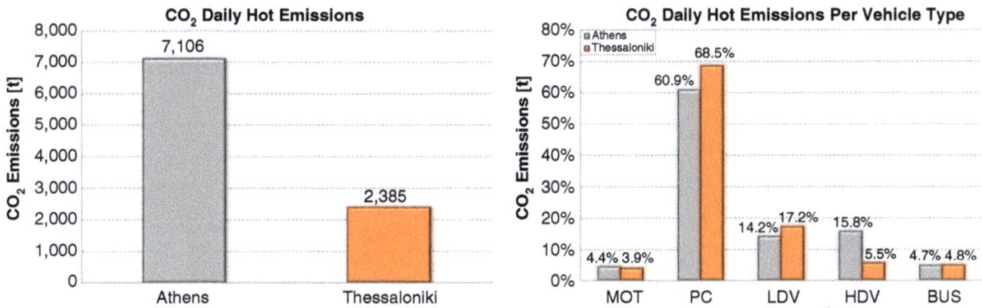

Fig. 58.6 Daily CO_2 hot emissions (*left*) and contribution of each vehicle category (*right*) for a typical weekday of October 2010 in Greater Athens Area and for a typical weekday of November 2014 in metropolitan area of Thessaloniki

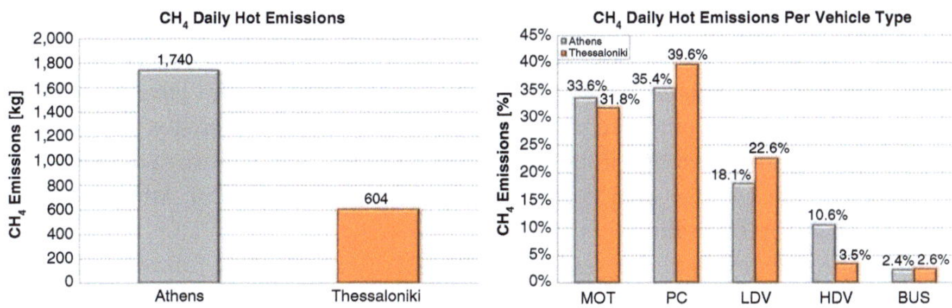

Fig. 58.7 Daily CH_4 hot emissions (*left*) and contribution of each vehicle category (*right*) for a typical weekday of October 2010 in Greater Athens Area and for a typical weekday of November 2014 in metropolitan area of Thessaloniki

duty vehicles (17.2 % and 5.5 % respectively). Finally, the two cities have almost the same level of CO_2 emissions coming from buses and two-wheel vehicles (mopeds and motorcycles).

On absolute level, the daily CO_2 hot emissions in Athens are three times higher than the corresponding ones in Thessaloniki. This is in line with the number of vehicle trips simulated with the traffic software used herein (about 3.8 and 1.3 million vehicle trips for Athens and Thessaloniki respectively). Since the vehicle trips in the case of Athens are about three times the trips in Thessaloniki, it is reasonable to expect that CO_2 emissions in Athens will be approximately three times the corresponding ones in Thessaloniki.

Regarding methane, the situation is similar to CO_2 in respect to absolute emission levels (Fig. 58.7). It is noteworthy that the share of two-wheel vehicles is quite high, almost similar to the share of passenger cars. On the other hand, the contribution of passenger cars in CH_4 hot emissions in both cities has been reduced substantially compared to CO_2—almost 42 % relative reduction (from 60.9 % for CO_2 to 35.4 % for CH_4 in the case of Athens and from 68.5 % to 39.6 % respectively for Thessaloniki).

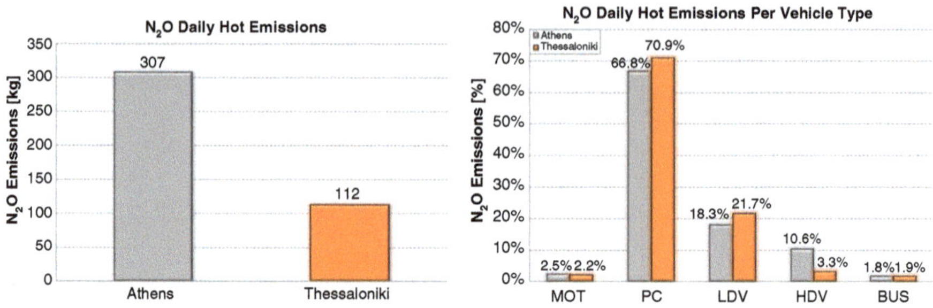

Fig. 58.8 Daily N_2O hot emissions (*left*) and contribution of each vehicle category (*right*) for a typical weekday of October 2010 in Greater Athens Area and for a typical weekday of November 2014 in metropolitan area of Thessaloniki

The two-wheel vehicles are responsible for almost one third of CH_4 hot emissions in each city. This is due to the fact that two-wheel vehicles have higher CH_4 hot emission factors compared with other vehicle types (Ntziachristos and Samaras 2014). Furthermore, another important factor is the relative high share of two-wheel vehicles on the average fleet composition, which for the case of Athens is about 19.8 %, and, for Thessaloniki is 12.3 %.

The absolute levels of the last GHG pollutant of interest-N_2O-follow the trend of CO_2 and CH_4 emissions (Fig. 58.8); N_2O hot emissions in Athens are about 2.7 times the respective ones in Thessaloniki. However, when daily N_2O hot emissions are split into the main vehicle categories, the percentages of passenger cars and light duty vehicles are higher than in the CO_2 case. The shares in both cities are relatively similar except for heavy duty vehicles, which in the case of Athens are much higher (10.6 % compared with 3.3 % in Thessaloniki).

The previous charts depict the recent situation in the two largest cities of Greece in respect to GHG emissions. Since the country is away from its Kyoto targets (EEA 2014), measures should be taken in order to reduce GHG emissions and meet these targets. In road transport the measures should focus on alleviating the traffic in main boulevards of the cities, since they constitute the major sources of pollution. Moreover, the adoption of stricter policies and cleaner technologies in passenger cars, light duty vehicles (i.e. new Euro 6 regulation), as well as in two-wheel vehicles will be beneficial for GHG emissions.

Conclusions

Climate change is already perceptible in Greece. There are clear indications of warming in the country from the early 1990s (Feidas and Lalas 2001); gradually strengthened and record-breaking hot summers days are an increasingly regular occurrence. Road transport in Greece constitutes a significant source of GHG emissions, and it consequently affects the local climate change.

In this paper, the hourly and daily GHG hot emissions were examined in both Athens and Thessaloniki, the two more heavily-populated cities in Greece. In both case studies the results at link level revealed the local pollution hot-spots and the high-emission links, which usually lie along the main urban highways. These hot-spots are not constant, but changing throughout the day, proportionally following transport activity. During the rush hour in the morning (8–9 A.M.), GHG emission attain their highest values, while the opposite occurs late at night (3–4 A.M.).

Passenger cars, and, secondarily, light duty vehicles are the vehicle categories which mainly contribute to GHG in the cities examined. Moreover, two-wheel vehicles should be also taken into account in respect of CH_4 emissions. Thus, apart from alleviating the traffic in the large arterials, the application of stricter policies and cleaner technologies in these vehicle categories should be also examined, considering that the country should meet its Kyoto targets.

References

Ayfadopoulou, G., Stamos, I., Mitsakis, E., & Salanova, J. M. (2012). Dynamic traffic assignment based evacuation planning for CBD areas. *Procedia: Social and Behavioral Sciences, 48*, 1078–1087.

Ewen, C., Anagnostopoulou, M., & Ward, N. (2009). Monitoring of heavy metal levels in roadside dusts of Thessaloniki, Greece in relation to motor vehicle traffic density and flow. *Environmental Monitoring and Assessment, 57*, 483–498.

European Environment Agency (EEA). (2014). *Annual European Union greenhouse gas inventory 1990–2012 and inventory report 2014*. Technical report No. 09/2014.

Feidas, H., Lalas, D. (2001). Climatic changes in Mediterranean and Greece: A critical review. In: *Seventh international conference on environmental science and technology*, Ermoupolis, Syros, Greece.

Friedrich, M., & Galster, M. (2009). Methods for generating connectors in transport planning models. In *TRB annual meeting*, 2009, Washington, DC.

Gentile, G., & Noekel, K. (2009). Linear user cost equilibrium: The new algorithm for traffic assignment in VISUM. In *Proceedings of European transport conference*, 2009. Leeuwenhorst Conference Centre. Netherlands.

Graus, W., & Blomen, E. (2008). *A low carbon vision for Greece in 2050*. Project number: PECSNL073688. Utrecht: Ecofys Netherlands BV.

Hare, W. (2003). *Assessment of knowledge on impacts of climate change—Contribution to the specification of Art. 2 of the UNFCCC*. WBGU (German Advisory Council on Global Change).

Hellenic Statistical Authority. (2014). Demographic and social characteristics of the Resident Population of Greece according to the 2011 Population—Housing Census revision of 20/3/2014.

IPCC (Intergovernmental Panel on Climate Change). (2015). Climate change 2014: Synthesis report.

Ministry of Environment, Energy and Climate Change (MEECC). (2012). *Annual inventory submission under the convention and the Kyoto protocol for greenhouse and other gases for the years 1990–2010*, Athens, Greece.

Ministry of Environment, Energy and Climate Change (MEECC). (2014). *Sixth national communication and 1st biennial report under the United Nations framework convention on climate change*, Athens, Greece.

Mitsakis, E., Stamos, I., Salanova Grau, J. M., Chrysochoou, E., Iordanopoulos, P., & Ayfadopoulou, G. (2013). Urban mobility indicators for Thessaloniki. *Journal of Traffic and Logistics Engineering, 1*, 148–152.

NOAA (National Oceanic and Atmospheric Administration). (2005). *NOAA reports December warmer than average, global temperature warmer than average.*

Ntziachristos, L., & Samaras, Z. (2014). Exhaust emissions from road transport. In *EMEP/EEA emission inventory guidebook 2013* (update September 2014). EMEP.

Ntziachristos, L., Gkatzoflias, D., Kouridis, C., & Samaras, Z. (2009). COPERT: A European Road Transport Emission Inventory. *Information Technologies in Environmental Engineering*, 491–504.

Progiou, A., & Ziomas, I. (2012). Twenty-year road traffic emissions trend in Greece. *Water, Air and Soil Pollution, 223*, 305–317.

Rosinowski, J. (1994). Entwicklung und Implementierung eines ÖPNV-Matrixkorrekturverfahrens mit Hilfe von Methoden der Theorie unscharfer Mengen (Fuzzy-Sets-Theorie). Master thesis, University of Karlsruhe.

Samaras, C., Ntziachristos, L., & Samaras, Z. (2014). *COPERT Micro: A tool to calculate the vehicle emissions in urban areas*. Paris, France: Transport Research Arena 2014.

Stamos, I., Salanova, J. M., Mitsakis, E., & Ayfadopoulou, G. (2011). *Large scale dynamic traffic assignment model for real-time traveler information services*. ITS 2011 "Innovation and Society", Patras, Greece.

Wardrop, J. G. (1952). Some theoretical aspects of road traffic research. IN *Proceedings of the Institute of civil engineers. Part II* (pp. 325–378).

Chapter 59
The Effect of Parameter Selection on Fume Formation Rate in SMAW of AH36 Shipbuilding Steel and Analysis with ANOVA Method

Tolga Mert, Levent Bilgili, Kutsi Mert Senoz, Uğur Buğra Çelebi, and Serkan Ekinci

Introduction

Shipbuilding consists of various types of production processes such as painting, blasting, cutting and welding, which cause considerable amount of deleterious emissions. Shipbuilding includes the characteristics of both manufacturing and construction processes. Due to the great amount of steel used during shipbuilding, welding, which is used to join steel sheet metals, is one of the main manufacturing methods in shipbuilding.

Welding Types

There are more than 80 different types of welding operations, including brazing, thermal cutting and gauging, in commercial use. In welding and similar operations, such as brazing, thermal cutting, and gauging, the most frequently used method for generating heat is obtained either from an electric arc or a gas-oxygen flame (Mener et al. 2001). Figure 59.1 presents the most common welding and cutting techniques used in shipbuilding.

Welding provides a powerful manufacturing tool for high quality joining of metallic components. Essentially, all metals and alloys can be welded; some with ease, others requiring special precautions. Of these processes, some of more common types include shielded metal arc welding (SMAW), gas metal arc welding (GMAW), flux-cored arc welding (FCAW), submerged arc welding (SAW), gas

T. Mert (✉) • L. Bilgili • K.M. Senoz • U.B. Çelebi • S. Ekinci
Yildiz Technical University, Barbaros Bulvari, Besiktas, Istanbul 34349, Turkey
e-mail: tmert@yildiz.edu.tr; lbilgili@yildiz.edu.tr; kmsenoz@yildiz.edu.tr; ucelebi@yildiz.edu.tr; ekinci@yildiz.edu.tr

P. Grammelis (ed.), *Energy, Transportation and Global Warming*, Green Energy and Technology, DOI 10.1007/978-3-319-30127-3_59

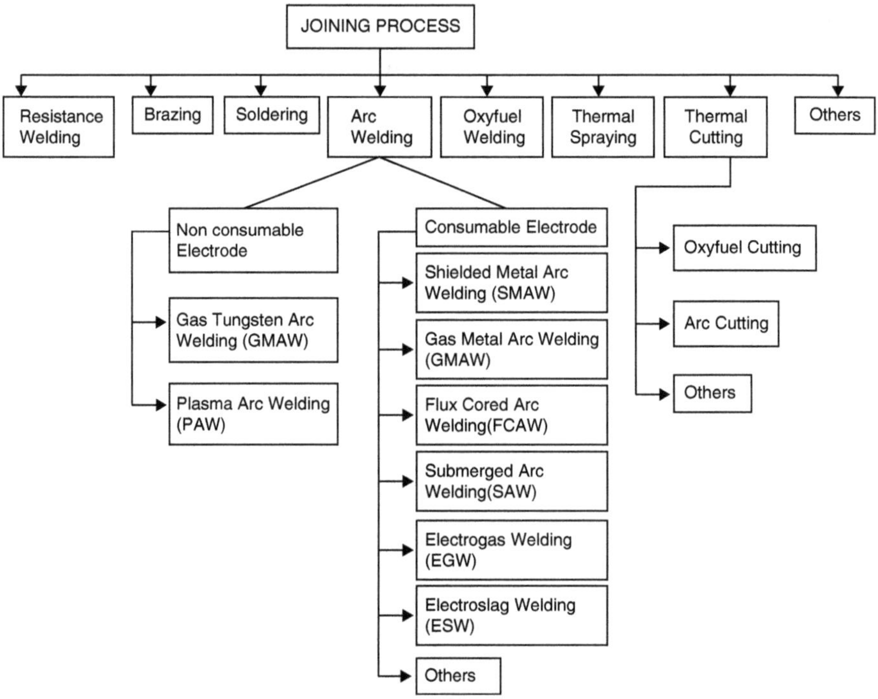

Fig. 59.1 Welding and cutting methods in shipbuilding (Celebi 2008)

tungsten arc welding (GTAW), and others such as, plasma arc welding, and oxy-gas welding (NIOSH 2002). Electric arc welding is the most frequently used process. Although electric arc welding method can be divided into various types, the welding process is more or less the same. A flow of electricity across the tip of the electrode and to the base metal creates the heat needed for melting and joining the metal parts. The electric current melts both the electrode and the base metal at the joint to form a molten pool, which solidifies upon cooling. In SMAW method, the core rod conducts electric current to form the arc and serves as a filler metal for the joint. The electrode cover provides stability to the arc and protects the molten metal. GTAW uses a non-consumable tungsten electrode that creates an arc between the electrode and the weld pool. An inert shielding gas is used in the process at no applied pressure. GMAW is a consumable electrode welding process that produces an arc between the base metal and a continuously supplied filler metal. Externally supplied gas is used to shield the arc. SAW forms an arc between a bare metal electrode and the work contained in a blanket of granular flux. The electrodes serve as the filler material, although a welding rod or metal granules may be added (EPA 1994).

Welding Emissions

Welding process is an important source for harmful emissions. Welding fumes and emissions, which include various numerous gases and particulates, are classified as dangerous for human health and environment.

Hazardous metals listed in the 1990 Clean Air Act Amendments that have been detected in welding fume include manganese, nickel, chromium, cobalt and lead. Additionally, the hexavalent form of chrome ($Chrome^{+6}$) is also found in some welding fume emissions. The emissions of toxic air contaminants during welding have potential adverse human health impacts (Mener et al. 2001). Both, the quantity and characteristics of air emissions can change from one welding process to another. The composition of base metals, welding electrodes, and operating variables of each process can influence welding emissions (Kura et al. 2009). The main pollutants of concern generated during welding operations are particulate matter and particulate phase hazardous air pollutants. Only electric arc welding generates pollutants in quantities of major concern. The welding fume is formed by the vaporization and recondensation of metallic elements upon cooling in ambient air. As such, the particulate matter produced is generally submicron in size with approximately 50–75 % of the particles having diameters in the range of 0.4–0.8 μm. The amount of the emissions generated can vary substantially from process to process (NSRP 2001). Welding fume particles less than 1 mm in diameter constitute the greatest health hazard because of their ability to penetrate deep into the lungs (Konarski et al. 2003). Welding is considered as a dangerous occupation because: (1) there are a multiplicity of factors that can endanger the health of a welder, such as heat, burns, radiation, noise, fumes, gases, and electrocution; and (2) the high variability in chemical composition of welding fumes which differ according to the work piece, method employed, and surrounding environment. The particulates and gases generated during welding are considered to be the most harmful in comparison with the other by-products of welding (NIOSH 2002). Metal fume fever is a possibility after exposure to manganese fume. Chronic manganese poisoning, characterized by a severe disorder of the nervous system, has been reported in welders working in confined spaces on high-manganese steels. Health effects can only occur if a worker is actually exposed to the hazard. The risk of injury or disease usually increases with the duration and frequency of exposure to the agent, and the intensity/concentration and toxicity of the agent (NIOSH 1990).

In addition to the fume, it is known that gas phase pollutants, such as carbon dioxide (CO_2), carbon monoxide (CO), nitrogen oxides (NO_x), sulfur dioxide (SO_2) and ozone (O_3), are generated during welding operations. CO may be formed by the incomplete combustion of the electrode cover or flux and by the use of CO_2. High concentrations of CO may be rapidly fatal without producing significant warning symptoms (Popović et al. 2014). It is also known that CO has an effect on global warming (AEA Energy & Environment 2008). CO_2 plays a major role in the atmosphere in global warming and anthropogenic climate change. CO_2 is the largest contributor among well-mixed long-lived greenhouse gases, accounting for more than 63 % of the total. Besides, inhaling CO_2 during welding may cause rapid breathing, rapid beating of the heart, headache, sweating, mental depression and death. SO_2 emissions contribute as a secondary particulate pollutant to formation of particulate matter (PM) in the atmosphere, an important air pollutant in terms of its adverse impact on human health (Popović et al. 2014). In some recent studies, it was indicated that while SO_2 has an effect on global cooling (AEA Energy & Environment 2008), PM has effects both on global warming and global cooling (Endresen et al. 2003). Ozone and nitrogen oxides are produced by the interaction of ultraviolet light (from the welding arc) with the surrounding air. These compounds are irritating to the eyes, nose and throat (Popović et al. 2014). In a recent study, it is also indicated that NO_x emissions may have more impact on global warming than CO_2 (Henningsen 2000).

Experimental Study

Fume emission experiments have been realized inside a fume chamber (Fig. 59.2). AH-36 grade ship building steel round plates with diameter of 290 mm have been used in the experiments and these plates have been welded with rutile (EN ISO 2560-A: E 42 0 RR 12) and basic (EN ISO 2560-A: E 46 6 B 42 H5) covered electrodes for 45 s. Welding power source was Lincoln Electric Invertec V260-S. Chemical composition of base metal and the electrodes are given in Table 59.1. Voltage values have been measured using Fluke 375 True Rms clamp meter. Fume was captured on Whatman GF/A glass fiber filters. Filters have been stored in a furnace at 110 °C for at least 1 h before and after the experiments. Shimadzu BL-320H electronic balance has been used to weigh filters and electrodes. Three different welding speeds (i.e. 25 cm min^{-1}, 35 cm min^{-1} and 45 cm min^{-1}) and three different current settings (i.e. 100 A, 120 A, 140 A) have been utilized in the experiments.

Design of Experiment (L18 Taguchi Design)

Taguchi method is one of the most robust and reliable design of experiment methods, whose aims are to analyze the effect of process parameters on process performance, define contribution of each variable and optimize the process. Table 59.2 shows three process factors (i.e. electrode type, welding speed and current) and three levels, which were used to form L18 orthogonal array for Taguchi design.

Fig. 59.2 Fume chamber

Table 59.1 Chemical composition of AH-36 steel, rutile and basic covered electrodes (% wt.)

Material/electrode	C	Mn	Si	P_{max}	S_{max}
AH-36 steel	0.18	1.30	0.45	0.035	0.035
Rutile electrode	0.07	0.5	0.3	–	–
Basic electrode	0.08	1.4	0.4	–	–

Table 59.2 Experimental factors and levels

Symbol	Factor	Level 1	Level 2	Level 3
A	Electrode type	Rutile	Basic	–
B	Welding speed (cm min^{-1})	25	35	45
C	Current (A)	100	120	140

Results and Discussion

Table 59.3 presents L18 orthogonal array experimental set and FFR measurements.

Experimental data were analyzed using signal to noise (S/N) ratio and the analysis of variance (ANOVA) methods with the help of Minitab 17 software. Since it is better to have smaller FFR in welding, smaller is better equation was chosen (Eq. 59.1). Here, n is the number of repeated experiments, Y is the measured value of the response variable.

Table 59.3 L18 experimental set and FFR measurements

	Factors and levels			Output
Run	Electrode type	Welding speed (cm min^{-1})	Current (A)	FFR (g min^{-1})
1	1	1	1	0.129
2	1	1	2	0.187
3	1	1	3	0.235
4	1	2	1	0.143
5	1	2	2	0.195
6	1	2	3	0.268
7	1	3	1	0.145
8	1	3	2	0.179
9	1	3	3	0.283
10	2	1	1	0.275
11	2	1	2	0.293
12	2	1	3	0.396
13	2	2	1	0.283
14	2	2	2	0.324
15	2	2	3	0.423
16	2	3	1	0.264
17	2	3	2	0.337
18	2	3	3	0.436

$$\text{The lower—the better:} \quad S/N = -10\log\left(\left(\sum Y^2\right)/n\right) \qquad (59.1)$$

Table 59.4 shows response table for S/N ratios. S/N ratios in Taguchi method are used to investigate the significance of the factors. Higher S/N ratios yield to the aim, which is lower FFR. S/N ratios in Table 59.4 and Fig. 59.3 depict electrode type and current are significant in formation of fume, unlike welding speed and selection of the covered electrode (rutile or basic) is the biggest contributor in fume formation. It is clearly understood that FFR is minimum at first level of electrode type (A1), first level of welding speed (B1) and first level of current (C1). Therefore, optimum design parameter combination was found to be A1B1C1.

ANOVA analysis was performed to investigate the significance of the factors on FFR. ANOVA results were based on general linear model. P-value should be equal or less than 0.05 with 95 % confidence level. Since P-values in Table 59.5 are less than 0.05, confidence level is assured. The percentile effects of each factor were also calculated. The percentile effect of electrode type, welding speed and current were found to be 60.10 %, 1.16 % and 37.16 %, respectively. Coefficient of determination (R-sq (adj)) in ANOVA analysis was 97.78 %.

Table 59.4 Response table for S/N ratios

Level	A	B	C
1	14.465	12.476	14.200
2	9.598	11.801	12.272
3	–	11.817	9.623
Delta	4.867	0.675	4.577
Rank	1	3	2

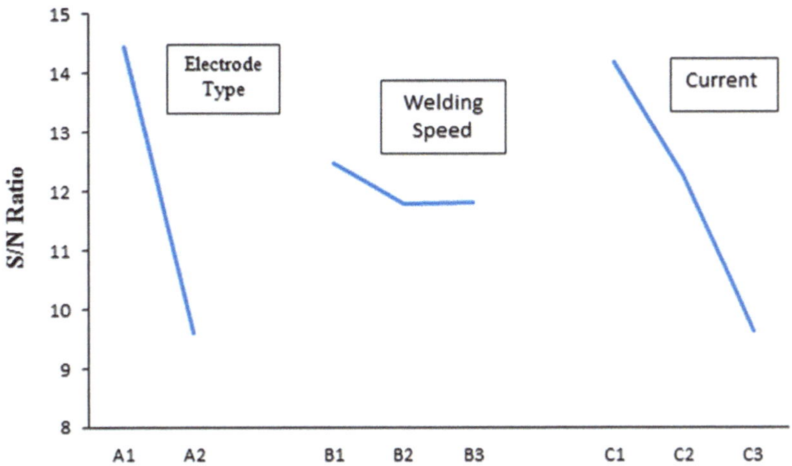

Fig. 59.3 S/N ratios of factor levels for FFR

Table 59.5 ANOVA results for FFR

Source	Degree of freedom	Adj SS	Adj MS	F-value	P-value	% Effect
A	1	0.089324	0.089324	460.88	0.000	60.10
B	2	0.001734	0.000867	4.47	0.035	1.16
C	2	0.055238	0.027619	142.51	0.000	37.16
Error	12	0.002326	0.000194			1.56
Total	17	0.148622				100

Conclusion

Due to the high preferability of welding in joining processes, it is one of the important sources of greenhouse gases. Although the production of greenhouse gas strongly depends on the welding conditions, such as electrode type, current value, type of the base material, welding speed, the ability of the welder etc., greenhouse gases (most particularly CO_2) are inevitable end products for most of the welding processes. Despite the fact that the harmful effects of welding fume are well documented, there is little research on hazardous gas emissions and further investigations of their effect on global warming must be realized.

Permissible exposure limits for the welding workers have a great importance for their health. Therefore, studies on welding fume and emissions are the first step to investigate and understand the limits. This study mainly focused on welding fume formation rate (FFR) and thus, it has the characteristics of a first stage of a long-term welding experiment. It is believed the effect of welding on global warming and worker exposure will be understood more comprehensively with the help of these studies.

Acknowledgement This experimental study was performed under the sponsorship of Turk Loydu (Turkish Lloyd) and Coordinatorship of Scientific Research Projects of Yildiz Technical University (Project No.: 2013-10-01-KAP05 and Project No.: 2013-06-01-GEP01).

References

AEA Energy & Environment. (2008). *Greenhouse gas emissions from shipping trends, projections and abatement potential*. Final report to the Committee on Climate Change, Restricted Commercial, ED 43808, Issue Number 4, 3 Sept 2008.

Celebi, U. B. (2008). Wastes in shipbuilding and painting emission estimation, Ph.D. thesis, Yildiz Technical University, İstanbul (in Turkish).

Endresen, Ø., Sørgard, E., Sundet, J. K., Dalsøren, S. B., Isaksen, I. S. A., Berglen, T. F., et al. (2003). Emission from international sea transportation and environmental impact. *Journal of Geophysical Research, 108*, 4560. doi:10.1029/2003JD003751.

EPA. (1994). *Development of particulate and hazardous emission factors for electric arc welding*. AP-42, Section 12.19, 20 May 1994.

Henningsen, R. F. (2000). Study of greenhouse gas emissions from ships. Final report to the International Maritime Organization, Norwegian Marine Technology Research Institute (MARINTEK), Trondheim, Norway, Mar 2000.

Konarski, P., Iwanejko, I., & Ćwil, M. (2003). Core–shell morphology of welding fume micro- and nanoparticles. *Vacuum, 70*, 385–389.

Kura, B., Jackens, J., & Keay, J. (2009). New weld fume chamber design to assess HAP emissions potential and promote cleaner production. In *Second international workshop/advances in cleaner production, key elements for a sustainable world: Energy, water and climate change*, São Paulo, Brazil, 20–22 May 2009.

Mener, W. C., Rosen, P. L., Austin, D. M., & ve Holt, W.S. (2001). Shipyard welding emission factor development. In *10th international emission inventory conference—"one atmosphere, one inventory, many challenges"*, May 1–3.

NIOSH. (1990). *Welding: Fumes and gases*. Australian Government Publishing Service, Canberra WAP 90/034GS.

NIOSH. (2002). *Nomination of welding fumes for toxicity studies*. National Institute for Occupational Safety and Health, 20 Feb 2002.

NSRP. (2001). *Emission factors for flux core rod used in gas shielded processes*. Project No. N1-98-1 Subtask 43, 16 May 2001.

Popović, O., Cvetković, R. P., Burzić, M., Lukić, U., & Beljić, B. (2014). Fume and gas emission during arc welding: Hazards and recommendation. *Renewable and Sustainable Energy Reviews, 37*, 509–516.

Chapter 60
An Online Visualization Tool for Assessing the Robustness of Multimodal Transport Networks in Case of Extreme Weather Events and Natural Hazards

Iraklis Stamos, Evangelos Mitsakis, and Georgia Aifadopoulou

Introduction

As climate change induced extreme weather events and natural hazards are expected to gain in severity and intensity in the following years, it is of central importance to quantify the potential impacts and effects, among others, in the transportation sector as well. In an effort to do so, this paper builds on a recently developed methodology for analysing the effects of EWE and NH on passenger flows and presents the development of a user-friendly visualization tool for depicting the outcomes of this analysis. The proposed tool addresses transport related entities/experts (e.g. planning and decision making authorities, operators, owners), policy entities, climate change related actors, meteorological agencies, related international organizations and entities, as well as researchers and individual users. The tool's aim is to provide visualized information on EWE and NH occurrence probabilities; climate change related projections; information on multimodal transport networks; and visualized representation of substitutability opportunities between modes.

A limited number of efforts have taken place in the field of visualizing climate related datasets. Such efforts include the Green and Blue Space adaptation for urban areas and eco towns (GRaBS) and the Diachronic Inventory of Forest fires and the mapping tool of the National Resources Defense Council (NRDC).

GRaBS (PPGIS 2015) has been developed in the framework of the Green and Blue Space Adaptation for Urban Areas and Eco Towns (GRaBS) project, dealing with the integration of climate change adaptation into regional planning and development. In an effort to aid the strategic planning of climate change adaptation

I. Stamos (✉) • E. Mitsakis • G. Aifadopoulou
Center for Research and Technology Hellas—Hellenic Institute of Transport,
6th Km Charilaou Thermis Road, 57001 Thessaloniki, Greece
e-mail: stamos@certh.gr; emit@certh.gr; gea@certh.gr

P. Grammelis (ed.), *Energy, Transportation and Global Warming*, Green Energy and Technology, DOI 10.1007/978-3-319-30127-3_60

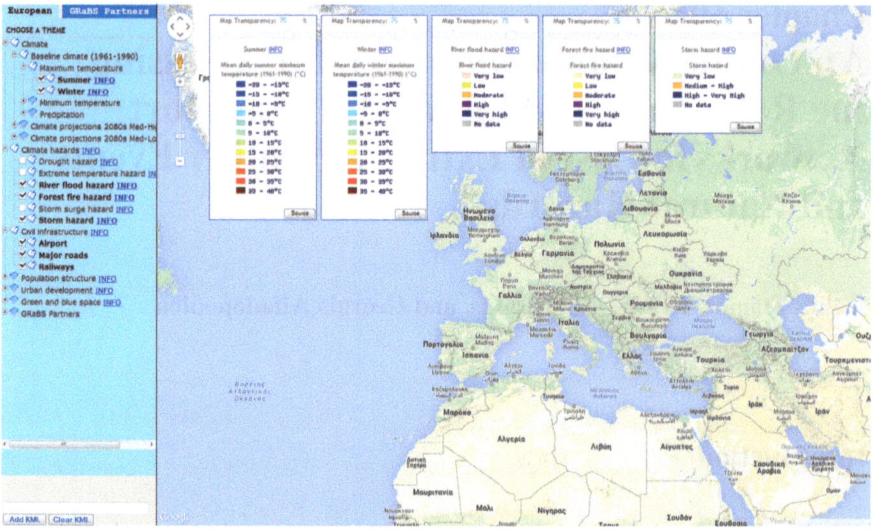

Fig. 60.1 View from the GRaBS adaptation Toolkit (source: http://www.ppgis.manchester.ac.uk/grabs/start.html)

responses, a risk and vulnerability assessment tool has been developed, with an emphasis on flooding and heat stress. The GRaBS tool assesses the vulnerability of urban areas to climate change impacts, with an additional assessment of relative patterns of spatial risk where suitable data is available (Fig. 60.1).

The National Observatory of Athens has developed an inventory of forest fires (NOA 2015), aiming to depict the results of the mapping of burned areas over Greece for the last 30 years (1984–2012). The services provide the user the ability to select and switch between a set of datasets (background layers, fire layers, burnt areas per year, areas which have been burnt mostly during this period, etc.) (Fig. 60.2).

The National Resources Defense Council of the USA has developed an Extreme Weather Event map for the year 2012 (NRDC 2012), when a series of "weather records" were broken, to the frequency and intensity of recorded extreme weather events and natural hazards attributed to the change of climate (Fig. 60.3).

Outline of the Visualization Tool

The proposed online visualization tool provides users with two services, as depicted in Fig. 60.4:

- Information on past and future frequency of selected weather phenomena in 134 locations in Europe, visualized on the European territory.

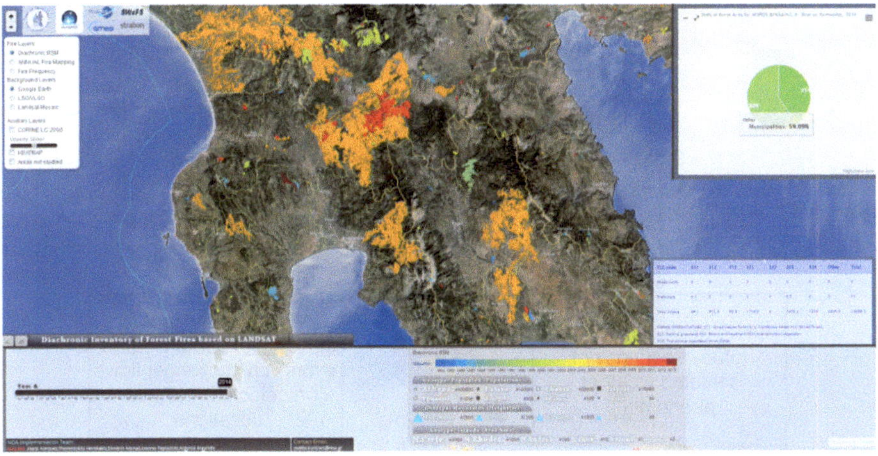

Fig. 60.2 View from the NOA tool—Diachronic Inventory of Forest fires (source: http://ocean.space.noa.gr/diachronic_bsm/)

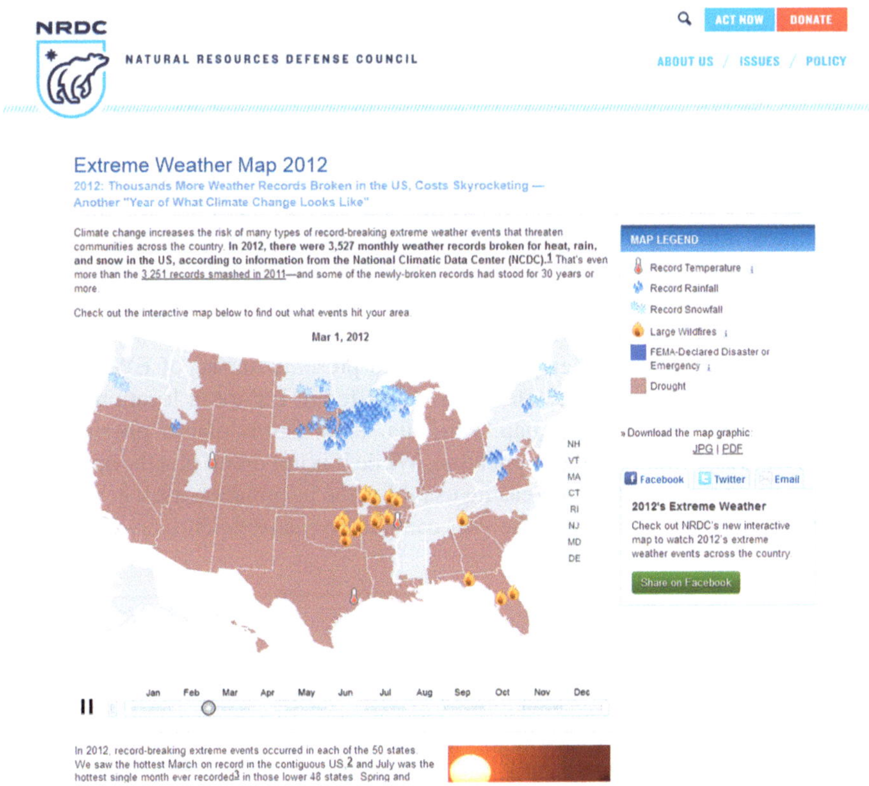

Fig. 60.3 View from the NRDC tool—Extreme Weather Map 2012 (source: http://www.nrdc.org/health/extremeweather/)

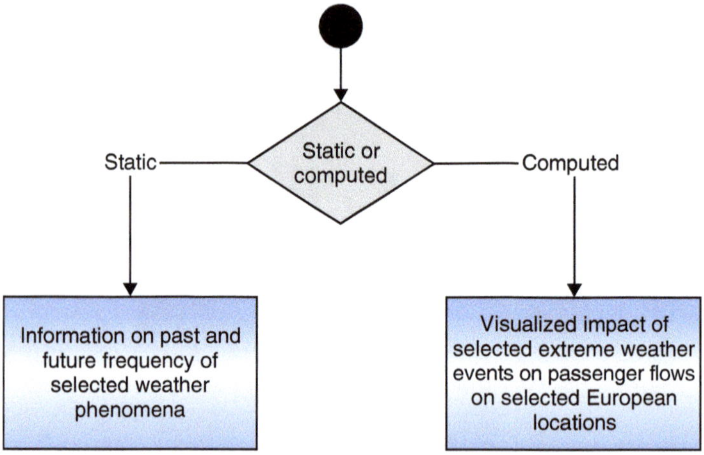

Fig. 60.4 Visualization tool end-user services

- Impacts of selected extreme weather events on passenger flows for surface and air transport modes on 14 locations in Europe, visualized on the European territory.

While the first service of the tool is only informative, providing to the end-users with a limited range of actions, as they are only able to see visualized outputs of a predefined dataset on preselected locations and as no computational process will take place, the second service is more complex. It provides its users with interaction capabilities, allowing own user's input that will determine the visualized outputs, which are based on computations of certain parameters. The following figure depicts the options for the end-user regarding the service of the visualization tool.

The visualization tool is a web-based application that provides information about past and future weather phenomena and presents impact of EWE and NH on passenger flows. Both cases are related to different European locations and the visualization is done using a map. Particular emphasis is placed on simplicity, usability and stylishness of the application. While it is important that the application is as simple as possible, thus increasing usability, the information itself is visualized in a comprehensive way so that the user can absorb it as efficiently as possible.

Data Needs

This section describes the data types and data formats needed for the development and operation of the proposed visualization tool. Each dataset is described individually in the next sections.

Table 60.1 Indicative format of locations and coordinates data

Location	Latitude	Longitude
Achleiten	48.583	13.416
Algeciras	36.167	−5.527
Amsterdam	52.427	4.762
Antwerp	51.328	4.313
Athens	381.543	238.588
Barcelona	414.946	21.417
Basel	47.565	7.594
Belfast	54.614	−5.925
Belgrad	447.412	204.889
Bergen	60.387	5.326
Berlin	525.098	134.367
Bilbao	433.274	−30.248

Geographical Data (GD)

Location coordinates (Latitude, Longitude) in WGS84 projection exist for all 134 locations that are used by the visualization tool. The 14 locations used for depicting the impacts of extreme weather events are a subset of those 134 locations (Table 60.1).

Temporal Periods (TP)

There are datasets for three periods in the visualization tool:

- 1971–2000
- 2001–2040
- 2041–2070

Weather Phenomena (WP)

The weather phenomena included in the databases of the visualization tool are the following:

1. Wind Gusts (WG)
2. Snowfall (SF)
3. Blizzard (BL)
4. Heavy Precipitation (HP)
5. Heat Waves (HW)
6. Cold Waves (CW)

Weather Phenomena Categories (WPC)

Each of the above defined weather phenomena (WP) has the following sub-categories:

1. Wind Gusts (WG)
   (a) 17 m/s
   (b) 25 m/s
   (c) 32 m/s
2. Snowfall (SF)
   (a) 10 cm/d
   (b) 10 cm/d
   (c) 20 cm/d
3. Blizzard (BL)
4. Heavy Precipitation (HP)
   (a) 30 mm/d
   (b) 100 mm/d
   (c) 150 mm/d
5. Heat Waves (HW)
   (a) >25 °C
   (b) >32 °C
   (c) >43 °C
6. Cold Waves (CW)
   (a) <0 °C
   (b) <−7 °C
   (c) <−20 °C

Weather Phenomena Categories Data (WPCD)

Each of the 134 locations receives one value for the above mentioned categories for all weather phenomena, representing the number of days that this event will occur in addition to the 1st Temporal Category (1971–2000). Tables, sized GD × WPC, for each WP and TP are available (i.e. five tables of 134 × 3 × 3 and one table of 134 × 1 × 3). Values within these Tables are ranged between: −60.7 and 196.6. An indicative layout format is depicted in the Table 60.2.

Additionally, these Tables will be also provided in such a way so as to represent the probability of the event occurring in one of the second (2001–2040) or third (2040–2070) temporal periods, with values ranging between 0 and 1.

Table 60.2 Indicative format of Weather Phenomena Categories Data

	1971–2000			2001–2040			2040–2070		
	Wind gust			Wind gust			Wind gust		
Location	17 m/s	25 m/s	32 m/s	17 m/s	25 m/s	32 m/s	17 m/s	25 m/s	32 m/s
Achleiten	14.43	0.38	0.00	14.43	0.38	0.00	14.43	0.38	0.00
Algeciras	14.14	0.24	0.00	14.14	0.24	0.00	14.14	0.24	0.00
Amsterda	44.29	2.38	0.10	44.29	2.38	0.10	44.29	2.38	0.10
Antwerp	35.00	1.14	0.05	35.00	1.14	0.05	35.00	1.14	0.05
Athens	22.76	0.62	0.00	22.76	0.62	0.00	22.76	0.62	0.00
Barcelona	2.81	0.05	0.00	2.81	0.05	0.00	2.81	0.05	0.00
Basel	14.10	0.29	0.05	14.10	0.29	0.05	14.10	0.29	0.05
Belfast	75.33	6.81	0.52	75.33	6.81	0.52	75.33	6.81	0.52
Belgrad	2.14	0.00	0.00	2.14	0.00	0.00	2.14	0.00	0.00
Bergen	35.43	1.29	0.05	35.43	1.29	0.05	35.43	1.29	0.05
Berlin	18.62	0.33	0.00	18.62	0.33	0.00	18.62	0.33	0.00
Bilbao	23.57	1.76	0.10	23.57	1.76	0.10	23.57	1.76	0.10

Source: MOWE-IT project (2014)

Demand Data (DD)

Demand data describe travel demand information (i.e. number of trips per mode) between certain nodes of a network. Demand data available for the proposed visualization tool refer to 14 selected cities' in-between connections. Demand data are classified in the three modes examined: rail, road and air transport. Therefore, three Tables (for each mode of transport), sized 14 × 14 are available. Values within these Tables range between 0 and 999,999,999. An indicative layout format is depicted in Table 60.3 (for road transport).

Supply Data (SD)

Supply data are classified in Cost, Time and Value of Time data.

Cost Data (CD)

Cost data describe the cost in € for the connections of the 14 selected cities for each three modes that are examined in MOWE-IT. Therefore, three tables (for each mode of transport), sized 14 × 14 are available. Values within these Tables are ranged between 0 and 999,999,999. An indicative layout format for road transport is depicted in Table 60.4.

Table 60.3 Indicative format of Demand Data (in passenger trips)

Demand Data—Road Transport														
	ATH	AMS	BCN	COP	FRA	LON	MAD	MAL	MOS	MUC	PAR	ROM	STO	ZUR
ATH	0	393	454	348	344	432	550	459	380	304	433	176	461	369
AMS	393	0	270	142	54	68	311	275	285	100	78	274	260	129
BCN	454	270	0	384	240	271	7	6	504	274	194	281	505	228
COP	348	142	385	0	148	197	440	391	255	112	207	356	121	224
FRA	344	5	240	146	0	95	343	246	272	49	98	225	266	79
LON	432	68	270	197	94	0	312	276	337	188	79	347	315	201
MAD	550	313	102	442	345	309	0	66	606	370	236	377	558	324
MAL	459	276	7	391	246	277	65	0	511	279	199	286	512	234
MOS	380	288	504	330	274	337	606	511	0	272	345	414	134	331
MUC	304	78	274	180	49	188	370	279	270	0	138	177	230	78
PAR	433	78	193	207	84	80	234	199	345	124	0	343	325	146
ROM	176	274	280	355	223	345	376	286	414	176	326	0	407	179
STO	461	260	505	121	266	315	558	512	134	230	325	407	0	340
ZUR	369	130	228	223	79	201	324	233	331	77	145	179	340	0

Source: ETISPlus project (2010)

Table 60.4 Indicative format of Cost Data (in €)

Input Data—Supply Data—Cost—Road Transport

	ATH	AMS	BCN	COP	FRA	LON	MAD	MAL	MOS	MUC	PAR	ROM	STO	ZUR
ATH	0	393	454	348	344	432	550	459	380	304	433	176	461	369
AMS	393	0	270	142	54	68	311	275	285	100	78	274	260	129
BCN	454	270	0	384	240	271	7	6	504	274	194	281	505	228
COP	348	142	385	0	148	197	440	391	255	112	207	356	121	224
FRA	344	5	240	146	0	95	343	246	272	49	98	225	266	79
LON	432	68	270	197	94	0	312	276	337	188	79	347	315	201
MAD	550	313	102	442	345	309	0	66	606	370	236	377	558	324
MAL	459	276	7	391	246	277	65	0	511	279	199	286	512	234
MOS	380	288	504	330	274	337	606	511	0	272	345	414	134	331
MUC	304	78	274	180	49	188	370	279	270	0	138	177	230	78
PAR	433	78	193	207	84	80	234	199	345	124	0	343	325	146
ROM	176	274	280	355	223	345	376	286	414	176	326	0	407	179
STO	461	260	505	121	266	315	558	512	134	230	325	407	0	340
ZUR	369	130	228	223	79	201	324	233	331	77	145	179	340	0

Source: MOWE-IT project (2014)

Time Data (TD)

Time data describe the cost in € for the connections of the 14 selected cities for all three modes of transport that are examined in MOWE-IT. Therefore, three tables (for each mode of transport), sized 14×14 are available. Values within these Tables are ranged between 0 and 999,999,999. An indicative layout format for air transport is depicted in Table 60.5.

Value of Time Data (VoTD)

This data category describes the Value of Time and refers to € per hour for all connections between the 14 selected cities, for all three modes of transport that are examined. Therefore, three tables (for each transport mode transport), sized 14×14 are available. Values within these tables range between 0 and 50. An indicative layout format for rail transport is depicted in Table 60.6.

Model Parameters (MP)

This data category describes certain parameters that are used in the functions "module" and concern:

- The weight of cost for every mode (WoC)
- The weight of time for every mode (WoT)
- The quality parameter for every mode (QP)
- A fixed scaling factor parameter for every mode (μ)

User Input

This section describes the inputs of the end-user to enable the visualization tool to compute and depict the impacts of extreme weather events on passenger flows per mode.

Type of Extreme Weather Event (WP)

The end-user chooses the type of the extreme weather event that will be examined by the tool. The extreme weather event types are similar to the weather phenomena described previously.

Table 60.5 Indicative format of Time Data (in hours)

Input Data—Supply Data—Time—Air Transport

	ATH	AMS	BCN	COP	FRA	LON	MAD	MAL	MOS	MUC	PAR	ROM	STO	ZUR
ATH	0	5	2	5	6	2	3	3	6	2	3	1	12	4
AMS	2	0	2	1	1	1	3	8	4	2	1	2	2	4
BCN	3	2	0	5	4	2	2	1	4	2	2	2	7	2
COP	6	1	3	0	2	2	8	5	7	2	2	5	1	4
FRA	4	1	4	1	0	4	4	8	8	1	1	2	2	3
LON	4	1	2	2	2	0	2	10	6	4	1	3	4	2
MAD	5	3	1	6	3	2	0	1	7	5	2	7	7	2
MAL	4	5	1	7	2	6	1	0	9	2	5	4	7	2
MOS	5	6	10	5	12	6	2	8	0	3	6	6	10	6
MUC	3	2	2	2	1	5	5	7	9	0	2	2	7	1
PAR	3	1	2	2	1	1	2	2	5	2	0	2	6	1
ROM	2	3	2	5	2	3	3	6	6	2	2	0	9	2
STO	7	2	3	1	4	2	6	12	1	10	7	6	0	5
ZUR	3	2	4	2	1	1	2	2	6	1	1	2	3	0

Source: MOWE-IT project (2014)

Table 60.6 Indicative format of Value of Time Data (in €/h)

Value of Time—Rail Transport														
	ATH	AMS	BCN	COP	FRA	LON	MAD	MAL	MOS	MUC	PAR	ROM	STO	ZUR
ATH	0	18	17	19	18	19	17	17	17	18	19	18	19	21
AMS	18	0	18	20	19	19	18	18	18	19	20	19	20	21
BCN	17	18	0	19	18	18	17	17	17	18	19	18	19	20
COP	19	20	19	0	20	21	19	19	19	20	21	20	21	22
FRA	18	19	18	20	0	19	18	18	18	19	20	19	20	21
LON	19	19	18	21	19	0	18	18	19	19	20	20	20	22
MAD	17	18	17	19	18	18	0	17	17	18	19	18	19	20
MAL	17	18	17	19	18	18	17	0	17	18	19	18	19	20
MOS	17	18	17	19	18	19	17	17	0	18	19	18	19	21
MUC	18	19	18	20	19	19	18	18	18	0	20	19	20	21
PAR	19	20	19	21	20	20	19	19	19	20	0	20	21	22
ROM	18	19	18	20	19	20	18	18	18	19	20	0	20	22
STO	19	20	19	21	20	20	19	19	19	20	21	20	0	22
ZUR	21	21	20	22	21	22	20	20	21	21	22	22	22	0

Source: EWENT project (2012)

Location of Occurrence (LO)

The end-user will have the ability to choose the exact location within the EU territory that an extreme weather event occurs. In order for the tool to compute the impacts of the chosen event, it will need to assign this event to the closest (based on proximity—direct distance) city within the network of the 14 pre-selected cities. The assignment of the event is a "one-to-one" relation with the city, as the end-user is able to choose another additional event, which again is assigned to the closest city (even if the city is the same).

Occurrence Probability (OP)

For each Weather Phenomenon identified previously, its probability of occurring in one or more locations (out of the 14 considered herein) is provided by the end-user. Values are ranged between 0 (when no event is present) to 1 (when an event is occurring now, i.e. at real-time). Additionally, the end-user will need to be able to choose the probability of the event based on the available Weather Phenomena Categories Data—WPCD.

Duration of Impact (DI)

For each Weather Phenomenon identified, the expected duration of its impact in hours in one or more locations (out of the 14 considered herein) is provided by the end-user. Values range between 0 and 24.

Operation Reduction (OR)

For each Weather Phenomenon identified and for each mode considered, the expected operation reduction of the performance of the respective network is provided by the end-user. Values range between 0 (no impact at all) and 1 (full operational impairment of the respective network). According to the process described before, after the end-user chooses the location of the event, the event is assigned to one of the 14 preselected cities. After this assignment, the operation reduction is considered to impact all connections that have either as origin or as destination the assigned city. For instance, if the end-user chooses that extreme heat waves will occur in the Netherlands (assumingly creating wildfires), having as impact that the road network operations will be affected by 50 %, then a table of the following form is created (Table 60.7).

Table 60.7 Exemplary output of the operation reduction in road transport

Heatwaves—Operation Reduced by (%) O/D—Road Transport														
	ATH	AMS	BCN	COP	FRA	LON	MAD	MAL	MOS	MUC	PAR	ROM	STO	ZUR
ATH	0.0	0.5	0.0	0.0	0.0	0.0	0.0	0.0	0.0	0.0	0.0	0.0	0.0	0.0
AMS	0.5	0.5	0.5	0.5	0.5	0.5	0.5	0.5	0.5	0.5	0.5	0.5	0.5	0.5
BCN	0.0	0.5	0.0	0.0	0.0	0.0	0.0	0.0	0.0	0.0	0.0	0.0	0.0	0.0
COP	0.0	0.5	0.0	0.0	0.0	0.0	0.0	0.0	0.0	0.0	0.0	0.0	0.0	0.0
FRA	0.0	0.5	0.0	0.0	0.0	0.0	0.0	0.0	0.0	0.0	0.0	0.0	0.0	0.0
LON	0.0	0.5	0.0	0.0	0.0	0.0	0.0	0.0	0.0	0.0	0.0	0.0	0.0	0.0
MAD	0.0	0.5	0.0	0.0	0.0	0.0	0.0	0.0	0.0	0.0	0.0	0.0	0.0	0.0
MAL	0.0	0.5	0.0	0.0	0.0	0.0	0.0	0.0	0.0	0.0	0.0	0.0	0.0	0.0
MOS	0.0	0.5	0.0	0.0	0.0	0.0	0.0	0.0	0.0	0.0	0.0	0.0	0.0	0.0
MUC	0.0	0.5	0.0	0.0	0.0	0.0	0.0	0.0	0.0	0.0	0.0	0.0	0.0	0.0
PAR	0.0	0.5	0.0	0.0	0.0	0.0	0.0	0.0	0.0	0.0	0.0	0.0	0.0	0.0
ROM	0.0	0.5	0.0	0.0	0.0	0.0	0.0	0.0	0.0	0.0	0.0	0.0	0.0	0.0
STO	0.0	0.5	0.0	0.0	0.0	0.0	0.0	0.0	0.0	0.0	0.0	0.0	0.0	0.0
ZUR	0.0	0.5	0.0	0.0	0.0	0.0	0.0	0.0	0.0	0.0	0.0	0.0	0.0	0.0

Source: MOWE-IT project (2014)

Architecture

The architecture of the visualization tool is comprised of two main components as depicted in Fig. 60.5 the server-side and the client-side. The server-side consists of a pre-processor and a database. The pre-processor is responsible for creating the client-side's user interface (UI) when the user starts the application. Data for the UI is retrieved from the database by the pre-processor during the initialization of the application. However, the pre-processor can provide additional data to the application also during runtime. The role of the client-side is mainly to visualize the available data to the user. However, the role includes also handling the pre-visualization calculations for the data as well as the previously mentioned communication with the server-side.

User Interface

The client-side User Interface (UI) is divided into two main functionalities based on the source of data. The user is required to choose whether she/he wants to visualize the static weather phenomena data or computed data. The latter are generated based on input provided by the user and contain information of how extreme weather events impact passenger flows.

Use of Static Data

In the case of static data, the UI consists of a map and a search field for locations. The map shows a small marker for each of the 134 European locations, for which existing weather phenomena data are stored in the tool. For displaying the data, the user can select a dot or type the desired location into the search field. This action leads to data being visualized with a tree-like structure on top of the selected location. Figure 60.6 depicts an indicative illustration of the data tree.

The data tree consists of location name, climatic region, temporal period as well as the actual data for each weather phenomenon. Each weather phenomenon has its own icon which acts as an on/off switch for showing a tree branch containing the corresponding weather phenomenon data. Thus, the user can decide which weather phenomena data are visible. If the temporal period is changed, the data shown is automatically changed, in order to correspond to the selected period.

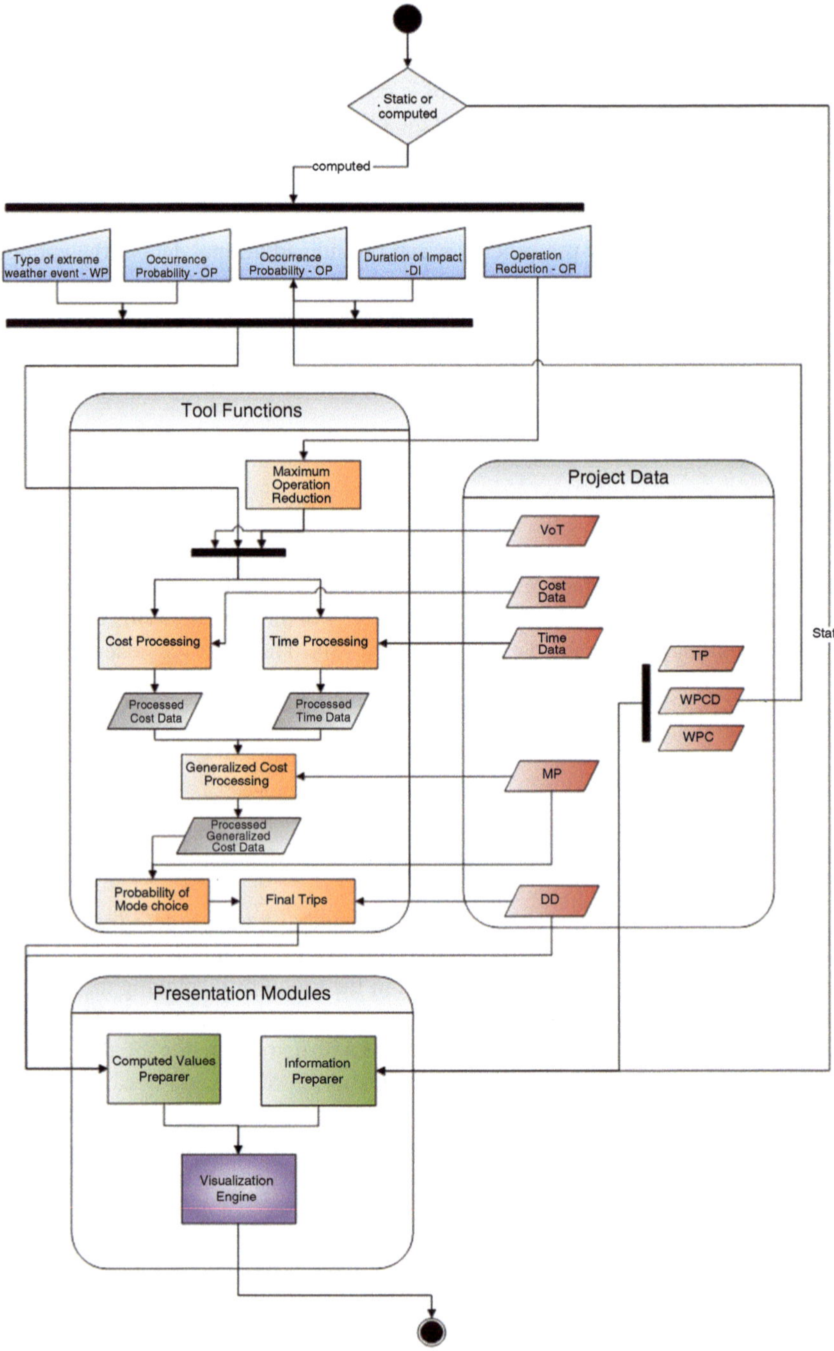

Fig. 60.5 Visualization tool architecture

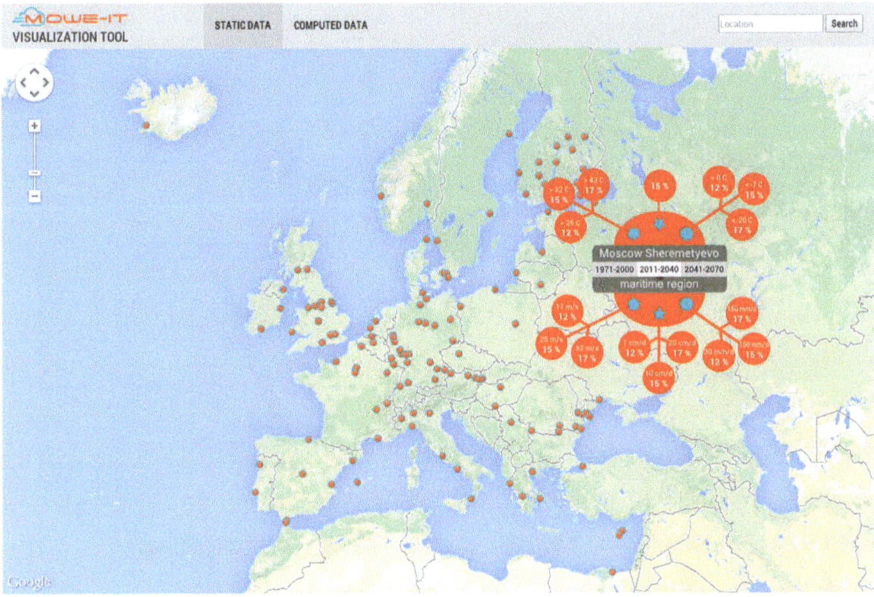

Fig. 60.6 Data tree visualization for static data

Use of Computed Data

In the case of computed data, the UI consists of a map and a list of extreme weather events added by the user. The user is able to add new weather events by using a form, which is opened by selecting a dedicated button/link alongside the weather event list. For each event type, subcategory, location, occurrence probability, duration, operation reduction and transport mode must be defined. Type and subcategory must be selected from predefined values, while location can be selected from predefined values or pointed from the map. If location is selected from the map, the tool will calculate which predefined location is closest to the selected location and assign the matched predefined location to the event. Occurrence probability can be chosen from predefined values or defined manually. Duration and operation reduction must be defined manually, and finally, transport mode must be selected from predefined values. Once all of these are defined for one event, and the user provides a confirmation, the event is added to the list of all events. The user can, in addition to adding new events, modify and remove existing events.

A marker is presented on the map for each location belonging to a group of 14 pre-selected locations in Europe. Before the user can interact with the map locations, she/he is required to add at least one weather event. The location to which the event is assigned to, will be the origin of the connection. At this point, the user is able to select (click on) one of the other 13 locations, which will then be assigned as the destination for the user selected connection. The connection data is then visualized with two parallel lines connecting the two locations. These lines present

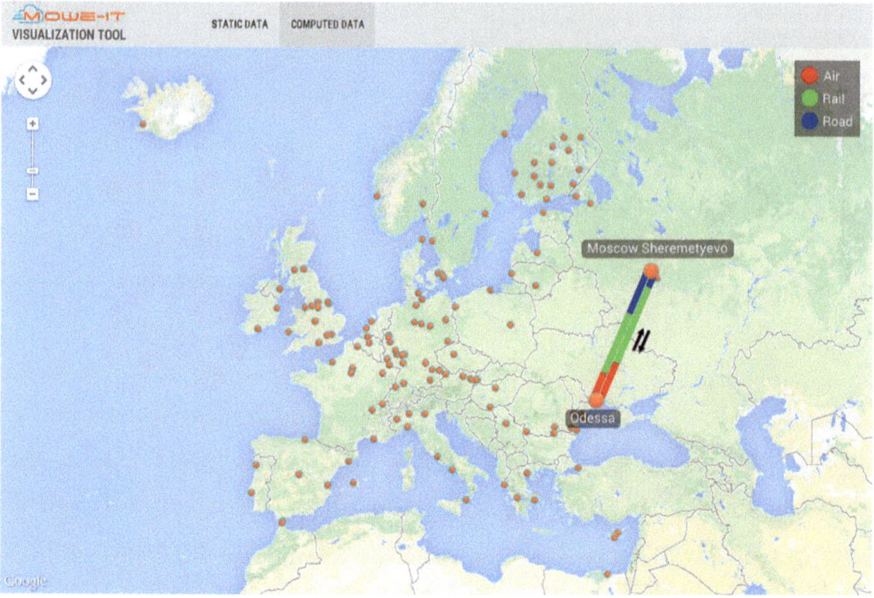

Fig. 60.7 Visualization of computed data

the data with and without the added weather event(s) respectively. Both lines are composed of three parts, each of which represents one of the three transport modes (air, rail and road). Each line is divided amongst these three modes by their percentage of the total number of trips combined from all three modes. An indicative visualization result is presented in the Fig. 60.7.

After the user has defined a connection with both origin and destination, she/he is able to reverse the direction of the connection by selecting (clicking on) a button alongside the lines. This leads to data being updated and presented respectively.

Conclusions

A data-driven method, which utilizes existing datasets for quantifying modal differentiated passenger trips in cases of extreme weather event, has been recently proposed, which enables the development of support tools for related entities and stakeholders, which are confronted with the assessment of the resilience of passenger transportation networks. The proposed visualization tool serves as a prototype for presenting weather data and mobility of people between major locations in Europe. The tool has been designed to be available for any additions in terms of data or subject area coverage, for instance visualization of European patterns of extreme weather events such as flooding etc. It is anticipated that such functionality is useful in future projects, where these types of analyses will be required.

Policy-makers, operators and transport system users are likely to find the information provided by the visualization tool useful in aiding choices and mapping of alternatives in the case of anticipated disruptions. Also, in the case of unexpected disruptions, the tool will also be useful as it allows for reacting to problems encountered with best possible response. A major research challenge with respect to passenger transport is the availability of substitution in quantity. There are supply constraints on road and rail networks that prevent infinite expansion of volumes, which is particularly relevant in high density routes. Also, carriage such as coaches and buses may not exist physically at right locations, which will add to limitations of substitution. One way to work with this problem would be to introduce disaster preparation into fleet management of operators to better target the areas in need, when disruptions take place. However, as the nature of extreme weather events suggests, predicting periods are short and anticipation difficult. Future research should lie in the field of obtaining data at real-time, so that the outcomes of the tool can be used at any given time. In addition, all input in the tool should be filtered and checked for inconsistencies in order to improve the accuracy and validity of presented results.

Acknowledgement This work was supported by the European Commission Seventh Framework Programme, under the "MOWE-IT: Management of Weather Events in the Transport System" project.

References

ETISPlus project. http://www.etisplus.eu/default.aspx.
EWENT Project. Retrieved March 15, 2015 from http://ewent.vtt.fi/.
MOWE-IT project, Retrieved March 18, 2015 from http://www.mowe-it.eu/.
Retrieved February 20, 2015 from http://www.ppgis.manchester.ac.uk/grabs/start.html.
Retrieved March 02, 2015 from http://ocean.space.noa.gr/diachronic_bsm/.
Retrieved March 14, 2015 from http://www.nrdc.org/health/extremeweather/.

Part XV
Securing Sustainable Mobility to Mitigate Climate Change

Climate Impact Effects of Air and Maritime Transport

Chapter 61
Manmade Changes in Cirrus Clouds from 1984 to 2007: A Preliminary Study

Kostas Eleftheratos, Gunnar Myhre, Patrick Minnis, Ioannis Kapsomenakis, and Christos Zerefos

Introduction

Aircraft flying at high altitudes form line shaped clouds which are called contrails. Contrails form if the ambient air is cold enough (Schumann 1996). In dry air the contrails dissipate quickly, and their impact is of minor importance, but in moist air which is super-saturated with respect to ice, the contrails spread and grow with the uptake of ambient water vapour, and become contrail-cirrus (Schumann 2005). Contrail-cirrus would not exist without the prior formation of contrails (i.e., Gierens 2007). This kind of cirrus formation occurs in regions with high air traffic (i.e., North America, North Atlantic and Europe). Contrail evolution into cirrus clouds

K. Eleftheratos (✉)
Laboratory of Climatology and Atmospheric Environment, Faculty of Geology and Geoenvironment, National and Kapodistrian University of Athens, Athens 15784, Greece

Biomedical Research Foundation of the Academy of Athens, Athens, Greece
e-mail: kelef@geol.uoa.gr

G. Myhre
Center for International Climate and Environmental Research Cicero-Oslo (CICERO), Oslo, Norway

P. Minnis
Atmospheric Sciences, NASA Langley Research Center, Hampton, VA, USA

I. Kapsomenakis
Research Centre for Atmospheric Physics and Climatology, Academy of Athens, Athens, Greece

C. Zerefos
Biomedical Research Foundation of the Academy of Athens, Athens, Greece

Research Centre for Atmospheric Physics and Climatology, Academy of Athens, Athens, Greece

Navarino Environmental Observatory (N.E.O.), Messinia, Greece

P. Grammelis (ed.), *Energy, Transportation and Global Warming*, Green Energy and Technology, DOI 10.1007/978-3-319-30127-3_61

has been observed both in North America and Europe (Minnis et al. 1998; Haywood et al. 2009).

The possible influence of air traffic emissions on cirrus cloud positive trends has been examined using various methods and different datasets for different periods of records (i.e., Zerefos et al. 2003; Minnis et al. 2004; Stubenrauch and Schumann 2005; Stordal et al. 2005; Mannstein and Schumann 2005; Krebs 2006). The major findings of those studies were summarized by Gierens (2007). For example, in central Europe it was found that the observed positive trends in cirrus due to aircraft are about 1–2 % cover per decade. Mannstein and Schumann (2005) had estimated ~3 % additional cloud coverage due to aircraft over Europe, which is about ten times higher than the coverage by linear contrails (~0.3 %) alone. However, their conclusion that the coverage by additional cirrus clouds in Europe is about 10 times higher than by linear contrails was no longer supported (Mannstein and Schumann 2007). As air traffic increases (~5 % per year), manmade (aviation) cirrus clouds are expected to increase. These additional clouds contribute to the greenhouse effect (i.e., Minnis et al. 2004; Sausen et al. 2005).

The objective of this study is to analyse long-term changes in cirrus clouds from ISCCP over congested air traffic corridors during the past two decades or so, in association with the changes in global air traffic. Radiative forcing calculations of the additional cirrus coverage due to aircraft are also provided.

Data Sources

The cloud data set analysed in this study was produced by the International Satellite Cloud Climatology Project (Rossow and Schiffer 1999). The data are based on observations from a suite of operational geostationary and polar orbiting satellites. Visible radiances are used to retrieve the optical thickness of clouds and infrared radiances to retrieve cloud top temperature and pressure. The D2 dataset used in this study has a spatial resolution of 280 km (2.5o at the equator) and provides monthly averages of cloud properties of 15 different cloud types. The cloud types are derived based on radiometric definitions that rely on cloud optical thickness and cloud top pressure. Cirrus clouds are defined as those with optical thickness less than 3.6 and cloud top pressure less than 440 hPa. In this study we made use of the cirrus cloud data for the period 1984–2007. The ISCCP cloud properties have been tested extensively both against other satellite cloud retrievals and against surface cloud observations (Rossow and Schiffer 1999). Further information can be found by Zerefos et al. (2003).

The analysis also includes the study of relative humidity at high-flying altitude levels from the independently produced NCEP-DOE and ERA-Interim Reanalysis projects. The 250 and 300 hPa levels were selected for this study as a compromise upper tropospheric level for the tropics, middle and high latitudes. Both projects provide mean monthly gridded values of various atmospheric and surface parameters on 2.5o × 2.5o grid boxes and on global scale. We used monthly daily means which are monthly means computed at base times 0000, 0600, 1200 and 1800 UTC. The NCEP-DOE Reanalysis-2 data were downloaded from the internet at the webpage http://www.esrl.noaa.gov/psd/data/gridded/data.ncep.reanalysis2.html. The ERA-Interim monthly mean data were downloaded from the webpage http://data-portal.ecmwf.int/data/d/interim_moda/levtype=pl/.

To overcome the effect of seasonal variations, all trends were calculated after removing variations related to the seasonal cycle of the data. Cirrus cloud data were deseasonalized by subtracting the long-term monthly mean (1984–2007) pertaining to the same calendar month. All trends were evaluated as to their statistical significance by applying the *t*-test of each trend against the null hypothesis of no-trend for the appropriate number of degrees of freedom. Cirrus cloud data for the period 1991–1993 were not used in our trend analysis in order to minimize artificial satellite cloud retrievals after the eruption of Mt. Pinatubo in 1991 (Rossow and Schiffer 1999; Luo et al. 2002; Zerefos et al. 2003).

Results and Discussion

Manmade Changes in Cirrus Clouds

Manmade changes in cirrus cloudiness were examined by comparing cirrus cloud cover over higher traffic regions and over lower traffic regions in the northern mid-latitudes (e.g. United States of America, Europe, North Atlantic and North Pacific Ocean). We have restricted the analysis to the northern mid-latitudes because in the tropics it is difficult to detect possible effects of regional persistent contrails on cirrus cloud trends. This is because trends in dynamics and thermodynamics mask possible man-made trends in cirrus clouds in those regions. Taking also into account that flight frequencies and fuel consumption are moderate over the tropical air traffic corridors, it makes it even more difficult to detect and quantify any possible anthropogenic effects (Zerefos et al. 2007). In addition, because we wanted to avoid the strong impact of El Nino/Southern Oscillation on cirrus clouds in the tropics (Eleftheratos et al. 2007).

Regions that correspond to different air traffic loading have been compared using air traffic statistics based on the emissions inventory of the QUANTIFY project. More specifically, two adjacent regions one with low and another with high air traffic over North America, Europe, North Atlantic and North Pacific have been selected, respectively. Analytical information on the spatial distribution of the selected areas is presented in Table 61.1. Table 61.2 shows the density of air traffic over the selected regions as fractions of the total air traffic. This analysis basically assumes that the differences in the trends are the result of differences in air traffic between the adjacent regions.

Figure 61.1 shows the annual mean cirrus coverage averaged over all high air traffic regions in North America, Europe, North Atlantic and North Pacific in

Table 61.1 Definition of high and adjacent low air traffic regions analysed in this study

	High air traffic	Low air traffic
North America	30°N-45°N, 60°W-130°W	45°N-60°N, 60°W-130°W
Europe	40°N-60°N, 10°W-40°E	20°N-40°N, 10°W-40°E
North Atlantic	40°N-60°N, 10°W-60°W	20°N-40°N, 10°W-60°W
North Pacific	45°N-60°N, 145°E-130°W	30°N-45°N, 145°E-130°W

Table 61.2 Million kilometres travelled by aviation over high and adjacent low air traffic regions as fraction of the global kilometres travelled in the year 2000 at 9760–11590 m height (from the QUANTIFY emissions inventory)

	High air traffic		Low air traffic	
	Million km travelled	Fraction (%) of global	Million km travelled	Fraction (%) of global
North America	4448	28.1	631	4.0
Europe	2322	14.7	634	4.0
North Atlantic	1019	6.4	321	2.0
North Pacific	347	2.2	152	1.0
All regions above	**8136**	**51.4**	**1738**	**11.0**
Global	15837			

comparison to the adjacent lower air traffic regions. Although both time series show the same variability from year to year, their long-term trends differ. From 1984 to 2004 cirrus cloud cover increased from 10.3 to 12.1 % in high air traffic corridors (=1.8 % increase), whereas no such change could be detected in low traffic areas. At the same time, global average annual mean fuel consumption rose from 139 to 222 Tg from 1984 to 2004 (=60 % increase). Air traffic increased much faster at 5–6 % on a global annual average, i.e. by 165–220 % in that period (Lee et al. 2009). The high air traffic corridor cirrus increased at least at the global average rate. Thus, we estimate an additional cloud coverage of up to 0.01 % over these corridors for every per cent increase in air traffic in the period 1984–2004.

Long-term trends in cirrus clouds for each high and low air traffic region are summarized in Table 61.3. Long-term trends were estimated as the increase or decrease of cirrus cloud cover in per cent per decade inferred from a least square linear fit to the data.

The differences between the trends over high air traffic relative to the low air traffic regions are statistically significant at the 99 % confidence level. The trend differences could provide indications for the additional cloud coverage that has been induced by aviation over heavy air traffic regions. The results show an increase in cirrus coverage in high air traffic areas of +0.6 % per decade, contrasted by a decrease of 0.3 % per decade in low traffic areas, which could be interpreted as a 0.9 %/decade change in cirrus due to air traffic. In all cases the differences have a positive trend which is consistent with the increasing trend in global air traffic during the past 20 years.

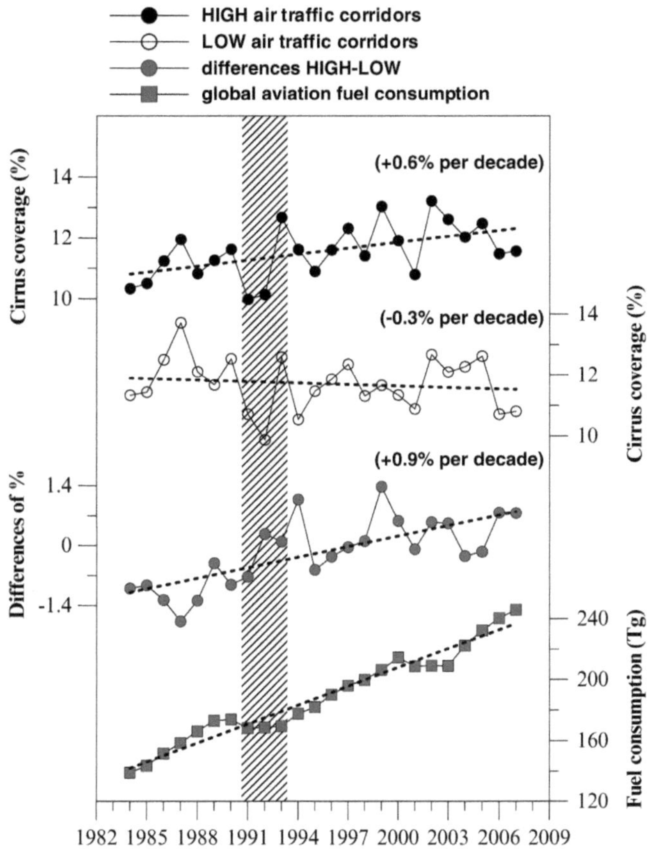

Fig. 61.1 Annual mean cirrus cloud cover averaged over high and over low air traffic regions in North America, Europe, North Atlantic and North Pacific in comparison to the global air traffic fuel consumption

Table 61.3 Trends in cirrus cloud cover (% per decade) and mean cirrus coverage (%) over high and low air traffic corridors

	High air traffic		Low air traffic		
1984–2007	Trend	Mean	Trend	Mean	Trend difference
North America	+0.7 (99 %)	18.7	+0.1 (–)	17.4	+0.6 (95 %)
Europe	+0.6 (99 %)	10.3	−0.3 (95 %)	7.7	+0.9 (99 %)
North Atlantic	+0.5 (99 %)	10.4	−0.4 (90 %)	12.6	+0.9 (99 %)
North Pacific	+0.3 (–)	7.1	−0.6 (99 %)	10.1	+0.9 (99 %)
All regions above	**+0.6 (99 %)**	**11.7**	**−0.3 (95 %)**	**12.0**	**+0.9 (99 %)**

Values in brackets refer to statistical significance of each trend

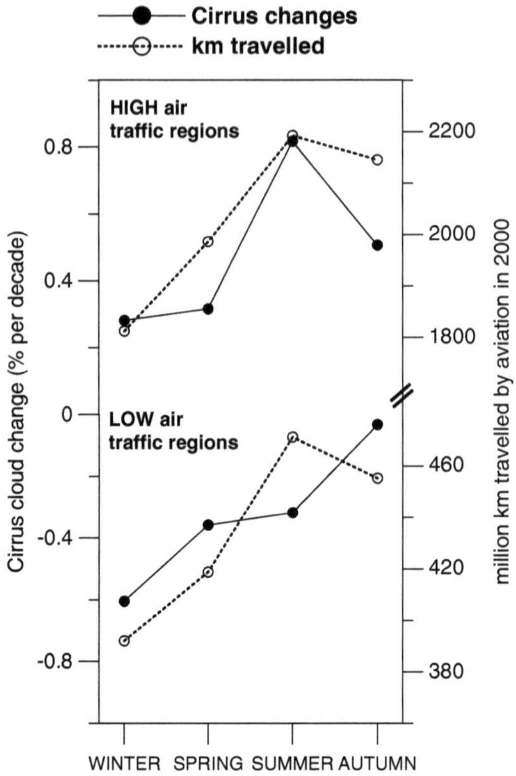

Fig. 61.2 Seasonal changes in cirrus clouds over the selected high and low air traffic regions

We also examined the seasonal changes of cirrus cloud cover over the high and low air traffic regions in North America, Europe, North Atlantic and North Pacific in comparison to the seasonal distribution of kilometres travelled by high-flying aircraft over these areas. The results are presented in Fig. 61.2.

It appears that there are positive changes in cirrus clouds over the high air traffic regions in all seasons and negative changes over the adjacent low air traffic counterparts. The trend differences are consistent with the differential increase in annual mean cirrus cloud amount. It should be mentioned here that the average percentage of cirrus cloud coverage over higher traffic areas in Europe and the US is higher in absolute numbers than the values of the nearest lower traffic continental regions. The difference is on the order of 2 % and the average is 18 %. The opposite is true over the oceans where the absolute average of cirrus cloud cover over higher traffic oceanic areas is about 2 % less compared to that over adjacent lower traffic oceanic regions.

The observed cloud changes could be related to changes in natural parameters. We have studied the temporal variability of relative humidity at 250 and 300 hPa and its relationship to changes in cirrus cloud cover. The results are presented in Fig. 61.3 which shows the changes in annual mean cirrus cloudiness over the examined high and low air traffic regions with respect to the changes in mean

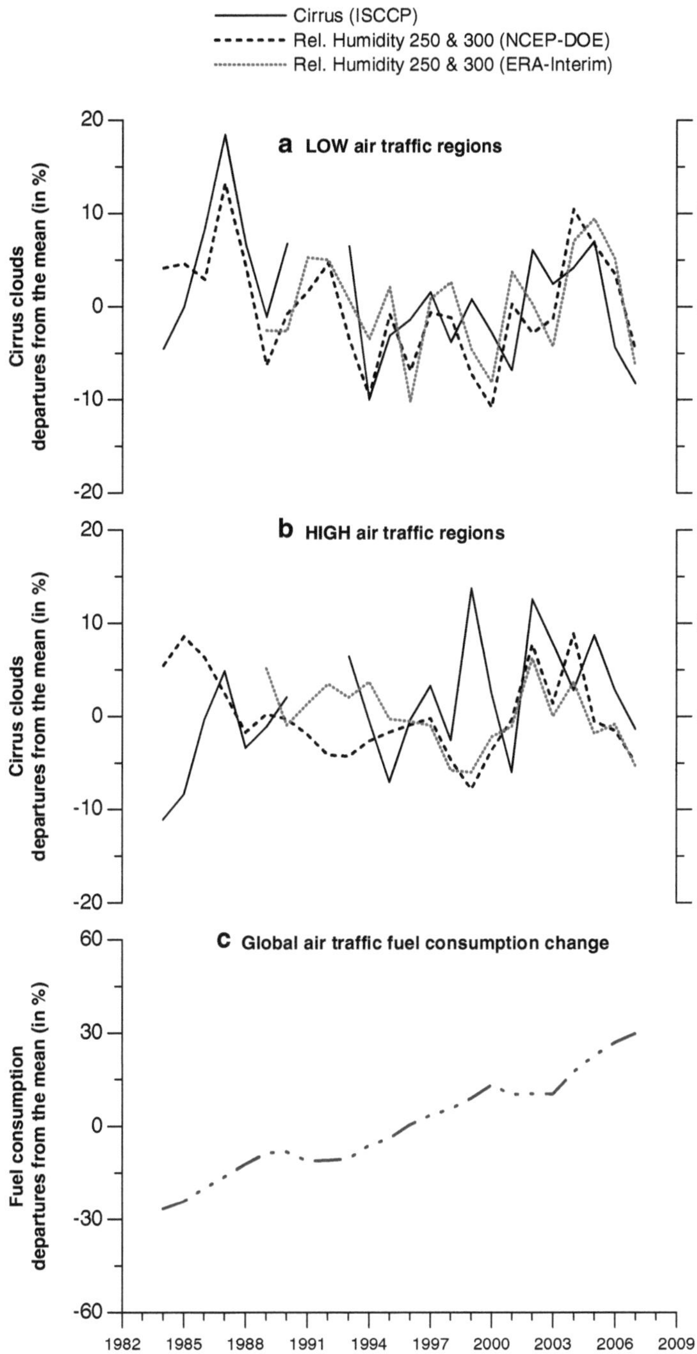

Fig. 61.3 Cirrus cloud changes (in % from the long-term mean) over low and high air traffic corridors with respect to corresponding changes in mean relative humidity at 250 and 300 hPa from the NCEP-DOE and ERA-Interim Reanalysis projects and in comparison to the global air traffic fuel consumption change

relative humidity at 250 and 300 hPa. The changes were calculated as percent departures from the long-term mean (1984–2007). It appears that over low air traffic regions there is general consistency between the observed changes in cirrus clouds and the changes in mean relative humidity. On the other hand, over the high air traffic regions the observed changes in cirrus clouds are less related to changes in relative humidity, and they follow the course of upward trends seen in global air traffic. The relative humidity changes at 250 and 300 hPa from the independently produced NCEP-DOE and ERA-Interim data sets are consistent with each other. While the greater disconnect between cirrus changes and relative humidity may in part be due to dehydration by contrails (Schumann et al. 2015), the increase in air traffic producing more contrails is most likely the primary cause.

Impact on Radiative Forcing

The result of manmade cirrus increase by 0.9 % per decade over the western air traffic corridors was expected to have a measurable effect in radiative forcing. We studied the effect using the DISORT radiative transfer code (Stamnes et al. 1988; Myhre and Stordal 2001). Meteorological data for temperature, water vapour, and clouds were taken from ECMWF analyses and are similar to the UiO model in Myhre et al. (2009). The cirrus optical properties are the same as in Myhre et al. (2009). The differences in cirrus cloud cover at the end of 24 years of increased air traffic are estimated to cause a radiative forcing of 17 mW/m^2 using an assumed optical depth of 0.4 for the additional cirrus cloud cover and 31 mW/m^2 using an assumed optical depth of 1.0. Table 61.2 shows that the four regions with highest aircraft traffic cover about 50 % of the global air traffic. Under the assumption that aircraft activity has the same impact in other regions as these four main aircraft regions we estimate a global radiative forcing of 34 mW/m^2 for optical depths of 0.4 and 62 mW/m^2 for optical depths of 1.0. These numbers are for contrail-induced cirrus for the period 1984–2007. Since the contrail induced cirrus optical depth is uncertain this causes a large uncertainty in our RF simulations. Our lower estimate is greater than the mean estimates of radiative forcing for the aviation-induced cirrus clouds from Minnis et al. (2004) for the period 1971–1995 (25 mW/m^2), from Sausen et al. (2005) for the year 2000 (30 mW/m^2), and from Lee et al. (2009) for the year 2005 (33 mW/m^2). Much of the difference could be due to the assumed optical depth, which exceeds that of the earlier estimates. The upper end of the range is far greater than any of the previous estimates and is probably unrealistic.

Conclusions

Manmade changes due to aviation were determined by comparing cirrus cloud trends inside and outside of high air traffic regions in North America, Europe, North Atlantic and North Pacific. Cirrus cloud cover increased by 0.9 % per decade in the period 1984–2007 over high air traffic regions, whereas no such change could be detected in low air traffic areas. This estimate is in agreement with earlier studies examining trends in cirrus cloudiness in relation to aviation (Zerefos et al. 2003; Minnis et al. 2004; Stordal et al. 2005; Stubenrauch and Schumann 2005) despite the differences among the examined datasets and periods of records. The resulting manmade cirrus increase of 0.9 % per decade over the western air traffic corridors was expected to have a measurable effect in radiative forcing. Radiative forcing calculations of the additional cloud cover due to aircraft yield a quite strong effect of approximately +34 mW/m^2 using an assumed optical depth of 0.4 for the additional cloud cover and +62 mW/m^2 for an optical depth of 1.0. Our estimates for the RF from aviation-induced cirrus clouds are greater than those from earlier studies (Minnis et al. 2004; Sausen et al. 2005; Lee et al. 2009). This could be due to the assumed optical depth, which exceeds that of the earlier estimates. As such, the upper end of the range is far greater than any of the previous estimates and is probably unrealistic.

Acknowledgement This study was conducted within the FP6 Integrated Project QUANTIFY (003893-GOCE) funded by EU.

References

Eleftheratos, K., Zerefos, C. S., Zanis, P., Balis, D. S., Tselioudis, G., Gierens, K., et al. (2007). A study on natural and manmade global interannual fluctuations of cirrus cloud cover for the period 1984-2004. *Atmospheric Chemistry and Physics, 7*, 2631–2642.

Gierens, K. (2007). Contrails, contrail cirrus, and ship tracks. In R. Sausen, A. Blum, D. S. Lee, & C. Brüning (Eds.), *Proceedings of an International Conference on Transport, Atmosphere and Climate (TAC)* (pp. 214–220). Oxford, UK. Retrieved June 26–29, 2006.

Haywood, J. M., Allan, R. P., Bornemann, J., Forster, P. M., Francis, P. N., Milton, S., et al. (2009). A case study of the radiative forcing of persistent contrails evolving into contrail-induced cirrus. *Journal of Geophysical Research, 114*, D24201. doi:10.1029/2009JD012650.

Krebs, K. (2006). Analyse des Einflusses des Flugverkehrs auf die natürliche Zirrusbewölkung über Europa, Nordafrika und dem Nordatlantik. *Dissertation*, Ludwing-Maximilians Universität München, 211pp (in German).

Lee, D. S., Fahey, D. W., Forster, P. M., Newton, P. J., Wit, R. C. N., Lim, L. L., et al. (2009). Aviation and global climate change in the 21th century. *Atmospheric Environment, 43*, 3520–3537.

Luo, Z., Rossow, W. B., Inoue, T., & Stubenrauch, C. J. (2002). Did the eruption of the Mt. Pinatubo volcano affect cirrus properties? *Journal of Climate, 17*, 2806–2820.

Mannstein, H., & Schumann, U. (2005). Aircraft induced contrail cirrus over Europe. *Meteorologische Zeitschrift, 14*, 549–554.

Mannstein, H., & Schumann, U. (2007). Corrigendum to "Aircraft induced contrail cirrus over Europe". *Meteorologische Zeitschrift, 16*, 131–132.

Minnis, P., Ayers, J. K., Palikonda, R., & Phan, D. (2004). Contrails, cirrus trends, and climate. *Journal of Climate, 17*, 1671–1685.

Minnis, P., Young, D. F., Garber, D. P., Nguyen, L., Smith, W. L., et al. (1998). Transformation of contrails into cirrus during success. *Geophysical Research Letters, 25*, 1157–1160.

Myhre, G., Kvalevag, M., Radel, G., Cook, J., Shine, K. P., Clark, H., et al. (2009). Intercomparison of radiative forcing calculations of stratospheric water vapour and contrails. *Meteorologische Zeitschrift, 18*, 585–596.

Myhre, G., & Stordal, F. (2001). On the tradeoff of the solar and thermal infrared radiative impact of contrails. *Geophysical Research Letters, 28*, 3119–3122.

Rossow, W. B., & Schiffer, R. A. (1999). Advances in understanding clouds from ISCCP. *Bulletin of the American Meteorological Society, 80*, 2261–2287.

Sausen, R., Isaksen, I., Grewe, V., Hauglustaine, D., Lee, D. S., Myhre, G., et al. (2005). Aviation radiative forcing in 2000: an update on IPCC (1999). *Meteorologische Zeitschrift, 14*, 555–561.

Schumann, U. (1996). On conditions for contrail formation from aircraft exhausts. *Meteorologische Zeitschrift, 5*, 4–23.

Schumann, U. (2005). Formation properties and climate effects of contrails. *C. R. Physique, 6*, 549–565.

Schumann, U., Penner, J. E., Chen, Y., Zhou, C., & Graf, K. (2015). Dehydration effects from contrails in a couple contrail-climate model. *Atmospheric Chemistry and Physics, 15*, 11179–11199.

Stamnes, K., Tsay, S. C., Wiscombe, W., & Jayaweera, K. (1988). Numerically stable algorithm for discrete-ordinate-method radiative-transfer in multiple-scattering and emitting layered media. *Applied Optics, 27*, 2502–2509.

Stordal, F., Myhre, G., Stordal, E. J. G., Rossow, W. B., Lee, D. S., Arlander, D. W., et al. (2005). Is there a trend in cirrus clouds cover due to aircraft traffic? *Atmospheric Chemistry and Physics, 5*, 2155–2162.

Stubenrauch, C. J., & Schumann, U. (2005). Impact of air traffic on cirrus coverage. *Geophysical Research Letters, 32*, L14813. doi:10.1029/2005GL022707.

Zerefos, C. S., Eleftheratos, K., Balis, D. S., Zanis, P., Tselioudis, G., & Meleti, C. (2003). Evidence of impact of aviation on cirrus cloud formation. *Atmospheric Chemistry and Physics, 3*, 1633–1644.

Zerefos, C. S., Eleftheratos, K., Zanis, P., Balis, D. S., & Tselioudis, G. (2007). Search for man-made cirrus contrails over Southeast Asia. *Terrestrial, Atmospheric and Oceanic Sciences, 18*, 459–474. doi:10.3319/TAO.2007.18.3.459(EA).

Chapter 62
Emission Routing in Maritime Transportation

Levent Bilgili and Ugur Bugra Celebi

Introduction

Worldwide shipping activities have a major role in trading and transportation across the globe. Beside the great benefits of ships, they play a considerable role on environmental issues due to their great potential to produce various types of solid, liquid and gaseous wastes such as rubbish, sludge, ballast water and air emissions.

According to International Maritime Organization (IMO), more than 90 % of global trade is carried by ships. Besides, a dramatic improvement was observed during the last century (IMO 2012). In a report, which is prepared by United Nations Conference on Trade and Development (UNCTAD), it is indicated that world seaborne trade and the world fleet grew by 3.8 and 4.1 % in 2013, respectively. Thus, the total trading volume and the total world fleet reach to 9.6 billion tons and 1.69 billion DWT, respectively (UNCTAD 2014).

Industrial revolution brought new technologies and fuels for different machines. After long centuries of human and wind power, ships began to use fossil fuels, first coal, then liquid fuels such as heavy fuel oil (HFO), marine diesel oil (MDO) and marine gas oil (MGO). Although these new energy sources are much stronger than natural sources, they have considerable side effects: Gas emissions. After the invention of internal combustion, liquid fossil fuels became more important.

L. Bilgili (✉) • U.B. Celebi
Yildiz Technical University, Barbaros Bulvari, Istanbul 34349, Turkey
e-mail: lbilgili@yildiz.edu.tr; ucelebi@yildiz.edu.tr

P. Grammelis (ed.), *Energy, Transportation and Global Warming*, Green Energy and Technology, DOI 10.1007/978-3-319-30127-3_62

Due to their carbon-based chemical structure, carbon-based gases are inevitable end products of combustion processes. Carbon dioxide (CO_2), carbon monoxide (CO), black carbon (BC) are the most common carbon-based exhaust gas emissions. Besides, due to the sulfur content of fossil fuels, sulfur-based by-products (sulfur oxides-SO_x) are produced during combustion. The amount of SO_x entirely depends on the amount of sulfur in the fossil fuel. Nitrogen oxides (NO_x) are other end products. They are formed due to air used during combustion process. NO_x amount strongly depends on the combustion characteristics and quality. Poor burning process results with the production of particulate matter (PM). PM consists of soot, ash and micro-sized carbon particles. Besides, due to the fossil fuels' contents, volatile organic compounds (VOC's) are produced during combustion.

Although it has been estimated that diesel exhaust contains almost 450 different compounds (Andreoni et al. 2008), generally some of them are considered and investigated due to the harmful impacts on environment and human health. Carbon dioxide, carbon monoxide, sulfur and nitrogen oxides and particulate matter are the main harmful emissions from internal combustion process of ships' diesel engines (Bilgili and Celebi 2014). CO_2, NO_x, SO_x, CO, PM, VOC's, methane (CH_4), black carbon (BC) and particulate organic matter (POM) are the most dangerous exhaust emissions for human health and environment, especially for global warming. Shipping related air emissions strongly depend on fuel type and engine type characteristics, characteristic combustion efficiency, characteristics of the lubricating oil, operation conditions of the ship, ship form and equipment efficiencies (Bilgili et al. 2014).

Air emissions from ships are investigated and documented well. In a recent study, it is estimated that there are almost 450 different types of components in ship exhaust emission (Andreoni et al. 2008).

There are several and comprehensive studies on ship emissions amount and their impacts on environment and human health. For nitrogen oxides Andreoni et al. (2008), MEPC (2005); for sulfur oxides Corbett et al. (2007); for particulate matter Schreier et al. (2007), Lauer et al. (2007), Eyring et al. (2010), Corbett et al. (2007); for carbon dioxide and carbon monoxide AEA Energy and Environment (2008), Miola et al. (2009), Haglind (2008) are the main and important studies.

According to the several recent studies, the estimated amount of CO_2 emission caused by shipping is 943.5 million tons (Psaraftis and Kontovas 2009) and around one billion tons, which corresponds the 2–4 % of global emissions (Tzannatos 2010). IMO estimated that the CO_2 emissions are about one billion tons in 2012 (IMO 2014). It is predicted that the ratio of shipping related CO_2 emissions in global amount will be increase to 12–18 % in 2050 (Heitmann and Khalilian 2010). Nitrogen oxides include nitrogen dioxide (NO_x) and nitrogen oxide (NO). Recent estimations presents that the amount of nitrogen oxide emissions from international shipping is 6.87 Tg (Corbett and Köhler 2003). It is also estimated that shipping related NO_x emissions are responsible for 15 % of global NO_x emissions (Eyring et al. 2005). Furthermore, in a recent study it is indicated that NO_x may have a considerable effect on global warming as well as, or more than, CO_2 (Henningsen 2000). Sulfur oxide emissions strongly depend on the sulfur content of fuel and shipping activities are responsible for 6.49 Tg of SO_x in terms of SO_2 which corresponds 5–8 % of global SO_x emissions (Corbett and Köhler 2003). It is known that SO_x emissions have an effect on global cooling (Andreoni et al. 2008). PM has different types, named accordingly to the diameter of the particles such as $PM_{2.5}$ and PM_{10} that are smaller than 2.5 and 10 μm, respectively. According to recent estimations, shipping activities are responsible for 0.9–1.7 million tons of PM, annually (Endresen et al. 2003). PM has both effects on global warming and global cooling. Especially on frequently used shipping routes, it is observed that lower layer clouds are affected by the PM emissions, and thus, changed their characteristics. These new clouds are called as cloud condensation nuclei and by reflecting the solar energy back to the space, they contribute to global cooling (Bilgili et al. 2014). VOC's have also positive effect on global warming (AEA Energy and Environment 2008).

Mediterranean Sea may be the most important and one of the most intense basins all over the world. It was estimated that NO_x, SO_2, PM and CO_2 emissions were 2, 1.45, 0.157 and 87.6 million tons, respectively, in 2005 (Cofala et al. 2007). Besides, in a recent study, it is estimated that 70, 44 and 36 % of the ship traffic occur within 200, 50 and 25 nautical miles from shore (IMO 2009). Thus, shipping emissions must be considered in detail for especially the population that settles in big cities near shores.

Application of the Models

In this study, it is aimed to calculate the total emissions produced during a total 80,000 DWT load by two different bulk carrier ships. While Ship A is 80,000 DWT, Ship Bis 40,000 DWT. Table 62.1 presents the main characteristics of ships.

Table 62.1 Main characteristics of ships

Main characteristics	Ship A	Ship B
L_{OA}	230 m	190 m
B	32 m	26 m
D	20 m	16 m
T	14.5 m	11.8 m
DWT	80,000 tons	40,000 tons
Main engine power	11060 kW	8825 kW
Auxiliary engine power	2160 kW	1776 kW
Speed	14.2 knots	12.75 knots

Table 62.2 Abbreviations and indices of equations

	Equation (62.1)	Equation (62.2)	Equation (62.3)
$\mathbf{E_{Trip}}$	Emission over a complete trip (tons)		
FC	Fuel consumption (tons)	–	Fuel consumption (tons)
EF	Emission factors (kg/tons)	Emission factors (kg/kWh)	Emission factors (Gg/tons)
i	Pollutant type		
j	Engine type		
m	Fuel type		
p	Different phases of trip		
LF	–	Engine load factor (%)	–
P	–	Engine nominal power (kW)	–
T	–	Time (hours)	–
e	–	Engine category	–
CV	–	–	Calorific value (TJ/tons)

There are three main estimation methods in order to calculate the emissions: Fuel consumption (FC) method, engine power (EP) method and energy (E) method. FC method is totally based on fuel consumption data which is gained from the noon reports of ships. Due to the real-time data, this method can be more reliable. EP method depends on engine power and cruising time. E method uses the fuel consumption data, fundamentally but focuses on the energy that is gained by combustion. Thus, the results may be different in three methods.

Trozzi (2010) offers the two following equations for FC and EP methods, respectively:

$$E_{Trip,\ i,j,m} = \sum \left(FC_{j,m,p} {}^{*} EF_{i,j,m,p} \right) \tag{62.1}$$

$$E_{Trip,i,j,m} = \sum \left[TP \sum \left(P_e {}^{*} LF_e {}^{*} EF_{e,i,j,m,p} \right) \right] \tag{62.2}$$

Equation (62.3) can be used for E method:

$$E_{Trip,i,j,m} = \sum FC_{j,m,p} {}^{*} CV_m {}^{*} EF_{i,j,m,p}) \tag{62.3}$$

Table 62.2 presents the meaning of abbreviations and indices of the equations.

The ships follow the route from Puerto la Cruz, Venezuela to Istanbul, Turkey with a 5 days stop for loading in Puerto la Cruz, 7 days stop in Thessaloniki, Greece for unloading and loading and 1 day stop in Istanbul for unloading. The route data is real and was taken from noon reports of the ships. The distance of the route is 5244 nm from Jose Terminal to Thessaloniki and 333 nm from Thessaloniki to Istanbul. The total distance is 5577 nm. Ship B follows the route twice to reach 80,000 DWT. Figure 62.1 presents the routes of the ships.

The emission factors that are used in this study are taken from the study of Cooper and Gustafsson (2004) and are shown in Table 62.3 for each type of

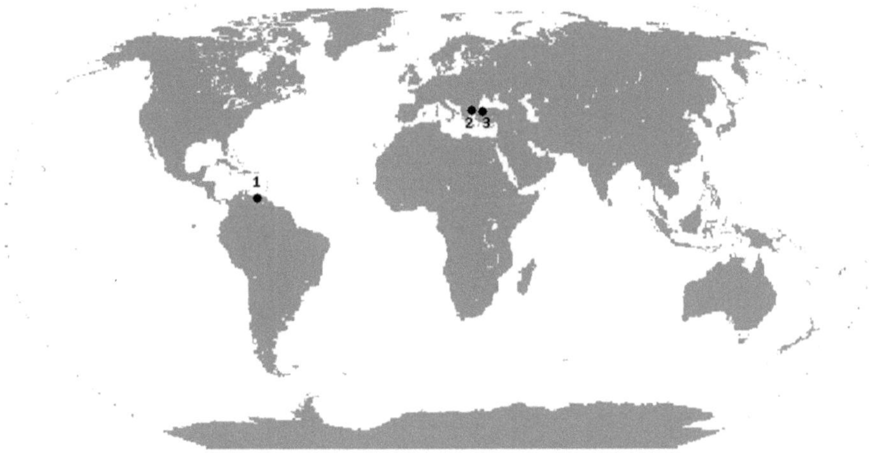

Fig. 62.1 The routes of the ships. [1]Puerto la Cruz; [2]Thessaloniki; [3]Istanbul

estimation methods. Both ships' main engines are classified as slow speed diesel engines. The auxiliary engines are classified as medium speed diesel engines. Thus, the emission factors were chosen properly to the speeds of the engines.

Discussion

The estimations, which are calculated by using Eqs. (62.1), (62.2) and (62.3), are shown in Tables 62.4 and 62.5 for Ship A and Ship B, respectively.

The emissions of Ship B must be multiplied by 2 due to reach the total 80,000 DWT load. In Fig. 62.2, the total emission amounts of total voyages are presented for FC method.

Figure 62.3 presents the total emissions according to the results of EP method.

In Fig. 62.3, the total emission estimations according to E method is shown.

As it can be seen in Figs. 62.2, 62.3 and 62.4, there is a considerable difference between the emissions of two ships when they carry the same amount of load in the same route. Thus, it can be said that bigger ships are more efficient in terms of emission/DWT ratio and environment issues.

In Figs. 62.5, 62.6 and 62.7, emission/DWT ratios of hip A and Ship B according to the results of the FC, EP and E methods are presented, respectively.

Table 62.3 Emission factors

	Fuel consumption method[a]				Engine power method[b]				Energy method[c]			
	Main engine		Aux. Engine		Main engine		Aux. Engine		Main engine		Aux. Engine	
	HFO	MDO	HFO	MDO	HFO	MDO	HFO	MDO	HFO	MDO	HFO	MDO
CO_2	3179	3179	3179	3179	620	588	722	690	78	74	78	74
CO	2.545	2.703	3.924	4.127	0.5	0.5	0.9	0.9	0.062	0.063	0.12	0.13
NO_x	87.136	91.561	61.728	62.440	18.1	17	14	13.5	2.13	2.14	1.51	1.48
SO_x	46	8	46	8	9	1.5	10.4	1.7	1.1	0.19	1.1	0.19
PM	6.667	1.081	2.203	0.922	1.3	0.2	0.5	0.2	0.16	0.025	0.057	0.023
CH_4	0.0308	0.0324	0.0176	0.0184	0.006	0.006	0.004	0.004	0.00075	0.00076	0.00045	0.00046
NMVOC	1.525	1.622	0.871	0.916	0.3	0.3	0.2	0.2	0.037	0.038	0.022	0.023

[a] kg/ton
[b] g/kWh
[c] t/TJ

Table 62.4 Emission amounts for Ship A (in tons)

	Fuel consumption method	Engine power method	Energy method
CO_2	2527.94	2363.55	2451.2
CO	2.22	1.98	2.29
NO_x	65.86	67.48	63.68
SO_x	34.90	32.40	32.99
PM	4.63	4.67	4.41
CH_4	0.02	0.02	0.02
NMVOC	1.12	1.11	1.08
Total	2636.72	2471.24	2555.69

Table 62.5 Emission amounts for Ship B (in tons)

	Fuel consumption Method	Engine power method	Energy method
CO_2	1605.61	2647.61	1557.71
CO	1.45	2.28	1.53
NO_x	41.17	74.39	39.86
SO_x	20.99	34.80	19.85
PM	2.78	5.00	2.65
CH_4	0.01	0.02	0.01
NMVOC	0.69	1.22	0.67
Total	1672.73	2765.35	1622.32

Conclusion

Although the ratio of total shipping activities in all kind of gaseous emissions is relatively low, especially the main and dense shipping routes, closed basins with high population such as Mediterranean and Aegean Seas and narrow water ways such as Turkish Straits are in high threat in terms of deleterious emissions. The share of shipping activities in GHG emission is low, as well. However, due to the great harmful effects of global warming, these emissions must be considered and minimized.

Ship emission estimation methods, which are mentioned in this study, provide the correct amounts to a certain degree. Due to the uncertainty, as well as differences between the results of the studies, of emissions factors, the amounts of estimated emissions may be totally different. Besides, obtaining the necessary data for calculations is considerably difficult. In this study, one of the emission minimization methods is investigated. Although the weather and sea conditions have remarkable effects on ship efficiency during voyage, they are ignored in this study. As it can be seen in Tables 62.3 and 62.4, ships produce similar amounts of emissions. However, when considering the load amount, Ship A is much more efficient in terms of emissions/DWT ratio, which can be used as an indicator for

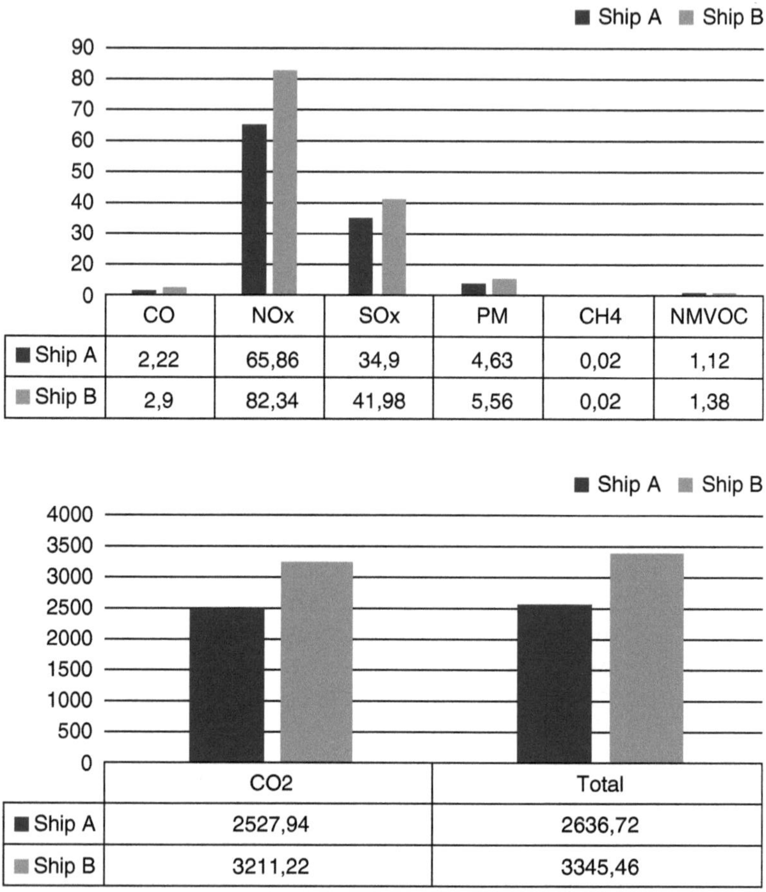

	CO	NOx	SOx	PM	CH4	NMVOC
Ship A	2,22	65,86	34,9	4,63	0,02	1,12
Ship B	2,9	82,34	41,98	5,56	0,02	1,38

	CO2	Total
Ship A	2527,94	2636,72
Ship B	3211,22	3345,46

Fig. 62.2 Ship emissions for fuel consumption method

capacity-environment relationship. Considering the difficulty to establish a reliable comparison between ships, indicators, such as developed in this study, are quite important. It is obvious that bigger ships are, generally, more environmental friendly. Although the bigger ships need more energy and material during manufacturing process, they may provide two-way benefits and solutions for carrying more goods with less energy and thus, emissions, in long-term.

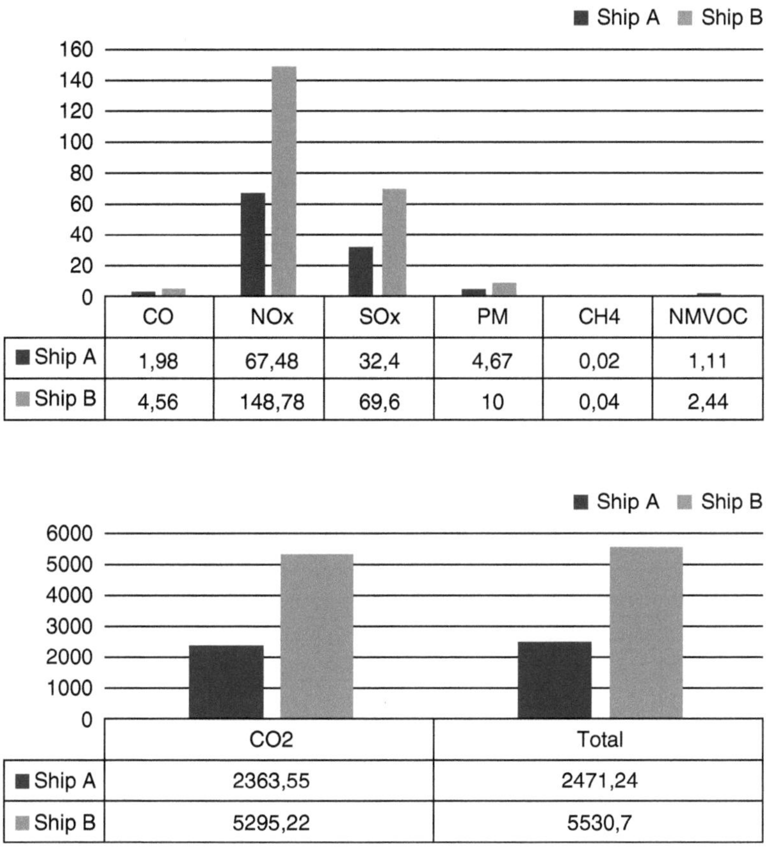

Fig. 62.3 Ship emissions for engine power method

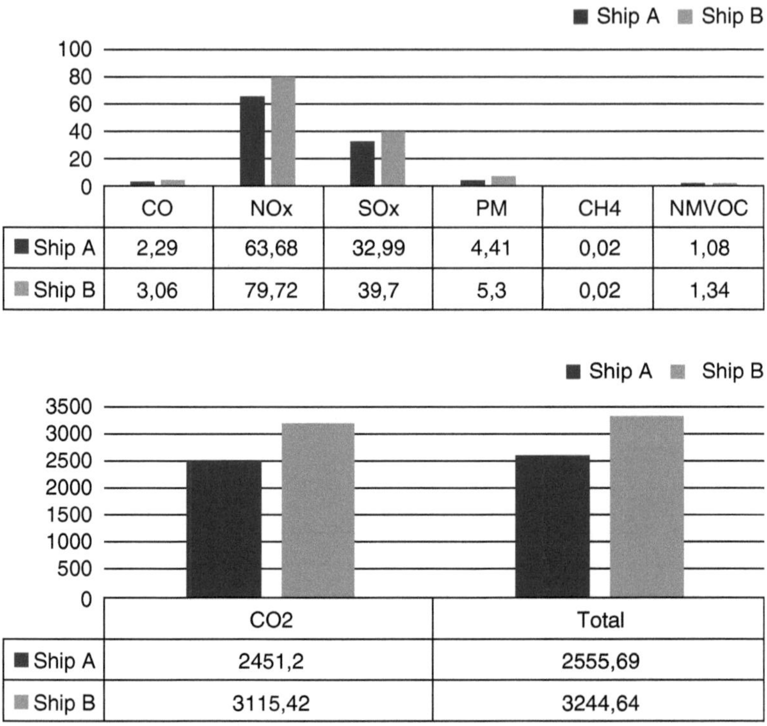

	CO	NOx	SOx	PM	CH4	NMVOC
Ship A	2,29	63,68	32,99	4,41	0,02	1,08
Ship B	3,06	79,72	39,7	5,3	0,02	1,34

	CO2	Total
Ship A	2451,2	2555,69
Ship B	3115,42	3244,64

Fig. 62.4 Ship emissions for energy method

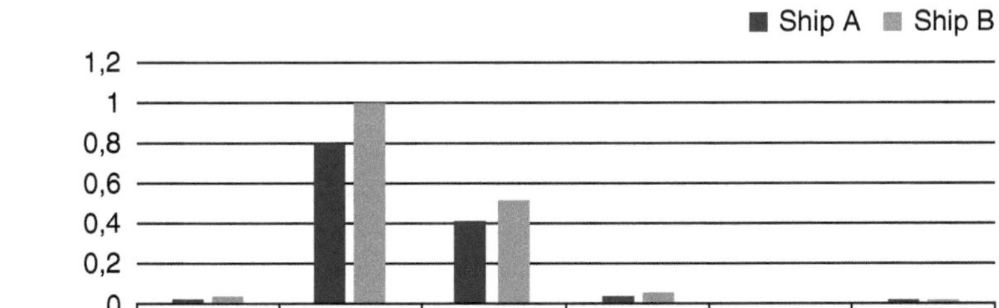

	CO	NOx	SOx	PM	CH4	NMVOC
Ship A	0,027	0,82	0,43	0,057	0,00025	0,014
Ship B	0,036	1	0,52	0,069	0,00025	0,017

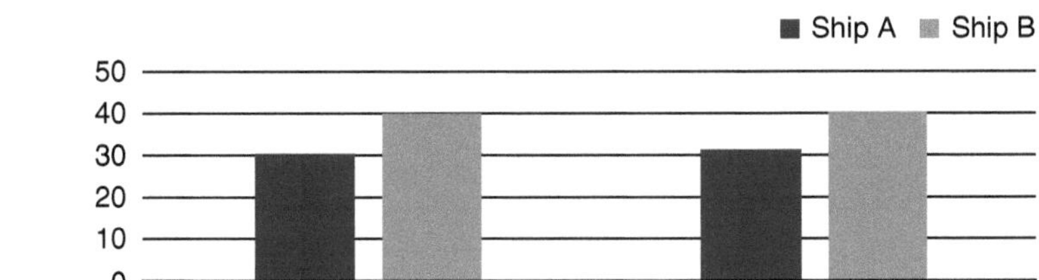

	CO2	Total
Ship A	31	32
Ship B	40	41

Fig. 62.5 (Emission/DWT)*1000 values for fuel consumption method

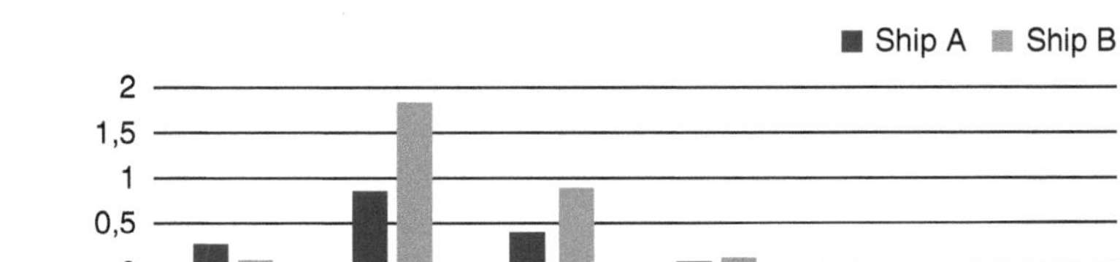

	CO	NOx	SOx	PM	CH4	NMVOC
Ship A	0,24	0,84	0,4	0,058	0,00025	0,013
Ship B	0,057	1,8	0,87	0,12	0,0005	0,03

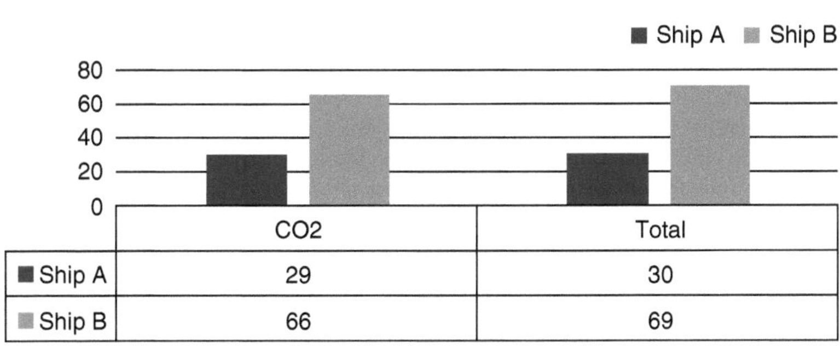

	CO2	Total
Ship A	29	30
Ship B	66	69

Fig. 62.6 (Emission/DWT)*1000 values for engine power method

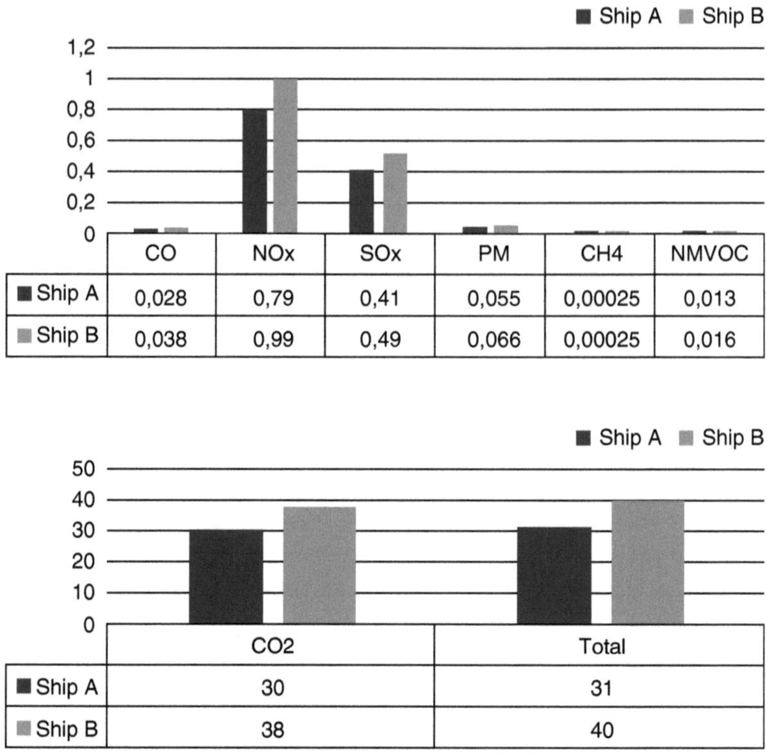

	CO	NOx	SOx	PM	CH4	NMVOC
Ship A	0,028	0,79	0,41	0,055	0,00025	0,013
Ship B	0,038	0,99	0,49	0,066	0,00025	0,016

	CO2	Total
Ship A	30	31
Ship B	38	40

Fig. 62.7 (Emission/DWT)*1000 values for energy method

References

AEA Energy & Environment. (2008). Green-house gas emissions from shipping: trends, projections and abatement potential, Final report to the Committee on Climate Change, Restricted Commercial, ED 43808, Issue Number 4, 3 September.

Andreoni, V., Miola, A., & Perujo, A. (2008). Cost effectiveness analysis of the emission abatement in the shipping sector emissions, European Commission Joint Research Center, Institute for Environment and Sustainability, ISBN: 978-92-79-11280-5.

Bilgili, L., & Celebi, U. B. (2014). Annual emission estimation comparison of two methods for a ship. In C. Guedes Soares & T. A. Santos (Eds.), *2nd International Conference on Maritime Technology and Engineering, MARTECH-2014, Maritime Technology and Engineering*, London: Taylor and Francis Group. ISBN: 978-1-138-02727-5.

Bilgili, L., Celebi, U. B., Alkan, G. B., Caglak, S. B., & Koralturk, G. (2014). Comparison of different ship emission estimation methodologies for annual emission footprint and reduction techniques of a bulk carrier. *Fresenius Environmental Bulletin, 23*(7), 1497–1509.

Buhaug, Ø., Corbett, J. J., Endresen, Ø., Eyring, V., Faber, J., Hanayama, S., et al. (2009). *Second IMO GHG Study*. London, UK: International Maritime Organization (IMO). April 009.

Cofala, J., Amann, M., Hezes, C., Wagner, F., Klimont, Z., Posch, M., Schopp, W., Tarasson, L., Jonson, J.E., Whall, C., & Stavrakaki, A. (2007). Analysis of policy measures to reduce ship emissions in the context of the revision of the national emissions ceiling directive, Final

Report, Submitted to the European Commission, DG Environment, Unit ENV/C1, Contract No. 070501/2005/419589/MAR/C1, IIASA Contract No. 06-107.7157-7160.
Cooper, D., & Gustafsson, T. (2004). Methodology for calculating emissions from ships: 1. Update of emission factors, Report for Swedish Methodology for Environmental Data, 2nd of February.
Corbett, J. J., & Köhler, H. W. (2003). Updated emissions from ocean shipping. *Journal of Geophysical Research, 108*(D20), 4650. doi:10.1029/2003JD003751.
Corbett, J. J., Winebrake, J. J., Green, E. H., Kasibhatla, P., Eyring, V., & Lauer, A. (2007). Mortality from ship emissions: A global assessment. *Environmental Science & Technology, 41* (24), 8233–8239.
Endresen, Ø., Sørgard, E., Sundet, J. K., Dalsøren, S. B., Isaksen, I. S. A., Berglen, T. F., et al. (2003). Emission from international sea transportation and environmental impact. *Journal of Geophysical Research, 108*, 4560. doi:10.1029/2002JD002898.
Eyring, V., Isaksen, I. S. A., Berntsen, T., Collins, W. J., Corbett, J. J., Endresen, Ø., et al. (2010). Assessment of transport impacts on climate and ozone: Shipping. *Atmospheric Environment, 44*(37), 4735–4771.
Eyring, V., Köhler, H. W., van Aardenne, J., & Lauer, A. (2005). Emissions from international shipping: 1. The last 50 years. *Journal of Geophysical Research, 110*(D17305), doi: 10.1029/ 2004JD005619.
Haglind, F. (2008). A review on the use of gas and steam turbine combined cycles as prime movers for large ships. Part I: Background and design. *Energy Conversion and Management, 49*, 3458–3467.
Heitmann, N., & Khalilian, S. (2010). Accounting for CO_2 emissions from international shipping: Burden sharing under different UNFCCC allocation options and regime scenarios, Kiel Working Paper No. 1665, October 2010.
Henningsen, R. F. (2000). Study of greenhouse gas emissions from ships, Final Report to the IMO, MARINTEK, Trondheim, Norway.
IMO. (2014). Reduction of GHG emissions from ships, Third GHG Study 2014-Final Report, Retrieved July 25, 2014.
IMO (International Maritime Organization). (2012). *International shipping facts and figures-information resources on trade, safety, security, environment*. Maritime Knowledge Center, Retrieved March 6, 2012.
Lauer, A., Eyring, V., Hendricks, J., Jöckel, P., & Lohmann, U. (2007). Effects on ocean-going shipping on aerosols and clouds. *Atmospheric Chemistry and Physics, 7*, 5061–5079.
MEPC. (2005). Prevention of air pollution from ships. Reduced shipping emissions of air pollution—feasible and cost-effective options. International Maritime Organization (IMO), Marine Environment Protection Committee (MEPC), 53rd session, 7 April.
Miola, A., Paccagnan, V., Mannino, I., Massarutto, A., Perujo, A., & Turvani, M. (2009). External costs of transportation case study: Maritime transport, EU-Report, Scientific and Technical Research Series. ISSN 1018-5593, Luxembourg.
Psaraftis, H. N., & Kontovas, C. A. (2009). CO_2 emission statistics for the world commercial fleet. *WMU Journal of Maritime Affairs, 8*, 1–25.
Schreier, M., Mannstein, H., Eyring, V., & Bovensmann, H. (2007). Global ship track distribution and radiative forcing from 1-year of AATSR-data. *Geophysical Research Letters, 34*, L17814. doi:10.1029/2007GL030664.
Trozzi, C. (2010). Emission estimation methodology for maritime navigation, *US EPA 19th International Emissions Inventory Conference*.
Tzannatos, E. (2010). Ship emissions and their externalities for Greece. *Atmospheric Environment, 44*, 2194–2202.
UNCTAD. (2014). In J. Rogers (Ed.), Review of maritime transport 2014. eISBN: 978-92-1-056861-6.

Chapter 63
Pollution Effects Onboard and Its Generated Solution for Minimized Pollution Effect

Munir Suner and Tankut Yildiz

Introduction

Mechanics and thermal energy proficiency is very important on board. The initial ones among the most important factors are viscosity, pollution and speed. These (pollution, speed and viscosity) cause mechanics and thermal energy losses Also the increase of these gives a rise to irregular turbulence and hydrodynamic roughness. The hydrodynamic roughness is not a problem in regular flow, but in turbulence flow besides increasing energy losses, it also causes vibration, voice pollution, cavitation and corrosion. In addition, due to flow of fluid it results in dynamic fatigue of materials, system failures and increasing amount of operational costs. In this section these so important terms will be mentioned about.

In the research conducted by S.D. Sharma and others in 2003, they gathered some useful information to help us comprehend the commencement of cavitation observed in propellers, measurement of sound of cavitation, number of cavitation, geometry of propellers and cavitation of turbulence much better (Sharma et al. 2003). During a survey conducted in 2010, the resistance of cast-iron against cavitation as well as the resistance of other metals except for iron against cavitation were examined and the obtained values were compared to the rates of carbon steel (Hattori and Kitagawa 2010). In a study conducted by X. Y. Li and others, the inspection and examination of using copper-manganese and aluminum alloys used for the method of welding to repair a vessel propeller exposed to cavitation was performed (Li et al. 2004). In another study conducted in 2010, sorts of cavitation

M. Suner (✉)
ITU Maritime Faculty Marine Engineering, Tuzla, Istanbul, Turkey
e-mail: msuner@itu.edu.tr

T. Yildiz
Alsancak Academy, Alsancak Izmir, Turkey
e-mail: yildiz.tankut@gmail.com

P. Grammelis (ed.), *Energy, Transportation and Global Warming*, Green Energy and Technology, DOI 10.1007/978-3-319-30127-3_63

and the cavitation which on the rudder tiller to the increase of cavitation in line with the acceleration of the speed of vessel were examined (Rhee et al. 2010). Ahmet Celek examined and inspected the effects of pollution, corrosion and cavitation which cause energy losses in his project titled “How to avoid energy losses on vessels” in 2012 (Çelek 2013). In one study Suner M. and Birdal O. illustrated that the impact of cavitation in ships on emissions. According to this study, the higher cavitation lead to higher energy losses and fuel consumption. This increase in energy loses and fuel consumption lead to higher CO_2 emissions, which results in temperature increases. Increase of CO_2 will be % 0.1155. If cavitation is prevented, it decrease consumption of fuel and this is approximately % CO_2 0.040425, and the others emission gases will also decrease. On other hand cavitation lead to deprecation of the systems, which lead to the wear of the materials? The worn materials pollute the water and leads to toxicity (Suner and Birdal 2013). During a survey conducted by Ballou and his friends (2008), Models represented powerful tools that could aid in fuel saving. An exemplary case is the voyage and vessel optimization system (VVOS), proven to reduce fuel consumption and related GHG emission through optimization of vessel operation reference to ocean environment parameters while recognizing ship performance criteria and safe operating limits (Ballou et al. 2008). In the study of Harrould-Kolieb and Savitz (2010), their forecasts demonstrated that the present IMO mitigation policy didn’t reduce emissions in the short-run. Policies regulating carbon dioxide emissions from ships were yet to see implementation by IMO regardless of their immense emission reduction potentials achievable via available operational and technical measures (Harrould-Kolieb and Savitz 2010). In study of Hochkirch and Bertram (2010), the energy consuming equipment on the ship could be reduced by adopting a number of approaches. Energy savings could be achieved by adopting efficient electronically controlled pumps, HVAC ventilation system, and energy saving lighting. The exhaust heat could be used for steam generation and hot coolant could be used for producing fresh water from sea water (Hochkirch and Bertram 2010). In one study, Mewis and Hollenbach (2007), illustrated that the typical fuel savings was 13 % for bulkers or tankers, 16–19 % for containerships, for a speed reduction of 5 % (Mewis and Hollenbach 2007). Hollenbach and Friesch (2007) illustrated that the high-efficiency rudders such as a twisted rudder with a Costa bulb might lower power consumption by 4 % (Hollenbach and Friesch 2007). In fact, 3 % of the world emissions reductions (44 Gt CO_2 per year) that is pivotal to stabilization of temperature under 2 °C must be provided by the Shipping sector (Vergara et al. 2012; Kontovas and Psaraftis 2011). Technologies revolving around refinement of mission, hull resistance reduction, prime mover and propulsion, and low carbon new fuels are crucial recipes for mitigating maritime emissions (Vergara et al. 2012; Kontovas and Psaraftis 2011). The design of vessels to achieve fuel efficiency is practicable for new ships as well as those yet to be constructed by manufactures. Apparently, historic vessels already in use prove difficult or most probably do not lend themselves easy to engineering modifications (Kontovas and Psaraftis 2011; Vergara et al. 2012). Adopting short-run simulations, Corbett et al. (2009) found out that speed reductions may result in emission reduction across a range of containership routes up to 70 % and lower speeds provide CO_2 reduction on most routes with higher costs when additional ships are added to maintain scheduled frequency (Corbett et al. 2009). Arteconi et al. (2010) in their study focused on assessing the emissions of GHG alone, comparing the use of LNG as a fuel in heavy-duty goods vehicles instead of the diesel normally used in such vehicles. Then they considered: (1) the life cycle of diesel fuel; (2) the life cycle of LNG brought to Europe by methane carrier ships and collected in liquid form at the regasification terminal (LNG-TER); (3) the life cycle of LNG produced at small-scale liquefaction plants (LNG-SSL) installed at service stations. In study was compared diesel and LNG fuels for use in heavy-duty vehicles in terms of GHG emissions throughout their life-cycle, in the setting of the European market (EU-15). For LNG, they considered two possible procurement strategies: its purchase directly from the regasification terminal (LNG-TER) and its small-scale production at the service stations (LNG-SSL). It emerged that LNG-TER afforded a 10 % reduction in GHG emissions by comparison with diesel, the majority of which was attributable to the combustion phase in the vehicle (14 %), while emissions relating to the production phase were much the same for the two fuels, and they were higher for LNG during the distribution phase because its procurement chain is less capillary and articulated (Arteconi et al. 2013). Gangolells et al. (2015) studied on an intelligent energy management system and applied on underground metro stations. They saw maintenance phase had very small positive impact (less than less than 0.5 %). In any case, the impact on resources was the largest (about 51 %), whereas the human health damage category amounted to approximately 35 % and the ecosystem quality damage category represented about 14 % of the total impact (Gangolells et al. 2015). Chua et al. (2013) reviewed the most recent developments concerning innovative cooling technologies and strategies found in literature that could potentially lower the kW/R ton of cooling systems. They presented intelligent air-control strategies and smart chiller sequencing methodologies that reduce the primary energy utilization for cooling. Finally, they illustrated through various literature sources how recent efforts had gradually but surely lowered the kW per R ton for air conditioning (Chua et al. 2015).

As it has been seen from the studies, as if in the literature, energy efficiency on vessel, pollution effects on the environmental (emission) there isn't any study pollution effect on injector, service pump, hydraulic cycles, Air Tower on the Heat Transfer, turbocharger vessel so.

In this study, the effects of turbulence, corrosion, pollution and vibration observed on vessels have been dealt with. The consideration of the significance of the matter with the samples experienced in life and presentation of suggestions for solving such problems are targeted. Afterwards, those events experienced on the vessel named M/V Infinity which was 30 years old in 2014 operated by Makro Maritime Lines, all of the incidents such as vibration, turbulence, corrosion, and pollution and their effects on vessels with the help of data and photographs taken and recorded on site will be examined thoroughly and those practiced suggestions for solutions for such incidents will be considered and dealt with.

Theory

Turbulence

As it is known; turbulence is irregular, in other words, mixed flow of fluid and transition among flow layers. The characteristic of flow is determined by Re number. Re number is shown in Eq. (63.1).

$$Re = \frac{\rho V.l_c}{\mu} \tag{63.1}$$

Here; ρ is density of fluid (kg/m^2), v is speed of fluid (m/s), l_c is characteristic length (m), μ is dynamics viscosity (Pa s), v is kinematic viscosity of fluid [$v = \mu/\rho$] (m^2/s)

On plate if $Re > 500000$, the flow is turbulent.

In internal flow if $Re > 4000$ the flow is turbulent.

Turbulence takes place inner and outer systems (pumps, pipe circuits, heat exchanger, air towers propeller) on vessels. The problem that occurs on board is basic structural element. Besides increasing energy losses; it also results in increasing of coefficient energy losses, fatigue of materials and failures in systems in time.

Cavitation

While fluid is passing through an enclosed space, if it is statics pressure decreases under the point pressure at heat, minerals and gases in fusion form in fluid turn into balloons. Later when is goes into this balloon explode and compact and fluid with high pressure effect, all these balloons explode. This matter is known as cavitation (Suner and Birdal 2013). On vessels it gives arise to big wearing, vibration and dynamic contraction, expansion, bottom area, screw propeller and valves. Also, increases pollutions.

Corrosion

Corrosion is being effected of material characteristics as a result of a chemical and electrochemical reaction of materials with environment in a negative way. Chemical reaction is a sort of reaction in which electron deals directly with another element in an environment where metal takes place. Metal usually gives electron to oxygen in environment and as a results of this reaction metal oxide oxidization is clear especially in high heats and corrosion product which appears in this situation is known as scaling in technical language (http://www.saritas.com.tr/ktlg/pdf/

Paslanmaz%20Celik%20Saritas%2006.pdf). Corrosion is observed on outer space (hull), especially when in contact o electric leakages with sea water, on metal and alloys in contact with seawater and machine, equipment and parts on vessels.

Pollution

As it is known, an important part of thermal and mechanic energy losses appearing materialize on border layer surface. During fluid flow both heat transfer and mechanic energy losses increase and also cause increase of resistance coefficient. This situation is an important problem in turbulence flow. Also pollution is a problem causing different difficulties. If amount of pollution is above regular below layer on border below layer; in other words if it is goes beyond, it takes the name of hydrodynamic pollution. Several sources such as cargo carried in engine, waste gas leakages in engines, seawater, oil and fuel taken are reason of pollution. Also corrosion and wearing out are sources of pollution and they are important problems. They cause failures such as clogging and being stuck of mechanic parts. One hundred percent of pollution prevention cannot be possible.

Vibration

Formation of vibration is observed during operation of ship engine, in structure of ship hull and in most mechanism. The source of this vibration is unbalanced forces and moments. Central inertial force of masses with rotary movement, moment of inertia forces, change of propeller trusts, engine rotation moment, insufficient balance of ring and intermediate shaft (Mocoşoglu 2013). Rupture and refraction appear in joining places as results of fatigue.

Applications M/V Infinity

IMO numbered vessel has been operated by Makro Shipping and Ship Management Ltd Since 2011. The deck of the vessel is seen in Fig. 63.1 and the Engine Room is seen in Fig. 63.2. M/V Infinity 1985 is a dry cargo vessel constructed in England. The Liberian Flaggedand 8115215.

In the following section of the study, some sample incidents such as turbulence, cavitation, corrosion, pollution and their effects experienced on the M/V Infinity Vessel from October 2013 to April 2014 will be examined and dealt with the photographs and data provided.

Fig. 63.1 M/V infinity

Fig. 63.2 M/V infinivity engine room

Cavitation Impeller

Hammering cause's energy and productivity losses at serious rates in pumping systems when experienced. Hammering creates worn outs Thus, it appears as an important hinder while performing fluidal substance transfer procedures. Its impact can be observed quite clearly on the impellers and the sockets of the impellers of the

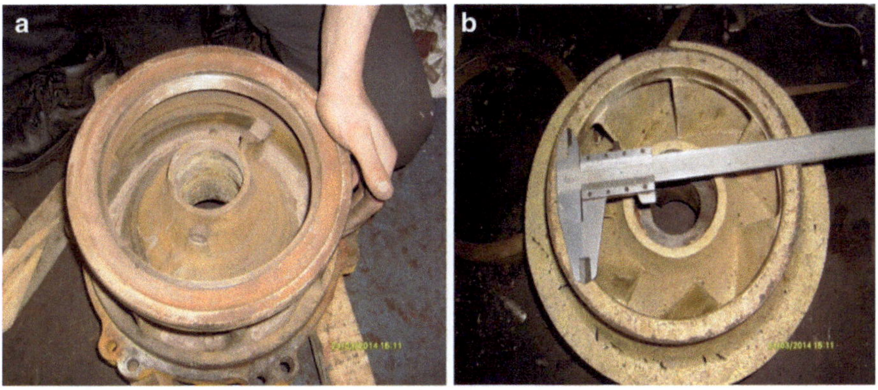

Fig. 63.3 (**a**) The impellers of the general service pump, (**b**) the socket of the ımpellers in the impellers and in the sockets of pumps

general service pump belonging to the M/V Infinity which has not gone through maintenance services for a long time. Those cracks and worn outs shown in Fig. 63.3a and those worn outs shown in Fig. 63.3b caused loss of pressure. After replacing the impeller with a new spare part and fulfilling the required maintenance, it was observed that the pressure at the pump outlet increased approximately 1 bar.

Corrosion in General Service Pump

Owing to of the effects the high rates of chlorine in the sea water and compounds of chlorine, high rates of humid and current oxygen content both in the air and water, corrosion becomes a significant issue to be dealt with delicately (Craig et al. 2006).

The cases of general service pumps as shown in Fig. 63.4 and besides the sea water of the main engine which belongs to the M/V Infinity as seen in Fig. 63.5 display an obvious sample to this situation. Both the leakages occurring in the packing gland and the sea water given to the throttle of packing gland for protection cause a considerable amount of corrosion in the long run and they also cause serious worn out on the material as well as some damages such as punctures and fractures here and there.

Therefore, it is necessary to avoid the contact of the risky parts of machines such as the cycle pump case with sea water and also paint containing additives or corrosion proof chemicals must be utilized in order to stop corrosion.

Fig. 63.4 The case of the main engine sea water pump close view

Fig. 63.5 The case of the general service pump

Corrosion on Outer Surface of Injector

Due to the misuse of the components and apparatus on ships or losing their edges such as the properties of sealing in time as a result of being worn out, several fresh water and sea water leakages are likely to occur and those leakages cause corrosion by affecting many parts of machines negatively.

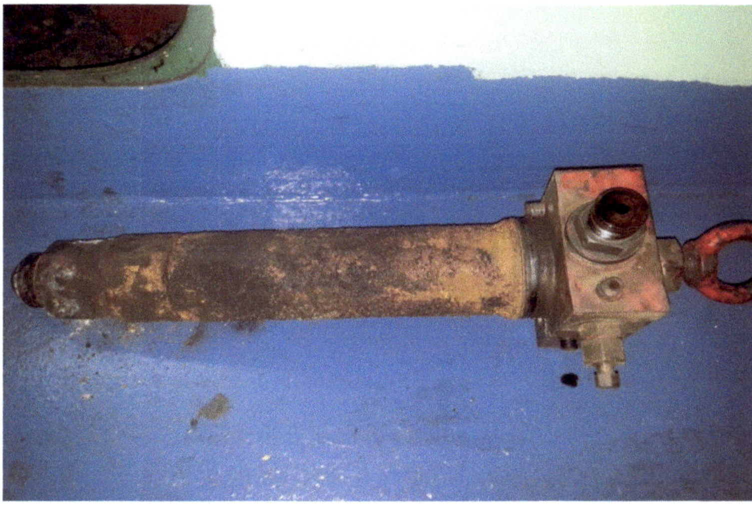

Fig. 63.6 Outer surface of the injector

As it is observed in Fig. 63.6, the leakage in the block of the main engine injector cooling water cycle belonging to the M/V Infinity caused serious amount of corrosion on the outer surface of the injector. Such corrosion incidents hinder performing disassembling parts of the injector to maintain and corrosion also gives damages on task performing parts of the injector in the long run and thus it creates many important problems. For this reason, it is necessary to avoid leakages and contact of machine parts with sea water.

Corrosion on the Deck of the Vessel

Corrosion effect of sea water causes many important problems on the deck of the vessel and the equipment on the deck. In Fig. 63.7, the hydraulic circuit of the hatch cover exposed to corrosion is observed. This corrosion impact exploded the links of the cycle valve and thus both the valve and the plastic hose had to be replaced with new ones.

Figure 63.8a, b present the effects of pollution and corrosion on the capstan located on the poop deck can be observed. The seal of capstan (Fig. 63.10) worn out, and resulted in oil leakage. Rust, pollution and corrosion occurred on the equipment of the capstan because they were constantly exposed to sea water and rain water yet no maintenance task was carried out for a long time, hydraulic oil leakages and gear jamming occurred as shown in Fig. 63.9.

Fig. 63.7 Corrosion impact on the hydraulic cycle of the hatch cover

Fig. 63.8 (**a**) Corrosion impact on the gear of the capstan, (**b**) different view

The Effect of Pollution in Air Tower on the Heat Transfer

Air Towers are utilized to cool the air of scavenge whereas turbochargers are used to boost the amount of air in scavenges both of which are adopted to increase the efficiency and productivity of the engine.

Those three diesel generators Model VTR-250 manufactured by Daihatsu Diesel Mfg. Co. Ltd which were installed on M/V Infinity have inter cooling systems and those and manufactured by BBC Brown Broveri VTR-250 model have turbochargers. Those turbo chargers own 22,500 rpm and maximum 630 °C working heat.

Fig. 63.9 Worn out oil seal on the capstan gear because of corrosion impact

Fig. 63.10 Gear jamming because of corrosion impact

As shown in Figs. 63.11, 63.12, 63.13, 63.14 and 63.15 those air towers, turbochargers, turbocharger socket, and close view of the socket as well as the scavengers of which cleaning and maintenance jobs were not performed for a long period of time resulted in decrease in their air pressures and consequently serious amounts of temperature increase occurred in their exhaust. Due to those

Fig. 63.11 Air tower

Fig. 63.12 A disassembled turbocharger

Fig. 63.13 The socket of a turbocharger

Fig. 63.14 Close view of the socket of a turbocharger

Fig. 63.15 Cleaning process of air tower

temperature increases, occasionally two generators had to be employed to perform jobs which can be carried out by only one generator in order to avoid probable harms and damages likely to occur because of excessive work load. In such cases, fuel expenses rose dramatically which caused monetary losses. Besides, another negative effect appeared since the life spans of the related machines were shortened considerably.

As seen in Fig. 63.15, after fulfilling all sorts of necessary cleaning and maintenance jobs of the air towers and turbochargers of the number 3 generator properly, the pressure of the scavenge went up from the rate of 0.4 bar to 1.2 bar and approximately 50 °C decrease was observed at the exhaust temperatures.

Conclusions

It has been investigated by examples that vibration, turbulence, cavitation, corrosion and pollution effects on board affect energy efficiency. Vibration local vibration throughout the vessel cause serious on board. On systems just after vibration it gives arise to material worn out, rupture and refraction in long term. Systems must be correctly fixed and hydraulic oils of vibration damper must be regularly checked and in the critical rotating specified during ship design voyage must be kept in minimum.

Turbulence is a problem which comes true in most of systems in which fluid flow inside vessel and around hull and propeller vessel. It is seen in pumps, pipelines and cantilevers where friction coefficient is high. It is reason of cavitation source, increase of energy losses and material fatigue. Just as the problems caused by vibration turbulence caused problems and decrease of energy. Pollution effect of the foremost plan here and must be prevented.

Corrosion is the important cause of increase of energy losses, decrease of heat transfer and material frazzling. Electric failures oxidization is important problems on board. Engine and its equipment exposed to corrosion must be maintained in the shortest time. On other hand, it has risk such as refraction, wearing and being pierced if they aren't maintained shortest time. These risks must be minimized projective paints must be used. Pollution is formed by external material such as powder, waste oil, exhaust gases, scaling, biological (sea organism and creatures), pollution, mud, and send. It is a problem that is most commonly faced with on board. It is a matter which concern the management directly. It is an important increase in energy losses, clogging, not carrying out function of systems and surface friction outsides vessels.

For instance the cleaning of evaporator, which is used for fresh water production, has provided 30 % increase in fresh water production. Again as pointed out in this study, due to clogging and stoppage of important equipment.

In other words hot carrying out its function, pollution give arise to accidents on board. Its solution is to maintain periodically. Serpentines, filters, pumps, valves, cooler, circuits must be regularly cleaned. This regular cleaning reduces risk of decrease of energy losses and fuel consumption and failure of systems as a results of energy losses pollution increases fuel amount.

Together with this situation because of use of fossil originated. Emissions increase. On vessels its effects can be reduced. Important rates as a result of appropriate operation and maintenance.

Acknowledgements The data, images and photographs used in this study are compiled by the author of this essay during the long term marine training voyage, actualized on board of the M/V Infinity Vessel, from October 2013 to April 2014. We would like to present our heartfelt gratitude to all of the seafarers and to the despondent owner, Makro Maritime Lines Ltd for their valuable contribution.

Nomenclature

P Pressure (N/m^2)
T Temperature (°C)

References

Arteconi, A., Brandoni, C., Evangelista, D., & Polonara, F. (2010). Life-cycle greenhouse gas analysis of LNG as a heavy vehicle fuel in Europe. *Applied Energy, 87*(6), 2005–2013.

Arteconi, A., Brandoni, C., Evangelista, D., & Polonara, F. (2013). Life-cycle greenhouse gas analysis of LNG as a heavy vehicle fuel in Europe. *Applied Energy, 87*(6), 2005–2013.

Ballou, P., Chen, H., & Horner, J. D. (2008). Advanced methods of optimizing ship operations to reduce emissions detrimental to climate change. OCEANS 2008. Kobe, Japan. Retrieved April 8–11, 2008.

Çelek, A. (2013). Gemilerde Meydana Gelen Enerji Kayıpları ve Önlenmesi İçin Yapılması Gerekenler. *Lisans Tezi* ITU (In Turkish).

Chua, K. J., Chou, S. K., Yang, W. M., & Yan, J. (2013). Achieving better energy-efficient air conditioning A review of technologies and strategies. *Applied Energy, 104*, 87–104

Chua, K. J., Chou, S. K., Yang, W. M., & Yan, J. (2015). Achieving better energy-efficient air conditioning—a review of technologies and strategies. *Applied Energy, 104*(2013), 87–104.

Corbett, J. J., Wang, H., & Winebrake, J. J. (2009). The effectiveness and costs of speed reductions on emissions from international shipping. *Transportation Research Part D: Transport and Environment, 14*(8), 593–598.

Craig, B. D., Lane, R. A., Rose, D. H. (2006). *Corrosion prevention and control: A program management guide for selecting materials*. AMMTIAC.

Gangolells, M., Casals, M., Forcada, N., Macarulla, M., & Giretti, A. (2015). Environmental impacts related to the commissioning and usage phase of an intelligent energy management system. *Applied Energy, 138*(15), 216–223.

Harrould-Kolieb, E., & Savitz, J. Shipping solutions (2010). Technological and operational methods available to reduce CO2. Oceana, pp. 6–18.

Hattori, S., & Kitagawa, T. (2010). *Analysis of cavitation erosion resistance of cast iron and nonferrous metals based on database and comparison with carbon steel data*. Graduate School of Engineering, University of Fukui Japan.

Hochkirch, K., & Bertram, V. (2010). *Engineering options for more fuel efficient ships*. Montreal, Canada: Green Tech.

Hollenbach, U., & Friesch, J. (2007) Efficient hull designs—what can be gained. *1st International Conference on Ship Efficiency*. Hamburg, Germany. Retrieved October 8–9, 2007.

http://www.saritas.com.tr/ktlg/pdf/Paslanmaz%20Celik%20Saritas%2006.pdf.

Kontovas, C., & Psaraftis, H. N. (2011). Reduction of emissions along the maritime intermodal container chain: Operational models and policies. *Maritime Policy & Management, 38*(4), 451–469.

Li, X. Y., Yan, Y. G., Ma, L., Xu, Z. M., & Li, J. G. (2004). *Cavitation erosion and corrosion behavior of copper–manganese–aluminum alloy weldment* (State Key Lab for Marine Corrosion and Protection). PR China: Luoyang Ship Materials Research Institute Qingdao Branch.

Mewis, F., & Hollenbach, U. (2007). Hydrodynamic measures to reduce energy consumption in ship operation. *Hansa, 144*(5), 40–58.

Mocoşoglu, O. (2013). Gemi titreşimlerinin enerji verimliliği açısından incelenmesi, ITU, Istanbul, Turkey design project. (In Turkish).

Rhee, S. H., Lee, C., Lee, H. B., & Jungkeun, O. (2010). *Rudder gap cavitation fundamental understanding and its suppression devices*. Republic of Korea: Research Institute of Marine Systems Engineering, Seoul National University, Department of Naval Architecture & Ocean Engineering.

Sharma, S. D., Mani, K., & Arakeri, V. H. (2003). *Cavitation noise studies on marine propellers*. Bangalore, India: Departments of Mechanical and Civil Engineering, Indian Institute of Science.

Suner, M., & Birdal, O. (2013). *Effect of cavitation in ships on the environment*. 957–973, New York: Springer.

Suner and Tankut Yildiz (2016). Loss of Energy due to Pollution and its Effects on Ship Main Engine and Auxiliary Machinery, March 2016 Applied Energy.

Vergara, J., McKesson, C., & Walczak, M. (2012). Sustainable energy for the marine sector. *Energy Policy, 49*, 333–345.

Chapter 64
Carbon-Mitigating Air Transport: Analysis of Current Policy Instruments

Ioanna Pagoni and Voula Psaraki-Kalouptsidi

Introduction

In response to the increasing evolution of aviation greenhouse gas emissions and to the current forecasts for further emissions' growth (in nominal terms) in the future, several environmental measures have been implemented worldwide. These measures are related to technological improvements, operational efficiency improvements, use of alternative fuels and economic instruments. This chapter provides an overall review of the environmental measures implemented on aviation sector focusing on the market-based measures for air transport discussing their characteristics, their geographic area of applicability and their potential effect on airline market and environment.

Economic instruments encompass a range of policy tools that allow internalization of the environmental external costs in the context of the principle "the polluter pays." There are various forms of economic instruments for air transport such as emission charges/fees, extra landing charges based on the amount of emissions, trading schemes and voluntary carbon offsetting schemes. The extent to which these economic policies could affect the air ticket price is an important issue and has been studied by various researchers. Scheelhaase et al. (2010) studied the economic impact of EU ETS on Lufthansa and Continental Airlines. The results indicated that Lufthansa's environmental costs are higher, due to its wide extent of operations within the EU region. Girardet and Spinler (2013) used dynamic programming to calculate the optimal percentage of ticket price change due to changes in kerosene's price and carbon-dioxide (CO_2) emissions' charges. Albers

I. Pagoni (✉) • V. Psaraki-Kalouptsidi
Department of Transportation Planning and Engineering, School of Civil Engineering, National Technical University of Athens, 5, Iroon Polythechniou, Zografou, Athens 15773, Greece
e-mail: ipagoni@central.ntua.gr; vpsaraki@civil.ntua.gr

P. Grammelis (ed.), *Energy, Transportation and Global Warming*, Green Energy and Technology, DOI 10.1007/978-3-319-30127-3_64

et al. (2007) examined the impact of EU ETS on the EU and non-EU carriers' profitability and route demand for selected routes. To capture the potential effects on air travel demand they used values for price elasticities calculated in other studies and concluded that on the whole the competitive implications for individual airlines are only marginal and could not threaten any airline's competitive position. Lu (2009) examined the impact of environmental charges on air passenger demand using the study of six intra-European short-haul routes in two city pairs. The results indicate that the potential demand reduction is higher for the low-cost carrier Easyjet compared to that of full service carriers, because of its lower fares. Miyoshi (2014) investigated the differences in terms of equity between Annex I and non-Annex I airlines and their passengers after the implementation of EU ETS. The author selected a specific route and constructed a logit model to estimate the impact of travel costs increase on the route's market share. The results demonstrated that the EU ETS is an effective instrument unless the carbon price is very low.

This chapter contributes to the above literature by considering the effects of a carbon scheme on a U.S. airline network encompassing the top ten air travel markets. The associated aircraft fuel burn and CO_2 emissions are computed and the results are presented at both a disaggregate level (flight-level analysis) and an aggregate level (market-level analysis). The computations are conducted for all flight modes based on data from the 1st quarter of 2012. A unit carbon price is applied to calculate the resulting carbon costs an airline has to pay in the context of the carbon scheme. Data from a combination of databases derived from ICAO, EUROCONTROL and U.S. Department of Transportation are used. We assume that the carbon cost is passed onto the passengers, resulting in air fare increase. The results indicate that the prevailing carbon price plays a critical role to the efficiency of a carbon fee scheme. Air ticket prices are found to increase to some extent. However, the results differ from flight to flight and from market to market, since flights' parameters, such as aircraft type, distance and flight profile, are differentiated.

Four-Pillar Approach for Sustainable Aviation

Most aviation organizations, aiming to achieve aviation environmental goals, rely their initiatives on a four-pillar approach (FAA 2015; IATA 2013a; Sgouridis et al. 2011) presented in Table 64.1. This approach is related to technological improvements (i.e. enhanced aircraft and engine performance efficiency), operational efficiency improvements (i.e. air traffic management, fuel optimized flight path), use of alternative fuels and deployment of market-based instruments.

Table 64.1 Four-pillar approach adopted by aviation stakeholders

I. Technological Improvements	II. Operational efficiency improvements
New composite lightweight materials New aircraft designs (improved aerodynamics and reduced drag) Wing configuration Newer and more fuel-efficient engines	Operational weight reduction Fuel optimized flight path Continuous descent approach More efficient air traffic management infrastructure
III. Alternative fuels	IV. Market-based measures
Fuel made from plant oil, animal fat or waste organic material such as Camelina, Jatropha, Algae, Halophytes etc	Emission charges Cap-and-trade programs Emission offsetting

Technological Improvements

Technological improvements are expected to significantly contribute to fuel burn and emissions reduction. The initiatives for improving aircraft fuel efficiency focus on reducing aircraft weight, improving aircraft aerodynamics to reduce drag and improving engine efficiency, to reduce fuel burn per unit thrust (Grote et al. 2014). The evolution of newer and more fuel-efficient airframes and engines, the use of new composite lightweight materials, the introduction of new aircraft designs are some of the future technologies considered by aircraft, engine and equipment manufacturers. According to IATA (2013b) the most promising airframe technology is the deployment of "Hybrid-Wing-Body aircraft" which may offer 10–25 % fuel reduction benefits and is estimated to be market available after 2026. Among engine technologies, the introduction of "New engine core concepts" is estimated to provide the greatest emission reduction of around 25–30 % (IATA 2013b). However, the success of such technologies strongly depends on uncertain factors such as development status, benefits, risk and research and development costs. For example, some technologies, such advanced airframes which are expected to have large potential on emissions reduction, may be years away from being available, and developing and adopting them is likely to face high development costs.

Operational Efficiency Improvements

Achieving more efficient aircraft operations is another pillar for reducing emissions from aviation. These improvements can be achieved through optimized airline operations, such as aircraft operational weight reductions by removing unnecessary onboard equipment, or optimized air traffic management operations, such as fuel optimized flight path, continuous descent approach and reduced flight delays. Both the United States and the European Union have taken important steps towards developing modernized air traffic management systems. In the United States, Federal Aviation Administration (FAA) is leading a multiagency effort to transform current U.S. air traffic control system to the Next Generation Air Transportation System (NextGen). NextGen is believed to increase safety, capacity and efficiency of the U.S. airspace system (FAA 2014). In Europe, the European Union launched the Single European Sky initiative which de-fragments the European airspace in order to provide air traffic conditions with lower delays, greater safety standards, lower aircraft emissions and lower costs related to service provision (Commission of the European Communities 2008). According to IATA (2013b), CO_2 emissions from commercial flights could be reduced by 28 million tones in 2020 through airline operational measures.

Alternative Fuels

Sustainable alternative fuels will play a critical role in the effort of achieving environmentally-friendly aircraft operations. Four drivers for the development of alternative jet fuel are recognized by Hileman and Stratton (2014): economic sustainability, environmental sustainability, energy supply diversity and competition for energy resources. Alternative fuels have lower lifecycle carbon dioxide (CO_2) emissions than conventional kerosene and thus offer the opportunity to reduce aviation's contribution to climate change and to air quality. However, the widespread production and development of alternative fuels may raise several concerns with the most important being the requirements on land and water usage (Hileman and Stratton 2014; Rojo et al. 2015). Since the approval for commercial use of biofuels, various airlines have experimented with using of biofuels on commercial flights which have demonstrated that alternative fuels can be safe and technically sound (IATA 2014a).

Market-Based Instruments

Technology, operations and alternative fuel measures are the long-term solution for aviation's sustainable growth. However, due to the time required for their implementation, market-based measures are viewed as a short-term promising option for reducing aviation emissions. Market-based instruments may take the form of an emission charge, a cap-and-trade program with allowance auctions or an emission offsetting (Carlsson and Hammar 2002; IMF 2011; GAO 2009). The aim of such policies is to establish a price per unit of emitted CO_2 and motivate airlines to adopt low-emissions technologies on their aircraft fleet in order to reduce their emissions.

Under an *emission charge* on carbon dioxide (or other greenhouse gas), a fee for every tonne of carbon dioxide emitted is levied to each polluter (airline). It can be implemented on any aircraft operation within a given airline network or an airport. Most known examples of emissions charges include the charges imposed by airports to airlines for their landing and take-off operations at the airports.

According to economic theory, an optimal emissions charge should be set at a level that represents the marginal damage cost of the emissions (GAO 2009).

Within a *cap-and-trade program* the amount of total emissions is capped to a predefined limit. Cap-and-trade programs are also known as emissions trading schemes. The airlines are issued emission permits up to the established cap. If they exceed the cap, they can "buy" additional emission allowances to cover the excess between what they emit and the cap. If an airline emits less CO_2 than the cap they can "sell" the emission allowances.

Carbon offset schemes allow individuals or companies to invest in environmental projects around the world in order to mitigate the carbon emissions from their activities. In the field of air transport, there is a variety of voluntary offset programs currently operating. At the time of ticket purchase, passengers are encouraged to pay for the amount of CO_2 emissions resulting from their travel by making charitable contributions to several environmental projects such as forest conservation and renewable energy (IATA 2008). The use of carbon offsets can be conceptually related to emissions trading. However, offsetting is a voluntary activity, undertaken by passengers in a largely unregulated environment.

The next section gives a detailed overview of the existing or under-consideration market-based instruments in the aviation industry.

Existing or Under-Consideration Market-Based Instruments for Air Transport

European Emissions Trading Scheme (EU ETS)

The most known cap-and-trade policy is the European Union Emissions Trading Scheme (EU ETS). In the context of the Kyoto Protocol's requirements, EU implemented the EU ETS in 2005 which included several sectors such as power and heat generation, mineral industries etc. (EU 2003). In 2008, the EU ETS was expanded so that, from 2012, it would include air traffic operations, by EU and non-EU airlines that depart from or arrive at European airports, i.e. both intra-EU flights and between EU and non-EU airports (EU 2008). This meant that, for example, an American air carrier operating a flight from New York to London would still have to comply with the EU ETS. Within the EU ETS, a cap is set on the carbon dioxide emissions from all covered flights by aircraft operators and, then, allowances are allocated, bought or sold among the airlines.

The inclusion of non-EU airlines in the EU ETS prompted strong objections from several countries and stakeholders in United States, China, Russia, and India. In light of these disagreements, in 2014, the EU ETS directive was amended in order to apply to flights that both originate and end within the European Economic Area (EU 2014). Thus, from 2014 to 2016 flights to and from countries outside the EU would benefit from a general exemption. All CO_2 emissions from flights between airports in the EU would continue to be covered, while all overflights are exempt (Arvanitakis and Hind 2014).

Emission Charges

Emission charges represent any charge imposed to an airline for the amount of CO_2 it emits either at an airport during the Landing and Take-Off (LTO) phase (airport charge) or during cruise (en-route charge). In other studies, emission charges may be referred to as "emission fees" or "emission taxes."

There are several examples of airport emission charges being applied in Europe, with Swiss and Swedish airports having the longest history. In 1997, Zurich airport became the first airport worldwide to introduce an NO_X emission-based charge in order to address air quality problems caused by air traffic. Later, Geneva airport followed this measure in 1998, Bern airport in 2001, Basel airport in 2003 and Lugano airport in 2007 (Zurich Airport n.d.). Other airports in Germany and United Kingdom have already implemented such a policy by imposing extra landing charges on airlines, based on their amount of emissions generated at the vicinity of the airports. Table 64.2 presents the emission charges applied to European airports.

Airport emission charges reflect cost externalities and aim to compensate society for the consequences of relevant emissions. The charge at Zurich airport was introduced to encourage airlines to use less polluting aircraft when using the airport. At the same time, the weight-based landing fee was reduced to ensure that the charge remained revenue-neutral for the airport (Zurich Airport n.d.).

Table 64.2 Airport emission charges in several European countries

Country	Airport	Emission charge per landing or per take-off (for 2014 or else as indicated)	Start year
Switzerland	Zurich	2.50 CHF per kg NO_X	1997
	Geneva	1.40 CHF per kg NO_X	1998
	Berne	3.30 CHF per kg NO_X (year 2010)	2001
	Lugano	3.40 CHF per kg NO_X (year 2010)	2007
United Kingdom	Heathrow	£8.82 per kg NO_X	2004
	Gatwick	£5.26 per kg NO_X (year 2013)	2005
Germany	Frankfurt	1.50€ per kg NO_X	2008
	Munich		2008
	Cologne Bonn		2008
	Hamburg		2010
	Dusseldorf		2011
Denmark	Copenhagen	DKK 16.60 per kg NOx	2010

An example of a CO_2 emission charge applied to the entire part of flight (and not only to the LTO phase at the vicinity of airports) is the Norwegian CO_2 tax on aviation fuel. It was implemented in 1999 for all domestic and international flights, although it later withdrew the tax relating to international flights due to violation of many bilateral aviation agreements which Norway had with other countries (OECD 2005). Along with the introduction of ETS for aviation on the 1st of January 2012, the CO_2 tax has been reduced in Norway (ICAO 2012).

Voluntary Carbon Offset Programs by Airlines

Carbon offset programs give the passengers the option to "neutralize" their proportion of aircraft emissions on a particular journey. Passengers who are willing to offset their emissions, pay an extra fee to the ticket price in order to invest in environmental projects. These projects are usually based in developing countries and most commonly designed to reduce future emissions.

Various airlines have already implemented voluntary carbon offset programs. Until 2014, over 35 airlines had launched their own voluntary carbon offset schemes (IATA 2014c); some of them are the following:

- *U.S. airlines*: Delta, JetBlue, United, Virgin Atlantic.
- *Canadian airlines*: Air Canada.
- *European airlines*: Austrian Airlines, Brussels Airlines, Easyjet, Lufthansa, SAS, TAP Portugal.
- *Australian airlines*: Cathay Pacific, Jetstar, Qantas Airways, Virgin Australia.
- *Asian airlines*: Japan Airlines, Qatar Airlines, Thai Airways.
- *African airlines*: Kenya Airways, South African Airways.

Given that the above carbon offset schemes are not mandatory, their success, in terms of passengers participating, may be uncertain. Reports by various airlines demonstrate a relatively satisfactory uptake of voluntary carbon offsets. For example, Jetstar and Virgin Australia indicate that 10 % of their domestic aviation passengers choose to offset their flights (Australian Government 2012).

Portuguese CO_2 Tax

In 2014, a CO_2 tax was proposed to be applied to flights departing from Portuguese airports. According to the proposal (IATA 2014b), the CO_2 levy would be imposed as follows:

- For domestic flights (between Portuguese airports) the levy would be 3€ per passenger per flight.

- For flights departing from Portugal to an airport outside of the European Economic Area, a 15€ per passenger would be collected for each flight.
- International flights departing from Portugal and arriving at airports outside Portugal and within the European Economic Area were excluded from the scheme.

However, the above proposal was not supported by the Portuguese Government and thus was not implemented. Given that EU ETS was implemented in 2012 for aviation, domestic flights within Portugal would be charged twice (both under the EU ETS and the Portuguese CO_2 tax).

ICAO Global Market-Based Measure

International Civil Aviation Organization (ICAO) recommends that as part of a broader package of measures including new technology, more efficient operations and better use of infrastructure, a single global market-based measure should be agreed to address aviation's climate impacts. Therefore, during its 38th session held in 2013, ICAO agreed on the development of a global market-based measure to mitigate carbon emissions from international aviation. This scheme will be finalized and adopted at the next ICAO Assembly in 2016 and will be implemented in 2020 (Arvanitakis and Hind 2014).

Trading schemes, emission-related airport charges and en-route charge/taxes provide regulated airlines an incentive to reduce their fuel consumption in the most cost-effective way since they are imposed an extra cost for polluting. This cost, which can be referred to as "carbon cost," is related to the amount of emissions under the scheme, the unit price of carbon and the structure of the scheme. An airline may decide to pass the carbon cost (either fully or partially) onto the passengers, which will lead to a ticket price increase. Ticket price increases also occur in the case when a passenger chooses to voluntarily offset the CO_2 emissions of his air travel, via the voluntary carbon offset schemes offered by the airlines.

The extent to which airlines would pass any carbon costs to passengers through higher fares and the way passengers would respond to these changes still need to be investigated. The next section examines the implementation of a carbon scheme in domestic air travel within United States and its impact on ticket prices.

Implementing an Emission Fee Scheme in United States

In this section we develop a carbon emission fee scheme for domestic flights operating between airports in the United States by American air carriers. Although there is currently no national market-based instrument to mitigate aviation emissions in the United States, we believe that emission costs on U.S. airline market have been and may continue to be a reality. This can be explained if we consider the impact of the existing and the planned economic instruments on the economy of U.S. air carriers and, in some cases, the ticket prices; for example, the proposed global market-based measure by ICAO and the existing voluntary carbon offset programs offered to passengers by American air carriers. Furthermore, until EU ETS amendment in 2014, American air carriers were obliged to comply with EU ETS requirements and were imposed the extra carbon cost for every flight between U.S. and Europe. Major U.S. airlines, including Delta, United and American Airlines, responded to the EU ETS by introducing a $3 surcharge per passenger for each flight under the EU ETS.

CO_2 emissions are the largest and most certain of the greenhouse gas emissions from the aviation sector. While non-CO_2 emissions are potentially significant, there currently exists a high degree of scientific uncertainty regarding their impact. A CO_2-based scheme is most likely to be compatible with other trading schemes (i.e. ICAO global scheme, EU ETS etc.) and so enhances the potential for future trading between schemes.

The examined emission fee scheme is implemented in a selected airline network within U.S. This encompasses the ten most popular origin and destination cities, in terms of passengers travelled, within the U.S. air travel market. The sample period is the first quarter of 2012 (2012:Q1). Air traffic data is derived from publicly available data from the U.S. Department of Transportation. The database of the Airline Origin and Destination Survey (DB1B) is used as it includes a 10 % sample of domestic airline tickets sold by U.S. airlines including detailed data on flight fares, itineraries (Origin, Connecting and Destination airports), air carriers and number of passengers.

The selected flight network is given in Table 64.3 and illustrated in Fig. 64.1. It should be noted that Fig. 64.1 illustrates not only the origin and destination airports, which form the ten most popular routes given in Table 64.3, but also the connecting airports of the non-direct flights. Table 64.4 presents the airports located in the origin and destination cities of the top ten routes.

For each flight in the selected network fuel consumption and related CO_2 emissions are calculated. The methodology to compute fuel burn and carbon emissions flight-by-flight is presented in a companion paper (Pagoni and Psaraki 2015). Fuel and emissions are computed separately for Landing/Take-off phase and the Climb-Cruise-Descent phase. The ICAO Engine Exhaust Emissions Databank (ICAO 2014) is used to compute the fuel consumption during the Landing/Take-off phase. For the remaining flight above 3000 feet (Climb-Cruise-Descent phase), the EUROCONTROL's Base of Aircraft Data (BADA) model (EUROCONTROL 2013) is used.

Then a unit carbon price is applied flight-by-flight to compute the extra carbon costs the regulated airline has to pay to cope with the scheme. The unit carbon price is determined based on historical price data from the well-known EU ETS. The average unit CO_2 price has fallen the last years due to the proposed EU ETS reform in 2012 and the economic recession in EU. The average price of one EU-allowance (one tonne of CO_2) in EU ETS was 22€ in 2008 and dropped to 7.3€ in 2012

Table 64.3 Top 10 markets for the sample period 2012:Q1

Market id	Origin city	Destination city	Passengers in our sample
1	San Francisco, CA	Los Angeles, CA	619,480
2	Los Angeles, CA	San Francisco, CA	616,950
3	New York City, NY	Miami, FL	440,330
4	Miami, FL	New York City, NY	435,280
5	New York City, NY	Orlando, FL	284,290
6	Orlando, FL	New York City, NY	282,530
7	Los Angeles, CA	New York City, NY	275,790
8	New York City, NY	Los Angeles, CA	275,020
9	Chicago, IL	New York City, NY	229,460
10	New York City, NY	Chicago, IL	228,120

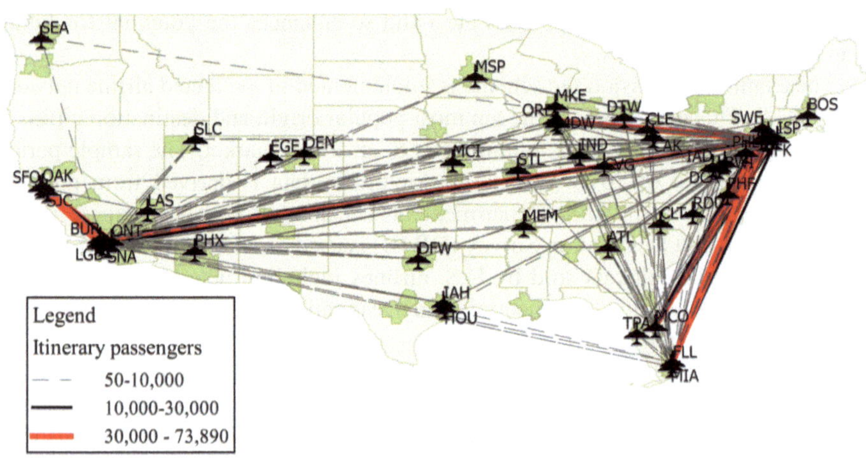

Fig. 64.1 Illustration of the U.S. network selected for the implementation of the carbon scheme

Table 64.4 Airports included in the sample

Origin or destination city	Airports included
San Francisco, CA	OAK, SFO, SJC
Los Angeles, CA	BUR, LAX, LGB, ONT, SNA
New York City, NY	EWR, HPN, ISP, JFK, LGA, SWF
Miami, FL	FLL, MIA
Orlando, FL	MCO
Chicago, IL	MDW, ORD

(SENDECO2 2014). In this paper, a carbon price of \$20 per tonne CO_2 is adopted as baseline scenario. However results are also presented for values of \$10 and \$50 per tonne CO_2 for sensitivity purposes. We assume that each airline would pass the 100 % of carbon cost onto the passengers. This is consistent with the profit maximizing behavior observed in competitive markets. The resulting fee on an airline is determined by the product of the unit carbon price times the total amount of CO_2 emitted by the aircraft of the airline.

Results and Discussion

Flight-Level Analysis

We first present our results on individual flight-level. Direct or one-stop flights between the top ten origin and destination cities are selected for this purpose. Each row of Table 64.5 represents a unique flight along with the origin and destination airports, the connecting airport (if existing), the ticketing airline, the average air fare for the sample period (2012:Q1) and the aircraft type (for each segment). Fuel consumption and CO_2 emissions are accordingly calculated for each flight. The estimations include all flight phases: take-off at the origin and the connecting (if existing) airports, landing at the connecting (if existing) and destination airports and cruise phase. At a final step, carbon emissions and carbon costs per passenger are calculated for the three scenarios: baseline ($20 per tn CO_2), low ($10 per tn CO_2) and high ($50 per tn CO_2).

Figure 64.2 presents the carbon costs per passenger and the percentage change in air fares in case the airline fully passes these costs onto the passengers. Computations refer to the baseline scenario. Figure 64.2 illustrates results derived from the online carbon calculators by Delta, JetBlue and United airlines. These are developed from the airlines in the context of their carbon offset programs. It should be noted that these carbon calculators do not explain their assumptions on aircraft type used and whether the flight is direct or one-stop. Thus, differences from these estimates are expected.

Market-Level Analysis

Fuel burn and emissions calculations are then aggregated for each of the ten selected origin and destination markets. For each market (see also Table 64.3) within the selected flight network, the associated CO_2 emissions and the impact of different carbon prices scenarios (low: $10, baseline: $20 and high: $50 per tonne CO_2) on average air fares and air travel demand is estimated. To compute the impact of carbon costs on demand, price elasticity of demand is required. The value of -1.601 is assumed based on Pagoni and Psaraki (2015).

Table 64.6 presents the results of the market-level analysis. Each row refers to the market (Origin and destination city) along with the distance flown between the

Table 64.5 Calculated fuel and CO_2 emissions and associated fee for individual flights

Market id	Origin	Conn.	Dest.	Airline	Fare [$]	Aircraft type [seg.1/seg.2]	Fuel burn [tn]	CO_2 [tn]	CO_2 [tn/pax]	CO_2 cost [$/pax]		
										Baseline	Low	High
2	LAX	–	OAK	DL	110	CRJ2	1.49	4.70	0.118	2.35	1.18	5.88
3	EWR	ATL	FLL	DL	192	A319/B752	10.44	32.95	0.265	5.30	2.65	13.26
6	MCO	–	LGA	B6	180	A321	6.16	19.46	0.147	2.95	1.47	7.37
7	ONT	IAH	EWR	UA	246	B738/B738	13.49	42.59	0.372	7.45	3.72	18.62
9	ORD	–	JFK	B6	133	E190	3.27	10.33	0.115	2.30	1.15	5.74

DL Delta Airlines, *B6* Jet Blue, *UA* United, *seg.1/2* flight segment 1 or 2, *Dest* destination airport, *Conn* connecting airport

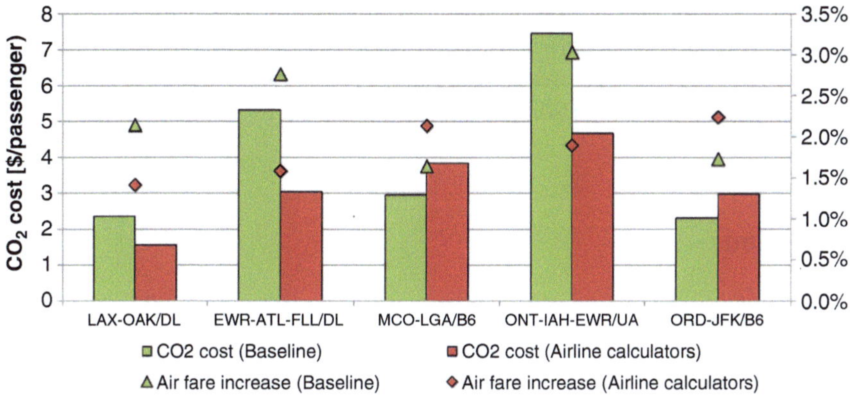

Fig. 64.2 Flight-level analysis: absolute and percentage change of air fare due to carbon fees

Table 64.6 Market-level analysis: CO_2 emissions, air fare increase and travel demand decrease

Market id	Distance [statute miles]	Total CO_2 emissions [tn]	Average price increase [$]			Average demand decrease [%] (assumed price elasticity = −1.601)		
			Low	Baseline	High	Low (%)	Baseline (%)	High (%)
1	340	69,073	0.91	1.81	4.53	−1.34	−2.68	−6.70
2	340	71,088	0.94	1.87	4.68	−1.38	−2.76	−6.90
3	1100	259,949	1.97	3.93	9.83	−1.78	−3.56	−8.91
4	1100	234,284	1.82	3.64	9.10	−1.64	−3.28	−8.20
5	975	133,901	1.73	3.46	8.65	−1.72	−3.43	−8.58
6	975	120,510	1.64	3.28	8.20	−1.63	−3.26	−8.14
7	2488	480,325	4.04	8.08	20.19	−2.60	−5.20	−13.00
8	2488	552,573	4.80	9.59	23.98	−3.08	−6.16	−15.41
9	735	71,193	1.41	2.81	7.03	−1.55	−3.09	−7.73
10	735	84,020	1.74	3.48	8.70	−1.92	−3.84	−9.60

cities, the total amount of CO_2 emitted, the average fare increase for the three scenarios and the corresponding average demand decrease. If the carbon price is set to 20$/tn CO_2 (baseline), one way ticket prices go up from $1.81 to $9.59, while demand falls from 2.68 to 6.16 %. More significant effects are noticed when the carbon price is set to 50 $/tn CO_2. Markets 7 and 8 will experience the greatest impact due to the extra carbon cost as they generate the largest amount of CO_2 emissions. In contrast, markets 1 and 2 impose the lowest carbon cost on the passengers and will thus have the lowest demand drop of 2.68 % and 2.76 % respectively (for the baseline 20$/tn CO_2).

The effectiveness of a carbon fee scheme program in balancing the benefits and costs of the emission reductions depend on factors originating from its structure and the prevailing carbon prices in the stock market. In general, a carbon fee which results in an extra cost for the airlines would provide them with an economic incentive to reduce their emissions. However, the results of this paper indicate that a carbon scheme could create strong incentives for CO_2 reduction only if the unit carbon price is set at a sufficiently high level.

Conclusions

This chapter examines the impact of a carbon scheme on air fares and air travel demand. As a case study, a network is designed which consists of the top ten markets, in terms of passengers travelled, within the United States. Carbon emissions and resulting carbon costs are calculated for every flight operating in the network. The results indicate that market ticket prices increase from $1.81 to $9.59 for the baseline carbon price of 20$/tn CO_2. The findings confirm that the carbon scheme will create incentives for CO_2 reduction only if carbon prices are set at a sufficiently high level. Further research examining the impact of strategic interactions on carbon trading schemes is under way.

References

Albers, S., Buhne, J. A., & Peters, H. (2007). Emission trading-strategy implications for airlines. *Proceedings of the 2007 European Transport Conference*. Leeuwenhorst, Netherlands.

Arvanitakis, A., & Hind, P. (2014). *Designing a Market-Based Measure for International Aviation Emissions*. FNI Climate Policy Perspectives 14, Fridtjof Nansen Institute, Norway.

Australian Government. (2012). *Managing the carbon footprint of Australian aviation*. Australia: Department of Infrastructure and Transport, Commonwealth of Australia.

Carlsson, F., & Hammar, H. (2002). Incentive-based regulation of CO_2 emissions from international aviation. *Journal of Air Transport Management, 8*, 365–372.

Commission of the European Communities. (2008). *Single European Sky II: Towards more sustainable and better performing aviation*. COM(2008) 389 final, Brussels.

EUROCONTROL. (2013). *User Manual for the Base of Aircraft Data (BADA)*. Revision 3.11, EEC Technical/Scientific Report No. 13/04/16-01, EUROCONTROL Experimental Centre.

European Union (EU). (2008). Directive 2008/101/EC of the European Parliament and of the Council of 19 November 2008 amending Directive 2003/87/EC so as to include aviation activities in the scheme for greenhouse gas emission allowance trading within the Community. *Official Journal of the European Union*.

European Union (EU). (2014). Regulation (EU) No421/2014 of the European Parliament and of the Council of 16 April 2014 amending Directive 2003/87/EC establishing a scheme for greenhouse gas emission allowance trading within the Community, in view of the implementation by 2020 of an international agreement applying a single global market-based measure to international aviation emissions. *Official Journal of the European Union*.

European Union. (EU). (2003). Directive 2003/87/EC of the European Parliament and of the Council of 13 October 2003 establishing a scheme for greenhouse gas emission allowance trading within the Community and amending Council Directive 96/61/EC. *Official Journal of the European Union*, L275.

Federal Aviation Administrator (FAA). (2014). *NextGen update: 2014*. Washington, DC: Office of NextGen.
Federal Aviation Administrator (FAA). (2015). *Aviation emissions, impacts and mitigation: A primer*. Office of Environment and Energy.
Girardet, D., & Spinler, S. (2013). Surcharge management of kerosene and CO_2 costs for airlines under the EU's emission trading. *Journal of Air Transport Management, 26*, 25–30.
Government Accountability Office (GAO). (2009). *Aviation and climate change: Aircraft emissions expected to grow, but technological and operational improvements and government policies can help control emissions*. Report to Congressional Committees, GAO-09-554, Washington, DC.
Grote, M., Williams, I., & Preston, J. (2014). Direct carbon dioxide emissions from civil aircraft. *Atmospheric Environment, 95*, 214–224.
Hileman, J. I., & Stratton, R. W. (2014). Alternative jet fuel feasibility. *Transport Policy, 34*, 52–62.
International Air Transport Association (IATA). (2008). *Aviation carbon offset programmes: IATA guidelines and toolkit*. Version 1, May 2008.
International Air Transport Association (IATA). (2013a). *Responsibly addressing climate change*.
International Air Transport Association (IATA). (2013b). *Technology roadmap* (4th ed.).
International Air Transport Association (IATA). (2014a). Fact sheet: Alternative fuels. http://www.iata.org/pressroom/facts_figures/fact_sheets/pages/alt-fuels.aspx. Retrieved March 17, 2015.
International Air Transport Association (IATA). (2014b). *Impact of proposed CO_2 tax on Portugal*. IATA Economic Briefing.
International Air Transport Association (IATA). (2014c). Fact sheet: Carbon offset. http://www.iata.org/pressroom/facts_figures/fact_sheets/pages/carbon-offsets.aspx. Retrieved March 17, 2015.
International Civil Aviation Organization (ICAO). (2012). Action plan of Norway. http://www.icao.int/environmental-protection/Documents/ActionPlan/Norway_AP_En.pdf. Retrieved March 10, 2015.
International Civil Aviation Organization (ICAO). (2014). *Engine exhaust emissions databank* (1st ed., Issue 20b). http://easa.europa.eu/document-library/icao-aircraft-engine-emissions-databank#1. Retrieved March 24, 2014.
International Monetary Fund (IMF). (2011). *Market-based instruments for international aviation and shipping as a source of climate finance*.
Lu, C. (2009). The implications of environmental costs on air passenger demand for different airline business models. *Journal of Air Transport Management, 15*, 158–165.
Miyoshi, C. (2014). Assessing the equity impact of the European Union Emission Trading Scheme on an African Airline. *Transport Policy, 33*, 56–64.
Organization for Economic Co-operation and Development (OECD). (2005). *The political economy of the Norwegian aviation fuel tax*. COM/ENV/EPOC/CTPA/CFA(2005)18/FINAL, Paris, France.
Pagoni, I., & Psaraki, V. (2015). Air travel demand in US: The effects of a carbon emissions fee. *Transportation Research Board 94th Annual Meeting*, Washington DC, Retrieved January 11–15, 2015.
Rojo, C., Vancassel, X., Mirabel, P., Ponche, J. L., & Garnier, F. (2015). Impact of alternative jet fuels on aircraft-induced aerosols. *Fuel, 144*, 335–341.
Scheelhaase, J., Grimme, W., & Schaefer, M. (2010). The inclusion of aviation into the EU emission trading scheme-Impacts on competition between European and non-European network airlines. *Transportation Research Part D, 15*(1), 14–25.
SENDECO2. (2014). *CO_2 prices: Graphic and historical data*. http://www.sendeco2.com/index-uk.asp. Retrieved June 11, 2014.

Sgouridis, S., Bonnefoy, P. A., & Hansman, R. J. (2011). Air transportation in a carbon constrained world: Long-term dynamics of policies and strategies for mitigating the carbon footprint of commercial aviation. *Transportation Research Part A, 45*, 1077–1091.

Zurich Airport. (n.d.). *Aircraft emission charges Zurich airport*. http://www.zurich-airport.com/~/media/FlughafenZH/Dokumente/Das_Unternehmen/Laerm_Politik_und_Umwelt/Luft/Emission_Charges_2010.pdf. Retrieved March 15, 2015.

Chapter 65
Ship Life Cycle Greenhouse Gas Emissions

Stefanos Chatzinikolaou, Nikolaos Ventikos, Levent Bilgili, and Ugur Bugra Celebi

Introduction

Ships play an indispensable role in trading and transportation across the world. Due to the great carrying capacity, they have a wide area of utilization. It is estimated and accepted that more than 90 % of global trade is carried by ships. Throughout the last century, the shipping industry has seen a general trend of increases in total trade volume (IMO 2012). According to the last annual review of United Nations Conference on Trade and Development (UNCTAD), world seaborne trade grew by 3.8 % in 2013, and the total trading volume reached to nearly 9.6 billion tons. Similarly, the world fleet grew by 4.1 % in 2013. Thus, the total world fleet reaches to 1.69 billion DWT in January 2014 (UNCTAD 2014). The growth in population and the global economy cause an intense demand for maritime transport. Since shipping activities are already contributing to some important environmental impacts, the projected growth of international shipping is expected to result in further environmental problems in the future.

Although CH_4, and N_2O_x gases have much more potential for global warming, CO_2, which is the major end product of internal combustion in marine engines should be the main focus (Helfre and Boot 2013; Edwards et al. 1996). In addition,

S. Chatzinikolaou (✉)
RINA SERVICES, 5, Aitolikou str., Piraeus 18545, Greece
e-mail: stefanos.chatzinikolaou@rina.org

N. Ventikos
National Technical University of Athens, 9, Herron Politechniou str., Zografou - Athens, Greece
e-mail: niven@mail.ntua.gr

L. Bilgili • U.B. Celebi
Yildiz Technical University, Yildiz Campus Barbaros Bulvari, 34349 Besiktas, Istanbul, Turkey
e-mail: lbilgili@yildiz.edu.tr; ucelebi@yildiz.edu.tr

P. Grammelis (ed.), *Energy, Transportation and Global Warming*, Green Energy and Technology, DOI 10.1007/978-3-319-30127-3_65

shipping activities cause numerous different air emissions such as sulfur oxides (SO_X), nitrogen oxides (NO_X), particulate matter (PM) and black carbon (BC) which have other considerable harmful effects on human health and the environment.

Shipping-Related GHG Emissions

The main GHG emission of shipping is CO_2. In 2012, it is estimated that the total CO_2 of international shipping was approximately 949 million tons. The total GHG emissions (CO_2, CH_4 and N_2O), is estimated at 972 million tons for this year. Therefore, it is revealed that approximately 97.63 % of the GHG emissions of shipping are attributed to CO_2. For the period 2007–2012, although CO_2 from shipping activities has increased, it accounted for only 3.1 % of the annual global CO_2 and approximately 2.8 % of the annual GHGs. It is also estimated that for shipping, the average annual (2007–2012) totals for NO_x (as NO_2) and SO_x (SO_2) are 20.9 and 11.3 million tons respectively. International shipping is estimated to produce approximately 18.6 and 10.6 million tons of NO_x (as NO_2) and SO_x (as SO_2). Global NO_x and SO_x emissions from shipping activities represent about 15 and 13 % of the global NO_x and SO_x anthropogenic emissions respectively. The total CH_4, N_2O and NMVOC emissions amount is estimated approximately at 300,000, 40,000 and 700 tons in 2012, respectively. Refrigerant and air conditioning gas releases account for the majority of HFC (and HCFC) emissions from ships. Use of SF_6 and PCFs in ships is documented as rarely being used in considerable quantities; hence these emissions are not covered in this study. Refrigerant and air conditioning gas releases from shipping result in an additional 15 million tons of CO_2 equivalent emissions. Inclusion of reefer container refrigerant emissions yields 13.5 and 21.8 million tons of CO_2 emissions. Combustion emissions of SO_x, NO_x, PM, CO and NMVOCs are also correlated with fuel consumption patterns, with some variability according to properties of combustion across engine types, fuel properties etc., which affect emissions substances differently (IMO 2014).

The Mediterranean Sea is an important area with respect to shipping activities which includes vital routes of international trade. Moreover, this basin is considered as one of the most congested sea areas of the world due to the heavy coastal and interantional sea transportation activities. For the Mediterranean Sea, it was estimated that NO_x, SO_2, PM and CO_2 emissions were 2, 1.45, 0.157 and 87.6 million tons, respectively, in 2005 (Cofala et al. 2007). Considering that 70, 44 and 36 % of the ship traffic occurs within 200, 50 and 25 nautical miles from shore (IMO 2009), it is reasonable for ship air emissions to be considered and explored in detail for their impacts to populations especially those located near the coast line.

Emission Abatement Technologies and Legislations

Although ship air emissions represent a small part of the total global emissions, the international shipping sector is putting effort to implement solutions for decreasing the environmental footprint of ships and specifically control both GHG and non GHG emissions. Despite the fact that it is impossible to eliminate the carbon-based emissions, reductions are possible with the implementation of conventional and innovative methods and technologies. These methods and technologies can be divided into sub-categories according to their focus. These sub-categories are (1) Ship Design, (2) Propulsion, (3) Machinery, (4) Operation, (5) Fuel Type.

Ship design refers more to the ship's hull form optimization. Increasing ship size or capacity may also reduce the fuel consumption (in terms of fuel consumption per ton-nm of cargo transported) by about 10–25 %. The main advantage of this method is that it is suitable for application to all kinds of ships (ABS 2013). Modern design methods may facilitate improvements in propeller design, especially if design methods progress to reliable prediction of full-scale wake fields and hull propeller interaction, considering speed and load case ranges instead of just a single operation point. Such improved propeller design procedures may be in place within the next 10 years. Potential savings of 0-4 % were reported in a survey among experts from ship model basins (MEPC 2011). Counter-rotating propellers (coupled counter-rotating propellers allow for the trailing aft propeller to recover some of the energy from the slipstream of the forward propeller); optimized propeller-hull interface (optimized design of the hull, protruding appendages and the propeller(s) to reduce frictional and residual resistance results in reduced hull propulsion system interference and improved fuel consumption); optimized propeller blade sections (propeller blades designed for reduced friction and cavitation reduce fuel consumption) can be used for all types of ships and provide 12; 4 and 2 % emission reduction, respectively. Using wind power with Flettner rotor and kites may have a reduction effect up to 30 and 20 %, respectively (Crist 2009). Propeller polishing, which provides a smooth surface and thus reduces the resistance, can decrease the emission by 3–8 % (Wang and Lutsey 2013). Machinery modifications are frequently used as effective methods for reducing emissions. Although the hybrid machinery systems (using diesel engine and fuel cells together) have a limited effect on general efficiency, CO_2 emissions can be reduced by 30 %. Internal engine modifications (IEM) (changing fuel valves with low-NO_x slide valves, retard injection, higher compression ratio, increased turbo efficiency, common rail injection, valve timing and different nozzles) have considerable effects on NO_x, VOCs and PM emissions. IEM methods can reduce the NO_x emissions up to 30–40 %. Exhaust Gas Recirculation (EGR) and Selective Catalytic Reduction (SCR) technologies are accepted solutions in the industry and can reduce the NO_x emission by 35 and 90–95 %, respectively (Andreoni et al. 2008). From the operational measures perspective, the reduction of the service speed of the ship has also considerable positive effects on fuel consumption and currently it is widely applied as a win-win solution for environmental and economic performance of the vessel.

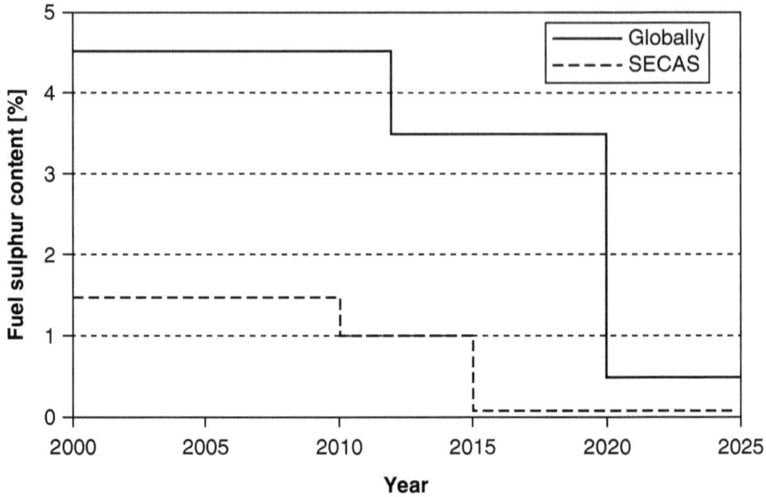

Fig. 65.1 Current and projected restrictions on fuel sulfur content according to IMO (Haglind 2008)

Reducing speed by 1 knot may reduce fuel consumption by 12–22 % (ABS 2013). Weather routing is a relatively easy and effective solution, which provides a reduction by 10 %. Cleaner and well-coated hulls may reduce the emission by 3–5 %. Besides, overall energy awareness of the crew have a positive effect on reducing the emission up to 10 % (Crist 2009). PM emission can be reduced around 18 and 20 % by reducing the sulfur content of the fuel to 1.5 and 0.5 %, respectively (ENTEC 2005).

The international law of the sea finds its highest authority in the IMO which is a body of the United Nations (UN) appointed to set international standards for safety and pollution. Regarding the protection of marine environment, MARPOL represents the most important IMO convention currently into force. Annex VI of MARPOL sets limits on SO_x, NO_x and other emissions from marine vessel operations and prohibits deliberate emissions of ozone-depleting substances (Miola et al. 2010). Figures 65.1 and 65.2 presents the SO_x and NO_x limits according to Regulation 14 and Regulation 13, respectively.

IMO has developed two major programs for shipping-related emissions: Energy Efficiency Design Index (EEDI) and Ship Energy Efficiency Management Plan (SEEMP). Both mechanisms, which fall under the category "energy efficiency/ emissions standards," are the first ever mandatory GHG regulations for the shipping industry. These mechanisms, which came into effect on 1 January 2013, apply to all ships of 400 tons gross tonnage and above. While the EEDI sets a minimum energy efficiency standard for new ships, the SEEMP enables ship owners to measure the fuel efficiency of existing ships and to monitor the effects of any changes in operation. The EEDI, which is probably the most important measure, will allow ship designers and builders to use the solutions that they consider most

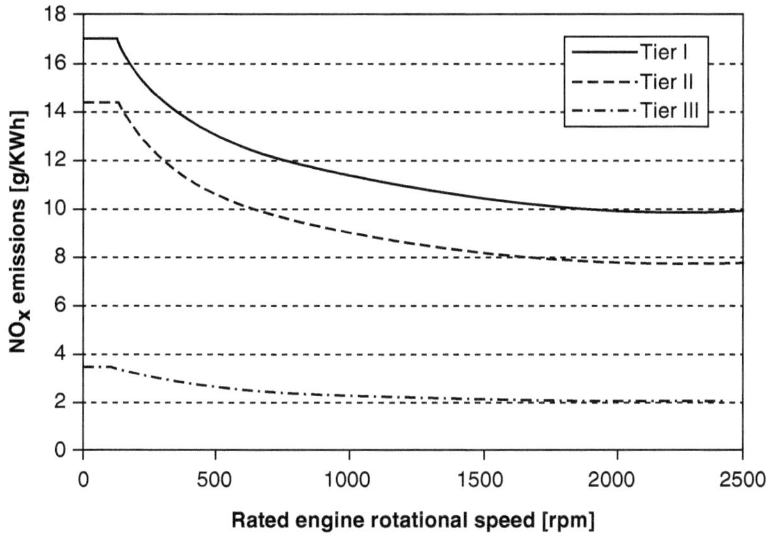

Fig. 65.2 Current and projected NO_x emission limits according to IMO (Haglind 2008)

cost-efficient to comply with the regulations. Based on the EEDI, the CO_2 reduction level (grams of CO_2 per ton mile) for the first phase (2015–2019) is set to 10 % and will be tightened every 5 years. The baseline is the average efficiency for ships built between 2000 and 2010. Note that developing countries will not have to implement the standards until 2017, allowing them time to develop shipbuilding capacity (Helfre and Boot 2013).

Results and Discussion

Emissions from shipping are produced throughout the ship life cycle which can be divided in four important phases: (1) Shipbuilding, (2) Operation (3) Maintenance, and (4) Dismantling/Recycling. The emissions and wastes of these phases might be similar in some cases but in general are different in terms of emission type and quantities. Various types of emissions and wastes are produced in solid, liquid and gaseous form during the life cycle phases of the ship.

Ship Life Cycle Assessment

According to the ISO framework, the LCA method is the compilation and evaluation of the inputs, outputs and potential environmental impacts of a product system throughout its life cycle (ISO 2006). LCA usually consists of four main components: goal definition and scoping, inventory analysis, impact assessment, and interpretation. After the identification of the system and its boundaries the inventory analysis is formulated in order to collect information on the quantities of energy and materials used and environmental releases (e.g., air emissions) throughout the life cycle of the system. The impact assessment step (usually referred to as Life Cycle Impact Assessment, LCIA) translates the inventory results into impacts and it is realized in three mandatory steps according to the ISO standards: (a) selection of impact categories, (b) classification, and (c) characterization. Selection of impact categories involves the identification of relevant categories of impact for the particular study's needs (i.e. climate change, eutrophication acidification etc.). Classification is the assignment of inventory results to impact categories. This should be done by assigning the inventory results that are not only exclusive to one impact category but also relate to other impact categories, including distinction between parallel mechanisms. Various characterization methods are available to address different environmental impacts (the impact categories) covered by the LCA methodology. Each one of these characterization methods uses a cause-effect pathway and impact indicators, to produce the so called characterization factors (CFs). Therefore, CFs are weighting factors that aggregate life cycle emissions (De Schryver 2010). There are available a number of impact assessment methods for use in the LCA framework which have been reported and analysed thoroughly in the literature (Hischier et al. 2010; EC-JRC 2011). Within the maritime transport sector, previous life cycle studies conducted (starting from the 1990s) have demonstrated that the LCA method may be well employed for the environmental life cycle evaluation of ships. Analysts agree that for a ship which is a complex system integrating multiple systems the LCA method might become more effective by detailing the ship system into subsystems (i.e. hull, machinery, equipment for cargo etc.). The ship LCA experimental studies have also highlighted the significance of the system boundaries selection which is a highly subjective process and may drastically affect the final results (Fet 2002; Chatzinikolaou and Ventikos 2013).

The usual life span of a cargo ship is 20–25 or more years and consists of four major stages i.e. the shipbuilding, operation, maintenance, and disposal/recycling. The Ship LCA framework accepts that the ship (the System) is a set of incorporated subsystems functioning throughout these life cycle stages. A subsystem is defined as an individual unit that is part of the defined total system. Subsystems are further detailed into system elements which themselves consist of sets of processes. The methodology accepts that processes have an environmental footprint at any life cycle stage which is viewed as inputs (material and energy use) and outputs (air emissions). Processes might be the same across life cycle stages but their footprint usually alters owing to different inputs which in effect result in different air emissions results. The initial goal is to identify important (in terms of emissions) processes and thoroughly explore their environmental input-output which will thereafter allow for emissions inventory and impact results to be provided at any other level (system element, subsystem or overall system) or life cycle stage (Chatzinikolaou and Ventikos 2015).

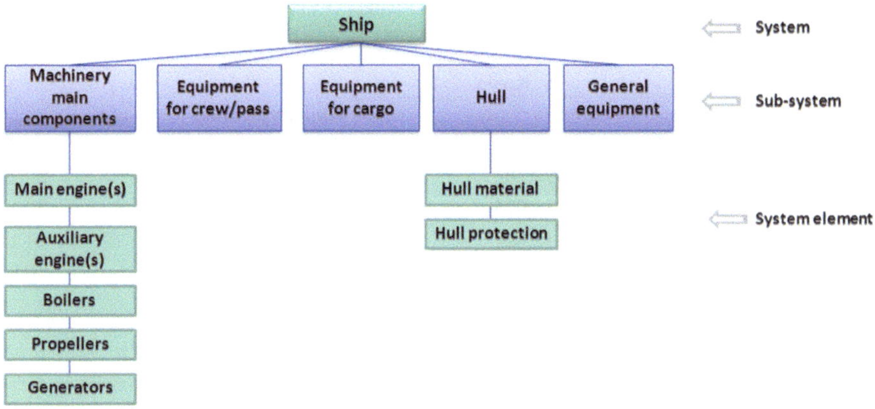

Fig. 65.3 Ship-related air emissions assessment from a life cycle perspective (Chatzinikolaou and Ventikos, 2014)

This paper places particular focus on two ship subsystems due to their importance in air emissions outcome, namely the machinery main components and the hull subsystem (Fig. 65.3).

The hull subsystem is divided in the hull material and hull protection system element for which corresponded processes are identified and modelled in an input-output context. The main machinery components subsystem is divided in system elements that correspond to the primary components of the engine room, for example main and auxiliary engines, boilers and generators.

The Ship LCA framework initial outcome is the development of an inventory with results of air emissions from all the identified process of the system. The emissions covered are: carbon dioxide (CO_2), carbon monoxide (CO), sulphur dioxide (SO_2), sulphur oxide (SO_x), nitrogen oxide (NO_x), particulate matter (PM), methane (CH4), volatile organic compound (VOC), and non-methane volatile organic compounds (NMVOC). Air emissions results can be provided for any ship system element, subsystem, and life cycle stage. Results are also available per year and total life cycle.

Case Study

The results of ship emissions are from previous work of Chatzinikolaou and Ventikos (2013), (2014), and (2015) on ship life cycle assessment. They have developed a framework for a ship LCA and conducted a case study in order to test this framework using a recently built ship, a Panamax type oil tanker (details of the ship are provided in Table 65.1).

The emissions produced during the life cycle of the case study ship are provided in the following figure. This table contains results for all emissions covered in the

Table 65.1 Case study for ship LCA: Panamax oil tanker (Chatzinikolaou and Ventikos 2014)

Year of built	2009		
Country	S. Korea		
General data		*Main particulars*	
Displacement (tonnes)	88,221	L_{BP} (m)	219.00
DWT (tonnes)	74,296	Breadth (m)	32.24
Lightship (tonnes)	13,925	Depth (m)	20.60
Steel weight (tonnes)	12,022	Draught (m)	14.17
Design speed (knots)	15.30	Block coefficient C_B	0.85
Machinery	*Main engine*	*Auxiliary engines (3)*	
MAN B&W	2 Stroke	4 Stroke	
Country of construction	S. Korea	S. Korea	
Unit power (kW)	12,240	740	
RPM	105	720	
Weight (tonnes)	368	19.7	

Table 65.2 Life cycle emission inventory—case study Panamax oil tanker

Emissions		Shipbuilding	Operation	Maintenance	Dismantling	Total life cycle
CO_2	Tons	2.29E+04	1.06E+06	9.62E+03	8.51E+03	1.10E+06
CO	Tons	4.53E+02	3.17E+03	8.16E+01	7.72E+02	4.48E+03
CH_4	Tons	4.06E+00	2.81E+01	1.48E+00	2.13E+00	3.58E+01
NO_x	Tons	1.28E+02	3.04E+04	9.20E+01	1.07E+02	3.07E+04
PM (all)	Tons	2.29E+01	2.45E+03	8.69E+00	2.25E+01	2.51E+03
SO_2	Tons	1.02E+02	1.57E+04	7.39E+01	1.28E+02	1.60E+04
VOC	Tons	2.00E+01	–	5.78E+01	2.99E-01	7.81E+01

LCA framework. The following results per life cycle stage are focusing specifically to the CO_2 emissions (Table 65.2).

For the hull subsystem the important life cycle stages are shipbuilding, maintenance and disposal/recycling. The operation life cycle stage has no important emissions contribution to this subsystem. The operation life cycle stage is very important for the machinery subsystem due to the process of combustion of fuels in main and auxiliary engines which is responsible for 98 % of the GHG emissions of the ship system.

GHG Emissions in Shipbuilding

The shipbuilding life cycle stage results reveal that the hull subsystem produces larger amount of emissions compared to the construction of the machinery sub-system. Looking deeper into the hull subsystem, the hull material system element produces larger amount of emissions than the hull protection system

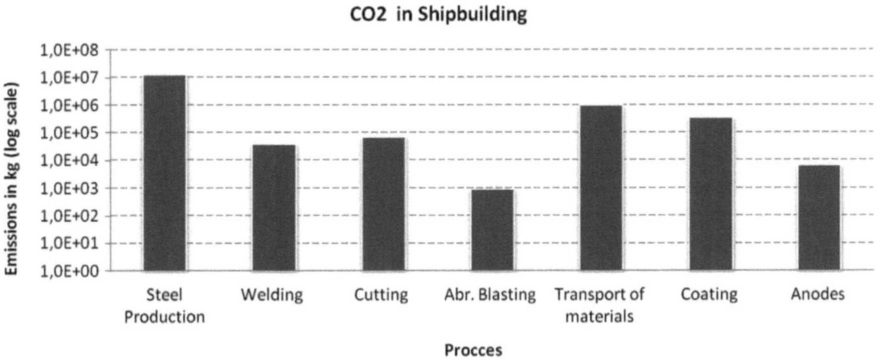

Fig. 65.4 CO_2 emissions in shipbuilding process—Panamax oil tanker vessel (Chatzinikolaou and Ventikos, 2014)

element. Steel production and elaboration is the main process with respect to air emissions. Other processes included in the hull material system element are: transport of steel to yard, cutting of steel, sandblasting and welding.

Coating is a major operation in any shipyard. It refers to all kinds of paint application performed at various stages from the initial priming of the steel, to the final paint application. Outdoor painting (layers of paint applied to the outer skin of a ship hull), is in general higher than indoor painting (layers applied to the inner parts of the ship). Emission calculations are based on the work of Celebi (2008). For the ship of the case study the distribution of CO_2 emissions per shipbuilding process is shown in the following figure (Fig. 65.4). Steel production process (which is included in the hull material system element) is responsible for nearly 90 % of the total CO_2 emissions during shipbuilding.

GHG Emissions in Operation

CO_2, NO_x, SO_x, CO, PM, VOCs, methane (CH_4), PM, black carbon (BC) and particulate organic matter (POM) are the most dangerous exhaust emissions for human health and environment, especially for global warming. Shipping related air emissions strongly depend on fuel type and engine type characteristics, characteristic combustion efficiency, characteristics of the lubricating oil, operation conditions of the ship, ship form and equipment efficiencies (Bilgili et al. 2014). The vast part of emissions in the phase of operation is produced from the machinery subsystem. Marine engines operation remains the largest source of environmental impacts in the life cycle of the ship, representing for example 98 % of the total life cycle impact for a for a cargo vessel according to Kameyama et al. (2004). In Fig. 65.5, emissions of CO_2 in the operation phase are shown in an annual basis (assuming a 25 years life cycle). This analysis of emissions per year demonstrates

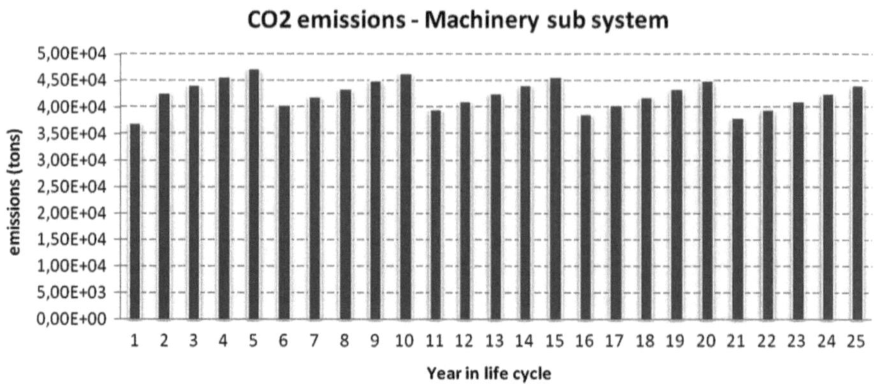

Fig. 65.5 Carbon Dioxide emissions per year of operation—machinery subsystem (Panamax oil tanker vessel) (Chatzinikolaou and Ventikos 2013)

that emissions are continuously increasing in intervals of 5 years which corresponds to the typical period between major maintenance works also known as dry-docking periods (Chatzinikolaou and Ventikos 2013). As a result, every 5 years there is an emission peak which reflects the impact of the deterioration of the hull of the ship due to marine growth. This phenomenon increases the frictional resistance of the ship which in effect increases the propulsion power demand to keep the same speed. The case study's scenario examined in this study assumes that the ballast and laden speeds remain constant throughout the ship's life cycle at the service speed of the ship (14 knots). This particular scenario is acknowledged that it is not reflecting the current shipping practice; however it is the scenario that matches better to the design specifications of this ship and can be used as a reference point for any other real scenario of operation. Overall, 96 % of the total CO_2 emissions are produced during the phase of operation (1.06 million tons) (Chatzinikolaou and Ventikos 2015).

GHG Emissions in Maintenance

Emissions in the phase of maintenance are mainly derived by the hull subsystem. No important maintenance processes have been identified for the machinery of the ship. The processes considered in the maintenance of the hull are identical to the shipbuilding processes but with different input–output (use of energy and materials and resulting emissions).

It is estimated that maintenance process is responsible for 0.01 % of total CO_2 emissions of a ship's life cycle (Fig. 65.6).

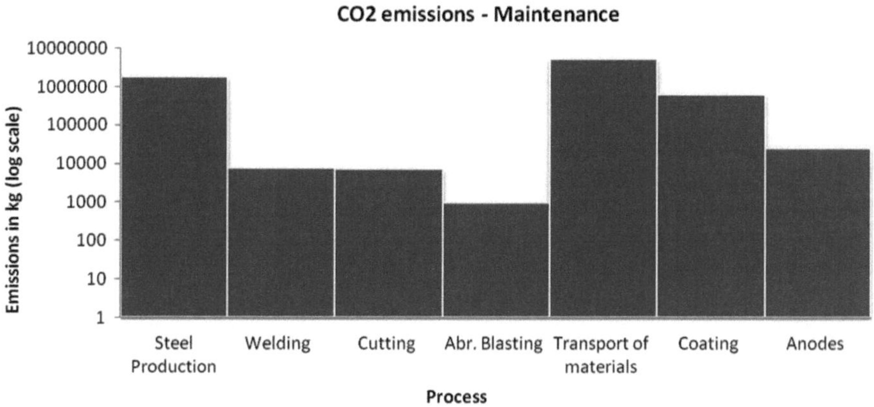

Fig. 65.6 Carbon Dioxide emissions during maintenance periods of the Panamax tanker (Chatzinikolaou and Ventikos 2014)

GHG Emissions in Disposal/Recycling

If there's no appropriate integrated system for the recycling or reusing of ship related steel, machines, auxiliaries and even furnishings, such materials will remain unused and useless at the end of the ship's life. In this respect, ship recycling facilities contribute to sustainable development and represent and environmentally friendly way of dismantling ships. At the end of a ship's life cycle, the ship contains not only various recyclable materials but also a range of hazardous and toxic substances. Many substances in a ship's structure and equipment may be defined as hazardous and toxic under the existing 1989 Basel Convention on the Control of Transboundary Movements of Hazardous Wastes and Their Disposal. The so called Hong Kong Convention, adopted by IMO in 2009, is the last and most important regulation on ship recycling wastes. Environmental information regarding the final life cycle stage of ships is generally unavailable. Owing to certain incentives (i.e. low labour costs, imperative need for steel and second hand material) the global ship dismantling centre is currently located in the so called Indian subcontinent (India, Bangladesh, Pakistan). Some 80 % or more of the end of life ships ends up in the Indian subcontinent dismantling sites and usually takes place by applying the beaching method which allows the demand for infrastructure (piers, sufficient depth of the harbour, cranes etc.) to be replaced by a mud flat and a huge labour force. The dismantling operations are based around the principle of maximum separation of the steel structure of the ship without making any use of technology similar to shipbuilding. Safety, environmental and human health aspects are critical in the majority of these sites. The ship recycling scenario which has been examined in the case study of the Panamax tanker has made use of information for the process or steel re-rolling which is common practice in these dismantling sites. The amount of steel of the ship treated and recovered with this specific method of re-rerolling for an oil tanker vessel ranges from 72 to 81 % of the total steel weight of the ship (Chatzinikolaou and Ventikos 2013). The estimated GHG emissions from the re-rolling of the case study ship are 8500 tons of CO_2 and 2.13 tons CH_4. It is acknowledged though, that by using only this information only an indication of the environmental footprint of ship recycling is provided. It is also acknowledged that the current mechanisms established in the majority of these ship recycling sites are forming severe threats to the environment and human health at the local level. In fact, these practices could be used as an illustrative example of the benefits of life cycle thinking as a holistic approach of studying the ship system (Chatzinikolaou and Ventikos 2014).

Conclusion

This work discusses important issues related to air emissions throughout the ship life cycle with particular focus on GHG of cargo ships. Literature information is provided as regards the contribution of the international maritime transport sector to GHG as well as non GHG emissions and available measures that are implemented for reducing these emissions have been also reported. This study also includes important elements of a framework which has been developed for assessing ship air emissions from a life cycle perspective, and some illustrative results from a case study conducted to evaluate and test this framework.

Overall, this work aims at providing assistance to the understanding, quantification and communication of knowledge related to shipping air emissions aspects. It is without any doubt that adding knowledge to these issues is part of the solution to continuously improving the efficiency of maritime transport and increasing its benefit to the society at large.

References

ABS. (2013). Ship energy efficiency measures. Retrieved June 4, 2013.

Andreoni, V., Miola, A., Perujo, A. (2008). Cost effectiveness analysis of the emission abatement in the shipping sector emissions, European Commission Joint Research Center, Institute for Environment and Sustainability, ISBN: 978-92-79-11280-5.

Bilgili, L., Celebi, U. B., Alkan, G. B., Caglak, S. B., & Koralturk, G. (2014). Comparison of different ship emission estimation methodologies for annual emission footprint and reduction techniques of a bulk carrier. *Fresenius Environmental Bulletin, 23*(7), 1497–1509.

Celebi, U. B. (2008). Wastes in shipbuilding and paint emission estimation, Ph.D. Thesis, Yildiz Technical University, Istanbul, Turkey (in Turkish).

Chatzinikolaou, S. D., & Ventikos, N. P. (2013). Assessment of ship emissions in a life cycle perspective. *3rd International Exergy, Life Cycle Assessment and Sustainability Workshop & Symposium*, Nisyros, Greece. Retrieved 7–9 July 2013.

Chatzinikolaou, S., & Ventikos, N. P. (2014). Assessing environmental impacts of ships from a life cycle perspective. In *Proceedings of the 2nd International Conference on Maritime Technology and Engineering, MARTECH 2014*, Lisbon, Portugal. Retrieved October 15–17, 2014.

Chatzinikolaou, S., & Ventikos, N. P. (2015). Critical analysis of air emissions from ships: Lifecycle thinking and results. In H. Psaraftis (Ed.), *Green transport logistics: the quest for win-win solutions* (International Series in Operations Research & Management Science). Switzerland: Springer. ISBN 978-3-319-17175-3.

Cofala, J., Amann, M., Hezes, C., Wagner, F., Klimont, Z., Posch, M., Schopp, W., Tarasson, L., Jonson, J. E., Whall, C., & Stavrakaki, A. (2007). Analysis of policy measures to reduce ship emissions in the context of the revision of the national emissions ceiling directive, Final Report, Submitted to the European Commission, DG Environment, Unit ENV/C1, Contract No. 070501/2005/419589/MAR/C1, IIASA Contract No. 06-107.7157-7160.

Crist, P. (2009). Greenhouse gas emissions reduction potential from international shipping, Joint Transport Research Centre of the OECD and the International Transport Forum, Retrieved May 2009.

De Schryver, A. M. (2010). Value choices in life cycle impact assessments, PhD-Thesis Radboud University, Nijmegen, ISBN: 978-94-91066-02-3.

EC-JRC. (2011). Recommendations based on existing environmental impact assessment models and factors for life cycle assessment in European context ILCD Handbook—International Reference Life Cycle Data System. European Union EUR24571EN. ISBN 978-92-79-17451-3.

Edwards, J. H., Galbally, I. E., Meyer, C. P., & Weeks, I. A. (1996). Lifecycle emissions and energy analysis of LNG, oil and coal, Final Report to Woodside Petroleum Pty Ltd.

ENTEC. (2005). Service contract on ship emissions: Assignment, abatement and market-based instruments, Task 2c-SO2 Abatement Final Report, Retrieved August, 2005.

Fet, A. M. (2002). Environmental reporting in marine transport based on LCA. *Journal of Marine Design and Operations B, B1*, 1476–1556.

Haglind, F. (2008). A review on the use of gas and steam turbine combined cycles as prime movers for large ships. Part I: Background and design. *Energy Conversion and Management, 49*, 3458–3467.

Helfre, J. F., & Boot, P. A. C. (2013). Emission reduction in the shipping industry: Regulations, exposure and solutions. *Sustainalytics*.

Hischier, R., et al. (2010). Implementation of life cycle impact assessment methods ecoinvent report No. 3, v2.2. Dübendorf: Swiss Centre for Life Cycle Inventories.

IMO. (2009). Prevention of ghg emissions from ships. *Second IMO GHG Study 2009 – Final Report*, MEPC 59/4/7. April 2009.

IMO. (2014). Reduction of GHG emissions from ships, Third GHG Study 2014-Final Report, Retrieved July 25, 2014.

IMO (International Maritime Organization). (2012). International shipping facts and figures-information resources on trade, safety, security, environment. Maritime Knowledge Center, Retrieved March 6, 2012.

ISO. (2006). ISO 14040 international standard. In Environmental management—life cycle assessment—principles and framework. Geneva, Switzerland: International Organization for Standardization.

Kameyama M. K., Hiraoka, A., Sakurai, T. Naruse & Tauchi H. (2004). In: *Proceedings of the Sixth International Conference on EcoBalance*. 25–27 October, Tsukuba, Japan, pp. 159–162.

MEPC. (2011). Air pollution and energy efficiency, GHG emissions-mitigating measures for oil tankers, Submitted by Oil Companies International Marine Forum (OCIMF), MEPC 63/INF.7. Retrieved November 25, 2011.

Miola, A., Ciuffo, B., Marra, M., & Giovine, E. (2010). Analytical framework to regulate air emissions from maritime transport, European Commission Joint Research Center, Institute for Environment and Sustainability.

UNCTAD. (2014). In J. Rogers (Ed.), Review of maritime transport 2014. eISBN: 978-92-1-056861-6.

Wang, H., & Lutsey, N. (2013). Long-term potential for increased shipping efficiency through the adoption of industry-leading practices, The International Council on Clean Transportation, Retrieved July, 2013.

MIX
Papier aus verantwortungsvollen Quellen
Paper from responsible sources
FSC® C105338

If you have any concerns about our products,
you can contact us on
ProductSafety@springernature.com

In case Publisher is established outside the EU,
the EU authorized representative is:
Springer Nature Customer Service Center GmbH
Europaplatz 3, 69115 Heidelberg, Germany

Printed by Libri Plureos GmbH
in Hamburg, Germany